READINGS IN THE PHILOSOPHY OF SCIENCE

READINGS IN THE PHILOSOPHY OF SCIENCE

Second Edition

Edited by

Baruch A. Brody

Richard E. Grandy

PRENTICE HALL, ENGLEWOOD CLIFFS, NEW JERSEY 07632

Library of Congress Cataloging-in-Publication Data

Readings in the philosophy of science.

Includes bibliographical references.
1. Science—Philosophy. 2. Science—Methodology.
I. Brody, Baruch A. II. Grandy, Richard E.
Q175.3.R415 1989 501 88-32202
ISBN 0-13-761065-3

Cover design: *George Cornell*
Manufacturing buyer: *Peter Havens*

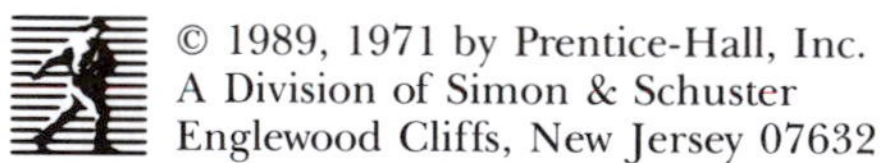

Printed in the United States of America

10 9 8 7 6 5 4 3 2 1

ISBN 0-13-761065-3

Prentice-Hall International (UK) Limited, *London*
Prentice-Hall of Australia Pty. Limited, *Sydney*
Prentice-Hall Canada Inc., *Toronto*
Prentice-Hall Hispanoamericana, S.A., *Mexico*
Prentice-Hall of India Private Limited, *New Delhi*
Prentice-Hall of Japan, Inc., *Tokyo*
Simon & Schuster Asia Pte. Ltd., *Singapore*
Editora Prentice-Hall do Brasil, Ltda., *Rio de Janeiro*

CONTENTS

PREFACE

It is eighteen years since the first edition of this anthology appeared. During those years, it was used as a text in many philosophy of science courses. Many of those who taught those courses have suggested that a new edition is required because of the extensive developments in the field in the last eighteen years. This second edition has been developed in response to those suggestions.

Several crucial features of the first edition have been retained. The anthology continues to be divided into three major sections, one on explanation and prediction, one on scientific theories, and one on the confirmation of scientific hypotheses. Moreover, each of those sections still begins with selections from the major logical empiricists (especially Rudolf Carnap and Carl G. Hempel) whose writings clearly constitute the point of departure for current discussions. Finally, we continue to provide both introductions to each section and an updated bibliographical essay, both of which were much-used features of the first edition.

There are, however, crucial changes in the second edition. To begin with, the philosophy of science continued to develop rapidly in the 1970s, and new post-empiricist figures have emerged to dominate the field. These include such diverse authors as Peter Achinstein, Nancy Cartwright, Paul Feyerabend, Clark Glymour, Ian Hacking, Thomas Kuhn, Imre Lakatos, Larry Laudan, Wesley Salmon, and Bas Van Fraassen. This edition reflects the significance of their contributions. Moreover, recent developments have directed considerable attention to the applications of general themes in the philosophy of science to methodological issues in the specific sciences. We have therefore included selections from those writings as a fourth section of the book. In short, the major changes in this second edition reflect the shifting nature of the field.

Baruch Brody
Richard Grandy

PREFACE TO THE FIRST EDITION

No area in philosophy has grown as rapidly in recent years as has the philosophy of science. There are an increasing number of specialists in this area, and a great many schools have added courses in the philosophy of science to their undergraduate and graduate curriculum. Unfortunately, there has not been a corresponding increase in the number of texts and anthologies suitable for these courses, and teachers have been forced to require their students to purchase numerous books and to spend a good deal of time in their library reserve rooms reading additional material. The purpose of this anthology is to alleviate this unsatisfactory situation by providing sufficient current material of high quality in one book so that a good undergraduate (or first year graduate) philosophy of science course can be built entirely around it.

This anthology is divided into three sections: Section I concerns scientific explanation and prediction; Section II, the structure and function of scientific theories; and Section III, the confirmation of scientific hypotheses. These problems have been chosen because they are clearly central to the philosophy of science; no worthwhile course in this field can possibly omit them. While further sections could have been added, it seemed more desirable to treat the central problems at some depth rather than to treat every problem superficially.

Three principles have guided me in choosing the selections: (a) as far as possible, every major position on a particular issue should be represented; (b) the various selections, besides presenting their own point of view, should also analyze other points of view; (c) selections should presuppose as little philosophy, logic, and science as is possible. It is hoped, therefore, that this book will give even a student relatively new to philosophy a genuine sense of the difficulties and potentialities inherent in a variety of positions, rather than a mere superficial knowledge of what several philosophers have said about a particular issue.

Each section begins with a substantial introduction that structures the section by explaining its problems, by stating the possible solutions to these problems that have been offered, and by briefly indicating the roles of the particular selections in the discussion of these problems. It is hoped that these introductions will enable the student to understand by himself what is said in the selections, thereby freeing the teacher from the task of exposition and enabling him to devote his time in class to the critical analysis and evaluation of the positions presented in the selections.

No book could reasonably hope to include all the worthwhile material that has been published in the philosophy of science in recent years. Since the serious student will certainly want access to this material, this book therefore concludes with a detailed bibliographical essay that introduces the student to material not included in this volume.

This book would probably never have appeared were it not for the constant aid and advice of Alan Lesure of Prentice-Hall. I would like to take this opportunity, therefore, to express my appreciation to him. I am also indebted to several of my colleagues, particularly to Sylvain Bromberger and Jerry Katz, for their advice on many points, and to my secretary, Martha Sullivan, for her invaluable aid in producing the manuscript for this book. Finally, in dedicating this book to my wife, I am merely expressing my awareness of the many debts that I owe to her, the least of which is for her encouragement that enabled me to complete this book.

Baruch Brody

GENERAL INTRODUCTION

Philosophers of science are primarily concerned with three kinds of questions. One kind deals with the implications of new scientific findings for traditional philosophical issues; thus, philosophers of science question whether the principle of indeterminacy in quantum mechanics shows that human actions are not entirely determined or whether recent research on computers and artificial intelligence supports the thesis that human beings are merely very complex machines. A second kind of question deals with the analysis of the fundamental concepts of the diverse scientific disciplines; thus, philosophers of science have analyzed such concepts as number, space, force, goal, and living organism. Finally, a third kind of question deals with the nature of the goals of the scientific enterprise and of the methods the scientist employs to attain these goals; thus, philosophers of science have argued about whether science should attempt to explain, or merely describe, the data observed in the laboratory or whether the scientist must postulate the existence of unobserved entities in order to deal effectively with his data. The selections in this anthology are directed primarily to this last type of problem, although in the process they frequently deal with the other types as well.

Some philosophers believe that this last type of problem, the nature of the goals and methods of science, cannot be dealt with since there are no permanent goals or methods of science. Much as scientists give up one scientific theory in favor of another, so they formulate new conceptions of their goals and of the methods to be employed in reaching them. Accordingly, the best we can hope for is a detailed account of the many methods for reaching these goals favored by scientists throughout history.

This view is clearly based upon the following assumptions: (a) different scientists at different points in the development of science have had different conceptions of the goals and methods of science; (b) the task of the philosopher of science is to describe the conception of the goals and methods of science actually held by scientists. Those who disagree with this view, while generally in agreement about (a), challenge (b) on the grounds that it confuses a normative discipline like the philosophy of science with descriptive disciplines like the history and sociology of science. After all, the existence of disciplines like the history and sociology of science that describe these many conceptions of the goals and methods of science in no way rules out the legitimacy of a discipline that is aimed at determining which of these conceptions are correct.

This normative conception of the philosophy of science goes back at least as far as the Renaissance philosophers of science, such as Bacon, Galileo, and Descartes, who claimed that earlier incorrect scientific methods caused the sterility of science before their time. Moreover, many of the authors represented here hold this conception. It is not our purpose at this point to decide whether this normative account, or the earlier descriptive account, of the philosophy of science is correct; we want merely to warn the reader that the following works were written by men with differing viewpoints about this fundamental issue.

The philosophy of science, as conceived of in this book, is not a new discipline. Both Plato and Aristotle wrote accounts of the goals and methods of the scientist. Extremely important and valuable work was produced in this area during the seventeenth century by Bacon, Galileo, Descartes, Newton, and Leibniz and during the nineteenth century by Herschel, Whewell, Mill, Jevons, Peirce, Mach, Herz, Poincare, Duhem, and Pearson. However, all the selections included in this book were written in the present century. These earlier writings, although still informative and provocative, contain many ideas which are today considered outmoded; it seemed, therefore, advisable to let the student begin his study of the philosophy of science with more recent writings. Those interested in acquiring a more thorough understanding of the subject must eventually become familiar with these classical texts. The bibilography contains several references that will enable the student to begin work on this task.

The most important school in twentieth-century philosophy of science is logical positivism (or logical empiricism). This movement began in Austria and Germany in the 1920's, spread to England and America in the 1930's, and continues to have a great influence on the philosophic community in these countries. This school held that most of the traditional philosophical issues are meaningless pseudo-problems and that, therefore, philosophers ought to turn their attention to more legitimate problems such as the nature of the goals and methods of the empirical sciences. Given this great interest in the philosophy of science and a variety of highly sophisticated analytical tools developed by modern logicians, these logical positivists produced a powerful and persuasive conception of the scientific enterprise that has been adopted by many philosophers and scientists who do not agree with the basic presuppositions of logical positivism. Nearly all the sections in this book begin with selections by important logical empiricists, like Professors Rudolf Carnap, Carl G. Hempel, and Ernest Nagel, who helped develop this conception of science.

In the last twenty-five years, many philosophers of science have argued that the logical empiricist view of science should be supplemented or supplanted. Some derive this conclusion from their analysis of the historical development of successful science, others from general logical, epistemological, and metaphysical assumptions. Most of the selections that follow the introductory readings from the empiricists contain these post-empiricist arguments. Although no text could possibly include all of the interesting material that has appeared in recent years, we have tried to include in this text a balanced sampling of the best material which has appeared so that the student can obtain a fair view of the present state of the subject.

PART I Theories

Introduction

The most obvious products of science are practical—microwave ovens, computers, plastics and vaccines. But these by-products are all by-products of the most distinctive product of science: theories which allow scientists to explain and predict observable events by relating them to unobservable objects and events.

Theories are the product of a mature stage of science, one that can be achieved only after empirical laws have first been discovered. The contrast between theories and empirical laws can be illustrated in the study of gases. A well-known example of an empirical law is Boyle's law, which states that when the temperature of a gas is constant, the pressure of the gas is inversely proportional to its volume. The kinetic theory of gases states that gases are composed of very small particles in motion whose collisions with the container are the pressure and whose average energy determines the temperature. There is an obvious difference between Boyle's law, which contains descriptive terms that refer to directly measurable properties of observable objects, and the kinetic theory, which makes reference to particles too small to observe and to their properties.

The logical positivists were led by this obvious difference to divide the vocabulary of science into two parts—the observational vocabulary and the theoretical. This distinction raises many questions: How do scientists understand and use the theoretical terms of a theory? Are the theoretical terms to be taken as really describing parts of the world or are they only convenient fictions? And does the distinction itself stand up under careful scrutiny in further cases?

The logical empiricists discussed all of these issues, and their views are described in the selections by Professor Carnap and Professor Hempel, which begin this section.

Professor Carnap presents the standard logical empiricist conception of the structure of a theory: A theory is a partially interpreted formal system. Its precisely formulated axioms contain both observational and theoretical primitive terms, and from these axioms one can deduce the consequences of the theory. Not all of the primitive terms are interpreted; in particular, the theoretical primitive terms, which refer to nonobservable entities, properties, or processes, cannot be directly interpreted. Nevertheless, the scientists understand the whole theory in the sense that they use it to derive non-theoretical consequences that are directly interpretable.

Consider a more detailed account of our example, the kinetic theory of gases. A set of axioms might be:

1. All gases are composed of many small molecules that are in a state of perpetual rapid motion.
2. At normal pressures, the volume of the molecules is extremely small compared to the total volume of the gas.
3. The molecules are spherical and no energy is lost when molecules collide.
4. The pressure exerted by the gas is the force-per-unit area due to the impact of the individual molecules on the wall of the containing vessel.
5. The temperature of a gas is directly proportional to the average kinetic energy of the individual molecules of the gas.

From these axioms one can deduce Boyle's Law and other experimental laws which are stated in purely observational terms ("temperature of the gas," "pressure of the gas") which can all be directly interpreted. But some of the primitive terms of the theory cannot be so interpreted because they refer to the nonobservable. Nevertheless, the physicists understand the whole theory because they know how to use it to derive empirical consequences.

Given this picture of the structure of a scientific theory, one can easily see its function. As in the case of an ordinary experimental law, one can deduce empirical consequences from a theory and so use the theory to explain data that have been observed and to predict the occurrence of future events.

Professor Hempel points out that some of the axioms play a central role in the interpretation of the theory. These are the axioms, like (4) and (5), that relate the observational and theoretical terms, thus enabling the axioms containing theoretical terms to have empirical consequences. For this reason, Hempel calls these axioms "interpretative sentences" (others have called them "correspondence rules").

As Hempel points out, this approach to the meaning of scientific terms is far less demanding than Professor P. W. Bridgman's "operationalist" approach according to which every descriptive term in science should be defined by a rule stating that the term applies in a given case if and only if the performance of a given operation yields a specific result. Operationalism runs into serious difficulties with theoretical terms (and also with dispositional terms) because the appropriate operations do not seem to exist. Therefore, the logical empiricists have modified the operationalist position to require merely that each term in science has some experimental import, even the weak one that appearance in a partially interpreted formal system affords.

In recent years, many philosophers have objected to part or all of the logical empiricist conception of theories, and the rest of this section is concerned with these objections. Perhaps the most fundamental of these is the very possibility of drawing an observational-theoretical distinction at all. Professor Grover Maxwell argues for example that the particular point at which one draws the distinction between the observational and the theoretical is a function of our physiological makeup, of our current state of knowledge, and of the instruments available. Since these latter two factors vary, the distinction will be drawn at different places at different times. Consequently, Maxwell argues, there can be no ontological significance to the observational-theoretical distinction, and the controversy over the reality of theoretical entities is therefore a pseudo-controversy.

Professor Hacking takes up the specific example of the microscope and counters Maxwell's position by elaborating on the nature of microscopes. Contrary to Maxwell's claim, the optical principles underlying all significant microscopes are quite different from those involved in the simple lenses used in eyeglasses or in magnifying glasses. Moreover, there are many different principles used in designing and constructing various types of microscopes. Thus, the argument that viewing through a microscope is a simple and direct extension of our normal mode of viewing is incorrect. However, this does not lead Hacking to be skeptical about the reality of what is seen with a microscope. He deploys two arguments for realism. The first is the high degree of agreement as to what is seen using microscopes based on quite distinct principles. The second is that by techniques of size reduction one can transform a normal visible grid with labels to a size at which it is invisible to the naked eye but can be seen in the same structural detail through microscopes.

Another line of criticism raised against the positivist account of theories is that by ignoring the role of models, it misrepresents the way in which scientists use and understand theories. For example, Professor N. R. Campbell argues that such models are necessary for the understanding of a theory and that they are more fundamental than the correspondence rules. In the case of the kinetic theory, the model would be the conception of a gas as like a collection of billiard balls, only smaller.

Professor R. B. Braithwaite countered this claim by trying to show that the empirical consequences of a theory depend only on the correspondence rules. Thus, the correspondence rules determine the entire empirical significance and the model is irrelevant. Professor Spector argues that some models, including the one mentioned for the kinetic theory, eventually become a standard part of the interpretation of the theory. This occurs when we discover an identity between the references of the observable terms in the theory and in the model. The whole theory is interpreted then, via the model, by extending the understanding of the theoretical terms as having the same reference in the theory as in the model.

Although most logical empiricists and some other philosophers agreed with Carnap's picture of the structure and meaning of a theory, there has been far less agreement about the scientist's commitment to the actual existence of theoretical entities. Does acceptance of the kinetic theory, for example, imply a commitment to the existence of unobservable molecules with the properties postulated by the theory? The view that it does is based upon the assumption that a correct scientific theory is a true description of some unobservable reality. This view is called "realism" while the alternative view is labelled "anti-realism"; the debate over this issue is the subject of section Ic.

Professor van Fraassen views the attempt to distinguish an observational component of science from a theoretical component as an important project that was carried out in the wrong way. He does not believe that we can make such a distinction on the basis of vocabulary, isolating observational vocabulary from the remainder on linguistic grounds. Instead he argues that we must make a distinction between observable and unobservable events, processes and objects based on our best current scientific theories. These theories, by virtue of informing us about the structure of the world and our sensory systems, will tell us what is observable and what must remain a matter of pure theory. Van Fraassen concludes that we have no reason to believe in the theoretical apparatus of a theory: if two theories are both empirically adequate, then there is no reason to worry about choosing between them.

Professor Cartwright offers a novel perspective regarding the debate about realism. She offers a number of examples in which several incompatible laws are offered as explanations of a single phenomenon. From this she concludes that we ought to be realistic not about the kinds of things mentioned in laws and theories but about the underlying causes of the phenomena to be explained. Her most detailed example involves the explanation of the fact that the lines in atomic spectra have an observable width. Her other examples—the radiometer (a set of vanes painted black on one side and white on the other side in a vacuum) and Brownian motion—both involve the explanation of observable effects in terms of molecular causes.

Professor Putnam rejects the positivist conception of theoretical terms according to which their meaning is given by their relation to the observational terms of the theory. On his approach the relation between the terms and the world are determined by the history of theorizing about that kind of thing. He is particularly concerned to reject the conclusions of Feyerabend and others that distinct theories have no point of contact because their expressions mean different things. Putnam rejects this conclusion, which he describes as "idealist," meaning that it construes the world as our creation, in favor of the realism according to which our theories may well talk about unobservable items even when the theories are radically wrong about their nature.

Professor Fine is dissatisfied with both realist and anti-realist positions. He is in agreement with the rejection of a Feyerabend position which regards truth as dependent upon theory; moreover, he finds van Fraassen's position unnatural, depending as it does on an expectation that science give an unequivocal and noncircular answer to what is observable but then *uniformly* suspend belief beyond that point. Fine's own "natural ontological attitude" does not, in his words, depend on interpreting science providing any philosophical interpretation of what science is doing. Whether this solves the problems or merely refuses to face them is a question we leave to the reader.

One apparent consequence of the positivist view of theories is that the meaning of theoretical terms is given via their relations in the axioms to observational terms. This leads to the conclusion that if the theory is changed, the meanings of the theoretical terms must also change, a conclusion that has seemed to many critics to create difficulties. Professor Paul Feyerabend has been one of the leading critics of the logical empiricist view that observational laws are deduced from theories. He claims that this presupposes that the laws and the theories never have conflicting empirical consequences and that the meaning of the terms in the laws does not change when the new theory is formulated. Feyerabend calls these two conditions the "consistency condition" and "the condition of meaning invariance," and he argues that neither is actually satisfied by scientific practice. He also argues that it would be unreasonable for the scientist to confine himself or herself to the formulation of theories that satisfy these conditions. Therefore, Feyerabend concludes that the logical empiricists were mistaken and that we should view new theories not as covering laws for earlier laws and theories but as attempts to replace these earlier laws with which they conflict.

One way of avoiding the problems which seem to follow from the change of meaning attendant on the change of theory is to argue that the reference of scientific terms does not change because it is determined by the causal history of the use of the term and by facts, perhaps as yet unknown, about the world. Professor Shapere discusses a version of this approach and argues that it is no better than the empiricist approach it is intended to supplant.

His main argument is that the causal history approach would commit scientists eventually to a dogmatic identification of a particular set of properties as the ultimate definition of a term, and thus would replace the apparent relativism of the earlier meaning change approach with an unacceptable dogmatism. His own solution to the problems, which he sketches at the end of his essay, is to argue that there is a continuity of reasons underlying the changes of meaning and that these continuities unite the apparently disparate discussions that occur over a period of time.

Professor Laudan argues against two dogmas that he believes have been widely accepted in discussions of theory change and theory evaluation. The first is that scientific progress requires that all problems solved by a previous theory must also be solved by any subsequent theory if we are to progress. The second is that to evaluate two rival theories we must be able to translate fully statements of the two theories into a neutral third language. Laudan argues that both of these dogmas are false; he cites numerous historical examples of theories which were accepted as successors and as constituting scientific progress even though they did not solve all of the problems solved by the theory they supplanted. With regard to evaluation, he points out that if we can simply count the number of problems solved by rival theories, the we would be able to compare their problem-solving abilities without either an exact matching of problems or a translation of the theories.

The Classical Approach

Rudolf Carnap

THEORIES AS PARTIALLY INTERPRETED FORMAL SYSTEMS

* * *

23. PHYSICAL CALCULI AND THEIR INTERPRETATIONS

The method described with respect to geometry can be applied likewise to any other part of physics: We can first construct a calculus and then lay down the interpretation intended in the form of semantical rules, yielding a physical theory as an interpreted system with factual content. The customary formulation of a physical calculus is such that it presupposes a logico—mathematical calculus as its basis, e.g., a calculus of real numbers in any of the forms discussed above (§18). To this basic calculus are added the specific primitive signs and the axioms, i.e., specific primitive sentences, of the physical calculus in question.

Thus, for instance, a calculus of mechanics of mass points can be constructed. Some predicates and functors (i.e., signs for functions) are taken as specific primitive signs, and the fundamental laws of mechanics as axioms. Then semantical rules are laid down stating that the primitive signs designate, say, the class of material particles, the three spatial coordinates of a particle x at the time t, the mass of a particle x, the class of forces acting on a particle x or at a space point s at the time t. (As we shall see later [§24], the interpretation can also be given indirectly, i.e., by semantical rules, not for the primitive signs, but for certain defined signs of the calculus. This procedure must be chosen if the semantical rules are to refer only to observable properties.) By the interpretation, the theorems of the calculus of mechanics become physical laws, i.e., universal statements describing certain features of events; they constitute physical mechanics as a theory with factual content which can be tested by observations. The relation of this theory to the calculus of mechanics is entirely analogous to the relation of physical

---

Reprinted by permission from Rudolf Carnap, *Foundations of Logic and Mathematics*, University of Chicago Press, 1939.

to mathematical geometry. The customary division into theoretical and experimental physics corresponds roughly to the distinction between calculus and interpreted system. The work in theoretical physics consists mainly in constructing calculi and carrying out deductions within them; this is essentially mathematical work. In experimental physics interpretations are made and theories are tested by experiments.

In order to show by an example how a deduction is carried out with the help of a physical calculus, we will discuss a calculus which can be interpreted as a theory of thermic expansion. To the primitive signs may belong the predicates "Sol" and "Fe", and the functors "lg", "te", and "th." Among the axioms may be A1 and A2. (Here, x, β and the letter with subscripts are real number variables; the parentheses do not contain explanations as in former examples, but are used as in algebra and for the arguments of functors.)

A1. For every $x, t_1, t_2, l_1, l_2, T_1, T_2$, β [if [$x$ is a Sol and lg $(x, t_1) = l_1$ and lg$(x, t_2) = l_{fi}2$ and te$(x, t_1) = T_1$ and te$(x, t_2) = T_2$ and th$(x) = \beta$] then $l_2 = l_1 \times (1 + \beta \times (T_2 - T_1))$].

A2. For every x, if [x is a Sol and x is a Fe] then th$(x) = 0.000012$.

The *customary interpretation*, i.e., that for whose sake the calculus is constructed, is given by the following semantical rules. lg(x,t) designates the length in centimeters of the body x at the time t (defined by the statement of a method of measurement); te(x, t) designates the absolute temperature in centigrades of x at the time t (likewise defined by a method of measurement); th(x) designates the coefficient of thermic expansion for the body x; Sol designates the class of solid bodies; Fe the class of iron bodies. By this interpretation, A1 and A2 become physical laws. A1 is the law of thermic expansion in quantitative form, A2 the statement of the coefficient of thermic expansion for iron. As A2 shows, a statement of a physical constant for a certain substance is also a universal sentence. Further, we add semantical rules for two signs occurring in the subsequent example: The name c designates the thing at such and such a place in our laboratory; the numerical variable t as time coordinate designates the time-point t seconds after August 17, 1938, 10:00 A.M.

Now we will analyze an example of a derivation within the calculus indicated. This derivation D_2 is, when interpreted by the rules mentioned, the deduction of a prediction from premises giving the results of observations. The construction of the derivation D_2 is however entirely independent of any interpretation. It makes use only of the rules of the calculus, namely, the physical calculus indicated together with a calculus of real numbers as basic calculus. We have discussed, but not written down, a similar derivation D_1 (§19), which, however, made use only of the mathematical calculus. Therefore the physical laws used had to be taken in D_1 as premises. But here in D_2 they belong to the axioms of the calculus (A1 and A2, occurring as [6] and [10]: Any axiom or theorem proved in a physical calculus may be used within any derivation in that calculus without belonging to the premisses of the derivation, in exactly the same way in which a proved theorem is used within a derivation in a logical or mathematical calculus, e.g., in the first example of a derivation in §19 sentence (7), and in D_1 (§19) the sentences which in D_2 are called (7) and (13). Therefore only singular sentences (not containing variables) occur as premisses in D_2. (For the distinction between premisses and axioms see the remark at the end of §19.)

Derivation D_2:

Premises

1. c is a Sol.
2. c is a Fe.
3. te$(c, 0) = 300$.
4. te$(c, 600) = 350$.
5. lg$(c, 0) = 1{,}000$.

Axiom A1	6. For every x, t_1, t_2, l_1, l_2, T_1, T_2, β [if [x is a Sol and lg(x, t_1) $= l_1$ and lg(x, t_2) $= l_2$ and te(x, t_1) $= T_1$ and te(x, t_2) $= T_2$ and th(x) $= \beta$] then $l_2 = l_1 \times (1 + \beta \times (T_2 - T_1))$].
Proved mathem. theorem:	7. For every l_1, l_2, T_1, T_2, β [$l_2 - l_1 = l_1 \times \beta \times (T_2 - T_1)$ if and only if $l_2 = l_1 \times (1 + \beta \times (T_2 - T_1))$].
(6)(7)	8. For every x, t_1, . . . (as in [6]) . . . [if [---] then $l_2 - l_1 = l_1 \times \beta \times (T_2 - T_1)$].
(1)(3)(4)(8)	9. For every l_1, l_2, β [if [th(c) $= \beta$ and lg(c, 0) $= l_1$ and lg(c, 600) $= l_2$] then $l_2 - l_1 \times \beta \times (350 - 300)$].
Axiom A2	10. For every x, if [x is a Sol and x is a Fe] then th(x) $=$ 0.000012.
(1)(2)(10)	11. th(c) $=$ 0.000012.
(9)(11)(5)	12. For every l_1, l_2, [if [lg(c, 0) $= l$, and lg(c, 600) $= l_2$] then $l_2 - l_1 = 1{,}000 \times 0.000012 \times (350 - 300)$].
Proved mathem. theorem:	13. $1{,}000 \times 0.000012 \times (350 - 300) = 0.6$.
(12)(13) *Conclusion:*	14. lg(c, 600) $-$ lg(c, 0) $=$ 0.6.

On the basis of the interpretation given before, the premisses are singular sentences concerning the body *c*. They say that *c* is a solid body made of iron, that the temperature of *c* was at 10:00 A.M. 300° abs. and at 10:10 A.M. 350° abs., and that the length of *c* at 10:00 A.M. was 1,000 cm. The conclusion says that the increase in the length of *c* from 10:00 to 10:10 A.M. is 0.6 cm. Let us suppose that our measurements have confirmed the premisses. Then the derivation yields the conclusion as a prediction which may be tested by another measurement.

Any physical theory, and likewise the whole of physics, can in this way be presented in the form of an interpreted system, consisting of a specific calculus (axiom system) and a system of semantical rules for its interpretation; the axiom system is, tacitly or explicitly, based upon a logico-mathematical calculus with customary interpretation. It is, of course, logically possible to apply the same method to any other branch of science as well. But practically the situtation is such that most of them seem at the present time to be not yet developed to a degree which would suggest this strict form of presentation. There is an interesting and successful attempt of an axiomatization of certain parts of biology, especially genetics, by Woodger (Vol. I, No. 10). Other scientific fields which may expect to be accessible soon to this method are perhaps chemistry, economics, and some elementary parts of psychology and social science.

Within a physical calculus the mathematical and the physical theorems, i.e., *C*-true formulas, are treated on a par. But there is a fundamental difference between the corresponding *mathematical* and the *physical propositions* of the physical theory, i.e., the system with customary interpretation. This difference is often overlooked. That physical theorems are sometimes mistaken to be of the same nature as mathematical theorems is perhaps due to several factors, among them the fact that they contain mathematical symbols and numerical expressions and that they are often formulated incompletely in the form of a mathematical equation (e.g., A1 simply in the form of the last equation occurring in it). A mathematical proposition may contain only logical signs, e.g., "for every m, n, $m + n = n + m$," or descriptive signs also, if the mathematical calculus is applied in a descriptive system. In the latter case the proposition, although it contains signs not belonging to the mathematical calculus, may still be provable in this calculus, e.g., lg(c) + lg(d) = lg(d) + lg(x) (lg designates length as before). A physical proposition always contains descriptive signs because otherwise it could not have factual content; in addition, it usually contains also logical signs. Thus the difference between mathematical theorems and physical theorems in the interpreted system does not depend upon the kinds of signs occurring but rather on the kind of truth of the theorems. The truth of a mathematical theorem, even if it contains descriptive signs, is not dependent upon any facts concerning the designata of these signs. We can determine its truth if we know only the semantical rules: Hence it is L-true. (In the example of the theorem just mentioned, we need not know the length of the body *c*). The truth of a physical theorem, on the other hand, depends upon the properties of the designata of the descriptive signs occuring. In order to determine its truth, we have to make observations concerning these designata; the knowledge of the semantical rules is not sufficient. (In the case of A2, e.g., we have to carry out experiments with solid iron bodies.) Therefore, a physical theorem, in contradistinction to a mathematical theorem, has factual content.

24. ELEMENTARY AND ABSTRACT TERMS

We find among the concepts of physics and likewise among those of the whole of empirical science—differences of abstractness. Some are more elementary than others, in the sense that we can apply them in concrete cases on the basis of observations in a more direct way than others. The others are more abstract; in order to find out whether they hold in a certain case, we have to carry out a more complex procedure, which however also rests finally on observations. Between quite elementary concepts and those of high abstraction there are many intermediate levels. We shall not try to give an exact definition for "degree of abstractness"; what is meant will become sufficiently clear by the following series of sets of concepts, proceeding from elementary to abstract concepts: bright, dark, red, blue, warm, cold, sour, sweet, hard, soft (all concepts of this first set are meant as properties of things, not as sense-data); coincidence; length; length of time; mass, velocity, acceleration, density, pressure; temperature, quantity of heat; electric charge, electric current, electric field; electric potential, electric resistance, coefficient of induction, frequency of oscillation; wave function.

Suppose that we intend to construct an interpreted system of physics—or of the whole of science. We shall first lay down a calculus. Then we have to state semantical rules of the kind SD for the specific signs, i.e., for the physical terms. (The SL-rules are presupposed as giving the customary interpretation of the logico-mathematical basic calculus.) Since the physical terms form a system, i.e., are connected with one another, obviously we need not state a semantical rule for each of them. For which terms, then, must we give rules, for the elementary or for the abstract ones? We can, of course, state a rule for any term, no matter what its degree of abstractness, in a form like this: "The term te designates temperature," provided the meta-language used contains a corresonding expression (here the word *temperature*) to specify the designatum of the term in question. But suppose we have in mind the following purpose for our syntactical and semantical description of the system of physics: The description of the system shall teach a layman to understand it, i.e., to enable him to apply it to his observations in order to arrive at explanations and predictions. A layman is meant as one who does not know physics but has normal senses and understands a language in which observable properties of things can be described (e.g., a suitable part of everyday nonscientific English). A rule like "the sign *P* designates the property of being blue" will do for the purpose indicated; but a rule like "the sign *Q* designates the property of being electrically charged" will not do. In order to fulfill the purpose, we have to give semantical rules for elementary terms only, connecting them with observable properties of things. For our further discussion we suppose the system to consist of rules of this kind, as indicated in the diagram on the next page.

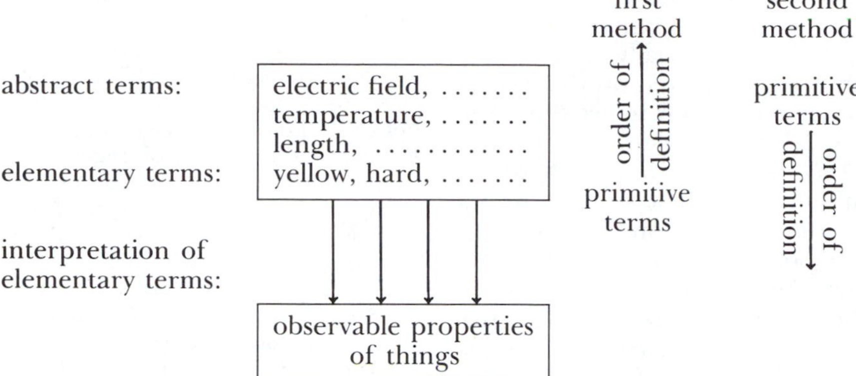

Now let us go back to the construction of the calculus. We have first to decide at which end of the series of terms to start the construction. Should we take elementary terms as primitive signs, or abstract terms? Our decision to lay down the semantical rules for the elementary terms does not decide this question. Either procedure is still possible and seems to have some reasons in its factor, depending on the point of view taken. The *first method* consists in taking elementary terms as primitive and then introducting on their basis further terms step by step, up to those of highest abstraction. In carrying out this procedure, we find that the introduction of further terms cannot always take the form of explicit definitions; conditional definitions must also be used (so-called reduction sentences[see Vol. I, No. 1, p. 50]). They describe a method of testing for a more abstract term, i.e., a procedure for finding out whether the term is applicable in particular cases, by referring to less abstract terms. The first method has the advantage of exhibiting clearly the connection between the system and observation and of making it easier to examine whether and how a given term is empirically founded. However, when we shift our attention from the terms of the system and the methods of empirical confirmation to the laws, i.e., the universal theorems, of the system, we get a different perspective. Would it be possible to formulate all laws of physics in elementary terms, admitting more abstract terms only as abbreviations? If so, we would have that ideal of a science in sensationalistic form which Goethe in his polemic against Newton, as well as some positivists, seems to have had in mind. But it turns out—this is an empirical fact, not a logical necessity—that it is not possible to arrive in this way at a powerful and efficacious system of laws. To be sure, historically, science started with laws formulated in terms of a low level of abstractness. But for any law of this kind, one nearly always later found some exceptions and thus had to confine it to a narrower realm of validity. The higher the physicists went in the scale of terms, the better did they succeed in formulating laws applying to a wide range of phenomena. Hence we understand that they are inclined to choose the *second method*. This method begins at the top of the system, so to speak, and then goes down to lower and lower levels. It consists in taking a few abstract terms as primitive signs and a few fundamental laws of great generality as axioms. Then further terms, less and less abstract, and finally elementary ones, are to be introduced by definitions; and here, so it seems at present, explicit definitions will do. More special laws, containing less abstract terms, are to be proved on the basis of the axioms. At least, this is the direction in which physicists have been striving with remarkable success, especially in the past few decades. But at the present time, the method cannot yet be carried through in the pure form indicated. For many less abstract terms no definition on the basis of abstract terms alone is as yet known; hence those terms must also be taken as primitive. And many more special laws, especially in biological fields, cannot yet be proved on the basis of laws in abstract terms only; hence those laws must also be taken as axioms.

Now let us examine the result of the interpretation if the first or the second method for the construction of the calculus is chosen. In both cases the semantical rules concern the elementary signs. In the first method these signs are taken as primitive. Hence, the semantical rules give a complete interpretation for these signs and those explicitly defined on their basis. There are, however, many signs, especially on the higher levels of abstraction, which can be introduced not by an explicit definition but only by a conditional one. The interpretation which the rules give for these signs is incomplete. This is due not to a defect in the semantical rules but to the method by which these signs are introduced; and this method is not arbitrary but corresponds to the way in which we really obtain knowledge about physical states by our observations.

If, on the other hand, abstract terms are taken as primitive—according to the second method, the one used in scientific physics—then the semantical rules have no direct relation to the primitive terms of the system but refer to terms introduced by long chains of definitions. The calculus is first constructed floating in the air, so to speak; the construction begins at the top and then adds lower and lower levels. Finally, by the semantical rules, the lowest level is anchored at the solid ground of the observable facts. The laws, whether general or special, are not directly interpreted, but only the singular sentences. For the more abstract terms, the rules determine only an *indirect interpretation*, which is—here as well as in the first method—in a certain sense incomplete. Suppose B is defined on the basis of A; then, if A is directly interpreted, B is, although indirectly, also interpreted completely; if, however, B is directly interpreted, A is not necessarily also interpreted completely (but only if A is also definable by B).

To give an example, let us imagine a calculus of physics constructed, according to the second method, on the basis of primitive specific signs like "electromagnetic field," "gravitational field," "electron," "proton," etc. The system of definitions will then lead to elementary terms, e.g., to Fe, defined as a class of regions in which the configuration of particles fulfills certain conditions, and Na-yellow as a class of space-time regions in which the temporal distribution of the electromagnetic fields fulfills certain conditions. Then semantical rules are laid down stating that Fe designates iron and Na-yellow designates a specified yellow color. (If "iron" is not accepted as sufficiently elementary, the rules can be stated for more elementary terms.) In this way, the connection between the calculus and the realm of nature to which it is to be applied is made for terms of the calculus which are far remote from the primitive terms.

Let us examine, on the basis of these discussions, the example of a derivation D_2 (§23). The premisses and the conclusion of D_2 are singular sentences, but most of the other sentences are not. Hence the premisses and the conclusion of this, as of all other derivations of the same type, can be directly interpreted, understood, and confronted with the results of observations. More of an interpretation is not necessary for a practical application of a derivation. If, in confronting the interpreted premisses with our observations, we find them confimed as true, then we accept the conclusion as a prediction and we may base a decision upon it. The sentences occurring in the derivation between premisses and conclusion are also interpreted, at least indirectly. But we need not make their interpretation explicit in order to be able to construct the derivation and to apply it. All that is necessary for its construction are the formal rules of the calculus. This is the advantage of the method of formalization, i.e., of the separation of the calculus as a formal system from the interpretation. If some persons want to come to an agreement about the formal correctness of a given derivation, they may leave aside all differences of opinion on material questions or questions of interpretation. They simply have to examine whether or not the given series of formulas fulfils the formal rules of the calculus. Here again, the function of calculi in empirical science becomes clear as instruments for transforming the expression of what we know or assume.

Against the view that for the application of a physical calculus we need an interpretation only for singular sentences, the following objection will perhaps be raised. Before we accept a derivation and believe its conclusion we must have accepted the physical calculus which furnishes the derivation and how can we decide whether or not to accept a physical calculus for application without interpreting and understanding its axioms? To be sure, in order to pass judgment about the applicability of a given physical calculus, we have to confront it in some way or other with observation, and for this purpose an interpretation is necessary. But we need no explicit interpretation of the axioms, nor even of any theorems. The empirical examination of a physical theory given in the form of a calculus with rules of interpretation is not made by interpreting and understanding the axioms and then considering whether they are true on the basis of our factual knowledge. Rather, the examination is carried out by the same procedure as that explained before for obtaining a prediction. We construct derivations in the calculus with premisses which are singular sentences describing the results of our observations, and with singular sentences which we can test by observations as conclusions. The physical theory is indirectly confirmed to a higher and higher degree if more and more of these predictions are confirmed and none of them is disconfirmed by observations. Only singular sentences with elementary terms can be directly tested; therefore, we need an explicit interpretation only for these sentences.

25. "UNDERSTANDING" IN PHYSICS

The development of physics in recent centuries, and especially in the past few decades, has more and more led to that method in the construction, testing, and application of physical theories which we call *formalization*, i.e., the construction of a calculus supplemented by an interpretation. It was the progress of knowledge and the particular structure of the subject matter that suggested and made practically possible this increasing formalization. In consequence, it became more and more possible to forego an "intuitive understanding" of the abstract terms and axioms and theorems formulated with their help. The possibility and even necessity of abandoning the search for an understanding of that kind was not realized for a long time. When abstract, nonintuitive formulas, as, e.g., Maxwell's equations of electromagnetism, were proposed as new axioms, physicists endeavored to make them "intuitive" by constructing a "model," i.e., a way of representing electromagnetic micro-processes by an analogy to known macro-processes, e.g., movements of visible things. Many attempts have been made in this direction, but without satisfactory results. It is important to realize that the discovery of a model has no more than an aesthetic or didactic or at best a heuristic value, but is not at all essential for a successful application of the physical theory. The demand for an intuitive understanding of the axioms was less and less fulfilled when the development led to the general theory of relativity and then to quantum mechanics, involving the wave function. Many people, including physicists, have a feeling of regret and disappointment about this. Some, especially philosophers, go so far as even to contend that these modern theories, since they are not intuitively understandable, are not at all theories about nature but "mere formalistic constructions," "mere calculi." But this is a fundamental misunderstanding of the function of a physical theory. It is true a theory must not be a "mere calculus" but possess an interpretation, on the basis of which it can be applied to facts of nature. But it is sufficient, as we have seen, to make this interpretation explicit for elementary terms; the interpretation of the other terms is then indirectly determined by the formulas of the calculus, either definitions or laws, connecting them with the elementary terms. If we demand from the modern physicist an answer to the question what he means by the symbol Ψ of his calculus, and are astonished that he cannot give an answer, we ought to realize that the situation was already the same in classical physics. There the physicist could not tell us what he meant by the symbol *E* in Maxwell's equations. Perhaps, in order not to refuse an answer, he would tell us that *E* designates the electric field vector. To be sure, this statement has the form of a semantical rule, but it would not help us a bit to understand the theory. It simply refers from a symbol in a symbolic calculus to a corresponding word expression in a calculus of words. We are right in demanding an interpretation for *E* but that will be given indirectly by semantical rules referring to elementary signs together with the formulas connecting them with *E*. This interpretation enables us to use the laws containing *E* for the derivation of predictions. Thus we understand *E*, if "understanding" of an expression, a sentence, or a theory means capability of its use for the description of known facts or the prediction of new facts. An "intuitive understanding" or a direct translation of *E* into terms referring to observable properties is neither necessary nor possible. The situation of the modern physicist is not essentially different. He knows how to use the symbol Ψ in the calculus in order to derive predictions which we can test by observations. (If they have the form of probability statements, they are tested by statistical results of observations.) Thus the physicist, although he cannot give us a translation into everyday language, understands the symbol Ψ and the laws of quantum mechanics. He possesses that kind of understanding which alone is essential in the field of knowledge and science.

Carl G. Hempel

A LOGICAL APPRAISAL OF OPERATIONISM

Operationism, in its fundamental tenets, is closely akin to logical empiricism. Both schools of thought have put much emphasis on definite experiential meaning or import as a necessary condition of objectively significant discourse, and both have made strong efforts to establish explicit criterions of experiential significance. But logical empiricism has treated experiential import as a characteristic of statements—namely, as their susceptibility to test by experiment or observation—whereas operationism has tended to construe experiential meaning as a characteristic of concepts or of the terms representing them—namely, as their susceptibility to operational definition.

BASIC IDEAS OF OPERATIONAL ANALYSIS

An operational definition of a term is conceived as a rule to the effect that the term is to apply to a particular case if the performance of specified operations in that case yields a certain characteristic result. For

Reprinted from *Scientific Monthly,* I, No. 79 (October 19, 1954), 215–20 by permission of the editor.

example, the term *harder than* might be operationally defined by the rule that a piece of mineral *x*, is to be called harder than another piece of mineral, *y*, if the operation of drawing a sharp point of *x* across the surface of *y* results in a scratch mark on the latter. Similarly, the different numerical values of a quantity such as length are thought of as operationally definable by reference to the outcomes of specified measuring operations. To safeguard the objectivity of science, all operations invoked in this kind of definition are required to be intersubjective in the sense that different observers must be able to perform "the same operation" with reasonable agreement in their results.[1]

P.W. Bridgman, the originator of operational analysis, distinguishes several kinds of operation that may be invoked in specifying the meanings of scientific terms.[2] The principal ones are (1) what he calls *instrumental operations*—these consist in the use of various devices of observation and measurement—and (2) paper-and-pencil operations, verbal operations, mental experiments, and the like—this group is meant to include, among other things, the techniques of mathematical and logical inference as well as the use of experiments in imagination. For brevity, but also by way of suggesting a fundamental similarity among the procedures of the second kind, I shall refer to them as *symbolic operations*.

The concepts of operation and of operational definition serve to state the basic principles of operational analysis, of which the following are of special importance:

(1) "Meanings are operational." To understand the meaning of a term, we must know the operational criterions of its application,[3] and every meaningful scientific term must therefore permit of an operational definition. Such definition may refer to certain symbolic operations and it always must ultimately make reference to some instrumental operation.[4]

(2) To avoid ambiguity, every scientific term should be defined by means of one unique operational criterion. Even when two different operational procedures (for instance, the optical and the tactual ways of measuring length) have been found to yield the same results, they must still be considered as defining different concepts (for example, optical and tactual length), and these should be distinguished terminologically because the presumed coincidence of the results is inferred from experimental evidence, and it is "not safe" to forget that the presumption may be shown to be spurious by new, and perhaps more precise, experimental data.[5]

(3) The insistence that scientific terms should have unambiguously specifiable operational meanings serves to insure the possibility of an objective test for the hypotheses formulated by means of those terms.[6] Hypotheses incapable of operational test or, rather, questions involving untestable formulations, are rejected as meaningless: "If a specific question has meaning, it must be possible to find operations by which an answer may be given to it. It will be found in many cases that the operations cannot exist, and the question therefore has no meaning."[7]

The emphasis on "operational meaning" in scientifically significant discourse has unquestionably afforded a salutary critique of certain types of procedure in philosophy and in empirical science and has provided a strong stimulus for methodological thinking. Yet, the central ideas of operational analysis as stated by their proponents are so vague that they constitute not a theory concerning the nature of scientific concepts but rather a program for the development of such a theory. They share this characteristic with the insistence of logical empiricism that all significant scientific statements must have experiential import, that the latter consists in testability by suitable data of direct observation, and that sentences which are entirely incapable of any test must be ruled out as meaningless "pseudo hypotheses." These ideas, too, constitute not so much a thesis or a theory as a program for a theory that needs to be formulated and amplified in precise terms.

An attempt to develop an operationist theory of scientific concepts will have to deal with a least two major issues: The problem of giving a more precise explication of the concept of operational definition; and the question whether operational definition in the explicated sense is indeed necessary for, and adequate to, the introduction of all non-observational terms in empirical science.

I wish to present here in brief outline some considerations that bear on these problems. The discussion will be limited to the descriptive, or extralogical, vocabulary of empirical science and will not deal, therefore, with Bridgman's ideas on the status of logic and mathematics.

A BROADENED CONCEPTION OF OPERATIONAL DEFINITION AND OF THE PROGRAM OF OPERATIONAL ANALYSIS

The terms "operational meaning" and "operational definition," as well as many of the pronouncements made in operationist writings, convey the suggestion that the criterions of application of any scientific term must ultimately refer to the outcome of some specified type of manipulation of the subject matter under investigation. Such emphasis would evidently be overly restrictive. An operational definition gives experiential meaning to the term it introduces because it enables us to decide on the applicability of that term to a given case by observing the response the case shows under specifiable test conditions. Whether these conditions can be brought about at will by "instrumental operations" or whether we have to wait for their occurrence is of great interest for the practice of scientific research, but it is inessential in securing experiental import for the defined term; what matters for this latter purpose is simply that the relevant test conditions and the requisite response be of such kind that different investigators can ascertain, by direct observation and with reasonably good agreement, whether, in a given case, the test conditions are realized and whether the characteristic response does occur.

Thus, an operational definition of the simplest kind—one that, roughly speaking, refers to instrumental operations only will have to be construed more broadly as introducing a term by the stipulation that it is to apply to all and only those cases which, under specified observable conditions *S*, show a characteristic observable response *R*.

However, an operational definition cannot be conceived as specifying that the term in question is to apply to a given case only if *S* and *R* actually occur in that case. Physical bodies, for example, are asserted to have masses, temperatures, charges, and so on, even at times when these magnitudes are not being measured. Hence, an operational definition of a concept—such as a property or a relationship, for example—will have to be understood as ascribing the concept to all those cases that *would* exhibit the characteristic response if the test conditions *should* be realized. A concept thus characterized is clearly not "synonymous with the corresponding set of operations."[8] It constitutes not a manifest but a potential character, namely, a disposition to exhibit a certain characteristic response under specified test conditions.

But to attribute a disposition of this kind to a case in which the specified test condition is not realized (for example, to attribute solubility-in-water to a lump of sugar that is not actually put into water) is to make a generalization, and this involves an inductive risk. Thus, the application of an operationally defined term to an instance of the kind here considered would have to be adjudged "not safe" in precisely the same sense in which Bridgman insists it is "not safe" to assume that two procedures of measurement that have yielded the same results in the past will continue to do so in the future. It is now clear that if we were to reject any procedure that involves an inductive risk, we would be prevented not only from using more than one operational criterion in introducing a given term but also from ever applying a disposition term to any case in which the characteristic manifest conditions of application are not realized;

thus, the use of dispositional concepts would, in effect be prohibited.

A few remarks might be added here concerning the non-instrumental operations countenanced for the introduction especially of theoretical terms. In operationist writings, those symbolic procedures have been characterized so vaguely as to permit the introduction, by a suitable choice of "verbal" or "mental" operations, of virtually all those ideas that operational analysis was to prohibit as devoid of meaning. To meet this difficulty, Bridgman has suggested a distinction between "good" and "bad" operations;[9] but he has not provided a clear criterion for this distinction. Consequently, this idea fails to plug the hole in the operationist dike.

If the principles of operationism are to admit the theoretical constructs of science but to rule out certain other kinds of terms as lacking experiential, or operational, meaning, then the vague requirement of definability by reference to instrumental and "good" symbolic operations must be replaced by a precise characterization of the kinds of sentences that may be used to introduce, or specify the meanings of, "meaningful" nonobservational terms on the basis of the observational vocabulary of science. Such a characterization would eliminate the psychologistic notion of mental operations in favor of a specification of the logicomathematical concepts and procedures to be permitted in the context of operational definition.

The reference just made to the observational vocabulary of science is essential to the idea of operational definition; for it is in terms of this vocabulary that the test conditions and the characteristic response specified in an operational definition are described and by means of which, therefore, the meanings of operationally defined terms are ultimately characterized. Hence, the intent of the original operationist insistence on intersubjective repeatability of the defining operations will be respected if we require that the terms included in the observational vocabulary must refer to attributes (properties and relationships) that are directly and publicly observable—that is, whose presence or absence can be ascertained, under suitable conditions, by direct observation, and with good agreement among different observers.[10]

In sum, then, a precise statement and elaboration of the basic tenets of operationism require an explication of the logical relationships between theoretical and observational terms, just as a precise statement and elaboration of the basic tenets of empiricism require an explication of the logical relationships connecting theoretical sentences with observation sentences describing potential data of direct observation.

SPECIFICATION OF MEANING BY EXPLICIT DEFINITION AND BY REDUCTION

Initially, it may appear plausible to assume that all theoetical terms used in science can be fully defined by means of the observational vocabulary. There are various reasons, however, to doubt this assumption.

First of all, there exists a difficulty concerning the definition of the scientific terms that refer to dispositions—and, as is noted in a foregoing paragraph, all the terms introduced by operational definition have to be viewed as dispositional in character. Recent logical studies strongly suggest that dispositions can be defined by reference to manifest characteristics, such as those presented by the observational vocabulary, only with the help of some "nomological modality" such as the concept of nomological truth, that is, truth by virtue of general laws of nature.[11] But a concept of this kind is presumably inadmissible under operationist standards, since it is neither a directly observable characteristic nor definable in terms of such characteristics.

Another difficulty arises when we attempt to give full definitions, in terms of observables, for quantitative terms such as "length in centimeters," "duration in seconds," "temperature in degrees Celsius." Within scientific theory, each of these is allowed to assume any real-number value within a certain interval; and the question therefore arises whether each of the infinitely many permissible values, say of length, is capable of an operational specification of meaning. It can be shown that it is impossible to characterize every one of the permissible numerical values by some truth-functional combination of observable characteristics, since the existence of a threshold of discrimination in all areas of observation allows for only a finite number of nonequivalent combinations of this kind.[12]

Difficulties such as these suggest the question whether it is not possible to conceive of methods more general and more flexible than definition for the introduction of scientific terms on the basis of the observational vocabulary. One such method has been developed in considerable detail by Carnap. It makes use of so-called reduction sentences, which constitute a considerably generalized version of definition sentences and are especially well suited for a precise reformulation of the intent of operational definitions. As we noted earlier, an operational definition of the simplest kind stipulates that the concept it introduces, say *C*, is to apply to those and only those cases which, under specified test conditions *S*, show a certain characteristic response *R*. In Carnap's treatment, this stipulation is replaced by the sentence

$$Sx \rightarrow (Cx \equiv Rx) \qquad (1)$$

or, in words: If a case *x* satisfies the test condition *S*, then *x* is an instance of *C* if and only if *x* shows the response *R*. Formula (1), called a bilateral reduction sentence, is not a full definition (which would have to be of the form *Cx* ≡ . . ., with *Cx* constituting the definiendum); it specifies the meaning of *Cx*, not for all cases, but only for those that satisfy the condition *S*. In this sense, it constitutes only a partial, or conditional, definition for *C*[13]. If *S* and *R* belong to the observational vocabulary of science, formula (1) schematizes the simplest type of operational definition, which invokes (almost) exclusively instrumental operations or, better, experiential findings. Operational definitions that also utilize symbolic operations would be represented by chains of reduction sentences containing logical or mathematical symbols. Some such symbols occur even in formula (1), however; and clearly, there can be no operational definition that makes use of no logical concepts at all.

INTERPRETATIVE SYSTEMS

Once the idea of a partial specification of meaning is granted, it appears unnecessarily restrictive, however, to limit the sentences effecting such partial interpretation to reduction sentences in Carnap's sense. A partial specification of the meanings of a set of nonobservational terms might be expressed, more generally, by one or more sentences that connect those terms with the observational vocabularly but do not have the form of reduction sentences. And it seems well to countenance, for the same purpose, even stipulations expressed by sentences containing only nonobservational terms; for example, the stipulation that two theoretical terms are to be mutually exclusive may be regarded as a limitation and, in this sense, a partial specification of their meanings.

Generally, then, a set of one or more theoretical terms, $t_1, t_2, \ldots, t_n$, might be introduced by any set *M* of sentences such that (i) *M* contains no extra logical terms other than $t_1, t_2, \ldots, t_n$, and observation terms, (ii) *M* is logically consistent, and (iii) *M* is not equivalent to a truth of formal logic. The last two of these conditions serve merely to exclude trivial extreme cases. A set *M* of this kind will be referred to briefly as an *interpretative system*, its elements as *interpretative sentences*.

Explicit definitions and reduction sentences are special types of interpretative sentences, and so are the meaning postulates recently suggested by Kemeny and Carnap.[14]

The interpretative sentences used in a given theory may be viewed simply as postulates of that theory,[15] with all the observation terms, as well as the terms introduced by the interpretative system, being treated as primitives. Thus construed, the specification of the meanings of nonobservational terms in science resembles what has sometimes been called the implicit definition of the primitives of an axiomatized theory by its postulates. In this latter procedure, the primitives are all uninterpreted, and the postulates then impose restrictions on any interpretation of the primitives that is to turn the postulates into true sentences. Such restrictions may be viewed as partial specifications of meaning. The use of interpretative systems as here envisaged has this distinctive peculiarity, however: The primitives include a set of terms—the observation terms—which are antecedently understood and thus not in need of any interpretation, and by reference to which the postulates effect a partial specification of meaning for the remaining, nonobservational, primitives. This partial specification again consists in limiting those interpretations of the nonobservational terms that will render the postulates true.

IMPLICATIONS FOR THE IDEA OF EXPERIENTIAL MEANING AND FOR THE DISTINCTION OF ANALYTIC AND SYNTHETIC SENTENCES IN SCIENCE

If the introduction of nonobservational terms is conceived in this broader fashion, which appears to accord with the needs of a formal reconstruction of the language of empirical science, then it becomes pointless to ask for the operational definition or the experiential import of any one theoretical term. Explicit definition by means of observables is no longer generally available, and experiential—or operational—meaning can be attributed only to the set of all the nonobservational terms functioning in a given theory.

Furthermore, there remains no satisfactory general way of dividing all conceivable systems of theoretical terms into two classes: Those that are scientifically significant and those that are not; those that have experiential import and those that lack it. Rather, experiential, or operational, significance appears as capable of gradations. To begin with one extreme possibility: The interpretative system *M* introducing the given terms may simply be a set of sentences in the form of explicit definitions that provide an observational equivalent for each of those terms. In this case, the terms introduced by *M* have maximal experiential significance, as it were. In another case, *M* might consist of reduction sentences for the theoretical terms; these will enable us to formulate, in terms of observables, a necessary and a (different) sufficient condition of application for each of the introduced terms. Again *M* might contain sentences in the form of definitions or reduction sentences for only some of the nonobservational terms it introduces. And finally, none of the sentences in *M* might have the form of a definition or of a reduction sentence; and yet, a theory whose terms are introduced by an interpretative system of this kind may well permit of test by observational findings, and in this sense, the system of its nonobservational terms may possess experiential import.[16]

Thus, experiential significance presents itself as capable of degrees, and any attempt to set up a dichotomy allowing only experientially meaningful and experientially meaningless concept systems appears as too crude to be adequate for a logical analysis of scientific concepts and theories.

The use of interpretative systems is a more inclusive method of introducing theoretical terms than the method of meaning postulates developed by Carnap and Kemeny. For although meaning postulates are conceived as analytic and, hence, as implying only analytic consequences, an interpretative system may imply certain sentences that contain observation terms but no theoretical terms and are neither formal truths of logic nor analytic in the customary sense. Consider, for example, the following two interpretative sentences, which form what Carnap calls a reduction pair, and which interpret C by means of observation predicates, R_1, S_1, R_2, S_2:

$$S_1x \rightarrow (R_1x \rightarrow Cx) \qquad (2.1)$$

$$S_2x \rightarrow (R_2x \rightarrow -Cx). \qquad (2.2)$$

Since in no case the sufficient conditions for C and for $-C$ (non-C) can be satisfied jointly, the two sentences imply the consequence[17] for every case x,

$$-(S_1x \cdot R_1x \cdot S_2x \cdot R_2x), \qquad (3)$$

that is, no case x exhibits the attributes S_1, R_1, S_2, R_2 jointly. Now, an assertion of this kind is not a truth of formal logic, nor can it generally be viewed as true solely by virtue of the meanings of its constituent terms. Carnap therefore treats this consequence of formulas (2.1) and (2.2) as empirical and as expressing the factual content of the reduction pair from which it was derived. Occurrences of this kind are by no means limited to reduction sentences, and we see that in the use of interpretative systems, specification of meaning and statement of empirical fact—two functions of language often considered as completely distinct—become so intimately bound up with each other as to raise serious doubt about the advisability or even the possibility of preserving that distinction in a logical reconstruction of science. This consideration suggests that we dispense with the distinction, so far maintained for expository purposes, between the interpretative sentences, included in M, and the balance of the sentences constituting a scientific theory: We may simply conceive of the two sets of sentences as constituting one "intepreted theory."

The results obtained in this brief analysis of the operationist view of significant scientific concepts are closely analogous to those obtainable by a similar study of the logical empiricist view of significant scientific statements, or hypotheses.[18] In the latter case, the original requirement of full verifiability or full falsifiability by experiential data has to give way to the more liberal demand for confirmability—that is, partial verifiability. This demand can be shown to be properly applicable to entire theoretical systems rather than to individual hypotheses—a point emphasized, in effect, already by Pierre Duhem. Experiential significance is then seen to be a matter of degree, so that the originally intended sharp distinction between cognitively meaningful and cognitively meaningless hypotheses (or systems of such) has to be abandoned; and it even appears doubtful whether the distinction between analytic and synthetic sentences can be effectively maintained in a formal model of the language of empirical science.

NOTES

1. P.W. Bridgman, "Some general principles of operational analysis" and "Rejoinders and second thoughts," *Psychol. Rev.*, LII (1945), 246; "The nature of some of our physical concepts," *Brit. J. Phil, Sci.*, I (1951), 258.

2. ——— "Operational analysis," *Phil, Sci.*, V (1938), 123; *Brit. J. Phil. Sci.*, I, (1951), 258.

3. ——— *Phil, Sci.*, V (1938), 116.

4. ——— *Brit. J. Phil. Sci.*, I (1951), 260.

5. ——— *The Logic of Modern Physics* (New York: Macmillan, 1927), pp. 6, 23-24; *Phil. Sci.*, V (1938), 121; *Psychol. Rev.*, LII (1945), 247; "The operational aspect of meaning," *Synthése*, VIII (1950-51), 255.

6. ——— *Psychol. Rev.*, LII (1945), 246.

7. ——— *The Logic of Modern Physics*, p. 28.

8. ——— *ibid.*, p. 5; qualified by Bridgman's reply [*Phil, Sci.*, V (1938), 117] to R.B. Lindsay, "A critique of operationalism in physics," *Phil. Sci.*, IV (1937), a qualification that was essentially on the ground, quite different from that given in the present paper, that operational meaning is only a necessary, but presumably not a sufficient characteristic of scientific concepts.

9. ——— *Phil. Sci.*, V (1938), 126; "Some

implications of recent points of view in physics," *Rev. intern. phil.*, III (1949), 484. The intended distinction between good and bad operations is further obscured by the fact that in Bridgman's discussion the meaning of "good operation" shifts from what might be described as "operation whose use in operational definition insures experimental meaning and testability" to "scientific procedure—in some very broad sense—which leads us to correct predictions."

10. The condition thus imposed upon the observational vocabulary of science is of a pragmatic character; it demands that each term included in that vocabulary be of such a kind that under suitable conditions, different observers can, by means of direct observation, arrive at a high degree of whether the term applies to a given situation. The expression *coincides with* as applicable to instrument needles and marks on scales of instruments is an example of a term meeting this condition. That human beings are capable of developing observational vocabularies that satisfy the given requirement is a fortunate circumstance: without it, science as an intersubjective enterprise would be impossible.

11. To illustrate briefly, it seems reasonable, prima facie, to define "x is soluble in water" by "if x is put in water then x dissolves." But if the phrase *if . . . then . . .* is here construed as the truth-functional, or "material," conditional, then the objects qualified as soluble by the definition include, among others, all those things that are never put in water—no matter whether or not they are actually soluble in water. This consequence—one aspect of the "paradoxes of material implication"—can be avoided only if the aforementioned definiens is construed in a more restrictive fashion. The idea suggests itself to construe "x is soluble in water" as short for "by virtue of some general laws of nature, x dissolves if x is put in water," or briefly, "it is nomologically true that if x is put in water then x dissolves." The phrase if . . . then . . . may now be understood in the truth-functional sense again. However, the acceptability of this analysis depends, of course, upon whether nomological truth can be considered as a sufficiently clear concept. For a fuller discussion of this problem complex, see especially R. Carnap, "Testability and meaning," *Phil. Sci.*, III (1936) and IV (1937) and N. Goodman, "The problem of counterfactual conditionals," *J. Phil.*, XLIV (1947).

12. In other words, it is not possible to provide, for every theoretically permissible value r of the length $l(x)$ of a rod x, a definition of the form

$$[l(x) = r] =_{df} C(P_1 x, P_2 x, \ldots, P_n x),$$

where $P_1, P_2 \ldots P_n$ are observable characteristics, and the definiens is an expression formed from $P_1 x, P_2 x, \ldots P_n x$ with help of the connective words *and, or*, and *not* alone.

It is worth noting, however, that if the logical constants allowed in the definiens include, in addition to truth-functional connectives, also quantifiers and the identity sign, then a finite observational vocabulary may permit the explicit definition of a denumerable infinity of further terms. For instance, if "*x* spatially contains *y*" and "*y* is an apple" are included in the observational vocabulary, then it is possible to define the expressions "*x* contains 0 apples," "*x* contains exactly 1 apple," "*x* contains exactly 2 apples," and so forth, in a matter familiar from the Frege-Russel construction of arithmetic out of logic. Yet even if definitions of this type are countenanced—and no doubt they are in accord with the intent of operationist analysis—there remain serious obstacles for an operationist account of the totality of real numbers which are permitted as theoretical values of length, mass, and so forth. On this point, see C.G. Hempel, *Fundamentals of Concept Formation in Empirical Science* (Univ. of Chicago Press, Chicago, 1952), Sec. 7 Gustav Bergman, in his contribution to the present symposium, deplores this argument—although he agrees with its point—on the ground that it focuses attention on a characteristic shared by all quantitative concepts instead of bringing out the differences between, say, length and the psi-function. He thinks this regrettable because, after all, as he puts it, "the real numbers are merely a part of the logical apparatus; concept formation is a matter of the descriptive vocabulary." I cannot accept the suggestion conveyed by this statement. To be sure, the theory of real numbers can be developed as a branch (or as an extension) of logic; however, my argument concerns not the definability of real numbers in logical terms, but the possibility of formulating an observational equivalent for each of the infinitely many permissible real-number values of length, temperature, and so forth. And this is clearly a question concerning the descriptive vocabulary rather than merely the logical apparatus of empirical science. I quite agree with Bergmann, however, that it would be of considerable interest to explicate whatever logical differences may obtain between quantitative concepts which, intuitively speaking, exhibit different degrees of theoretical abstractness, such as length on the one hand and the psi-function on the other.

13. The use of reduction sentences circumvents one of the difficulties encountered in the the attempt to give explicit and, thus, complete definitions of disposition terms: The conditional and biconditional signs occuring in formula (1) may be construed truth-functionally without giving rise to undesirable consequences of the kind characterized in n. 11. For details, see R. Carnap, "Testability and meaning," *Phil. Sci.* (1936-37), Part II; also C. G. Hempel, *Fundamentals of Concept Formation in Empirical Science*, Secs. 6 and 8. Incidentally, the use of nomological concepts is not entirely avoided in Carnap's procedure; the reduction sentences that are permitted for the introduction of new terms are required to satisfy certain conditions of logical or of nomological validity. See R. Carnap. *Phil. Sci.*, III and IV (1936-37), 442-43.

14. J. G. Kemeny, "Extension of the methods of inductive logic," *Philosophical Studies*, III (1952); R. Carnap, "Meaning postulates," *ibid*, III (1952).

15. For the case of Carnap's reduction sentences, the postulational interpretation was suggested to me by N. Goodman and by A. Church.

16. This is illustrated by the following simple model case: The theory *T* consists of the sentence $(x)((C_1x \cdot C_2x) \rightarrow C_3x)$ and its logical consequences; the three "theoretical" terms occuring in it are introduced by the interpretative set *M* consisting of the sentences $0_1x \rightarrow (C_1x \cdot C_2x)$ and $(C_1x \cdot C_3x) \rightarrow (0_2x v 0_3x)$, where 0_1, 0_2, 0_3, belong to the observational vocabulary. As is readily seen, *T* permits, by virtue of *M*, the "prediction" that if an object has the observable property 0_1 but lacks the observable property 0_2, then it will have the observable property 0_3. Thus *T* is susceptible to experiental test, although *M* provides for none of its consitutent terms both a necessary and a sufficient observational, or operational criterion of application.

17. Carnap calls it the representative sentence of the pair of formulas (2.1) and (2.2). See R. Carnap, *Phil. Sci.*, III and IV (1936-37), pp. 444 and 451. Generally, when a term is introduced by several reduction sentences representing different operational criterions of application, then the agreement among the results of the corresponding procedures, which must be presupposed if the reduction sentences are all to be compatible with one another, is expressed by the representative sentence associated with the given set of reduction sentences. The representative sentence reflects, therefore, the inductive risk which, as Bridgman has stressed, is incurred by using more than one operational criterion for a given term.

18. C.G. Hempel, "Problems and changes in the empiricist criterion of meaning," *Rev. intern. phil.*, IV (1951), and the "The concept of cognitive significance: a reconsideration," *Proc. Am. Acad. Arts Sci.*, LXXX (1951); W. V. Quine, "Two dogmas of empiricism," *Phil. Rev.* XL (1951).

The Observational–Theoretical Distinction

Grover Maxwell

THE ONTOLOGICAL STATUS OF THEORETICAL ENTITIES

That anyone today should seriously contend that the entities referred to by scientific theories are only convenient fictions, or that talk about such entities is translatable without remainder into talk about sense contents or everyday physical objects, or that such talk should be regarded as belonging to a mere calculating device and, thus, without cognitive content—such contentions strike me as so incongruous with the scientific and rational attitude and practice that I feel this paper *should* turn out to be a demolition of straw men. But the instrumentalist views of outstanding physicists such as Bohr and Heisenberg are too well known to be cited, and in a recent book of great competence, Professor Ernest Nagel concludes that "the opposition between [the realist and the instrumentalist] views [of theories] is a conflict over preferred modes of speech" and "the question as to which of them is the 'correct position' has only terminological interest."[1] The phoenix, it seems, will not be laid to rest.

The literature on the subject is, of course, voluminous, and a comprehensive treatment of the problem is far beyond the scope of one essay. I shall limit myself to a small number of constructive arguments (for a radically realistic interpretation of theories) and to a critical examination of some of the more crucial assumptions (sometimes tacit, sometimes explicit) that seem to have generated most of the problems in this area.[2]

From Grover Maxwell, "The Ontological Status of Theoretical Entities", *Minnesota Studies in the Philosophy of Science*, Vol. III. Herbert Feigl and Grover Maxwell, eds., University of Minnesota Press, Minneapolis.

THE PROBLEM

Although this essay is not comprehensive, it aspires to be fairly self-contained. Let me, therefore, give a pseudohistorical introduction to the problem with a piece of science fiction (or fictional science).

In the days before the advent of microscopes, there lived a Pasteur-like scientist whom, following the usual custom, I shall call Jones. Reflecting on the fact that certain diseases seemed to be transmitted from one person to another by means of bodily contact or by contact with articles handled previously by an afflicted person, Jones began to speculate about the mechanism of the transmission. As a "heuristic crutch," he recalled that there is an obvious *observable* mechanism for transmission of certain afflictions (such as body lice), and he postulated that all, or most, infectious diseases were spread in a similar manner but that in most cases the corresponding "bugs" were too small to be seen and, possibly, that some of them lived inside the bodies of their hosts. Jones proceeded to develop his theory and to examine its testable consequences. Some of these seemed to be of great importance for preventing the spread of disease.

After years of struggle with incredulous recalcitrance, Jones managed to get some of his preventative measures adopted. Contact with or proximity to diseased persons was avoided when possible, and articles which they handled were "disinfected" (a word coined by Jones) either by means of high temperatures or by treating them with certain toxic preparations which Jones termed "disinfectants." The results were spectacular: Within ten years the death rate had declined 40 percent. Jones and his theory received their well-deserved recognition.

However, the "crobes" (the theoretical term coined by Jones to refer to the disease-producing organisms) aroused considerable anxiety among many of the philosophers and philosophically inclined scientists of the day. The expression of this anxiety usually began something like this: "In order to account for the facts, Jones must assume that his crobes are too small to be seen. Thus the very postulates of his theory preclude their being observed; they are *unobservable in principle.*" (Recall that no one had envisaged such a thing as a microscope.) This common prefatory remark was then followed by a number of different "analyses" and "interpretations" of Jones's theory. According to one of these, the tiny organisms were merely convenient fictions—*façons de parler*—extremely useful as heuristic devices for facilitating (in the "context of discovery") the thinking of scientists but not to be taken seriously in the sphere of cognitive knowledge (in the "context of justification"). A closely related view was that Jones's theory was merely an instrument, useful for organizing observation statements and (thus) for producing desired results, and that, therefore, it made no more sense to ask what was the nature of the entities to which it referred than it did to ask what was the nature of the entities to which a hammer or any other tool referred.[3] "Yes," a philosopher might have said, "Jones's theoretical expressions are just meaningless sounds or marks on paper which, when correlated with observation sentences by appropriate syntactical rules, enable us to predict successfully and otherwise organize data in a convenient fashion." These philosophers called themselves "instrumentalists."

According to another view (which, however, soon became unfashionable), although expressions containing Jones's theoretical terms were genuine sentences, they were translatable without remainder into a set (perhaps infinite) of observation sentences. For example, "There are crobes of disease *X* on this article" was said to translate into something like this: "If a person handles this article without taking certain precautions, he will (probably) contract disease *X;* and if this article is first raised to a high temperature, then if a person handles it at any time afterward, before it comes into contact with another person with disease *X,* he will (probably) not contract disease *X;* and. . . ."

Now virtually all who held any of the views so far noted granted, even insisted, that theories played a useful and legitimate role in the scientific enterprise. Their concern was the elimination of "pseudo problems" which might arise, say, when one began wondering about the "reality of supraempirical entities," etc. However, there was also a school of thought, founded by a psychologist named Pelter, which differed in an interesting manner from such positions as these. Its members held that while Jones's crobes might very well exist and enjoy "full-blown reality," they should not be the concern of medical research at all. They insisted that if Jones had employed the correct methodology, he would have discovered, even sooner and with much less effort, all of the observation laws relating to disease contraction, transmission, etc. without introducting superfluous links (the crobes) into the causal chain.

Now, lest any reader find himself waxing impatient, let me hasten to emphasize that this crude parody is not intended to convince anyone, or even to cast serious doubt upon sophisticated varieties of any of the reductionistic positions caricatured (some of them not too severely, I would contend) above. I am well aware that there are theoretical entities and theoretical entities, some of whose conceptual and theoretical statuses differ in important respects from Jones's crobes. (I shall discuss some of these later.) Allow me, then, to bring the Jonesean prelude to our examination of observability to a hasty conclusion.

Now Jones had the good fortune to live to see the invention of the compound microscope. His crobes were "observed" in great detail, and it became possible to identify the specific kind of *microbe* (for so they began to be called) which was responsible for each different disease. Some philosophers freely admitted error and were converted to realist positions concerning theories. Others resorted to subjective idealism or to a thoroughgoing phenomenalism, of which there were two principal varieties. According to one, the one "legitimate" observation language had for its descriptive terms only those which referred to sense data. The other maintained the stronger thesis that *all* "factual" statements were *translatable* without remainder into the sense-datum language. In either case, any two non-sense data (e.g., a theoretical entity and what would ordinarily be called an "observable physical object") had virtually the same status. Others contrived means of modifying their views much less drastically. One group maintained that Jones's crobes actually never had been unobservable in principle, for, they said, the theory did not imply the impossibility of finding a mean (e.g., the microscope) of observing them. A more radical contention was that the crobes were not observed at all; it was argued that what was seen by means of the microscope was just a shadow or an image rather than a corporeal organism.

THE OBSERVATIONAL-THEORETICAL DICHOTOMY

Let us turn from these fictional philosophical positions and consider some of the actual ones to which they roughly correspond. Taking the last one first, it is interesting to note the following passage from Bergmann: "But it is only fair to point out that if this . . . methodological and terminological analysis [for the thesis that there are no atoms] . . . is strictly adhered to, even stars and microscopic objects are not physical things in a literal sense, but merely by courtesy of language and pictorial imagination. This might seem awkward. But when I look through a microscope, all I see is a patch of color which creeps through the field like a shadow over a wall. And a shadow, though real, is certainly not a physical thing."[4]

I should like to point out that it is also the case that if this analysis is strictly adhered to, we cannot observe physical things through opera glasses, or even through ordinary spectacles, and one begins to wonder about the status of what we see through an ordinary windowpane. And what about distortions due to temperature gradients—however small and, thus, always present—in the ambient air? It really *does* "seem awkward" to say that when people who wear glasses describe what they see they are talking about shadows, while those who employ unaided vision talk about physical things—or that when we look through a windowpane, we can only *infer* that it is raining, while if we raise the window, we may "observe directly" that it is. The point I am making is that there is, in principle, a continuous series beginning with looking through a vacuum and containing these as members: looking through a windowpane, looking through glasses, looking through binoculars, looking through a low-power microscope, looking through a high-power microscope, etc., in the order given. The important consequence is that, so far, we are left without criteria which would enable us to draw a nonarbitary line between "observation" and "theory." Certainly, we will often find it convenient to draw such a to-some-extent-arbitrary line; but its position will vary widely from context to context. (For example, if we are determining the resolving characteristics of a certain microscope, we would certainly draw the line beyond ordinary spectacles, probably beyond simple magnifying glasses, and possibly beyond another microscope with a lower power of resolution.) But what ontological ice does a mere methodologically convenient observational-theoretical dichotomy cut? Does an entity attain physical thinghood and/or "real existence" in one context only to lose it in another? Or, we may ask, recalling the continuity from observable to unobservable, is what is seen through spectacles a "little bit less real" or does it "exist to a slightly less extent" than what is observed by unaided vision?[5]

However, it might be argued that things seen through spectacles and binoculars look like ordinary physical objects, while those seen through microscopes and telescopes look like shadows and patches of light. I can only reply that this does not seem to me to be the case, particularly when looking at the moon, or even Saturn, through a telescope or when looking at a small, though "directly observable," physical object through a low-power microscope. Thus, again, a continuity appears.

"But," it might be objected, "theory tells us that what we see by means of a microscope is a real image, which is certainly distinct from the object on the stage." Now first of all, it should be remarked that it seems odd that one who is espousing an austere empiricism which requires a sharp observational-language/theoretical-language distinction (and one in which the former language has a privileged status) should need a theory in order to tell him what is observable. But, letting this pass, what is to prevent us from saying that we still observe the object on the stage, even though a "real image" may be involved? Otherwise, we shall be strongly tempted by phenomenalistic demons, and at this point we are considering a physical-object observation language rather than a sense-datum one. (Compare the traditional puzzles: Do I see one physical object or two when I punch my eyeball? Does one object split into two? Or do I see one object and one image? Etc.)

Another argument for the continuous transition from the observable to the unobservable (theoretical) may be adduced from theoretical considerations themselves. For example, contemporary valency theory tells us that there is a virtually continuous transition from very small molecules (such as those of hydrogen) through "medium-sized" ones (such as those of the fatty acids, polypeptides, proteins, and viruses) to extremely large ones (such as crystals of the salts, diamonds, and lumps of polymeric plastic). The molecules in the last-mentioned group are macro, "directly observable" physical objects but are, nevertheless, genuine, single molecules; on the other hand, those in the first mentioned group have the same perplexing properties as subatomic particles (de Broglie waves, Heisenberg indeterminacy, etc.). Are we to say that a large protein molecule (e.g., a virus) which can be "seen" only with an electron microscope is a little less real or exists to somewhat less an extent than does a molecule of a polymer which can be seen with an optical microscope? And does a hydrogen molecule partake of only an infinitesimal portion of existence or reality? Although there certainly is a continuous transition from observability to unobservability, any talk of such a continuity from full-blown existence to nonexistence is, clearly, nonsense.

Let us now consider the next to last modified position which was adopted by our fictional philosophers. According to them, it is only those entities which are *in principle* impossible to observe that present special problems. What kind of impossibility is meant here? Without going into a detailed discussion of the various types of impossibility, about which there is abundant literature with which the reader is no doubt familiar, I shall assume what usually seems to be granted by most philosophers who talk of entities which are unobservable in principle, i.e., that the theory(s) itself (coupled with a physiological theory of perception, I would add) entails that such entities are unobservable.

We should immediately note that if this

analysis of the notion of unobservability (and, hence, of observability) is accepted, then its use as a means of delimiting the observation language seems to be precluded for those philosophers who regard theoretical expressions as elements of a calculating device—as meaningless strings of symbols. For suppose they wished to determine whether or not "electron" was a theoretical term. First, they must see whether the theory entails the sentence "Electrons are unobservable." So far, so good, for their calculating devices are said to be able to select genuine sentences, provided they contain no theoretical terms. But what about the selected "sentence" itself? Suppose that "electron" is an observation term. It follows that the expression is a genuine sentence and asserts that electrons are unobservable. But this entails that "electron" is not an observation term. Thus, if "electron" is an observation term, then it is *not* an observation term. Therefore, it is not an observation term. But then it follows that "Electrons are unobservable" is not a genuine sentence and does not assert that electrons are unobservable, since it is a meaningless string of marks and does not assert anything whatever. Of course, it could be stipulated that when a theory "selects" a meaningless expression of the form "*X*'s are unobservable," then *X* is to be taken as a theoretical term. But this seems rather arbitrary.

But, assuming that well-formed theoretical expressions are genuine sentences, what shall we say about unobservability in principle? I shall begin by putting my head on the block and argue that the present-day status of, say, electrons is in many ways similar to that of Jones's crobes before microscopes were invented. I am well aware of the numerous theoretical arguments for the impossibility of observing electrons. But suppose new entities are discovered which interact with electrons in such a mild manner that if an electron is, say, in an eigenstate of position, then, in certain circumstances, the interaction does not disturb it. Suppose also that a drug is discovered which vastly alters the human perceptual apparatus—perhaps even activates latent capacities so that a new sense modality emerges. Finally, suppose that in our altered state we are able to perceive (not necessarily visually) by means of these new entities in a manner roughly analogous to that by which we now see by means of photons. To make this a little more plausible, suppose that the energy eigenstates of the electrons in some of the compounds present in the relevant perceptual organ are such that even the weak interaction with the new entities alters them and also that the cross sections, relative to the new entities, of the electrons and other particles of the gases of the air are so small that the chance of any interaction here is negligible. Then we might be able to "observe directly" the position and possibly the approximate diameter and other properties of some electrons. It would follow, of course, that quantum theory would have to be altered in some respects, since the new entities do not conform to all its principles. But however improbable this may be, it does not, I maintain, involve any logical or conceptual absurdity. Furthermore, the modification necessary for the inclusion of the new entities would not necessarily change the meaning of the term *electron*.[6]

Consider a somewhat less fantastic example, and one which does not involve any change in physical theory. Suppose a human mutant is born who is able to "observe" ultraviolet radiation, or even X rays, in the same way we "observe" visible light.

Now, I think that it is extremely improbable that we will ever observe electrons directly (i.e., that it will ever be reasonable to assert that we have so observed them). But this is neither here nor there; it is not the purpose of this essay to predict the future development of scientific theories, and hence, it is not its business to decide what actually is observable or what will become observable (in the more or less intuitive sense of "observable" with which we are now working). After all, we are operating, here, under the assumption that it is theory, and thus science itself, which tells us what is or is not, in this sense, observable (the "in principle" seems to have become superflous). And this is the heart of the matter; for it follows that, at least for this sense of "observable," there are no a priori or philosophical criteria for separating the observable from the unobservable. By trying to show that we can talk about the *possibility* of observing electrons without committing logical or conceptual blunders, I have been trying to support the thesis that any (nonlogical) term is a *possible* candidate for an observation term.

There is another line which may be taken in regard to delimitation of the observation language. According to it, the proper term with which to work is not "observ*able*" but, rather, "observ*ed*." There immediately comes to mind the tradition beginning with Locke and Hume (No idea without a preceding impression!), running through Logical Atomism and the Principle of Acquaintance, and ending (perhaps) in contemporary positivism. Since the numerous facets of this tradition have been extensively examined and criticized in the literature, I shall limit myself here to a few summary remarks.

Again, let us consider at this point only observation languages which contain ordinary physical-object terms (along with observation predicates, etc., of course). Now, according to this view, all descriptive terms of the observation language must refer to that which has been observed. How is this to be interpreted? Not too narrowly, presumably, otherwise each language user would have a different observation language. The name of my Aunt Mamie, of California, whom I have never seen, would not be in my observation language, nor would "snow" be an observation term for many Floridians. One could, of course, set off the observation language by means of this awkward restriction, but then, obviously, not being the referent of an observation term would have no bearing on the ontological status of Aunt Mamie or that of snow.

Perhaps it is intended that the referents of observation terms must be members of a *kind,* some of whose members have been observed, or instances of a *property,* some of whose instances have been observed. But there are familiar difficulties here. For example, given any entity, we can always find a kind whose only member is the entity in question; and surely expressions such as "men over 14 feet tall" should be counted as observational even though no instances of the "property" of being a man over 14 feet tall have been observed. It would seem that this approach must soon fall back upon some notion of simples or determinables vs. determinates. But is it thereby saved? If it is held that only those terms which refer to observed simples or observed determinates are observation terms, we need only remind ourselves of such instances as Hume's notorious missing shade of blue. And if it is contended that in order to be an observation term an expression must at least refer to an observed determinable, then we can always find such a determinable which is broad enough in scope to embrace any entity whatever. But even if these difficulties can be circumvented, we see (as we knew all along) that this approach leads inevitably into phenomenalism, which is a view with which we have not been concerning ourselves.

Now it is not the purpose of this essay to give a detailed critique of phenomenalism. For the most part, I simply assume that it is untenable, at least in any of its translatability varieties.[7] However, if there are any unreconstructed phenomenalists among the readers, my purpose, insofar as they are concerned, will have been largely achieved if they will grant what I suppose most of them would stoutly maintain anyway, i.e., that theoretical entities are no worse off than so-called observable physical objects.

Nevertheless, a few considerations concerning phenomenalism and related matters may cast some light upon the observational-theoretical dichotomy and, perhaps, upon the nature of the "observation language." As a preface, allow me some overdue remarks on the latter. Although I have contended that the line between the observable and the unobservable is diffuse, that it shifts from one scientific problem to another and that it is constantly being pushed toward the "unobservable" end of the spectrum as we develop better means of observation—better instruments—it would, nevertheless, be fatuous to minimize the importance of the observation base, for it is absolutely necessary as a confirmation base for statements which do refer to entities which are unobservable at a given time. But we should take as its basis and its unit not the "observational term" but, rather, the quickly decidable sentence. (I am indebted to Feyerabend, *loc. cit.,* for this terminology.) A quickly decidable sentence (in the technical sense employed here) may be defined as a singular, nonanalytic sentence such that a reliable, reasonably sophisticated language user can very quickly decide[8] whether to assert it or deny it when he is reporting on an occurrent situation. *Observation term* may now be defined as a "descriptive (non-logical) term which may occur in a quickly decidable sentence," and *observation sentence* as a "sentence whose only descriptive terms are observation terms."

Returning to phenomenalism, let me emphasize that I am not among those philosophers who hold that there are no such things as sense contents (even sense data), nor do I believe that they play no important role in our perception of "reality." But the fact remains that the referents of most (not all) of the statements of the linguistic framework used in everyday life and in science are not sense contents but, rather, physical objects and other publicly observable entities. Except for pains, odors, "inner states," etc., *we do not usually observe sense contents;* and although there is good reason to believe that they play an indispensable role in observation, *we are usually not aware of them when we* (visually or tactilely) *observe physical objects.* For example, when I observe a distorted, obliquely reflected image in a mirror, I may seem to be seeing a baby elephant standing on its head; later I discover it is an image of Uncle Charles taking a nap with his mouth open and his hand in a peculiar position. Or, passing my neighbor's home at a high rate of speed, I observe that he is washing a car. If asked to report these observations I could quickly and easily report a baby elephant and a washing of a car; I probably would not, without subsequent observations, be able to report what colors, shapes, etc. (i.e., what sense data), were involved.

Two questions naturally arise at this point: How is it that we can (sometimes) quickly decide the truth or falsity of a pertinent observation sentence? and, What role do sense contents play in the appropriate tokening of such sentences? The heart of the matter is that these are primarily scientific-theoretical questions rather than "purely logical," "purely conceptual," or "purely epistemological." If theoretical physics, psychology, neurophysiology, etc., were sufficiently advanced, we could give satisfactory answers to these questions, using, in all likelihood, the physical-thing language as our observation language and *treating sensations, sense contents, sense data, and "inner states" as theoretical* (yes, theoretical!) *entities.*[9]

It is interesting and important to note that, even before we give completely satisfactory answers to the two questions considered above, we can, with due effort and reflection, train ourselves to "observe directly" what were once theoretical entities—the sense contents (color sensations, etc.)—involved in our perception of physical things. As has been pointed out before, we can also come to observe other kinds of entities which were once theoretical. Those which most readily come to mind involve the use of instruments as aids to observation. Indeed, using our painfully acquired theoretical knowledge of the world, we come to see that we "directly observe" many kinds of so-called theoretical things. After listening to a dull speech while sitting on a hard bench, we begin to become poignantly aware of the presence of a considerably strong gravitational field, and as Professor Feyerabend is fond of pointing out, if we were carrying a heavy suitcase in a changing gravitational field, we could observe the changes of the $G\mu\nu$ of the metric tensor.

I conclude that our drawing of the observational-theoretical line at any given point is an accident and a function of our physiological make-up, our current state of knowledge, and the instruments we happen to have available and, therefore, that it has no ontological significance whatever.

* * *

NOTES

1. E. Nagel, *The Structure of Science* (New York: Harcourt, Brace, and World, 1961), Chap. vi.

2. For the genesis and part of the content of some of the ideas expressed herein, I am indebted to a number of sources; some of the more influential are H. Feigl, "Existential Hypotheses," *Philosophy of Science,* XVII (1950), 35–62; P. K. Feyerabend, "An Attempt at a Realistic Interpretation of Experience," Proceeding of the Aristotelian Society, LVIII (1958), 144–70; N. R. Hanson, *Patterns of Discovery* (Cambridge: Cambridge University Press, 1958); E. Nagel, *loc. cit.;* Karl Popper, *The Logic of Scientific Discovery* (London: Hutchinson, 1959); M. Scriven, "Definitions, Explanations, and Theories," in *Minnesota Studies in the Philosophy of Science,* eds. H. Feigl, M. Scriven, and G. Maxwell, Vol. II (Minneapolis: University of Minnesota Press, 1958); Wilfrid Sellars, "Empiricism and the Philosophy of Mind," in *Minnesota Studies in the Philosophy of Science,* eds. H. Feigl and M. Scriven, Vol. I (Minneapolis: University of Minnesota Press, 1956), and "The Language of Theories," in *Current Issues in the Philosophy of Science,* eds. H. Feigl and G. Maxwell (New York: Holt, Rinehart, and Winston, 1961).

3. I have borrowed the hammer analogy from E. Nagel, "Science and [Feigl's] Semantic Realism," *Philosophy of Science,* XVII (1950), 174–81 but it should be pointed out that Professor Nagel makes it clear that he does not necessarily subscribe to the view which he is explaining.

4. G. Bergmann, "Outline of an Empiricist Philosophy of Physics," *American Journal of Physics,* II (1943), 248–58, 335–42, reprinted in *Readings in the Philosophy of Science,* eds. H. Feigl and M. Brodbeck (New York: Appleton-Century-Crofts, 1953), pp. 262–87.

5. I am not attributing to Professor Bergmann the absurd views suggested by these questions. He seems to take a sense-datum language as his observation language (the base of what he called "the empirical hierarchy"), and, in some ways, such a position is more difficult to refute than one which purports to take an "observable-physical-object" view. However, I believe that demolishing the straw men with which I am now dealing amounts to desirable preliminary "therapy." Some nonrealist interpretations of theories which embody the presupposition that the observable-theoretical distinction is sharp and ontologically crucial seem to me to entail positions which correspond to such straw men rather closely.

6. For arguments that it is possible to alter a theory without altering the meaning of its terms, see my "Meaning Postulates in Scientific Theories," in *Current Issues in the Philosophy of Science,* eds. Feigl and Maxwell.

7. The reader is no doubt familiar with the abundant literature concerned with this issue. See, for example, Sellars's "Empiricism and the Philosophy of Mind," which also contains references to other pertinent works.

8. We may say "noninferentially" decide, provided this is interpreted liberally enough to avoid starting the entire controversy about observability all over again.

9. Cf. Sellars, "Empiricism and the Philosophy of Mind." As Professor Sellars points out, this is the crux of the "other-minds" problem. Sensations and inner states (relative to an intersubjective observation language, I would add) are theoretical entities (and they "really exist") and *not* merely actual and/or possible behavior. Surely it is the unwillingness to countenance theoretical entities—the hope that every sentence is translatable not only into some observation language but into the physical-thing language—which is responsible for the "logical behaviorism" of the neo-Wittgensteinians.

Ian Hacking

DO WE SEE THROUGH A MICROSCOPE?

A couple of years ago I was discussing scientific realism with Dr. Jal Parakh, a biologist from Western Washington University. We had talked about many of the things that philosophers find important. He diffidently added that, from his point of view, a main reason for believing in the existence of entities postulated by theory is that we have evolved better and better ways of actually seeing them. I began to protest against this naive instinct that bypasses the philosophical issues, but I had to stop. Isn't what he says right?

Last fall, during a lecture in Stanford University's "Microscopy for Biologists" course, the professor, Dr. Paul Green, casually remarked that "X-ray diffraction microscopy is now the main interface between atomic structure and the human mind." Dr. Green is a nuts and bolts man, not given to philosophizing. Philosophers of science who discuss realism and anti-realism must needs know a little about the instruments that inspire such eloquence. What follows is a first start, which limits itself to biology and which hardly gets beyond the light microscope. Even that is a marvel of marvels which, I suspect, not many philosophers well understand. Microscopes do not work in the way that most untutored people suppose. But why, it may be asked, should a philosopher care how they work? Because a correct understanding is necessary to elucidate problems of scientific realism as well as answering the question posed by my title.

Reprinted by permission from *Pacific Philosophical Quarterly*, Vol. 62, no. 4, October 1981.

Our philosophical literature is full of intricate accounts of causal theories of perception, yet they have curiously little to do with real life. We have fantastical descriptions of aberrant causal chains which, Gettier-style, call in question this or that conceptual analysis. But the modern microscopist has far more amazing tricks than the most imaginative of armchair students of perception. What we require in philosophy is better awareness of the truths that are stranger than fictions. We ought to have some understanding of those astounding physical systems "by whose augmenting power we now see more/than all the world has ever done before."[1]

THE GREAT CHAIN OF BEING

Philosophers have written dramatically about telescopes. Galileo himself invited philosophizing when he claimed to see the moons of Jupiter, assuming that the laws of vision in the celestial sphere are the same as those on earth. Paul Feyerabend has used that very case to urge that great science proceeds as much by propaganda as by reason: Galileo was a con man, not an experimental reasoner. Pierre Duhem used the telescope to present his famous thesis that no theory need ever be rejected, for phenomena that don't fit can always be accommodated by changing auxiliary hypotheses (if the stars aren't where theory predicts, blame the telescope, not the heavens). By comparison the microscope has played a humble role, seldom used to generate philosophical paradox. Perhaps this is because everyone expected to find worlds within worlds here on earth. Shakespeare is merely an articulate poet of the great chain of being when he writes of Queen Mab and her minute coach "drawn with a team of little atomies . . . her waggoner, a small grey coated gnat not half so big as a round little worm prick'd from the lazy finger of a maid."[2] One expected tinies beneath the scope of human vision. When dioptric glasses were to hand, the laws of direct vision and refraction went unquestioned. That was a mistake. I suppose no one understood how a microscope works before Ernst Abbe (1840–1905). One immediate reaction, by a president of the Royal Microscopical Society, and quoted for years in many editions of the standard American textbook on microscopy, was that we do not, after all, see through a microscope. The theoretical limit of resolution

> [A] Becomes explicable by the research of Abbe. It is demonstrated that microscopic vision is *sui generis.* There is and there can be *no* comparison between microscopic and macroscopic vision. The images of minute objects are not delineated microscopically by means of the ordinary laws of refraction; they are not *dioptical* results, but depend entirely on the laws of *diffraction.*[3]

I think that means that we do not see, in any ordinary sense of the word, with a microscope.

PHILOSOPHERS OF THE MICROSCOPE

Every twenty years or so a philosopher has said something about microscopes. As the spirit of logical positivism came to America, one could read Gustav Bergman telling us that as he used philosophical terminology,

> . . . microscopic objects are not physical things in a literal sense, but merely by courtesy of language and pictorial imagination. . . . When I look through a microscope, all I see is a patch of color which creeps through the field like a shadow over a wall.[4]

In due course Grover Maxwell, denying that there is any fundamental distinction between observational and theoretical entities, urged a continuum of vision: "looking through a window pane, looking through glasses, looking through binoculars, looking through a low power microscope, looking through a high power microscope, etc."[5] Some entities may be invisible at one time and later, thanks to a new trick of technology, they become observable. The distinction between the observable and the merely theoretical is of no interest for ontology.

Grover Maxwell was urging a form of scientific realism. He rejected any anti-realism that holds that we are to believe in the existence of only the observable entities that are entailed by our theories. In his new anti-realist book *The Scientific Image,* Bas van Fraassen strongly disagrees. He calls his philosophy constructive empiricism. He holds that *"Science aims to give us theories which are empirically adequate; and acceptance of a theory involves as belief only that it is empirically adequate."*[6] Six pages later he attempts this gloss: "To accept a theory is (for us) to believe that it is empirically adequate—that what the theory says *about what is observable* (by us) is true." Clearly then it is essential for van Fraassen to restore the distinction between observable and unobservable. But it is not essential to him, exactly where we should draw it. He grants the "observable" is a vague term whose extension itself may be determined by our theories. At the same time, he wants the line to be drawn in the place which is, for him, most readily defensible, so that even if he should be pushed back a bit in the course of debate, he will still have lots left on the "unobservable" side of the fence. He distrusts Grover Maxwell's continuum and tries to stop the slide from seen to inferred entities as early as possible. He quite rejects the idea of a continuum.

There are, says van Fraassen, two quite distinct kinds of case arising from Grover Maxwell's list. You can open the window and see the fir tree directly. You can walk up to at least some of the objects you see through binoculars and see them in the round, with the naked eye. (Evidently he is not a bird watcher.) But there is no way to see a blood platelet with the naked eye. The passage from a magnifying glass to even a low-powered microscope is the passage from what we might be able to observe with the eye unaided, to what we could not observe except with instruments. Van Fraassen concludes that we do not see through a microscope. Yet we see through some telescopes. We can go to Jupiter and look at the moons, but we cannot shrink to the size of a paramecium and look at it. He also compares the vapor trail made by a jet and the ionization track of an electron in a cloud chamber. Both result from similar physical processes, but you can point ahead of the trail and spot the jet, or at least wait for it to land, but you can never wait for the electron to land and be seen.

Taking van Fraassen's view to the extreme you would say that you have observed or seen something by the use of an optical instrument only if human beings with fairly normal vision could have seen that very thing with the naked eye. The ironist will retort: "What's so great about 20-20 human vision?" It is doubtless of some small interest to know the limits of the naked eye, just as it is a challenge to climb a rock face without pitons or Everest without oxygen. But if you care chiefly to get to the top you will use all the tools that are handy. Observation, in my book of science, is not passive seeing. Observation is a skill. Any skilled artisan cares for new tools. I elsewhere use Caroline Herschel to illustrate the supremely skilled observer.[7] She discovered more comets than anyone, using a rather simple tool, a sky sweeper, and was backed up by the telescopes of her brother William Herschel. Our confidence that she saw comets has, contrary to van Fraassen, nothing to do with a fiction of getting up close and seeing that they are indeed comets—that's still impossible. To understand whether she was seeing, or whether one sees through the microscope, one needs to know quite a lot about the tools.

DON'T JUST PEER: INTERFERE

Philosophers tend to look on microscopes as black boxes with a light source at one end and a hole to peer through at the other. There are, as Grover Maxwell puts it, low power and high power microscopes, more and more of the same kind of thing. That's not right, nor are microscopes just for looking through. In fact a philosopher will certainly not see through a microscope until he has learned to use several of them. Asked to draw what he sees he may, like James Thurber, draw his own reflected eyeball, or, like Gustav Bergman, see only "a patch of color which creeps through the field like a shadow over a wall." He will certainly not be able to tell a dust particle from a fruit fly's salivary gland until he has started to dissect a fruit fly under a microscope of modest magnification.

That is the first lesson: you learn to see through a microscope by doing, not just by looking. There is a parallel to Berkeley's *New Theory of Vision,* according to which we have three-dimensional vision only after learning what it is like to move around in the world and intervene in it. Tactile sense is correlated with our allegedly two dimensional retinal image, and this learned cueing produces three-dimensional perception. Likewise a scuba diver learns to see in the new medium of the oceans only by swimming around. Whether or not Berkeley was right about primary vision, new ways of seeing, acquired after infancy, involve learning by doing, not just passive looking. The conviction that a particular part of a cell is there as imaged is, to say the least, reinforced when, using straightforward physical means, you microinject a fluid into just that part of the cell. We see the tiny glass needle—a tool that we have ourselves hand crafted under the microscope—jerk through the cell wall. We see the lipid oozing out of the end of the needle as we gently turn the micrometer screw on a large, thoroughly macroscopic, plunger. Blast! Inept as I am, I have just burst the cell wall, and must try again on another specimen. John

Dewey's jeers at the "spectator theory of knowledge" are equally germane for the spectator theory of microscopy.

This is not to say that practical microscopists are free from philosophical perplexity. I quote from the most thorough of available textbooks intended for biologists:

[B] The microscopist can observe a familiar object in a low power microscope and see a slightly enlarged image which is "the same as" the object. Increase of magnification may reveal details in the object which are invisible to the naked eye; it is natural to assume that they, also, are "the same as" the object. (At this stage it is necessary to establish that detail is not a consequence of damage to the specimen during preparation for microscopy.) But what is actually implied by the statement that "the image is the same as the object"?

Obviously the image is a purely optical effect. . . . The "sameness" of object and image in fact implies that the physical interactions with the light beam that render the object visible to the eye (or which would render it visible, if large enough) are identical with those that lead to formation of an image in the microscope. . . .

Suppose however, that the radiation used to form the image is a beam of ultraviolet light, x-rays, or electrons, or that the microscope employs some device which converts differences in phase to changes in intensity. The image then cannot possibly be "the same" as the object, even in the limited sense just defined! The eye is unable to perceive ultraviolet, x-ray, or electron radiation, or to detect shifts of phase between light beams. . . .

This line of thinking reveals that the image must be *a map of interactions between the specimen and the imaging radiation.*[8]

The author goes on to say that all of the methods she has mentioned, and more, "can produce 'true' images which are, in some sense, 'like' the specimen." She also remarks that in a technique like the radioautogram "one obtains an 'image' of the specimen . . . obtained exclusively from the point of view of the location of radioactive atoms. This type of 'image' is so specialized as to be, generally, uninterpretable without the aid of an additional image, the photomicrograph, upon which it is superposed."

This microscopist is happy to say that we see through a microscope only when the physical interactions of specimen and light beam are "identical" for image formation in the microscope and in the eye. Contrast my quotation [A] from an earlier generation, which holds that since the ordinary light microscope works by diffraction, it is not the same as ordinary vision but is *sui generis.* Can microscopists [A] and [B] who disagree about the simplest light microscope possibly be on the right philosophical track about "seeing"? The scare quotes around "image" and "true" suggest more ambivalence in [B]. One should be especially wary of the word "image" in microscopy. Sometimes it denotes something at which you can point, a shape cast on a screen, a micrograph, or whatever. But on other occasions it denotes as it were the input to the eye itself. The conflation results from geometrical optics, in which one diagrams the system with a specimen in focus and an "image" in the other focal plane, where the "image" indicates what you will see if you place your eye there. I do resist one inference that might be drawn even from quotation [B]. It may seem that any statement about what is seen with a microscope is theory-loaded: loaded with the theory of optics or other radiation. I disagree. One needs theory to make a microscope. You do not need theory to use one. Theory may help to understand why objects perceived with an interference-contrast microscope have asymmetric fringes around them, but you can learn to disregard that effect quite empirically. Hardly any biologists know enough optics to satisfy a physicist. Practice—and I mean in general doing, not looking—creates the ability to distinguish between visible artefacts of the preparation or the instrument, and the real structure that is seen with the microscope. This practical ability breeds conviction. The ability may require some understanding of biology, although one can find first class technicians who don't even know biology. At any rate physics is simply irrelevant to the biologist's sense of microscopic reality. His observations and manipulations seldom bear any load of physical theory at all.

BAD MICROSCOPES

I have encountered the impression that Leeuwenhoek invented the microscope, and that since then people have gone on to make better and better versions of the same kind of thing. I would like to correct that idea.

Leeuwenhoek, hardly the first microscopist, was a technician of genius. His scopes had a single lens, and he made a lens for each specimen to be examined. The object was mounted upon a pin at just the right distance. We don't quite know how he made such marvellously accurate drawings of his specimens. The most representative collection of his lenses-plus-specimen was given to the Royal Society in London, which lost the entire set after a century or so in what are politely referred to as suspicious circumstances. But even by that time the glue for his specimens had lost its strength and the objects had begun to fall off their pins. Almost certainly Leeuwenhoek got his marvelous results thanks to a secret of illumination rather than lens manufacture, and he seems never to have taught the public his technique. Perhaps Leeuwenhoek invented dark field illumination, rather than the microscope. That guess should serve as the first of a long series of possible reminders that many of the chief advances in microscopy have had nothing to do with optics. We have needed microtomes to slice specimens thinner, aniline dyes for staining, pure light sources, and, at more modest levels, the screw micrometer for adjusting focus, fixatives and centrifuges.

Although the first microscopes did create a terrific popular stir by showing worlds within worlds, it is important to note that after Hooke's compound microscope, the technology did not markedly improve. Nor did much new knowledge follow after the excitement of the initial observations. The microscope became a toy for English ladies and gentlemen. The toy would consist of a microscope and a box of mounted specimens from the plant and animal kingdom. Note that a box of mounted slides might well cost more than the purchase of the microscope itself. You did not just put a drop of pond water on a slip of glass and look at it. All but the most expert would require a ready mounted slide to see *anything*. Indeed considering the optical aberrations it is amazing that anyone ever did see anything through a compound microscope, although in fact, as always in experimental science, a really skillful technician can do wonders with awful equipment.

There are about eight chief aberrations in bare-bones light microscopy. Two important ones are spherical and chromatic. The former is the result of the fact that you polish a lens by random rubbing. That, as can be proven, gives you a spherical surface. A light ray travelling at a small angle to the axis will not focus at the same point as a ray closer to the axis. For angles i for which sin i differs at all from i we get no common focus of the light rays, and so a point on the specimen can be seen only as a smear through the microscope. This was well understood by Huygens, who also knew how to correct it in principle, but practical combinations of concave and convex lenses to avoid spherical aberration were a long time in the making.

Chromatic aberrations are caused by differences in wavelength between light of different colors. Hence red and blue light emanating from the same point on the specimen will come to focus at different points. A sharp red image is superimposed on a blue smear or vice versa. Although rich people liked to have a microscope about the house for entertainments, it is no wonder that serious science had nothing to do with the instrument. We often regard Bichat as the founder of histology, the study of living tissues. In 1800 he would not allow a microscope in his lab.

> When people observe in conditions of obscurity each sees in his own way and according as he is affected. It is, therefore, observation of the vital properties that must guide us rather than the blurred images provided by the best of microscopes.[9]

No one tried very hard to make achromatic microscopes, because Newton had written that they are physically impossible. They were made possible by the advent of flint glass, with refractive indices different from that of ordinary glass. A doublet of two lenses of different refractive indices can be made to cancel out the aberration perfectly for a given pair of red and blue wavelengths, and although the solution is imperfect over the whole spectrum, it is pretty negligible and can be improved by a triplet of lenses. The first person to get the right ideas was so secretive that he sent the specifications for the lenses of different kinds of glass to two different contractors. They both subcontracted with the same artisan, who then formed a shrewd guess that the lenses were for the same device. In due course, in 1758, the idea was pirated. A court case for the patent rights was decided in favor of the pirate, John Doland. The High Court Judge ruled:

> It was not the person who locked the invention in his scritoire that ought to profit by a patent for such an invention, but he who brought it forth for the benefit of the public.[10]

The public did not benefit all that much. Even up in to the 1860s there were serious debates as to whether globules seen through a microscope were artefacts of the instrument or genuine elements of living material. (They were artefacts.) Microscopes did get better and aids to microscopy improved at rather a greater rate. If we draw a graph of development we get a first high around 1660, then a slowly ascending plateau until a great leap around 1870; the next great period, which is still with us, commences about 1945. An historian has plotted this graph with great precision, using as a scale the limits of resolution of surviving instruments of different epochs.[11] Making a subjective assessment of great applications of the microscope, we would draw a similar graph, except that the 1870/1660 contrast would be greater. Few truly memorable facts were found out with a microscope until after 1860. The surge of new microscopy is partly due to Abbe, but the most immediate cause of advance was the availability of aniline dyes for staining. Living matter is mostly transparent. The new aniline dyes made it possible for us to see microbes and much else.

ABBE AND DIFFRACTION

How do we "normally" see? Mostly we see reflected light. But if we are using a magnifying glass to look at a specimen illumined from behind, then it is transmission, or absorption, that we are "seeing." So we have the following idea: to see something through a light microscope is to see patches of dark and light corresponding to the proportions of light transmitted or absorbed. We see changes in the amplitude of light rays. I think that even Huygens knew there is something wrong with this conception, but not until 1873 could one read in print how a microscope works.[12]

Ernst Abbe provides the happiest example of a rags-to-riches story. Son of a spinning-mill workman, he yet learned mathematics and was sponsored through the Gymnasium. He became a lecturer in mathematics, physics, and astronomy. His optical work led him to associate. He was taken on by the small firm of Carl Zeiss in Jena, and when Zeiss died he became an owner; he retired to a life of philanthropy. Innumerable mathematical and practical innovations by Abbe turned Carl Zeiss into the greatest of optical firms. Here I consider only one.

Abbe was interested in resolution. Magnification is worthless if it "magnifies" two distinct dots into one big blur. One needs to resolve the dots into two distinct images. G. B. Airy, the English Astronomer Royal, had seen the point already when considering the properties of a telescope needed to distinguish twin stars. It is a matter of diffraction. The most familiar example of diffraction is the fact that shadows of objects with sharp boundaries are fuzzy. This is a consequence of the wave character of light. When light travels between two narrow slits, some of the beam may go straight through, but some of it will bend off at an angle to the main beam, and some more will bend off at a larger angle: these are the first-order, second-order, etc., diffracted rays.

Abbe took as his problem how to resolve (i.e., visibly distinguish) parallel lines on a diatom. These lines are very close together and of almost uniform separation and width. He was soon able to take advantage of even more regular artificial diffraction gratings. His analysis is an interesting example of the way in which pure science is applied, for he worked out the theory for the pure case of looking at a diffraction grating, and inferred that this represents the infinite complexity of the physics of seeing a heterogeneous object with a microscope.

When light hits a diffraction grating most of it is diffracted rather than transmitted. It is emitted from the grating at the angle of first, second, or third order diffractions, where the angles of the diffracted rays are in part a function of the distances between the lines on the grating. Abbe realized that in order to see the slits on the grating, one must pick up not only the transmitted light, but also at least the first order diffracted ray. What you see, in fact, is best represented as a Fourier synthesis of the transmitted and the diffracted rays. Thus according to Abbe the image of the object is produced by the interference of the light waves emitted by the principle image, and the secondary images of the light source which are the result of diffraction.

Practical applications abound. Evidently you will pick up more diffracted rays by having a wider aperture for the objective lens, but then you obtain vastly more spherical aberration as well. Instead you can change the medium between the specimen and the lens. With something denser than air, as in the oil immersion microscope, you capture more of the diffracted rays within a given aperture and so increase the resolution of the microscope.

Even though the first Abbe-Zeiss microscopes were good, the theory was resisted for a number of years, particularly in England and America, who had enjoyed a century of dominating the market. Even by 1910 the very best English microscopes, built on purely empirical experience, although stealing a few ideas from Abbe, could resolve as well or better than the Zeiss equipment. The expensive craftsmen with trial-and-error skills were doomed. It was not, however, only commercial or national rivalry which made some people hesitate to believe Abbe. In an American textbook of 1916 I find it stated that an alternative (and more "common sense") theory of "ordinary" vision is now once again in the ascendant and will soon scuttle Abbe![13] Resistance arose partly from surprise at what Abbe asserted, with the apparent consequence that, as quotation [A] has it, "there is and can be *no* comparison between microscopic and macroscopic vision."

If you hold (as my more modern quotation [B] still seems to hold), that what we see is essentially a matter of a certain sort of physical processing in the eye, then everything else must be more in the domain of optical illusion or at best of mapping. On that account the systems of Leeuwenhoek and of Hooke do allow you to see. After Abbe even the conventional light microscope is essentially a Fourier synthesizer of first or even second order diffractions. Hence you must modify your notion of seeing or hold that you never see through a serious microscope. Before reaching a conclusion on this question, we had best examine some more recent instruments.

A PLETHORA OF SCOPES

We move on to after World War II. Most of the ideas had been around during the interwar years, but did not get beyond prototypes until later. One invention is a good deal older, but it was not properly exploited for a while.

The first practical problem for the cell biologist is that most living material does not show up under an ordinary light microscope because it is transparent. To see anything you have to stain the specimen. Aniline dyes are the world's number one poison, so what you will see is a very dead cell, which is also quite likely to be a structurally damaged cell, exhibiting structures that are an artefact of the preparation. However it turns out that living material varies in its birefringent (polarizing) properties. So let us incorporate into our scope a polarizer and an analyzer. The polarizer transmits to the specimen only polarized light of certain properties. In the simplest case, let the analyzer be placed at right angles to the polarizer, so as to transmit only light of polarization opposite to that of the polarizer. The result is total darkness. But suppose the specimen is itself birefringent; it may then change the plane of polarization of the incident light, and so a visible image may be formed by the analyzer. "Transparent" fibers of striated muscle may be observed in this way, without any staining, and relying solely on certain properties of light that we do not normally "see."

Abbe's theory of diffraction, augmented by the polarizing microscope, leads to something of a conceptual revolution. We do not have to see using the "normal" physics of seeing in order to perceive structures in living material. In fact we never do. Even in the standard case we synthesize diffracted rays rather than seeing the specimen by way of "normal" visual physics. Then the polarizing microscope reminds us that there is more to light than refraction, absorption and diffraction. *We could use any property of light that interacts with a specimen in order to study the structure of the specimen.* Indeed we could use any property of *any kind of wave* at all.

Even when we stick to light there is lots to do. Ultraviolet microscopy doubles resolving power, although its chief interest lies in noting the specific ultraviolet absorptions that are typical of certain biologically important substances. In fluorescence microscopy the incident illumination is cancelled out, and one observes only light re-emitted at different wave lengths by natural or induced phosphorescence or fluorescence. This is an invaluable histological technique for certain kinds of living matter. More interesting, however, than using unusual modes of light transmission or emission are the games we can play with light itself: the Zelnicke phase-contrast microscope and the Nomarski interference microscope.

A specimen that is transparent is uniform with respect to light absorption. It may still possess invisible differences in refractive index in various parts of its structure. The phase contrast microscope converts these into visible differences of intensity in the image of the specimen. In an ordinary microscope the image is synthesized from the diffracted waves *D* and the directly transmitted waves *U*. In the phase contrast microscope the *U* and *D* waves are physically separated in an ingenious although physically simple way, and one or the other kind of wave is then subject to a standard phase delay which has the effect of producing in focus phase contrast corresponding to the differences in refractive index in the specimen.

The interference contrast microscope is perhaps easier to understand. The light source is simply split by a half-silvered mirror, and half the light goes through the specimen while half is kept as an unaffected reference wave to be recombined for the output image. Changes in optical path due to different refractive indices within the specimen thus produce interference effects with the reference beam.

The interference microscope is attended by illusory fringes but is particularly valuable because it provides a quantitative determination of refractive indices within the specimen. Naturally once we have such devices in hand, endless variations may be constructed, such as polarizing interference microscopes, multiple beam interference, phase modulated interference, and so forth.

TRUTH IN MICROSCOPY

The differential interference-contrast technique is distinguished by the following characteristics: Both clearly visible outlines (edges) within the object and continuous structures (striations) are imaged in their true profile.

So says a Carl Zeiss sales catalogue to hand. What makes the enthusiastic sales person suppose that the images produced by these several optical systems are "true"? Of course, the images are "true" only when one has learned to put aside distortions. There are many grounds for the conviction that a perceived bit of structure is real or true. One of the most natural is the most important. I shall illustrate it with my own first experience in the laboratory.[14] Low-powered electron microscopy reveals small dots in red blood cells. These are called dense bodies: that means simply that they are electron dense, and show up on a transmission electron microscope without any preparations or staining whatsoever. On the basis of the movements and densities of these bodies in various stages of cell development or disease, it is guessed that they may have an important part to play in blood biology. On the other hand they may simply be artefacts of the electron microscope. One test is obvious: can one see these selfsame bodies using quite different physical techniques? In this case the problem is fairly readily solved. The low resolution electron microscope is about the same power as a high resolution light microscope. The dense bodies do not show up under every technique, but are revealed by fluorescent staining and subsequent observation by the fluorescent microscope.

Slices of red blood cell are fixed upon a microscopic grid. This is literally a grid: when seen through a microscope one sees a grid each of whose squares is labelled with a capital letter. Electron micrographs are made of the slices mounted upon such grids. Specimens with particularly striking configurations of dense bodies are then prepared for fluorescence microscopy. Finally one compares the electron micrographs and the fluorescence micrographs. One knows that the micrographs show the same bit of the cell, because this bit is clearly in the square of the grid labelled *P*, say. In the fluorescence micrographs there is exactly the same arrangement of grid, general cell structure, and of the seven "bodies" seen in the electron micrograph. It is inferred that the bodies are not an artefact of the electron microscope.

Two physical processes—electron transmission and fluorescent re-emission—are used to detect the bodies. These processes have virtually nothing in common between them. They are essentially unrelated chunks of physics. It would be a preposterous coincidence if, time and again, two completely different physical processes produced identical visual configurations which were, however, artefacts of the physical processes rather than real structures in the cell.

Note that no one actually produces this "argument from coincidence" in real life. One simply looks at the two (or preferably more) sets of micrographs from different physical systems, and sees that the dense bodies occur in exactly the same place in each pair of micrographs. That settles the matter in a moment. My mentor, Dr. Richard Skaer, had in fact expected to prove that dense bodies are artefacts. Five minutes after examining his completed experimental micrographs he knew he was wrong.

Note also that no one need have any ideas what the dense bodies *are*. All we know is that there are some structural features of the cell rendered visible by several techniques. Microscopy itself will never tell all about these bodies (if indeed there is anything important to tell). Biochemistry must be called in. Also, instant spectroscopic analysis of the dense body into constitutent elements is now available, by combining an electron microscope and a spectroscopic analyzer. This works much like spectroscopic analyses of the stars.

COINCIDENCE AND EXPLANATION

Arguments from coincidence have been put to more general use in discussions of scientific realism. In particular J. J. C. Smart notes that good theories are used to explain diverse phenomena. It would, he says, be a cosmic coincidence if the theory were false and yet correctly predicted all the phenomena:

> One would have to suppose that there were unnumerable lucky accidents about the behavior mentioned in the observational vocabulary, so that they behaved miraculously *as if* they were brought about by the non-existent things ostensibly talked about in the theoretical vocabulary.[15]

Van Fraassen challenges this and related arguments for realism that deploy what Gilbert Harman calls "inference to the best explanation," or what Hans Reichenbach and Wesley Salmon call the "common cause" argument. So it may seem as if my talk of coincidence puts me in the midst of an ongoing feud. Not so! My argument is much more localized, and commits me to none of the positions of Smart or Salmon.

First of all, we are not concerned with an observational and theoretical vocabulary. There may well be no theoretical vocabulary for the things seen under the microscope—"dense body" means nothing else than something dense, i.e., that shows up under the electron microscope without any staining or other preparation. Secondly we are not concerned with explanation. We see the same constellations of dots whether we use an electron microscope or fluorescent staining, and it is no "explanation" of this to say that some definite kind of thing (whose nature is as yet unknown) is responsible for the persistent arrangement of dots. Thirdly we have no theory which predicts some wide range of phenomena. The fourth and perhaps most important difference is this: we are concerned to distinguish artefacts from real objects. In the metaphysical disputes about realism, the contrast is between "real although unobservable entity" and "not a real entity, but rather a tool of thought." With the microscope we know there are dots on the micrograph. The question is, are they artefacts of the physical system or are they structures present in the specimen itself? My argument from coincidence says simply that it would be a preposterous coincidence if two totally different kinds of physical systems were to produce exactly the same arrangements of dots on micrographs.

THE ARGUMENT OF THE GRID

I now venture a philosopher's aside on the topic of scientific realism. Van Fraassen says we can see through a telescope because although we need the telescope to see the moons of Jupiter when we are positioned on earth, we could go out there and look at the moons with the naked eye. Perhaps that fantasy is close to fulfillment, but it is still science fiction. The microscopist avoids fantasy. Instead of flying to Jupiter he shrinks the visible world. Consider the grid that we used for re-identifying dense bodies. The tiny grids are made of metal; they are barely visible to the naked eye. They are made by drawing a very large grid with pen and ink. Letters are neatly inscribed by a draftsman at the corner of each square on the grid. Then the grid is reduced photographically. Using what are now standard techniques, metal is deposited on the resulting micrograph. Grids are sold in packets, or rather tubes, of 100, 250, and 1,000. The procedures for making such grids are entirely well understood, and as reliable as any other high-quality mass production system.

In short, rather than disporting ourselves to Jupiter in an imaginary space ship, we are routinely shrinking a grid. Then we look at the tiny disc through almost any kind of microscope and see exactly the same shapes and letters as were drawn in the large by the first draftsman. It is impossible seriously to entertain the thought that the minute disc, which I am holding by a pair of tweezers, does not in fact have the structure of a labelled grid. I know that what I see through the microscope is veridical because we *made* the grid to be just that way. I know that the process of manufacture is reliable, because we can check the results with the microscope. Moreover we can check the results with any kind of microscope, using any of a dozen unrelated physical processes to produce an image. Can we entertain the possibility that, all the same, this is some gigantic coincidence? Is it false that the disc is, in fine, in the shape of a labelled grid? Is it a gigantic conspiracy of thirteen totally unrelated physical processes that the large scale grid was shrunk into some non-grid which when viewed using twelve different kinds of microscopes still looks like a grid? To be an anti-realist about that grid you would have to invoke a malign Cartesian demon of the microscope.

The argument of the grid probably requires a healthy recognition of the disunity of science, at least at the phenomenological level. Light microscopes, trivially, all use light, but interference, polarizing, phase contrast, direct transmission, fluorescence, and so forth all exploit essentially unrelated phenomenological aspects of light. If the same structure can be discerned using many of these different aspects of light waves, we cannot seriously suppose that the structure is an artefact of all the different physical systems. Moreover I emphasize that all these physical systems are made by people. We as it were purify some aspect of nature, isolating, say, the phase interference character of light. We design an instrument knowing in principle exactly how it will work, just because optics is so well understood a science. We spend a number of years debugging several prototypes, and finally have an off-the-shelf instrument, through which we discern a particular structure. Several other off-the-shelf instruments, built upon entirely different principles, reveal the same structure. No one, short of the Cartesian sceptic can suppose that the structure is made by the instruments rather than inherent in the specimen.

It was once not only possible but perfectly sensible to ban the microscope from the histology lab on the plain grounds that it chiefly revealed artefacts of the optical system rather than the structure of fibers. That is no longer the case. It is always a problem in innovative microscopy to become convinced that what you are seeing is really in the specimen rather than an artefact of the preparation of the optics. But by 1981, as opposed to 1800, we have a vast arsenal of ways of gaining such conviction. I emphasize only the "visual" side. Even there I am simplistic. I say that if you can see the same fundamental features of structure using several different physical systems, you have excellent reason for saying, "that's real" rather than, "that's an artefact." It is not conclusive reason. But the situation is no different from ordinary vision. If black patches on the tarmac road are seen, on a hot day, from a number of different perspectives, but always in the same location, one concludes that one is seeing puddles rather than the familiar illusion. One may still be wrong. One is wrong, from time to time, in microscopy, too. Indeed the sheer similarity of the kinds of mistakes made in macroscopic and microscopic perception may increase the inclination to say, simply, that one sees through a microscope.

I must repeat that just as in large scale vision, the actual "images" or micrographs are only one small part of the confidence in reality. In a recent lecture the molecular biologist G. S. Stent recalled that in the late forties or early fifties *Life* magazine had a full color cover of an electron micrograph, labelled, excitedly, "the first photograph of the gene."[16] Given the theory, or lack of theory, of the gene at that time, said Stent, the title did not make any sense. Only a greater understanding of what a gene is can bring the conviction of what the micrograph shows. We become convinced of the reality of bands and interbands on chromosomes not just because we see them, but because we formulate conceptions of what they do, what they are for. But in this respect, too, microscopic and macroscopic visions are not different: a Laplander in the Congo won't see much in the bizarre new environment until he starts to get some idea what is in the jungle.

Thus I do not advance the argument from coincidence as the sole basis of our conviction that we see true through the microscope. It is one element, a compelling visual element, that combines with more intellectual modes of understanding, and with other kinds of experimental work. Biological microscopy without practical biochemistry is as blind as Kant's intuitions in the absence of concepts.

THE ACOUSTIC MICROSCOPE

I here avoid the electron microscope. There is no more "the" electron microscope than "the" light microscope: all sorts of different properties of electron beams are used. A simple but comprehensive explanation requires another essay. In case, however, we have in mind too slender a diet of examples based upon the properties of visible light, let us briefly consider the most disparate kind of radiation imaginable: sound.[17]

Radar, invented for aerial warfare, and sonar, invented for war at sea, remind us that longitudinal and transverse wave fronts can be put to the same kinds of purpose. Ultrasound is "sound" of very high frequency. Ultrasound examination of the foetus *in vitro* has recently won well deserved publicity. Over forty years ago Soviet scientists suggested a microscope using sound of frequency 1000 times greater than audible noise. Technology has only recently caught up to this idea. Useful prototypes are just now in operation.

The acoustic part of the microscope is relatively simple. Electric signals are converted into sound signals and then, after interaction with the specimen, are reconverted into electricity. The subtlety of present instruments lies in the electronics rather than the acoustics. The acoustic microscope is a scanning device. It produces its images by converting the signals into a spatial display on a television screen, a micrograph, or, when studying a large number of cells, a videotape.

As always a new kind of scope is interesting because of the new aspects of a specimen that it may reveal. Changes in refractive index are vastly greater for sound than for light. Moreover sound is transmitted through objects that are completely opaque. Thus one of the first applications of the acoustic microscope is in metallurgy, and also in detecting defects in silicon chips. For the biologist, the prospects are also striking. The acoustic microscope is sensitive to density, viscosity, and flexibility of living matter. Moreover the very short bursts of sound used by the scanner do not immediately damage the cell. Hence one may study the life of a cell in a quite literal way: one will be able to observe changes in viscosity and flexibility as the cell goes about its business.

The rapid development of acoustic microscopy leaves us uncertain where it will lead. A couple of years ago the research reports carefully denied any competition with electron microscopes; they were glad to give resolution at about the level of light scopes. Now, using the properties of sound in supercooled solids, one can emulate the resolution of electron scopes, although that is not much help to the student of living tissue!

Do we see with an acoustic microscope?

LOOKING WITH A MICROSCOPE

Do we see through a microscope? Let us first do away with the anachronistic word *through*. Looking through a lens was the first step in technology, then came peering through the tube of a compound microscope. The micrograph is more to the point: we study photographs taken with a microscope. Thanks to the enormous focal length of an electron microscope it is natural to view the image on a large flat surface so everyone can stand around and point to what's interesting. Scanning microscopes necessarily constitute the image on a screen or plate. Any image can be digitized and retransmitted on a television display or whatever. Moreover digitization is marvellous for censoring noise and even reconstituting lost information. Do not, however, become awed by technology. In the study of crystal structure, one good way to get rid of noise is to cut up a micrograph in a systematic way, paste it back together, and rephotograph it for interference contrast.

We do not in general see *through* a microscope; we see with one. But do we *see* with a microscope? It would be silly to debate the ordinary use of the word *see,* a word already put to innumerable uses of an entirely intellectual sort. "Now I see the point," and kindred employments in mathematics. Or consider how the physicist writes of the hypothetical entities. I quote from a lecture listing twelve fermions, or fundamental constituents of matter, including electron neutrinos, deuterons, etc. We are told that "of these fermions, only the t quark is yet unseen. The failure to observe tt′ states in $e+e-$ anihilation at PETRA remains a puzzle. . . ."[18] Seeing and observing for this high energy physicist are a long way from the eye. (Probably seeing acquired its peculiar association with ocular vision only at the start of the nineteenth century, as is manifested in the twin doctrines called positivism and phenomenology, the philosophies that say seeing is with the eye, not the mind.)

Consider a device for low-flying jet planes, laden with nuclear weapons, skimming a few dozen yards from the surface of the earth in order to evade radar detection. The vertical and horizontal scale are both of interest to the pilot; he needs both to see a few hundred feet down and miles and miles away. So the visual information is digitized, processed, and cast on a head-up display on the windscreen. The distances are condensed and the altitude is expanded. Does the pilot *see* the terrain? I should say so. It would be foolish to put in some unnatural word like *perceive* to indicate that the seeing employs an instrument. Note that this case is not one in which the pilot could have seen the terrain by getting off the plane and taking a good look. There is no way of getting a look at that much landscape without an instrument.

Consider the electron diffraction microscope with which I produce images either in conventional space or in reciprocal space. Reciprocal space is, roughly speaking, conventional space turned inside out; near is far and far is near. Crystallographers often find it most natural to study their specimens in reciprocal space. Do they see them in reciprocal space? They certainly say so, and thereby call in question the Kantian doctrine of the uniqueness of perceptual space.

How far could one push the concept of seeing? Suppose I take an electronic paint brush and paint on a television screen, an accurate picture (I) of a cell that I have previously studied, say, by using a digitized and reconstituted image (II). Even if I am "looking at the cell" in case (II), in (I) I am only looking at a drawing of the cell. What is the difference? The important feature is that in (II) there is a direct interaction between a wave source, an object, and a series of physical events that end up in an image of the object. To use quotation [B] once again, in case (II) we have a map of interactions between the specimen and the imaging radiation. If the map is a good one, then (II) is seeing with a microscope.

This is doubtless a liberal extension of the notion of seeing. We see with an acoustic microscope. We see with television, of course. We do not say that we saw an attempted assassination *with* the television, but *on* the television. That is mere idiom, inherited from "I heard it on the radio." We distinguish between seeing the television broadcast live or not. We have endless distinctions to be made with various adverbs, adjectives, and even prepositions. I know of no confusion that will result from talk of seeing with a microscope.

SCIENTIFIC REALISM

When an image is a map of interactions between the specimen and the image of radiation, and the map is a good one, then we are seeing with a microscope. What is a good map? After discarding or disregarding aberrations or artefacts, the map should represent some structure in the specimen in essentially the same two- or three-dimensional set of relationships as are actually present in the specimen.

Does this bear on scientific realism? First let us be clear that it can bear in only the modest way. I do not even argue here for the reality of objects and structure that can be discerned only by the electron microscope (That calls for another essay). I have spoken chiefly of light microscopy. Now imagine a reader initially attracted by van Fraassen and who thought that objects seen only with light microscopes do not count as observable. That reader could change his mind, and admit such objects into the class of observable entities. This would still leave intact all the main philosophical positions of van Fraassen's anti-realism.

But if we conclude that we see with the light microscopes, does it follow that the objects we report seeing are real? No. For I have said only that we should not be stuck in the nineteenth century rut of positivism-cum-phenomenology, and that we should allow ourselves to talk of seeing with a microscope. Such a recommendation implies a strong commitment to realism about microscopy, but it begs the question at issue. This is clear from my quotation from high-energy physics, with its cheerful talk of our having seen electron neutrinos, deuterons, and so forth. The physicist is a realist, too, and he shows this by using the word *see,* but his usage is no *argument* that there are deuterons. Here perhaps is one source of the philosophers' scepticism of Dr. Parakh's suggestion that one can become a convinced realist because of advances in microscopy.

Does microscopy then beg the question of realism? No. On closer inspection, Parakh's suggestion is right. We *are* convinced of the structures that we observe using various kinds of microscopes. Our conviction arises partly from our success at systematically removing aberrations and artefacts. In 1800 there was no such success. Bichat banned the microscope from his dissecting rooms, for one did not, then, observe structures that could be confirmed to exist in the specimens. But now we have by and large got rid of aberrations; we have removed many artefacts, disregard others, and are always on the lookout for undetected frauds. We are convinced about the structures we seem to see because we can interfere with them in quite physical ways, say by microinjecting. We are convinced because instruments using entirely different physical principles lead us to observe pretty much the same structures in the same specimen. We are convinced by our clear understanding of most of the physics used to build the instruments that enable us to see, but this theoretical conviction plays a relatively small part. We are more convinced by the admirable intersections with biochemistry, which confirm that the structures that we discern with the microscope are individuated by distinct chemical properties, too. We are convinced not by a high powered deductive theory about the cell—there is none—but because of a large number of interlocking low-level generalizations that enable us to control and create phenomena in the microscope. In short, we learn to move around in the microscopic world. Berkeley's *New Theory of Vision* may not be the whole truth about infantile binocular three-dimensional vision, but is surely on the right lines when we enter the new worlds within worlds that the microscope reveals to us.

NOTES

1. From a poem, "In Commendation of the Microscope," by Henry Powers, 1664. Quoted in Saville Bradbury, *The Microscope, Past and Present* (Oxford: Pergammon, 1968).

2. W. Shakespeare, *Romeo and Juliet,* 1.4.58, 66–67.

3. William B. Carpenter, *The Microscope and Its Revelations,* 8th ed., revised by W. H. Dallinger, London and Philadelphia, 1899. Quoted in S. H. Gage, *The Microscope,* 9th ed. (Ithaca: Comstock), 21. Gage contrasts the alternative theory that microscopic vision "is with the unaided eye, the telescope and the photographic camera. This is the original view, and the one which many are favoring at the present day."

4. G. Bergmann, "Outline of an Empiricist Philosophy of Physics," *American Journal of Physics,* 11 (1943):248–58, 335–42. Reprinted in *Readings in the Philosophy of Science,* ed. H. Feigl and M. Brodbeck (New York: Appleton-Century-Crofts, 1953).

5. G. Maxwell, "The Ontological Status of Theoretical Entities," in *Minnesota Studies in the Philosophy of Science,* vol. 3, ed. H. Feigl and G. Maxwell (Minneapolis: University of Minnesota Press, 1962), 3–27.

6. B. C. van Fraassen, *The Scientific Image* (Oxford: Clarendon Press, 1980), 12.

7. I. Hacking, "Spekulation, Berechnung und die Erschaffung von Phänomemen," in *Versuchungen: Aufsätze zur Philosophie Paul Feyerabends,* ed. P. Duerr (Frankfurt: Suhrkamp 1981), 2 Band, 126–58, esp. p. 134.

8. E. M. Slayter, *Optical Methods in Biology* (New York: Wiley, 1970), 261–63.

9. X. Bichat, *Anatomie générale appliquéé á la physiologie et à la médecine* (Paris: Brosson, Gaber et cie, 1801), 51.

10. Quoted in Bradbury (note 1), 130.

11. S. Bradbury and G. L'E. Turner, eds., *Historical Aspects of Microscopy* (Cambridge: Heffer, 1967).

12. E. Abbe, "Beitrage zur Theorie des Mikroscop und der mikroskopische Wahrnemung."

13. S. H. Gage (note 3), 11th ed., 1916. The direct quotation from Carpenter and Dallinger is dropped in the 12th edition of 1917, but the spirit is retained, including the *"sui generis."* Gage does admit that "Certain very striking experiments have been devised to show the accuracy of Abbe's hypothesis, but as pointed out by many, the ordinary use of the microscope never involves the conditions realized in these experiments" (page 301). How Imre Lakatos would have delighted in this degenerating programme of preserving the naive picture of vision, complete with "monster-barring" of the striking experiments! This passage remained unchanged in essentials even in the 17th edition of 1941.

14. I owe a particular debt of gratitude to my friend R. J. Skaer of Peterhouse, Cambridge, who allowed me to spend a good deal of time in his cell biology laboratory in the Department of Haematological Medicine, Cambridge University.

15. J. J. C. Smart, *Between Science and Philosophy* (New York: Random House, 1968), 150.

16. I think Stent must have been referring to *Life,* 17 March 1947, p. 83.

17. C. F. Quate, "The Acoustic Microscope," *Scientific American* 241 (Oct. 1979), 62–69. R. N. Johnston, A. Atalar, J. Heiserman, V. Jipson, and C. F. Quate, "Acoustic Microscopy: Resolution of Subcellular Detail," *Proceedings of the National Academy of Sciences U.S.A.,* 76 (1979): 3325–29.

18. C. Y. Prescott, "Prospects for Polarized Electrons at High Energies," Stanford Linear Accelerator, SLAC-PUB-2630, Oct. 1980: 5.

Marshall Spector

MODELS AND THEORIES

1. INTRODUCTION

In this paper I will attempt to show that an important currently held view as to the nature of a model for a physical theory is in error. Indirectly, I shall be concerned to show that a certain related thesis about the nature of physical theories themselves is infected with serious difficulties. This will be done in the context of a modification of the former view which I shall offer.

It will be essential to have before us a careful statement of the latter thesis before we begin, for it seems to me that much of its apparent plausibility rests on the fact that presentations of it are not always very clear. (In fact, I believe that several influential attempts to *refute* this analysis of physical theories also rest on unclear and misleading presentations of it.)

According to this thesis, a physical theory is to be analysed as an empirically interpreted hypothetico-deductive system or formal calculus—in Rudolf Carnap's terms, a "semantical system."[1] Its most basic assumption is that a general distinction can be drawn between two types of terms occurring in physical theories—*observation* terms and *theoretical* terms.[2] The former, terms like "green," "desk," "longer than," refer to observable objects, properties, relations, and events, and can be understood independently of any physical theory. The latter, expressions like "electron," "magnetic field," "spin angular momentum," refer to unobservable (= theoretical) objects, properties, etc., and can be understood only in the context of the theories in which they occur. The attractiveness of the view I shall be examining lies in its claim to present a general, well-articulated schema for showing just how we are to understand theories which seem to talk about objects which we have never observed, and perhaps never will directly observe. In outline, the analysis is as follows:

It is maintained that we can distinguish two components in any theory. The first is its *calculus,* which is the logical skeleton of the theory, considered as devoid of any empirical meaning. This will consist of a set of primitive formulas, i.e., sentences which are taken as postulates of the calculus, and other formulas which are obtained by derivation from the postulates in accordance with specified rules of transformation. Two types of terms appearing in the calculus may be distinguished: the primitive terms, i.e., terms which are not defined on the basis of other terms within the calculus; and non-primitive terms, i.e., those which are introduced on the basis of the primitives. (This distinction is not identical with that between observation terms and theoretical terms.) When this calculus, or syntactical system, is given an empirical interpretation, or meaning, it becomes a system of empirical statements having the structure of a hypothetico-deductive system. The primitive formulas become empirical hypotheses, and the derived formulas become empirical statements which will be true if the hypotheses are true.

To illustrate some of these ideas, consider the kinetic theory of gases. The postulates of this theory will contain such expressions as "molecule," "mass of a molecule," and "position of a molecule." These might be considered as primitives. Other expressions, such as "momentum of a mole-

Reprinted from the *British Journal for the Philosophy of Science* (1965) by permission of the author and the publishers, Cambridge University Press.

cule" and "mean kinetic energy of a group of molecules," will be introduced on the basis of the primitives. One of the postulates might be "All gases are composed of molecules." A typical derived formula might be, "If the pressure of a gas is increased while its temperature remains constant, its volume will decrease."

The second component of a theory, the empirical interpretation, is given to the calculus by *semantical rules* for terms of the calculus, i.e., rules which are formulated in a suitable metalanguage (usually "ordinary" English, or German, etc.) and provides the meaning of the terms by stating what properties, relations, or individuals the terms designate.[3] An example is, "The term *P* of the calculus designates the pressure of a sample of gas."

Now the authors we shall be considering maintain that:

(a) not all terms of the calculus of a theory *need* be given semantical rules, and
(b) not all terms of the calculus *can* be given semantical rules.

(It has not been clearly recognised that these are *distinct* claims. The importance of noting this distinction will become clear as we proceed.) Only the observation terms of the unanalysed theory are thus "directly interpreted." That is, if we look upon the calculus of a semantical system as the uninterpreted logical skeleton of a theory, for which the semantical system provides a *reconstruction,* only those terms in the calculus which represent the observation terms of the unreconstructed theory will be given semantical rules in the completed reconstruction. Theoretical terms of the unanalysed theory will not be given semantical rules in the semantical system reconstruction. It is claimed that such terms *cannot* be "understood in themselves," but must be understood—given their meaning—in an "indirect" manner, through the role they play in the theory. Such terms obtain an empirical meaning if and only if they appear in sentences in the calculus which also contain terms which *are* given semantical rules—the observation terms. Such sentences are known as *correspondence rules.*

An example of a correspondence rule is the postulate stated earlier: "All gases are composed of molecules." Symbolically, this would read $(x)\ (Gx \supset Qx)$. The term *G* is given the semantical rule "*G* designates the property of being a sample of gas," and the term *Q* (which is a symbolic translation of the theoretical expression "is composed of molecules") is not given a semantical rule, but is said to obtain a partial meaning indirectly by virtue of its occurrence in a sentence which contains a term (*G*) whose meaning is given directly and completely by a semantical rule. (It is important to notice that *all* the terms of a theory are found in the calculus, including the observation terms. If this is not kept in mind, there arises a tendency to confuse semantical rules with correspondence rules.)

This, in outline, is the analysis of physical theories which underlies the analysis of the concept of a model for a theory which I shall be examining. For reasons which should be apparent, I shall refer to it as the *partial interpretation thesis.*[4]

The clearest and most precise explication of the concept of a model for a physical theory offered by a proponent of the partial interpretation thesis is that given by Braithwaite,[5] so the burden of my analysis will be directed towards his position. A model for a theory, according to Braithwaite, is to be understood as *another interpretation* of the theory's calculus, in which the *theoretical* terms are directly interpreted (by semantical rules).[6] It is sometimes also stated that these direct interpretations must be in terms of familiar concepts. But this need not be stated as a separate condition if we remember Braithwaite's (and Carnap's) injunction to the effect that this is the only way that direct interpretations can usefully be given.[7] Now if we use the term *model* to designate not a domain of non-linguistic entities, but rather the statements about such entities, we can say that a theory and a model for the theory are two sets of statements which share the same calculus, but with the epistemological order of the two reversed. As Braithwaite puts it:

> A theory and a model for it . . . have the same formal structure, since theory and model are both represented by the same calculus. . . . But the theory and the model have different epistemological structures; in the model the logically prior premisses determine the meaning of the terms occurring in the representation in the calculus of the conclusions; in the theory the logically posterior consequences determine the meaning of the theoretical terms occurring in the representation in the calculus of the premisses.[8]

Braithwaite offers this explication of the concept of a model not as an arbitrary definition, but rather as "an attempt to make more precise the notion of a model for a scientific theory widely current in discussions of the philosophy of science." [*ibid.*]

Braithwaite also points out, quite emphatically, what models allegedly are *not*. Braithwaite claims that one of the chief "dangers" in the use of models is the tendency that: "The theory will be identified with a model for it, so that the objects with which the model is concerned . . . will be supposed actually to be the same as the theoretical concepts of the theory." [*op. cit.* p. 93.]

Nagel also warns against confusing the model with the theory itself. After commenting on the possibility that a model may be "an obstacle to the fruitful development of a theory," he writes: "The only point that can be affirmed with confidence is that a model for a theory is not the theory itself."[9] Braithwaite goes on to say that:

> Thinking of scientific theories by means of models is always *as-if* thinking; hydrogen atoms behave (in certain respects) as if they were solar systems each with an electronic planet revolving round a protonic sun. But hydrogen atoms are not solar systems; it is only useful to think of them as if they were such systems if one remembers all the time that they are not.[10]

According to the explication offered by Braithwaite, then, the objects of the model cannot be identified with the theoretical objects of the theory. Such an identification would be a logical error; the possibility of this identification (i.e., the question of the "reality" of the objects of the model) cannot even arise.[11]

However, as I shall argue, these questions *do* legitimately arise for some systems which physicists would recognise as models; and since Braithwaite's explication of the concept of a model cannot allow for this, it is inadequate. To show this, I shall begin by distinguishing four types of domains. They will be quite different in important respects; yet Braithwaite's explication will not be able to distinguish them. Then I will offer a modification of Braithwaite's explication which will be able to distinguish these types of domains. I shall conclude with an analysis of the effects of this modification on the partial interpretation thesis itself.

2. FOUR "BRAITHWAITEAN" MODELS

(1) Suppose we have a geometrical interpretation of the calculus of some physical theory. (Braithwaite frequently uses geometrical interpretations of his "factor-theories" as examples of models.) Such an interpretation would be a model for the theory in Braithwaite's sense, and yet the possible identification of the "objects" of this system with the theoretical objects of the theory would not even arise for the physicist. There would be no question of the lines, triangles, circles, etc., being identified with, or being similar to, the theoretical objects of the theory. Only the formal structure would be relevant. If there are, in fact, triangles, etc., which satisfy the postulates of a given physical theory (its calculus), then the objects of this system are "real"; but this does not mean that the theoretical objects of the theory are triangles, etc. This sort of identification would indeed be a mistake, but a strange sort of mistake that only a modern-day Pythagorean might make. (It is certainly quite different from the mistake of

identifying the theoretical objects of the kinetic theory of gases with, say, billiard balls.) In this sense, a model has nothing whatsoever to do with the *domain* of the theory. Here, the system is another interpretation of the theory's calculus in a very strong sense of "another." The objects of the "model" are of a different logical type from the objects of the domain of the theory. A physicist would probably not call this a model at all.

(2) It has been recognised that there is a rather thoroughgoing analogy between some of the laws of acoustical theory and those of electric circuit theory. Put another way, there is an important correspondence between acoustical systems and electric circuits. Let us consider a specific instance of this. If we have a series circuit consisting of a resistance R, a capacitance C, and an inductance L, with a periodic electromotance ϵ equal to $E \cos(wt)$, the charge on the capacitor, q, will satisfy the equation

$$L\frac{d^2q}{dt^2} + R\frac{dq}{dt} + \frac{1}{C}q = E \cos(wt).$$

Now, consider a Helmholtz resonator, which is a "flask" of volume V with a neck of length d, radius a, and cross-sectional area $S = \pi a^2$. A sound wave of amplitude P impinges on the resonator opening, so that the driving force at the neck is given by $SP \cos(wt)$. The air displacement z at the neck will then satisfy the equation

$$(pd'S)\frac{d^2z}{dt^2} + \left(\frac{pck^2S2}{2\pi}\right)\frac{dz}{dt} + \left(\frac{pc_2S^2}{V}\right)z = SP \cos(wt),$$

where $d' = d + 16a/2\pi$, p = mean air density, c = sound wave velocity, and $k = w/c$. These two equations are of the same form:

$$a\frac{d^2\zeta}{dt^2} + \beta\frac{d\zeta}{dt} + \gamma\zeta = \delta \cos(wt).$$

For other acoustical systems, it is also possible to find electric circuit "analogues." In fact, this can be done in a general way, and certain combinations of acoustical parameters are given names taken from electric circuit theory. In the system just described, for example, the quantity $pck^2/2\pi$ is called the "acoustical resistance," R; V/pc^2 is called the "compliance," C; and pd'/S is called the "inertance," M. Moreover, these names are not given merely on the basis of the expressions appearing in the same place in the acoustical equation as the corresponding expressions in the equation for the series circuit; the analogy is more than formal. Thus, the inductance of an electric circuit is a measure of the tendency of the circuit to resist changes in current ($= dq/dt$), while the inertance of the acoustical system is a measure of the tendency of the system to resist changes in air velocity ($= dz/dt$). Similarly, the capacitance of a circuit is a measure of the ability of charge to "pile up," so to speak, in one place in the circuit, while the compliance of an acoustical system, e.g., the resonator, is a measure of just how far back the air in the neck will "allow itself" to be pushed. (If it is claimed that the analogy is nevertheless a formal one only, this would strengthen the point I shall presently make.)

Finally, the analogy is actually used to solve practical problems. It is not merely a heuristic aid—allowing one theory to be taught in terms of the other. To quote from a textbook in acoustics:

> Consideration of the equivalent electric circuit offers many advantages in solving practical engineering problems of applied acoustics. For example, many acoustical systems are so complicated that their mathematical analysis is very difficult, if not impossible, and their design by a cut-and-try [sic] experimental method is extremely tedious, as each change involves constructing a new part. On the other hand, if it is possible to set up an equivalent electric circuit, the electrical constants of this network may readily be varied [actually—not just in the mathematical analysis]

to obtain the desired experimental characteristics, and the constants of the mechanical system may then be calculated from their electrical equivalents. This technique has been used in the design of loudspeaker systems and other acoustical devices.[12]

Now, given all of this, is electric circuit theory a model for acoustical theory? (Or, are electric circuits models for acoustical theory?) Usually, the term *analogue* is used here, though one would be understood if the above relation were referred to as a modelling relation. The important point is, again, that there is no question of the identification of the subject matter of the two theories. Acoustics deals with small vibrations in air (generally, elastic fluids), whereas electric circuit theory deals with the movement of electric charge in certain types of systems. The analogy is perhaps not completely formal, but the substantive similarities are rather weak.

(3) Here I have in mind systems which physicists would certainly recognise as models—systems for which the question of "reality" *does* arise. But here, the physicist knows that the objects of the model cannot be identified with the theoretical objects of the theory for *definite physical reasons,* although there is some *substantive similarity.* As an example of this, consider the following passage from the arch-modellist, William Thompson (Lord Kelvin). (I have italicised the parts to which I wish attention drawn.)

> To think of ponderable matter, imagine for a moment that we make a *rude* mechanical *model.* Let this be . . . [Here Kelvin describes a rather elaborate contraption, as Duhem might have called it] . . . you will have a *crude model,* as it were, of what Helmholtz makes the subject of his paper on anomalous dispersion. . . . If we had only dispersion to deal with there would be no difficulty in getting a full explanation by putting this *not* in a *rude* mechanical model form, but in a form which would commend itself to our judgment as presenting the *actual* mode of action of the particles, of gross matter, whatever they may be upon the luminiferous ether. . . . It seems that there must be something in this molecular hypothesis, and that as a mechanical symbol, it is certainly *not a mere* hypothesis, but a *reality.* But alas for the difficulties of the undulatory theory of light. . . .[13]

Kelvin has offered a model, which is only a "rude" model, *because* it will not work in certain important, fully specified, situations. If it were not for these other circumstances, which contradict the phenomena that could be expected on the basis of the model, this would not be a rude model, but a "reality." Consider also the following passage:

> The luminiferous ether we must imagine to be a substance which *so far as* luminiferous vibrations are concerned moves *as if* it were an elastic solid. That it moves *as if* it were an elastic solid *in respect to* the luminiferous vibrations is the fundamental assumption of the wave theory of light. [*loc. cit.* p. 9 (Italics mine).]

Kelvin does not identify his models with the domain of the relevant theory for *empirical* reasons—not because of a general philosophical point about the relation between models and theories. For Kelvin goes on to show that there are *other specific* phenomena with respect to which the ether does *not* behave like an elastic solid (for example, the fact that material bodies move through it, showing that it sometimes behaves as if it were a fluid).[14] I think it is quite clear from the quoted passages that if it were not for these other, specific, intransigent phenomena, Kelvin would have dropped the as-if, rude-model, terminology.[15]

Notice that there is no question of identifying the ether with the actual gadgetry in a constructed model, just as for a nineteenth-century physicist there is no question of identifying the molecules of the kinetic theory of gases with actual billiard balls (which are made of ivory, have numbers on them, etc.). The proposed identity would be between the molecules and elastic spheres, which are exemplified by billiard balls; or between the ether and an array of springs, etc., which is exemplified by the gadget on the laboratory bench.

Failure to make this distinction is what lends plausibility to Braithwaite's statement about hydrogen atoms and solar systems quoted near the end of the introduction to this paper. Of course hydrogen atoms are not solar systems; the nucleus of the hydrogen atom is not a star and the electron is not a planet—just as gas molecules are not billiard balls. But is it as obvious that the hydrogen atom is not a system in which something is going around something else? (Unfortunately there is no standard general term for such systems, analogous to the term "elastic sphere" to describe billiard balls and other such objects.) Bohr believed this to be the case; we no longer do for definite physical reasons, although the similarity is still there—hence the "as-if."

(4) Finally, there are systems which are no longer *merely* models. Here, the question of the reality of the model (i.e., the identification of the objects of the model with the theoretical objects of the theory) is not only significant, but is answered in the affirmative. A system may originally be introduced as a model in sense (3), but may eventually be modified to such an extent that we will finally speak of the identity of the objects of this system with the theoretical objects of the theory. When this happens, we may or may not continue to speak of the system as being a *model.* Actual usage may depend on factors such as the historical development of the theory. For example, we may still speak of the discrete particle model for the kinetic theory of gases, even though we have identified the concept of a molecule from the kinetic theory with the concept of a discrete particle (or "object"). Here, the model is not *another* interpretation of the theory's calculus, but a *filling out of the original interpretation.* (This point will be made clearer in the following sections.)

Now, the first serious problem with Braithwaite's explication of the concept of a model as simply another interpretation of the theory's calculus is that it cannot distinguish between the four types of systems just discussed—and these *are* distinguished by physicists. For Braithwaite, all four would be models. However, we found that the first would not be recognised as such by physicists. The second would usually be described as an analogue. The fourth would probably not be described as a model either, but for a quite different reason—it is too good, as it were. "Questions of reality" arise for some of these systems, but not for others; whereas for Braithwaite, it is claimed that they should not arise for any of them. In other words, Braithwaite is correct in saying that identity of logical structure alone is not enough to be able to claim identity between the objects of the model and the theoretical objects of the theory. But I take this as showing that Braithwaite's explication of the concept of a model is inadequate, and not as an insight into the nature of models for physical theories.

It could be objected at this point that reliance on what physicists say in such situations is not sufficient for establishing a point about the logic of the situation (or for refuting such a point, as I am attempting to do). Perhaps not; I shall have more to say about this below. But suppose that we could emend Braithwaite's explication of the concept of a model in such a way as to take account of what physicists say—which *would* leave a place for the distinctions drawn above. And suppose that this emendation would be such that it could be accepted by the partial interpretation theorists as being within the letter and spirit of their analysis of the structure of physical theories. Clearly this would be a gain in understanding. I shall, in the next section, offer such a modification of Braithwaite's explication. I shall also attempt to show that this modification will be able to overcome certain further difficulties inherent in Braithwaite's explication.

3. AN EMENDATION OF BRAITHWAITE'S EXPLICATION WITH AN EXAMPLE

As long as only identity of formal structure is required between a theory and a model for it, there can be no relevant connection between the domain or subject matter of the theory and the domain of the model. A geometrical model for the calculus of a physical theory provides a striking example of this, and the electric circuit analogue for acoustical theory provides an example where the domains are both of physical objects, yet quite different. In each case, we do not have a model in the physicist's sense, and there is no question of the identification of the objects of the model with the theoretical objects of the theory.

If, however, the observable properties of the domain of the theory—the designata of the observation terms—are similar to the properties of the model represented by these same terms when the calculus is interpreted in the domain of the model, then the possibility arises of comparing the properties of the model represented by the theoretical terms of the calculus with the theoretical objects of the theory. That is, *we can argue by analogy to the nature of the theoretical properties.* A good example of this is provided by the elementary kinetic theory of gases and its usual model. It will be useful to compare in some detail the interpretations of the calculus of the kinetic theory of gases in the theory itself and in the elastic sphere model for the theory. (Notice that this distinction itself sounds somewhat strange, for this theory is usually presented by means of the model. The reasons for this will become clear as we proceed.)

In the theory, certain *defined* terms of the calculus are interpreted as designating the observable properties of a sample of gas in a container at equilibrium—volume, pressure, and temperature.[16] In the model, on the other hand, certain *primitive* terms of the calculus are interpreted as designating a group of elastic spheres in a container and some of their properties. Some of the primitive formulas express propositions describing these elastic spheres (their masses, for example), and others are interpreted as expressing the laws of classical dynamics (the laws governing their motion).

Now we notice that the designata of some of the defined terms of the calculus, when used to represent the theory, are similar (in this case, identical) to the corresponding designata of these terms, when the calculus is used to represent statements about the model. Thus, for example, there is an expression in the calculus, which by the interpretation given the calculus in the model, represents the total rate of momentum transfer per unit area to the walls of the container in which the elastic spheres are moving. This expression, when the calculus is used to represent the theory, appears in a correspondence rule of biconditional form (a definition, in Carnap's sense) with the observation term P, which designates the pressure of the gas. But according to classical dynamics, rate of momentum transfer per unit area is equal to force per unit area, which is (by definition) the pressure on the wall of the container. Thus we not only have a formal identity—a shared calculus, but also a *substantive* identity of two properties, one from the domain of the theory and one from the domain of the model.[17]

The same is trivially true in the case of volume, thus giving us two sets of *identical properties*—out of a possible three.[18] We are left with temperature in the theory and mean kinetic energy of the elastic spheres in the model as the designata of a defined term (T) in the calculus when used in the theory and in the model, respectively. Are these last two properties also similar? Reasoning by analogy on the basis of the first two similarities (identities, in this case), we would expect so, which would amount to suspecting that gases are, in fact, *composed of* elastic spheres (suspecting that molecules are elastic spheres—that the model "corresponds with reality"). At first sight, however, it would seem that the temperature of a gas and the mean kinetic energy of a swarm of elastic spheres are quite dissimilar. But we are not concerned here with temperature as a felt quality of bodies or as a sensation (hot-cold). Rather, we are concerned with the level of a column of mercury (for example—assuming that we are using a mercury thermometer) in a capillary tube. Now we notice that if liquids were also composed of elastic spheres,[19] they would expand if placed in contact with a gas, so composed, if (and only if) the mean kinetic energy of the particles of the gas were higher than in the liquid immediately before contact. In fact, an equilibrium state would soon be reached in which the mean kinetic energies of the two would be the same, at which time the liquid would cease expanding. All of this follows from classical dynamics. But this is just what is observed. When a mercury thermometer is inserted in a gas, the column will rise (or fall) for a time, and reach a stationary level.

On the basis of this, and the two identities mentioned earlier, we can conclude at least that gases behave just as we would expect them to behave if they were in fact composed of small elastic spheres in incessant motion. That is, we can tentatively identify the objects of the model with the theoretical objects of the theory. Gases behave as if they were composed of elastic spheres. The "as-if" does not here indicate that we have contrary information (cf. Kelvin) but rather that we may still feel that there are other tests to which we would like to put this hypothesis before committing ourselves. If we had not the slightest idea of what other tests would be relevant, or if we could satisfy ourselves that other tests were impossible, there would be no point in the "as-if." In this case, there are such further tests, and they follow directly from a consideration of the elastic sphere model. (The theory itself—considered as the partially interpreted calculus with semantical rules for the observation terms only—affords no reason whatsoever to expect these other phenomena.) I have in mind the so-called "molecular beam" experiments.

We open a slit in one wall of the container of gas, and by a suitable experimental arrangement[20] we can see whether the results are those which would be expected (quantitatively as well as qualitatively) if a stream of small (unobserved) particles having a certain mean kinetic energy were to issue from the slit in accordance with the dynamics of such particles. As is well known, such experiments do in fact confirm the hypothesis that gases are composed as described—that we can identify the theoretical objects of the theory with the objects of the model. The measured velocities of the particles are what they would have to be in order to identify the temperature of the gas with the mean kinetic energies of these particles, thus completing the analogy.

Imagine a physicist who now says: "Well, it has seemed as if gases might be composed of these small elastic spheres, considering the similarities you mentioned; now I'm convinced that they really are. After all, the results of this experiment too were exactly what could be expected if they were." Has he committed an error? Has he succumbed to one of the "dangers" involved in model thinking? If he had relied only on the *formal* characteristics of the model, he would indeed have made a mistake, although in this case a fortunate one. But the argument was based on the *substantive* characteristics of the model, and as such is perfectly good analogical reasoning. And it would still have been good reasoning even if the molecular beam experiment had given wholly different results, thus refuting the identification of the theoretical objects with the objects of the model. (Although we might then doubt the reliability of the apparatus rather than the model!) His mistake would then merely be that he had chosen a false hypothesis—but not a meaningless one, or an improbable one, or one which indicated a basic misunderstanding of the very use of models. The identification, on the basis of the first two sets of identical properties alone, is at least probable, and the results of the molecular beam experiment increase the probability.

Notice that it is not necessary at this point to produce a satisfactory theory of confirmation. Any explication of the notion of the probability of a scientific hypothesis which did not reflect the above would have to be rejected, as the reasoning sketched is as good a paradigm as can be found of one type of sound scientific reasoning.[21]

Thus, for a model to qualify as a "candidate for reality" (or, for a system to qualify as a model in the physicist's sense) it must be more than just some other interpretation of the theory's calculus. There must be a substantive similarity between the designata of the defined terms of the calculus when used to represent the observable properties in the domain of the theory and when used to represent the (formally) corresponding properties in the domain of the model. If there is, moreover, an identity of these properties, the model is no longer "merely" a model; the theoretical objects and properties simply *are*. . . .

This, then, is the emendation of Braithwaite's explication which will allow one to distinguish between the four types of domains sketched earlier, and which will accommodate an important type of reasoning based on models which physicists do in fact use. This emended form of Braithwaite's explication can also account for some of the uses to which models are put by physicists in modifying and extending a theory. I shall attempt to show this in the next section. But at the same time, it will become apparent that this emendation cannot be accepted by one who holds the partial interpretation thesis. We shall see that the modification which I have suggested is in contradiction with one of the most basic assumptions of this type of analysis of the structure of physical theories.

4. MODELS AND MODIFICATIONS OF A THEORY

The laws deduced from the simplified kinetic theory of gases sketched above are not as accurate as we would like them to be, and we would like to modify the theory to remedy this situation. In this case, the elastic sphere model "points to its own extension,"[22] providing leads as to how it can reasonably be modified. For example, we might argue as follows:

If gases are composed of elastic spheres, perhaps we should take their radii into account (i.e., they are not just point masses). Also, they may exert forces on each other when not in actual contact. If we make these assumptions, and modify the calculus of the theory accordingly, we will be able to derive a formula which, when interpreted term by term in the domain of the theory, is the van der Waals equation of state,

$$\left(P + \frac{a}{V^2}\right) (V - b) = cT,$$

where a is a constant for a particular type of gas depending on the force law, and b is a constant for each type of gas depending on the radii of the spheres. This is a more accurate description of the behaviour of gases than the original ideal gas law, $PV = kT$.

The new primitive formulas of the calculus (or, the primitive formulas of the new calculus) which allowed for the derivation of this law were essentially "read off" from the model. By this I mean that a domain of objects was described which obeys the laws of another familiar theory—the dynamics of rigid spheres attracting each other in accordance with a stated force law (involving a). The statements of the description, together with the statements of the laws of this other theory, upon disinterpretation, become the primitive formulas of the modified calculus. (In the original model, the other theory was the dynamics of mass-points interacting only by contact. This was an idealisation, in an obvious sense, hence the name "ideal gas law" for the equation of state derived from the associated theory. Note also how this is a reduction of part of thermodynamics to classical dynamics.)

In this modification of the model, the identity among derived properties in the model and in the theory, spoken of earlier, still obtains. Thus if we had started with the van der Waals equation of state, we could have used the same sort of reasoning by analogy described earlier to establish, or make probable, the identification of the primitive properties of this model with the theoretical properties of the theory. And once again, this sort of reasoning would not be valid if we had considered only the formal similarity (as Braithwaite correctly recognises when he argues against identifying the model with the theory). Moreover, if we had started with the van der Waals equation, it would have been impossible to *construct* the calculus of the kinetic theory without thinking of the model. (The very form in which the equation is stated betrays its origin in the model.) And if perchance someone hit upon the right modification of the original calculus without considering the substantive aspects of the model, he would have no justification for believing it probable.

This difference between what can be done with the "theory itself" and with the model is even more striking when we consider the partial interpretation thesis view that theoretical terms obtain their meaning by being connected with observation terms via correspondence rules. For now we have a much more complicated calculus just in order that the directly interpreted formula relating *P, V,* and *T* can be more accurate for a certain range of conditions. Now suppose (contrary to fact) that the ideal gas law were more accurate than the van der Waals law. Before dropping the modified theory in favour of the simpler one, physicists would, I think, look for experimental errors, because the modified model, associated with the modified theory, appears to be a *more plausible* representation of the theoretical objects than is the original model. But on the basis of the partial interpretation thesis, and its associated explication of the notion of a model, this would be an irrational procedure, stemming from a "misunderstanding" of what a model is.

Also, the derivation of the results of the molecular beam experiment could not be carried out with either the simplified or the full theory without a change in the observation language itself. We would need new terms in the calculus designating the distance between the container of gas and a sheet of film located inside a rotating cylinder, the angular velocity of the cylinder, the degree of blackening of the film, etc.[23] But if we introduce these new terms and the new correspondence rules needed for them, and at the same time accept the partial interpretation thesis as to how theoretical terms obtain their empirical meaning, we will have the very strange result that the meanings of the terms *molecule, mass, velocity,* etc., have changed. After all, the theoretical terms are supposed to obtain their meaning from the observation terms through correspondence rules, and we have added new observation terms and correspondence rules. In the model, on the other hand, there is no such change in meaning. "Elastic sphere" means the same when the model is first conceived as it does after we realise that this model points to a new *test* of the theory in the molecular beam experiment. But we found that we could identify the objects of the model with the theoretical objects of the theory, through analogical reasoning based on substantive similarities in the two domains. Therefore, on the basis of this identification we are forced to conclude (what is more plausible on its own merit) that the addition of the new test with its corresponding set of new observation terms and correspondence rules does not change the meaning of the theoretical terms of the theory, but rather gives new ways to test the truth of the theoretical statements (whose meanings are given in some other way—directly, as will soon become apparent).

The view that the meanings of the theoretical terms change with the addition of new observation terms and corresondence rules has a similar strange consequence when we consider the sort of unification of physical theories that is accomplished by reasoning based on the use of a common model. For example, Niels Bohr, in his 1913 paper, "On the Constitution of Atoms and Molecules,"[24] using reasoning based on a model of the atom (notice that it is not called a model for atomic theory) which can be considered as a detailing of the elastic sphere model of kinetic theory, explains why certain lines in the spectrum of hydrogen gas had been missing, and predicts under what conditions we can expect to observe them. According to the partial interpretation thesis, even though the Bohr theory contains the terms *atom, mass,* etc., these terms have a different meaning from that in the kinetic theory of gases. The observation language of the Bohr theory is different from that of the kinetic theory, as are (of course) the correspondence rules; and his reasoning, which depends on the identity of the meanings of these terms in the two theories, would be invalid.

In cases like this, it is the model which is the heart of the physicist's investigations. If we disinterpret the calculus that is read off from the model, and reinterpret it "from the bottom up," and call the result the theory itself, looking now upon the model as merely another possible interpretation of the theory's calculus, we must accept the conclusion that arguments such as Bohr's are colossal logical blunders—fortunate blunders—showing a lack of understanding of how theoretical terms get their meaning, and a lack of understanding of what a physical theory is.

Now it may well be the case that much of what physicists *say* about the methodology of their science is of dubious value—I do not wish to argue this point here. But the partial interpretation thesis, coupled with Braithwaite's explication of the concept of a model (which leaves out substantive similarities), is not only in contradiction with this; it is also in contradiction with actual theory construction and theory modification—what physicists *do.* Now Braithwaite, Carnap, and Nagel say that they are not interested in constructing a logic of discovery. Rather, they are interested in analysing the final product. But their position is not neutral with respect to the former; it instead declares certain types of reasoning which are paradigms of physical genius to be (fortunate) logical mistakes of a very basic sort. (The parenthetical adjective alone should make one suspicious.)

We could not even describe the kinetic theory of gases as a reduction of thermodynamics to (statistical) dynamics. We could only notice that there is a formal similarity between some of the primitive formulas of the kinetic theory and the laws of classical dynamics. But these laws would not have the same meaning in each case. In the kinetic theory, the meanings of the terms *force, mass, momentum,* etc., would have to be analysed in terms of the observation terms *pressure, temperature,* and *volume.*

What has happened here? I think the basic point is that the suggested emendation of Braithwaite's explication of the notion of a model, i.e., taking into account substantive features of the model, as plausible as it may seem, cannot be accepted by one who holds the partial interpretation thesis. That is, even though my suggested emendation eliminates the difficulties I have been pointing out, and apparently does so while remaining within the letter and spirit of the programme of the partial interpretation theorists, the emendation contradicts a basic assumption of the partial interpretation thesis. For analogical reasoning from *substantive* similarities in the designata of terms in the derived formulas in the model and in the theory to substantive similarities in the theoretical (primitive) properties amounts to a direct interpretation of the theoretical terms in the theory. Thus, in the case of an *identity* of derived properties, the completion of the analogy is tantamount to *giving semantical rules for the theoretical terms.* (If there is only a similarity, in which case we have one type of "as-if thinking," we are giving qualified semantical rules, so to speak—cf. the earlier quotations from Kelvin.) But according to the partial interpretation thesis, we *cannot* give a consistent direct interpretation (semantical rules) for the theoretical terms of the calculus when it is used to express the theory, for "we could not understand them" (Carnap); the theoretical terms are "not understood in themselves," but only as part of the whole system (Braithwaite). Theoretical terms gain what empirical meaning they have *only* "from below"; the "zipper" moves from the bottom up in giving meaning to the theoretical terms. Thus, if this thesis is held as to how theoretical terms become meaningful, one is forced to consider a model as some *other* interpretation of the calculus. To accept our results is to accept the possibility of giving meaning directly for theoretical terms.

5. CONCLUSIONS

Our suggested emendation of Braithwaite's explication of the concept of a model has implied that theoretical terms *can* be given direct interpretations—semantical rules, where the metalanguage used to state them *is* "understood." It is in fact the same metalanguage used to interpret the observation terms of the calculus. That is, if we wish to analyse physical theories in terms of interpreted formal calculi, we *can* give semantical rules for theoretical terms (thus, for *all* terms of the calculus). Notice that this still gives the meanings of the theoretical terms "on the basis of observations," but by statements outside the calculus rather than by correspondence rules.

I have also shown that theoretical terms *must* be given semantical rules, to allow for certain types of physical reasoning. Certain extensions and modifications of a theory were based on the use of a model *in the emended sense,* which we saw is tantamount to interpreting the theoretical terms of the theory. But this refutes what I referred to as assumptions (a) and (b) of the partial interpretation thesis in the introduction to this paper.[25]

All of this also implies that theoretical or unobservable *objects* may (in some cases) be described by observational *predicates.* Let us see why this is so. If a semantical rule is to be successful in giving meaning to a theoretical term of the calculus, it must be stated in a metalanguage which is "already understood." That is, it must supply a designatum which is understood independently of the theory being reconstructed. Otherwise it would be, as Carnap has put it, a useless transcription "from a symbol in a symbolic calculus to a corresponding word expression in a calculus of words."[26] This requirement has usually been put in the form of demanding that the properties, things, and relations which are to be the designata of the terms of the calculus must be observable, i.e., the metalanguage must be a theory-uninfected observation language. (This observation language is not to be confused with the observation language of the theory's calculus, which is part of the object language being analysed.) It would seem that familiarity should be sufficient; but let us grant the stronger requirement of observability for the present. Now, if we give semantical rules for the theoretical predicates (say) of the calculus, and do so in accordance with the just mentioned condition on the metalanguage, we will have the result that *unobservable objects may be characterised by observational predicates.* More accurately, we will have a sentence of the form $P(a)$, with a designating an unobservable (= theoretical) object and P designating a property of which it is possible to observe instances applying to observable objects (= "having weight" as applied to an electron or atom).

This is the same conclusion reached earlier from a consideration of the concept of a model for a theory. We saw that the identification of the objects (and properties) of the model with the designata of the theoretical terms of "the theory itself" was tantamount to giving semantical rules for theoretical terms and thus applying observational predicates to unobservable objects. In the example of the kinetic theory of gases, this amounted to maintaining that the unobservable atoms could have observational properties—or at least familiar properties (mass, velocity) from another theory (classical dynamics).

Moreover, we saw that there is nothing unintelligible about unobservable objects being characterised by observational predicates. The reasoning that leads a physicist to impute observational properties to unobservable objects was seen to be perfectly acceptable analogical reasoning.

These conclusions, however, conflict with the most basic presupposition of the proponents of the semantical system approach. This is the assumption that we can distinguish in a general manner two types of terms or concepts (and statements) in physical theories: The observation terms, which designate observable properties, relations, events, and objects *only;* and the theoretical terms, which purport to designate unobservable properties, relations, events, and objects *only.* This "dual language model" of the vocabulary of science has been expressed by Carnap as follows:

> [We accept] the customary and useful [division of] the language of science into two parts, the observation language and the theoretical language, [where] the observation language uses terms designating observable properties for the description of observable things and events, [and] the theoretical language [uses] terms which may refer to unobservable aspects or features of events, e.g., to microparticles . . .[27]

We have seen, therefore, that what are perhaps the three most basic assumptions of the partial interpretation thesis are in error.

Now, these assumptions involve two crucial notions—that of an observation term, and that of a theoretical term. In this paper, I have treated these notions as if they were clear and unproblematic. They are not, however, and a full evaluation of the partial interpretation thesis—and the semantical system approach itself—requires a careful analysis of them. This I hope to do in future papers.

NOTES

1. The oldest and most precise formulation of this position is found in Rudolf Carnap, "Foundations of Logic and Mathematics," I, No. 3, of the *International Encyclopedia of Unified Science* (Chicago, 1939). Later statements of this view are found in other writings of Carnap; in various papers by Carl Hempel; in Ernest Nagel's *The Structure of Science* (New York, 1961); in R. B. Braithwaite's *Scientific Explanation* (Cambridge, 1953); Arthur Pap's *Introduction to the Philosophy of Science* (Glencoe, Ill., 1962); Ernest Hutten's *The Language of Modern Physics* (London, 1956); Peter Caws's *The Philosophy of Science* (Princeton, 1965). Further references may be found in the works just cited.

2. See Carnap, *loc. cit.* p. 203; Carnap, "The Methodological Character of Theoretical Concepts," in *Minnesota Studies in the Philosophy of Science,* I, 38; Nagel, *op. cit.* pp. 81 ff.; Braithwaite, *op. cit.* p. 51. The same assumption is made by the other authors mentioned in n. 1. Most of them hasten to say that the distinction may not be a sharp one.

3. See Carnap, "Foundations of Logic and Mathematics," *loc. cit.* p. 153; Carnap, *Meaning and Necessity* (Chicago, 1946), pp. 4 f. See also the introductory sections of Alonzo Church, *Introduction to Mathematical Logic* (Princeton, 1956).

4. I have borrowed this expression from Professor Peter Achinstein of the Johns Hopkins University.

5. See his book, *Scientific Explanation,* as well as his more recent paper, "Models in the Empirical Sciences," in *Logic, Methodology, and Philosophy of Science,* eds. Nagel, Suppes, and Tarski (Stanford, 1960).

6. It will not usually matter whether we use the term *model* to refer to the domain of objects which the interpreted calculus makes statements about, or to the interpretative statements themselves.

7. Cf. "Foundations of Logic and Mathematics," p. 204. The metalanguage must be already understood. The familiarity of the concepts need not entail their observational character. They may be from another, "better understood" theory; but eventually, the chain must end with observational concepts. (See Braithwaite, "Models in the Empirical Sciences," *loc. cit.* p. 227.)

8. *Scientific Explanation,* p. 90.

9. *The Structure of Science,* p. 116.

10. *Scientific Explanation,* p. 31.

11. Compare Nagel, *op. cit.* p. 116.

12. L. E. Kinsler and A. R. Frey, *Fundamentals of Acoustics* (New York, 1950), p. 33.

13. *Baltimore Lectures on Molecular Dynamics and the Wave Theory of Light* (London, 1904), pp. 12–14.

14. Thus it was much like the product once on the market in the United States known as "Silly Putty." This was a substance which could be

moulded like clay, but which would also bounce if dropped several feet. It behaved like a fluid under some circumstances (slowly applied force), but like an elastic solid under others (rapidly applied force). Ice is another example; glaciers flow, although ice is "usually" brittle.

15. See also Edmund Whittaker, "Models of the Aether," *A History of Theories of Aether and Electricity* (New York, 1960), I, Chap. ix. Whittaker describes a series of (mechanical) models of the ether, and it is apparent that the "question of reality" *did* arise for these models and was taken quite seriously. In each of these cases, there was some *substantive* similarity between the domain of the model and the domain of the theory (electrodynamics). And in each case, the model was considered only as an "as-if" because it didn't work, i.e., certain phenomena (though not all) expected on the basis of the model were not observed.

It is also interesting to note that in many of the models cited by Whittaker, part of the reason for failure was a lack of complete *formal* identity. Thus, according to Braithwaite's explication, certain paradigm cases of models employed by physicists would have to be described as *not* being models at all! But I shall let this pass.

16. If pressure and temperature are not considered as sufficiently elementary or observable, the analysis could be carried out in terms of the observed heights of mercury columns; but this would be at the expense of clarity without changing the results.

17. Here, the identity is established on the basis of classical dynamics, rather than being an observed identity. But, as Braithwaite has noted, the familiarity need not necessarily involve observability. (See n. 7, p. 292 supra.)

18. We are at present interested in only three "observable" properties of the theory—those which enter into the ideal gas law, $PV = kT$.

19. But where the spheres also exert an attractive force on one another. This is admittedly crude, but putting in all of the details would only complicate matters without helping (or hindering) the point I wish to make.

20. See F. W. Sears, *Thermodynamics* (New York, 1955), p. 241—or almost any other elementary book on thermodynamics.

21. This type of reasoning is fruitfully compared with C. S. Peirce's "abductive reasoning"; "The surprising fact, *C*, is observed; but if *A* were true, *C* would be a matter of course. Hence, there is reason to suspect *A* is true." [*Collected Papers of C. S. Peirce* (Cambridge, Mass., 1935), V, para. 189.]

Einstein's explanation of the Brownian motion on the basis of the kinetic theory of gases is a more striking example of this. It was this which convinced many doubters that the atomic theory was not just a "convenient fiction."

22. This is Braithwaite's phrase in "Models in the Empirical Sciences," *loc. cit.* p. 229.

23. See Sears, *op. cit.*

24. *Philosophical Magazine*, XXVI, 1913, pp. 9–10.

25. I should point out here that in one sense the conclusions I have so far drawn, and those I shall presently draw, are not entirely new. [See, for example, R. Harré, *An Introduction to the Logic of the Sciences* (London, 1960) Chap. iv; M. B. Hesse, *Models and Analogies in Science* (London, 1963); N. R. Campbell, *Physics, The Elements* (Cambridge, 1920) Chap. vi (recently reprinted in a paper-back edition by Dover entitled *Foundations of Science*); H. Putnam, "What Theories Are Not," in *Proceedings of the Congress of Logic, Methodology, and Philosophy of Science* (Stanford, 1962) pp. 240–51.]

However, I believe that I have argued for these conclusions in an importantly novel way. I have been concerned to show how these conclusions can be *generated out of* the "partial interpretation" analysis of the structure of physical theories and models. I do not believe that this has been done by the authors cited above. (Hesse comes the closest to this sort of undertaking.) My *general* point of view, which may be called, roughly, "realistic," is of course not new at all.

26. "Foundations of Logic and Mathematics," p. 210.

27. "The Methodological Character of Theoretical Concepts," in *Minnesota Studies in the Philosophy of Science*, I (Minneapolis, 1956), 38.

Realism

Bas C. van Fraassen

TO SAVE THE PHENOMENA*

After the demise of logical positivism, scientific realism has once more returned as a major philosophical position. I shall not try here to criticize that position, but rather attempt to outline a comprehensive alternative.[1]

I

What exactly is scientific realism? Naively stated, it is the view that the picture science gives us of the world is true, and the entities postulated really exist. (Historically, it added that there are real necessities in nature; I shall ignore that aspect here.[2]) But that statement is too naive; it attributes to the scientific realist the belief that today's scientific theories are (essentially) right.

The correct statement, it seems to me, must indeed be in terms of epistemic attitudes, but not so directly. The aim of science is to give us *a literally true story of what the world is like;* and the proper form of acceptance of a theory is to believe that it is true. This is the statement of scientific realism: "To have good reason to accept a theory is to have good reason to believe that the entities it postulates are real," as Wilfrid Sellars has expressed it. Accordingly, an anti-realism is a position according to which the aims of science can well be served without giving such a literally true story, and acceptance of a theory may properly involve something less (or other) than belief that it is true.

The idea of a literally true account has two aspects: the language is to be literally construed; and, so construed, the account is true. This divides the anti-realists into two sorts. The first sort holds that science is or aims to be true, properly (but not literally) construed. The second holds that the language of science should be literally construed, but its theories need not be true to be good. The anti-realism I advocate belongs to the second sort.

II

When Newton wrote his *Mathematical Principles of Natural Philosophy* and *System of the World,* he carefully distinguished the phenomena to be saved from the reality he postulated. He distinguished the "absolute magnitudes" that appear in his axioms from their "sensible measures" which are determined experimentally. He discussed carefully the ways in which, and extent to which, "the true motions of particular bodies [may be determined] from the apparent," via the assertion that "the apparent motions . . . are the differences of true motions."[3]

---

*To be presented in an APA symposium on Scientific Realism, December 28, 1976. Richard N. Boyd and Clark Glymour will comment; see this JOURNAL, this issue, pp. 633–635 and 635–637, respectively.

The research for this paper was supported by Canada Council Grant S74-0590. An earlier version was presented at the Western Division of the Canadian Philosophical Association (Calgary, October 1975).

I want to acknowledge my debt to Clark Glymour, Princeton University, for the challenge of his critiques of conventionalism in his dissertation and unpublished manuscripts.

Reprinted by permission from the author and *The Journal of Philosophy,* Vol. 73, No. 18, October 21, 1976.

The "apparent motions" form relational structures defined by measuring relative distances, time intervals, and angles of separation. For brevity, let us call these relational structures *appearances.* In the mathematical model provided by Newton's theory, bodies are located in Absolute Space, in which they have real or absolute motions. But within these models we can define structures that are meant to be exact reflections of those appearances and are, as Newton says, identifiable as differences between true motions. These structures, defined in terms of the relevant relations between absolute locations and absolute times, which are the appropriate parts of Newton's models, I shall call *motions,* borrowing Simon's term.[4]

When Newton claims empirical adequacy for his theory he is claiming that his theory has some model such that *all actual appearances are identifiable with* (*isomorphic to*) motions in that model.

Newton's theory goes a great deal further than this. It is part of his theory that there is such a thing as Absolute Space, that absolute motion is motion relative to Absolute Space, that absolute acceleration causes certain stresses and strains and thereby deformations in the appearances, and so on. He offered, in addition, the hypothesis (his term) that the center of gravity of the solar system is at rest in Absolute Space. But, as he himself noted, the appearances would be no different if that center were in any other state of constant absolute motion.

Let us call Newton's theory (mechanics and gravitation) *TN,* and *TN(v)* the theory *TN* plus the postulate that the center of gravity of the solar system has constant absolute velocity. By Newton's own account, he claims empirical adequacy for *TN*(0); and also claims that, if *TN*(0) is empirically adequate, then so are all the theories *TN(v).*

Recalling what it was to claim empirical adequacy, we see that all the theories *TN(v)* are empirically equivalent exactly *if all the motions in a model of TN(v) are isomorphic to motions in a model TN(v + w),* for all constant velocities v and w. For now, let us agree that these theories are empirically equivalent, referring objections to a later section.

III

What exactly is the "empirical import" of *TN*(0)? Let us focus on a fictitious and anachronistic philosopher Leibniz*, whose only quarrel with Newton's theory is that he does not believe in the existence of Absolute Space. As a corollary, of course, he can attach no "physical significance" to statements about absolute motion. Leibniz* believes, like Newton, that *TN*(0) is empirically adequate; but not that it is true. For the sake of brevity, let us say that Leibniz* *accepts* the theory but that he does not *believe* it; when confusion threatens we may expand that idiom to say that he *accepts the theory as empirically adequate,* but does not *believe it to be true.* What does Leibniz* believe, then?

Leibniz* believes that *TN*(0) is empirically adequate, and hence, equivalently, that all the theories *TN(v)* are empirically adequate. Yet we cannot identify the theory that Leibniz* holds about the world—call it *TNE*—with the common part of all the theories *TN(v).* For each of the theories *TN(v)* has such consequences as that the earth has *some* absolute velocity, and that Absolute Space exists. In each model of each theory *TN(v)* there is to be found something other than motions, and there is the rub.

To believe a theory is to believe that one of its models correctly represents the world. A theory may have isomorphic models; that redundancy is easily removed. If it has been removed, then to believe the theory is to believe that exactly one of its models correctly represents the world. Therefore, if we believe of a family of theories that all are empirically adequate, but each goes beyond the phenomena, then we are still free to believe that each is false, and hence their common part is false. For that common part is phrasable as: one of the models of one of those theories correctly represents the world.

IV

It may be objected that theories will seem empirically equivalent only so long as we do not consider their possible extensions.[5] The equivalence may generally, or always, disappear when we consider their implications for some further domain of application. The usual example is Brownian motion; but this is imperfect, for it was known that phenomenological and statistical thermodynamics disagreed even on macroscopic phenomena over sufficiently long periods of time. But there is a good, *fictional* example: the combination of electromagnetism with mechanics, if we ignore the unexpected null results that led to the replacement of classical mechanics.

Maxwell's theory was not developed as part of mechanics, but it did have mechanical models. This follows from a result of Koenig, as detailed by Poincaré in the preface of his *Electricité et Optique* and elsewhere. But the theory had the strange new feature that velocity itself, not just its derivative, appears in the equations. A spate of thought experiments was designed to measure absolute velocity, the simplest perhaps that of Poincaré:

> Consider two electrified bodies; though they seem to us at rest, they are both carried along by the motion of the earth; . . . therefore, equivalent to two parallel currents of the same sense and these two currents should attract each other. In measuring this attraction, we shall measure the velocity of the earth; not its velocity in relation to the sun or the fixed stars, but its absolute velocity.[6]

The null outcome of all experiments of this sort led to the replacement of classical by relativistic mechanics. But let us imagine that values *were* found for the absolute velocities; specifically for that of the center of the solar system. Then, surely, one of the theories *TN*(v) *would be confirmed and the others falsified?*

This reasoning is spurious. Newton made the distinction between true and apparent motions without presupposing more than that basic mechanics in which Maxwell's theories has models. Each motion in a model of *TN*(v) is isomorphic to one in some model of *TN*($v + w$), for all constant velocities v and w. Could this assertion of empirical equivalence possibly be controverted by those nineteenth-century reflections? The answer is *no.* The thought experiment, we may imagine, confirmed the theory that added to *TN* the hypothesis:

HO. The center of gravity of the solar system is at absolute rest.
EO. Two electrified bodies moving with absolute velocity v attract each other with force $F(v)$.

This theory has a consequence strictly about appearances:

CON. Two electrified bodies, moving with velocity v relative to the center of gravity of the solar system attract each other with force $F(v)$.

However, the same consequence can be had by adding to *TN* the two alternative hypotheses:

Hw. The center of gravity of the solar system has absolute velocity w.
Ew. Two electrified bodies moving with absolute velocity $v + w$ attract each other with force $F(v)$.

More generally, for each theory *TN*(v) there is an electromagnetic theory $E(v)$ such that $E(0)$ is Maxwell's and all the combined theories *TN*(v) plus $E(v)$ are empirically equivalent.

There is no originality in this observation, of which Poincaré discusses the equivalent immediately after the passage I cited above. Only familiar examples, but rightly stated, are needed, it seems, to show the feasibility of concepts of empirical adequacy and equivalence. In the remainder of this paper I shall try to generalize these considerations, while showing that the attempts to explicate those concepts *syntactically* had to reduce them to absurdity.

V

The idea that theories may have hidden virtues by allowing successful extensions to new kinds of phenomena, is too pretty to be left. Nor is it a very new idea. In the first lecture of his *Cours de philosophie positive,* Comte referred to Fourier's theory of heat as showing the emptiness of the debate between partisans of calorific matter and kinetic theory. The illustrations of empirical equivalence have that regrettable tendency to date; calorifics lost. Federico Enriques seemed to place his finger on the exact reason when he wrote: "The hypotheses which are indifferent in the limited sphere of the actual theories acquire significance from the point of view of their possible extension."[7] To evaluate this suggestion, we must ask what exactly is an extension of a theory.

Suppose that experiments really had confirmed the combined theory *TN*(0) plus *E*(0). In that case mechanics would have won a *victory.* The claim that *TN*(0) was empirically adequate would have been borne out by the facts. But such victorious extensions could never count for a theory as against one of its empirical equivalents.

Therefore, if Enriques' idea is to be correct, there must be another sort of extension, which is really a defeat—but qualified. For a theory *T* may have an easy or obvious modification which is empirically adequate, while another theory empirically equivalent to *T* does not. One example may be the superiority of Newton's celestial mechanics over the variant produced by Brian Ellis; Ellis himself seems to be of this opinion.[8] This is a *pragmatic* superiority and cannot suggest that theories, empirically equivalent in the sense explained, can nevertheless have different empirical import.

VI

We still need a general account of empirical adequacy and equivalence. It is here that the syntactic approach has most conspicuously failed. A theory was conceived as identifiable with the set of its theorems in a specified language. This language has a vocabulary, divided into two classes—the observational and theoretical terms. Let the first class be *E;* then the empirical import of theory *T* was said to be its subtheory *T*/*E*—those theorems expressible in that subvocabulary. *T* and *T*′ were declared empirically equivalent if *T*/*E* was the same as *T*′/*E*.

Obvious questions were raised and settled. Craig showed that, under suitable conditions, *T*/*E* is axiomatizable in the vocabulary *E*. Logicians attached importance to questions about restricted vocabularies, and this was apparently enough to make philosophers think them important too. The distinction between observational and theoretical terms was more debatable, and some changed the division into "old" and "newly introduced" terms.[9] But all this is mistaken. Empirical import cannot be isolated in this syntactic fashion. If that could be done, then *T*/*E* would say exactly what *T* says about what is observable, and nothing else. But consider: the quantum theory, Copenhagen version, says that there are things which sometimes have a position in space and sometimes do not. This consequence I have just stated without using theoretical terms. Newton's theory *TN* implies that there is something (to wit, Absolute Space) which neither has a position nor occupies volume. As long as unobservable entities differ systematically from observable entities with respect to observable characteristics, *T*/*E* will say that there are such things if *T* does.

The reduced theory *T*/*E* is not a description of the observable part of the world of *T;* rather, it is a hobbled and hamstrung version of *T*'s description of everything. Empirical equivalence fares as badly. In section II, $TN(0)$ and *TNE must* be empirically equivalent, but the above remark about *TN* shows that $TN(0)/E$ is not TNE/E. To eliminate such embarrassments, extensions of theories were considered in attempts to redefine empirical equivalence.[10] But these have similar absurd consequences.

The worst consequence of the syntactic approach was surely the way it focused philosophical attention on irrelevant technical questions. The expressions 'theoretical object' and 'observational predicate' mark category mistakes. Terms may be theoretical, but 'observable' classifies putative entities. Hence there cannot be a "theoretical/observable distinction." It is true surely that elimination of all theory-laden terms would leave no usable language; also that 'observable' is as vague as 'bald'. But these facts imply not at all that 'observable' marks an unreal distinction. It refers quite clearly to our limitations, the limits of observation, which are not incapacitating, but also not negligible.

VII

The phenomena are saved when they are exhibited as fragments of a larger unity. For that very reason it would be strange if scientific theories described the phenomena, the observable part, in different terms from the rest of the world they describe. And so an attempt to draw the conceptual line between phenomena and the transphenomenal by means of a distinction of vocabulary, must always have looked too simple to be good.

Not all philosophers who discussed unobservables, by any means, did so in terms of vocabulary. But there was a common assumption: that the distinction marked is philosophical. Hence it must be drawn, if at all, by philosophical analysis and, if attacked, by philosophical arguments. This attitude needs a Grand Reversal. If there are limits to observation, these are empirical, and must be described by empirical science. The classification marked by "observable" must be of entities in the world of science. And science, in giving content to the distinction, will reveal how much we believe when we accept it as empirically adequate.

A future Unified Science may detail the limits of observation exactly; meanwhile, extant theories are not silent on them. We saw Newton's delineation; for relativity theory, we have two revealing studies by Clark Glymour. The first shows that local (hence, I should think, measurable) quantities do not uniquely determine global features of space-time.[11] The second shows that these features also are not uniquely determined by structures each lying wholly within some absolute past cone—hence, I should think, by observable structures. It is the theory of relativity itself, after all, that places *an absolute* limit on the information we can gather, through the limiting function of the speed of light.

In the foundations of quantum mechanics much more attention has been given to measurement. Much of the discussion is about necessary limitations: the role of noise in amplification, the distinction between macro- and micro-observables.[12] Yet we have no such clarity as Glymour gave us for relativity theory, concerning the extent to which macro-structure determines micro-structure. The debate over scientific realism may at least have the virtue of directing attention to such questions.

Science itself distinguishes the observable that it postulates from the whole it postulates. The distinction, being in part a function of the limits science discloses on human observation, is anthropocentric. But, since science places human observers among the physical systems it means to describe, it also gives itself the task of describing anthropocentric distinctions. It is in this way that even the scientific realist must observe a distinction between the phenomena and the transphenomenal in the scientific world picture.

VIII

I have laid some philosophical misfortunes at the door of a mistaken orientation toward syntax. The alternative is to say that theories are presented directly by describing their models. But does this really introduce a new element? When you give the theorems of *T*, you give the set of models of *T*—namely, all those structures which satisfy the theorems. And, if you give the models, you give at least the set of theorems of *T*—namely, all those sentences which are satisfied in all the models. Does it not follow that we can as advantageously identify *T* with its theorems as with its models?

But there is an ellipsis in the argument. It is being assumed that there is a specific language *L* which is the one language that belongs to *T*. And indeed, the theorems of *T* in *L* determine, and are determined by, the set of model structures of *L* (that is, structures in which *L* is interpreted) in which those theorems are satisfied. However, the assumption that there is a language *L* which plays this role for *T* places important restrictions on what the set of models of *T* can be like.

A theory provides, among other things, a specification (more or less complete) of the parts of its models that are to be direct images of the structures described in measurement reports. In the case of Newton's mechanics, I called those parts *motions;* in general, let us call them *empirical substructures.* The structures described in measurement reports we may continue to call *appearances.* A theory is *empirically adequate* exactly if all appearances are isomorphic to empirical substructures in at least one of its models. Theory *T* is *empirically no stronger* than theory *T′* exactly if, for each model *M* of *T*, there is a model *M′* of *T′* such that all empirical substructures of *M* are isomorphic to empirical substructures of *M′*. Theories *T* and *T′* are *empirically equivalent* exactly if neither is empirically stronger than the other. In that case, as an easy corollary, each is empirically adequate if and only if the other is.

In section v, I distinguished two kinds of extensions, the first a sort of victory and the second a sort of defeat. Let us call the first a *proper extension:* this simply narrows the class of models. We may call a theory *empirically minimal* if it is not empirically equivalent to any of its proper extensions. Glymour has convincingly argued, in the work cited above, that General Relativity is not empirically minimal. The reason is, in my present terms, that only local properties of space-time enter the descriptions of the appearances, but models may differ in global properties. This is a further nontrivial example of empirical equivalence.

The second sort of extension I shall not try to define precisely. The idea is that models of the theory may differ in structure other than that of the empirical substructures. In that case the theory is not empirically minimal, but this may put it in the advantageous position of offering modeling possibilities when radically new phenomena come to light. An example may yet be offered by hidden-variable theories in quantum mechanics.[13]

In terms of the concepts now at our disposal, and the examples given, we can conclude that there are indeed nontrivial cases of empirical equivalence, non-uniqueness, and extendability, both proper and improper. Such cases are now seen to be quite possible *even if the formulation of the theory has not a single term that cannot be called observational, in some way.* And now it should be possible to state the issue of scientific realism, which concerns our epistemic attitude toward theories rather than their internal structure.

All the results of measurements are not in; they are never all in. Therefore we cannot know what all the appearances are. We can say that a theory is empirically adequate, that all the appearances will fit (the empirical substructures of) its models. Though we cannot know this with certainty, we can reasonably believe it. All this is the case not only for empirical adequacy but for truth as well. Yet there are two distinct epistemic attitudes that can be taken: we can *accept* a theory (accept it as empirically adequate) or *believe* the theory (believe it to be true). We can take it to be the aim of science to produce a literally true story about the world, or simply to produce accounts that are empirically adequate. This is the issue of scientific realism versus its (divided) opposition. The intrascientific distinction between the observable and the unobservable is an anthropocentric distinction; but it is reasonable that the distinction should be drawn in terms of *us,* when it is a question of *our* attitudes toward theories.

NOTES

1. For some criticisms, see my "Theoretical Entities: The Five Ways," *Philosophia* 4 (1974): 95–109, and "Wilfrid Sellars on Scientific Realism," *Dialogue,* XIV, 4 (December 1975): 606–616.

2. Cf. my "The Only Necessity Is Verbal Necessity," forthcoming in this JOURNAL, LXXIV, 2 (February 1977).

3. F. Cajori, ed., *Sir Isaac Newton's Mathematical Principles of Natural Philosophy and His System of the World* (Berkeley: University of California Press, 1960), p. 12.

4. Herbert A. Simon, "The Axiomatization of Classical Mechanics," *Philosophy of Science,* XXI, 4 (October 1954): 340–343.

5. See, for example, Richard N. Boyd, "Realism, Undetermination, and a Causal Theory of Evidence," *Noûs,* VII, 1 (March 1973): 1–12.

6. Henri Poincaré, *The Value of Science,* B. Halsted, tr. (New York: Dover, 1958), p. 98.

7. *Historical Development of Logic,* J. Rosenthal, tr. (New York: Holt, 1929), p. 230.

8. "The Origins and Nature of Newton's Laws of Motion," in R. Colodny, ed., *Beyond the Edge of Certainty* (Englewood Cliffs, N.J.: Prentice-Hall, 1965), pp. 29–68.

9. For example, David Lewis, "How to Define Theoretical Terms," this JOURNAL, LXVII, 13 (July 9, 1970): 427–446. This paper is not subject to my criticisms here; on the contrary, it provides independent reasons to conclude that the empirical import of a theory cannot be syntactically isolated.

10. See fn 5 above. We could say that Boyd's paper, like Lewis's, provides independent evidence that empirical import cannot be syntactically isolated. But Boyd concludes also that there can be no distinction between truth and empirical adequacy for scientific theories.

11. "Cosmology, Convention, and the Closed Universe," *Synthese,* XXIV, ½ (July/August 1972): 195–218; discussed in my "Earman on the Causal Theory of Time," *op. cit.,* pp. 87–95 (referred to therein by an earlier title).

12. See, for example, N. D. Cartwright, "Superposition and Macroscopic Observation," *Synthese,* XXIX (December 1974): 229–242, and references therein.

13. See Stanley Gudder, "Hidden Variables in Quantum Mechanics Reconsidered," *Review of Modern Physics,* XL (1968): 229–231; and section III of my "Semantic Analysis of Quantum Logic," in C. A. Hooker, ed., *Contemporary Research in the Foundations and Philosophy of Quantum Theory* (Dordrecht: Reidel, 1973), pp. 80–113.

Nancy Cartwright

THE REALITY OF CAUSES IN A WORLD OF INSTRUMENTAL LAWS

INTRODUCTION

Empiricists are notoriously suspicious of causes. They have not been equally wary of laws. Hume set the tradition when he replaced causal facts with facts about generalizations. Modern empiricists do the same. But nowadays Hume's generalizations are the laws and equations of high-level scientific theories. On current accounts, there may be some question about where the laws of our fundamental theories get their neces-

Reprinted by permission from the author and from *PSA 1980,* Vol. 2, P. Asquith and R. Giere, eds., 1981, pp. 38–48.

sity; but it is no question that these laws are the core of modern science. Bertrand Russell is well known for this view:

> The law of gravitation will illustrate what occurs in any exact science . . . Certain differential equations can be found, which hold at every instant for every particle of the system . . . But there is nothing that could be properly called 'cause' and nothing that could be properly called 'effect' in such a system.[1]

For Russell, causes 'though useful to daily life and in the infancy of a science, tend to be displaced by quite different laws as soon as a science is successful.'

It is convenient that Russell talks about physics, and that the laws he praises are its fundamental equations—Hamilton's equations or Schroedinger's, or the equations of general relativity. That is what I want to discuss too. But I hold just the reverse of Russell's view. I am in favour of causes and opposed to laws. I think that, given the way modern theories of mathematical physics work, it makes sense only to believe their causal claims and not their explanatory laws.

1. EXPLAINING BY CAUSES

Following Bromberger, Scriven, and others, we know that there are various things one can be doing in explaining. Two are of importance here: in explaining a phenomenon one can cite the causes of that phenomenon; or one can set the phenomenon in a general theoretical framework. The framework of modern physics is mathematical, and good explanations will generally allow us to make quite precise calculations about the phenomena we explain. Rene Thom remarks the difference between these two kinds of explanations, though he thinks that only the causes really explain: 'Descartes with his vortices, his hooked atoms, and the like explained everything and calculated nothing; Newton, with the inverse square of gravitation, calculated everything and explained nothing.'[2]

Unlike Thom, I am happy to call both explanation, so long as we do not illicitly attribute to theoretical explanation features that apply only to causal explanation. There is a tradition, since the time of Aristotle, of deliberately conflating the two. But I shall argue that they function quite differently in modern physics. If we accept Descartes's causal story as adequate, we must count his claims about hooked atoms and vortices true. But we do not use Newton's inverse square law as if it were either true or false.

One powerful argument speaks against my claim and for the truth of explanatory laws—the *argument from coincidence*. Those who take laws seriously tend to subscribe to what Gilbert Harman has called inference to the best explanation. They assume that the fact that a law *explains* provides evidence that the law is true. The more diverse the phenomena that it explains, the more likely it is to be true. It would be an absurd coincidence if a wide variety of different kinds of phenomena were all explained by a particular law, and yet were not in reality consequent from the law. Thus the argument from coincidence supports a good many of the inferences we make to best explanations.

The method of inference to the best explanation is subject to an important constraint, however—the requirement of non-redundancy. We can infer the truth of an explanation only if there are no alternatives that account in an equally satisfactory way for the phenomena. In physics nowadays, I shall argue, an acceptable causal story is supposed to satisfy this requirement. But exactly the opposite is the case with the specific equations and models that make up our theoretical explanations. There is redundancy of theoretical treatment, but not of causal account.

There is, I think, a simple reason for this: causes make their effects happen. We begin with a phenomenon which, relative to our other general beliefs, we think would not occur unless something peculiar brought it about. In physics we often mark this belief by labelling the phenomena as effects—the Sorbet effect, the Zeeman effect, the Hall effect. An effect needs something to bring it about, and the peculiar features of the effect depend on the particular nature of the cause, so that—in so far as we think we have got it right—we are entitled to infer the character of the cause from the character of the effect.

But equations do not bring about the phenomenological laws we derive from them (even if the phenomenological laws are themselves equations). Nor are they used in physics as if they did. The specific equations we use to treat particular phenomena provide a way of casting the phenomena into the general framework of the theory. Thus we are able to treat a variety of disparate phenomena in a similar way, and to make use of the theory to make quite precise calculations. For both of these purposes it is an advantage to multiply theoretical treatments.

Pierre Duhem used the redundancy requirement as an argument against scientific realism, and recently Hilary Putnam uses it as an argument against realism in general. Both propose that, in principle, for any explanation of any amount of data there will always be an equally satisfactory alternative. The history of science suggests that this claim may be right: we constantly construct better explanations to replace those of the past. But such arguments are irrelevant here; they do not distinguish between causal claims and theoretical accounts. Both are likely to be replaced by better accounts in the future.

Here I am not concerned with alternatives that are at best available only in principle, but rather with the practical availability of alternatives within theories we actually have to hand. For this discussion, I want to take the point of view that Putnam calls 'internal realism'; to consider actual physical theories which we are willing to account as acceptable, even if only for the time being, and to ask, 'Relative to that theory, which of its explanatory claims are we to deem true?' My answer is that causal claims are to be deemed true, but to count the basic explanatory laws as true is to fail to take seriously how physics succeeds in giving explanations.

I will use two examples to show this. The first—quantum damping and its associated line broadening—is a phenomenon whose understanding is critical to the theory of lasers. Here we have a single causal story, but a fruitful multiplication of successful theoretical accounts. This contrasts with the unacceptable multiplication of causal stories in the second example.

There is one question we should consider before looking at the examples, a question pressed by two colleagues in philosophy of science, Dan Hausman and Robert Ennis. How are we to distinguish the explanatory laws, which I argue are not to be taken literally, from the causal claims and more pedestrian statements of fact, which are? The short answer is that there is no way. A typical way of treating a problem like this is to find some independent criterion—ideally syntactical, but more realistically semantical—which will divide the claims of a theory into two parts. Then it is argued that claims of one kind are to be taken literally, whereas those of the other kind function in some different way.

This is not what I have in mind. I think of a physics theory as providing an explanatory scheme into which phenomena of interest can be fitted. I agree with Duhem here. The scheme simplifies and organizes the phenomena so that we can treat similarly happenings that are phenomenologically different, and differently ones that are phenomenologically the same. It is part of the nature of this organizing activity that it cannot be done very well if we stick too closely to stating what is true. Some claims of the theory must be literally descriptive (I think the claims about the mass and charge of the electron are a good example) if the theory is to be brought to bear on the phenomena; but I suspect that there is no general independent way of characterizing which these will be. What is important to realize is that if the theory is to have considerable explanatory power, most of its fundamental claims will not state truths, and that this will in general include the bulk of our most highly prized laws and equations.

2. EXAMPLES: QUANTUM DAMPING

In radiative damping, atoms de-excite, giving off photons whose frequencies depend on the energy levels of the atom. We know by experiment that the emission line observed in a spectroscope for a radiating atom is not infinitely sharp, but rather has a finite linewidth; that is, there is a spread of frequencies in the light emitted. What causes this natural linewidth? Here is the standard answer which physicists give, quoted from a good textbook on quantum radiation theory by William Louisell:

> There are many interactions which may broaden an atomic line, but the most fundamental one is the reaction of the radiation field on the atom. That is, when an atom decays spontaneously from an excited state radiatively, it emits a quantum of energy into the radiation field. This radiation may be reabsorbed by the atom. The reaction of the field on the atom gives the atom a linewidth and causes the original level to be shifted as we show. This is the source of the natural linewidth and the Lamb shift.[3]

Following his mathematical treatment of the radiative decay, Louisell continues:

> We see that the atom is continually emitting and reabsorbing quanta of radiation. The energy level shift does not require energy to be conserved while the damping requires energy conservation. Thus damping is brought about by the emission and absorption of real photons while the photons emitted and absorbed which contribute to the energy shift are called virtual photons.[4]

This account is universally agreed upon. Damping, and its associated line broadening, are brought about by the emission and absorption of real photons.

Here we have a causal story; but not a mathematical treatment. We have not yet set line broadening into the general mathematical framework of quantum mechanics. There are many ways to do this. One of the Springer Tracts by G. S. Agarwal[5] summarizes the basic treatments which are offered. He lists six different approaches in his table of contents: (1) Weisskopf-Wigner method; (2) Heitler-Ma method; (3) Goldberger-Watson method; (4) Quantum Statistical method: master equations; (5) Langevin equations corresponding to the master equation and a *c*-number representation; and (6) neoclassical theory of spontaneous emission.

Before going on to discuss these six approaches, I will give one other example. The theory of damping forms the core of current quantum treatments of lasers. Figure 4.1 is a diagram from a summary article by H. Haken on 'the' quantum theory of the laser.[6] We see that the situation I described for damping theory is even worse here. There are so many different treatments that Haken provides a 'family tree' to set straight their connections. Looking at the situation Haken himself describes it as a case of 'theory overkill.' Laser theory is an extreme case, but I think there is no doubt that this kind of redundancy of treatment, which Haken and Agarwal picture, is common throughout physics.

Agarwal describes six treatments of line broadening. All six provide precise and accurate calculations for the shape and width of the broadened line. How do they differ? All of the approaches employ the basic format of quantum mechanics. Each writes down a Schroedinger equation; but it is a *different* equation in each *different* treatment. (Actually among the six treatments there are really just three different equations.) The view that I am attacking takes theoretical explanations to provide, as best they can, statements of objective laws. On this view the six approaches that Agarwal lists compete with one another; they offer different laws for exactly the same phenomena.

But this is not Agarwal's attitude. Different approaches are useful for different purposes; they complement rather than compete. The Langevin and Master equations of (4) and (5), for instance, have forms borrowed from statistical mechanics. They were introduced in part because the

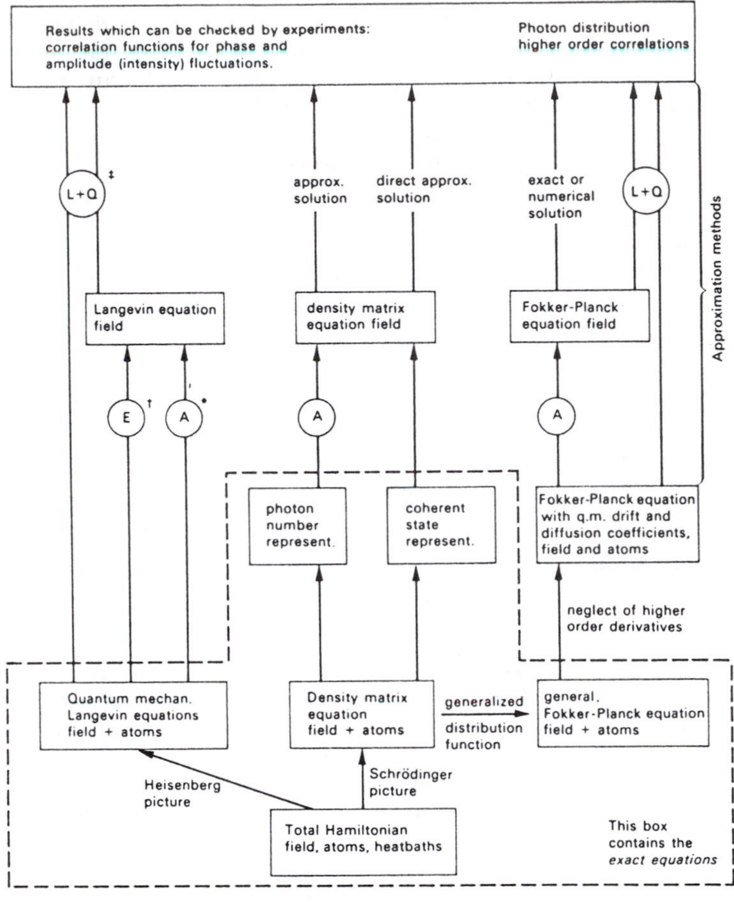

*A: adiabatic elimination of atomic variables
†E: exact elimination of atomic variables
‡L + Q: linearization and quantum mechanical quasilinearization

FIG. 4.1. Family tree of the quantum theory of the laser. (*Source:* Haken, 'The Semiclassical and Quantum Theory of the Laser'.)

development of lasers created an interest in photon correlation experiments. Clearly, if we have statistical questions, it is a good idea to start with the kind of equations from which we know how to get statistical answers.

Let us consider an objection to the point of view I have been urging. We all know that physicists write down the kinds of equations they know how to solve; if they cannot use one approximation, they try another; and when they find a technique that works, they apply it in any place they can. These are commonplace observations that remind us of the pragmatic attitude of physicists. Perhaps, contrary to my argument, the multiplication of theoretical treatments says more about this pragmatic orientation than it does about how explanatory laws ought to be viewed. I disagree. I think that it does speak about laws, and in particular, shows how laws differ from causes. We do not have the same pragmatic tolerance of causal alternatives. We do not use first one causal story in explanation, then another, depending on the ease of calculation, or whatever.

The case of the radiometer illustrates. The radiometer was introduced by William Crookes in 1873, but it is still not clear what makes it work. Recall from the Introduction to these essays that there are three plausible theories. The first attributes the motion of the vanes to light pressure. This explanation is now universally rejected. As M. Goldman remarks in 'The Radiometer Revisited',

> A simple calculation shows that on a typical British summer day, when the sky is a uniform grey (equally luminous all over) the torque from the black and silver faces exactly balance, so that for a perfect radiometer [i.e., a radiometer with a perfect vacuum] no motion would be possible.[7]

Two explanations still contend. The first is the more standard, textbook account, which is supported by Goldman's calculations. It supposes that the motion is caused by the perpendicular pressure of the gas in the perfect vacuum against the vanes. But as we have seen, on Maxwell's account the motion must be due to the tangential stress created by the gas slipping around the edge of the vanes. There is a sense in which Maxwell and Goldman may both be right: the motion may be caused by a combination of tangential and perpendicular stress. But this is not what they claim. Each claims that the factor he cites is the single significant factor in bringing about the motion, and only one or the other of these claims can be accepted. This situation clearly contrasts with Agarwal's different theoretical treatments. In so far as we are interested in giving a causal explanation of the motion, we must settle on one account or the other. We cannot use first one account, then the other, according to our convenience.

I know of this example through Francis Everitt, who thinks of building an experiment that would resolve the question. I mention Everitt's experiment again because it argues for the difference in objectivity which I urge between theoretical laws and causal claims. It reminds us that unlike theoretical accounts, which can be justified only by an inference to the best explanation, causal accounts have an independent test of their truth: we can perform controlled experiments to find out if our causal stories are right or wrong. Experiments of these kinds in fact play an important role in an example from which Wesley Salmon defends inferences to the best explanation.

3. THE ARGUMENT FROM COINCIDENCE

In a recent paper,[8] Salmon considers Jean Perrin's arguments for the existence of atoms and for the truth of Avogadro's hypothesis that there are a fixed number of molecules in any gram mole of a fluid. Perrin performed meticulous experiments on Brownian motion in colloids from which he was able to calculate Avogadro's number quite precisely. His 1913 tract, in which he summarizes his experiments and recounts the evidence for the existence of atoms, helped sway the community of physicists in favour of these hypotheses. Besides Brownian motion, Perrin lists thirteen quite different physical situations which yield a determination of Avogadro's number. So much evidence of such a variety of kinds all pointing to the same value must surely convince us, urges Perrin, that atoms exist and that Avogadro's hypothesis is true.

For many, Perrin's reasoning is a paradigm of inference to the best explanation; and it shows the soundness of that method. I think this misdiagnoses the structure of the argument. Perrin does not make an inference to the best explanation, where explanation includes anything from theoretical laws to a detailed description of how the explanandum was brought about. He makes rather a more restricted inference—an inference to the most probable cause.

A well-designed experiment is constructed to allow us to infer the character of the cause from the character of its more readily observable effects. Prior to Perrin, chemists focused their attention on the size and velocities of the suspended particles. But this study was unrewarding; the measurements were difficult and the results did not signify much. Perrin instead studied the height distribution of the Brownian granules at equilibrium. From his results, with a relatively simple model for the collision interactions, he was able to calculate Avogadro's number. Perrin was a brilliant experimenter. It was part of his genius that he was able to find quite specific effects which were peculiarly sensitive to the exact character of the causes he wanted to study. Given his model, the fact that the carrier fluids had just 6×10^{23} atoms for every mole made precise and calculable differences to the distribution he observed.

The role of the model is important. It brings out exactly what part coincidence plays in the structure of Perrin's argument. Our reasoning from the character of the effect to the character of the cause is always against a background of other knowledge. We aim to find out about a cause with a particular structure. What effects appear as a result of that structure will be highly sensitive to the exact nature of the causal processes which connect the two. If we are mistaken about the processes that link cause and effect in our experiment, what we observe may not result in the way we think from the cause under study. Our results may be a mere artefact of the experiment, and our conclusions will be worthless.

Perrin explicitly has this worry about the first of the thirteen phenomena he cites: the viscosity of gases, which yields a value for Avogadro's number via Van der Waal's equation and the kinetic theory of gases. In his *Atoms* he writes that 'the probable error, for all these numbers is roughly 30 per cent, owing to the approximations made in the calculations that lead to the Clausius-Maxwell and Van der Waal's equations.' He continues: 'The Kinetic Theory justly excites our admiration. [But] it fails to carry complete conviction, because of the many hypotheses it involves.' (I take it he means 'unsubstantiated hypotheses'.) What sets Perrin's worries to rest? He tells us himself in the next sentence: 'If by entirely independent routes we are led to the same values for the molecular magnitudes, we shall certainly find our faith in the theory considerably strengthened.'[9]

Here is where coincidence enters. We have thirteen phenomena from which we can calculate Avogadro's number. Any one of these phenomena—if we were sure enough about the details of how the atomic behaviour gives rise to it—would be good enough to convince us that Avogadro is right. Frequently we are not sure enough; we want further assurance that we are observing genuine results and not experimental artefacts. This is the case with Perrin. He lacks confidence in some of the models on which his calculations are based. But he can appeal to coincidence. Would it not be a coincidence if each of the observations was an artefact, and yet all agreed so closely about Avogadro's number? The convergence of results provides reason for thinking that the various models used in Perrin's diverse calculations were each good enough. It thus reassures us that those models can legitimately be used to infer the nature of the cause from the character of the effects.

In each of Perrin's thirteen cases we infer a concrete cause from a concrete effect. We are entitled to do so because we assume that causes make effects occur in just the way that they do, via specific, concrete causal processes. The structure of the cause physically determines the structure of the effect. Coincidence enters Perrin's argument, but not in a way that supports inference to the best explanation in general. There is no connection analogous to causal propagation between theoretical laws and the phenomenological generalizations which they bring together and explain. Explanatory laws summarize phenomenological laws; they do not make them true. Coincidence will not help with laws. We have no ground for inferring from any phenomenological law that an explanatory law must be just so; multiplying cases cannot help.

I mentioned that Gilbert Harman introduced the expression 'inference to the best explanation'. Harman uses two examples in his original paper.[10] The first is the example that we have just been discussing: coming to believe in atoms. The second is a common and important kind of example from everyday life: inferring that the butler did it. Notice that these are both cases in which we infer facts about concrete causes: they are not inferences to the laws of some general explanatory scheme. Like Perrin's argument, these do not vindicate a general method for inferring the truth of explanatory laws. What they illustrate is a far more restrictive kind of inference: inference to the best cause.

4. CONCLUSION

Perrin did not make an inference to the best explanation, only an inference to the most probable cause. This is typical of modern physics. 'Competing' theoretical treatments—treatments that write down different laws for the same phenomena—are encouraged in physics, but only a single causal story is allowed. Although philosophers generally believe in laws and deny causes, explanatory practice in physics is just the reverse.

NOTES

1. Bertrand Russell, 'On the Notion of Cause,' *Mysticism and Logic* (London: Allen & Unwin, 1917), p. 194.

2. Rene Thom, *Structural Stability and Morphogenesis,* trans. C. H. Waddington (Reading, Mass.: W. A. Benjamin, 1972), p. 5.

3. William H. Louisell, *Quantum-Statistical Properties of Radiation* (New York: John Wiley & Sons, 1973), p. 285.

4. Ibid., p. 289.

5. See G. S. Agarwal, *Quantum-Statistical Theories of Spontaneous Emission and their Relation to Other Approaches* (Berlin: Springer-Verlag, 1974).

6. H. Haken, 'The Semiclassical and Quantum Theory of the Laser', in S. M. Kay and A. Maitland (eds.), *Quantum Optics* (London: Academic Press, 1970), p. 244.

7. M. Goldman, 'The Radiometer Revisited', *Physics Education* 13 (1978), p. 428.

8. See fn 21, Introduction.

9. Jean Perrin, *Atoms,* trans. D. L. Hammick (New York: D. Van Nostrand Co., 1916), p. 82.

10. G. H. Harman, 'Inference to the Best Explanation', *Philosophical Review* 74 (1965), pp. 88–95.

Paul M. Churchland

THE ANTI-REALIST EPISTEMOLOGY OF VAN FRAASSEN'S *THE SCIENTIFIC IMAGE**

At several points in the reading of van Fraassen's book, I feared I would no longer be a realist by the time I completed it. Fortunately, sheer doxastic inertia has allowed my convictions to survive its searching critique, at least temporarily, and as of today, van Fraassen and I still hold different views. I am a scientific realist, of unorthodox persuasion, and van Fraassen is a constructive empiricist, whose persuasions currently define the doctrine. I assert that global excellence of theory is the ultimate measure of truth and ontology at all levels of cognition, even at the observational level. Van Fraassen asserts that descriptive excellence at the observational level is the only genuine measure of any theory's truth, and that one's acceptance of a theory should create no ontological commitments whatever beyond the observational level.

Against his first claim I will maintain that observational excellence or 'empirical adequacy' is only one epistemic virtue among others, of equal or comparable importance. And against his second claim I will maintain that the ontological commitments of any theory are wholly blind to the idiosyncratic distinction between what is and what is not humanly observable, and so should be our own ontological commitments. Criticism will be directed primarily at van Fraassen's *selective* scepticism in favor of observable ontologies over unobservable ontologies; and against his view that the superempirical theoretical virtues (simplicity, coherence, explanatory power) are merely pragmatic virtues, irrelevant to the estimate of a theory's truth. My aims are not merely critical, however. Scientific realism does need reworking, and there are good reasons for moving it in the direction of van Fraassen's constructive empiricism, as will be discussed in the closing section of this paper. But those reasons do not support the sceptical theses at issue.

* * *

Before pursuing our differences, it will prove useful to emphasize certain convictions we share. Van Fraassen is already a scientific realist in the minimal sense that he interprets theories literally and he concedes them a truth-value. Further, we agree that the observable/unobservable distinction is entirely distinct from the non-theoretical/theoretical distinction, and we agree as well that all observation sentences are irredeemably laden with theory.

Additionally, I absolutely reject many sanguine assumptions common among realists. I do not believe that on the whole our beliefs must be at least roughly true; I do not believe that the terms of 'mature' sci-

*This paper is an expanded version of a presentation delivered at the annual meetings of the Canadian Philosophical Association (in Halifax, May, 1981), in a session devoted to a critical discussion of van Fraassen [5], and Churchland [1]. Van Fraassen's contribution to that session has appeared, as a critical notice of [1], in [6].

I should like to thank Hartry Field, Michael Stack, Bas van Fraassen, Clark Glymour, Barney Keaney, Stephen Stich, and Patricia Churchland for helpful discussion of the issues here addressed.

Reprinted by permission from *Pacific Philosophical Quarterly*, Vol. 63, no. 2, July 1982.

ences must typically refer to real things; and I very much doubt that the Reason of *homo sapiens,* even at its best and even if allowed infinite time, would eventually encompass all and/or only true statements.

This scepticism is born partly from an historical induction: so many past theories, rightly judged excellent at the time, have since proved to be false. And their current successors, though even better .founded, seem but the next step in a probably endless and not obviously convergent journey. (For a most thorough and insightful critique of typical realist theses, see the recent paper by Laudan [4].)

Evolutionary considerations also counsel a healthy scepticism. Human reason is a hierarchy of heuristics for seeking, recognizing, storing, and exploiting information. But those heuristics were invented at random, and they were selected for within a very narrow evolutionary environment, cosmologically speaking. It would be miraculous if human reason were completely free of false strategies and fundamental cognitive limitations, and doubly miraculous if the theories we accept failed to reflect those defects.

Thus some very realistic reasons for scepticism with respect to any theory. Why then am I still a scientific realist? Because these reasons fail to discriminate between the integrity of observables and the integrity of unobservables. If anything is compromised by these considerations, it is the integrity of theories generally. That is, of *cognition* generally. Since our observational concepts are just as theory-laden as any others, and since the integrity of those concepts is just as contingent on the integrity of the theories that embed them, our observational ontology is rendered *exactly as dubious* as our non-observational ontology.

This parity should not seem surprising. Our history reveals mistaken ontological commitments in both domains. For example, we have had occasion to banish phlogiston, caloric, and the luminiferous aether from our ontology, but we have also had occasion to banish witches, and the starry sphere that turns about us daily. And these latter items were as 'observable' as you please.

Since these sceptical considerations are indifferent to the distinction between what is and is not observable, they provide no reason for resisting a commitment to unobservable ontologies *while allowing* a commitment to observable ontologies. The latter appear as no better off than the former. For me then, the empirical success of a theory remains a reason for thinking the theory to be true, and for accepting its overall ontology. The inference from success to truth should no doubt be tempered by the sceptical considerations adduced, but the inference to *un*observable ontologies is not rendered *selectively* dubious. Thus I remain a scientific realist. My realism is highly circumspect, but the circumspection is uniform for unobservables and observables alike.

Perhaps I am wrong in this. Perhaps we should be selectively sceptical in the fashion van Fraassen recommends. Does he have other arguments for refusing factual belief and ontological commitment beyond the observational domain? Indeed, he does. In fact, he does not appeal to historical induction or evolutionary humility at all. These are *my* reasons for scepticism (and they will remain, even if I manage to undermine van Fraassen's). They have been introduced here to show that, while there are some powerful reasons for scepticism, those reasons do not place unobservables at a selective disadvantage.

Very well, what are van Fraassen's reasons for scepticism? They are very interesting. To summarize quickly, he does a compelling job of deflating certain standard realist arguments (from Smart, Sellars, Salmon, Boyd, and others) to the effect that, given the aims of science, we have no alternative but to bring unobservables (not just into our calculations, but) into our literal ontology. He also argues rather compellingly that the superempirical virtues, such as simplicity and comprehensive explanatory power, are at bottom merely pragmatic virtues, having nothing essential to do with any theory's truth. This leaves only empirical adequacy as a genuine measure of any theory's truth. Roughly, a theory is empirically adequate if and only if everything it says about *observable* things is true. Empirical adequacy is thus a necessary condition on a theory's truth.

However, claims van Fraassen, the truth of any theory whose ontology includes unobservables is always radically underdetermined by its empirical adequacy since a great many logically incompatible theories can all be empirically equivalent. Accordingly, the inference from empirical adequacy to truth now appears presumptuous in the extreme, especially since it has just been disconnected from additional selective criteria such as simplicity and explanatory power, criteria which might have reduced the arbitrariness of the particular inference drawn. Fortunately, says van Fraassen, we do not need to make such wanton inferences since we can perfectly well understand science as an enterprise that never really draws them. Here we arrive at his positive conception of science as an enterprise whose sole intellectual aims are empirical adequacy and the satisfaction of certain human intellectual needs.

The central element in this argument is the claim that, in the case of a theory whose ontology includes unobservables, its empirical adequacy underdetermines its truth. (We should notice that in the case of a theory whose ontology is completely free of unobservables, its empirical adequacy does not underdetermine its truth: in that case, truth and empirical adequacy are obviously identical. Thus van Fraassen's *selective* scepticism with respect to unobservables.) That is, for any theory T inflated with unobservables, there will always be many other such theories incompatible with T, but empirically equivalent to it.

In my view, the notions of "empirical adequacy" and its cognate relative term "empirically equivalent" are extremely thorny notions of doubtful integrity. If we attempt to explicate a theory's "empirical content" in terms of the observation sentences it entails (or entails-if-conjoined-with available background information, or with possible future background information, or with possible future theories), we generate a variety of notions which are variously empty, context-relative, ill-defined, or incompatible with the claim of underdetermination. Van Fraassen is entirely aware of these difficulties and proposes to avoid them by giving the notions at issue a model-theoretic rather than a syntactic explication. I am unconvinced that this improves matters decisively (on this issue see Wilson [7]). But let me sidestep the issue for now, since the matter is difficult and there is a simpler objection to be voiced.

The empirical adequacy of any theory is itself something that is radically underdetermined by any evidence conceivably available to us. Recall that, for a theory to be empirically adequate, what it says about observable things must be true—*all* observable things, in the past, in the indefinite future, and in the most distant corners of the cosmos. But since any actual data possessed by us must be finite in its scope, it is plain that we here suffer an underdetermination problem no less serious than that claimed above. This is Hume's problem, and the lesson is that even observation-level theories suffer radical underdetermination by the evidence. Accordingly, theories about observables and theories about unobservables appear on a par again, so far as scepticism is concerned.

Van Fraassen thinks there is an important difference between the two cases, and one's first impulse is to agree with him. We are all willing to concede the existence of Hume's problem—the problem of justifying the inference to unobserv*ed* entities. But the inference to entities that are downright unobserv*able* appears as a different and additional problem.

I do not see that it is. Consider the different reasons why entities or processes may go unobserved by us. First, they may go unobserved because, relative to our natural sensory apparatus, they fail to enjoy an appropriate spatial or temporal *position.* They may exist in the Upper Jurassic Period, for example, or they may reside in the Andromeda Galaxy. Second, they may go unobserved because, relative to our natural sensory apparatus, they fail to enjoy the appropriate spatial or temporal *dimensions.* They may be too small, or too brief, or too large, or too protracted. Third, they may fail to enjoy the appropriate *energy,* being too feeble, or too powerful, to permit useful discrimination. Fourth and fifth, they may fail to have an appropriate *wavelength,* or an appropriate *mass.* Sixth, they may fail to 'feel' the relevant fundamental forces our sensory apparatus exploits, as with our inability to observe the background neutrino flux, despite the fact that its energy density exceeds that of light itself.

This list could be lengthened, but it is long enough to suggest that being spatially or temporally distant from our sensory apparatus is only one undistinguished way, one among many ways, in which an entity or process can fall outside the compass of human observation.

There is perhaps some point to calling a thing "observ*able*" if it fails only the first test (spatio-temporal proximity), and "unobservable" if it fails any of the others. But that is only because of the contingent practical fact that one generally has somewhat more *control* over the spatio-temporal perspective of one's sensory systems than one has over their size, or reaction time, or mass, or wavelength sensitivity, or chemical constitution. Had we been less mobile—rooted to the earth like Douglas Firs, say—yet been more voluntarily plastic in our sensory constitution, the distinction between the 'merely unobserved' and the 'downright unobservable' would have been very differently drawn. It may help to imagine here a suitably rooted arboreal philosopher named (what else?) Douglas van Fiirrsen, who, in his sedentary wisdom, urges an anti-realist scepticism concerning the spatially very *distant* entities postulated by his fellow trees.

Admittedly, for any distant entity one can in principle always change the relative spatial position of one's sensory apparatus so that the entity is observed: one can go to it. But equally, for any microscopic entity one can in principle always change the relative spatial *size* or *configuration* of one's sensory apparatus so that the entity is observed. Physical law imposes certain limitations on such plasticity, but so also does physical law limit how far one can travel in a lifetime.

The point of all this is that there is no special or additional problem about inferences to the existence of entities commonly called "unobservables." Such entities are merely those that go unobserved by us for reasons *other* than their spatial or temporal distance from us. But whether the 'gap' to be bridged is spatio-temporal, or one of the many other gaps, the logical/epistemological problem is the same in all cases: ampliative inference and underdetermined hypotheses. I therefore fail to see how van Fraassen can justify tolerating an ampliative inference when it bridges a gap of spatial distance, while refusing to tolerate an ampliative inference when it bridges a gap of, for example, spatial size. Hume's problem and van Fraassen's problem (or Duhem's problem) collapse into one.

Van Fraassen attempts to meet such worries about the inescapable ubiquity of speculative activity by observing that ". . . it is not an epistemological principle that one may as well hang for a sheep as for a lamb" ([5], p. 72). Agreed. But it is a principle of *logic* that one may as well hang for a sheep as for a sheep, and van Fraassen's lamb (empirical adequacy) is just another sheep.

Let me summarize. As van Fraassen sets it up, and as the instrumentalists set it up before him, the realist looks more gullible than the non-realist, since the realist is willing to extend belief beyond the observable, while the non-realist insists on confining belief within that domain. I suggest, however, that it is really the non-realists who are being the more gullible in this matter, since they suppose that the epistemic situation of our beliefs about observables is in some way superior to that of our beliefs about unobservables. But in fact their epistemic situation is not superior. They are exactly as dubious as their non-observational cousins. Their causal history is different perhaps, but not their credibility.

Simply to hold *fewer* beliefs from a given set is of course to be less adventurous, but it is not necessarily to be applauded. One might decide to relinquish all one's beliefs save those about objects weighing less than 500 kg, and perhaps one would then be logically safer. But in the absence of some relevant epistemic difference between one's beliefs about such objects and one's beliefs about other objects, this is perversity, not parsimony.

* * *

Let me now try to address the question of whether the theoretical virtues such as simplicity, coherence, and explanatory power are *epistemic* virtues genuinely relevant to the estimate of a theory's truth, as tradition says, or merely *pragmatic* virtues, as van Fraassen urges. His view promotes empirical adequacy, or evidence of empirical adequacy, as the only genuine measure of a theory's truth, the other virtues (insofar as they are distinct from these) being cast as purely pragmatic virtues, to be valued only for the human needs they satisfy. Despite certain compelling features of the account of explanation that van Fraassen provides, I remain inclined towards the traditional view.

My reason is simplicity itself. Since there is no way of conceiving or representing "the empirical facts" that is completely independent of speculative assumptions, and since we will occasionally confront theoretical alternatives on a scale so comprehensive that we must also choose between competing modes of conceiving what the empirical facts before us *are*, then the epistemic choice between these global alternatives cannot be made by comparing the extent to which they are adequate to some common touchstone, 'the empirical facts.' In such a case, the choice must be made on the comparative global virtues of the two global alternatives, T_1-plus-the-observational-evidence-therein-construed, versus T_2-plus-the-observational-evidence-therein-(differently)-construed. That is, it must be made on superempirical grounds such as relative coherence, simplicity, and explanatory unity. In such cases, "empirical adequacy" becomes just one dimension of coherence.

Such cases as these are reminiscent of Carnap's 'external' questions, and it may be that van Fraassen, like Carnap, does not regard them as factual questions, but as essentially pragmatic questions. I would disagree, since I regard so-called 'external' questions as arrayed on a smooth continuum with 'internal' (i.e., factual) questions. The arguments are presented elsewhere ([1], sections 7 and 10), however, so I shall not repeat them here.

As I see it then, values such as ontological simplicity, coherence, and explanatory power are some of the brain's criteria for recognizing information, for distinguishing information from noise. And I think they are even more fundamental values than is 'empirical adequacy,' since collectively they can overthrow an entire conceptual framework for representing the empirical facts. Indeed, they even dictate how such a framework is constructed by the questing infant in the first place. One's observational taxonomy is not 'read off' the world directly; rather, one comes to it piecemeal, and by stages, and one settles on that taxonomy which finds the greatest coherence and simplicity in the world, and most and the simplest lawful connections.

I can bring together my protective concerns for unobservables and for the superempirical virtues by way of the following thought experiment. Consider a man for whom absolutely *nothing* is observable. All of his sensory modalities have been surgically destroyed, and he has no visual, tactile, or other sensory experience of any kind. Fortunately, he has mounted on top of his skull a microcomputer fitted out with a variety of environmentally-sensitive transducers. The computer is connected to his association cortex (or perhaps the frontal lobe, or Wernicke's area) in such a way as to cause in him a continuous string of singular beliefs about his local environment. These "intellectual intuitions" are not infallible, but let us suppose that they provide him with much the same information that our perceptual judgments provide us.

For such a person, or for a society of such persons, the *observable* world is an empty set. There is no question, therefore, of their evaluating any theory by reference to its 'empirical adequacy,' as characterized by van Fraassen (i.e., isomorphism between some observable features of the world and some 'empirical substructure' of one of the theory's models). But such a society is still capable of science, I assert. They can invent theories, construct explanations of the facts-as-represented-in-past-spontaneous-beliefs, hazard predictions of the facts-as-represented-in-future-spontaneous-beliefs, and so forth. In principle, there is no reason they could not learn as much as we have. (cf. Feyerabend [3])

But it is plain in this case that the global virtues of simplicity, coherence, and explanatory unification are what *must* guide the continuing evolution of their collected beliefs. And it is plain as well that their ontology, whatever it is, must consist entirely of *un*observable entities. To invite a van Fraassenean disbelief in unobservable entities is in this case to invite the suspension of all beliefs beyond tautologies! Surely reason does not require them to be so abstemious.

It is time to consider the objection that those aspects of the world which are successfully monitored by the transducing microcomputer should count as 'observables' for the folk described, despite the lack of any appropriate field of internal sensory qualia to mediate the external circumstance and the internal judgment it causes. Their tables-and-chairs ontology, as expressed in their spontaneous judgments, could then be conceded legitimacy.

I will be the first to accept such an objection. But if we do accept it, then I do not see how we can justify van Fraassen's selective scepticism with respect to the wealth of 'unobservable' entities and properties reliably monitored by *our* transducing measuring instruments (electron microscopes, cloud chambers, chromatographs, etc.). The spontaneous singular judgments of the working scientist, at home in his theoretical vocabulary and deeply familiar with the measuring instruments to which his conceptual system is responding, are not worse off, causally or epistemologically, than the spontaneous singular judgments of our transducer-laden friends. If scepticism is to be put aside above, it must be put aside here as well.

My concluding thought experiment is a complement to the one just outlined. Consider some folk who observe, not less of the world than we do, but more of it. Suppose them able to observe a domain normally closed to us: the micro-world of virus particles, DNA strands, and large protein molecules. Specifically, suppose a race of humanoid creatures each of whom is born with an electron microscope permanently in place over his left 'eye.' The scope is biologically constituted, let us suppose, and it projects its image onto a human-style retina, with the rest of their neurophysiology paralleling our own.

Science tells us, and I take it that van Fraassen would agree, that virus particles, DNA strands, and most other objects of comparable dimensions count as observable entities for the humanoids described. The humanoids, at least, would be justified in so regarding them and in including them in their ontology.

But we humans may not include such entities in our ontology, according to van Fraassen's position, since they are not observable with our unaided perceptual apparatus. We may not include such entities in our ontology *even though we can construct and even if we do construct electron microscopes of identical function, place them over our left eyes, and enjoy exactly the same microexperience as the humanoids.*

The difficulty for van Fraassen's position, if I understand it correctly, is that his position requires that a humanoid and a scope-equipped human must embrace *different* epistemic attitudes towards the microworld, even though their causal connections to the world and their continuing experience of it be identical: the humanoid is required to be a realist with respect to the microworld, and the human is required to be an anti-realist (i.e., an agnostic) with respect to the microworld. But this distinction between what we and they may properly embrace as real seems to me to be highly arbitrary and radically under-motivated. For the only difference between the humanoid and a scope-equipped human lies in the *causal origins* of the transducing instruments feeding information into their respective brains. The humanoid's scope owes its existence to information coded in his genetic material. The human's scope owes its existence to information coded in his cortical material, or in technical libraries. I do not see why this should make any difference in their respective ontological commitments, whatever they are, and I must decline to embrace any philosophy of science which says that it must.

* * *

I now turn from critic of van Fraassen's position to advocate. One of the most central elements in his view seems to me to be well-motivated and urgently deserving of further development. As he explains in his introductory chapter, his aim is to reconceive the relation of theory to world, and the units of scientific cognition, and the virtue of those units when successful. He says,

> I use the adjective 'constructive' to indicate my view that scientific activity is one of construction rather than discovery: construction of models that must be adequate to the phenomena, and not discovery of truth concerning the unobservable. ([5], p. 5)

The traditional view of human knowledge is that the unit of cognition is the sentence or proposition, and the cognitive virtue of such units is truth. Van Fraassen rejects this overtly linguistic guise for his empiricism. He invites us to reconceive a theory as a set of models (rather than as a set of sentences), and he sees empirical adequacy (rather than truth) as the principal virtue of such units.

Though I reject his particular reconception, and the selective scepticism he draws from it, I think the move away from the traditional conception is entirely correct. The criticism to which I am inclined is that van Fraassen has not moved quite far enough. Specifically, if we are to reconsider truth as the aim or product of cognitive activity, I think we must reconsider its applicability right across the board, and not just in some arbitrarily or idiosyncratically segregated domain of 'unobservables.' That is, if we are to move away from the more naive formulations of scientific realism, we should move in the direction of *pragmatism* rather than in the direction of a positivistic instrumentalism. Let me elaborate.

When we consider the great variety of cognitively active creatures on this planet—sea slugs and octopi, bats, dolphins, and humans; and when we consider the ceaseless reconfiguration in which their brains or central ganglia engage—adjustments in the response potentials of single neurons made in the microsecond range, changes in the response characteristics of large systems of neurons made in the seconds-to-hours range, dendritic growth and new synaptic connections and the selective atrophy of old connections effected in the day-upwards range; then van Fraassen's term 'construction' begins to seem highly appropriate. There is endless construction and reconstruction, both functional and structural. Further, it is far from obvious that truth is either the primary aim or the principal product of this activity. Rather, its function would appear to be the ever more finely tuned administration of the organism's *behavior*. Natural selection does not care whether a brain has or tends toward true beliefs, so long as the organism reliably exhibits reproductively advantageous behavior. Plainly there is going to be *some* connection between the faithfulness of the brain's 'world-model' and the propriety of the organism's behavior. But just as plainly, the connection is not going to be direct.

While we are considering cognitive activity in biological terms and in all branches of the phylogenetic tree, we should note that it is far from obvious that sentences or propositions or anything remotely like them constitute the basic elements of cognition in creatures generally. Indeed, as I have argued at length elsewhere ([1], chapter 5; [2]), it is highly unlikely that the sentential kinematics embraced by folk psychology and orthodox epistemology represents or captures the basic parameters of cognition and learning even in humans. That framework is part of a common-sense theory that threatens to be either superficial or false. If we are ever to understand the *dynamics* of cognitive activity, therefore, we may have to reconceive our basic unit of cognition as something other than the sentence or proposition, and reconceive its virtue as something other than truth.

Success of this sort on the descriptive/explanatory front would likely have normative consequences. Truth, as currently conceived, might cease to be an aim of science. Not because we had lowered our sights and reduced our epistemic standards, as van Fraassen's constructive empiricism would suggest, but because we had raised our sights, in pursuit of some epistemic goal even *more* worthy than truth. I cannot now elucidate such goals, but we should be sensible of their possible existence. The notion of "truth," after all, is but the central element in a normative *theory*, and *praxis* makes progress no less than *theoria*.

The notion of truth is suspect on purely metaphysical grounds anyway. It suggests straightaway the notion of the Complete and Final True Theory: at a minimum, the infinite set of all true sentences. Such a theory would be, by epistemic criteria, the best theory possible. But nothing whatever guarantees the existence of such a unique theory. Just as there is no largest positive integer, it may be that there is no best theory. It may be that for any theory whatsoever, there is always an even better theory, and so ad infinitum. If we were thus unable to speak of the set of all true sentences, what sense could we make of truth sentence-by-sentence?

These considerations do invite a 'constructive' conception of cognitive activity, one in which the notion of truth plays at best a highly derivative role. The formulation of such a conception, adequate to all of our epistemic criteria, is the outstanding task of epistemology. I do not think we will find that conception in a model-theoretic version of positivistic instrumentalism, nor do I think we will find it quickly. But the empirical brain begs unravelling, and we have plenty of time.

Finally, there is a question put to me by Stephen Stich. If ultimately my view is even more sceptical than van Fraassen's concerning the relevance or applicability of the notion of truth, why call it scientific *realism* at all? For at least two reasons. The term "realism" still marks the principal contrast with its traditional adversary, positivistic instrumentalism. Whatever the integrity of the notion of truth, theories about unobservables have just *as much* a claim to truth, epistemologically and metaphysically, as theories about observables. Second, I remain committed to the idea that there exists a world, independent of our cognition, with which we interact, and of which we construct representations: for varying purposes, with varying penetration, and with varying success. Lastly, our best and most penetrating grasp of the real is still held to reside in the representations provided by our best theories. Global excellence of theory remains the fundamental measure of rational ontology. And that has always been the central claim of scientific realism.

REFERENCES

1. Churchland, Paul M. (1979), *Scientific Realism and the Plasticity of Mind* (Cambridge University Press).

2. ______ (1981), "Eliminative Materialism and the Propositional Attitudes," *The Journal of Philosophy,* vol. 78, no. 2, February.

3. Feyerabend, Paul K. (1969), "Science Without Experience," *The Journal of Philosophy,* vol. 66, no. 22.

4. Laudan, Larry (1981), "A Confutation of Convergent Realism," *Philosophy of Science,* vol. 48, no. 1, March.

5. van Fraassen, Bas C. (1980), *The Scientific Image* (Oxford University Press).

6. ______ (1981), "Critical Notice of Paul Churchland: Scientific Realism and the Plasticity of Mind," *Canadian Journal of Philosophy,* vol. 11, no. 3, September.

7. Wilson, Mark (1980), "The Observational Uniqueness of Some Theories," *The Journal of Philosophy,* vol. 77, no. 4, April.

Hilary Putnam

EXPLANATION AND REFERENCE*

I. GENERAL SIGNIFICANCE OF THE TOPIC

In this paper I try to contrast realist theories of meaning with what may be called 'idealist' theories of meaning. But a word of explanation is clearly in order.

There is no Marxist 'theory of meaning' but there are a series of remarks on the correspondence between concepts and things, on concepts, and on the impossibility of *a priori* knowledge in the writings of Engels (cf. Engels, 1959) which clearly bear on problems of meaning and reference. In particular, there is a passage† in which Engels makes the point that a concept may contain elements which are not correct. A contemporary scientific characterization of fish would include, Engels says, such properties as life under water and breathing through gills; yet lungfish and other anomalous species which lack these properties are classified as fish for scientific purposes. And Engels argues, I think correctly, that to stick to the letter of the 'definition' in applying the concept *fish* would be bad science. In short, Engels contends that:

(1) Our scientific conception (I would say 'stereotype') of a fish includes the property 'breathing through gills', but
(2) 'All fish breath through gills' is not true! (and, *a fortiori,* not analytic).

I do not wish to ascribe to Engels an anachronistic sophistication about contemporary logical issues, but without doing this it is fair to say on the basis of this argument that Engels *rejects* the model according to which such a concept as *fish* provides anything like analytically necessary and sufficient conditions for membership in a natural kind. Two further points are of importance: (1) The fact that the concept 'natural kind *all* of whose members live under water, breath through gills, etc.' does not strictly fit the natural kind Fish does not mean that the concept does not *correspond* to the natural kind Fish. As Engels puts it, the

*First published in G. Pearce and P. Maynard (eds.), *Conceptual Change* (Dordrecht-Reidel 1973) pp. 199–221.

†In a letter written to Conrad Schmidt in 1895; cf. Marx (1942), pp. 527–30. My agreement is with Engels' realism, not his 'dialectical materialism'.

concept is not exactly correct (as a description of the corresponding natural kind) but that does not make it a *fiction.* (2) The concept is continually changing as a result of the impact of scientific discoveries, but that does not mean that it ceases to correspond to the same natural kind (which is itself, of course, also changing). Again, without attributing to Engels a sophisticated theory of meaning and reference, it is fair, I think, to restate the essential gist of these two points in the following way: concepts which are not strictly true of anything may yet refer to something; and concepts in different theories may refer to the same thing. Of these two points, the second is obvious for most realists; with a few possible exceptions (e.g., Paul Feyerabend), realists have held that there are successive scientific theories about the *same* things: about heat, about electricity, about electrons, and so forth; and this involves treating such terms as 'electricity' as *trans-theoretical* terms, as Dudley Shapere has called them (cf. Shapere, 1969), i.e., as terms that have the same reference in different theories. The first point is more controversial; the idea that concepts provide necessary and sufficient conditions for class membership has often been attacked but, nonetheless, constantly reappears. Without it, however, the other point is moot. Bohr assumed in 1911 that there are (at every time) numbers p and q such that the (one dimensional) position of a particle is q and the (one dimensional) momentum is p; if this was part of the meaning of 'particle' for Bohr, and in addition, 'part of the meaning' means 'necessary condition for membership in the extension of the term', then electrons are *not* particles in Bohr's sense, and, indeed, there are *no* particles 'in Bohr's sense'. (And no 'electrons' in Bohr's sense of 'electron', etc.) None of the terms in Bohr's 1911 theory referred! It follows on this account that we cannot say that present electron theory is a better theory of the same particles that Bohr was referring to. I take it that this is the line of thinking that Paul Feyerabend represents. On an account like Shapere's, however, Bohr would have been referring to electrons when he used the word 'electron', notwithstanding the fact that some of his beliefs about electrons were mistaken, and *we* are referring to those same particles notwithstanding the fact that some of our beliefs—even beliefs included in our scientific 'definition' of the term 'electron'—may very likely turn out to be equally mistaken. This seems right to me. The main technical contribution of this paper will be a sketch of a theory of meaning which supports Shapere's insights.

An 'idealist' theory of meaning, as I am using the term, might go like this (in its simplest form): the meaning of such a sentence as 'electrons exist' is a function of certain *predictions* that can be derived from it (in a pure idealist theory, these would have to be predictions about *sensations*); these predictions are clearly a function of the *theory* in which the sentence occurs; thus 'electrons exist' has no meaning apart from this, that or the other theory, and it has a different meaning in different theories.

The question of 'reference' is a harder one for an idealist: the essence of idealism is to view scientific theories and concepts as instruments for predicting sensations and not as representatives of real things and magnitudes. But a sophisticated idealist is likely to say that the question of reference is 'trivial':† if one has a scientific language L containing the term 'electron', then one can certainly construct a metalanguage ML over it *á la* Tarski, and define 'reference' in such a way that ' "electron" refers to electrons' is a trivial theorem. But if different scientific theories T_1 and T_2 are associated with different formal languages L_1 and L_2 (as they must be if the words have different meanings in T_1 and T_2), then they will be associated with different *meta*-languages ML_1 and ML_2. In ML_1 we can say ' "electron" refers to electrons', meaning that 'electron' in the sense of T_1 refers to elec-

---

*See, for example, the discussion by Hempel, in *Aspects of Scientific Explanation*, (Free Press, New York), (1965), pp. 217–18. A contrasting view is sketched in chapter 13, volume I of these papers.

trons *in the sense of* T_1, and in ML_2 we can say '"electron" refers to electrons' meaning that 'electron' in the sense of T_2 refers to electrons *in the sense of* T_2; but there is no *ML* in which we can even express the statement that 'electron' refers to the same entities in T_1 and T_2—or, at least, no prescription for constructing such an *ML* has been provided by Positivist philosophers of science. In short, just as the idealist regards 'electron' as *theory dependent,* so does he regard the semantical notions of reference and truth as theory dependent; just as the realist regards 'electron' as *trans-theoretical,* so does he regard truth and reference as trans-theoretical.

II. THE MEANING OF PHYSICAL MAGNITUDE TERMS

A. A causal account of meaning

My purpose here is to sketch an account of the meaning of physical magnitude terms (e.g., 'temperature', 'electrical charge'); not an account of meaning in general, although I will try to indicate similarities between what is said here about these terms and what Kripke has said about proper names and what I have said elsewhere about natural kind words. (Kripke's work has come to me second hand; even so, I owe him a large debt for suggesting the idea of causal chains as the mechanism of reference.)

On a traditional view, any term has an intension and an extension. 'Knowing the meaning' is having knowledge of the intension; what it is to 'know' an intension (construed, usually, as an abstract entity of some kind) is never explained. The extension of the color term 'red', for example, is the class of red things; the intension, according to Carnap, is the property Red. Carnap spoke of 'grasping' the intension of terms; what it would be to 'grasp' the property Red was never explained; probably Carnap would have equated it with knowing how to verify sentences of the form '*x* is red', but this comes from his theory of knowledge, not his writings on semantics. In any case, understanding words is a matter of having knowledge. Full linguistic competence in connection with a word may require more knowledge than just the intension; for example, syntactical knowledge, knowledge of cooccurrence regularities, etc.; but linguistic competence, like understanding, is a matter of *knowledge*—not necessarily explicit knowledge—knowledge in the wide sense, implicit as well as explicit, 'knowing how' as well as 'knowing that', skills and abilities as well as facts, but all *knowledge* none the less.

According to the theory I shall present this is fundamentally wrong. Linguistic competence and understanding are not just *knowledge.* To have linguistic competence in connection with a term it is not sufficient, in general, to have the full battery of usual linguistic knowledge and skills; one must, in addition, be in the right sort of relationship to certain distinguished situations (normally, though not necessarily, situations in which the *referent* of the term is present). It is for this reason that this sort of theory is called a 'causal theory' of meaning.

Coming to physical magnitude terms, what every user of the term 'electricity' knows is that electricity is a magnitude of some sort—and, in fact, not even that: electricity was thought at one time to possibly be a sort of substance, and so was heat. At any rate, speakers know that 'electricity' and 'heat' are putative physical *quantities*—capable of more and less, and capable of location. (I do not think that even these statements are *analytic,* but I think they have a kind of *linguistic* association with the terms in question.) In a developed semantic theory one might introduce a special semantic marker, e.g., 'physical quantity', for terms of this sort. I cannot, however, think of anything that *every* user of the term 'electricity' *has* to know except that electricity is (associated with the notion of being) a physical magnitude of some sort, and, possibly, that 'electricity' (or electrical charge or charges) is capable of flow or motion. Benjamin Franklin knew that 'electricity' was manifested in the form of sparks and lightning bolts; someone else might know about currents and electromagnets; someone else might know about atoms consisting of positively and negatively charged particles. They could all use the term 'electricity' without there being a discernible 'intension' that they all share. I want to suggest that what they do have in common is this: that each of them is connected by a certain kind of causal chain to a situation in which a *description* of electricity is given, and generally a *causal* description—that is, one which singles out electricity as *the* physical magnitude *responsible* for certain effects in a certain way.

Thus, suppose I were standing next to Ben Franklin as he performed his famous experiment. Suppose he told me that 'electricity' is a physical quantity which behaves in certain respects like a liquid (if he were a mathematician he might say 'obeys an equation of continuity'); that it collects in clouds, and then, when a critical point of some kind is reached, a large quantity flows from the cloud to the earth in the form of a lightning bolt; that it runs along (or perhaps 'through') his metal kite string; etc. He would have given me an *approximately correct definite description* of a physical magnitude. I could now use the term 'electricity' myself. Let us call this event—my acquiring the ability to use the term 'electricity' in this way—an *introducing event.* It is clear that each of my later uses will be causally connected to this introducing event, as long as those uses exemplify the ability I acquired in that introducing event. Even if I use the term so often that I forgot when I first learned it, the intention to refer to the same magnitude that I referred to in the past by using the word links my present use to those earlier uses, and indeed the word's being in my present vocabulary at all is a causal product of earlier events—ultimately of the introducing event. If I teach the word to someone else by telling him that the word 'electricity' is the name of a physical magnitude, and by telling him certain facts about it which do not constitute a causal description—e.g., I might tell him that like charges repel and unlike charges attract, and that atoms consist of a nucleus with one kind of charge surrounded by satellite electrons with the opposite kind of charge—even if the facts I tell him do not constitute a definite description of any kind, let alone a causal description—still, the word's being in his vocabulary will be causally linked to its being in my vocabulary, and hence, ultimately, to an introducing event.

I said before that different speakers use the word 'electricity' without there being a discernible 'intension' that they all share. If an 'intension' is anything like a necessary and sufficient condition, then I think that this is right. But it does not follow that there are no ideas about electricity which are in some way linguistically associated with the word. Just as the idea that tigers are striped is linguistically associated with the word 'tiger', so it seems that some idea that 'electricity' (i.e., electric charge or charges) is capable of flow or motion *is* linguistically associated with 'electricity'. And perhaps this is all—apart from being a physical magnitude or quantity in the sense described before—that is linguistically associated with the word.

Now then, if anyone knows that 'electricity' is the name of a physical quantity, and his use of the word is connected by the sort of causal chain I described before to an introducing event in which the causal description given was, in fact, a causal description of electricity, then we have a clear basis for saying that he uses the word to refer to electricity. Even if the causal description failed to describe electricity, if there is good reason to treat it as a misdescription *of electricity* (rather than as a description of nothing at all)—for example, if electricity was described as the physical magnitude with such-and-such properties which is responsible for such-and-such effects, where in fact electricity is responsible for the effects in question, and the speaker intended to refer to the magnitude responsible for those effects, but mistakenly added the incorrect information 'electricity has such-and-such properties' because he mistakenly thought that the magnitude responsible for those effects had those further properties—we still have a basis for saying that both the original speaker and the persons to whom he teaches the word use the word to refer to electricity.

If a number of speakers use the word 'electricity' to refer to electricity, and, in addition, they have the standard sorts of associations with the word—that it refers to a magnitude which can move or flow—then, I suggest, the question of whether it has 'the same meaning' in their various idiolects simply does not arise. If a word is linguistically associated with a necessary and sufficient condition in the way that 'bachelor' is, then that sort of question *can* arise; but it does not arise, for example, in the case of proper names, and it does not arise, for a similar reason, in the case of physical magnitude terms. Thus if you know that 'Quine' is a name and I know that 'Quine' is a name and, in addition, we both refer to the same person when we use the word (even if the causal chains linking us to the referent are quite different) then the question of whether 'Quine' has the same meaning in my idiolect and in yours does not arise. More precisely: if the referent is the same, and we both associate the same minimal linguistic information with the word 'Quine', namely, that it is a person's name, then the word is treated as the same word whether it occurs in your idiolect or in mine. Similarly, 'electricity' is the same word in Ben Franklin's idiolect and in mine. Of course, if you had wrong linguistic ideas about the name 'Quine'—for example, if you thought 'Quine' was a female name (not just that Quine was a woman, but that the name was restricted to females)—then there would be a difference in meaning.

This account stresses causal descriptions because physical magnitudes are invariably discovered through their effects, and so the natural way to first single out a physical magnitude is as the magnitude responsible for certain effects. Of course, the words 'responsible', 'causes', etc., do not literally have to occur in the description: *spin,* for example, was introduced by describing it as a physical magnitude having half-integral values characteristic of certain elementary particles, and giving a *law* connecting it with magnitudes previously introduced; I intend the notion of a causal description to include this case. And it is not a 'necessary truth' that the description introducing a new physical magnitude should involve a notion of cause or law; but I am not trying in this paper to state 'necessary truths'.

Once the term 'electricity' has been introduced into someone's vocabulary (or into his 'idiolect', as the dialect of a single speaker is called) whether by an introducing event, or by his learning the word from someone who learned it via an introducing event, or by his learning the word from someone linked by a chain of such transmissions to an introducing event, the referent in that person's idiolect is also fixed, even if no knowledge that that person has fixed it. And once the referent is fixed, one can use the word to formulate any number of theories about that referent (and even to formulate theoretical definitions of that referent which may be correct or incorrect scientific characterizations of that referent), without the word's being in any sense a different word in different theories. Thus the account just given fulfils the desideratum with which we started—it makes such terms as 'electricity' trans-theoretical. The 'operational criteria' you can give for the presence of electricity will depend strongly on what theory you accept; but, without the illicit identification of meaning with operational criteria, it does not follow at all that *meaning* depends on the theory you accept.

The possibility of formulating definite descriptions (or even misdescriptions) of physical magnitudes depends upon the availability in our language of such 'broad spectrum' notions as *physical magnitude* and *causes;* that these play a crucial role in the introduction of physical magnitude terms was argued in chapter 13, volume I. In that paper, however, I did not distinguish between *defining* what I then called theoretical terms and *introducing* them. Of course, if we have available a language in which we can formulate descriptions of the referents of our various physical magnitude terms, then we can consider the various theories that we have containing those terms as so many different systems of sentences in that one language. To the extent that we can do this, we can treat the notions of reference and truth appropriate to that language as trans-theoretical notions also.

B. Kripke's theory of proper names

I have already acknowledged a heavy indebtedness to Kripke's (unpublished) work on proper names. Since I have heard mainly secondhand reports of that work, I shall not attempt to describe it here in any great detail. But, as it has come down to me, the key idea is that a person may use a proper name to refer to a thing or person *X* even though he has *no* true beliefs about *X*. For example, suppose someone asks me who Quine is, and I falsely tell him that Quine was a Roman emperor. If he believes me, and if he goes on to use the word 'Quine' with the intention of referring to the person to whom *I* refer as Quine, then he will say such things as 'Quine was a Roman emperor'—and he will be referring to a contemporary logician. Of course, he still has some true beliefs about Quine (beyond the belief that Quine is or was a person); for example, that Quine is or was named 'Quine'; but Kripke has more elaborate examples to show that even this is not always the case. On Kripke's view, the essential thing is this: that the use of a proper name to refer involves the existence of a causal chain of a certain kind connecting the user of the name (and the particular event of his using the name) to the bearer of the name.

Now then, I do not feel that one should be quite as liberal as Kripke is with respect to the causal chains one allows. I do not see much point, for example, in saying that someone is referring to Quine when he uses the name 'Quine' if he thinks that 'Quine' was a Roman emperor, and that is all he 'knows' about Quine; unless one has *some* beliefs about the bearer of the name which are true or approximately true, then it is at best idle to consider that the name refers to that bearer in one's idiolect. But what seems right about Kripke's account is that the knowledge an individual user of a language has, need not at all fix the reference of the proper names in that individual's idiolect; the reference is fixed by the fact that that individual is causally linked to other individuals who were in a position to pick out the bearer of the name, or of some names from which the name descended. Indeed, what is important about Kripke's theory is not that the use of proper names is 'causal'—what is not?—but that the use of proper names is *collective*. Anyone who uses a proper name to refer is, in a sense, a member of a collective which had 'contact' with the bearer of the name: if it is surprising that a particular member of the collective need not have had such contact, and need not even have any good idea of the bearer of the name, it is only surprising because we think of language as private property.

The relationship of this theory of Kripke's to the above theory of physical magnitude terms should be obvious. Indeed, one might say that physical magnitude terms *are* proper names: they are proper names of *magnitudes* not *things*—however, this would be wrong, I think, since some physical magnitude terms (e.g., 'heat') are linguistically associated with rather rich information about the referent. The important thing about proper names is that it would be ridiculous to think that having linguistic competence can be equated in their case with knowledge of a necessary and sufficient condition—thus one is led to search for something other than the knowledge of the speaker which fixes the referent in their case.

It will be noted that I required a causal chain from the use of the physical magnitude term back to an introducing event—

not back to an event in which the physical magnitude played a significant role. The reason is that, although no one in practice is going to be in a position to give a definite description of a physical magnitude unless he is causally connected to such an event, the nature of *that* causal chain seems not to matter. As long as one is in a position to give a definite description (or even a misdescription), one is in a position to introduce the term; and the chain from there on is something about which much more definite statements can be made. (In my opinion, it would be good to make a similar modification in Kripke's theory of proper names.)

C. Natural kind words

In chapter 8 of this volume I presented an account of natural kind words (e.g., 'lemon') which has some relation to the present account of physical magnitude terms. I suggested that anyone who has linguistic competence in connection with 'lemon' satisfies three conditions: (1) He has implicit knowledge of such facts as the fact that 'lemon' is a concrete noun, that it is the 'name of a fruit', etc.—information given by classifying the word under certain natural syntactic and semantic 'markers'. I criticized Jerrold Katz for the view that natural systems of semantic markers can enable us to give the exact meaning of each term (or of *any* natural kind term); but *some* of the information associated with a word can naturally be represented by classifying the word under such familiar headings as 'noun', 'concrete', etc. (2) He associates the word with a certain 'stereotype'—yellow color, tart taste, thick peel, etc. (3) He uses the word to *refer* to a certain natural kind—say, a natural kind of fruit whose most essential feature, from a biologist's point of view, might be a certain kind of DNA.

Two points were most important in the argument of that paper. The first was that the properties mentioned in the stereotype (and, I would add, the properties indicated by the semantic markers) are not being analytically predicated of each member of the extension, or, indeed, of any members of the extension. It is not analytic that all tigers have stripes, nor that some tigers have stripes; it is not analytic that all lemons are yellow, nor that some lemons are yellow; it is not even analytic that tigers are animals or that lemons are fruits. The stereotype is *associated* with the word; it is not a necessary and sufficient condition for membership in the corresponding class, nor even for being a normal member of the corresponding class. Engels' example of the word 'fish' fits right in here: what Engels was pointing out was precisely that the stereotype associated with the term 'fish' even in scientific, as opposed to lay, usage is not a necessary and sufficient condition. The second point was that speakers must be referring to a particular natural kind for us to treat them as using the same word 'lemon', or 'aluminum', or whatever. The weakness of that paper, apart from being very poorly organized and presented, is that nothing positive is said about the conditions under which a speaker who uses a word (say 'aluminum' or 'elm tree') is referring to one set of things rather than another. Clearly, the speaker who uses the word 'aluminum' need not be able to tell aluminum from molybdenum, and the speaker who use the term 'elm tree' cannot tell elm trees from beech trees if he happens to be me. But then what does determine the reference of the terms 'aluminum', and 'molybdenum' in my idiolect? In the previous papers, I suggested that the reference is fixed by a test known to experts; it now seems to me that this is just a special case of my use being causally connected to an introducing event. For natural kind words too, then, linguistic competence is a matter of knowledge plus causal connection to introducing events (and ultimately to members of the natural kind itself). And this is so far the same reason as in the case of physical magnitude terms; namely, that the use of a natural kind word involves in many cases membership in a 'collective' which has contact with the natural kind, which knows of tests for membership in the natural kind, etc., only as a collective. The idea that linguistic competence in connection with a natural kind word involves more than just having the right extension or reference (where this is now explained via a causal account), but also associating the right stereotype seems to me to carry over to physical magnitude words. Natural kind words can be associated with 'strong' stereotypes (stereotypes that give a strong picture of a stereotypical member—even to the point of enabling one to tell, in most cases, if something belongs to the natural kind), as in the case of 'lemon' or 'tiger', or with 'weak' stereotypes (stereotypes that give no idea of what a sufficient condition for membership in the class would be), as in the case of 'molybdenum' or (unless I am a very atypical speaker) 'elm'. Similarly, it seems to me that the physical magnitude term 'temperature' is associated with a very strong stereotype, and 'electricity' with a weak one.

D. Objections and questions

It is obvious that the account presented here must face certain hard questions. Without attempting to think of all of them myself, I should like to list a few that may help to launch discussion.

(1) One question that must be faced by all causal theories of meaning is how to make more precise the notion of a causal chain of the appropriate kind. How precisely can we describe the sorts of causal chains that must exist from one use of a word to a later use of the same word if we are to say that the referent or referents are the same in the two cases? And how much of a defect in these sorts of theories is it if one cannot be more precise on this point?

(2) It may seem counterintuitive that a natural kind word such as 'horse' is sharply distinguished from a term for a fictitious or nonexistent natural kind such as 'unicorn', and that a physical magnitude term such as 'electricity' is sharply distinguished from a term for a fictitious or nonexistent physical magnitude or substance such as 'phlogiston'. Indeed, I myself believe that if unicorns were found to exist and people began to discover facts about them, give nonobvious definite descriptions or approximately correct descriptions of the class of unicorns, etc., then the linguistic character of the word 'unicorn' would change; and similarly with 'phlogiston'; but this is certain to be controversial.

(3) Some people will argue that definitions of such terms as 'electricity' (or, more precisely, 'charge') are crucial in the exact sciences, and further that such definitions should be regarded as *meaning stipulations.* I agree with the first part of this—that definitions are important in science, provided one remembers what Quine has pointed out, that 'definition' is relative to a particular text or presentation, and that there is no such thing, in general, as the definition of a term 'in physics' or 'in biology'—only the definition in *X*, *Y*, or *Z*'s presentation or axiomatization. I disagree with the last part—that 'definitions' in science are meaning stipulations—but, again, this is certain to be controversial.

(4) Finally, there will be objections to my use of causal notions, from Humeans who expect them to be reduced away, and to my use of the term 'physical magnitude' from extensionalists and nominalists. Here I can only plead guilty to the belief that talk about what causes what, or what the laws of nature are, or what would happen if other things happen is *not* highly derived talk about mere regularities, and to the belief that the real world requires for its description not only reference to things but reference to physical magnitudes (cf. chapter 19, volume I of these papers)—in a sense of 'physical magnitude' in which physical magnitudes exist contingently, not as a matter of logical necessity, and in which magnitudes can be synthetically identical (e.g., temperature is the same magnitude as mean molecular kinetic energy).

III. WHY POSITIVISTIC THEORY OF SCIENCE IS WRONG

My contention in this paper is not that what is wrong with positivist theory of science is positivist theory of meaning. What is wrong with positivist theory of science is that it is based on an idealist or idealist-tending world view, and that that view does not correspond to reality. However, the idealist element in contemporary positivism enters precisely through the theory of meaning; thus part of any realist critique of positivism has to include at least a sketch of rival theory. In the present section, I want to turn from the task of sketching such a rival theory, which was just completed, to the task of showing that positivistic theory of explanation broadly construed—that is, positivist theory of scientific theory—does not correspond to reality any better than the older and less sophisticated idealist theories to which it is historically the successor.

Let us for a moment review some of those older theories. The oldest theory is Bishop Berkeley's. Here one already meets what might be called the *adequacy claim:* that is, the claim that a convinced Berkelian is *entitled* to accept standard scientific theory and practice, that Berkeley can give an account of the scientific method which would justify this. Indeed, I have heard philosophers argue that acceptance of Berkeley's metaphysics would not make any difference to the scientific theories one would accept. Here one already meets an important ambiguity. One can be claiming that a Berkelian can make the move of 'accepting' scientific theory in some sense other than accepting it as true or approximately-true: say, accepting it as a useful prediction heuristic. If this is what one means, then the claim is trivial. To be sure, Berkeley can 'accept' Newtonian physics in the Pickwickian sense of 'accept' as a useful scheme for making predictions. But Berkeley, to do him justice, was interested in much more: what he claimed was that an idealist could *reinterpret* (only he would not consider it *re*-interpretation, but rather *correct* interpretation) the notion of object so as to square both the layman's and the scientist's talk of objects with the idealist claim that reality consists of minds and their sensations ('spirits' and their 'ideas'). The difference between the two claims is the difference between accepting the idea that social practice is the test of truth and rejecting it, between accepting the idea that the overwhelming success of scientific theory offers some reason for accepting that theory as true or approximately-true, and claiming that success in practice is *no* indication of truth. Machian positivism fails for the same reason that Berkelian idealism does: although Mach makes the claim that his construction of the world out of sensations ('Empfindungen') is compatible with lay and scientific object-talk, no demonstration at all is given that this is so. The first philosopher to both precisely state and to undertake the task of *translating* thing-language into phenomenalistic language was Carnap (in *Logische Aufbau der Welt*). And what does Carnap do? He devotes the entire book to *preliminaries,* to 'reconstructions' *within* sensationalistic language (i.e., reductions of some sensation-concepts to others, not of thing-concepts to sensation-concepts), and then in the last chapter gives a sketch of the relation of thing-language to sensation-language which is *not* a translation, and which, indeed, amounts to no more than the old claim that we pick the thing-theory that is 'simplest' and most useful. In short, no demonstration is given at all that the positivist is entitled to quantify over (or refer to) material things.

It is with the failure of the phenomenalist translation enterprise, that is, with the failure to find *any* interpretation of object-concepts under which the prima facie incompatibility between an idealist world-view and a materialist world-view, between a world consisting of 'spirits and their ideas', or of 'Empfindungen', or of total experience-slices in one 'specious present', and a world consisting of fields and particles, simply *disappears*—it is with this failure that contemporary positivistic philosophy of science begins. Basically, two moves were made by the positivists after the failure of phenomenalist translation. The first was to give up construing scientific theories as systems of statements each of which had to have an intelligible interpretation (intelligible from the standpoint of what was taken as 'completely understood' or 'fully interpreted') and to construe them rather as mere calculi, whose objective was to give successful predictions and otherwise to be as 'simple' as possible. 'Scientific theories are partially interpreted calculi' (chapter 13, volume I of these papers). The second move was to shift from phenomenalist language to 'observable thing language' as one's reduction-base, i.e., to say that one was seeking an interpretation or 'partial interpretation' of physical theory in 'observable thing language', not in 'sensationalistic language'.

The second move may make it appear questionable whether positivism is still correctly characterized as an 'idealist' tendency—i.e., as a tendency which regards or tends to regard the 'hard facts' as just facts about actual and potential *experiences,* and all other talk as somehow just highly derived talk about actual and potential experiences. I, myself, think this characterization *is* still fundamentally correct despite the shift to 'observable thing predicates' for two reasons: (1) The cut between observable things and 'theoretical entities' was historically introduced as a substitute for the thing/sensation dichotomy. Indeed, the reduction of 'theoretical entities' to 'observable things and qualities' would hardly seem to be a natural problem to someone who did not have in the back of his head the older problem of reduction to *sensations.* The reduction of things to sensations is both a historically motivated problem and one which rests upon the sharpness of the distinction between a material thing and a sensation (of course, even this sharpness is partly an illusion, in a materialist view—substitute 'material process' for 'material thing'!), as well as the supposed 'certainty' one has concerning one's own sensations. But the reduction of electrons to tables and chairs, or, more generally, of 'unobservable' things to 'observable' things is not historically motivated, the distinction is not sharp (Grover Maxwell asked years ago if a dust mote is something 'given' when it is just big enough to see and a 'construct' when it is just too small to see—can the distinction between data and construct be a matter of size?), and one is not supposed to have certainty concerning observable things. (2) The positivists themselves frequently say that one could carry their analysis back down to the level of sensations, and that stopping with 'observable thing predicates' is a matter of *convenience.*†

In the remainder of this section I want to show that the first move—construing scientific theories as partially interpreted calculi—does not solve the adequacy problem at all. The positivist today is no more entitled than Berkeley was to accept scientific theory and practice—that is, his own story leads to no reason to think either that scientific theory is true, or that scientific practice tends to discover truth. In a sense, this is immediate. The positivist does not claim that scientific theory is 'true' in any trans-theoretic sense of 'true'; the only trans-theoretic notions he has are of the order of 'leads to successful prediction' and 'is simple'. Like the Berkelian, he has to fall back on the position that scientific theory is *useful* rather than true or approximately-true. But he does try to provide some account of the acceptability of scientific theories, even some account of their 'interpretation'. And he wants to maintain that in some sense the principle on which realist philosophy of science rests—that social practice is the test of truth, that the success of scientific theories is reason to think they are true or approximately-true—is right. What I want to show is that the notion of 'truth' that the positivist can give us is not

†E.g., Carnap says this on p. 63 in Carnap, R. "The Methodological Character in Theoretical Entities", in H. Feigl and M. Scriven (eds.) *Minnesota Studies in the Philosophy of Science* (Minneapolis: U. of Minnesota Press, 1956).

the one on which scientific practice is based.

A. Truth

When a realistically minded scientist—that is to say, a scientist *whose practice* is realistic, not one whose official 'philosophy of science' is realistic—accepts a theory, he accepts it as true (or probably true, or approximately-true, or probably approximately-true). Since he also accepts *logic*† he knows that certain moves *preserve truth.* For example, if he accepts a theory T_1 as true and he accepts a theory T_2 as true, then he knows that T_1 & T_2—the *conjunction* of T_1 and T_2—is also true, by logic, and so he accepts T_1 & T_2. If we talk about probability, we have to say that if T_1 is very highly probably true and T_2 is very highly probably true, then the conjunction T_1 & T_2 is also highly probable (though not *as* highly as the conjuncts separately), provided that T_1 is not negatively relevant to T_2, i.e., provided that T_2 is not only highly probable on the evidence, but also no less probable on the added assumption of T_1 (this is a judgment that must be made on the basis of what T_1 *says* and of background knowledge, of course). If we talk about approximate-truth, then we have to say that the approximations probably involved in T_1 and T_2 need to be compatible for us to pass from the approximate-truth of T_1 and T_2 to the approximate-truth of their conjunction. None of these matters is at all deep, from a realist point of view. But even if we confine ourselves to the simplest case, the case in which we can neglect the chances of error and the presence of approximations, and treat the acceptance of T_1 and T_2 as simply the acceptance of them as true, I want to suggest that the move from this acceptance to the acceptance of the conjunction is one to which one is not entitled on positivist philosophy of science. One of the simplest moves that scientists daily make, a move they make as a matter of propositional logic, a move which is central if scientific inquiry is to have any *cumulative* character at all, is totally *arbitrary* if positivist philosophy of science is right.

†The role of logic in empirical science is discussed in Putnam, Philosophy of Logic (1971).

The difficulty is very simple. Acceptance of T_1, for a positivist, means acceptance of the calculus T_1 as leading to successful predictions (i.e., all *observation sentences* which are theorems of T_1 are true; not all *sentences* which are theorems of T_1 are 'true' in any fixed trans-theoretic sense). Similarly, the acceptance of T_2 means the acceptance of T_2 as leading to successful predictions. But from the fact that T_1 leads to successful predictions and the fact that T_2 leads to successful predictions it does not follow at all that the conjunction T_1 & T_2 leads to successful predictions. The difficulty, in a nutshell, is that the predicate which plays the *role* of truth—the predicate 'leads to successful predictions'—does not have the *properties* of truth. The positivist may teach in his philosophy seminar that acceptance of a scientific theory is acceptance of it as 'simple and leading to true predictions', and then go out and do science (or his students may go out and do science) by verifying theories T_1 and T_2, conjoining theories which have been previously verified, etc.—but then there is just as great a discrepancy between what he teaches in his philosophy seminar and his *practice* as there was between Berkeley's teaching that the world consisted of spirits and their ideas and continuing in practice to daily rely on the material object conceptual system.

Nor does it help to bring in 'simplicity'. It is not obvious that the conjunction of simple theories is simple; and even if simplicity is preserved, the conjunction of simple theories which separately lead to no false predictions may even be *inconsistent* (examples are easy to construct). More sophisticated moves have indeed been made. Thus, for Carnap, truth of a theory is the same as truth of its 'Ramsey sentence' (for details see Hempel, *Aspects of Scientific Explanation*, 1965). But exactly the same objection applies: 'truth of the Ramsey sentence' does not have the properties of truth: if T_1 has a true Ramsey sentence and T_2 has a true Ramsey sentence, it does not at all follow that the conjunction does.

(For those readers familiar with Carnap's use of the Hilbert epsilon-symbol, it may be pointed out that the difficulty comes out in very sharp form in Carnap's symbolization of his interpretations of individual theoretical terms. Thus let $T_1(P)$, $T_2(P)$ be two theories containing exactly one theoretical term P. On Carnap's own symbolization of his view, what P means in T_1 is $\epsilon P T_1(P)$; what P means in T_2 is $\epsilon P T_2(P)$; and what P means in T_1 & T_2 is $\epsilon P[T_1(P) \,\&\, T_2(P)]$; this makes it explicit that P has different meanings in T_1 and T_2 and *yet a third meaning* in their conjunction.)

B. Simplicity

It is easy to construct a 'theory' in the positivist sense (a calculus containing some observation terms) which leads to no false predictions but which no scientist would dream of accepting. This is usually handled by saying that scientists only choose 'simple' theories. Also, a simple theory may mess up science as a whole: so it is said that scientists are trying to maximize the simplicity of 'total science'. 'Theory' means, then, 'formalization of total science, or of some piece which is independent of the rest of total science'. Unfortunately, no one has ever written down or ever will write down a 'theory' in this sense. The fact is, that positivist philosophy of science depends on a constant slide between giving the impression that one is talking about 'theories' in the customary sense—Newton's theory, Maxwell's theory, Darwin's theory, Mendel's theory—and saying, at key points of difficulty such as the one just alluded to, that one is *really* talking about a 'formalization of total science', or some such thing.

The difficulty with the rule 'choose the simplest theory compatible with the evidence' is that it is probably not *right,* or would probably not be right, even if one *could* formalize 'total science' (at a given time). Scientists are not trying to maximize some formal property of 'simplicity'; they are trying to maximize *truth* (or improve their approximation to truth, or increase the amount of approximate-truth they know without decreasing the goodness of the approximation, and so forth).

Of course, a realist might accept the rule 'choose the simplest hypothesis', if it could be shown that the simplest hypothesis is always the most *probable* on the basis of the rest of his knowledge. But this is not so on any usual measure of simplicity. For example, suppose I know just three points on interstate highway 40, and those three points lie on a straight line. Suppose also that the statement 'IS 40 is straight' is logically consistent with my total knowledge. Then accepting 'IS 40 is straight' would, on the usual simplicity metrics, be accepting the simplest hypothesis. Yet I would not in fact accept 'IS 40 is straight', nor would anyone with our background knowledge. Given that every other interstate highway has curves, and given the enormous length of IS 40 and the enormous impracticality of making a straight highway across the entire United States, it is overwhelmingly probable that IS 40 is *not* straight.

Can we not say that my *total* 'knowledge' is less simple if I accept 'IS 40 is straight'? Not, it seems to me, on the basis of any criterion of *simplicity* that I know of. What is obviously involved here is not *simplicity* but plausibility: what introducing the word 'simplicity' does is make it look as if a calculation which is in fact the calculation of the probability of a state of affairs is in reality just a calculation of a formal property (such as number of argument places, number of primitive symbols, length and number of the axioms, perhaps shape of the curves mentioned) of an uninterpreted or semi-interpreted *calculus,* even if the property of being the most probable hypothesis on background knowledge could be *represented* syntactically, omitting to mention that the representing property was the syntactic representation of a *probability measure*, and pretending that it was *just* a formal property (like having simple axioms), would be a way of disguising rather than revealing what was going on.

C. Confirmation

Indeed, positivist philosophers of science have made attempts at formalizing the logic of confirmation. These attempts are interesting (though so far unsuccessful) researches on *any* philosophy of science. But not only do they have nothing to do with positivist theory of meaning; they are in fact *incompatible* with it. Thus when they write about meaning, positivists tell us that 'theoretical terms' have different meanings in different theories; when they formalize confirmation theory, they invariably treat theories as systems of sentences in *one* language, and assume that all semantical concepts are *trans*-theoretic. Thus the positivists are engaged in formalizing *realistic* confirmation theory: not the confirmation theory (if there is one!) to which their own theory of meaning should lead.

What is going on here should be evident from Carnap's work on the foundations of mathematics. Carnap has a consistent tendency to *identify* concepts with their syntactic representations: thus, mathematical truth with theoremhood (after the discovery of Gödel's theorem, he either allowed 'non-constructive rules of proof', or simply assumed set theory, and took 'logical consequence' rather than derivability as the basic notion, although this trivialized the 'analysis' of mathematical truth). In the same way he would have liked to identify a state of affairs having a probability of, say, 0.9, with the corresponding sentence's having a c-value of 0.9 (where 'c' would be a syntactically defined measure on sentences in a formalized language). Even if Carnap had found a successful 'c-function', the fact is that it would have been successful because it corresponded to a reasonable probability measure over some collection of states of affairs; but this is just what Carnap's positivism did not allow him to say.

D. Auxiliary hypotheses

Sometimes, as we mentioned, the positivists make it explicit that the 'theories' to which their theory of science applies are 'formalizations of total science', and not theories in the usual sense; but their readers do, I think, tend to come away with the impression that their model *is* a model of a scientific theory in the usual sense—especially, a physical theory. Believing this involves believing that a physical theory is a calculus, or could easily be formalized as a calculus, and that its predictions are *self-contained*—that they are deduced from the explicitly stated assumptions of the theory itself. This leads to a comparison with social sciences which is derogatory to the social sciences—for the classic social science theories are clearly *not* self-contained in this sense. In short, the positivist attitude tends to be that social science is science only when and to the extent that it apes *physics*. And this for the reason that the mathematical model of a scientific theory provided by the positivists is thought to clearly fit *physical* theories.

But, in fact, it fits physical theories very badly, and this for the reason that even physical theories in the usual sense—e.g., Newton's Theory of Universal Gravitation, Maxwell's theory—lead to no predictions at all without a host of auxiliary assumptions, and moreover without auxiliary assumptions that are not at all law-like, but that are, in fact, assumptions about boundary conditions and initial conditions in the case of particular systems. Thus, if the claim that the term 'gravitation', for example, had a meaning which depended on the theory were true, and the theory included such auxiliary assumptions as that 'space is a hard vacuum', and 'there is no tenth planet in the solar system', then it would follow that discovery that space is *not* a hard vacuum or even that there is a tenth planet would change the meaning of 'gravitation'. I think one has to be pretty idealistic in one's intuitions to find this at all plausible! It is not so implausible that knowledge of the meaning of the term 'gravitation' involves some knowledge of the theory (although I think that this is wrong: the stereotype associated with 'gravitation' is not nearly as strong as a particular theory of gravitation), and this is probably what most readers think of when they encounter the claim that physical magnitude terms (usually called 'theoretical terms' to prejudge just the issue this paper discusses) are 'theory loaded'; but the actual meaning-dependence required by positivist meaning theory would be a dependence not just on the *laws* of the theory, but on the particular auxiliary assumptions—for, if these are not counted as part of the theory, then the whole theory-prediction scheme collapses at the outset.

Finally, neglect of the role that auxiliary assumptions actually play in science leads to a wholly incorrect idea of how a scientific theory is confirmed. Newton's theory of gravitation was not confirmed by checking predictions derived from it plus some set of auxiliary statements fixed in advance; rather the auxiliary assumptions had to be continually modified and expanded in the history of Celestial Mechanics. That scientific problems as often have the form of finding auxiliary hypotheses as they do of finding and checking predictions is something that has been too much neglected in philosophy of science;† this neglect is largely the result of the acceptance of the positivist model and its uncritical application to actual physical theories.

†I discuss this in chapter 16, volume I, of these papers.

Arthur Fine

AND NOT ANTI-REALISM EITHER*

EPIGRAPHS

1. *Realism:* "Out yonder there was this huge world, which exists independently of us human beings. . . . The mental grasp of this extra-personal world hovered before me as the highest goal. . . ."

Albert Einstein, "Autobiographical Notes"

2. *Anti-Realism:* "to get at something absolute without going out of your own skin!"

William James, letter to Tom Ward, October 9, 1868

*This paper was written during the tenure of a Guggenheim Fellowship. I want to thank the Foundation for their support. Thanks, too, to Micky Forbes for struggling with me through the ideas and their expression.

Reprinted by permission of the author and of the editor of *Noûs*, Vol. 18 (1984): pp. 51–66.

INTRODUCTION

As my title suggests, this paper is another episode in a continuing story. In the last episode the body of realism was examined, the causes of its death identified, and then the project of constructing a suitable successor for these post-realist times was begun ([3]). I called that successor the "natural ontological attitude" or NOA, for short, and I shall return to it below. In today's episode, however, the subject of criticism becomes anti-realism, and this is a live and, therefore, a shiftier target. For the death of realism has revived interest in several anti-realist positions and, appropriately enough, recent philosophical work has explored modifications of these anti-realisms to see whether they can be refurbished in order to take over, from realism, as the philosophy of science "of choice." My first object here will be to show that just as realism will not do for this choice position, neither will anti-realism. That job accomplished, I shall then sing some more in praise of NOA.

To understand anti-realism we have first to backtrack a bit and re-examine realism. Given the diverse array of philosophical positions that have sought the "realist" label, it is probably not possible to give a sketch of realism that will encompass them all. Indeed, it may be hopeless to try, even, to capture the essential features of realism. Yet that is indeed what I hope to do in identifying the core of realism with the following ideas. First, realism holds that there exists a definite world; that is, a world containing entities with relations and properties that are to a large extent independent of human acts and agents (or the possibilities thereof). Secondly, according to realism, it is possible to obtain a substantial amount of reliable and relatively observer-independent information concerning this world and its features, information not restricted, for example, to just observable features. I shall refer to these components of realism as (1) belief in a definite world-structure and (2) belief in the possibility of substantial epistemic access to that structure. This realism becomes "scientific" when we add to it a third component, namely, (3) the belief that science aims at (and, to some extent, achieves) all the epistemic access to the definite world structure that realism holds to be possible.

This sketch of realism highlights the ontological features that seem to me characteristic of it. But there is a semantical aspect as well. For in order to see science as working toward the achievement of the realist goal of substantial access to features of the definite world-structure, the theories and principles of science must be understood to be *about* that world-structure. Thus the *truth* of scientific assertions gets a specifically realist interpretation; namely, as a *correspondence* with features of the definite world structure.

I can put it very succinctly this way. The realist adopts a standard, model-theoretic, correspondence theory of truth, where the model is just the definite world structure posited by realism and where correspondence is understood as a relation that reaches right out to touch the world. (See [4] and [10].)

The "anti-realisms" that I want to examine and reject here all oppose the three tenets of realism understood as above (in spirit, if not always in words). They also reject the characteristically realist picture of truth as external-world correspondence. They divide among themselves over the question of whether or not that realist picture of truth ought to be replaced by some other picture. But they agree (again in spirit, if not in words) that although the realist has the aim of science wrong, in his third tenet above, it is important for us to understand what the correct aim of science is. This agreement is the mark of what I shall call "scientific" anti-realism. And the disagreement over offering truth pictures, then, divides the scientific anti-realists into those who are truthmongers and those who are not.

TRUTHMONGERS

The history of philosophy has witnessed a rather considerable trade in truth; including wholesale accounts like correspondence and coherence theories, or consensus and pragmatic theories, or indexical and relativist theories. There have also been special reductions available including phenomenalisms and idealisms. Among scientific anti-realists the wholesalers, recently, have tried to promote some kind of consensus-cum-pragmatic picture. I will try to give this picture a canonical representation so that we can identify the features that these particular anti-realisms have in common. So represented, it portrays the truth of a statement P as amounting to the fact that a certain class of subjects would accept

P under a certain set of circumstances. If we let the subjects be "perfectly rational" agents and the circumstances be "ideal" ones for the purposes of the knowledge trade (perhaps those marking the Peircean limit?), then we get the picture of truth as ideal rational acceptance, and this is the picture that Hilary Putnam paints for his "internal realism."[1] If the subjects are not perfectly rational and yet conscientious and well-intentioned about things, and we let the circumstances be those marking a serious dialogue of the kind that makes for consensus, where consensus is attainable, then we get the Wittgensteinean position that Richard Rorty calls "epistemological behaviorism."[2] Finally, if our subjects are immersed in the matrix of some paradigm and the circumstances are those encompassed by the values and rules of the paradigm, then we get the specifically paradigm-relative concept of truth (and of reference) that is characteristic of Thomas Kuhn's anti-realism ([8]). With these three applications in mind, I want to examine the merits—which is to say, to point out the demerits—of this sort of acceptance theory of truth.

Let us first be clear that these acceptance pictures of truth make for an anti-realist attitude towards science. That is a somewhat subtle issue, for the old Machian debates over the reality of molecules and atoms might suggest that realism turns on the putative truth (or not) of certain existence claims, especially claims about the existence of "unobservables." Since acceptance theories of truth, of the sort outlined above, might very well issue in the truth of such existence claims, one might be tempted to suggest, as well, that holders of acceptance theories could be realists. While there is no doubt a distinction to be drawn between those who do and those who do not believe in the existence, let us say, of magnetic monopoles; I think it would be a mistake to take that as distinguishing the realists from the others.[3] For it is not the *form* of a claim held true that marks off realism, it is rather the significance or content of the claim. The realist, say, wants to know whether there *really* are magnetic monopoles. He understands that in the way explained above, so that a positive answer here would signify a sort of reaching *out* from electrodynamic discourse *to* the very stuff of the world. The fact that scientific practice involves serious monopole talk, including what is described as manipulating monopoles and intervening in their behavior, does not even begin to address the issue of realism. For what realism is after is a very particular interpretation of that practice. This is exactly the interpretation that the picture of truth-as-acceptance turns us away from.

The special sort of correspondence that is built into the realist conception of truth orients us to face "*out* on the world," striving in our science to grab hold of significant chunks of its definite structure. The idea of truth as acceptance, however, turns us right around again to look back at our own collective selves, and at the interpersonal features that constitute the practice of the truth-game. (Compare the two epigraphs.) This turnabout makes for a sort of Ptolemaic counterrevolution. We are invited to focus on the mundane roots of truth-talk, and its various mundane purposes and procedures. Concepts having to do with acceptance provide a rich setting for all these mundane happenings. If we then take truth just to *be* the right sort of acceptance we reap a bonus for, when we bring truth down to earth in this way, we obtain insurance against the inherent, metaphysical aspects of realism.

I can well understand how the sight of realism unveiled might bring on disturbing, metaphysical shudders. And it's understandable, I think, that we should seek the seeming security provided by sheltering for awhile in a nest of interpersonal relations. But it would be a mistake to think that we will find truth there. For the anti-realism expressed in the idea of truth-as-acceptance is just as metaphysical and idle as the realism expressed by a correspondence theory.

I have not been able to locate a significant line of argument in the recent literature that moves to supply the warrant for an acceptance theory of truth. Rather, as I have noted, these anti-realists seem to have taken shelter in that corner mainly in reaction to realism. For when one sees that the realist conception of truth creates a gap that keeps the epistemic access one wants always just beyond reach, it may be tempting to try to refashion the idea of truth in epistemic terms in order, literally, to make the truth accessible. What allows the truthmongers to think that this is feasible, so far as I can tell, is a common turning toward behaviorism. In one way or another, these anti-realists seem sympathetic to the behaviorist idea that the working practices of conceptual exchange exhaust the meaning of that exchange, giving it its significance and providing it with its content. Thus we come to the idea that if the working practices of the truth exchange are the practices of acceptance, then acceptance is what truth is all about, and nothing but acceptance.

I do not have any new critique to offer concerning the flaws in behaviorism. Just about everyone recognizes that various special applications of behaviorism are wrong; for example, operationalism, or Watson-Skinnerism. So too, just about everyone has a sense of the basic error; namely, that behaviorism makes out everything it touches to be less than it is, fixing limits where none exist. Such, indeed, is the way of these anti-realisms: they fix the concept of truth, pinning it down to acceptance. One certainly has no more warrant for imposing this constraint on the basic concept of truth, however, than the operationalist has for imposing his constraints on more derivative concepts (like length or mass).

In fact, I think the warrant for behaviorism with regard to truth is considerably more suspicious than anything the operationalist ever had in mind. For whatever might possibly warrant the behaviorist conception of truth-as-acceptance should at least make that a conception we can take in and understand. Even if, as some maintain, truth is merely a regulative ideal, it must still be an ideal we can understand, strive for, believe in, glimpse—and so forth. But if, as the behaviorist holds, judgments of truth are judgments of what certain people would accept under certain circumstances, what are the ground rules for arriving at those judgments, and working with them as required? Naively, it looks like what we are called upon to do is to extrapolate from what *is* the case with regard to actual acceptance behavior to what *would be* the case under the right conditions. But how are we ever to establish what *is* the case, in order to get this extrapolation going, when that determination itself calls for a prior, successful round of extrapolation? It appears that acceptance locks us into a repeating pattern that involves an endless regress. Moreover, if we attend to the counterfactuality built into the "would accept" in the truth-as-acceptance formula, then I think we encounter a similar difficulty. To understand this conception of truth we must get a sense of how things would be were they different in certain respects from what they are now. Whatever your line about counterfactuals, this understanding involves at least either the idea of truth in altered circumstances, or the idea of truth in these actual circumstances. But each alternative here folds in upon itself, requiring in turn further truths. I believe there is no grounding for this process unless we turn away from the acceptance picture at some point.

It seems to me that the acceptance idea never *can* get off the ground, and that we cannot actually understand the picture of truth that it purports to offer. If we think otherwise that is probably because we are inclined to read into the truthmongers' project some truths (or ideas of truth) not having to do with acceptance at all—perhaps, even, some truths via correspondence! Thus, with respect to warrant and intelligibility, the acceptance picture emerges as quite on a par with the correspondence picture.[4]

There is, as I have noted, a very close connection between these two conceptions.

It is a typical dialectic that binds the metaphysics of realism to the metaphysics of behaviorism. Realism reaches out for *more* than can be had. Behaviorism reacts by pulling back to the "secure" ground of human behavior. In terms of that it tries to impose a limit, short of what realism has been searching for. The limit imposed by behaviorism, however, is simply *less* than what we require. So realism reacts by positing something more, and then reaches out for it again. What we can learn from this cycle is just what makes it run, and how to stop it.

Both the scientific realist and the scientific anti-realist of the acceptance sort share an attitude toward the concept of truth They think it is appropriate to give a theory, or account, or perhaps just a "picture" of truth. As Hilary Putnam pleads,

> But if all notions of rightness, both epistemic and (metaphysically) realist are eliminated, then what are our statements but noisemakings? What are our thoughts but *mere* subvocalizations? . . . Let us recognize that one of our fundamental self-conceptualizations, in Rorty's phrase, is that we are *thinkers,* and that *as* thinkers we are committed to there being *some* kind of truth, some kind of correctness which is substantial and not merely "disquotational." ([14]: 20–21)

Of course we are all committed to there being some kind of truth. But need we take that to be something like a "natural" kind? This essentialist idea is what makes the cycle run, and we can stop it if we stop conceiving of truth as a substantial something, something for which theories, accounts, or even pictures are appropriate. To be sure, the anti-realist is quite correct in his diagnosis of the disease of realism, and in his therapeutic recommendation to pay attention to how human beings actually operate with the family of truth concepts. Where he goes wrong is in trying to fashion out of these practices a completed concept of truth as a substantial something, one that will then act as a limit for legitimate human aspirations. If we do not join him in this undertaking and if we are also careful not to replace this anti-realist limit on truth by something else that goes beyond practice, then we shall have managed to avoid both realism and these truthmongering anti-realisms as well.

EMPIRICISM

But there are other anti-realisms to contend with. One well-known brand is empiricism, and this has made some notable progress in the sophisticated version that Bas van Fraassen calls "constructive empiricism" ([18]). This account avoids the reductive and foundationalist tendency of earlier empiricisms that sought to ground all truths in a sense-data or phenomenalist base. It also avoids the modification of this idea that ensnared logical-empiricism: the conception of a theory as a deductively closed logical system on the vocabulary of which there is imposed an epistemologically significant distinction between observables and unobservables. Instead, constructive empiricism takes a semantical view of a scientific theory; it views it as a family of models. And it lets science itself dictate what is or is not observable, where the "able" part refers to us and our limitations according to science. As for truth, it does not engage in trade but plumps for a literal construal.[5] The important concept for this brand of empiricism is the idea of empirical adequacy. This idea applies to theories, conceived of as above. Such a theory is empirically adequate just in case it has some model in which all truths about observables are represented. If truths about observables are called "phenomena," then a theory is empirically adequate just in case it saves the phenomena, *all* the phenomena. The distinctively anti-realist thesis of constructive empiricism is twofold: (1) that science aims only to provide theories that are empirically adequate and (2) that acceptance of a theory involves as belief only that it *is* empirically adequate. The intended contrast is with a realism that posits true theories as the goal of science and that takes acceptance of a theory to be belief in the truth of the theory. Since truth here is to be taken literally, the realist could well be committed to believing in the existence of unobservable entities literally, but never the constructive empiricist.

Indeed this brand of empiricism, along with its ancestors, involves a strong limitation on what it is legitimate for us to believe (in the sense of believe to be true). Where science is taken as the legitimating basis, we are allowed to believe that the scientific story about observables is true, and no more than that. It seems to me that there are two obvious testing points to probe with regard to any stance that seeks to impose limits on our epistemic attitudes. The first is to see whether the boundary can be marked off in a way that does not involve suspicious or obnoxious assumptions. The second is to see what the rationale is for putting the boundary just there, and to what extent that implacement is arbitrary. Let me take these in order.

A difficulty of the first sort begins to show up as soon as we ask why an attitude of belief is appropriate for the scientific judgment that something is observable. After all, that is supposed to be just another bit of science, and so our empiricism says that it is a candidate for affirmative belief (as opposed to agnostic reserve) just in case it is itself a judgment entirely within the realm of observables (according to science). What does that mean? Well, one might suppose that since the judgment that something is observable has a simple subject/predicate form, then both the subject of the judgment and the predicate must refer to what science holds to be observable. So, for example, the judgment that carrots are mobile would be a candidate for belief if, as we suppose, science classifies both carrots and mobility as observables. What then of the judgment that carrots are observable? In order for *it* to be a candidate for belief, we must suppose that science classifies both carrots and observability as observables. But now I think we ought to come to a full stop.

For if we accept the moves made so far, then we see that the combination of first, limiting belief to the observable and, second, letting science determine what counts as observable, has a terribly odd consequence. Namely, in order to believe in any scientific judgment concerning what is observable, we must take as a presupposition that the "property" or "characteristic" (or whatever) of "being observable" is itself an observable, *according to science.* Thus when we go down the list of entities supposedly using our science to determine which ones are observable and which not, the property of "being observable" must be classified as well, and indeed it must come out as observable. But this is surely something forced on us *a priori* by this empiricist philosophical stance. If there actually were such a property as "being observable," and science did actually classify it, who is to say how it must come out—or even whether it must come out at all as observable or not. *Science* is supposed to speak here, not philosophy. Thus if we accept the moves in the argument, this empiricism is suspiciously near to an inconsistency: It forces the hand of science exactly where it is supposed to follow it.

What then if we try to reject some move in the argument? What shall we question? Surely the requirement that we respect grammar and ask separately of subject and predicate whether it refers to an observable is not a necessary one. After all, to speak somewhat realistically, who can tell how a judgment confronts the world? Let us then give up the grammatical requirement and think again how to deal with the judgment that something is an observable; that is, how to construe it as a judgment entirely within the realm of observables. If I judge, scientifically, that carrots are observable, then I suppose I would have to identify some properties or features of carrots and show that these would induce the right sort of effects in an interaction, one party to which is a human being, *qua* observing instrument. To back up the counterfactual, here, (what effects would be induced) certainly several laws would enter the argument, very likely connecting entities that may themselves not be observable. Now, according to the empiricism at issue, I do not have to believe this whole theoretical story, only its observational part. *That* I do have to believe if it is to warrant my belief in the observability of carrots. But since the question here was precisely how to identify the observational part of a simple judgment (that carrots are observable) I think I am stuck. I do not know what to believe in my scientific story that issues in the observability of carrots, unless I can pick out its observational parts. And I cannot identify a part of the story as observational unless I can support that identification by means of beliefs based on observational parts of still other covering stories. I really think that we cannot break out of this cycle—or rather break in to get it going—without some external stipulations, or the like, as to what to believe to be observable. Thus an aprioristic resolution of the philosophical squabble over what to take as observable seems required by this empiricism, just as it was by the older ones.

There is, however, a deft maneuver that could get things going again. It is simply not to raise the question of observability where what is at stake is itself a judgment of observability. Thus we could exempt those special judgments from the test of observability and allow ourselves to believe them in just the way that we would if they had actually passed the test. Indeed if we allow this exception for judgments of observability, then no difficulties seem to arise by way of beliefs being sanctioned that are not really warranted. But if we try to avoid obnoxious assumptions concerning what is observable by granting exemptions from the general empiricist rule in certain special cases, then why—we must ask—should that rule be necessary for the others? This brings us to the second testing point for a philosophy that seeks to impose limits on one's epistemic attitudes; namely, to examine the rationale for the limit—especially to see how arbitrary it is.

We can push this question hard if we recognize that there is a loosely graded vocabulary concerning observability. We do, after all, draw a distinction between what is *observable,* which is rather strict, and what is *detectable,* which is somewhat looser. To get a feel for the distinction, we might, for instance, picture the difference here as between what we would "observe," in the right circumstances, with our sense organs as they are, and what we would "detect" in those same circumstances were our eyes, for example, replaced by electron microscopes. In this grading system atoms . . . would count as detectable but not (strictly) observable. It seems to me that distinctions of this sort are, in fact, at work in the vocabulary of observation, and van Fraassen certainly recognizes some such ([18]: 16–17). With this in mind . . . I think we can make the question of observability, as a warrant for belief, very acute by asking why restrict the realm of belief to what is observable, as opposed, say, to what is detectable?

I think the question is acute, because I cannot imagine any answer that would be compelling. Are we supposed to refrain from believing in atoms, and various truths about them, because we are concerned over the possibility that what the electron microscope reveals is merely an artifact of the machine? If this is our concern, then we can address it by applying the cautious and thorough procedures and analyses involved in the use and construction of that machine, as well as the cross-checks from other detecting devices, to evaluate the artifactuality (or not) of the atomic phenomena. If we can do this satisfactorily according to tough standards, are we then still not supposed to frame beliefs about atoms, and why not now? Surely the end product of such inquiries, when each one pursues a specific area of uncertainty or possible error, can only be a very compelling scientific documentation of the grounds for believing that we are, actually, detecting atoms.[6] Faced with such substantial reasons for believing that we are detecting atoms what, except purely a priori and arbitrary conventions, could possibly dictate the empiricist conclusion that, nevertheless, we are unwarranted actually to engage in *belief* about atoms? What holds for detectability holds as well for the other information-bearing modalities, ones that may be even more remotely connected with strict observability. The general lesson is that, in the context of science, adopting an attitude of belief has as warrant precisely that which science itself grants, nothing more but certainly nothing less.[7]

The stance of empiricism, like that of the truthmongers, is (in part) a moral stance. They both regard metaphysics, and in particular the metaphysics of realism, as a sin. They both move in the direction of their anti-realism in order to avoid that sin. But the behaviorism to which the truthmongers turn, as we have seen, locks them into a comic dance with realism, a *pas de deux* as wickedly metaphysical as ever there was. The empiricist, I think, carries a comparable taint. For when he sidesteps science and moves into his courtroom, there to pronounce his judgments of where to believe and where to withhold, he avoids metaphysics only by committing, instead, the sin of epistemology. We ought not to follow him in this practice. Indeed, I think courtesy requires, at this point, a discreet withdrawal.

NOA: THE NATURAL ONTOLOGICAL ATTITUDE

The "isms" of this paper each derive from a philosophical program in the context of which they seek to place science. The idea seems to be that when science is put in that context its significance, rationality, and purpose, as it were, just click into place. Consequently, the defense of these "isms," when a defense is offered, usually takes the form of arguing that the favorite one is better than its rivals because it makes better sense of science than do its rivals.[8]

What are we to conclude from this business of placing science in a context, supplying it with an aim, attempting to make better sense of it, and so forth? Surely, it is that realism and anti-realism alike view science as susceptible to being set in context, provided with a goal, and being made sense of. And what manner of object, after all, could show such susceptibilities other than something that could not or did not do these very things for itself? What binds realism and anti-realism together is this. They see science as a set of practices in need of an interpretation, and they see themselves as providing just the right interpretation.

But science is not needy in this way. Its history and current practice constitute a rich and meaningful setting. In that setting questions of goals or aims or purposes occur spontaneously and *locally*. For what purpose is a particular instrument being used, or why use a tungsten filament here rather than a copper one? What significant goals would be accomplished by building accelerators capable of generating energy levels in excess of 10^4 GeV? Why can we ignore gravitational effects in the analysis of Compton scattering? Etc. These sorts of questions have a teleological cast and, most likely, could be given appropriate answers in terms of ends, or goals, or the like. But when we are asked what is the aim of science itself, I think we find ourselves in a quandary, just as we do when asked "What is the purpose of life?" or indeed the corresponding sort of question for any sufficiently rich and varied practice or institution. As we grow up I think we learn that such questions really do not require an answer, but rather they call for an empathetic analysis to get at the cognitive (and temperamental) sources of the question, and then a program of therapy to help change all that.

Let me try to collect up my thoughts by means of a metaphor (or is it an allegory?). The realisms and anti-realisms seem to treat science as a sort of grand performance, a play, or opera, whose production requires interpretation and direction. They argue among themselves as to whose "reading" is best.[9] I have been trying to suggest that if science is a performance, then it is one where the audience and crew play as well. Directions for interpretation are also part of the act. If there are questions and conjectures about the meaning of this or that, or its purpose, then there is room for those in the production too. The script, moreover, is never finished, and no past dialogue can fix future action. Such a performance is not susceptible to a reading or interpretation in any global sense, and it picks out its own interpretations, locally, as it goes along.

To allow for such an open conception of science, the attitude one adopts must be neither realist nor anti-realist. It is the attitude I want to call your attention to under the name of NOA, the natural ontological attitude. The quickest way to get a feel for NOA is to understand it as undoing the idea of interpretation, and the correlative idea of invariance (or essence).

The attitude that marks NOA is just this: Try to take science on its own terms, and try not to read things into science. If one adopts this attitude, then the global interpretations, the "isms" of scientific philosophies, appear as idle overlays to science: not necessary, not warranted and, in the end, probably not even intelligible. It is fundamental to NOA that science has a history, rooted indeed in everyday thinking. But there need not be any aspects invariant throughout that history, and hence, contrary to the isms, no necessary uniformity in the overall development of science (including projections for the future). NOA is, therefore, basically at odds with the temperament that looks for definite boundaries demarcating science from pseudo-science, or that is inclined to award the title "scientific" like a blue ribbon on a prize goat. Indeed the anti-essentialist aspect of NOA is intended to be very comprehensive, applying to all the concepts used in science, even the concept of truth.

Thus NOA is inclined to reject *all* interpretations, theories, construals, pictures, etc. of truth, just as it rejects the special correspondence theory of realism and the acceptance pictures of the truthmongering anti-realisms. For the concept of truth is the fundamental semantical concept. Its uses, history, logic, and grammar are sufficiently definite to be partially catalogued, at least for a time. But it cannot be "explained" or given an "account of" without circularity. Nor does it require anything of the sort. The concept of truth is open-ended, growing with the growth of science. Particular questions (Is this true? What reason do we have to believe in the truth of that? Can we find out whether it is true? Etc.) are addressed in well-known ways. The significance of the answers to those questions is rooted in the practices and logic of truth-judging (which practices, incidentally, are by no means confined to acceptance, or the like), but that significance branches out beyond current practice along with the growing concept of truth. For, present knowledge not only redistributes truth-values among past judgments, present knowledge also re-evaluates the whole character of past practice. There is no saying, in advance, how this will go. Thus there is no projectible sketch now of what truth signifies, nor of what areas of science (e.g. "fundamental laws") truth is exempt from—nor ever will there be. Some questions, of course, are not settled by the current practices of truth judging. Perhaps some never will be settled.

NOA is fundamentally a heuristic attitude, one that is compatible with quite different assessments of particular scientific investigations; say, investigations concerning whether or not there are magnetic monopoles. At the time of this writing the scientific community is divided on this issue. There is a long history of experimental failure to detect monopoles, and one recent success—maybe. I believe that there are a number of new experiments under way, and considerable theoretical work that might narrow down the detectable properties of monopoles.[10] In this context various ways of putting together the elements that enter into a judgment about monopoles will issue in various attitudes toward them, ranging from complete agnosticism to strong belief. NOA is happy with any of these attitudes. All that NOA insists is that one's ontological attitude towards monopoles, and everything else that might be collected in the scientific zoo (whether observable or not), be governed by the very same standards of evidence and inference that are employed by science itself. This attitude tolerates all the differences of opinion, and all the varieties of doubt and skepticism, that science tolerates. It does not, however, tolerate the prescriptions of empiricism, or of other doctrines that externally limit the commitments of science. Nor does it overlay the judgment say, that monopoles do exist, with the special readings of realism or of the truthmongering anti-realisms. NOA tries to let science speak for itself, and it trusts in our native ability to get the message without having to rely on metaphysical or epistemological hearing aids.

I promised to conclude these reflections by singing in praise of NOA. The refrain I had in mind is an adaptation of a sentiment that Einstein once expressed concerning Mozart. Einstein said that the music of Mozart (read "NOA") seems so natural that, by contrast, the music of other composers (read "realism" or "anti-realism") sounds artificial and contrived.

NOTES

1. Putnam [11] is an extended discussion, usefully supplemented by Putnam [13] and [14]. Originally, rational acceptability was merely offered as a "picture" of truth. But later it emerged as a "characterization," and as providing "the only sense in which we have a vital and working notion of it." ([14]:5)

2. Rorty [15]. In his symposium talk for the March, 1983 Pacific Division, APA meetings (in Berkeley), Rorty announced a new position that he called "revisionary pragmatism." This new stance pulls back from various of Rorty's earlier commitments, including some of his ideas about truth. I have not been able to figure out, however, just what it rejects or what it retains.

3. Cartwright [1] and Hacking [5] and [6] adopt this way of distinguishing a significant form of realism. See my [3] and [4] for a critique. The only background that seems to me to support the idea that the truth of certain existence claims makes for realism is an account of truth as external-world correspondence. I do not believe that Hacking adopts such a view. I do not know about Cartwright. I might mention here that Putnam's tactic of calling his position a kind of realism (an "internal" kind), while also seeing in it a "transcendental idealism" ([12]: 6), seems founded on nothing more than the amusing idea that whatever is not solipsism is, *ipso facto*, a realism. See ([13]: 162) and ([14]: 13).

4. Other ways of displaying the gap between "truth" and the favored version of acceptance would be to ask whether the acceptance formula is true, or what "accept as true" comes to under the formula, or what now would guarantee the idempotency of "is true." Pursuing such lines of inquiry, along with the ones in the text, will show that the sense and grammar of truth is not that of acceptance. But, of course, it does not follow that truth is not acceptance (really!). Nor could such lines of inquiry really subvert the program of replacing "truth" by acceptance, if one were determined to carry on with the program. One can always dodge the arguments and, where that fails, bite the argumentative bullets. In philosophy, as in other areas of rational discourse, inquiry must end in judgments. One can try to inform and tutor good judgment, but it cannot be compelled—not even by good-looking reasons.

5. Although van Fraassen ([18]: Esp. 9–11) is quite explicit about taking truth literally, he also seems tempted by the interpretative metaphor of realist-style correspondence. "A statement is true exactly if the actual world accords with the statement." ([18]: 90) "I would still identify the truth of a theory with the condition that there is an exact correspondence between reality and one of its models." ([18]: 197) If van Fraassen is taken literally, in these passages, then for him truth *is*, literally, real-world correspondence. If this were correct, then van Fraassen's empiricism would appear to be a restricted version of realism, a version where the epistemic access is restricted to observables. This makes his "anti-realism" seem considerably less radical than one might have thought. I, at any rate, had thought his idea of literal truth included the notion that "truth" was not to be further interpreted. On this understanding I thought that if he were persuaded out

of his attachment to observables, then his ideas would fit right in with NOA (see below). But now I think that may be wrong. If we were to make constructive empiricism lose its attachment to observables, we would (it seems) merely have regained realism, full blown.

6. The themes just touched on, especially the insistence on the specificity of scientific doubt and on following the scientific rationale that informs the vocabulary of observation, are forcefully elaborated by Shapere [16]. Part B ("intervening") of Hacking [6] is also required reading here.

7. See Hellman ([7]: esp. 247–248) for some cosmological "unobservables" in which we might have good scientific grounds for belief. But do not forget more familiar sorts of objects either, like unconscious (or "subliminal") causal factors in our behavior or, even, the nightly activity we call "dreaming"!

8. "However, there is also a positive argument for constructive empiricism—it makes better sense of science, and of scientific activity, than realism does and does so without inflationary metaphysics." ([18]: 73) I think van Fraassen speaks here for all the anti-realists. While I cannot recommend this defense of anti-realism, I think van Fraassen's own critique of the explanationist defenses of realism is very incisive, especially if complemented with the attack of Laudan [9]. My [3] contains a meta-theorem showing why such explanationist (or coherentist) defenses of realism are bound to fail.

9. This way of putting it suggests that the philosophies of realism and anti-realism are much closer to the hermeneutical tradition than (most of) their proponents would find comfortable. Similarly, I think the view of science that has emerged from these "isms" is just as contrived as is the shallow, mainline view of the hermeneuts (science as control and manipulation, involving only dehumanized and purely imaginery models of The World). In opposition to this, I do not suggest that science is hermeneutic-proof, but rather that in science, as elsewhere, hermeneutical understanding has to be gained *from the inside.* It should not be prefabricated to meet external, philosophical specifications. There is, then, no legitimate hermeneutical *account* of science, but only a hermeneutical activity that is a lively part of science itself.

10. See [17] for a review.

REFERENCES

1. N. Cartwright, *How the Laws of Physics Lie* (Oxford: Oxford University Press, 1983).

2. A. Fine, "The Young Einstein and the Old Einstein," in R. Cohen et al. (eds.), *Essays in Memory of Imre Lakatos* (Dordrecht, Holland: Reidel, 1976): 145–159.

3. ———, "The Natural Ontological Attitude" in J. Leplin (ed.), *Scientific Realism* (Berkeley, CA: University of California Press, 1984).

4. ———, "Is Scientific Realism Compatible with Quantum Physics?," *Noûs,* 1984.

5. I. Hacking, "Experimentation and Scientific Realism" in J. Leplin (eds.), *Scientific Realism* (Berkeley, CA: University of California Press, 1984).

6. ———, *Representing and Intervening,* (Cambridge, MA: Cambridge University Press, 1983).

7. G. Hellman, "Realist Principles," *Philosophy of Science* 50(1983): 227–249.

8. T. S. Kuhn, *The Structure of Scientific Revolutions,* 2nd Edition (Chicago: University of Chicago Press, 1970).

9. L. Laudan, "A Confutation of Convergent Realism," *Philosophy of Science* 48(1981): 19–49.

10. ———, "Realism without the Real," *Philosophy of Science,* 1984.

11. H. Putnam, *Reason, Truth and History* (Cambridge: Cambridge University Press, 1981).

12. ———, "Quantum Mechanics and the Observer," *Erkenntnis* 16(1981): 193–220.

13. ———, "Why There Isn't a Ready-Made World," *Synthese* 51(1982): 141–167.

14. ———, "Why Reason Can't Be Naturalized," *Synthese* 52(1982): 3–23.

15. R. Rorty, *Philosophy and the Mirror of Nature* (Princeton: Princeton University Press, 1979).

16. D. Shapere, "The Concept of Observation in Science and Philosophy," *Philosophy of Science* 49(1982): 485–525.

17. W. P. Trower and B. Cabrera, "Magnetic Monopoles: Evidence since the Dirac Conjecture," *Foundations of Physics* 13(1983): 195–216.

18. B. van Fraassen, *The Scientific Image* (Oxford: Clarendon Press, 1980).

Theory Change and Change of Meaning

P. K. Feyerabend

HOW TO BE A GOOD EMPIRICIST—A PLEA FOR TOLERANCE IN MATTERS EPISTEMOLOGICAL*

"Facts?" he repeated. "Take a drop more grog, Mr. Franklin, and you'll get over the weakness of believing in facts! Foul play, Sir!"

Wilkie Collins, *Moonstone*

1. CONTEMPORARY EMPIRICISM LIABLE TO LEAD TO ESTABLISHMENT OF A DOGMATIC METAPHYSICS

Today empiricism is the professed philosophy of a good many intellectual enterprises. It is the core of the sciences, or so at least we are taught, for it is responsible both for the existence and for the growth of scientific knowledge. It has been adopted by influential schools in aesthetics, ethics, and theology. And within philosophy proper the empirical point of view has been elaborated in great detail and with even greater precision. This predilection for empiricism is due to the assumption that only a thoroughly observational procedure can exclude fanciful speculation and empty metaphysics as well as to the hope that an empiristic attitude is most liable to prevent stagnation and to further the progress of knowledge. It is the purpose of the present paper to show that empiricism in the form in which it is practiced today cannot fulfill this hope.

Putting it very briefly, it seems to me that the contemporary doctrine of empiricism has encountered difficulties, and has created contradictions which are very similar to the difficulties and contradictions inherent in some versions of the doctrine of democracy. The latter are a well-known phenomenon. That is, it is well known that essentially totalitarian measures are often advertised as being a necessary consequence of democratic principles. Even worse—it not so rarely happens that the totalitarian character of the defended measures is not explicitly stated but covered up by calling them "democratic," the word *democratic* now being used in a new, and somewhat misleading, manner. This method of (conscious or unconscious) verbal camouflage works so well that it has deceived some of the staunchest supporters of true democracy. What is not so well-known is that modern empiricism is in precisely the same predicament. That is, some of the methods of modern empiricism which are introduced in the spirit of anti-dogmatism and progress are bound to lead to the establishment of a dogmatic metaphysics and to the construction of defense mechanisms which make this metaphysics safe from refutation by experimental inquiry. It is true that in the process of establishing such a metaphysics the words

*Revised paper originally appearing in *Inquiry*.

From *Philosophy of Science: The Delaware Seminar, II.* Reprinted by permission of the University of Delaware Press.

empirical or *experience* will frequently occur; but their sense will be as distorted as was the sense of "democratic" when used by some concealed defenders of a new tyranny.[1] This, then, is my charge: Far from eliminating dogma and metaphysics and thereby encouraging progress, modern empiricism has found a new way of making dogma and metaphysics respectable, viz., the way of calling them "well-confirmed theories," and of developing a method of confirmation in which experimental inquiry plays a large though well controlled role. In this respect, modern empiricism is very different indeed from the empiricism of Galileo, Faraday, and Einstein, though it will of course try to represent these scientists as following its own paradigm of research, thereby further confusing the issue.[2]

From what has been said above it follows that the fight for tolerance in scientific matters and the fight for scientific progress must still be carried on. What has changed is the denomination of the enemies. They were priests, or "school-philosophers," a few decades ago. Today they call themselves "philosophers of science," or "logical empiricists."[3] There are also a good many scientists who work in the same direction. I maintain that all these groups work against scientific progress. But whereas the former did so openly and could be easily discerned, the latter proceed under the flag of progressivism and empiricism and thereby deceive a good many of their followers. Hence, although their presence is noticeable enough they may almost be compared to a fifth column, the aim of which must be exposed in order that its detrimental effect be fully appreciated. It is the purpose of this paper to contribute to such an exposure.

I shall also try to give a positive methodology for the empirical sciences which no longer encourage dogmatic petrification in the name of experience. Put in a nutshell, the answer which this method gives to the question in the title is: You can be a good empiricist only if you are prepared to work with many alternative theories rather than with a single point of view and "experience." This plurality of theories must not be regarded as a preliminary stage of knowledge which will at some time in the future be replaced by the One True Theory. Theoretical pluralism is assumed to be an *essential feature* of all knowledge that claims to be objective. Nor can one rest content with a plurality which is merely abstract and which is created by denying now this and now that component of the dominant point of view. Alternatives must rather be developed in such detail that problems already "solved" by the accepted theory can again be treated in a new and perhaps also more detailed manner. Such development will of course take time, and it will not be possible, for example, at once to construct alternatives to the present quantum theory which are comparable to its richness and sophistication. Still, it would be very unwise to bring the process to a standstill in the very beginning by the remark that some suggested new ideas are undeveloped, general, metaphysical. *It takes time to build a good theory* [a triviality that seems to have been forgotten by some defenders of the Copenhagen point of view of the quantum theory]; and it also takes time to develop an alternative to a good theory. The *function* of such concrete alternatives is, however, this: They provide means of criticizing the accepted theory in a manner which goes *beyond* the criticism provided by a comparison of that theory "with the facts": However closely a theory seems to reflect the facts, however universal its use, and however necessary its existence seems to be to those speaking the corresponding idiom, its factual adequacy can be asserted only *after* it has been confronted with alternatives *whose invention and detailed development must therefore precede any final assertion of practical success and factual adequacy.* This, then, is the methodological justification of a plurality of *theories:* Such a plurality allows for a much sharper criticism of accepted ideas than does the comparison with a domain of "facts" which are supposed to sit there independently of theoretical considerations. The function of unusual *metaphysical* ideas which are built up in a nondogmatic fashion and which are then developed in sufficient detail to give an (alternative) account even of the most common experimental and observational situations is defined accordingly: They play a decisive role in the criticism and in the development of what is generally believed and "highly confirmed"; and they have therefore to be present at *any* stage of the development of our knowledge.[4] A science that is free from *metaphysics* is on the best way to become a *dogmatic* metaphysical system. So far the summary of the method I shall explain, and defend, in the present paper.

It is clear that this method still retains an essential element of *empiricism:* The decision between alternative theories is based upon *crucial experiments.* At the same time it must *restrict* the range of such experiments. Crucial experiments work well with theories of a low degree of generality whose principles do not touch the principles on which the ontology of the chosen observation language is based. They work well if such theories are compared with respect to a much more general background theory which provides a stable meaning for the observation sentences. However, this background theory, like any other theory, is itself in need of criticism. Criticism must use alternatives. Alternatives will be the more efficient the more radically they differ from the point of view to be investigated. It is bound to happen, then, that the alternatives do not share a single statement with the theories they criticize. Clearly, a crucial experiment is now impossible. It is impossible, not because the experimental device is too complex, or because the calculations leading to the experimental prediction are too difficult; it is impossible because there is no statement capable of expressing what emerges from the observation. This consequence, which severely restricts the domain of empirical discussion, cannot be circumvented by any of the methods which are currently in use and which all try to work with relatively stable observation languages. It indicates that the attempt to make empiricism a universal basis of all our factual knowledge cannot be carried out. The discussion of this situation is beyond the scope of the present paper.

On the whole, the paper is a concise summary of results which I have explained in a more detailed fashion in the following essays: "Explanation, Reduction, and Empiricism"; "Problems of Microphysics"; "Problems of Empiricism"; "Linguistic Philosophy and the Mind-Body Problem."[5] All the relevant acknowledgements can be found there. Let me only repeat here that my general outlook derives from the work of K. R. Popper (London) and David Bohm (London) and from my discussions with both. It was severely tested in discussion with my colleague, T. S. Kuhn (Berkeley). It was the latter's skillful defense of a scientific conservatism which triggered two papers, including the present one. Criticism by A. Naess (Oslo), D. Rynin (Berkeley), Roy Edgley (Bristol), and J. W. N. Watkins (London) have been responsible for certain changes I made in the final version.

2. TWO CONDITIONS OF CONTEMPORARY EMPIRICISM

In this section I intend to give an outline of some assumptions of contemporary empiricism which have been widely accepted. It will be shown in the sections to follow that these apparently harmless assumptions which have been explicitly formulated by some logical empiricists, but which also seem to guide the work of a good many physicists, are bound to lead to exactly the results I have outlined above: Dogmatic petrification and the establishment, on so-called "empirical grounds," of a rigid metaphysics.

One of the cornerstones of contemporary empiricism is its *theory of explanation.* This theory is an elaboration of some simple and very plausible ideas first proposed by Popper[6] and it may be introduced as follows: Let T and T' be two different scientific theories, T' the theory to be explained, or the explanandum, T the explaining theory, or the explanans. Explanation (of T') consists in the *derivation* of T' from T and initial conditions which specify the domain D' in which T' is applicable. Prima facie, this demand of derivability seems to be a very natural one to make for "otherwise the explanans would not constitute adequate grounds for the explanation" (Hempel[7]). It implies two things: First, that the consequences of a satisfactory explanans, T, inside D' must be compatible with the explanadum, T'; and secondly, that the main descriptive terms of these consequences must either coincide, with respect to their meanings, with the main descriptive terms of T', or at least they must be related to them via an empirical hypothesis. The latter result can also be formulated by saying that the meaning of T' must be unaffected by the explanation. "It is of the utmost importance," writes Professor Nagel,[8] emphasizing this point, "that the expressions peculiar to a science will possess meanings that are fixed by its *own* procedures, and are therefore intelligible in terms of its own rules of usage, whether or not the science has been, or will be [explained in terms of] the other discipline."

Now if we take it for granted that more general theories are always introduced with the purpose of explaining the existent successful theories, then every new theory will have to satisfy the two conditions just mentioned. Or, to state it in a more explicit manner:

(1) Only such theories are then admissible in a given domain which either *contain* the theories already used in this domain, or which are at least *consistent* with them inside the domain,[9] and

(2) meanings will have to be invariant with respect to scientific progress; that is, all future theories will have to be phrased in such a manner that their use in explanations does not affect what is said by the theories, or factual reports to be explained.

The two conditions I shall call the *consistency condition* and the *condition of meaning invariance,* respectively.

Both conditions are *restrictive* conditions and therefore bound profoundly to influence the growth of knowledge. I shall soon show that the development of actual science very often violates them and that it violates them in exactly those places where one would be inclined to perceive a tremendous progress of knowledge. I shall also show that neither condition can be justified from the point of view of a tolerant empiricism. However, before doing so I would like to mention that both conditions have occasionally entered the domain of the sciences and have been used here in attacks against new developments and even in the process of theory construction itself. Especially today, they play a very important role in the construction as well as in the defense of certain points of view in microphysics.

Taking first an earlier example, we find that in his *Wärmelehre,* Ernst Mach[10] makes the following remark:

> Considering that there is, in a purely mechanical system of absolutely elastic atoms no real analogue for the *increase of entropy,* one can hardly suppress the idea that a violation of the second law . . . should be possible if such a mechanical system were the *real* basis of thermodynamic processes.

And referring to the fact that the second law is a highly confirmed physical law, he insinuates (in his *Zwei Aufsaetze*[11]) that for this reason the mechanical hypothesis must not be taken too seriously. There were many similar objections against the kinetic theory of heat.[12] More recently, Max Born has based his arguments against the possibility of a return to determinism upon the consistency condition and the assumption which we shall here take for granted, that wave mechanics is incompatible with determinism:

> If any future theory should be deterministic it cannot be a modification of the present one, but

must be entirely different. How this should be possible without sacrificing a whole treasure of well established results [i.e., without contradicting highly confirmed physical laws and thereby violating the consistency condition] I leave the determinist to worry about.[13]

Most members of the so-called Copenhagen school of quantum theory would argue in a similar manner. For them the idea of complementarity and the formalism of quantization expressing this idea do not contain any hypothetical element as they are "uniquely determined by the facts."[14] Any theory which contradicts this idea is factually inadequate and must be removed. Conversely, an explanation of the idea of complementarity is acceptable only if it either contains this idea, or is at least consistent with it. This is how the consistency condition is used in arguments against theories such as those of Bohm, de Broglie, and Vigier.[15]

The use of the consistency condition is not restricted to such general remarks, however. A decisive part of the existing quantum theory *itself*, viz., the projection postulate,[16] is the result of the attempt to give an account of the definiteness of macro-objects and macro-events that is in accordance with the consistency condition. The influence of the condition of meaning invariance goes even further.

The Copenhagen-interpretation of the quantum theory [writes Heisenberg[17]] starts from a paradox. Any experiment in physics, whether it refers to the phenomena of daily life or to atomic events is to be described in the terms of classical physics . . . *We cannot and should not replace these concepts by any others* [my italics]. Still the application of these concepts is limited by the relation of uncertainty. We must keep in mind this limited range of applicability of the classical concepts while using them, but we cannot, and should not, try to improve them.

This means that the meaning of the classical terms must remain invariant with respect to any future explanation of microphenomena. Microtheories have to be formulated in such a manner that this invariance is guaranteed. The principle of correspondence and the formalism of quantization connected with it were explicitly devised for satisfying this demand. Altogether, the quantum theory seems to be the first theory after the downfall of the Aristotelian physics that has been quite explicitly constructed with an eye both on the consistency condition and the condition of (empirical) meaning invariance. In this respect it is very different indeed from, say, relativity which violates both consistency and meaning invariance with respect to earlier theories. Most of the arguments used for the defense of its customary interpretation also depend on the validity of these two conditions and they will collapse with their removal. An examination of these conditions is therefore very topical and bound deeply to affect present controversies in microphysics. I shall start this investigation by showing that some of the most interesting developments of physical theory in the past have violated both conditions.

3. THESE CONDITIONS NOT INVARIABLY ACCEPTED BY ACTUAL SCIENCE

The case of the consistency condition can be dealt with in a few words: It is well known (and has also been shown in great detail by Duhem[18]) that Newton's theory is inconsistent with Galileo's law of the free fall and with Kepler's laws; that statistical thermodynamics is inconsistent with the second law of the phenomenological theory; that wave optics is inconsistent with geometrical optics; and so on. Note that what is being asserted here is *logical* inconsistency; it may well be that the differences of prediction are too small to be detectable by experiment. Note also that what is being asserted is not the inconsistency of, say, Newton's theory and Galileo's law, but rather the inconsistency of *some consequences* of Newton's theory in the domain of validity of Galileo's law, and Galileo's law. In this last case the situation is especially clear. Galileo's law asserts that the acceleration of the free fall is a constant, whereas application of Newton's theory to the surface of the earth gives an acceleration that it not a constant but *decreases* (although imperceptibly) with the distance from the center of the earth. Conclusion: If actual scientific procedure is to be the measure of method, then the consistency condition is inadequate.

The case of meaning invariance requires a little more argument, not because it is intrinsically more difficult, but because it seems to be much more closely connected with deep-rooted prejudices. Assume that an explanation is required, in terms of the special theory of relativity, of the classical conservation of mass in all reactions in a closed system S. If $m', m'', m''', \ldots, m^i, \ldots$ are the masses of the parts $P', P'', P''', \ldots, P^i, \ldots$ of S, then what we want is an explanation of

$$\Sigma\, m^i = \text{const.} \qquad (1)$$

for all reactions inside S. We see at once that the consistency condition cannot be fulfilled: According to special relativity $\Sigma\, m^i$ will vary with the velocities of the parts relative to the coordinate system in which the observations are carried out, and the total mass of S will also depend on the relative potential energies of the parts. However, if the velocities and the mutual forces are not too large, then the variation of $\Sigma\, m^i$ predicted by relativity will be so small as to be undetectable by experiment. Now let us turn to the *meanings* of the terms in the relativistic law and in the corresponding classical law. The first indication of a possible change of meaning may be seen in the fact that in the classical case the mass of an aggregate of parts equals the sum of the masses of the parts:

$$M\,(\Sigma\, P^i) = \Sigma\, M(P^i).$$

This is not valid in the case of relativity where the relative velocities and the relative potential energies contribute to the mass balance. That the relativistic concept and the classical concept of mass are very different indeed becomes clear if we also consider that the former is a *relation,* involving relative velocities, between an object and a coordinate system, whereas the latter is a *property* of the object itself and independent of its behavior in coordinate systems. True, there have been attempts to give a relational analysis even of the classical concept (Mach). None of these attempts, however, leads to the relativisic idea with its velocity dependence on the coordinate system, which idea must therefore be added even to a *relational* account of classical mass. The attempt to identify the classical mass with the relativistic rest mass is of no avail either. For although both may have the same numerical value, the one is still dependent on the coordinate system chosen (in which it is at rest and has that specific value), whereas the other is not so dependent. We have to conclude, then, that $(m)_c$ and $(m)_r$ mean very different things and that $(\Sigma\, m^i)_c = \text{const.}$ and $(\Sigma\, m^i)_r = \text{const.}$ are very different assertions. This being the case, the derivation from relativity of either equation (1) or of a law that makes slightly different quantitative predictions with $\Sigma\, m^i$ used in the classical manner, will be possible only if a further premise is added which establishes a relation between the $(m)_c$ and the $(m)_r$. Such a "bridge law"—and this is a major point in Nagel's theory of reduction—is a hypothesis

> according to which the occurrence of the properties designated by some expression in the premises of the [explanans] is a sufficient, or a necessary and sufficient condition for the occurrence of the properties designated by the expressions of the [explanandum].[19]

Applied to the present case this would mean the following: Under certain conditions the occurrence of relativistic mass of a given magnitude is accompanied by the occurrence of classical mass of a corresponding magnitude; this assertion is inconsistent with another part of the explanans, viz., the theory of relativity. After all, this theory asserts that there are no invariants which are directly connected with mass measurements and it thereby asserts that $(m)_c$ does not express real features of physical systems. Thus we inevitably arrive at the conclusion that mass conservation cannot be explained in terms of relativity (or "reduced" to relativity) without a violation of meaning invariance. And if one retorts, as has been done by some critics of the ideas expressed in the present paper,[20] that meaning invariance is an essential part of both reduction and explanation, then the answer will simply be that equation (1) can neither be explained by, nor reduced to relativity. Whatever the *words* used for describing the situation, the *fact* remains that actual science does not observe the requirement of meaning invariance.

This argument is quite general and is independent of whether the terms whose meaning is under investigation are observable or not. It is therefore stronger than may seem at first sight. There are some empiricists who would admit that the meaning of theoretical terms may be changed in the course of scientific progress. However, not many people are prepared to extend meaning *variance* to observational terms also. The idea motivating this attitude is, roughly, that the meaning of observational terms is uniquely determined by the procedures of observation such as looking, listening, and the like. These procedures remain unaffected by theoretical advance.[21] Hence, observational meanings, too, remain unaffected by theoretical advance. What is overlooked, here, is that the "logic" of the observational terms is not exhausted by the procedures which are connected with their application "on the basis of observation." As will turn out later, it also depends on the more general ideas that determine the "ontology" (in Quine's sense) of our discourse. These general ideas may change without any change of observational procedures being implied. For example, we may change our ideas about the nature, or the ontological status (property, relation, object, process, etc.) of the color of a self-luminescent object without changing the methods of ascertaining that color (looking, for example). Clearly, such a change is bound profoundly to influence the meanings of our observational terms.

All this has a decisive bearing upon some contemporary ideas concerning the interpretation of scientific theories. According to these ideas, theoretical terms receive their meanings via correspondence rules which connect them with an observational language *that has been fixed in advance* and independently of the structure of the theory to be interpreted. Now, our above analysis would seem to show that *if we interpret scientific theories in the manner accepted by the scientific community*, then most of these correspondence rules will be either false, or nonsensical. They will be *false* if they *assert* the existence of entities denied by the theory; they will be *nonsensical* if they *presuppose* this existence. Turning the argument around, we can also say that the attempt to interpret the calculus of some theory that has been voided of the meaning assigned to it by the scientific community with the help of the double language system will lead to a very different theory. Let us again take the theory of relativity as an example: It can be safely assumed that the physical thing language of Carnap, and any similar language that has been suggested as an observation language, is not Lorentz-invariant. The attempt to interpret the *calculus* of relativity on *its* basis therefore cannot lead to the *theory* of relativity as it was understood by Einstein. What we shall obtain will be at the very most *Lorentz's interpretation* with its inherent asymmetries. This undesirable result cannot be evaded by the *demand* to use a different and more adequate observation language. The double language system assumes that theories which are not connected with some observation language do not possess an interpretation. The demand assumes that they do, and asks to choose the observation language most suited to it. It reverses the relation between theory and experience that is characteristic for the double language method of interpretation, which means, it gives up this method. Contemporary empiricism, therefore, has not led to any satisfactory account of the meanings of scientific theories.[22]

What we have shown so far is that the two conditions of section 2 are frequently violated in the course of scientific practice and especially at periods of scientific revolution. This is not yet a very strong argument. True: There are empirically inclined philosophers who have derived some satisfaction from the assumption that they only make explicit what is implicitly contained in scientific practice. It is therefore quite important to show that scientific practice is not what it is supposed to be by them. Also, strict adherence to meaning invariance and consistency would have made impossible some very decisive advances in physical theory such as the advance from the physics of Aristotle to the physics of Galileo and Newton. However, how do we know (independently of the fact that they do exist, have a certain structure, and are very influential—a circumstance that will have great weight with opportunists only[23]) that the sciences are a desirable phenomenon, that they contribute to the advancement of knowledge, and that their analysis will therefore lead to reasonable methodological demands? And did it not emerge in the last section that meaning invariance and the consistency condition *are* adopted by some scientists? Actual scientific practice, therefore, cannot be our last authority. We have to find out whether consistency and meaning invariance are *desirable* conditions, and this quite independently of who accepts and praises them and how many Nobel prizes have been won with their help.[24] Such an investigation will be carried out in the next sections.

4. INHERENT UNREASONABLENESS OF CONSISTENCY CONDITION

Prima facie, the case of the consistency condition can be dealt with in very few words. Consider for that purpose a theory T' that successfully describes the situation in the domain D'. From this we can infer (a) that T' agrees with a *finite* number of observations (let their class be F); and (b) that it agrees with these observations inside a margin M of error only.[25] Any alternative that contradicts T' outside F and inside M is supported by exactly the same observations and therefore acceptable if T' was acceptable (we shall assume that F are the only observations available). The consistency condition is much less tolerant. It eliminates a theory not because it is in disagreement with the *facts;* it eliminates it because it is in disagreement with *another theory*, with a theory, moreover, whose confirming instances it shares. *It thereby makes the as yet untested part of that theory a measure of validity.* The only difference between such a measure and a more recent theory is age and familiarity. Had the younger theory been there first, then the consistency condition would have worked in its favor. In this respect the effect of the consistency condition is rather similar to the effect of the more traditional methods of transcendental deduction, analysis of essences, phenomenological analysis, linguistic analysis. It contributes to the preservation of the old and familiar not because of any inherent advantage in it—for example, not because it has a better foundation in observation than has the newly suggested alternative, or because it is more elegant—but just because it is old and familiar. This is not the only instance where on closer inspection a rather surprising similarity emerges between modern empiricism and some of the school philosophies it attacks.

Now it seems to me that these brief considerations, although leading to an interesting *tactical* criticism of the consistency condition, do not yet go to the heart of the matter. They show that an alternative of the accepted point of view which shares its confirming instances cannot be *eliminated* by factual reasoning. They do not show that such an alternative is *acceptable;* and even less do they show that it *should be used.* It is bad enough, so a defender of the consistency condition might point out, that the accepted point of view does not possess full empirical support. Adding new theories *of an equally unsatisfactory character* will not improve the situation; nor is there much sense in trying to *replace* the accepted theories by some of their possible alternatives. Such replacement will be no easy matter. A new formalism may have to be learned and familiar problems may have to be calculated in a new way. Textbooks must be rewritten, university curricula readjusted, experimental results reinterpreted. And what will be the result of all the effort? Another theory which, from an empirical point of view, has no advantage whatever over and above the theory it replaces. The only real improvement, so the defender of the consistency condition will continue, derives from the *addition of new facts*. Such new facts will either support the current theories, or they will force us to modify them by indicating precisely where they go wrong. In both cases they will precipitate real progress and not only arbitrary change. The proper procedure must therefore consist in the confrontation of the accepted point of view with as many relevant facts as possible. The exclusion of alternatives is then required for reasons of expediency: Their invention not only does not help, but it even hinders progress by absorbing time and manpower that could be devoted to better things. And the function of the consistency condition lies precisely in this. It eliminates such fruitless discussion and it forces the scientist to concentrate on the facts which, after all, are the only acceptable judges of a theory. This is how the practicing scientist will defend his concentration on a single theory to the exclusion of all empirically possible alternatives.[26]

It is worthwhile repeating the reasonable core of this argument: Theories should not be changed unless there are pressing reasons for doing so. The only pressing reason for changing a theory is disagreement with facts. Discussion of incompatible facts will therefore lead to progress. Discussion of incompatible alternatives will not. Hence, it is sound procedure to increase the number of relevant facts. It is not sound procedure to increase the number of factually adequate, but incompatible alternatives. One might wish to add that formal improvements such as increase of elegance, simplicity, generality, and coherence should not be excluded. But once these improvements have been carried out, the collection of facts for the purpose of test seems indeed to be the only thing left to the scientist.

5. RELATIVE AUTONOMY OF FACTS

And this it is—provided these facts *exist, and are available independently of whether or not one considers alternatives to the theory to be tested.* This assumption on which the validity of the argument in the last section depends in a most decisive manner I shall call the assumption of the relative autonomy of facts, or the autonomy principle. It is not asserted by this principle that the discovery and description of facts is independent of *all* theorizing. But it *is* asserted that the facts which belong to the empirical content of some theory are available whether or not one considers alternatives to *this* theory. I am not aware that this very important assumption has ever been explicitly formulated as a separate postulate of the empirical method. However, it is clearly implied in almost all investigations which deal with questions of confirmation and test. All these investigations use a model in which a *single* theory is compared with a class of facts (or observation statements) which are assumed to be "given" somehow. I submit that this is much too simple a picture of the actual situation. Facts and theories are much more intimately connected than is admitted by the autonomy principle. Not only is the description of every single fact dependent on *some* theory (which may, of course, be very different from the theory to be tested). There exist also facts which cannot be unearthed except with the help of alternatives to the theory to be tested, and which become unavailable as soon as such alternatives are excluded. This suggests that the methodological unit to which we must refer when discussing questions of test and empirical content is constituted by a *whole set of partly overlapping, factually adequate, but mutually inconsistent theories.* In the present paper only the barest outlines will be given of such a test model. However, before doing this I want to discuss an example which shows very clearly the function of alternatives in the discovery of facts.

As is well known, the Brownian particle is a perpetual motion machine of the second kind and its existence refutes the phenomenological second law. It therefore belongs to the domain of relevant facts for this law. Now, could this relation between the law and the Brownian particle have been discovered in a *direct* manner, i.e., could it have been discovered by an investigation of the observational consequences of the phenomenological theory that did not make use of an alternative account of heat? This question is readily divided into two: (1) Could the *relevance* of the Brownian particle have been discovered in this manner? (2) Could it have been demonstrated that it actually *refutes* the second law? The answer to the first question is that we do not know. It is impossible to say what would have happened had the kinetic theory not been considered by some physicists. It is my guess, however, that in this case the Brownian particle would have been regarded as an oddity much in the same way in which some of the late Professor Ehrenhaft's astounding effects[27] are regarded as an oddity, and that it would not have been given the decisive position it assumes in contemporary theory. The answer to the second question is simply—No. Consider what the discovery of the inconsistency between the Brownian particle and the second law would have required! It would have required (a) measurement of the exact *motion* of the particle in order to ascertain the changes of its kinetic energy plus the energy spent on overcoming the resistance of the fluid; and (b) it would have required precise measurement of temperature and heat transfer in the surrounding medium in order to ascertain that any loss occurring here was indeed compensated by the increase of the energy of the moving particle and the work done against the fluid. Such measurements are beyond experimental possibilities.[28] Neither is it possible to make precise measurements of the heat transfer; nor can the path of the particle be investigated with the desired precision. Hence a "direct" refutation of the second law that considers only the phenomenological theory and the "facts" of Brownian motion is impossible. And, as is well known, the actual refutation was brought about in a very different manner. It was brought about via the kinetic theory and Einstein's utilization of it in the calculation of the statistical properties of the Brownian motion.[29] In the course of this procedure the phenomenological theory (T') was incorporated into the wider context of statistical physics (T) *in such a manner that the consistency condition was violated;* and *then* a crucial experiment was staged (investigations of Svedberg and Perrin).

It seems to me that this example is typical for the relation between fairly general theories, or points of view, and "the facts." Both the relevance and the refuting character of many very decisive facts can be established only with the help of other theories which, although factually adequate, are yet not in agreement with the view to be tested. This being the case, the production of such refuting facts may have to be preceded by the invention and articulation of alternatives to that view. Empiricism demands that the empirical content of whatever knowledge we possess be increased as much as possible. Hence *the invention of alternatives in addition to the view that stands in the center of discussion constitutes an essential part of the empirical method.* Conversely, the fact that the consistency condition eliminates alternatives now shows it to be in disagreement with empiricism and not only with scientific practice. By excluding valuable tests it decreases the empirical content of the theories which are permitted to remain (and which, as we have indicated above, will usually be the theories which have been there first); and it especially decreases the number of those facts which could show their limitations. This last result of a determined application of the consistency condition is of very topical interest. It may well be that the refutation of the quantum-mechanical uncertainties presupposes just such an incorporation of the present theory into a wider context which is no longer in accordance with the idea of complementarity and which therefore suggests new and decisive experiments. And it may also be that the insistence, on the part of the majority of contemporary physicists, on the consistency condition will, if successful, forever protect these uncertainties from refutation. This is how modern empiricism may finally lead to a situation where a certain point of view petrifies into dogma by being, in the name of experience, completely removed from any conceivable criticism.

6. THE SELF-DECEPTION INVOLVED IN ALL UNIFORMITY

It is worthwhile to examine this apparently empirical defense of a dogmatic point of view in somewhat greater detail. Assume that physicists have adopted, either consciously or unconsciously, the idea of the uniqueness of complementarity and that they therefore elaborate the orthodox point of view and refuse to consider alternatives. In the beginning such a procedure may be quite harmless. After all, a man can do only so many things at a time and it is better when he pursues a theory in which he is interested rather than a theory he finds boring. Now assume that the pursuit of the theory he chose has led to successes and that the theory has explained in a satisfactory manner circumstances that had been unintelligible for quite some time. This gives empirical support to an idea which to start with seemed to possess only this advantage: It was interesting and intriguing. The concentration upon the theory will now be reinforced, the attitude towards alternatives will become less tolerant. Now if it is true, as has been argued in the last section, that many facts become available only with the help of such alternatives, then the refusal to consider them *will result in the elimination of potentially refuting facts.* More especially, it will eliminate facts whose discovery would show the complete and irreparable inadequacy of the theory.[30] Such facts having been made inaccessible, the theory will appear to be free from blemish and it will seem that "all evidence points with merciless definiteness in the . . . direction . . . [that] all the processes involving . . . unknown interactions conform to the fundamental quantum law" (n. 14, p. 44). This will further reinforce the belief in the uniqueness of the current theory and in the complete futility of any account that proceeds in a different manner. Being now very firmly convinced that there is only one good microphysics, the physicists will try to explain even adverse facts in its terms, and they will not mind when such explanations are sometimes a little clumsy. By now the success of the theory has become public news. Popular science books (and this includes a good many books on the philosophy of science) will spread the basic postulates of the theory; applications will be made in distant fields. More than ever the theory will appear to possess tremendous empirical support. The chances for the consideration of alternatives are now very slight indeed. The final success of the fundamental assumptions of the quantum theory and of the idea of complementarity will seem to be assured.

At the same time it is evident, on the basis of the considerations in the last section, that this appearance of success *cannot in the least be regarded as a sign of truth and correspondence with nature.* Quite the contrary, the suspicion arises that the absence of major difficulties is a result of the decrease of empirical content brought about by the elimination of alternatives, and of facts that can be discovered with the help of these alternatives only. In other words, *the suspicion arises that this alleged success is due to the fact that in the process of application to new domains the theory has been turned into a metaphysical system.* Such a system will of course be very "successful" not, however, because it agrees so well with the facts, but because no facts have been specified that would constitute a test and because some such facts have even been removed. Its "success" *is entirely manmade.* It was decided to stick to some ideas and the result was, quite naturally, the survival of these ideas. If now the initial decision is forgotten, or made only implicitly, then the survival will seem to constitute independent support, it will reinforce the decision, or turn it into an explicit one, and in this way close the circle. This is how empirical "evidence" may be *created* by a procedure which quotes as its justification the very same evidence it has produced in the first place.

At this point an "empirical" theory of the kind described (and let us always remember that the basic principles of the present quantum theory and especially the idea of complementarity are uncomfortably close to forming such a theory) becomes almost indistinguishable from a myth. In order to realize this, we need only consider that on account of its all-pervasive character a myth such as the myth of witchcraft and of demonic possession will possess a high degree of confirmation on the basis of observation. Such a myth has been taught for a long time; its content is enforced by fear, prejudice, and ignorance as well as by a jealous and cruel priesthood. It penetrates the most common idiom, infects all modes of thinking and many decisions which mean a great deal in human life. It provides models for the explanation of any conceivable event, conceivable, that is, for those who have accepted it.[31] This being the case, its key terms will be fixed in an unambiguous manner and the idea (which may have led to such a procedure in the first place) that they are copies of unchanging entities and that change of meaning, if it should happen, is due to human mistake—this idea will now be very plausible. Such plausibility reinforces all the maneuvres which are used for the preservation of the myth (elimination of opponents included). The conceptual apparatus of the theory and the emotions connected with its application having penetrated all means of communication, all actions, and indeed the whole life of the community, such methods as transcendental deduction, analysis of usage, phenomenological analysis which are means for further solidifying the myth will be extremely successful (which shows, by the way, that all these methods which have been the trademark of various philosophical schools old and new, have one thing in common: They tend to *preserve* the *status quo* of the intellectual life).[32] Observational results too, will speak in favor of the theory as they are formulated in its terms. It will seem that at last the truth has been arrived at. At the same time, it is evident that all contact with the world has been lost and that the stability achieved, the semblance of absolute truth, *is nothing but the result of an absolute conformism.*[33] For how can we possibly test, or improve upon, the truth of a theory if it is built in such a manner that any conceivable event can be described, and explained, in terms of its principles? The *only* way of investigating such all-embracing principles is to compare them with a different set of *equally all-embracing* principles—but this way has been excluded from the very beginning. The myth is therefore of no objective relevance, it continues to exist solely as the result of the effort of the community of believers and of their leaders, be these now priests or Nobel prize winners. *Its "success" is entirely manmade.* This, I think, is the most decisive argument against any method that encourages uniformity, be it now empirical or not. Any such method is in the last resort a method of deception. It enforces an unenlightened conformism, and speaks of truth; it leads to a deterioration of intellectual capabilities, of the power of imagination, and speaks of deep insight; it destroys the most precious gift of the young, their tremendous power of imagination, and speaks of education.

To sum up: *Unanimity of opinion may be fitting for a church, for the frightened victims of some (ancient, or modern) myth, or for the weak and willing followers of some tyrant; variety of opinion is a feature necessary for objective knowledge; and a method that encourages variety is also the only method that is compatible with a humanitarian outlook.* To the extent to which the consistency condition (and, as will emerge, the condition of meaning invariance) delimits variety, it contains a theological element (which lies, of course, in the worship of "facts" so characteristic for nearly all empiricism).

7. INHERENT UNREASONABLENESS OF MEANING INVARIANCE

What we have achieved so far has immediate application to the question whether the meaning of certain key terms should be kept unchanged in the course of the development and improvement of our knowledge. After all, the meaning of every term we use depends upon the theoretical context in which it occurs. Hence, if we consider two contexts with basic principles which either contradict each other, or which lead to inconsistent consequences in certain domains, it is to be expected that some terms of the first context will not occur in the second context with exactly the same meaning. Moreover, if our methodology demands the use of mutually inconsistent, partly overlapping, and empirically adequate theories, then it thereby also demands the use of conceptual systems which are mutually *irreducible* (their primitives cannot be connected by bridge laws which are meaningful *and* factually correct) and it demands that meanings of terms be left elastic and that no binding commitment be made to a certain set of concepts.

It is very important to realize that such a tolerant attitude towards meanings, or such a change of meaning in cases where one of the competing conceptual systems has to be abandoned, need not be the result of directly accessible observational difficulties. The law of inertia of the so-called *impetus theory* of the later Middle Ages[34] and Newton's own law of inertia are in perfect quantitative agreement: Both assert that an object that is not under the influence of any outer force will proceed along a straight line with constant speed. Yet despite this fact, the adoption of Newton's theory entails a conceptual revision that forces us to abandon the inertial law of the impetus theory, not because it is quantitatively incorrect but *because it achieves the correct predictions with the help of inadequate concepts.* The law asserts that the *impetus* of an object that is beyond the reach of outer forces remains constant.[35] The impetus is interpreted as an inner *force* which pushes the object along. Within the impetus theory such a force is quite conceivable as it is assumed here that forces determine *velocities* rather than accelerations. The concept of impetus is therefore formed in accordance with a law (forces determine velocities), and this law is inconsistent with the laws of Newton's theory and must be abandoned as soon as the latter is adopted. This is how the progress of our knowledge may lead to conceptual revisions for which no direct observational reasons are available. The occurrence of such changes quite obviously refutes the contention of some philosophers that the invariance of *usage* in the trivial and uninteresting contexts of the private lives of not too intelligent and inquisitive people indicates invariance of *meaning* and the superficiality of all scientific changes. It is also a very decisive objection against any crudely operationalistic account of both observable terms and theoretical terms.

What we have said applies even to singular statements of observation. Statements which are empirically adequate, and which are the result of observation (such as "here is a table") may have to be reinterpreted, not because it has been found that they do not adequately express what is seen, heard, felt, but because of some changes in sometimes very remote parts of the conceptual scheme to which they belong. Witchcraft is again a very good example. Numerous eyewitnesses claim that they have actually *seen* the devil; or *experienced* demonic influence. There is no reason to suspect that they were lying. Nor is there any reason to assume that they were sloppy observers, for the phenomena leading to the belief in demonic influence are so obvious that a mistake is hardly possible (possession; split personality; loss of personality; hearing voices; etc.). These phenomena are well known today.[36] In the conceptual scheme that was the one generally accepted in the fifteenth and sixteenth centuries, the only way of describing them, or at least the way that seemed to express them most adequately, was by reference to demonic influences. Large parts of this conceptual scheme were changed for philosophical reasons and also under the influence of the evidence accumulated by the sciences. Descartes's materialism played a very decisive role in discrediting the belief in spatially localizable spirits. The language of demonic influences was no part of the new conceptual scheme that was created in this manner. It was for this reason that a reformulation was needed, and a reinterpretation of even the most common "observational" statements. Combining this example with the remarks at the beginning of the present section, we now realize that according to the method of classes of alternative theories a lenient attitude must be taken with respect to the meanings of all the terms we use. We must not attach too great an importance to "what we mean" by a phrase, and we must be prepared to change whatever little we have said concerning this meaning as soon as the need arises. Too great concern with meanings can only lead to dogmatism and sterility. Flexibility, and even sloppiness in semantical matters is a prerequisite of scientific progress.[37]

8. SOME CONSEQUENCES

Three consequences of the results so far obtained deserve a more detailed discussion. The first consequence is an evaluation of *metaphysics* which differs significantly from the standard empirical attitude. As is well known, there are empiricists who demand that science start from observable facts and proceed by generalization, and who refuse the admittance of metaphysical ideas at any point of this procedure. For them, only a system of thought that has been built up in a purely inductive fashion can claim to be genuine knowledge. Theories which are partly metaphysical, or "hypothetical," are suspect, and are best not used at all. This attitude has been formulated most clearly by Newton[38] in his reply to Pardies's second letter concerning the theory of colors:

> If the possibility of hypotheses is to be the test of truth and reality of things, I see not how certainty can be obtained in any science; since numerous hypotheses may be devised, which shall seem to overcome new difficulties.

This radical position, which clearly depends on the demand for a theoretical monism, is no longer as popular as it used to be. It is now granted that metaphysical considerations may be of importance when the task is to *invent* a new physical theory; such invention, so it is admitted, is a more or less irrational act containing the most diverse components. Some of these components are, and perhaps must be, metaphysical ideas. However, it is also pointed out that as soon as the theory has been developed in a formally satisfactory fashion and has received sufficient confirmation to be regarded as empirically successful, it is pointed out that in the very same moment it can *and must* forget its metaphysical past; metaphysical speculation must *now* be replaced by empirical argument.

> On the one side I would like to emphasize [writes Ernst Mach on this point[39]] that *every and any*

idea is admissible as a means for research, provided it is helpful; still, it must be pointed out, on the other side, that it is very necessary from time to time to free the presentation of the *results* of research from all inessential additions.

This means that empirical considerations are still given the upper hand over metaphysical reasoning. Especially in the case of an inconsistency between metaphysics and some highly confirmed empirical theory it will be decided, *as a matter of course,* that the theory or the result of observation must stay, and that the metaphysical system must go. A very simple example is the way in which materialism is being judged by some of its opponents. For a materialist the world consists of material particles moving in space, of collections of such particles. Sensations, as introspected by human beings, do not look like collections of particles, and their observed existence is therefore assumed to refute and thereby to remove the metaphysical doctrine of materialism. Another example which I have analyzed in "Problems of Microphysics" is the attempt to eliminate certain very general ideas concerning the nature of microentities on the basis of the remark that they are inconsistent "with an immense body of experience" and that "to object to a lesson of experience by appealing to metaphysical preconceptions is unscientific."

The methodology developed in the present paper leads to a very different evaluation of metaphysics. Metaphysical systems are scientific theories in their most primitive stage. If they *contradict* a well-confirmed point of view, then this indicates their usefulness as an alternative to this point of view. Alternatives are needed for the purpose of criticism. Hence, metaphysical systems which contradict observational results of well-confirmed theories *are most welcome* starting points of such criticism. Far from being misfired attempts at anticipating, or circumventing, empirical research which were deservedly exposed by a reference to experience, they are the only means at our disposal for examining those parts of our knowledge which have already become observational and which are therefore inaccessible to a criticism "on the basis of observation."

A second consequence is that a new attitude has to be adopted with respect to the *problem of induction.* This problem consists in the question of what justification there is for asserting the truth of a statement S given the truth of another statement, S', whose content is smaller than the content of S. It may be taken for granted that those who want to justify the truth of S also assume that after the justification the truth of S will be *known.* Knowledge to the effect that S implies the *stability* of S (we must not change, remove, criticize, what we know to be true). The method we are discussing at the present moment cannot allow such stability. It follows that the problem of induction, at least in some of its formulations, is a problem whose solution leads to undesirable results. It may therefore be properly termed a pseudoproblem.

The third consequence, which is more specific, is that *arguments from synonymy* (or from coextensionality), far from being that measure of adequacy as which they are usually introduced, are liable severely to impede the progress of knowledge. Arguments from synonymy judge a theory or a point of view not by its capability to mimic the world but rather by its capability to mimic the descriptive terms of another point of view which for some reason is received favorably. Thus for example, the attempt to give a materialistic, or else a purely physiological, account of human beings is criticized on the grounds that materialism, or physiology, cannot provide synonyms for "mind," "pain," "seeing red," "thinking of Vienna," in the sense in which these terms are used either in ordinary English (provided there is a well-established usage concerning these terms, a matter which I doubt) or in some more esoteric mentalistic idiom. Clearly, such criticism silently assumes the principle of meaning invariance, that is, it assumes that the meanings of at least some fundamental terms must remain unchanged in the course of the progress of our knowledge. It cannot therefore be accepted as valid.[40]

However, we can, and must, go still further. The ideas which we have developed above are strong enough not only to *reject* the demand for synonymy, wherever it is raised, but also to *support* the demand for irreducibility (in the sense in which this notion was used at the beginning of section 7). The reason is that irreducibility is a presupposition of high critical ability on the part of the point of view shown to be irreducible. An outer indication of such irreducibility which is quite striking in the case of an attack upon commonly accepted ideas is the feeling of *absurdity:* We deem absurd what goes counter to well-established linguistic habits. The absence, from a newly introduced set of ideas, of synonymy relations connecting it with part of the accepted point of view; the feeling of absurdity therefore indicate that the new ideas are fit for the purpose of criticism, i.e., that they are fit for either leading to a strong *confirmation* of the earlier theories, or else to a very revolutionary *discovery:* Absence of synonymy, clash of meanings, absurdity are desirable. Presence of synonymy, intuitive appeal, agreement with customary modes of speech, far from being *the* philosophical virtue, indicate that not much progress has been made and that the business of investigating what is commonly accepted *has not even started.*

9. HOW TO BE A GOOD EMPIRICIST

The final reply to the question put in the title is therefore as follows. A good empiricist will not rest content with the theory that is in the center of attention and with those tests of the theory which can be carried out in a direct manner. Knowing that the most fundamental and the most general criticism is the criticism produced with the help of alternatives, he will try to invent such alternatives.[41] It is, of course, impossible at once to produce a theory that is formally comparable to the main point of view and that leads to equally many predictions. His first step will therefore be the formulation of fairly general assumptions which are not yet directly connected with observations; this means that his first step will be the invention of a new *metaphysics.* This metaphysics must then be elaborated in sufficient detail in order to be able to compete with the theory to be investigated as regards generality, details of prediction, precision of formulation.[42] We may sum up both activities by saying that a good empiricist must be a critical metaphysician. Elimination of all metaphysics, far from increasing the empirical content of the remaining theories, is liable to turn these theories into dogmas. The consideration of alternatives together with the attempt to criticize each of them in the light of experience also leads to an attitude where meanings do not play a very important role and where arguments are based upon assumptions of fact rather than analysis of (archaic, although perhaps very precise) meanings. The effect of such an attitude upon the development of human capabilities should not be underestimated either. Where speculation and invention of alternatives is encouraged, bright ideas are liable to occur in great number and such ideas may then lead to a change of even the most "fundamental" parts of our knowledge, i.e., they may lead to a change of assumptions which either are so close to observation that their truth seems to be dictated by "the facts," or which are so close to common prejudice that they seem to be "obvious," and their negation "absurd." In such a situation it will be realized that neither "facts" nor abstract ideas can ever be used for defending certain principles, come what may. Wherever facts play a role in such a dogmatic defense, we shall have to suspect foul play (see the opening quotation)—the foul play of those who try to turn good science into bad, because unchangeable, metaphysics. In the last resort, therefore, being a good empiricist means being critical, and basing one's criticism not just on an abstract principle of skepticism but upon *concrete suggestions* which indicate in every single case how the accepted point of view might be further tested and further investigated and which thereby prepare the next step in the development of our knowledge.

For support of research the author is indebted to the National Science Foundation and the Minnesota Center for the Philosophy of Science.

NOTES

1. K. R. Popper, *The Open Society and Its Enemies* (Princeton, N.J.: Princeton University Press, 1953).

2. It is very interesting to see how many so-called empiricists, when turning to the past, completely fail to pay attention to some very obvious facts which are incompatible with their empiristic epistemology. Thus, Galileo has been represented as a thinker who turned away from the empty speculations of the Aristotelians and who based his own laws upon facts which he had carefully collected beforehand. Nothing could be further from the truth. *The Aristotelians could quote numerous observational results in their favor.* The Copernican idea of the motion of the earth, on the other hand, did not possess independent observational support, at least not in the first 150 years of its existence. Moreover, it was inconsistent with facts and highly confirmed physical theories. And *this* is how modern physics started: not as an observational enterprise, *but as an unsupported speculation that was inconsistent with highly confirmed laws.* For details and further references, see my "Realism and Instrumentalism," to appear in *The Critical Approach: Essays in Honor of Karl Popper.* (subnote 2*.)

* P. K. Feyerabend, "Realism and Instrumentalism," in *The Critical Approach: Essays in Honor of Karl Popper,* ed. M. Bunge (Glencoe, Illinois: The Free Press, to be published).

3. One might be inclined to add those who base their pronouncements upon an analysis of what they call "ordinary language." I do not think they deserve to be honored by a criticism. Paraphrasing Galileo, one might say that they "deserve not even that name, for they do not talk plainly and simply but are content to adore the shadows, philosophizing not with due circumspection but merely from having memorized a few ill-understood principles."

4. It is nowadays frequently assumed that "if one considers the history of a special branch of science, one gets the impression that non-scientific elements . . . relatively frequently occur in the earlier stages of development, but that they gradually retrogress in later stages and even tend to disappear in such advanced stages which become ripe for more or less thorough formalization." (H. J. Groenewold, *Synthese* (1957), p. 305). Our considerations in the text would seem to show that such a development is very undesirable and can only result in a well-formalized, precisely expressed, and completely petrified metaphysics.

5. These essays were published in Vol. III of the *Minnesota Studies in the Philosophy of Science;* in Vols. I and II of the *Pittsburgh Studies in the Philosophy of Science;* and in *Problems of Philosophy, Essays in Honor of Herbert Feigl,* respectively.

6. See subnote 6*. The decisive feature of Popper's theory, a feature which was not at all made clear by earlier writers on the subject of explanation, is the emphasis he puts on the initial conditions and the implied possibility of two kinds of laws, viz., (1) laws concerning the temporal sequence of events; and (2) laws concerning the space of initial conditions. In the case of the quantum theory, the laws of the second kind provide very important information about the nature of the elementary particles and it is to *them* and *not* to the laws of motion that reference is made in the discussions concerning the interpretation of the uncertainty relations. In general relativity, the laws formulating the initial conditions concern the structure of the universe at large and only by overlooking them could it be believed that a purely relational account of space would be possible. For the last point, cf. Hill, subnote 6†.

* K. R. Popper, *Logic of Scientific Discovery* (New York, 1959), Sec. 12. This is a translation of his *Logik der Forschung* published in 1935.

† E. L. Hill, "Quantum Physics and the Relativity Theory," in *Current Issues in the Philosophy of Science,* eds. H. Feigl and G. Maxwell (New York: Holt, Rinehart and Winston, 1961).

7. C. G. Hempel, "Studies in the Logic of Explanation," reprinted in *Readings in the Philosophy of Science,* eds. H. Feigl and M. Brodbeck (New York, 1953), p. 321.

8. E. Nagel, "The Meaning of Reduction in the National Sciences," reprinted in *Philosophy of Science,* eds. A. C. Danto and S. Morgenbesser (New York, 1960), p. 301.

9. It has been objected to this formulation that theories which are consistent with a given explanandum may still contradict each other. This is quite correct, but it does not invalidate my argument. For as soon as a single theory is regarded as sufficient for explaining all that is known (and represented by the other theories in question), it will have to be consistent with all these other theories.

10. E. Mach, *Wärmelehre* (Leipzig, 1897), p. 364.

11. E. Mach, *Zwei Aufsaetze* (Leipzig, 1912).

12. For a discussion of these objections, cf. ter Haar's review article in *Reviews of Modern Physics* (1957).

13. M. Born, *Natural Philosophy of Cause and Chance* (New York: Oxford University Press, 1948), p. 109.

14. L. Rosenfeld, "Misunderstandings about the Foundations of the Quantum Theory," in *Observation and Interpretation* (London, 1957), p. 42.

15. Cf. the discussions in *Observation and Interpretation* (See n. 14).

16. For details and further literature, cf. Sec. 11 of my paper "Problems of Microphysics."

17. W. Heisenberg, *Physics and Philosophy* (New York, 1958), p. 44.

18. P. Duhem, *La Théorie Physique: Son Objet, Sa Structure* (Paris, 1914), Chaps. ix and x. See also K. R. Popper, "The Aim of Science," *Ratio*, Vol. I (1957).

19. E. Nagel, n. 8, p. 302.

20. Cf. Sec. 4.7 of M. Scriven's paper "Explanations, Predictions, and Laws," in Vol. III of the *Minnesota Studies in the Philosophy of Science*. Similar objections have been raised by Kraft (Vienna) and Rynin (Berkeley).

21. For an exposition and criticism of this idea, cf. my "Attempt at a Realistic Interpretation of Experience," *Proceedings of the Aristotelian Society*, New Series, LVIII (1958), 143–70.

22. It must be admitted, however, that Einstein's original interpretation of the special theory of relativity is hardly ever used by contemporary physicists. For them the theory of relativity consists of two elements: (1) the Lorentz transformations; and (2) mass-energy equivalence. The Lorentz transformations are interpreted purely formally and are used to make a selection among possible equations. This interpretation does not allow to distinguish between Lorentz's original point of view and the entirely different point of view of Einstein. According to it Einstein achieved a very minor *formal* advance [this is the basis of Whittaker's attempt to "debunk" Einstein]. It is also very similar to what application of the double language model would yield. Still, an undesirable philosophical procedure is not improved by the support it gets from an undesirable procedure in physics. [The above comment on the contemporary attitude towards relativity was made by E. L. Hill in discussions at the Minnesota Center for the Philosophy of Science.]

23. In about 1925 philosophers of science were bold enough to stick to their theses even in those cases where they were inconsistent with actual science. They meant to be *reformers* of science, and not *imitators*. (This point was explicitly made by Mach in his controversy with Planck. Cf. again his *Zwei Aufsaetze*, n. 11.) In the meantime they have become rather tame (or beat) and are much more prepared to change their ideas in accordance with the latest discoveries of the historians, or the latest fashion of the contemporary scientific enterprise. This is very regrettable, indeed, for it considerably decreases the number of the rational critics of the scientific enterprise. And it also seems to give unwanted support to the Hegelian thesis (which is now implicitly held by many historians and philosophers of science) that what exists has a "logic" of its own and is for that very reason reasonable.

24. Even the most dogmatic enterprise allows for discoveries (cf. the "discovery" of so-called "white Jews" among German physicists during the Nazi period). Hence, before hailing a so-called discovery, we must make sure that the system of thought which forms its background is not of a dogmatic kind.

25. The indefinite character of all observations has been made very clear by Duhem, n. 18, Chap. ix. For an alternative way of dealing with this indefiniteness, cf. S. Körner, *Conceptual Thinking* (New York, 1960).

26. More detailed evidence for the existence of this attitude and for the way in which it influences the development of the sciences may be found in Kuhn's book *Structure of Scientific Revolutions*, subnote 26*. The attitude is extremely common in the contemporary quantum theory. "Let us enjoy the successful theories we possess and let us not waste our time with contemplating what *would* happen if *other* theories were used"—this seems to be the motto of almost all contem-

porary physicists (cf. Heisenberg, n. 17, pp. 56, 144) and philosophers (cf. Hanson, subnote 26†). It may be traced back to Newton's papers and letters (to Hooke, and Pardies) on the theory of color. See also n. 23.

* T. Kuhn, *Structure of Scientific Revolutions* (Chicago: University of Chicago Press, 1962).

† N. R. Hanson, "Five Cautions for the Copenhagen Critics," *Philosophy of Science,* XXVI (1959), 325–37.

27. Having witnessed these effects under a great variety of conditions, I am much more reluctant to regard them as mere curiosities than is the scientific community of today. Cf. also my edition of Ehrenhaft's lectures, *Einzelne Magnetische Nord- und Südpole und deren Auswirkung in den Naturwissenschaften* (Vienna, 1947).

28. R. Fürth, *Zeitschrift für Physik,* 81 (1933), 143–62.

29. For these investigations, cf. A. Einstein, *Investigations on the Theory of the Brownian Motion,* subnote 29*, which contains all the relevant papers by Einstein and an exhaustive bibliography by R. Fürth. For the experimental work, cf. J. Perrin, *Die Atome,* subnote 29†. For the relation between the phenomenological theory and the kinetic theory, cf. also Smoluchowski, subnote 29** and Popper, subnote 29‡. Despite Einstein's epoch-making discoveries and von Smoluchowski's splendid presentation of their effect (for the latter cf. also subnote 29§), the present situation in thermodynamics is extremely unclear, especially in view of the continued presence of the ideas of reduction which we criticized in the text above. To be more specific, it is frequently attempted to determine the entropy balance of a complex *statistical* process by reference to the (refuted) *phenomenological* law after which procedure fluctuations are superimposed in a most artificial fashion. For details cf. Popper, *loc. cit.*

* A. Einstein, *Investigations on the Theory of the Brownian Motion* (New York, 1956).

† J. Perrin, *Die Atome,* (Leipzig, 1920).

** M. v. Smoluchowski, 'Experimentell nachwiesbare, der üblichen Thermodynamik widersprechende Molekularphanomene,' *Physikalische Zeitschrift,* XIII (1912), 1069.

‡ K. R. Popper, "Irreversibility, or, Entropy since 1905," *British Journal for the Philosophy of Science,* VIII (1957), 151.

§ *Oeuvres de Marie Smoluchowski,* II (Cracouvie, 1927), 226 ff., 316 ff., 462 ff., and 530 ff.

30. The quantum theory can be adapted to a great many difficulties. It is an open theory in the sense that apparent inadequacies can be accounted for in an *ad hoc* manner, by *adding* suitable operators, or elements in the Hamiltonian, rather than by recasting the whole structure. A refutation of its basic formalism (i.e., of the formalism of quantization, and of non-commuting operators in a Hilbert space or a reasonable extension of it) would therefore demand proof to the effect that *there is no conceivable adjustment of the Hamiltonian, or of the operators used* which makes the theory conform to a given fact. It is clear that such a general statement can only be provided by an *alternative theory* which of course must be detailed enough to allow for independent and crucial tests.

31. For a very detailed description of a once very influential myth, cf. C. H. Lea, *Materials for a History of Witchcraft,* 3 Vols. (New York, 1957), as well as *Malleus Malleficarum,* translated by Montague Summers (who, by the way, counts it "among the most important, wisest [sic!], and weightiest books of the world") (London, 1928).

32. Quite clearly, analysis of usage, to take only one example, presupposes certain regularities concerning this usage. The more people differ in their fundamental ideas, the more difficult will it be to uncover such regularities. Hence, analysis of usage will work best in a closed society that is firmly held together by a powerful myth such as was the philosophy in the Oxford of about ten years ago.

33. Schizophrenics very often hold beliefs which are as rigid, all-pervasive, and unconnected with reality, as are the best dogmatic philosophies. Only such beliefs come to them naturally whereas a professor may sometimes spend his whole life in attempting to find arguments which create a similar state of mind.

34. For details and further references, cf. Sec. 6 of my "Explanation, Reduction, and Empiricism," *loc. cit.*

35. We assume here that a dynamical rather than a kinematic characterization of motion has been adopted. For a more detailed analysis, cf. again the paper referred to in the previous footnote.

36. For very vivid examples, cf. K. Jaspers, *Allgemeine Psychopathologie* (Berlin, 1959), pp. 75–123.

37. Mae West is by far preferable to the preci-

sionists: "I ain't afraid of pushin' grammar around so long as it sounds good" (*Goodness Had Nothing to do With It,* New York, 1959, p. 19).

38. I. B. Cohen, ed., *Isaac Newton's Papers & Letters on Natural Philosophy* (Cambridge, Mass.: Harvard University Press, 1958), p. 106.

39. "Der Gegensatz zwischen der mechanischen und der phaenomenologischen Physik," *Wärmelehre* (Leipzig, 1896), pp. 362 ff.

40. For details concerning the mind-body problem, cf. my "Materialism and the Mind-Body Problem," *Review of Metaphysics* (Sept. 1963).

41. In my paper "Realism and Instrumentalism," subnote 2*, I have tried to show that this is precisely the method which has brought about such spectacular advances of knowledge as the Copernican Revolution, the transition to relativity and to quantum theory.

42. Cf. Sec. 13 of my "Realism and Instrumentalism."

Dudley Shapere*

REASON, REFERENCE, AND THE QUEST FOR KNOWLEDGE

This paper examines the "causal theory of reference", according to which science aims at the discovery of "essences" which are the objects of reference of natural kind terms (among others). This theory has been advanced as an alternative to traditional views of "meaning", on which a number of philosophical accounts of science have relied, and which have been criticized earlier by the present author. However, this newer theory of reference is shown to be equally subject to fatal internal difficulties, and to be incompatible with actual science as well. Indeed, it rests on assumptions which it shares with the purportedly opposing theory of meaning. Behind those common assumptions is the supposition that the nature of science can be illuminated by an examination of alleged necessities of language which are independent of the results and methods of scientific inquiry. An alternative view of science is proposed, according to which the goals and language of science develop as integral parts of the process of demarcating science from non-science, a process in which the notion of a "reason" gradually assumes a decisive role. On this view, the comparability, competition, and development of scientific ideas are understood without reliance on either common "meanings" or common "references" as fundamental tools of analysis.

I

There is a traditional theory of meaning, or, more exactly, a class of theories of meaning, which has been much discussed in recent years both within and outside philosophy of science. Briefly, the idea is the following: we choose certain features as defining a type of thing, as giving the "meaning" of the term we choose to use for that type of thing. From then on, those features will serve as "criteria", in the sense of necessary and, or perhaps, or sufficient conditions for anything's being that sort of thing, for deserving to be referred to by that term. If the

*I am grateful for the opportunity to have visited at the Institute for Advanced Study, Princeton, New Jersey, in 1981, during which time this paper was written. An earlier and longer version was presented at the International Symposium on Philosophy, Querétaro, Mexico, in August, 1980.

Reprinted by permission from the author and from *Philosophy of Science,* Vol. 49, no. 1, March 1982, pp. 1–23.

criteria are not satisfied (or, for that version known as the "cluster" theory, if "enough" of them are not satisfied), we will not apply the term in question to it. According to such theories, nothing we find out empirically about the things in question can be relevant to changing those criteria; for if a thing did not satisfy them, it would not even be that sort of thing, and hence could not be a counterexample affecting the criteria. One aspect of this type of theory that has been the focus of much attention among philosophers of science is that it, if anything, was supposed to account for continuity of discussion between different, and particularly successive, theories: theories would be "talking about the same thing", and thus really in agreement or competition with one another, if and only if the terms used in those theories have the same "meaning".

Saul Kripke and Hilary Putnam have raised two major objections against this traditional theory of meaning, in any of its forms. First, the properties used originally in identifying a substance or natural kind (or, more generally, specified as constituting the meaning of the term in question) need not belong to the substance or kind; they may—even all of them—be found not to be true of that substance or kind. And second, other substances or things may be found which have all those properties and yet are not that kind of substance or thing. Hence we must reject the view that the properties initially assigned to kinds of things and substances give the "meanings" of the terms we use to refer to those kinds and substances, in the sense either of a set of necessary and/or sufficient conditions for applying the term or a set (cluster) of conditions "enough" of which must be satisfied. "*A priori*", Kripke asserts, "all we can say is that it is an empirical matter whether the characteristics originally associated with the kind apply to its members universally, or even ever, and whether they are in fact jointly sufficient for membership in the kind" (Kripke 1980, p. 137; *cf.*, also, Kripke 1977; Putnam 1977, 1978, 1979).[1]

It should be clear from my previous writings that I agree fundamentally with these criticisms. I have argued in a number of papers that, for science, there are no conditions governing the application of a term which are immune from revision in the light of further experience. And I have diagnosed a number of difficulties and unacceptable positions in the interpretation of science—difficulties such as those revolving around the subject of "meaning change", and views such as that of the alleged "incommensurability" of at least some scientific theories—as due to confusions engendered in considerable part by reliance on traditional doctrines of meaning. I have urged that such doctrines should be abandoned as inadequate tools for the attempt to forge a satisfactory interpretation of science and scientific change. (Shapere 1964, 1966, 1971, 1974b, 1977, 1980, 1981). In the remainder of this paper, I will provide still more support for these views. However, I will argue that the alternative view which Kripke and Putnam offer in place of those doctrines must also be rejected, at least as it applies to science. My purpose will nevertheless go far beyond mere criticism of that specific view; for I will try to show that both the Kripke-Putnam approach to the interpretation of science in terms of reference, and the opposing approach in terms of meaning, share certain fundamental assumptions, and that, despite their deep differences, those approaches stem from the same tradition. And I shall argue that those assumptions constitute a violation of the spirit of science, and that the tradition of philosophical analysis from which the two views flow must be banished from the philosophy of science if we are to make true progress in understanding the scientific enterprise. But more importantly, as far as I am able within the confines of this brief paper, I will use my discussion of these matters to bring out, in ways that further what I have said in previous writings, some central features of the approach and views which I have come to believe are appropriate and correct.

The alternative with which Kripke and Putnam would replace the traditional theory of meaning makes reference, rather than meaning, the guarantor that we are "talking about the same thing" despite any changes in the descriptive criteria associated with the thing. The properties by which we originally identify cats or gold or water do not establish conditions which anything must satisfy in order to count as a cat or gold or water. Cats and gold and water are what cats and gold and water *really are,* and the properties by which we initially identify them may have nothing to do with what they really are; those properties do not constitute the irrevocable "meaning" of the term, but only serve, as Kripke says, to "fix the reference". It is the latter, not the meaning, which remains "rigid"—that is, which, in being passed down historically from a hypothetical initial baptism, remains unaltered throughout any vicissitudes, no matter how radical, in what we believe about and attribute to that thing or kind of thing. What we are talking about—referring to—all along, throughout the history of our use of the term after the initial baptism, is what the object or kind really is; and our aim in science is to discover what the thing or kind of thing really is, its very nature, its essence. According to Kripke, therefore, "In general, science attempts, by investigating basic structural traits, to find the nature, and thus the essence (in the philosophical sense) of the kind." (1980, p. 138) The discovery of that essence would be an *a posteriori* one. Of course we might be wrong in thinking that we have arrived at that discovery—that a certain property is indeed an essential property of the substance or kind in question. But *if* we are right—if we have examined a certain substance and found what it "really is"—then from that point on, we will call by the name of that substance or kind all and only those things, in this or any possible world, that have that essential property. If we encounter something in some hitherto-unexamined region of the universe that resembles that substance in every respect except that of possessing what we have found to be the "very nature" of the substance, we will not call that new thing by the name of that substance. Our referential practice—what we say or would say when we discover what a substance or kind really is, and what we are talking about pending that discovery—thus conforms to the existence of metaphysically necessary truths, of essences; and it is in view of the existence of those metaphysical truths and the linguistic practices which manifest them that the so-called "causal theory of reference" claims that science is a search for "essences" in the "philosophical sense". My argument against this thesis will fall into two parts. First I shall argue that the alleged linguistic practice of calling something by the name of a substance (once we know what that substance really is) if and only if it has the essential property of that substance neither would nor should nor does take place. Having so argued, I shall then ask what aspects of science could remain to be illuminated by the doctrine that there are metaphysical truths about essences which it is the object of science to discover. (It is at this point that I shall explore the view that we refer to the essence of a thing or kind even before we have discovered what that essence is.) And I shall conclude that none remain; for an understanding of science, we have no need of that doctrine.

There are certain difficulties in the formulation of the Kripke-Putnam thesis which I will not discuss here. I will simply mention one of these, because it lurks in the background of some things I will talk about. Nothing satisfactory is said about how we are to decide what is to count as an "essential" property. In particular cases, this is by no means an easy matter. There are properties which are true of a substance which Kripke and Putnam would not want to consider "essential" to it; and on the other hand, an essential property apparently need not be a fundamental one (being an element of atomic number 79 is not a "fundamental" property of gold, at least in the sense that it is explainable in deeper quantum-mechanical terms). Talking about "what a substance really is" is thus quite ambiguous in the absence of any discussion of criteria of essentiality. Much could and should be said about this point, but I shall have to pass over it with only one brief comment: we must not suppose that universality is a *criterion* of essentiality—that is, that in order to decide whether a property is an essential property of a certain kind or substance, we must first ascertain (perhaps in addition to other things) whether it belongs to all instances of that kind or substance. On the contrary, for Kripke and Putnam, we discover, from an examination of things of that kind in our spatiotemporal region, what the essence is, and *from then on* refuse to consider anything to be that kind of thing unless it has that property.

II

With regard to the case of gold, then, the thesis before us is the following. *Given* that we have found gold to be an element of atomic number 79, we will thenceforth call something gold if and only if it has that property. In this paper, I will examine just the "only if" portion of this thesis. But that will, I think, be enough to establish the points I want to make about the Kripke-Putnam thesis.

Let us grant, then, uncritically for the moment, that scientists have found, in their examination of our spatiotemporal region of the universe, that gold has certain fundamental or essential properties: it is an element of atomic number 79. And let us assume—still leaving aside any "epistemic" questions of how they find out—that they are right: the claim is true, and furthermore they know it. But later they find some other region in which there is no substance which is an element of atomic number 79; let us even suppose that in that region there are no "elements", and nothing corresponding to "atomic numbers". Is it clear that, if we were to come across such a region, we would not call something in it gold because that substance is not an element of atomic number 79—that we would call a substance gold *only if* it had that characteristic, and that, since nothing in that region had that characteristic, nothing in it would be (i.e., be called by us) gold? Certainly it is not clear; in fact, I shall argue that if certain conditions prevail in that region, precisely the contrary will be in accord with the scientific spirit.

Suppose there is a substance in that region having all the characteristics of gold except of being composed of an atomic nucleus of atomic number 79. We might come to understand this in terms like the following: there exists in that region a peculiar field which smears out the discrete particles of the nucleus and the characteristics of it relevant to its atomic number, but leaves the more "superficial" physical and chemical properties intact; the field also alters the ratio of strength of the electromagnetic to the strong force, leaving that substance easily splittable into two other substances. If the field were removed (or the substance removed from the region), the nuclear characteristics and the ratio we know between the two forces would be restored. (Perhaps, on the other hand, we might conclude on various grounds that *ours* is an exceptional region in which a peculiar field warps substances into the form of elemental nuclei.) Our physics has such patterns of explanation within it; and should we find such an explanation of the situation, it seems clear that the existence of that explanation, together with the similarities between our gold and the particular substance in question, would lead us to call that substance gold *despite* its not having the characteristics Kripke says it must have if we are to call it gold. Even if we had only a range of alternative applicable explanations, without knowing which if any was correct, we would still undoubtedly call the substance gold on the ground that we would understand how the situation *could* arise. Nor need we wait until the substance has been transported out of that region (or put into our own peculiar field, if that is the way things are) before deciding to call it gold;

the reasons for accepting the distorting-field explanation might be quite convincing in themselves. Even if we were unable to remove the substance from that region or to bring it into our own (suppose such transport always resulted in its destruction), we might still have abundant reason for calling it gold.

But the case is stronger yet. For although having such an explanation (or such a range of possible explanations) might give us grounds for calling the substance gold, the existence of such an explanation, or of any other explanation, is not a precondition of our deciding to call it gold. Extending our usage of the term 'gold' is not dependent on a presupposed explanation or even a presupposed set of alternative possible explanations. It is not necessary to assume that *understanding* of the situation always be present as a condition of extending our use of the term to the stuff in the new region, as long as other considerations are sufficiently compelling. For example, if *all*, or even *many*, of the substances in the new region had all or most of the same characteristics as certain corresponding substances in our region, but lacked in each case the property found to be "essential" in our region, our suspicions would, to say the least, be aroused. Our hypothesis would be that the new substances were counterparts of substances in our region (there is absolutely no problem here about what are the counterparts), despite the fact that they do not have the characteristics we have found (truly, as we have assumed for the sake of argument) to belong to those substances in our region, and also despite the fact that we have no explanation or even possible explanation of that fact as yet; and we would *then* begin looking for explanations of that difference. That is, our question would be: Why is gold here not an element of atomic number 79? If the total evidence were somewhat weaker—if, for example, not so many, though still several, substances in the new region shared all or most properties of their earthly counterparts except for the essential property—the question would be correspondingly more tentatively stated: Is this substance perhaps gold, and if so, why is it not an element of atomic number 79? It is only if the total evidence were quite weak that we would withhold the term 'gold' from the substance in the new region—though reasons might accumulate later which would strengthen the case that it is gold. And on the other hand, no matter how strong the evidence might be, our attribution of the term 'gold' to the counterpart substance is defeasible: we might decide, in the light of our investigation, that despite the initial evidence, the substance was not gold after all.

Finally, note that a return to our earthly laboratories to re-examine our prototypical instances of gold would be irrelevant in the example I have given. For, since they are in our field, those instances would show the same nature we had already found for them, and would shed no light on the behavior of the corresponding substances in the other region. Nor would we need to transport our prototypical instances to the other region to see what would happen to them there (again, suppose even that we *could* not so transport them). Under the circumstances I have envisaged, our reasons for calling the new substance gold would be strong enough, independent of such investigations (though such investigations might, of course, strengthen them yet further).

The upshot of our discussion so far is, of course, that *it is not just one property or set of properties—the "essential" ones—that determines or affects how scientists will apply terms in new situations; all the (true) properties may, as in this example, play a role, and furthermore, the properties and behavior of other entities (substances, etc.) may also play a role—as is again the case in this example.* No doubt, too, the possession of certain properties (more fundamental ones) will play a more important role than the possession of others; but importance will be balanced against other factors, such as, in the example given, sheer numbers. Still further, the availability in current science of applicable explanatory patterns, or the actual existence of such an explanation, may also play a role, though, as I have argued, it need not.

I have focused on Kripke's arguments; but Putnam has advanced very similar ones. In a particularly imaginative and influential discussion, he asks us to consider a "Twin Earth" in which is found a substance indistinguishable from water at normal temperatures and pressures, but whose chemical formula is not H_2O, but something else, XYZ. Then, he claims, that liquid is not water; *we* would not call it water, even if the inhabitants of Twin Earth did. His conclusion is the same as Kripke's: ". . . once we have discovered the nature of water, nothing counts as a possible world in which water doesn't have that nature. Once we have discovered that water (in the actual world) is H_2O, *nothing counts as a possible world in which water isn't* H_2O". (Putnam 1977, p. 130; 1979, p. 233; italics his) But the reply to Putnam's example is the same as to Kripke's: if the substance had all, or even a great many, of the properties of our water except that of being H_2O, and especially if other substances on Twin Earth similarly resembled substances on Earth, scientists would suspect strongly that it *was* water under circumstances that, from our point of view, were extraordinary.

The Kripke-Putnam reply to my argument would undoubtedly be that our encounter with the substance in the new region only shows that we were *mistaken* in our belief that gold or water was "essentially" what we had supposed; all that is shown by my example, they would say, is that being an element of atomic number 79 turned out not to be the "very nature" of gold, as we had previously thought it to be. And in general, according to this response, *any* particular claim as to what is true—or rather, essentially true—of something may turn out to be false. But, the response concludes, the point is that *if* the claim is true, and also satisfies further conditions of essentiality, then certain consequences will follow and will be reflected in the linguistic practices of scientists.

The first problem with this reply is that it seems impossible to see how, on the Kripke-Putnam view, scientists could ever come to the conclusion that they were mistaken, once they have come to know, or even to suppose they know, that gold is essentially an element of atomic number 79. For in consequence of coming to that knowledge they will, according to them, thenceforth refuse to call anything gold unless it has that property, and so *cannot* come to the conclusion that the substance in the other region was gold, and that therefore they were mistaken in thinking gold was, in its very nature, an element of atomic number 79. Whether they are right or not, once they commit themselves at any stage—*whether at an initial baptism or after long investigation*—to supposing they have found the essence of gold, they are thenceforth stuck with that conclusion, and it can never transpire that they will find themselves wrong; there can *be* no counterexamples. The gold-like substance in the new region would be *dismissed out of hand* as not being gold, just as, on the old theory of meaning, one would dismiss out of hand the offer of this married man as a counterexample to the thesis that all bachelors are unmarried. The common ground of the old theory of meaning and the new theory of reference begins to be revealed.

Putnam is aware of the difficulty, but only in passing. He admits that ". . . we can perfectly well imagine having experiences that would convince us (and that could make it rational to believe that) water *isn't* H_2O. In that sense, it is conceivable that water isn't H_2O. It is conceivable but it isn't possible! Conceivability is no proof of possibility". (1977, p. 130; 1979, p. 233) With that remark, however, Putnam simply changes the subject, leaving his and Kripke's theory with a paradox that should have been its epitaph. It is impossible to hold *both* that scientists commit themselves in the way Kripke and Putnam suggest, *and* that they might yet find themselves to have been mistaken. Given the doctrine of rigid reference, merely to *assert* that we might be mistaken is not enough. For, having found what we suppose to be the essence, if we then engage in the practice of rigid reference, there is no way to discover that we might have been wrong; but if we admit that we *could* be wrong, we are not practicing rigid reference.

Suppose, though, that we waive this objection; suppose, that is, that the Kripke-Putnam view has been successfully supplemented by a theory that shows how we can be mistaken as to the essence of a substance despite our rigidly refusing to consider anything an instance of that substance that does not have that essential property. Then if one of the Kripke-Putnam examples is shown not to support the thesis of rigid reference, it would be open to a defender of the thesis to say that it just wasn't a good example, that the property previously held to be essential turned out not to be essential after all. Even then there would remain a serious flaw in the proposed reply to my argument. For the real import of this possibility lies in the fact that *no examples at all* would ever be admitted as having the credentials required to bring about the freezing of usage that Kripke and Putnam suggest does or might take place. If nothing will be admitted as a telling counterexample, it is equally clear that neither are there, nor even could there be, any supporting instances. And if examples can throw no light on the theory, the theory can likewise throw no light on the examples. The proposed reply, by divorcing the Kripke-Putnam position from its own examples, only succeeds in alienating that position still further from actual science.

Behind these two difficulties of the proposed reply lies the root of the failure of the Kripke-Putnam position as regards science, at least as I have discussed it so far. For it is not only that (as I have argued) scientists would not, when they have found what a substance truly ("essentially") is, apply the name of that substance "rigidly" to all and only those things having that essential property. It is also that, if they did, they would be justified in doing so only if they had knowledge of an extent that we do not have; at best (i.e., granting the certainty of the conclusions we draw from what we have observed), the knowledge that we do have does not countenance the commitment Kripke and Putnam envision. At best—questions about the adequacy of *that* claimed knowledge aside—we know only the things we have observed thus far. This "epistemic" consideration *cannot* be dismissed as irrelevant to the Kripke-Putnam thesis about linguistic commitment; for what it brings out is that that thesis is irrelevant to real science, which does not, even at best, obtain the sort of knowledge required for such commitment. To base rigid commitment on the knowledge that we do have would be sheer dogmatism, incompatible with the spirit of science which is reflected in what scientists clearly would do in hypothetical cases like those of gold-like and water-like substances in a new region of the universe. And thus, not only is it the case that scientists *would not* observe the kind of linguistic commitment that Kripke and Putnam allege they do or might; as I have now argued, they *should not,* because to do so would violate the spirit of science.

III

But it is also the case that science, at least when it is most characteristically scientific, *does not* observe such linguistic commitment. The examples given by Kripke and Putnam are science fiction stories, and what I have done is simply to turn those examples against them. Science fiction stories have the disadvantage that they often permit the construction of arguments too vague and general to be critically evaluated. But what is important about the use to which I have put the Kripke-Putnam stories is that the kind of *attitude* exemplified in my version, far from being an artificial construction, is characteristic of the scientific approach. For we found in my revision of their examples the following four ingredients: the attempt to find relationships and differences; the use of those relationships as bases for classification; the shaping of vocabulary in the light of the relationships and differences we have found; and the refusal to be bound irrevocably to some prior categorization, or to some prior vocabulary or concepts, however well-founded. These four features are typically found in the dynamics of forming and revising the subject-matter investigated by a scientific field—in what I have elsewhere referred to as the formation and change of scientific domains. They can be given a generalized systematic treatment, but it is not necessary to do that here. What is pertinent here is the way those four features operate in the case of actual linguistic practice in science.

For what happens in the sorts of cases relevant to the present issue is the following. We find reason to believe that things are of a certain sort. Particularly when those beliefs approach or achieve the ideal of being successful and free from specific doubt, we accept them and use them as bases for our further thinking and talking about the world. These processes are well illustrated by the chemical revolution of the eighteenth century. There we find the language in terms of which we talk, scientifically at least, about the world or certain aspects of it, reformulated, and on a very broad scale. The reformulation was based on an incorporation of new beliefs, based on reasons, into the vocabulary for naming and describing the objects of chemical study. In particular, in that case, the beliefs so incorporated were a specific version of the idea that material substances are to be understood in terms of their constituents—what they are made of. When a more mature chemistry added two further ideas—that material substances are to be understood not only in terms of their constituents, but also in terms of the arrangement of those constituents and the forces holding them together—the view amounted to what may be called the *compositional* approach to the understanding of material substances. (Shapere 1974a) Here I will use that term in reference to the first of these three ideas.

Whatever one might suppose, the view that material things can be understood in terms of their constituents is by no means an *a priori* or necessary truth. While it was present from early times, it was rivalled by, among others, a view according to which material entities—at least "earthy" entities—are all of one or at most a few basic types, whose similarities and differences are to be understood in terms of their various degrees of fulfillment or perfection of that one type, and not, or at least not simply, in terms of the intermingling of more basic constituents in them. And laboratory manipulation of those substances was to consist of bringing the substances to their proper perfection. That trend of thought, so foreign to the compositional approach, was more powerful than the latter in the alchemical tradition. In the sixteenth through eighteenth centuries, when it underwent a revival, the compositional approach was confronted with a serious objection. What, precisely, does fire (then the primary means of operating on material substances) do to the substance on which it acts? To the compositionalist, fire merely separates substances into their constituent parts, while not altering those parts in any way. But how could we know that the fire did not act on the substance so as to *alter* it, so as to *produce* the breakdown products which did not exist as constituents of the substance prior to the application of the fire? Things might have been that way. But the compositional approach gradually gained credence through the discovery that a set of substances could be found which were breakdown products in a great many reactions, which themselves could not be broken down by any means at the disposal of the chemistry of the time, and which could be reassembled to form the original substance. But in order to develop that compositional view in a consistent way, more had to be involved in the view than that. The concept of an element had to be given a new interpretation; further, a positive (at first, positive or zero) weight had to be shown to be of central relevance chemically; and it had to be argued that "airs" (gases) were true constituents of material substances, i.e., that they participated in chemical reactions, despite the fact that many, including many compositionalists, had believed that they did not. The totality of such considerations was not sufficient to establish the compositional approach in any final way, or even anything near that; it still had to prove its worth in a fuller way. But those circumstances in which Lavoisier's version of a compositional approach was successful were coupled with another, largely independent consideration to justify a radical revision of scientific linguistic practice. For in the process of discovering new substances and new techniques for handling them, the chemists of the seventeenth and eighteenth centuries had come more and more to the view that sensory qualities, which had hitherto served as the primary bases for classifying and naming material substances, were inadequate for that purpose. On their basis, substances by then known to be different had been confused; and the same substance, if produced by different methods, was often considered to be two or more different substances. The group centered around Lavoisier therefore proposed a revision of the vocabulary of material substances to reflect a, and more specifically his, compositional understanding of those substances: the names of compound substances (like sodium chloride) embodied the theory in terms of which they were understood. Had the compositional view, in a form at least descended from Lavoisier, proved fruitless, no doubt that reform of language would have been rescinded, those beliefs peeled away from the vocabulary in which they had been incorporated—as, indeed, the vocabulary of the phlogiston theory was so discarded, and as, in a parallel case, classifications of stars and consequent naming of types of stars in the light of theories of stellar evolution were withdrawn when those evolutionary theories had to be discarded. (Shapere 1977) As it was, however, the reform of chemical nomenclature was vindicated in its essentials in succeeding decades; and the new vocabulary contributed significantly to suggesting new problems and lines of research.

Though not ordinarily so sweeping, similar extensions and shifts of vocabulary are characteristic of many stages of the scientific enterprise. The development of science consists, in part, of such a shifting of associations of items into new classifications, and of a constant redescribing, and often a renaming, of the items which which it deals. Change is by no means limited merely to renaming of previously-conceived kinds; the pattern of division into kinds is also altered, old kinds being split or united, and new ones introduced. Classifications into kinds of things or substances is not something given, there for us to understand, brought to science to investigate but not to alter, whose essences we are trying to ascertain. Rather, our classifications are shaped and reshaped in the light of what we learn. And the objects to be classified, too, are not simply given; what objects there are is something to be found out, if indeed we find out that there are objects (and kinds) at all in any fundamental sense. Finally, however apt it may be in some cases to talk about the shifts as wholesale, kaleidoscopic, revolutionary, they are not, in the most characteristically scientific cases at least, capricious. There are cases where the considerations on which the revisions are based are abundant and compelling; where they are less strong, as in the case of the reform of chemical nomenclature, fuller justification may have to await further developments. That we may have to await such after-the-fact justification should not be surprising, especially when one considers relatively early cases of scientific thought. For as science is, among other things, a search for better ways of thinking and talking about the world, and for an understanding of the relations between what is talked about, it should be clear that at stages when its objects and relations are but little understood, there will be correspondingly little guidance. Then it may have to rely on ill-grounded hypotheses, obtained, sometimes, from what are judged, then or later, to be extraneous considerations. With the accumulation of beliefs that have proved successful and relatively free from reasons for specific doubt—beliefs in the form of both vocabulary and understanding of relationships—the reasoning by which science advances becomes more autonomous, more self-generating. It becomes more able to generate problems, lines of research, hypotheses, and so forth, from its own structure.

But not only do shifts of the sorts we are considering take place *as a result* of beliefs that have been found successful and free from doubt (and perhaps satisfy other conditions as well); as we have seen, science also *incorporates* such beliefs into its conception of the subject-matter of its inquiry. Such utilization of prior well-grounded belief for determining the strategies of science, and the incorporation of such beliefs into the structure and language of science, is an instance of a more general process that has come to characterize science, a process which we may call the internalizing of relevant considerations. It is a process of gradual reformulation, in the light of what we have learned, of the scientific enterprise—its goals, problems, patterns of explanation, and indeed all its aspects, in addition to its subject-matter. The aim is to make all these so tightly bound by considerations of relevance, and so comprehensive of such relations, that all reasoning about a subject-matter and its problems can be based solely on a consideration of that subject-matter and the well-grounded information known to be relevant to it. It is in that respect that we may say that, in science, we not only learn, but also learn how to learn; and part of that learning consists in learning how to talk and think about the world. Further, this internalizing of relevant considerations exemplifies a third factor, in addition to "success" and "freedom from doubt", that must go into any analysis of what a "scientific reason" is: namely, the intuition that in any argument concerning a subject-matter, those considerations will count as reasons which have to do with that specific subject-matter. (Shapere 1981)

Viewed from another perspective, that process through which scientific reasoning becomes more "internal" is essentially one of gradually distinguishing the scientifically relevant from the irrelevant, of gradually separating what will be counted as internal and what as external to science, of gradually demarcating science from nonscience.[2] It is an ideal we have learned to seek; but it is far from being fully achieved, simply because we do not know everything. It should therefore again not be surprising that "internal considerations" are not yet always adequate for formulating or dealing with scientific problems. Nevertheless, the distinction between scientifically relevant and scientifically irrelevant considerations is far better delineated now than it was in the days, say, of Kepler or Lavoisier.

The process of internalization includes, as we have seen, the building of well-founded information into our descriptive vocabulary. But here as elsewhere, science is constantly open to the possibility that doubt may (though it need not) arise, that our present views, including the ways we "conceptualize" objects and kinds, and name and describe them, may have to be revised or rejected and replaced. Despite the fact that compositionalist views were built into the very language in terms of which we name and describe material substances, that view may even yet have to be rejected; today, indeed, it again faces potential crisis, at least where fundamental theory is concerned. For if quark confinement is accepted, the notion of an independent elementary particle may come to be viewed as a high-energy, short-range approximation, and a truly fundamental physical theory might no longer be characterizable in terms of the particles it postulates. Our reasons for belief—and that includes naming, describing, and classifying—are not in any case conclusive; it is always possible, at least as far as our present understanding is concerned, that reason for doubt may arise.

Even in that majority of cases of scientific change where reform of language does not take place, the properties or criteria associated with the application of a term may be altered. We may indeed *call* by 'gold' only those things which satisfy our latest or most fundamental well-grounded beliefs about nature and about a specific kind of substance in particular; to this extent Kripke and Putnam are right. But we are also always prepared to find reasons for changing what we attribute to (or call) gold no matter how fundamental and well-established that attribution may be. Kripke-Putnam commitment never comes.

IV

In the account I have given, our scientific linguistic practice has been described wholly within the framework of a conception in which, in the light of new findings, we arrive at new beliefs and incorporate them into the ways we think and talk about nature. Is there any remaining place for the concept of "essence", of "metaphysically necessary truth"? As we have seen, Kripke and Putnam hold that if scientists were to discover what a substance in our region "really is", they would thenceforth call something by the name of that substance only if (and if) it has that property. Against this doctrine, I have argued that scientists would not so behave were they to make such discoveries, and furthermore that they should not and do not so behave. The examples adduced by Kripke and Putnam were turned against them, and it was shown that in fact no specific examples of scientific reasoning can support the Kripke-Putnam thesis. If there is to be anything to their claims, then—at least with regard to science—the point cannot be exhibited in the behavior of scientists subsequent to any discovery, even of the sort they have in mind. Can it then be manifested in the behavior of scientists—in the way they talk—in the very process of seeking discoveries? More specifically, can we say that the discovery of essences is the *aim* of science, an aim which is reflected in our (here, scientists') alleged referential practice of "talking about" what gold, water, and so forth "really are", no matter what specific beliefs we may hold regarding those substances, and whether or not we do or even can ever achieve that aim?

The trouble with this view is that, unless the notion of "essence", of the "very nature" of a substance or kind, is expanded to the point of emptiness, it cannot describe the aim of science without limiting the options of science in an unacceptable way. The problem is not just that Kripke and Putnam sometimes talk in ways that suggest that understanding the "very natures" of substances would be understanding those substances in a compositional way; the validity of the latter approach, if indeed it is valid, is contingent. But the notion of "essence" could, I imagine, be expanded to cover alternatives to compositional understanding. Nor is the problem merely that there is no guarantee that there are well-circumscribed boundaries between substances or kinds, and well-defined sets of essential properties for them, after the fashion of Aristotelian essences; perhaps the difference between kinds is rather like the difference between metals and non-metals. That, too, is a contingent matter; but no doubt the concept of essence could be expanded to take in that contingency too. Nor is the difficulty simply that fundamental understanding may have nothing to do with distinct kinds, which might ultimately be understood in far more general terms; that possibility, I am willing to grant, might be covered by some appropriate distinction between knowledge of essential truth and knowledge of fundamental truth. The real difficulty is that there are such a variety of possibilities, and it is folly to suppose that we are aware of all of them that might arise in future science. Even to cover all those of which we are aware, we would have to include the possibility that we might arrive at the conclusion that we cannot understand what things are "really like", and even that the very notion of things being "really like" anything may be inapt. One might think that the Copenhagen Interpretation of quantum mechanics is wrong; but can it be ruled out as impossible? Yet in at least some versions, it suggests that certain properties (e.g., position and momentum) tell us nothing about "entities" themselves, but only about the system consisting of object plus instrument. True, some properties remain attributable to entities (e.g., mass and charge); but is it impossible that a scientific theory might involve—for definite reasons, as in the case of position and momentum—the conclusion that *no* properties are so ascribable? If that view, or anything like it, were to prevail, we might say that we had found a kind of truth; but in that case, it would be only in a very stretched sense that we would have found out something about the essences of kinds and substances. Yet we would certainly not want to conclude that such a theory had failed to fulfill the aims of science, any more than we would want to say that about quantum theory. Furthermore, in such a case, there might even be reason, as some have argued, for concluding that there are no such things

The aim of science, then, cannot be restricted in any way in advance of the investigation of nature. Earlier, we found that the Kripke-Putnam thesis about the linguistic commitment of scientists after their discovery of essences can receive no support from any specific scientific belief; rather, it must be compatible with any outcome of scientific investigation, with any scientific belief. It is hopeless to try to view this compatibility as an advantage of the thesis—as showing the thesis to be independent of the mere contingencies of scientific belief. For now, in considering whether the discovery of essences can be the aim of science, we see that such compatibility can be purchased only at the cost of interpreting the concept of "essence" so broadly that it is without content. Like its ancestors in the history of philosophy, the new doctrine of essences can be true only if it is empty; to whatever extent it is specific, it is false.

But we must be clear as to just where that falsity lies. Note that I am not objecting to the idea that specific sense might be made of the notion of "essence"; nor do I even object to the idea that someday, somehow, we might be in a position to obtain knowledge which is certain, for which there is not even an in-principle possibility of doubt, and which satisfies whatever sensible conditions of essentiality philosophers might lay down. What I have been calling the "spirit of science" is, after all, itself not an *a priori* or necessary condition of science, but is based on a particular view about the way things are and the way we can find out about them that we have developed through encounter with the world. Roughly speaking, we have come to accept the possibility that specific doubt may always arise because we have come (for good reasons) to believe that space and time are of incomparably vaster extent, and the entities existing therein of incomparably vaster number and internal complexity, than we can examine adequately in our experience, and also that the only way to learn about the way things are is to examine the things existing in space and time intensively and serially. And though we have no specific reason to doubt that view, there may still be profound surprises in store regarding it. In short, given a better analysis of essentiality than Kripke and Putnam have provided, I am willing to admit that the discovery of essences, in some specific sense, might be an eventual achievement of science. But whatever may be the case in non-scientific contexts, in science we have learned not to impose inviolable restrictions on possible outcomes of our investigations: we do not search exclusively for "essences" in any specific (i.e., restrictive) sense of that term; we search for whatever we find. If—or to the extent that—scientists talk as though they seek to find out what things are "really (essentially) like", they do so because, in grappling with experience, they have found reason to talk that way, and not because such talk is imposed on them as a necessary consequence of the nature of referential language. And thus—my earlier objections aside—what I am objecting to is the claim that science *must* always aim at discovery of "essences", *where that claim is supposed to have any specific content.* What we strive for in science are views which will be as successful and free of specific doubt as possible, if any such views can be arrived at; and even that is something we have learned to seek. But nothing can be specified, in advance of the actual study of nature, about the particular sort of view we may arrive at. If the concept of essence is understood broadly enough to cover *any* outcome of scientific investigation, well and good—though saying that science is a search for essences then becomes highly misleading. The trouble is that the Kripke-Putnam view *seems* to claim more: that it constitutes a profound comment on the aims of science. This is what I object to.

Everything I have said is compatible with "realism" to at least the following degree: that we *may* eventually arrive at a view which is fully successful in all respects; regarding which no specific reason for doubt ever arises which is not removed satisfactorily; and to which no equally successful alternative view is ever found. My account does not even rule out the possibility that we may arrive at beliefs that we could know fulfilled those conditions. But such eventualities are by no means guaranteed: whatever one says about "truth", the "truth" which science seeks must be compatible with *either* "realism" *or* the discovery that realism cannot be accepted. The Kripke-Putnam view, at least if it is interpreted as having some specific content, limits science to seeking a "realist" account. What I have argued is that that is too limiting. And in any case, it cannot be their theory of reference that explains our putative success in achieving knowledge; for that view is, or at any rate must be, compatible with *any* beliefs we may hold. Nothing else is needed, nor is anything else available, to explain the success of science than the reasons we have for accepting the beliefs we do; and nothing else is available or needed for providing truth-conditions for claims about gold and so forth except the beliefs we have best reason to accept.[3]

The issue here is a deeply important one for the attempt to understand science. For on the Kripke-Putnam view, the aim of science is specified in a way that is independent of the content of science. In this respect, that view is a descendant of the idea that science must make "metaphysical presuppositions", though that old idea is now given a new twist in terms of the philosophy of language. Or from another, broader perspective, their view, like the old theory of meaning, offers a bifurcated account of the scientific enterprise. On the one hand, we are told, there is the content of scientific belief, established on the basis of "epistemic" considerations; and on the other hand, there is the aim of science, which is independent of that content, and is established on grounds independent of "epistemic" considerations. And so our question is this: Can the scientific enterprise be understood without appealing to concepts which transcend the content of scientific belief and the methods by which we arrive at it? Can we provide an account of science in which such concepts as reference, and as the aims of science, are integrated with an account of the content of scientific belief itself, or must that account be bifurcated in the way I have described? My answer should by now be clear: I have argued that we neither would, nor should, nor do, nor need, employ the concepts of "essence" and "metaphysical necessity" in the ways Kripke and Putnam allege we do, unless those concepts are understood so broadly as to be empty. More positively, my suggestion is that the process of "internalization" which characterizes science is in part a process directed toward making scientific reasoning autonomous, making it grounded fully in the investigation of the experience it tries to understand; and that included in that process is the shaping—the internalization—of its own goals in the light of what we learn.

If a realistic account is indicated by the successes of present science (and in many aspects of science it is), that account is a conclusion from our experience, not the product of an *a priori* or transcendental argument in the philosophy of language. Even more than I object to the idea that science aims at the discovery of essences, I protest against this latter view—the view, namely, that what science must aim at can be established by an examination of the nature of language. Here my criticisms of the Kripke-Putnam view flow together, and join also with my objections to the traditional theory of meaning. I remarked earlier that the arguments raised by Kripke and Putnam against the traditional theory of meaning were sound and correct: any properties (criteria) we originally use in identifying some thing or kind may later be found not to hold of that thing or kind. My agreement can now be seen to be simply a manifestation of my view that *all* of what we say about the thing or kind may, in principle, be subject to doubt and revision or rejection (barring certain sorts of new discoveries about how to learn, discoveries that we have no specific reason whatever to expect). But from this point of view, my disagreement with Kripke and Putnam is a manifestation of the very same point. For in saying that there *is* something at whose discovery we might arrive (and to which we have been referring all along) and whose discovery will lead to practices which are in principle (because of the nature of reference) immune to revision, they have abandoned the very point which they so incisively raised against the traditional theory of meaning. Their view remains in agreement with that theory on one fundamental point: that there is something about science that is in principle immune to revision in the face of experience. The difference between them is merely that, for the traditional theory of meaning, that which is immune to revision is something that is established at an inception, while for Kripke and Putnam it is something that is aimed at and perhaps arrived at after long empirical investigation. I reject the fundamental assumption common to both views.

Both the traditional theory of meaning and the Kripke-Putnam theory of reference stem from the same source: the idea that, by an analysis of features of language, we should be able to lay down inviolable conditions on the knowledge-seeking enterprise. Having rejected the principle of inviolability with regard to *any* aspect of that enterprise, I reject it in this instance also. Neither meaning nor reference, at least as understood by the views considered in this paper, can serve as tools in the attempt to understand science.[4]

V

Yet though we may thus reject the ideas of meaning and reference which are the technical artifacts of philosophers, those notions have their source in homely observations for which place must be made, if not as tools, then as products. After all, we do talk about things, and we do attribute meanings to what we say. But the work that needs to be done with regard to those observations can be done in the context of the picture of science and its changes that I have drawn. Consider, for example, the notorious problem of explaining how (or whether) continuity of discussion is possible between successive theories in science. To account for such continuity, it is unnecessary to assume *either* a common "meaning" *or* a common ("rigid") reference of terms. That Stoney, Thomson, and Feynman were, in certain aspects of their work, all "talking about" electrons is not guaranteed by a shared set of necessary and sufficient conditions for applying the term 'electron'; there *may* be shared ascriptions[5] (though not in the sense of necessary and sufficient conditions), but there is nothing to prevent the ultimate abandonment of all of them; they are not irrevocable. To assume that there *must* be shared and irrevocable "criteria" is simply a Platonic fallacy. Nor is continuity guaranteed by a baptismal act by Stoney and a "causal" handing-down of the term; that view, as I hope I have shown, results in the assumption of a something-I-know-not-what that does even less explanatory work, and certainly more mischief, than its Lockean ancestor. In particular (and in addition to my earlier objections), that view is no help at all in resolving the problem of continuity or comparability of scientific theories: for to say that continuity is guaranteed by the fact that we are talking about (referring to) the same "essence", where we do not or cannot know what that essence is, is merely to give a name to the bald assertion of continuity. Nor can it help to look at the "causal" connections between successive usages: as adherents of incommensurability might well argue by pointing to methods of education in science, a new generation may be conditioned—caused—to *think* it is referring to the same thing by 'electron' as its predecessors.

An emphasis on *causal* connections is thus futile for escaping the usual incommensurability arguments. (Here, at least, the old theory of meaning has the edge over its rival: for, its other flaws aside, it *could* answer those arguments *if* there were "meanings" in the sense it claimed, and *if* those meanings were indeed shared by successive members of a scientific tradition. But those conditions are not satisfied, and in any case the other objections to the theory make it unable to do any better than its rival in accounting for scientific change.) Rather than either common meanings or common references, what serve, and what alone can serve, in science at least, as providers of continuity, are the *reasons* connecting the successor idea to its ancestors. Where it occurs in science, continuity is achieved by what may be called a "chain-of-reasoning connection." An example is the chain which led in successive stages from Stoney's views about the electron, through Thomson's, and on to Feynman's, there having been reasons at each stage for the dropping or modification of some things that were said about electrons and the introduction of others. The meaning of the term 'electron', in the only sense that can be made or is needed, then, is this: a family of criteria related by a chain-of-reasoning connection. This is sufficient warrant for the claim that the theories concerned were saying competing things about electrons.

Thus, what Feynman and his associates "meant" in the homely sense in discussing electrons is understood in terms of what Feynman and his associates *said* electrons were—the set of properties they attributed to electrons.[6] What they were talking *about* is similarly determined by those properties, or, if you prefer, criteria. But those properties or criteria were in turn related to Thomson's and Stoney's by specifiable connections. The considerations forming those connections had been determined to count as "reasons" as a result of the process of "internalization" described in Part III, above. We thus need not say that Feynman and the others were all really referring to a common something-I-know-not-what; all the work that concept tries and fails to do can be done simply through this analysis, together with the recognition, as part of our analysis, that Feynman's or the others' criteria—what they said about electrons, how they used the term 'electron'—might become subject to doubt or replacement. Electrons can thus be understood as "trans-theoretical", something about which we can have competing theories, without assuming that there is either a common meaning or a common reference of the term 'electron'.[7] A somewhat more subtle analysis must be given for the competition of theories at least one of which is rejected and thus does not belong to a chain of theories each of which (or at least the later ones in the series) was *accepted* in turn. But even there, there will be found, on proper analysis, a common subject-matter, common problems, and usually at least some common set of other relevant information from the scientific storehouse.

As I hope I have shown in this and other papers, the technical concepts of meaning and reference stemming from the philosophy of language have failed to clarify the scientific enterprise. On the contrary, they have only succeeded either in introducing hopeless confusion or in contradicting some of the most fundamental aspects and achievements of that enterprise. Their vagaries, confusions, and paradoxes, their arbitrary presuppositions and apriorisms, their epistemological relativisms and metaphysical absolutes, must all be avoided. The only way of doing this is to abandon those technical concepts themselves, as philosophers and others have understood them, and to exorcise completely the error of supposing that scientific reasoning is subservient to certain alleged necessities of language, and that the study of the latter is therefore deeper than the study of the former. The situation, I have argued, is rather the reverse. I have tried to show how this is so, and how its being so can be recognized in a more adequate understanding of the scientific enterprise.

NOTES

1. In speaking of Putnam in this paper, I refer only to the views he expressed in the abovementioned works (with the exception of the final chapter of [1978]). More recently, he has abandoned some but not all of those ideas, and for this reason I have focused here more on Kripke's than on Putnam's views. But I have included discussion of some of the latter, both because they remain influential, and because they often throw additional light on the topics at issue. But there are many aspects even of Putnam's earlier views that I do not discuss here.

2. Note that the idea of an "internalizing of relevant considerations" provides a resolution of the debate, among philosophers and historians of science, as to whether scientific change is governed primarily by "internal" or "external" factors.

3. Margalit has put the arguments I am rejecting in a way that exhibits my point glaringly: "In determining the extension of 'salt', say—which is part of the task of determining the reference conditions of the sentence constituents, which is, in turn, part of the task of determining its truth conditions—the chemical structure of salt is crucial. The reasoning here is clear. Our ability to say something true about the world depends not just on what we *believe* about the world, but also on its actual structure. In order to explain the success of our language in saying truthful things about salt we need information about the constitution of salt . . . The chemical structure of salt is, therefore, essential for determining the truth conditions of sentences where 'salt' occurs in a non-empty way . . . one of the enterprises undertaken by the theory of language is to explain the connection between language and the world. The explanation of this connection requires that we understand both sides of it, *i.e.*, language and the world. The world, however, is what *science* tells us that is the world." (Margalit 1979, p. 22) But the contrast made here between "beliefs" and the "actual structure" of the world plays no role in determining the truth-conditions *we use*. Scientific information is not something distinct from "beliefs", but itself consists of (well-grounded) beliefs. It is the final sentence in this quotation from Margalit, not the preceding discussion, that carries the real weight.

4. In this paper, I have not addressed the contention that Kripke's views on reference and essences stem from certain technical features of (current) modal logic. If those views follow necessarily from the apparatus of modal logic, then if my criticisms are justified, that apparatus does not do justice to actual science. But it seems to me questionable whether his views of reference and essences *are* so required by modal logic. If they do not so follow, then of course it remains an open question whether, and to what extent and in what ways, modal logic can illuminate the structure and reasoning of science. In that case, as with all formal systems, its applicability cannot be taken for granted, but is something that must be established.

5. We must of course distinguish between "criteria for (properties and behavior used in) identifying something as an *x*" and "criteria for being an *x*"; but my not doing so in the present context is unimportant. It should be remembered that, in the usual situation in mature areas

of science, the connection between the two is tight, since the former—e.g., shapes of tracks of electrons—are explained in terms of the properties ascribed to x's, together with a knowledge of the ways x's interact with other entities and a knowledge of the particular circumstances of observation.

6. We thus need not suppose that there is a "concept" of "electron" consisting of a set of necessary and sufficient conditions remaining unchanged in all uses of the term 'electron'. If the ascription of some particular subset of properties happens to be common to all uses of the term 'electron', that is a contingent fact: For people to be "talking about the same thing" when they use that term, there need be no such set, the "concept of electron" being simply the family (ancestors and descendants) of reason-related criteria each "generation" of which consists of what is attributed to electrons at a given stage in the succession.

When combined with the earlier discussion of the "internalizing of relevant considerations" (in the final four paragraphs of Section III, above), this concept of 'concept' suggests an analysis of what it is to be a "reason" in science (and in other areas of human thought as well). As I remarked in that earlier section, for something to be a "reason" is for it to be (among other things) *relevant,* in a sense broader than is capturable by formal logic, to that for which it is alleged to be a reason. This implies that, in the process of discovering what is and what is not relevant to given claims—that is, in delineating the distinction between "internal" and "external" considerations, and increasing its ability to arrive at and test ideas on the basis of internal considerations alone—science gradually clarifies what is to count as a scientific reason in a given domain of inquiry. The discussion given now, in the present section, implies that such clarification of what *counts* as a reason constitutes at the same time a clarification of the *concept* of "(a) reason" itself, in the only sense of 'concept' that can ultimately make sense. This suggested way of understanding 'reason' as inextricably bound to what is "internal" to a subject points the way toward understanding how it is possible for science to progress on the basis of considerations that are "objective" and "rational". These remarks are developed in detail in a forthcoming work, "The Concept of Observation in Science and Philosophy", where it is shown that even allegedly "metascientific concepts" undergo a process of revision in response to new scientific findings.

7. For the notion of a "transtheoretical term", see Shapere (1969, Part II). Putnam (1973) has appealed to this idea in developing his own theory of reference; as should be clear, his use of it was not what I had in mind.

REFERENCES

Kripke, S. (1977), "Identity and Necessity", in Schwartz, S. (ed.), *Naming, Necessity, and Natural Kinds:* 66–101. Ithaca: Cornell University Press.

Kripke, S. (1980), *Naming and Necessity.* Cambridge, Mass.: Harvard University Press.

Margalit, A. (1979), "Sense and Science", in Saarinen, E., *et al.* (eds.), *Essays in Honour of Jaakko Hintikka:* 17–47. Dordrecht: Reidel.

Putnam, H. (1973), "Explanation and Reference", in Pearce, G., and Maynard, P. (eds.), *Conceptual Change:* 199–221. Dordrecht: Reidel.

Putnam, H. (1977), "Meaning and Reference", in Schwartz, S. (ed.), *Naming, Necessity, and Natural Kinds:* 119–132. Ithaca: Cornell University Press.

Putnam, H. (1978), *Meaning and the Moral Sciences.* London: Routledge and Kegan Paul.

Putnam, H. (1979), *Philosophical Papers,* Vol. II: *Mind, Language, and Reality.* Cambridge, Eng.: Cambridge University Press.

Shapere, D. (1964), "The Structure of Scientific Revolutions", *Philosophical Review 73:* 383–394.

Shapere, D. (1966), "Meaning and Scientific Change", in Colodny, R. (ed.), *Mind and Cosmos:* 41–85. Pittsburgh: University of Pittsburgh Press.

Shapere, D. (1969), "Notes Toward a Post-Positivistic Interpretation of Science", in Achinstein, P., and Barker, S. (eds.), *The Legacy of Logical Positivism:* 115–160. Baltimore: Johns Hopkins University Press.

Shapere, D. (1971), "The Paradigm Concept", *Science 172:* 706–709.

Shapere, D. (1974a), "On the Relations Between Compositional and Evolutionary Theories", in Ayala, F., and Dobzhansky, T. (eds.), *Studies in the Philosophy of Biology:* 187–202.

Shapere, D. (1974b), "Scientific Theories and Their Domains", in Suppe, F. (ed.), *The Structure of Scientific Theories:* 518–565. Urbana: University of Illinois Press.

Shapere, D. (1977), "The Influence of Knowledge on the Description of Facts", in Suppe, F., and Asquith, P. (eds.) *PSA 1976:* 281–298. East Lansing: Philosophy of Science Association.

Shapere, D. (1980), "The Character of Scientific Change", in Nickles, T. (ed.), *Scientific Discovery, Logic, and Rationality:* 61–116. Dordrecht: Reidel.

Shapere, D. (1981), "The Scope and Limits of Scientific Change", in Cohen, L. J., *et. al.* (eds.), *Logic, Methodology and Philosophy of Science VI,* forthcoming. Amsterdam: North-Holland.

Shapere, D. (forthcoming), "The Concept of Observation in Science and Philosophy".

Larry Laudan*

TWO DOGMAS OF METHODOLOGY

For a very long time, much work in the philosophy of science has been predicated on the existence of two crucial dependences. The first assumed dependence is between cognitive progress or growth, on the one hand, and the cumulative retention of explanatory success, on the other. The second dependence makes the appraisal of the cognitive merits of different theories crucially contingent upon the existence of rules for translating one theory into another. These linkages have become such matters of faith that entire philosophies of science have been predicated upon them, in spite of the fact that neither of these connections has been cogently argued for (thus the reference to 'dogmas' in the title).

It will be the claim of this short essay that neither the necessity, nor even the advisability, of either of these connections has been convincingly established.

*I am grateful to my colleagues at the University of Pittsburgh for their helpful comments on an earlier version of this paper, and especially to Adolf Grünbaum whose criticisms saved me from some egregious errors.

Reprinted by permission from the author and from *Philosophy of Science,* Vol. 43, no. 4, December 1976, pp. 585–597.

1. PROGRESS WITHOUT CUMULATION

It is a widely-shared conviction—accepted alike by thinkers as diverse as Whewell, Collingwood, Popper, Lakatos, Post, Stegmüller and others—that a necessary (and for some authors, even a sufficient) condition for scientific progress and growth is what I shall call *the cumulativity postulate* (*CP*). In brief, *CP* states that the replacement of one theory by another is progressive—or represents cognitive growth—if and only if the successor can explain everything successfully explained by the predecessor *and something else as well.* Specific versions of *CP* focus on different units of analysis (problems, questions, and content have been the most common), but they all share a conviction that progress can occur only if we preserve all the "successes" (whether defined as solved problems, answered questions, explained facts or true entailments) of previous theories.

Collingwood, for instance, formulates *CP* in terms of problems and their solutions:

> If thought in its first phase, after solving the initial problems of that phase, is then, through solving these, brought up against others which defeat

it; and if the second [attempt] solves these further problems *without losing its hold on the solution of the first,* so that *there is gain without any corresponding loss,* then there is progress. And there can be progress on no other terms. *If there is any loss, the problem of setting loss against gain is insoluble.* ([2], p. 329)

Popper, too, sometimes speaks of progress at problem-solving, but he most usually articulates *CP* using the language of either "empirical content" or "explanatory success." Most recently, he has put it this way:

a new theory, however revolutionary, must always be able to explain *fully* the success of its predecessor. In *all* those cases in which its predecessor was successful, it must yield results at least as good . . . ([12], p. 83)

Lakatos' espousal of *CP* occurs in the context of his discussion of the progress of "research programmes." In order for any series of theories, T_1, T_2, T_3, . . . to be progressive (and even 'scientific'), Lakatos insists that every later member of the series must contain its predecessor, along with added "auxiliary clauses" which allow it (among other things) to explain some anomaly of its predecessor.[1] Any series whose latest member, *no matter how successful,* failed to explain *everything* its predecessors could, does not qualify as progressive. Another recent version of *CP* has been formulated by Heinz Post, who claims not only that scientists should strive after cumulativity, but also that:

as a matter of empirical historical fact L [i.e., new] theories [have] always explained the *whole* of [the well-confirmed part of their predecessors] . . . contrary to Kuhn, *there is never any loss of successful explanatory power.* [my italics] ([13], p. 229)[2]

By contrast with these philosophers, both Kuhn and Feyerabend have insisted that if progress is construed as involving cumulativity, then there are many episodes in the history of science which must be classified as non-progressive or even regressive. Put starkly, their claim is that we must either assert that the bulk of science is irrational (because not progressive), or else accept that progress has nothing to do with scientific rationality.[3]

The aim of this section is to show:

(a) that there are, indeed, many non-cumulative episodes in the history of science;
(b) that many of these episodes involved the *rationally well-founded* replacement of one theory by a *non-cumulative* successor;
(c) that a defensible notion of cognitive progress can be retained by dropping *CP;*
(d) that this modified conception of progress fits those historical cases (in (a) above) where the cumulative requirement breaks down.

Before I develop these claims, however, we need to be clearer about what *CP,* in its various formulations, amounts to.

In its more extreme versions, which I shall call *objective* or absolute *cumulativity* (CP_s), the postulate requires that *T_2 is progressive with respect to T_1 if and only if T_2 entails all the true consequences of T_1 plus some additional true consequences.* The chief difficulty with CP_s, of course, is that we cannot enumerate the entire class of true consequences of any two universal theories, T_1 and T_2, and so any *direct* comparison of their true consequences is out of the question. However, if T_2 entails T_1, and not *vice versa,* then we are entitled to say, even in the absence of such direct comparison, that all the true consequences of T_1 must be properly included in the true consequences of T_2. But it is presumably *only* in such cases that we could be warranted in saying that objective progress had occurred. Sadly, *none* of the major cases of theory change in history of science seems to satisfy the stringent conditions of entailment between theories required to decide whether absolute cumulativity has occurred. Copernican astronomy does not entail Ptolemaic astronomy; Newton's mechanics does not entail the mechanics of Galileo, Kepler or Descartes; the special theory of relativity does not entail the Lorentz-modified aether theory; Darwin's theory does not entail Lamarck's; Lavoisier's does not entail Priestley's; statistical mechanics does not entail classical thermodynamics, which in turn does not entail its predecessor, the caloric theory of heat, which finally does not entail the earlier kinetic theory of heat; wave optics does not entail corpuscularian optics, etc.

Perhaps aware of such difficulties with CP_s, some authors (e.g., Collingwood and Popper) have offered a weaker version of *CP*, which I shall call *epistemic cumulativity* (CP_w). This requires that T_2 *is progressive with respect to* T_1 *if and only if all the facts thus far explained* (or all the questions thus far answered, or all the problems thus far solved) *by* T_1 *can be explained* (*or answered or solved*) *by* T_2, *and* T_2 *can also be shown to explain some fact* (*etc.!*) *not explained by* T_1. What is initially attractive about CP_w is that it neither limits progress to cases of theory entailment, nor does it require an enumeration of all the consequences of a theory. Suppose, for example, that one theory, T_a, *has so far solved problems m, n* and *o*, while T_b can solve *m, n, o* and *p*. The weaker cumulativity postulate would warrant us in saying that T_b was progressive with respect to T_a, even if T_b does not entail T_a. By making the relatively innocuous assumption that we can identify those problems which two theories have *already* solved, the epistemic version of *CP* seems to offer large scope for talking about actual scientific cases; or so it has been widely assumed.

Unfortunately, however, the weaker version, CP_w, turns out to be far too strong to capture our pre-analytic judgments about the progressiveness of many of the most important transformations in the history of scientific ideas. As we shall see shortly, the conditions demanded by CP_w, every bit as much as those of CP_s, are generally violated by what we know about the course of scientific development. The chief flaw in both versions of *CP* is their commitment to what we might call the "Russian-doll model of cognitive progress." According to this model, every acceptable theory includes, encapsulated within itself, *all* the explanatory successes of its predecessors. It is this requirement which ultimately proves to be the undoing of both CP_s and CP_w.

One way of exposing the weaknesses in the Russian-doll model has been explored by Paul Feyerabend and Adolf Grünbaum, who have shown that some of the problems which some theories solve can not be formulated within the ontology presupposed by their competitors.[4] When one such theory is replaced by a significantly different one, many of the previously solved problems cannot even be formulated, let alone solved. Within the caloric theory of heat, for instance, a commonly posed (and often solved) problem was represented by the question: "What is the character (attractive or repulsive) of the force relations between heat particles?" Once substantive theories of heat were abandoned, such problems vanished; they literally became meaningless. Similarly, problems about the fine structure of the electromagnetic aether—tackled by (among others) Kelvin, Larmor and Boltzmann—received neither formulation nor solution at the hands of relativity theorists.

In the face of examples such as those discussed by Feyerabend and Grünbaum, a proponent of *CP* might argue that the cumulativity postulate could still be retained, provided that it was applied only to that sub-set of problems whose formulation is possible within the langauge of *all* the relevant theories being compared. The rationale for such a restriction might be that if some problem, p_a, was solved by a theory T_1, but inexpressible within T_2, then p_a was—at least so far as proponents of T_2 are concerned—*a mere pseudo-problem* which T_2 should not be expected to resolve. Equally, some of the apparent successes of T_2 (specifically, those which cannot be represented in the framework of T_1) would not be viewed as genuine explanatory successes by defenders of T_1. Considering such circumstances, a liberal defender of cumulativity might propose a modified version of CP_w along these lines: any successor theory, T_2, is progressive with respect to its predecessor T_1 if and only if both (a) T_2 can solve all the problems already solved by T_1 which are well-formed within the framework of T_2 and (b) T_2 can solve some additional problems (which are well-formed within T_1).

Although this line of argument may provide an answer to some of Grünbaum's challenges, it manifestly does not save *CP* from historical refutation; for it is easy to show that many of the problems solved by predecessor theories have *not* been solved by successor theories, even when those problems can be coherently formulated within the language of the successors. If this claim is true, and if we wish to preserve a sense of 'progress' which is applicable to paradigm cases of progressive change, then CP_w, along with CP_s, will have to be abandoned.

I shall mention briefly four well-known cases in the history of science. In each, there is substantial evidence to support the claim that the successor theory notably failed to deal with certain problems solved by its predecessor:

(i) *The celestial mechanics of Newton and Descartes.* By the 1670s, the celestial mechanics of Descartes was widely accepted (in spite, incidentally, of its failure to offer any explanation for the precession of the equinoxes—which had been a solved problem since antiquity). One of the core problems for Descartes, as for Kepler before him, was that of explaining why all the planets move in the same direction around the sun. Descartes theorized that the planets were carried by a revolving vortex which extended from the sun to the periphery of the solar system. The motion of this vortex would entail that all the planets moved in the same direction. Newton, on the other hand, proposed no machinery whatever for explaining the uniform direction of revolution. It was perfectly compatible with Newton's laws for the planets to move in quite different directions. It was acknowledged by both critics and defenders of the newer Newtonian system that it failed to solve this problem which had been explained by the earlier Cartesian system.

(ii) *The electrical theories of Nollet and Franklin.* During the 1730s and 1740s, one of the central explanatory tasks of electrical theory was the explanation of the fact, discovered by Dufay, that like-charged bodies repelled one another. In the early 1740s, the Abbé Nollet explained this in terms of an electrical vortex action, and his theories were widely accepted. In the late 1740s, Franklin produced his one-fluid theory of electricity which offered a very convincing account of the electrical condenser. That theory assumed that electrical particles repelled one another, and that positive electrification arose from an excess of the fluid and negative electrification from a deficiency of the fluid. With this model, Franklin—by his own admission—could offer no coherent account of the repulsion between negatively-charged bodies. Nonetheless, Franklin's theory quickly displaced Nollet's as the dominant electrical theory.

(iii) *Caloric and early kinetic theories of heat.* One of the major explanatory triumphs of the proto-kinetic theories of heat in the 17th century was their ability to explain the generation of heat by friction. By the late 18th century, however, kinetic theories were confounded by such anomalies as specific and latent heats, endothermic and exothermic chemical reactions and animal heat—phenomena which could all be explained on a substantial theory of heat. For these reasons, substantial theories of heat displaced kinetic ones, in spite of the failure (stressed by Rumford and Davy) of substantial theories to explain frictional heat.

(iv) *The geology of Lyell and his predecessors.*[5] Prior to Hutton and Lyell, geological theories had been concerned with a wide range of empirical problems, among them: how aqueous deposits are consolidated into rocks, how the earth originated from celestial matter and slowly acquired its present form, when and where the various plants and animals originated, how the earth retains its heat, the subterraneous origins of volcanos and hot springs, how and when various mineral veins were formed, etc. Solutions, of varying degress of adequacy, had been offered to each of these problems in the 18th century. The system of Lyell, and similar ones which largely displaced these earlier geological theories by the mid-19th century, did not offer *any* explanation for *any* of the problems cited above.

In none of these cases, as well as numerous others, does CP_s or CP_w allow us to call such theory transitions progressive. What all these cases have in common is a gain *and* a loss of problem-solving success, which makes talk of cumulativity—in any usual sense of the term—inappropriate. (It is perhaps worth stressing, since some defenders of *CP* utilize a "limited case" approach, that in none of the historical examples I have cited can the predecessor theory be shown to be an approximation to, or a limiting case of, its successor.)[6]

Confronted by the wholesale discrepancy between the demands of *CP* and the realities of history, several choices are open to us. We could, for instance, decide that the history of science is, in fact, non-progressive, because non-cumulative. Alternatively, we could ask whether some objective notion of scientific progress, which dispenses with cumulativity, could be articulated, which would do more justice to our historical intuitions. My inclination is to suggest that the examples above indicate that we were probably mistaken in the first place in linking progress to any form of strict cumulativity and to claim further that an adequate characterization of progress can be given without any reference to *CP*.

To see how this might be done, we should begin with some preliminary remarks about what a theory of progress is designed to do. In its usual signification, progress is a goal- or aim-theoretic concept. To say that "*x* represents progress" is an elliptical way of saying "*x* represents progress towards goal *y*."[7] If our concern is with cognitive (as opposed to material, spiritual or other forms of) progress, then *y* must be some cognitive aim or aims, such as truth, greater predictive power, greater problem-solving capacity, greater coherence, or greater content. With respect to at least some of these cognitive goals, we can meaningfully speak of progress without reference to *CP*.

To take one obvious example, suppose that our cognitive aim is to possess theoretical machinery which allows us to solve the greatest number of problems. Cognitive progress would then be represented by any choice which produced a theory which solved more problems than its predecessor (but *not* necessarily all the problems of its predecessor). Suppose, for instance, that T_1 has already solved problems *a, b* and *c,* while T_2 solved *a, b, d, f* and *g*. If our cognitive aim is to possess solutions to the largest number of problems, then clearly T_2 is progressive with respect to T_1. If one allows for the possibility that the problems which a theory has solved can be individuated and counted, then we have the makings of a measure of progress.[8] (Indeed, all the historical cases discussed above could be shown to be progressive in precisely this sense.)

Consider a different cognitive aim, such

as maximizing the probability of our beliefs. Given two competing theories, the more progressive would be that one with the higher probability. Assuming that non-zero probabilities can be found for theories, then it could presumably be shown that T_2 is more probable than T_1, even if some of the confirming instances of T_1 are not also confirming instances of T_2. Similar remarks could be made about progress if we took coherence, consistency or any number of other cognitive features, singly or in combination, as constitutive of the aim of science. *In none of these cases need we postulate the cumulativity of facts explained* (*or problems solved*) in order to have a workable measure of scientific progress.

Thus far we have argued that *CP* is not a necessary condition for many types of scientific progress. It is equally not a *sufficient* condition for progress. One can readily imagine, for instance, that some T_2, when compared with T_1, satisfies CP_w and yet generates greater incoherence or conceptual unclarity than T_1. In so far as coherence is a cognitive goal of science, T_2 might well be judged unprogressive, even though it solved all the problems of its predecessor and some others as well. Similarly, turning to CP_s, it is conceivable that some T_2 might entail T_1 and yet so violate our cognitive ambitions that it was regarded as non-progressive.

My object here is not to defend any one of these possible cognitive goals to the exclusion of the others.[9] I have sought, rather, to show that many of the cognitive aims which have been proposed for science would allow us to make definitive judgments about the progressiveness of theoretical changes without committing ourselves to the overly restrictive demands of cumulativity.

Those who would deny that these determinations of progressiveness are possible in the absence of cumulativity must show that it is, in principle, impossible to rank-order non-cumulative theories with respect to the degree to which they exemplify our cognitive ambitions. So long as such a rank-ordering is possible, progress—even without cumulativity of content—may prove to be a viable methodological tool for comparative theory appraisal.

Before leaving this topic, an important rider should be added: in suggesting that cumulativity is neither necessary nor sufficient for scientific progress, I am not denying that cumulative changes, when they can be obtained, are cognitively valuable. To the extent that cumulativity is closely related to explanatory scope it is unquestionably a worthwhile desideratum. But it does not deserve the pride of place it has been accorded in those philosophies of science which make it into a *sine qua non* for progress.

2. EVALUATION WITHOUT TRANSLATION

It has been widely assumed that rational choice between two theories is only possible when the statements of one can be fully translated into the other, or when both can be translated into some "neutral" third language. Indeed, one major source of anxiety concerning the Kuhn-Feyerabend theses on incommensurability may be located in the conviction of many philosophers that an absence of full or partial translatability between the object languages of theories being compared precludes any possibility of objective, rational evaluation of the relative merits of those theories.

I do not wish to defend the doctrine of non-translatability, for it is confronted with several difficulties, which are as well-known as they are acute. I do intend to show, however, that both proponents and critics of that doctrine have mistakenly assented to the conditional claim that if object language incommensurability is once conceded, then the rational appraisal of competing theories is the first casualty. Kuhn summarizes what seems to be the dominant view when he writes: "[the] comparison of two successive theories demands a language into which at least the empirical consequences of both can be translated without loss or change" ([8], p. 266). More recently, Feyerabend has claimed that "the phenomena of *incommensurability* . . . creates [sic] problems for all theories of rationality," ([4], p. 214) and has gone so far as to suggest that the choice between incommensurable theories may be only a subjective "matter of taste" ([3], p. 228).[10]

Presupposed by this linkage between comparability and translatability is the assumption that *every* parameter relevant to the comparative appraisal of theories requires a translation of the object-level claims of one theory into the object-level language of another theory. It will be my claim, on the contrary, that there are many features of theories—features which are relevant to their epistemic appraisal—which can be determined without any machinery for translating the substantive claims of the theory into any other language.

Of course, it is true that certain approaches to the logic of comparative theory appraisal are badly undermined by non-translatability. Consider a pair of examples:

(i) *Content Analysis.* Among the followers of Popper (including Lakatos and Watkins), it is taken for granted that the basic measure for comparing theories is content. Unless we can compare the content of two competing theories, then we have no way of rank-ordering them on a preference scale. Content comparisons are inevitably comparisons of the object languages of theories; without a high (perhaps even a complete) degree of translatability, the Popperian machinery for theory appraisal is sterile. Similarly, in their demand that successful theories must explain the anomalies of their predecessors, the Popperians require a high measure of inter-theoretic translation; for without such translations, one could never establish that a later theory solved the anomalies of its predecessors.

(ii) *Bayesian Inference.* In any account of the evolution of scientific theories which utilizes an analogue of Bayes' theorem, object language comparisons are inevitably presupposed. Assessments of prior probabilities, for instance, are usually seen as involving background knowledge assumptions in order to obtain an initial probability for some empirical proposition, M. Unless M can be expressed both in the language of the theory whose posterior probability is being assessed *and* in the language of the background knowledge, then such assessments are impossible. Even more damaging is the fact that the Bayesian schema requires it to be meaningful to speak of the probability of M-like statements within theories which are contraries of the theory whose probability is being computed. Unless we can show that M is a well-formed statement within the contraries to T, then T's posterior probability is, again, indeterminate.

Hence, there can be no doubt that certain tools which have been widely touted as instruments for the comparative appraisal of theories are incompatible with the thesis of radical non-translatability. But it has not been sufficiently appreciated that this situation reflects a *contingent* feature of those particular tools and that there are other methods for objectively discussing the relative cognitive merits of different theories which do not require *any* inter-theoretic translations of object-level statements.

To see why, we must return to our earlier discussion of cognitive goals. As we pointed out there, numerous proposals have been made concerning the aims of science: among them, maximum problem-solving ability, maximum internal coherence, simplicity and minimum anomalies. With any one of these goals in mind, we could (and some philosophers already have) generate machinery for comparing theories. None of this machinery need entail any object-level comparability between the theories under analysis. If, for instance, we wish to maximize the number of problems our theories solve, we need only compute—for each theory—the number of problems it has solved. Assuming, as I think we can, that proponents of different theories may be able to agree upon objective criteria for individuating problems, then we can determine how many problems a theory solves without going beyond the theory itself. All extant theories could then be rank-ordered with respect to their problem-solving indices, without once asking whether they can be translated into one another's languages.

Similarly, coherence or simplicity can be so defined that the determination of the degree of coherence or simplicity of a theory does not necessitate any reference to the substantive claims made in other theories. Here again, we could in principle determine that (say) T_a was more coherent that T_b even if no statement in T_a was translatable into the language of T_b. Whether the aims I have mentioned are ones to which we should subscribe exclusively is beside the point. I have mentioned them here only to illustrate the fact that appraisals of the relative merits of different theories do not necessarily depend upon the existence of object-level translation procedures between the theories being compared. The fact (if it is a fact) that scientists who subscribe to radically different theories about the natural world may not understand one another does not preclude the possibility that—if they share certain meta-level goals—they may nonetheless be able to agree on the relative cognitive successes and failures of their respective theories.

We can sum up the argument of this section quite simply: incommensurability of theories at the object-level does not entail incomparability at the meta-level. *The widely-held assumption that non-translatability leads inevitably to cognitive relativism is simply mistaken.* If we are disposed to attack the Kuhn-Feyerabend doctrine of incommensurability for its own deficiencies, that is fine; but it is not fair play to dismiss the incommensurability thesis by invoking the supposed *reductio ad absurdum* that an acceptance of that thesis makes any objective evaluation of different theories impossible.

NOTES

1. See Lakatos ([9], p. 118). Cf. also the Lakatos-Zahar claim that "one research program supercedes another only if it has excess truth content over its rival, in the sense that it predicts progressively all that its rival truly predicts and some more besides" ([10], p. 369).

2. See also Koertge [6].

3. In a well-known passage, Kuhn asserts that "new paradigms seldom or never possess all the [problem-solving] capabilities of their predecessors" ([7] p. 168). He also writes: "there are [explanatory] losses as well as gains in scientific revolutions" ([7], p. 166). (Cf. also Kuhn [8], p. 277.)

4. Cf. Grünbaum [5] and Feyerabend [4].

5. I am grateful to R. Laudan for this example.

6. Such a view is represented in Born's comment that "the continuity of our science has not been affected by all these turbulent happenings, as the older theories have always been included as limiting cases in the new ones," ([1], p. 122). Post ([14], p. 228) espouses a similar view.

7. Strictly, we should unpack it as "*x* represents progress towards goal *y, with respect to competitor z.*"

8. A defense and fuller treatment of a problem-solving approach to progress can be found in Laudan [12], upon which this paper is based. That study moves well beyond the simplifying assumption used here (i.e., that all problems are of equal importance).

9. It should be added, parenthetically, that one major approach to the problem of progress—namely, the self-corrective thesis of Reichenbach, Peirce and Salmon—is not committed to *CP*. On their account, theories could conceivably "move closer to the truth" (and thus show progress) *without* explaining all the successes of their predecessors. Unfortunately, there are other well-known and acute difficulties facing proponents of this approach. (For a brief discussion of some of those difficulties, see Laudan [11].)

10. Feyerabend later insists that between incommensurable theories, choice can only be based upon "our subjective wishes" ([4], p. 285).

REFERENCES

1. Born, M. *Physics in My Generation.* New York: Dover, 1960.

2. Collingwood, R. *The Idea of History.* New York: Oxford University Press, 1956.

3. Feyerabend, P. "Consolations for the Specialist." In *Criticisms and the Growth of Knowledge.* Edited by I. Lakatos and A. Musgrave. Cambridge: Cambridge University Press, 1970. Pages 197–230.

4. Feyerabend, P. *Against Method.* New Jersey: Humanities Press, 1975.

5. Grünbaum, A. "Can a Theory Answer more Questions than One of Its Rivals?" *British Journal for the Philosophy of Science,* 27 (1976): Page 1–23.

6. Koertge, N. "Theory Change in Science." In *Conceptual Change.* Edited by G. Pearce and P. Maynard. Dordrecht: Reidel, 1973. Pages 167ff.

7. Kuhn, T. S. "Reflections on My Critics." In *Criticism and the Growth of Knowledge.* Edited by I. Lakatos and A. Musgrave. Cambridge: Cambridge University Press, 1970. Pages 231–278.

8. Lakatos, I. "The Methodology of Scientific Research Programmes." In *Criticism and the Growth of Knowledge.* Edited by I. Lakatos and A. Musgrave. Cambridge: Cambridge University Press, 1970. Pages 91–195.

9. Lakatos, I. and Zahar E. "Why did Copernicus' Research Program Supercede Ptolemy's?" In *The Copernican Achievement.* Edited by R. Westman. Berkeley: University of California Press, 1975. Pages 354–383.

10. Laudan, L. "C. S. Peirce and the Trivialization of the Self-Corrective Thesis." In *Foundations of Scientific Method in the 19th Century.* Edited by R. Giere and R. Westfall. Bloomington: Indiana University Press, 1973. Pages 275–306.

11. Laudan, L. *Progress and Its Problems.* Berkeley: University of California Press, forthcoming 1977.

12. Popper, K. "The Rationality of Scientific Revolutions." In *Problems of Scientific Revolutions.* Edited by R. Harré. Oxford: 1975. Pages 72–101.

13. Post, H. "Correspondence, Invariance and Heuristics." *Studies in History and Philosophy of Science,* 2 (1971): 213–255.

PART II Explanation and Causality

Introduction

Many people believe that the scientist's major concern is with collecting information about particular objects and events. Such a view is not compatible with the fact that scientific books and articles are focused on laws and theories rather than on items of particular information. Why do scientists focus on the formulation of laws and theories? The standard answer is that these laws and theories are needed to fulfill two major goals of the scientific enterprise: the explanation of what has been observed and the prediction of what will be observed. This answer, which presupposes that explanation and prediction necessarily involve laws and theories, is defended by Hempel and Oppenheim in their classic article "Studies in the Logic of Explanation."

In that article, Hempel and Oppenheim

formulate what has come to be called the "covering-law" model of the explanation and prediction of events. This model is called the covering-law model because it involves laws and theories. In the deductive-nomological version of this model, the version which is central to this article, the *explanandum,* the sentence describing what has to be explained, is deduced from the *explanans,* the sentences describing that which is being offered as an explanation; moreover, the explanans must contain at least one non-statistical law or theory.

Hempel and Oppenheim claim that all explanations and predictions fit the covering-law model. What type of evidence do they offer for that claim? To begin with, they give several examples of explanations and predictions and show that they fit the model. Secondly, they argue that many explanations and predictions which do not seem to fit this model actually do fit it. Finally, they argue that only explanations which fit this model meet the requirement that the explanans constitute adequate grounds for the explanandum.

As Wesley Salmon asserts in his account of statistical explanation, the deductive-nomological version of the covering-law model needs to be supplemented by some account of the statistical explanation of particular events by reference to laws and theories of a statistical nature. He points out that, with the rise of classical statistical mechanics during the latter half of the nineteenth century, such explanations have been firmly entrenched in science. Hempel had attempted to formulate a second version of the covering-law model of explanation for statistical explanations, a version he called the inductive-statistical version. In this account, the explanans of an inductive-statistical explanation (explanans which contain a statistical law or theory) offer strong inductive support to the explanandum, even though the explanandum cannot be deduced from the explanans. Because of this difference, Hempel had to add the additional requirement of maximal specificity.

Salmon argues that this account of statistical explanation must be inadequate. There are some purported statistical explanations in which the explanans give high inductive support to the explanandum but which are not acceptable statistical explanations and there are acceptable statistical explanations in which the explanans do not give strong inductive support to the explanandum. Salmon advocates instead a statistical relevance model of statistical explanation, a model in which the explanans contain an assembly of facts statistically relevant to the explanandum, regardless of the degree of probability which results. The rest of the selection from Salmon is a more detailed presentation of that model.

The next three authors focus primarily on the covering-law model in general, rather than on the inductive-statistical version of it. Nancy Cartwright argues that, on the whole, the true covering-laws required by the model (and by many other models of explanation) do not exist. Most phenomena are covered at best by *ceterus parabus* generalizations, generalizations that only hold under ideal conditions which rarely obtain. Explanations using these generalizations serve the function of explanations, but they are not, contra Hempel, explanations from true covering laws.

Baruch Brody argues that the covering-law model is inadequate because it allows as acceptable many purported explanations that really are not explanations. He claims that the difficulty posed by these purported explanations was already known by Aristotle and that we can only solve it by reference to metaphysical notions of causes and essences which Aristotle employed in his theory of explanation. In particular, Brody argues that a deductive-nomological explanation must contain either a description of the event which is the cause of the event described in the explanandum or it must attribute to a class of objects which contains at least one object in the event described in the explanandum a property had essentially by that class of objects.

Peter Achinstein argues that there is a fundamental objection to most models of scientific explanation. Like the covering-law model of explanation, most models of explanation insist that singular sentences in the explanans cannot by themselves entail the explanandum (the NES requirement). This is in part why the covering-law model requires that the explanans contain laws essential for the explanation. Secondly, like the covering-law model of explanation, most models of explanation insist that, apart from the question of the truth of the explanans, the satisfactoriness of the explanation is an a priori matter. Achinstein claims that many models (including Hempel's, Salmon's, and Brody's) can be saved from counterexamples only by violating the a priori requirement. Models violating that requirement, he argues, will also violate the NES requirement or will entail as a condition of adequacy that there is some true singular sentence which entails the explanandum. All of this casts doubt upon the original need for covering-laws.

Two final critics of the covering-law model argue that no theory of explanation can be adequate unless it recognizes the essentially pragmatic character of scientific explanations. Michael Scriven argues that causality is indispensable in contemporary scientific practice and that the notion of causality is essentially pragmatic, referring to the solution of a range of particular problems raised in a particular context. Bas van Fraassen argues that explanations involve identifying certain salient factors and that salience is determined by pragmatic concepts of relevance and of contrasting explanations. Consequently, what is or is not an explanation is determined by pragmatic features.

It is clear from this survey of the critics of the original covering-law model of explanation that the classical model advocated by Hempel and Oppenheim is no longer generally accepted and that no alternative model has taken its place as the standard account of explanation. Quite a number of authors seem to feel that a theory of explanation requires invoking concepts of causality and that this concept itself is in need of further explication. Contrast, for example, Brody's metaphysical account of causality with Scriven's very pragmatic account. We turn therefore to an examination of the concept of causality.

Hempel and Oppenheim claim that a set of conditions described in the explanans can be said to be the cause of the event described in the explanandum just in case there are laws which imply that whenever the conditions of that type occur, an event of that kind will also occur. In short, for Hempel and Oppenheim, causes are conditions which are sufficient for the occurrence of the effect. This contrasts with views that see causes as conditions which are necessary for the occurrence of the effect.

J. L. Mackie claims that causes are neither necessary nor sufficient for the occurrence of their effects, but that a cause is a condition of a sort related to necessary and sufficient conditions. He gives a concrete example of a cause to show that it is by itself neither necessary nor sufficient for the effect: It is not necessary because the effect could have been produced in other ways and not sufficient because it would not have produced the effect by itself. Using the notions of necessary condition and of sufficient condition, Mackie defines the notion of an INUS condition, and he then defines the notion of a cause in light of an INUS condition. The reader should carefully examine the details of Mackie's theory of causality.

Hans Reichenbach, who accepts the type of regularity analysis of causality common to both Hempel and Oppenheim and Mackie, argues in a classic essay reprinted in this section that modern science requires that we accept the idea that probability must take the place of causality. He suggests, however, that it should be possible to formulate a regularity approach to all of these concepts. This suggestion has been developed by a number of authors including I. J. Good, P. Suppes, and W. Salmon.

The Classical Approach

Carl G. Hempel and Paul Oppenheim*

STUDIES IN THE LOGIC OF EXPLANATION

§1. Introduction

To explain the phenomena in the world of our experience, to answer the question "Why?" rather than only the question "What?", is one of the foremost objectives of all rational inquiry; and especially, scientific research in its various branches strives to go beyond a mere description of its subject matter by providing an explanation of the phenomena it investigates. While there is rather general agreement about this chief objective of science, there exists considerable difference of opinion as to the function and the essential characteristics of scientific explanation. In the present essay, an attempt will be made to shed some light on these issues by means of an elementary survey of the basic pattern of scientific explanation and a subsequent more rigorous analysis of the concept of law and of the logical structure of explanatory arguments.

The elementary survey is presented in part I of this article; part II contains an analysis of the concept of emergence; in part III, an attempt is made to exhibit and to clarify in a more rigorous manner some of the peculiar and perplexing logical problems to which the familiar elementary analysis of explanation gives rise.

PART I. ELEMENTARY SURVEY OF SCIENTIFIC EXPLANATION

§2. Some Illustrations

A mercury thermometer is rapidly immersed in hot water; there occurs a temporary drop of the mercury column, which is then followed by a swift rise. How is this phenomenon to be explained? The increase in temperature affects at first only the glass tube of the thermometer; it expands and thus provides a larger space for the mercury inside, whose surface therefore drops. As soon as by heat conduction the rise in temperature reaches the mercury, however, the latter expands, and as its coefficient of expansion is considerably larger than that of glass, a rise of the mercury level results. This account consists of statements of two kinds. Those of the first kind indicate certain conditions which are realized prior to, or at the same time as, the phenomenon to

*This paper represents the outcome of a series of discussions among the authors; their individual contributions cannot be separated in detail. The technical developments contained in Part IV, however, are due to the first author, who also put the article into its final form. [Parts II and IV omitted in this reprinting.]

We wish to express our thanks to Dr. Rudolf Carnap, Dr. Herbert Feigl, Dr. Nelson Goodman, and Dr. W. V. Quine for stimulating discussions and constructive criticism.

be explained; we shall refer to them briefly as antecedent conditions. In our illustration, the antecedent conditions include, among others, the fact that the thermometer consists of a glass tube which is partly filled with mercury, and that it is immersed into hot water. The statements of the second kind express certain general laws; in our case, these include the laws of the thermic expansion of mercury and of glass, and a statement about the small thermic conductivity of glass. The two sets of statements, if adequately and completely formulated, explain the phenomenon under consideration: They entail the consequence that the mercury will first drop, then rise. Thus, the event under discussion is explained by subsuming it under general laws, i.e., by showing that it occurred in accordance with those laws, by virtue of the realization of certain specified antecedent conditions.

Consider another illustration. To an observer in a row boat, that part of an oar which is under water appears to be bent upwards. The phenomenon is explained by means of general laws—mainly the law of refraction and the law that water is an optically denser medium than air—and by reference to certain antecedent conditions—especially the facts that part of the oar is in the water, part in the air, and that the oar is practically a straight piece of wood. Thus, here again, the question "*Why* does the phenomenon happen?" is construed as meaning "According to what general laws, and by virtue of what antecedent conditions, does the phenomenon occur?"

So far, we have considered exclusively the explanation of particular events occurring at a certain time and place. But the question "Why?" may be raised also in regard to general laws. Thus, in our last illustration, the question might be asked: Why does the propagation of light conform to the law of refraction? Classical physics answers in terms of the undulatory theory of light, i.e., by stating that the propagation of light is a wave phenomenon of a certain general type and that all wave phenomena of that type satisfy the law of refraction. Thus, the explanation of a general regularity consists in subsuming it under another, more comprehensive regularity, under a more general law. Similarly, the validity of Galileo's law for the free fall of bodies near the earth's surface can be explained by deducing it from a more comprehensive set of laws, namely Newton's laws of motion and his law of gravitation, together with some statements about particular facts, namely the mass and the radius of the earth.

§3. The Basic Pattern of Scientific Explanation

From the preceding sample cases let us now abstract some general characteristics of scientific explanation. We divide an explanation into two major constituents, the *explanandum* and the *explanans.*[1] By the *explanandum,* we understand the sentence describing the phenomenon to be explained (not that phenomenon itself); by the *explanans,* the class of those sentences which are adduced to account for the phenomenon. As was noted before, the explanans falls into two subclasses; one of these contains certain sentences $C_1, C_2, \ldots, C_k$ which state specific antecedent conditions; the other is a set of sentences $L_1, L_2, \ldots, L_r$ which represent general laws.

If a proposed explanation is to be sound, its constituents have to satisfy certain conditions of adequacy, which may be divided into logical and empirical conditions. For the following discussion, it will be sufficient to formulate these requirements in a slightly vague manner; in part III, a more rigorous analysis and a more precise restatement of these criteria will be presented.

I. *Logical conditions of adequacy*

(R_1) The explanandum must be a logical consequence of the explanans; in other words, the explanandum must be logically deducible from the information contained in the explanans, for otherwise, the explanans would not constitute adequate grounds for the explanandum.

(R_2) The explanans must contain general laws, and these must actually be required for the derivation of the explanandum. We shall not make it a necessary condition for a sound explanation, however, that the explanans must contain at least one statement which is not a law; for, to mention just one reason, we would surely want to consider as an explanation the derivation of the general regularities governing the motion of double stars from the laws of celestial mechanics, even though all the statements in the explanans are general laws.

(R_3) The explanans must have empirical content; i.e., it must be capable, at least in principle, of test by experiment or observation. This condition is implicit in (R_1); for since the explanandum is assumed to describe some empirical phenomenon, it follows from (R_1) that the explanans entails at least one consequence of empirical character, and this fact confers upon it testability and empirical content. But the point deserves special mention because, as will be seen in section 4, certain arguments which have been offered as explanations in the natural and in the social sciences violate this requirement.

II. *Empirical condition of adequacy*

(R_4) The sentences constituting the explanans must be true. That in a sound explanation, the statements constituting the explanans have to satisfy some condition of factual correctness is obvious. But it might seem more appropriate to stipulate that the explanans has to be highly confirmed by all the relevant evidence available rather than that it should be true. This stipulation, however, leads to awkward consequences. Suppose that a certain phenomenon was explained at an earlier stage of science, by means of an explanans which was well supported by the evidence then at hand, but which had been highly disconfirmed by more recent empirical findings. In such a case, we would have to say that originally the explanatory account was a correct explanation, but that it ceased to be one later, when unfavorable evidence was discovered. This does not appear to accord with sound common usage, which directs us to say that on the basis of the limited initial evidence, the truth of the explanans, and thus the soundness of the explanation, had been quite probable, but that the ampler evidence now available made it highly probable that the explanans was not true, and hence that the account in question was not—and had never been—a correct explanation. (A similar point will be made illustrated, with respect to the requirement of truth for laws, in the beginning of section 6.)

Some of the characteristics of an explanation which have been indicated so far may be summarized in the following schema:

Logical deduction	$C_1, C_2, \ldots, C_k$	Statements of antecedent conditions	Explanans
	$L_1, L_2, \ldots, L_r$	General Laws	
→	E	Description of the empirical phenomenon to be explained	Explanandum

Let us note here that the same formal analysis, including the four necessary conditions, applies to scientific prediction as well as to explanation. The difference between the two is of a pragmatic character. If E is given, i.e., if we know that the phenomenon described by E has occurred, and a suitable set of statements $C_1, C_2, \ldots, C_k, L_1, L_2, \ldots, L_r$ is provided afterwards, we speak of an explanation of the phenomenon in question. If the latter statements are given and E is derived prior to the occurrence of the phenomenon it describes, we speak of a prediction. It may be said, therefore, that an explanation is not fully adequate unless its explanans, if taken account of in time, could have served as a basis for predicting the phenomenon under consideration.[2] Consequently, whatever will be said in this article concerning the logical characteristics of explanation or prediction will be applicable to either, even if only one of them should be mentioned.

It is this potential predictive force which gives scientific explanation its importance: Only to the extent that we are able to explain empirical facts can we attain the major objective of scientific research, namely not merely to record the phenomena of our experience, but to learn from them, by basing upon them theoretical generalizations which enable us to anticipate new occurrences and to control, at least to some extent, the changes in our environment.

Many explanations which are customarily offered, especially in prescientific discourse, lack this predictive character, however. Thus, it may be explained that a car turned over on the road "because" one of its tires blew out while the car was travelling at high speed. Clearly, on the basis of just this information, the accident could not have been predicted, for the explanans provides no explicit general laws by means of which the prediction might be effected, nor does it state adequately the antecedent conditions which would be needed for the prediction. The same point may be illustrated by reference to W. S. Jevons's view that every explanation consists in pointing out a resemblance between facts, and that in some cases this process may require no reference to laws at all and "may involve nothing more than a single identity, as when we explain the appearance of shooting stars by showing that they are identical with portions of a comet."[3] But clearly, this identity does not provide an explanation of the phenomenon of shooting stars unless we presuppose the laws governing the development of heat and light as the effect of friction. The observation of similarities has explanatory value only if it involves at least tacit reference to general laws.

In some cases, incomplete explanatory arguments of the kind here illustrated suppress parts of the explanans simply as "obvious"; in other cases, they seem to involve the assumption that while the missing parts are not obvious, the incomplete explanans could at least, with appropriate effort, be so supplemented as to make a strict derivation of the explanandum possible. This assumption may be justifiable in some cases, as when we say that a lump of sugar disappeared "because" it was put into hot tea, but it is surely not satisified in many other cases. Thus, when certain peculiarities in the work of an artist are explained as outgrowths of a specific type of neurosis, this observation may contain significant clues, but in general it does not afford a sufficient basis for a potential prediction of those peculiarities. In cases of this kind, an incomplete explanation may at best be considered as indicating some positive correlation between the antecedent conditions adduced and the type of phenomenon to be explained, and as pointing out a direction in which further research might be carried on in order to complete the explanatory account.

The type of explanation which has been considered here so far is often referred to as causal explanation. If E describes a particular event, then the antecedent circumstances described in the sentences C_1, C_2,

. . . , C_k may be said jointly to "cause" that event, in the sense that there are certain empirical regularities, expressed by the laws $L_1, L_2, \ldots, L_r$, which imply that whenever conditions of the kind indicated by $C_1, C_2 \ldots, C_k$ occur, an event of the kind described in E will take place. Statements such as $L_1, L_2, \ldots, L_r$, which assert general and unexceptional connections between specified characteristics of events, are customarily called causal, or deterministic laws. They are to be distinguished from the so-called statistical laws which assert that in the long run, an explicitly stated percentage of all cases satisfying a given set of conditions are accompanied by an event of a certain specified kind. Certain cases of scientific explanation involve "subsumption" of the explanandum under a set of laws of which at least some are statistical in character. Analysis of the peculiar logical structure of that type of subsumption involves difficult special problems. The present essay will be restricted to an examination of the causal type of explanation, which has retained its significance in large segments of contemporary science, and even in some areas where a more adequate account calls for reference to statistical laws.[4]

§4. Explanation in the Non-Physical Sciences. Motivational and Teleological Approaches

Our characterization of scientific explanation is so far based on a study of cases taken from the physical sciences. But the general principles thus obtained apply also outside this area.[5] Thus, various types of behavior in laboratory animals and in human subjects are explained in psychology by subsumption under laws or even general theories of learning or conditioning; and while frequently, the regularities invoked cannot be stated with the same generality and precision as in physics or chemistry, it is clear, at least, that the general character of those explanations conforms to our earlier characterization.

Let us now consider an illustration involving sociological and economic factors. In the fall of 1946, there occurred at the cotton exchanges of the United States a price drop which was so severe that the exchanges in New York, New Orleans, and Chicago had to suspend their activities temporarily. In an attempt to explain this occurrence, newspapers traced it back to a large-scale speculator in New Orleans who had feared his holdings were too large and had therefore begun to liquidate his stocks; smaller speculators had then followed his example in a panic and had thus touched off the critical decline. Without attempting to assess the merits of the argument, let us note that the explanation here suggested again involves statements about antecedent conditions and the assumption of general regularities. The former include the facts that the first speculator had large stocks of cotton, that there were smaller speculators with considerable holdings, that there existed the institution of the cotton exchanges with their specific mode of operation, etc. The general regularities referred to are—as often in semi-popular explanations—not explicitly mentioned; but there is obviously implied some form of the law of supply and demand to account for the drop in cotton prices in terms of the greatly increased supply under conditions of practically unchanged demand; besides, reliance is necessary on certain regularities in the behavior of individuals who are trying to preserve or improve their economic position. Such laws cannot be formulated at present with satisfactory precision and generality, and therefore, the suggested explanation is surely incomplete, but its intention is unmistakably to account for the phenomenon by integrating it into a general pattern of economic and socio-psychological regularities.

We turn to an explanatory argument taken from the field of linguistics.[6] In Northern France, there exists a large variety of words synonymous with the English "bee," whereas in southern France, essentially only one such word is in existence. For this discrepancy, the explanation has been suggested that in the Latin epoch, the South of France used the word "apicula," the North the word "apis." The latter, because of a process of phonologic decay in northern France, became the monosyllabic word "é"; and monosyllables tend to be eliminated, especially if they contain few consonantic elements, for they are apt to give rise to misunderstandings. Thus, to avoid confusion, other words were selected. But "apicula," which was reduced to "abelho," remained clear enough and was retained, and finally it even entered into the standard language, in the form "abbeille." While the explanation here described is incomplete in the sense characterized in the previous section, it clearly exhibits reference to specific antecedent conditions as well as to general laws.[7]

While illustrations of this kind tend to support the view that explanation in biology, psychology, and the social sciences has the same structure as in the physical sciences, the opinion is rather widely held that in many instances, the causal type of explanation is essentially inadequate in fields other than physics and chemistry, and especially in the study of purposive behavior. Let us examine briefly some of the reasons which have been adduced in support of this view.

One of the most familiar among them is the idea that events involving the activities of humans singly or in groups have a peculiar uniqueness and irrepeatability which makes them inaccessible to causal explanation because the latter, with its reliance upon uniformities, presupposes repeatability of the phenomena under consideration. This argument which, incidentally, has also been used in support of the contention that the experimental method is inapplicable in psychology and the social sciences, involves a misunderstanding of the logical character of causal explanation. Every individual event, in the physical sciences no less than in psychology or the social sciences, is unique in the sense that it, with all its peculiar characteristics, does not repeat itself. Nevertheless, individual events may conform to, and thus be explainable by means of, general laws of the causal type. For all that a causal law asserts is that any event of a specified kind, i.e., any event having certain specified characteristics, is accompanied by another event which in turn has certain specified characteristics; for example, that in any event involving friction, heat is developed. And all that is needed for the testability and applicability of such laws is the recurrence of events with the antecedent characteristics, i.e., the repetition of those characteristics, but not of their individual instances. Thus, the argument is inconclusive. It gives occasion, however, to emphasize an important point concerning our earlier analysis: When we spoke of the explanation of a single event, the term "event" referred to the occurrence of some more or less complex characteristic in a specific spatio-temporal location or in a certain individual object, and not to *all* the characteristics of that object, or to all that goes on in that space-time region.

A second argument that should be mentioned here[8] contends that the establishment of scientific generalizations—and thus of explanatory principles—for human behavior is impossible because the reactions of an individual in a given situation depend not only upon that situation, but also upon the previous history of the individual. But surely, there is no a priori reason why generalizations should not be attainable which take into account this dependence of behavior on the past history of the agent. That indeed the given argument "proves" too much, and is therefore a *non sequitur*, is made evident by the existence of certain physical phenomena, such as magnetic hysteresis and elastic fatigue, in which the magnitude of a specific physical effect depends upon the past history of the system involved, and for which nevertheless certain general regularities have been established.

A third argument insists that the explanation of any phenomenon involving purposive behavior calls for reference to motivations and thus for teleological rather than causal analysis. Thus, for example, a fuller statement of the suggested explanation for the break in the cotton prices would have to indicate the large-scale speculator's motivations as one of the factors determining the event in question. Thus, we have to refer to goals sought, and this, so the argument runs, introduces a type of explanation alien to the physical sciences. Unquestionably, many of the—frequently incomplete—explanations which are offered for human actions involve reference to goals and motives; but does this make them essentially different from the causal explanations of physics and chemistry? One difference which suggests itself lies in the circumstance that in motivated behavior, the future appears to affect the present in a manner which is not found in the causal explanations of the physical sciences. But clearly, when the action of a person is motivated, say, by the desire to reach a certain objective, then it is not the as yet unrealized future event of attaining that goal which can be said to determine his present behavior, for indeed the goal may never be actually reached; rather—to put it in crude terms—it is (a) his desire, present before the action, to attain that particular objective, and (b) his belief, likewise present before the action, that such and such a course of action is most likely to have the desired effect. The determining motives and beliefs, therefore, have to be classified among the antecedent conditions of a motivational explanation, and there is no formal difference on this account between motivational and causal explanation.

Neither does the fact that motives are not accessible to direct observation by an outside observer constitute an essential difference between the two kinds of explanation; for also the determining factors adduced in physical explanations are very frequently inaccessible to direct observation. This is the case, for instance, when opposite electric charges are adduced in explanation of the mutual attraction of two metal spheres. The presence of those charges, while eluding all direct observation, can be ascertained by various kinds of indirect test, and that is sufficient to guarantee the empirical character of the explanatory statement. Similarly, the presence of certain motivations may be ascertainable only by indirect methods, which may include reference to linguistic utterances of the subject in question, slips of the pen or of the tongue, etc.; but as long as these methods are "operationally determined" with reasonable clarity and precision, there is no essential difference in this respect between motivational explanation and causal explanation in physics.

A potential danger of explanation by motives lies in the fact that the method lends itself to the facile construction of *ex post facto* accounts without predictive force. It is a widespread tendency to "explain" an action by ascribing it to motives conjectured only after the action has taken place. While this procedure is not in itself objectionable, its soundness requires that (1) the motivational assumptions in question be capable of test, and (2) that suitable general laws be available to lend explanatory power to the assumed motives. Disregard of these requirements frequently deprives alleged motivational explanations of their cognitive significance.

The explanation of an action in terms of the motives of the agent is sometimes considered as a special kind of teleological explanation. As was pointed out above, motivational explanation, if adequately formulated, conforms to the conditions for causal explanation, so that the term "teleological" is a misnomer if it is meant to imply either a non-causal character of the explanation or peculiar determination of the present by the future. If this is borne in mind, however, the term "teleological" may be viewed, in this context, as referring to causal explanations in which some of the antecedent conditions are motives of the agent whose actions are to be explained.[9]

Teleological explanations of this kind

have to be distinguished from a much more sweeping type, which has been claimed by certain schools of thought to be indispensable especially in biology. It consists in explaining characteristics of an organism by reference to certain ends or purposes which the characteristics are said to serve. In contradistinction to the cases examined before, the ends are not assumed here to be consciously or subconsciously pursued by the organism in question. Thus, for the phenomenon of mimicry, the explanation is sometimes offered that it serves the purpose of protecting the animals endowed with it from detection by its pursuers and thus tends to preserve the species. Before teleological hypotheses of this kind can be appraised as to their potential explanatory power, their meaning has to be clarified. If they are intended somehow to express the idea that the purposes they refer to are inherent in the design of the universe, then clearly they are not capable of empirical test and thus violate the requirement (R_3) stated in section 3. In certain cases, however, assertions about the purposes of biological characteristics may be translatable into statements in non-teleological terminology which assert that those characteristics function in a specific manner which is essential to keeping the organism alive or to preserving the species.[10] An attempt to state precisely what is meant by this latter assertion—or by the similar one that without those characteristics, and other things being equal, the organism or the species would not survive—encounters considerable difficulties. But these need not be discussed here. For even if we assume that biological statements in teleological form can be adequately translated into descriptive statements about the life-preserving function of certain biological characteristics, it is clear that (1) the use of the concept of purpose is not essential in these contexts, since the term "purpose" can be completely eliminated from the statements in question, and (2) teleological assumptions, while now endowed with empirical content, cannot serve as explanatory principles in the customary contexts. Thus, e.g., the fact that a given species of butterflies displays a particular kind of coloring cannot be inferred from—and therefore cannot be explained by means of—the statement that this type of coloring has the effect of protecting the butterflies from detection by pursuing birds, nor can the presence of red corpuscles in the human blood be inferred from the statement that those corpuscles have a specific function in assimilating oxygen and that this function is essential for the maintenance of life.

One of the reasons for the perseverance of teleological considerations in biology probably lies in the fruitfulness of the teleological approach as a heuristic device: Biological research which was psychologically motivated by a teleological orientation, by an interest in purposes in nature, has frequently led to important results which can be stated in non-teleological terminology and which increase our scientific knowledge of the causal connections between biological phenomena.

Another aspect that lends appeal to teleological considerations is their anthropomorphic character. A teleological explanation tends to make us feel that we really "understand" the phenomenon in question, because it is accounted for in terms of purposes, with which we are familar from our own experience of purposive behavior. But it is important to distinguish here understanding in the psychological sense of a feeling of empathic familiarity from understanding in the theoretical, or cognitive, sense of exhibiting the phenomenon to be explained as a special case of some general regularity. The frequent insistence that explanation means the reduction of something unfamiliar to ideas or experiences already familiar to us is indeed misleading. For while some scientific explanations do have this psychological effect, it is by no means universal: The free fall of a physical body may well be said to be a more familiar phenomenon than the law of gravitation, by means of which it can be explained; and surely the basic ideas of the theory of relativity will appear to many to be far less familiar than the phenomena for which the theory accounts.

"Familiarity" of the explicans is not only not necessary for a sound explanation (as we have just tried to show), but it is not sufficient either. This is shown by the many cases in which a proposed explicans sounds suggestively familiar, but upon closer inspection proves to be a mere metaphor, or an account lacking testability, or a set of statements which includes no general laws and therefore lacks explanatory power. A case in point is the neovitalistic attempt to explain biological phenomena by reference to an entelechy or vital force. The crucial point here is not—as it is sometimes made out to be—that entelechies cannot be seen or otherwise directly observed; for that is true also of gravitational fields, and yet, reference to such fields is essential in the explanation of various physical phenomena. The decisive difference between the two cases is that the physical explanation provides (1) methods of testing, albeit indirectly, assertions about gravitational fields, and (2) general laws concerning the strength of gravitational fields, and the behavior of objects moving in them. Explanations by entelechies satisfy the analogue of neither of these two conditions. Failure to satisfy the first condition represents a violation of (R_3); it renders all statements about entelechies inaccessible to empirical test and thus devoid of empirical meaning. Failure to comply with the second condition involves a violation of (R_2). It deprives the concept of entelechy of all explanatory import; for explanatory power never resides in a concept, but always in the general laws in which it functions. Therefore, notwithstanding the flavor of familiarity of the metaphor it invokes, the neovitalistic approach cannot provide theoretical understanding.

The preceding observations about familiarity and understanding can be applied, in a similar manner, to the view held by some scholars that the explanation, or the understanding, of human actions requires an empathic understanding of the personalities of the agents.[11] This understanding of another person in terms of one's own psychological functioning may prove a useful heuristic device in the search for general psychological principles which might provide a theoretical explanation; but the existence of empathy on the part of the scientist is neither a necessary nor a sufficient condition for the explanation, or the scientific understanding, of any human action. It is not necessary, for the behavior of psychotics or of people belonging to a culture very different from that of the scientist may sometimes be explainable and predictable in terms of general principles even though the scientist who establishes or applies those principles may not be able to understand his subjects empathically. And empathy is not sufficient to guarantee a sound explanation, for a strong feeling of empathy may exist even in cases where we completely misjudge a given personality. Moreover, as the late Dr. Zilsel has pointed out, empathy leads with ease to incompatible results; thus, when the population of a town has long been subjected to heavy bombing attacks, we can understand, in the empathic sense, that its morale should have broken down completely, but we can understand with the same ease also that it should have developed a defiant spirit of resistance. Arguments of this kind often appear quite convincing; but they are of an *ex post facto* character and lack cognitive significance unless they are supplemented by testable explanatory principles in the form of laws or theories.

Familiarity of the explanans, therefore, no matter whether it is achieved through the use of teleological terminology, through neovitalistic metaphors, or through other means, is no indication of the cognitive import and the predictive force of a proposed explanation. Besides, the extent to which an idea will be considered as familiar varies from person to person and from time to time, and a psychological factor of this kind certainly cannot serve as a standard in assessing the worth of a proposed explanation. The decisive requirement for every sound explanation remains that it subsume the explanandum under general laws.

* * *

PART III. LOGICAL ANALYSIS OF LAW AND EXPLANATION

§6. Problems of the Concept of General Law

From our general survey of the characteristics of scientific explanation, we now turn to a closer examination of its logical structure. The explanation of a phenomenon, we noted, consists in its subsumption under laws or under a theory. But what is a law? What is a theory? While the meaning of these concepts seems intuitively clear, an attempt to construct adequate explicit definitions for them encounters considerable difficulties. In the present section, some basic problems of the concept of law will be described and analyzed; in the next section, we intend to propose, on the basis of the suggestions thus obtained, definitions of law and of explanation for a formalized model language of a simple logical structure.

The concept of law will be construed here so as to apply to true statements only. The apparently plausible alternative procedure of requiring high confirmation rather than truth of a law seems to be inadequate: It would lead to a relativized concept of law, which would be expressed by the phrase, "Sentence *S* is a law relatively to the evidence *E*." This does not seem to accord with the meaning customarily assigned to the concept of law in science and in methodological inquiry. Thus, for example, we would not say that Bode's general formula for the distance of the planets from the sun was a law relatively to the astronomical evidence available in the 1770's, when Bode propounded it, and that it ceased to be a law after the discovery of Neptune and the determination of its distance from the sun; rather, we would say that the limited original evidence had given a high probability to the assumption that the formula was a law, whereas more recent additional information reduced that probability so much as to make it practically certain that Bode's formula is not generally true, and hence not a law.[12]

Apart from being true, a law will have to satisfy a number of additional conditions. These can be studied independently of the factual requirement of truth, for they refer, as it were, to all logically possible laws, no matter whether factually true or false. Adopting a convenient term proposed by Goodman,[13] we will say that a sentence is lawlike if it has all the characteristics of a general law, with the possible exception of truth. Hence, every law is a lawlike sentence, but not conversely.

Our problem of analyzing the concept of law thus reduces to that of explicating the meaning of *lawlike sentence*. We shall construe the class of lawlike sentences as including analytic general statements, such as "A rose is a rose," as well as the lawlike sentences of empirical science, which have empirical content.[14] It will not be necessary to require that each lawlike sentence permissible in explanatory contexts be of the second kind; rather, our definition of explanation will be so constructed as to guarantee the factual character of the totality of the laws—though not of every single one of them—which function in an explanation of an empirical fact.

What are the characteristics of lawlike sentences? First of all, lawlike sentences are statements of universal form, such as "All robins' eggs are greenish-blue," "All metals are conductors of electricity," "At constant pressure, any gas expands with increasing temperature." As these examples illustrate, a lawlike sentence usually is not only of universal, but also of conditional form; it makes an assertion to the effect that universally, if a certain set of conditions, *C*, is realized, then another specified set of conditions, *E*, is realized as well. The standard form for the symbolic expression of a lawlike sentence is therefore the universal conditional. However, since any conditional statement can be transformed into a non-conditional one, conditional form will not be considered as essential for a lawlike sentence, while universal character will be held indispensable.

But the requirement of universal form is not sufficient to characterize lawlike sentences. Suppose, for example, that a certain basket, *b*, contains at a certain time *t* a number of red apples and nothing else.[15] Then the statement

(S_1) Every apple in basket *b* at time *t* is red.

is both true and of universal form. Yet the sentence does not qualify as a law; we would refuse, for example, to explain by subsumption under it the fact that a particular apple chosen at random from the basket is red. What distinguishes (S_1) from a lawlike sentence? Two points suggest themselves, which will be considered in turn, namely, finite scope, and reference to a specified object.

First, the sentence (S_1) makes, in effect, an assertion about a finite number of objects only, and this seems irreconcilable with the claim to universality which is commonly associated with the notion of law.[16] But are not Kepler's laws considered as lawlike although they refer to a finite set of planets only? And might we not even be willing to consider as lawlike a sentence such as the following?

(S_2) All the sixteen ice cubes in the freezing tray of this refrigerator have a temperature of less than 10 degrees centigrade.

This point might well be granted; but there is an essential difference between (S_1) on the one hand and Kepler's laws as well as (S_2) on the other: The latter, while finite in scope, are known to be consequences of more comprehensive laws whose scope is not limited, while for (S_1) this is not the case.

Adopting a procedure recently suggested by Reichenbach,[17] we will therefore distinguish between fundamental and derivative laws. A statement will be called a derivative law if it is of universal character and follows from some fundamental laws. The concept of fundamental law requires further clarification; so far, we may say that fundamental laws, and similarly fundamental lawlike sentences, should satisfy a certain condition of non-limitation of scope.

It would be excessive, however, to deny the status of fundamental lawlike sentence to all statements which, in effect, make an assertion about a finite class of objects only, for that would rule out also a sentence such as "All robins' eggs are greenish-blue," since presumably the class of all robins' eggs—past, present, and future—is finite. But again, there is an essential difference between this sentence and, say, (S_1). It requires empirical knowledge to establish the finiteness of the class of robins' eggs, whereas, when the sentence (S_1) is construed in a manner which renders it intuitively unlawlike, the terms "basket *b*" and "apple" are understood so as to imply finiteness of the class of apples in the basket at time *t*. Thus, so to speak, the meaning of its constitutive terms alone—without additional factual information—entails that (S_1) has a finite scope. Fundamental laws, then, will have to be construed so as to satisfy what we have called a condition of non-limited scope; our formulation of that condition however, which refers to what is entailed by "the meaning" of certain expressions, is too vague and will have to be revised later. Let us note in passing that the stipulation here envisaged would bar from the class of fundamental lawlike sentences also such undesirable candidates as "All uranic objects are spherical," where "uranic" means the property of being the planet Uranus; indeed, while this sentence has universal form, it fails to satisfy the condition of non-limited scope.

In our search for a general characterization of lawlike sentences, we now turn to a second clue which is provided by the sentence (S_1). In addition to violating the condition of non-limited scope, this sentence has the peculiarity of making reference to a particular object, the basket *b;* and this, too, seems to violate the universal character of a law.[18] The restriction which seems indicated here should however again be applied to fundamental lawlike sentences only; for a true general statement about the free fall of physical bodies on the moon, while referring to a particular object, would still constitute a law, albeit a derivative one.

It seems reasonable to stipulate, therefore, that a fundamental lawlike sentence must be of universal form and must contain no essential—i.e., uneliminable—occurrences of designations for particular objects. But this is not sufficient; indeed, just at this point, a particularly serious difficulty presents itself. Consider the sentence:

(S_3) Everything that is either an apple in basket *b* at time *t* or a sample of ferric oxide is red.

If we use a special expression, say "*x* is ferple," as synonymous with "*x* is either an apple in *b* at *t* or a sample of ferric oxide," then the content of (S_3) can be expressed in the form:

(S_4) Everything that is ferple is red.

The statement thus obtained is of universal form and contains no designations of particular objects, and it also satisfies the condition of non-limited scope; yet clearly, (S_4) can qualify as a fundamental lawlike sentence no more than can (S_3).

As long as "ferple" is a defined term of our language, the difficulty can readily be met by stipulating that after elimination of defined terms, a fundamental lawlike sentence must not contain essential occurrences of designations for particular objects. But this way out is of no avail when "ferple," or another term of the kind illustrated by it, is a primitive predicate of the language under consideration. This reflection indicates that certain restrictions have to be imposed upon those predicates, i.e., terms for properties or relations, which may occur in fundamental lawlike sentences.[19]

More specifically, the idea suggests itself of permitting a predicate in a fundamental lawlike sentence only if it is purely universal, or, as we shall say, purely qualitative, in character; in other words, if a statement of its meaning does not require reference to any one particular object or spatio-temporal location. Thus, the terms "soft," "green," "warmer than," "as long as," "liquid," "electrically charged," "female," "father of" are purely qualitative predicates, while "taller than the Eiffel Tower," "medieval," "lunar," "arctic," "Ming" are not.[20]

Exclusion from fundamental, lawlike sentences of predicates which are not purely qualitative would at the same time ensure satisfaction of the condition of non-limited scope; for the meaning of a purely qualitative predicate does not require a finite extension; and indeed, all the sentences considered above which violate the condition of non-limited scope make explicit or implicit reference to specific objects.

The stipulation just proposed suffers, however, from the vagueness of the concept of purely qualitative predicate. The question whether indication of the meaning of a given predicate in English does or does not require reference to some one specific object does not always permit an unequivocal answer since English as a natural language does not provide explicit definitions or other clear explications of meaning for its terms. It seems therefore reasonable to attempt definition of the concept of law not with respect to English or any other natural language, but rather with respect to a formalized language—let us call it a model language, *L*,—which is governed by a well-determined system of logical rules, and in which every term either is characterized as primitive or is introduced by an explicit definition in terms of the primitives.

This reference to a well-determined system is customary in logical research and is indeed quite natural in the context of any attempt to develop precise criteria for certain logical distinctions. But it does not by itself suffice to overcome the specific difficulty under discussion. For while it is now readily possible to characterize as not purely qualitative all those among the defined predicates in *L* whose definiens contain an essential occurrence of some individual name, our problem remains open for the primitives of the language, whose meanings are not determined by definitions within the language, but rather by semantical rules of interpretation. For we want to permit the interpretation of the primitives of *L* by means of such attributes as blue, hard, solid, warmer, but not by the properties of being a descendant of Napoleon, or an arctic animal or a Greek statue; and the difficulty is precisely that of stating rigorous criteria for the distinction between the permissible and the non-permissible interpretations. Thus, the problem of setting up an adequate definition for purely qualitative attributes now arises again; namely for the concepts of the meta-language in which the semantical interpretation of the primitives is formulated. We may postpone an encounter with the difficulty by presupposing formalization of the semantical meta-language, the meta-meta-language, and so forth; but somewhere, we will have to stop at a non-formalized meta-language, and for it a characterization of purely qualitative predicates will be needed and will present much the same problems as non-formalized English, with which we began. The characterization of a purely qualitative predicate as one whose meaning can be made explicit without reference to any one particular object points to the intended meaning but does not explicate it precisely, and the problem of an adequate definition of purely qualitative predicates remains open.

There can be little doubt, however, that there exists a large number of property and relation terms which would be rather generally recognized as purely qualitative in the sense here pointed out, and as permissible in the formulation of fundamental lawlike sentences; some examples have been given above, and the list could be readily enlarged. When we speak of purely qualitative predicates, we shall henceforth have in mind predicates of this kind.

* * *

NOTES

1. These two expressions, derived from the Latin *explanare,* were adopted in preference to the perhaps more customary terms "explicandum" and "explicans" in order to reserve the latter for use in the context of explication of meaning, or analysis. On explication in this sense, cf. Carnap [Concepts], p. 513. Abbreviated titles in brackets refer to the bibliography at the end of this article.

2. The logical similarity of explanation and prediction, and the fact that one is directed towards past occurrences, the other towards future ones, is well expressed in the terms "postdictability" and "predictability" used by Reichenbach in [Quantum Mechanics], p. 13.

3. [Principles], p. 533.

4. The account given above of the general characteristics of explanation and prediction in science is by no means novel; it merely summarizes and states explicitly some fundamental points which have been recognized by many scientists and methodologists.

Thus, e.g., Mill says: "An individual fact is said to be explained by pointing out its cause, that is, by stating the law or laws of causation of which its production is an instance," and "a law of uniformity in nature is said to be explained when another law or laws are pointed out, of which that law itself is but a case, and from which it could be deduced." ([Logic], Book III, Chap. xii, Sec. 1). Similarly, Jevons, whose general characterization of explanation was critically discussed above, stresses that "the most important process of explanation consists in showing that an observed fact is one case of a general law or tendency." ([Principles], p. 533.) Ducasse states the same point as follows: "Explanation essentially consists in the offering of a hypothesis of fact, standing to the fact to be explained as case of antecedent to case of consequent of some already known law of connection." ([Explanation], pp. 150–51.) A lucid analysis of the fundamental structure of explanation and prediction was given by Popper in [Forschung], Sec. 12, and, in an improved version, in his work [Society], especially in Chap. xxv and in n. 7 referring to that chapter. For a recent characterization of explanation as subsumption under general theories, cf., for example, Hull's concise discussion in [Principles], Chap. 1. A clear elementary examination of certain aspects of explanation is given in Hospers [Explanation], and a concise survey of many of the essentials of scientific explanation which are considered in the first two parts of the present study may be found in Feigl [Operationism], pp. 284 ff.

5. On the subject of explanation in the social sciences, especially in history, cf. also the following publications, which may serve to supplement and amplify the brief discussion to be presented here: Hempel ["Laws"]; Popper [*Society*]; White ["Explanation"]; and the articles "Cause" and "Understanding" in Beard and Hook [Terminology].

6. The illustration is taken from Bonfante [Semantics], Sec. 3.

7. While in each of the last two illustrations, certain regularities are unquestionably relied upon in the explanatory argument, it is not possible to argue convincingly that the intended laws, which at present cannot all be stated explicitly, are of a causal rather than a statistical character. It is quite possible that most or all of the regularities which will be discovered as sociology develops will be of a statistical type. Cf., on this point, the suggestive observations by Zilsel in [Empiricism] Sec. 8, and [Laws]. This issue does not affect, however, the main point we wish to make here, namely that in the social no less than in the physical sciences, subsumption under general regularities is indispensable for the explanation and the theoretical understanding of any phenomenon.

8. Cf., for example, F. H. Knight's presentation of this argument in [Limitations], pp. 251–52.

9. For a detailed logical analysis of the character and the function of the motivation concept in psychological theory, see Koch [Motivation]. A stimulating discussion of teleological behavior from the standpoint of contemporary physics and biology is contained in the article [Teleology] by Rosenblueth, Wiener and Bigelow. The authors propose an interpretation of the concept of purpose which is free from metaphysical connotations, and they stress the importance of the concept thus obtained for a behavioristic analysis of machines and living organisms. While our formulations above intentionally use the crude terminology frequently applied in philosophical arguments concerning the applicability of causal explanation to purposive behavior, the analysis presented in the article referred to is couched in behavioristic terms and avoids reference to "motives" and the like.

10. An analysis of teleological statements in biology along these lines may be found in Woodger [Principles], especially pp. 432 ff.; essentially the same interpretation is advocated by Kaufmann in [Methodology], Chap. 8.

11. For a more detailed discussion of this view on the basis of the general principles outlined above, cf. Zilsel [Empiricism], Secs. 7 and 8, and Hempel [Laws], Sec. 6.

12. The requirement of truth for laws has the consequence that a given empirical statement S can never be definitely known to be a law; for the sentence affirming the truth of S is logically equivalent with S and is therefore capable only of acquiring a more or less high probability, or degree of confirmation, relatively to the experimental evidence available at any given time. On this point, cf. Carnap [Remarks]. For an excellent non-technical exposition of the semantical concept of truth, which is here applied, the reader is referred to Tarski [Truth].

13. [Counterfactuals], p. 125.

14. This procedure was suggested by Goodman's approach in [Counterfactuals]. Reichenbach, in a detailed examination of the concept of law, similarly construes his concept of nomological statement as including both analytic and synthetic sentences; cf. [Logic], Chap. viii.

15. The difficulty illustrated by this example was stated concisely by Langford [Review], who referred to it as the problem of distinguishing between universals of fact and causal universals. For further discussion and illustration of this point, see also Chisholm [Conditional], especially pp. 301 f. A systematic analysis of the problem was given by Goodman in [Counterfactuals], especially Part III. While not concerned with the specific point under discussion, the detailed examination of counterfactual conditionals and their relation to laws of nature, in Chap. viii of Lewis's work [Analysis], contains important observations on several of the issues raised in the present section.

16. The view that laws should be construed as

not being limited to a finite domain has been expressed, among others, by Popper [Forschung], Sec. 13 and by Reichenbach [Logic], p. 369.

17. [Logic], p. 361. Our terminology as well as the definitions to be proposed later for the two types of law do not coincide with Reichenbach's, however.

18. In physics, the idea that a law should not refer to any particular object has found its expression in the maxim that the general laws of physics should contain no reference to specific space-time points, and that spatio-temporal coordinates should occur in them only in the form of differences or differentials.

19. The point illustrated by the sentences (S_3) and (S_4) above was made by Goodman, who has also emphasized the need to impose certain restrictions upon the predicates whose occurrence is to be permissible in lawlike sentences. These predicates are essentially the same as those which Goodman calls projectible. Goodman has suggested that the problems of establishing precise criteria for projectibility, of interpreting counterfactual conditionals, and of defining the concept of law are so intimately related as to be virtually aspects of a single problem. (Cf. his articles [Query] and [Counterfactuals].) One suggestion for an analysis of projectibility has recently been made by Carnap in [Application]. Goodman's note [Infirmities] contains critical observations on Carnap's proposals.

20. That laws, in addition to being of universal form, must contain only purely universal predicates was clearly argued by Popper ([Forschung], Sec. 14, 15). Our alternative expression "purely qualitative predicate" was chosen in analogy to Carnap's term "purely qualitative property" (cf. [Application]). The above characterization of purely universal predicates seems preferable to a simpler and perhaps more customary one, to the effect that a statement of the meaning of the predicate must require no reference to particular objects. For this formulation might be too exclusive since it could be argued that stating the meaning of such purely qualitative terms as "blue" or "hot" requires illustrative reference to some particular object which has the quality in question. The essential point is that no one specific object has to be chosen; any one in the logically unlimited set of blue or of hot objects will do. In explicating the meaning of "taller than the Eiffel Tower," "being an apple in basket *b* at the time *t*," "medieval," etc., however, reference has to be made to one specific object or to some one in a limited set of objects.

BIBLIOGRAPHY

Throughout the article, the abbreviated titles in brackets are used for reference.

Beard, Charles A., and Hook, Sidney, [Terminology], "Problems of terminology in historical writing." Chap. iv of Theory and practice in historical study: A report of the Committee on Historiography. New York: Social Science Research Council, 1946.

Bergmann, Gustav, [Emergence], "Holism, historicism, and emergence." *Philosophy of Science* II, (1944), 209–21.

Bonfante, G., [Semantics], Semantics, language. An article in P. L. Harriman, ed., *The Encyclopedia of Psychology.* Philosophical Library, New York, 1946.

Broad, C. D., [Mind], *The mind and its place in nature.* New York, 1925.

Carnap, Rudolf, [Semantics], *Introduction to Semantics.* Harvard University Press, 1942.

———. [Inductive Logic], "On Inductive Logic." *Philosophy of science,* XII (1945), 72–97.

———. [Concepts], "The Two Concepts of Probability." *Philosophy and Phenomenological Research,* V (1945), 513–32.

———. [Remarks], "Remarks on Induction and Truth." *Philosophy and Phenomenological Research,* VI (1946), 590–602.

———. [Application], "On the Application of Inductive Logic." *Philosophy and Phenomenological Research,* VIII (1947), 133–47.

Chisholm, Roderick M., [Conditional], "The Contrary-to-Fact Conditional." *Mind,* IV (1946), 289–307.

Church, Alonzo, [Logic], "Logic, Formal." An article in Dagobert D. Runes, ed. *The Dictionary of Philosophy.* Philosophical Library, New York, 1942.

Ducasse, C. J., [Explanation], "Explanation, Mechanism, and Teleology." *The Journal of Philosophy,* XXII (1925), 150–55.

Feigl, Herbert, [Operationism], "Operationism and Scientific Method." *Psychological Review,* LII (1945), 250–59, 284–88.

Goodman, Nelson, [Query], "A Query on

Confirmation." *The Journal of Philosophy,* XLIII (1946), 383–85.

———. [Counterfactuals], "The Problem of Counterfactual Conditionals." *The Journal of Philosophy,* XLIV (1947), 113–28.

———. [Infirmities], "On Infirmities of Confirmation Theory." *Philosophy and Phenomenological Research,* VIII (1947), 149–51.

Grelling, Kurt, and Oppenheim, Paul, [Gestaltbegriff], "Der Gestaltbegriff im Lichte der neuen Logik." *Erkenntnis,* VII (1937–38), 211–25, 357–59.

Grelling, Kurt, and Oppenheim, Paul, [Functional Whole], "Logical Analysis of Gestalt as Functional Whole." Preprinted for distribution at Fifth Internat. Congress for the Unity of Science, Cambridge, Mass., 1939.

Helmer, Olaf, and Oppenheim, Paul, [Probability], "A Syntactical Definition of Probability and of Degree of Confirmation." *The Journal of Symbolic Logic,* X (1945), 25–60.

Hempel, Carl G., [Laws], "The Function of General Laws in History." *The Journal of Philosophy,* XXXIX (1942), 35–48.

———. [Studies], "Studies in the Logic of Confirmation." *Mind,* LIV (1945); Part I: 1–26, Part II: 97–121.

Hempel, Carl G., and Oppenheim, Paul, [Degree], "A Definition of Degree of Confirmation." *Philosophy of Science,* XII (1945), 98–115.

Henle, Paul, [Emergence], "The Status of Emergence." *The Journal of Philosophy,* XXXIX (1942), 486–93.

Hospers, John, [Explanation], "On Explanation." *The Journal of Philosophy,* XLIII (1946), 337–56.

Hull, Clark L., [Variables], "The Problem of Intervening Variables in Molar Behavior Theory." *Psychological Review,* L (1943), 273–91.

———. [Principles] *Principles of Behavior.* New York, 1943.

Jevons, W. Stanley, [Principles], *The Principles of Science.* London, 1924. (1st ed. 1874.)

Kaufmann, Felix, [Methodology], *Methodology of the Social Sciences.* New York, 1944.

Knight, Frank H., [Limitations], "The Limitations of Scientific Method in Economics." In Tugwell, R., ed., *The Trend of Economics.* New York, 1924.

Koch, Sigmund, [Motivation], "The Logical Character of the Motivation Concept." *Psychological Review,* XLVIII (1941). Part I: 15–38, Part II: 127–54.

Langford, C. H., [Review], "Review" *in The Journal of Symbolic Logic,* VI (1941), 67–68.

Lewis, C. I., [Analysis], *An Analysis of Knowledge and Valuation.* La Salle, Ill., 1946.

McKinsey, J. C. C., [Review], "Review of Helmer and Oppenheim" [Probability]. *Mathematical Reviews,* VII (1946), 45.

Mill, John Stuart, [Logic], *A System of Logic.*

Morgan, C. Lloyd, *Emergent Evolution.* New York, 1923.

———. *The Emergence of Novelty.* New York, 1933.

Popper, Karl, [Forschung], *Logik der Forschung.* Wien, 1935.

———. [Society], *The Open Society and its Enemies.* London, 1945.

Reichenbach, Hans, [Logic], *Elements of Symbolic Logic.* New York, 1947.

———. [Quantum mechanics], *Philosophic Foundations of Quantum Mechanics.* University of California Press, 1944.

Rosenblueth, A., Wiener, N., and Bigelow, J., [Teleology], "Behavior, Purpose, and Teleology." *Philosophy of Science,* X (1943), 18–24.

Stace, W. T., [Novelty], "Novelty, Indeterminism and Emergence." *Philosophical Review,* XLVIII (1939), 296–310.

Tarski, Alfred, [Truth], "The Semantical Conception of Truth, and the Foundations of Semantics." *Philosophy and Phenomenological Research,* IV (1944), 341–76.

Tolman, Edward Chase, [Behavior], "Purposive Behavior in Animals and Men." New York, 1932.

White, Morton G., [Explanation], Historical Explanation. *Mind,* LII (1943), 212–29.

Woodger, J. H., [Principles], *Biological Principles.* New York, 1929.

Zilsel, Edgar, [Empiricism], "Problems of Empiricism." In *International Encyclopedia of Unified Science,* II, No. 8. The University of Chicago Press, 1941.

———. [Laws], "Physics and the Problem of Historico-Sociological Laws." *Philosophy of Science,* VIII (1941), 567–79.

Alternative Models

Wesley C. Salmon

STATISTICAL EXPLANATION AND ITS MODELS

THE PHILOSOPHICAL THEORY of scientific explanation first entered the twentieth century in 1962, for that was the year of publication of the earliest bona fide attempt to provide a systematic account of statistical explanation in science.[1] Although the need for some sort of inductive or statistical form of explanation had been acknowledged earlier, Hempel's essay "Deductive-Nomological vs. Statistical Explanation" (1962) contained the first sustained and detailed effort to provide a precise account of this mode of scientific explanation. Given the pervasiveness of statistics in virtually every branch of contemporary science, the late arrival of statistical explanation in philosophy of science is remarkable. Hempel's initial treatment of statistical explanation had various defects, some of which he attempted to rectify in his comprehensive essay "Aspects of Scientific Explanation" (1965a). Nevertheless, the earlier article did show unmistakably that the construction of an adequate model for statistical explanation involves many complications and subtleties that may have been largely unanticipated. Hempel never held the view—expressed by some of the more avid devotees of the D-N model—that *all* adequate scientific explanations must conform to the deductive-nomological pattern. The 1948 Hempel-Oppenheim paper explicitly notes the need for an inductive or statistical model of scientific explanation in order to account for some types of legitimate explanation that actually occur in the various sciences (Hempel, 1965, pp. 250–251). The task of carrying out the construction was, however, left for another occasion. Similar passing remarks about the need for inductive or statistical accounts were made by other authors as well, but the project was not undertaken in earnest until 1962—a striking delay of fourteen years after the 1948 essay.

One can easily form the impression that philosophers had genuine feelings of ambivalence about statistical explanation. A vivid example can be found in Carnap's *Philosophical Foundations of Physics* (1966), which was based upon a seminar he offered at UCLA in 1958.[2] Early in the first chapter, he says:

> The general schema involved in *all explanation* can be expressed symbolically as follows:
>
> 1. $(x)\ (Px \supset Qx)$
> 2. Pa
> 3. Qa
>
> The first statement is the universal law that applies to any object x. The second statement asserts that a particular object a has the property

Wesley C. Salmon, *Scientific Explanation and the Causal Structure of the World.* Copyright © 1984 by Princeton University Press. Excerpt, pp. 24–47, reprinted with permission of Princeton University Press.

> *P*. These two statements taken together enable us to derive logically the third statement: object *a* has the property *Q* (1966, pp. 7–8, italics added).

After a single intervening paragraph, he continues:

> At times, in giving an explanation, the only *known* laws that apply are statistical rather than universal. In such cases, we must be content with a statistical explanation. (1966, p. 8, italics added)

Farther down on the same page, he assures us that "these are genuine explanations," and on the next page he points out that "In quantum theory . . . we meet with statistical laws that may not be the result of ignorance; they may express the basic structure of the world." I must confess to a reaction of astonishment at being told that all explanations are deductive-nomological, but that some are not, because they are statistical. This lapse was removed from the subsequent paperback edition (Carnap, 1974), which appeared under a new title.

Why did it take philosophers so long to get around to providing a serious treatment of statistical explanation? It certainly was not due to any absence of statistical explanations in science. In antiquity, Lucretius (1951, pp. 66–68) had based his entire cosmology upon explanations involving spontaneous swerving of atoms, and some of his explanations of more restricted phenomena can readily be interpreted as statistical. He asks, for example, why it is that Roman housewives frequently become pregnant after sexual intercourse, while Roman prostitutes to a large extent avoid doing so. Conception occurs, he explains, as a result of a collision between a male seed and a female seed. During intercourse the prostitutes wiggle their hips a great deal, but wives tend to remain passive; as everyone knows, it is much harder to hit a moving target (1951, p. 170).[3] In the medieval period, St. Thomas Aquinas asserted:

> The majority of men follow their passions, which are movements of the sensitive appetite, in which movements of heavenly bodies can cooperate: but few are wise enough to resist these passions. Consequently astrologers are able to foretell the truth in the majority of cases, especially in a general way. But not in particular cases; for nothing prevents man resisting his passions by his free will. (1947, 1:Qu. 115, a. 4, *ad* Obj. 3)

Astrological explanations are, therefore, of the statistical variety. Leibniz, who like Lucretius and Aquinas was concerned about human free will, spoke of causes that incline but do not necessitate (1951, p. 515; 1965, p. 136).

When, in the latter half of the nineteenth century, the kinetic-molecular theory of gases emerged, giving rise to classical statistical mechanics, statistical explanations became firmly entrenched in physics. In this context, it turns out, many phenomena that *for all practical purposes* appear amenable to strict D-N explanation—such as the melting of an ice cube placed in tepid water—must be admitted *strictly speaking* to be explained statistically in terms of probabilities almost indistinguishable from unity. On a smaller scale, Brownian motion involves probabilities that are, both theoretically and practically, definitely less than one. Moreover, two areas of nineteenth-century biology, Darwinian evolution and Mendelian genetics, provide explanations that are basically statistical. In addition, nineteenth-century social scientists approached such topics as suicide, crime, and intelligence by means of "moral statistics" (Hilts, 1973).

In the present century, statistical techniques are used in virtually every branch of science, and we may well suppose that most of these disciplines, if not all, offer statistical explanations of some of the phenomena they treat. The most dramatic example is the statistical interpretation of the equations of quantum mechanics, provided by Max Born and Wolfgang Pauli in 1926–1927; with the aid of this interpretation, quantum theory explains an impressive range of physical facts.[4] What is even more important is that this interpretation brings in statistical considerations at the most basic level. In nineteenth-century science, the use of probability reflected limitations of human knowledge; in quantum mechanics, it looks as if probability may be an ineluctable feature of the physical world. The Nobel laureate physicist Leon Cooper expresses the idea in graphic terms: "Like a mountain range that divides a continent, feeding water to one side or the other, the probability concept is the divide that separates quantum theory from all of physics that preceded it" (1968, p. 492). Yet it was not until 1962 that any philosopher published a serious attempt at characterizing a statistical pattern of scientific explanation.

INDUCTIVE-STATISTICAL EXPLANATION

When it became respectable for empirically minded philosophers to admit that science not only describes and predicts, but also explains, it was natural enough that primary attention should have been directed to classic and beautiful examples of deductive explanation. Once the D-N model had been elaborated, either of two opposing attitudes might have been taken toward inductive or statistical explanation by those who recognized the legitimacy of explanations of this general sort. It might have been felt, on the one hand, that the construction of such a model would be a simple exercise in setting out an analogue to the D-N model or in relaxing the stringent requirements for D-N explanation in some straightforward way. It might have been felt, on the other hand, that the problems in constructing an appropriate inductive or statistical model were so formidable that one simply did not want to undertake the task. Some philosophers may unreflectingly have adopted the former attitude; the latter, it turns out, is closer to the mark.

We should have suspected as much. If D-N explanations are deductive arguments, inductive or statistical explanations are, presumably, inductive arguments. This is precisely the tack Hempel took in constructing his inductive-statistical or I-S model. In providing a D-N explanation of the fact that this penny conducts electricity, one offers an explanans consisting of two premises: the particular premise that this penny is composed of copper, and the universal law-statement that all copper conducts electricity. The explanandum-statement follows deductively. To provide an I-S explanation of the fact that I was tired when I arrived in Melbourne for a visit in 1978, it could be pointed out that I had been traveling by air for more than twenty-four hours (including stopovers at airports), and almost everyone who travels by air for twenty-four hours or more becomes fatigued. The explanandum gets strong inductive support from those premises; the event-to-be-explained is thus subsumed under a statistical generalization.

It has long been known that there are deep and striking disanalogies between inductive and deductive logic.[5] Deductive entailment is transitive; strong inductive support is not. Contraposition is valid for deductive entailments; it does not hold for high probabilities. These are *not* relations that hold in some approximate way if the probabilities involved are high enough; once we abandon strict logical entailment, and turn to probability or inductive support, they break down entirely. But much more crucially, as Hempel brought out clearly in his 1962 essay, the deductive principle that permits the addition of an arbitrary term to the antecedent of an entailment does not carry over at all into inductive logic. If A entails B, then $A.C$ entails B, whatever C may happen to stand for. However, no matter how high the probability of B given A, there is no constraint whatever upon the probability of B given both A and C. To take an extreme case, the probability of a prime number being odd is one, but the probability that a prime number smaller than 3 is odd has the value zero. For those who feel uneasy about applying probability to cases of this arithmetical sort, we can readily supply empirical examples. A thirty-year-old Australian with an advanced case of lung cancer has a low probability of surviving for five more years, even though the probability of surviving to age thirty-five for thirty-year-old Australians in general is quite high. It is *this* basic disanalogy between deductive and inductive (or probabilistic) relations that gives rise to what Hempel called *the ambiguity of inductive-statistical explanation*—a phenomenon that, as he emphasized, has no counterpart in D-N explanation. His *requirement of maximal specificity* was designed expressly to cope with the problem of this ambiguity.

Hempel illustrates the ambiguity of I-S explanation, and the need for the requirement of maximal specificity, by means of the following example (1965, pp. 394–396). John Jones recovers quickly from a streptococcus infection, and when we ask why, we are told that he was given penicillin, and that almost all strep infections clear up quickly after penicillin is administered. The recovery is thus rendered probable relative to these explanatory facts. There are, however, certain strains of streptococcus bacteria that are resistant to penicillin. If, in addition to the above facts, we were told that the infection is of the penicillin-resistant type, then we would have to say that the prompt recovery is rendered *improbable* relative to the available information. It would clearly be scientifically unacceptable to ignore such relevant evidence as the penicillin-resistant character of the infection; the requirement of maximal specificity is designed to block statistical explanations that thus omit relevant facts. It says, in effect, that when the class to which the individual case is referred for explanatory purposes—in this instance, the class of strep infections treated by penicillin—is chosen, we must not know how to divide it into subsets in which the probability of the fact to be explained differs from its probability in the entire class. If it has been ascertained that this particular case involved the penicillin-resistant strain, then the original explanation of the rapid recovery would violate the requirement of maximal specificity, and for that reason would be judged unsatisfactory.[6]

Hempel conceived of D-N explanations as valid deductive arguments satisfying certain additional conditions. Explanations that conform to his inductive-statistical or I-S model are correct inductive arguments also satisfying certain additional restrictions. Explanations of both sorts can be characterized in terms of the following four conditions:

1. The explanation is an argument with correct (deductive or inductive) logical form,
2. At least one of the premises must be a (universal or statistical) law,
3. The premises must be true, and
4. The explanation must satisfy the requirement of maximal specificity.

This fourth condition is automatically satisfied by D-N explanations by virtue of the fact that their explanatory laws are universal generalizations. If all A are B, then obviously there is no subset of A in which the probability of B is other than one. This condition has crucial importance with respect to explanations of the I-S variety. In general, according to Hempel (1962a, p. 10), an explanation is an argument (satisfying these four conditions) to the effect that the event-to-be-explained was to be expected by virtue of certain explanatory facts. In the case of I-S explanations, this means that the premise must lend high inductive probability to the conclusion—that is, the explanandum must be highly probable with respect to the explanans.

Explanations of the D-N and I-S varieties can therefore be schematized as follows (Hempel, 1965, pp. 336, 382):

$$\text{(D-N)}\quad \begin{array}{cl} C_1, C_2, \ldots, C_j & \text{(particular explanatory conditions)} \\ \underline{L_1, L_2, \ldots, L_k} & \text{(general laws)} \\ E & \text{(fact-to-be-explained)} \end{array}$$

The single line separating the premises from the conclusion signifies that the argument is deductively valid.

$$\text{(I-S)}\quad \begin{array}{ll} C_1, C_2, \ldots, C_j & \text{(particular explanatory conditions)} \\ \underline{\underline{L_1, L_2, \ldots, L_k}}\ [r] & \text{(general laws, at least one statistical)} \\ E & \text{(fact-to-be-explained)} \end{array}$$

The double lines separating the premises from the conclusion signifies that the argument is inductively correct, and the number r expresses the degree of inductive probability with which the premises support the conclusion. It is presumed that r is fairly close to one.[7]

The high-probability requirement, which seems such a natural analogue of the deductive entailment relation, leads to difficulties in two ways. First, there are arguments that fulfill all of the requirements imposed by the I-S model, but that patently do not constitute satisfactory scientific explanations. One can maintain, for example, that people who have colds will probably get over them within a fortnight if they take vitamin C, but the use of vitamin C may not explain the recovery, since almost all colds clear up within two weeks regardless. In arguing for the use of vitamin C in the prevention and treatment of colds, Linus Pauling (1970) does not base his claims upon the high probability of avoidance or quick recovery; instead, he urges that massive doses of vitamin C have a bearing upon the probability of avoidance or recovery—that is, the use of vitamin C is relevant to the occurrence, duration, and severity of colds. A *high* probability of recovery, given use of vitamin C, does not confer explanatory value upon the use of this drug with respect to recovery. An *enhanced* probability value does indicate that the use of vitamin C may have some explanatory force. This example, along with a host of others which, like it, fulfill all of Hempel's requirements for a correct I-S explanation, shows that fulfilling these requirements does not constitute a sufficient condition for an adequate statistical explanation.

At first blush, it might seem that the type of relevance problem illustrated by the foregoing example was peculiar to the I-S model, but Henry Kyburg (1965) showed that examples can be found which demonstrate that the D-N model is infected with precisely the same difficulty. Consider a sample of table salt that dissolves upon being placed in water. We ask why it dissolves. Suppose, Kyburg suggests, that someone has cast a dissolving spell upon it—that is, someone wearing a funny hat waves a wand over it and says, "I hereby cast a dissolving spell upon you." We can then 'explain' the phenomenon by mentioning the dissolving spell—without for a moment believing that any actual magic has been accomplished—and by invoking the true universal generalization that all samples of table salt that have been hexed in this manner dissolve when placed in water. Again, an argument that satisfies all of the requirements of Hempel's model patently fails to qualify as a satisfactory scientific explanation because of a failure of relevance. Given Hempel's characterizations of his D-N and I-S models of explanation, it is easy to construct any number of 'explanations' of either type that invoke some irrelevancy as a purported explanatory fact.[8] This result casts serious doubt upon the entire epistemic conception of scientific explanation, as outlined in the previous chapter, insofar as it takes all explanations to be arguments of one sort or another.

The diagnosis of the difficulty can be stated very simply. Hempel's requirement of maximal specificity (RMS) guarantees that *all* known relevant facts must be included in an adequate scientific explanation, but there is no requirement to insure that *only* relevant facts will be included. The foregoing examples bear witness to the need for some requirement of this latter sort. To the best of my knowledge, the advocates of the 'received view' have not, until recently, put forth any such additional condition, nor have they come to terms with counterexamples of these types in any other way.[9] James Fetzer's *requirement of strict maximal specificity,* which rules out the use in explanations of laws that mention nomically irrelevant properties (Fetzer, 1981, pp. 125–126), seems to do the job. In fact, in (Salmon, 1979a, pp. 691–694), I showed how Reichenbach's theory of nomological statements could be used to accomplish the same end.

The second problem that arises out of the high-probability requirement is illustrated by an example furnished by Michael Scriven (1959, p. 480). If someone contracts paresis, the straightforward explanation is that he was infected with syphilis, which had progressed through the primary, secondary, and latent stages without treatment with penicillin. Paresis is one form of tertiary syphilis, and it never occurs except in syphilitics. Yet far less than half of those victims of untreated latent syphilis ever develop paresis. Untreated latent syphilis is the explanation of paresis, but it does not provide any basis on which to say that the explanandum-event was to be expected by virtue of these explanatory facts. Given a victim of latent untreated syphilis, the odds are that he will *not* develop paresis. Many other examples can be found to illustrate the same point. As I understand it, mushroom poisoning may afflict only a small percentage of individuals who eat a particular type of mushroom (Smith, 1958, Introduction), but the eating of the mushroom would unhesitatingly be offered as the explanation in instances of the illness in question. The point is illustrated by remarks on certain species in a guide for mushroom hunters (Smith, 1958, pp. 34, 185):

Helvella infula, "Poisonous to some, but edible for most people. Not recommended."
Chlorophyllum molybdites, "Poisonous to some but not to others. Those who are not made ill by it consider it a fine mushroom. The others suffer acutely."

These examples show that high probability does not constitute a necessary condition for legitimate statistical explanations. Taking them together with the vitamin C example, we must conclude—provisionally, at least—that a high probability of the explanandum relative to the explanans is neither necessary nor sufficient for correct statistical explanations, even if all of Hempel's other conditions are fulfilled. Much more remains to be said about the high-probability requirement, for it raises a host of fundamental philosophical problems, but I shall postpone further discussion of it until chapter 4.

Given the problematic status of the high-probability requirement, it was natural to attempt to construct an alternative treatment of statistical explanation that rests upon different principles. As I argued in (Salmon, 1965), statistical relevance, rather than high probability, seems to be the key explanatory relationship. This starting point leads to a conception of scientific explanation that differs fundamentally and radically from Hempel's I-S account. In the first place, if we are to make use of statistical relevance relations, our explanations will have to make reference to at least two probabilities, for statistical relevance involves a difference between two probabilities. More precisely, a factor C is statistically relevant to the occurrence of B under circumstances A if and only if

$$P(B|A.C) \neq P(B|A) \qquad (1)$$

or

$$P(B|A.C) \neq P(B|A.C). \qquad (2)$$

Conditions (1) and (2) are equivalent to one another, provided that C occurs with a nonvanishing probability within A; since we shall not be concerned with the relevance of factors whose probabilities are zero, we may use either (1) or (2) as our definition of statistical relevance. We say that C is positively relevant to B if the probability of B is greater in the presence of C; it is negatively relevant if the probability of B is smaller in the presence of C. For instance, heavy cigarette smoking is positively relevant to the occurrence of lung cancer, at some later time, in a thirty-year-old Australian male; it is negatively relevant to survival to the age of seventy for such a person.

In order to construct a satisfactory statistical explanation, it seems to me, we need a *prior probability* of the occurrence to be explained, as well as one or more *posterior probabilities.* A crucial feature of the explanation will be the comparison between the prior and posterior probabilities. In Hempel's case of the streptococcus infection, for instance, we might begin with the probability, in the entire class of people with streptococcus infections, of a quick recovery. We realize, however, that the administration of penicillin is statistically relevant to quick recovery, so we compare the probability of quick recovery among those who have received penicillin with the probability of quick recovery among those who have not received penicillin. Hempel warns, however, that there is another relevant factor, namely, the existence of the penicillin-resistant strain of bacteria. We must, therefore, take that factor into account as well. Our original reference class has been divided into four parts: (1) infection by non-penicillin-resistant bacteria, penicillin given; (2) infection by non-penicillin-resistant bacteria, no penicillin given; (3) infection by penicillin-resistant bacteria, penicillin given; (4) infection by penicillin-resistant bacteria, no penicillin given. Since the administration of penicillin is irrelevant to quick recovery in case of penicillin-resistant infections, the subclasses (3) and (4) of the original reference class should be merged to yield (3′) infection by penicillin-resistant bacteria. If John Jones is a member of (1), we have an explanation of his quick recovery, according to the S-R approach, not because the probability is high, but, rather, because it differs significantly from the probability in the original reference class. We shall see later what must be done if John Jones happens to fall into class (3′).

By way of contrast, Hempel's earlier high-probability requirement demands only that the posterior probability be sufficiently large—whatever that might mean—but makes no reference at all to any prior probability. According to Hempel's abstract model, we ask, "Why is individual x a member of B?" The answer consists of an inductive argument having the following form:

$$\begin{array}{l} P(B|A) = r \\ x \text{ is an } A \\ \hline\hline x \text{ is a } B \end{array} \quad [r]$$

As we have seen, even if the first premise is a statistical law, r is high, the premises are true, and the requirement of maximal specificity has been fulfilled, our 'explanation' may be patently inadequate, due to failure of relevancy.

In (Salmon, 1970, pp. 220–221), I advocated what came to be called the statistical-relevance or S-R model of scientific explanation. At that time, I thought that anything that satisfied the conditions that define that model would qualify as a legitimate scientific explanation. I no longer hold that view. It now seems to me that the statistical relationships specified in the S-R model constitute the *statistical basis* for a bona fide scientific explanation, but that this basis must be supplemented by certain *causal factors* in order to constitute a satisfactory scientific explanation. In chapters 5–9 I shall discuss the causal aspects of explanation. In this chapter, however, I shall confine attention to the statistical basis, as articulated in terms of the S-R model. Indeed, from here on I shall speak, not of the S-R model, but, rather, of the *S-R basis.*[10]

Adopting the S-R approach, we begin with an explanatory question in a form somewhat different from that given by Hempel. Instead of asking, for instance, "Why did x get well within a fortnight?" we ask, "Why did this person with a cold get well within a fortnight?" Instead of asking,

"Why is x a B?" we ask, "Why is x, which is an A, also a B?" The answer—at least for preliminary purposes—is that x is also a C, where C is *relevant* to B within A. Thus we have a prior probability $P(B|A)$—in this case, the probability that a person with a cold (A) gets well within a fortnight (B). Then we let C stand for the taking of vitamin C. We are interested in the posterior probability $P(B|A.C)$ that a person with a cold who takes vitamin C recovers within a fortnight. If the prior and posterior probabilities are equal to one another, the taking of vitamin C can play no role in explaining why this person recovered from the cold within the specified period of time. If the posterior probability is not equal to the prior probability, then C may, under certain circumstances, furnish part or all of the desired explanation. A large part of the purpose of the present book is to investigate the way in which considerations that are statistically relevant to a given occurrence have or lack explanatory import.

We cannot, of course, expect that every request for a scientific explanation will be phrased in canonical form. Someone might ask, for example, "Why did Mary Jones get well in no more than a fortnight's time?" It might be clear from the context that she was suffering from a cold, so that the question could be reformulated as, "Why did this person who was suffering from a cold get well within a fortnight?" In some cases, it might be necessary to seek additional clarification from the person requesting the explanation, but presumably it will be possible to discover what explanation is being called for. This point about the form of the explanation-seeking question has fundamental importance. We can easily imagine circumstances in which an altogether different explanation is sought by means of the same initial question. Perhaps Mary had exhibited symptoms strongly suggesting that she had mononucleosis; in this case, the fact that it was only an ordinary cold might constitute the explanation of her quick recovery. A given why-question, construed in one way, may elicit an explanation, while otherwise construed, it asks for an explanation that cannot be given. "Why did the Mayor contract paresis?" might mean, "Why did this adult human develop paresis?" or, "Why did this syphilitic develop paresis?" On the first construal, the question has a suitable answer, which we have already discussed. On the second construal, it has no answer—at any rate, we cannot give an answer—for we do not know of any fact in addition to syphilis that is relevant to the occurrence of paresis. Some philosophers have argued, because of these considerations, that scientific explanation has an unavoidably pragmatic aspect (e.g., van Fraassen, 1977, 1980). If this means simply that there are cases in which people ask for explanations in unclear or ambiguous terms, so that we cannot tell what explanation is being requested without further clarification, then so be it. No one would deny that we cannot be expected to supply explanations unless we know what it is we are being asked to explain. To this extent, scientific explanation surely has pragmatic or contextual components. Dealing with these considerations is, I believe, tantamount to choosing a suitable reference class with respect to which the prior probabilities are to be taken and specifying an appropriate sample space for purposes of a particular explanation. More will be said about these two items in the next section. In chapter 4—in an extended discussion of van Fraassen's theory—we shall return to this issue of pragmatic aspects of explanation, and we shall consider the question of whether there are any others.

THE STATISTICAL-RELEVANCE APPROACH

Let us now turn to the task of giving a detailed elaboration of the S-R basis. For purposes of initial presentation, let us construe the terms $A, B, C, \ldots$ (with or without subscripts) as referring to classes, and let us construe our probabilities in some sense as relative frequencies. This *does not mean* that the statistical-relevance approach is tied in any crucial way to a frequency theory of probability. I am simply adopting the heuristic device of picking examples involving frequencies because they are easily grasped. Those who prefer propensities, for example, can easily make the appropriate terminological adjustments, by speaking of chance setups and outcomes of trials where I refer to reference classes and attributes. With this understanding in mind, let us consider the steps involved in constructing an S-R basis for a scientific explanation:

1. We begin by selecting an appropriate reference class A with respect to which the prior probabilities $P(B_i|A)$ of the B_is are to be taken.
2. We impose an *explanandum-partition* upon the initial reference class A in terms of an exclusive and exhaustive set of attributes $B_1, \ldots, B_m$; this defines a sample space for purposes of the explanation under consideration. (This partition was not required in earlier presentations of the S-R model.)
3. Invoking a set of statistically relevant factors $C_1, \ldots, C_s$, we partition the initial reference class A into a set of mutually exclusive and exhaustive cells $A.C_1, \ldots, A.C_s$. The properties $C_1, \ldots, C_s$ furnish the *explanans-partition.*
4. We ascertain the associated probability relations:
   prior probabilities

   $$P(B_i|A) = p_i \quad \text{for all } i \ (1 \leq i \leq m)$$

   posterior probabilities

   $$P(B_i|A.C_j) = p_{ij} \quad \text{for all } i \text{ and } j \ (1 \leq i \leq m) \text{ and } (1 \leq j \leq s)$$

5. We require that each of the cells $A.C_j$ be homogeneous with respect to the explanandum-partition $\{B_i\}$; that is, none of the cells in the partition can be further subdivided in any manner relevant to the occurrence of any B_i. (This requirement is somewhat analogous to Hempel's requirement of maximal specificity, but as we shall see, it is a much stronger condition.)
6. We ascertain the relative sizes of the cells in our explanans-partition in terms of the following marginal probabilities:

   $$P(C_j|A) = q_j$$

   (These probabilities were not included in earlier versions of the S-R model; the reasons for requiring them now will be discussed later in this chapter.)
7. We require that the explanans-partition be a maximal homogeneous partition, that is—with an important exception to be noted later—for $i \neq k$ we require that $p_{ji} \neq p_{jk}$. (This requirement assures us that our partition in terms of $C_1, \ldots, C_m$ does not introduce any irrelevant subdivision in the initial reference class A.)
8. We determine which cell $A.C_j$ contains the individual x whose possession of the attribute B_i was to be explained. The probability of B_i within the cell is given in the list under 4.

Consider in a rather rough and informal manner the way in which the foregoing pattern of explanation might be applied in a concrete situation; an example of this sort was offered by James Greeno (1971a, pp. 89–90). Suppose that Albert has committed a delinquent act—say, stealing a car, a major crime—and we ask for an explanation of that fact. We ascertain from the context that he is an American teen-ager, and so we ask, "Why did this American teen-ager commit a serious delinquent act?" The prior probabilities, which we take as our point of departure, so to speak, are simply the probabilities of the various degrees of juvenile delinquency (B_i) among American teen-agers (A)—that is, $P(B_i|A)$. We will need a suitable explanandum-partition; Greeno suggests B_1 = no criminal convictions, B_2 = conviction for minor infractions only, B_3 = conviction for a major offense. Our sociological theories tell us that such factors as sex, religious background, marital status of parents, type of residential community, socioeconomic status, and several others are relevant to delinquent behavior. We therefore take the initial reference class of American teen-agers and divide it into males and females; Jewish, Protestant, Roman Catholic, no religion; parents married, parents divorced, parents never married; urban, suburban, rural place of residence; upper, middle, lower class; and so forth. Taking such considerations into account, we arrive at a large number *s* of cells in our partition. We assign probabilities of the various degrees of delinquent behavior to each of the cells in accordance with 4, and we ascertain the probability of a randomly selected American teen-ager belonging to each of the cells in accordance with 6. We find the cell to which Albert belongs—for example, male, from a Protestant background, parents divorced, living in a suburban area, belonging to the middle class. If we have taken into account all of the relevant factors, and if we have correctly ascertained the probabilities associated with the various cells of our partitions, then we have an S-R basis for the explanation of Albert's delinquency that conforms to the foregoing schema. If it should turn out (contrary to what I believe actually to be the case) that the probabilities of the various types of delinquency are the same for males and for females, then we would not use sex in partitioning our original reference class. By condition 5 we must employ *every* relevant factor; by condition 7 we must employ *only* relevant factors. In many concrete situations, including the present examples, we know that we have not found all relevant considerations; however, as Noretta Koertge rightly emphasized (1975), that is an ideal for which we may aim. Our philosophical analysis is designed to capture the notion of a fully satisfactory explanation.

Nothing has been said, so far, concerning the rationale for conditions 2 and 6, which are here added to the S-R basis for the first time. We must see why these requirements are needed. Condition 2 is quite straightforward; it amounts to a requirement that the sample space for the problem at hand be specified. As we shall see when we discuss Greeno's information-theoretic approach in chapter 4, both the explanans-partition and the explanandum-partition are needed to measure the information transmitted in any explanatory scheme. This is a useful measure of the explanatory value of a theory. In addition, as we shall see when we discuss van Fraassen's treatment of why-questions in chapter 4, his contrast class, which is the same as our explanandum-partition, is needed in some cases to specify precisely what explanation is being sought. In dealing with the question "Why did Albert steal a car?" we used Greeno's suggested explanandum-partition. If, however, we had used different partitions (contrast classes), other explanations might have been called forth. Suppose that the contrast class included: Albert steals a car, Bill steals a car, Charlie steals a car, and so forth. Then the answer might have involved no sociology whatever; the explanation might have been that, among the members of his gang, Albert is most adept at hot-wiring. Suppose, instead, that the contrast class had included: Albert steals a car, Albert steals a diamond ring, Albert steals a bottle of whiskey, and so forth. In that case, the answer might have been that he wanted to go joyriding.

The need for the marginal probabilities mentioned in 6 arises in the following way. In many cases, such as the foregoing delinquency example, the terms C_j that furnish the explanans-partition of the initial reference class are conjunctive. A given cell is determined by several distinct factors: sex *and* religious background *and* marital status of parents *and* type of residential community *and* socioeconomic status *and* . . . which may be designated D_k, E_n, F_r, . . . These factors will be the probabilistic contributing causes and counteracting causes that tend, respectively, to produce or prevent delinquency. In attempting to understand the phenomenon in question, it is important to know how each factor is relevant—whether positively or negatively, and how strongly—both in the population at large and in various subgroups of the population. Consider, for example, the matter of sex. It may be that within the entire class of American teen-agers (A) the probability of serious delinquency (B_3) is greater among males than it is among females. If so, we would want to know by how much the probability among males exceeds the probability among females and by how much it exceeds the probability in the entire population. We also want to know whether being male is always positively relevant to serious delinquency, or whether in combination with certain other factors, it may be negatively relevant or irrelevant. Given two groups of teen-agers—one consisting entirely of boys and the other entirely of girls, but alike with respect to all of the other factors—we want to know how the probabilities associated with delinquency in each of the two groups are related to one another. It might be that in each case of two cells in the explanandum-partition that differ from one another only on the basis of gender, the probability of serious delinquency in the male group is greater than it is in the female group. It might turn out, however, that sometimes the two probabilities are equal, or that in some cases the probability is higher in the female group than it is in the corresponding male group. Relationships of all of these kinds are logically possible.

It is a rather obvious fact that each of two circumstances can individually be positively relevant to a given outcome, but their conjunction can be negatively relevant or irrelevant. Each of two drugs can be positively relevant to good health, but taken together, the combination may be detrimental—for example, certain antidepressive medications taken in conjunction with various remedies for the common cold can greatly increase the chance of dangerously high blood pressure (Goodwin and Guze, 1979). A factor that is a contributing cause in some circumstances can be a counteracting cause in other cases. Problems of this sort have been discussed, sometimes under the heading of "Simpson's paradox," by Nancy Cartwright (1983, essay 1) and Bas van Fraassen (1980, pp. 108, 148–151). In (Salmon, 1975c), I have spelled out in detail the complexities that arise in connection with statistical relevance relations. The moral is that we need to know not only how the various factors D_k, E_n, F_r, . . . , are relevant to the outcome, B_i, but how the relevance of each of them is affected by the presence or absence of the others. Thus, for instance, it is possible that being female might in general be negatively relevant to delinquency, but it might be positively relevant among the very poor.

Even if all of the prior probabilities $P(B_i|A)$ and all of the posterior probabilities $P(B_i|A.C_j)$ furnished under condition 4 are known, it is not possible to deduce the conditional probabilities of the B_i's with respect to the individual conjuncts that make up the C_j's or with respect to combinations of them. Without these conditional probabilities, we will not be in a position to ascertain all of the statistical relevance relations that are required. We therefore need to build in a way to extract that information. This is the function of the marginal probabilities $P(C_j|A)$ required by condition 6. If these are known, such conditional probabilities as $P(B_i|A.D_k)$, $P(B_i|A.E_n)$, and $P(B_i|A.D_k.E_n)$ can be derived.[11] When 2 and 6 are added to the earlier characterization of the S-R model (Salmon et al., 1971), then, I believe, we have gone as far as possible in characterizing scientific explanations at the level of statistical relevance relations.

Several features of the new version of the S-R basis deserve explicit mention. It should be noted, in the first place, that conditions 2 and 3 demand that the entire initial reference class A be partitioned, while conditions 4 and 6 require that *all* of the associated probability values be given. This is one of several respects in which it differs from Hempel's I-S model. Hempel requires only that the individual mentioned in the explanandum be placed within an appropriate class, satisfying his requirement of maximal specificity, but he does not ask for information about any class in either the explanandum-partition or the explanans-partition to which that individual does not belong. Thus he might go along in requiring that Bill Smith be referred to the class of American male teen-agers coming from a Protestant background, whose parents are divorced, and who is a middle-class suburban dweller, and in asking us to furnish the probability of his degree of delinquency within that class. But why, it may surely be asked, should we be concerned with the probability of delinquency in a lower-class, urban-American, female teen-ager from a Roman Catholic background whose parents are still married? What bearing do such facts have on Bill Smith's delinquency? The answer, I think, involves serious issues concerning scientific generality. If we ask why this American teen-ager becomes a delinquent, then, it seems to me, we are concerned with *all* of the factors that are relevant to the occurrence of delinquency, and with the ways in which these factors are relevant to that phenomenon (cf. Koertge, 1975). To have a satisfactory scientific answer to the question of why this A is a B_i—to achieve full scientific understanding—we need to know the factors that are relevant to the occurrence of the various B_is for *any* randomly chosen or otherwise unspecified member of A. It was mainly to make good on this desideratum that requirement 6 was added. Moreover, as Greeno and I argued in *Statistical Explanation and Statistical Relevance,* a good measure of the value of an S-R basis is the gain in information furnished by the complete partitions and the associated probabilities. This measure cannot be applied to the individual cells one at a time.

A fundamental philosophical difference between our S-R basis and Hempel's I-S model lies in the interpretation of the concept of homogeneity that appears in condition 5. Hempel's requirement of maximal specificity, which is designed to achieve a certain kind of homogeneity in the reference classes employed in I-S explanations, is *epistemically relativized.* This means, in effect, that we must not *know* of any way to make a relevant partition, but it certainly does not demand that no possibility of a relevant partition can exist unbeknown to us. As I view the S-R basis, in contrast, condition 5 demands that the cells of our explanans-partition be *objectively* homogeneous; for this model, homogeneity is not epistemically relativized. Since this issue of epistemic relativization versus objective homogeneity is discussed at length in chapter 3, it is sufficient for now merely to call attention to this complex problem.[12]

Condition 7 has been the source of considerable criticism. One such objection rests on the fact that the initial reference class A, to which the S-R basis is referred, may not be maximal. Regarding Kyburg's hexed salt example, mentioned previously, it has been pointed out that the class of samples of table salt is not a maximal homogeneous class with respect to solubility, for there are many other chemical substances that have the same probability—namely, unity—of dissolving when placed in water. Baking soda, potassium chloride, various sugars, and many other compounds have this property. Consequently, if we take the maximality condition seriously, it has been argued, we should not ask, "Why does this sample of table salt dissolve in water?" but, rather, "Why does this sample of matter in the solid state dissolve when placed in water?" And indeed, one can argue, as Koertge has done persuasively (1975), that to follow such a policy often leads to significant scientific progress. Without denying her important point, I would nevertheless suggest, for purposes of elaborating the formal schema, that we take the initial reference class A as given by the explanation-seeking why-question, and look for relevant partitions within it. A significantly different explanation, which often undeniably represents scientific progress, may result if a different why-question, embodying a broader initial reference class, is posed. If the original question is not presented in a form that unambiguously determines a reference class *A,* we can reasonably discuss the advantages of choosing a wider or a narrower class in the case at hand.

Another difficulty with condition 7 arises if 'accidentally'—so to speak—two different cells in the partition, $A.C_i$ and $A.C_j$, happen to have equal associated probabilities p_{ki} and p_{kj} for all cells B_k in the explanandum-partition. Such a circumstance might arise if the cells are determined conjunctively by a number of relevant factors, and if the differences between the two cells cancel one another out. It might happen, for example, that the probabilities of the various degrees of delinquency—major offense, minor offense, no offense—for an upper-class, urban, Jewish girl would be equal to those for a middle-class, rural, Protestant boy. In this case, we might want to relax condition 7, allowing $A.C_i$ and $A.C_j$ to stand as separate cells, provided they differ with respect to at least two of the terms in the conjunction, so that we are faced with a fortuitous canceling of relevant factors. If, however, $A.C_i$ and $A.C_j$ differed with respect to only one conjunct, they would have to be merged into a single cell. Such would be the case if, for example, among upper-class, urban-dwelling, American teen-agers whose religious background is atheistic and whose parents are divorced, the probability of delinquent behavior were the same for boys as for girls. Indeed, we have already encountered this situation in connection with Hempel's example of the streptococcus infection. If the infection is of the penicillin-resistant variety, the probability of recovery in a given period of time is the same whether penicillin is administered or not. In such cases, we want to say, there is no relevant difference between the two classes—not that relevant factors were canceling one another out. I bring this problem up for consideration at this point, but I shall not make a consequent modification in the formal characterization of the S-R basis, for I believe that problems of this sort are best handled in the light of causal relevance relations. Indeed, it seems advisable to postpone detailed consideration of the whole matter of regarding the cells $A.C_j$ as being determined conjunctively until causation has been explicitly introduced into the discussion. As we shall see in chapter 9, (Humphreys, 1981, 1983) and (Rogers, 1981) provide useful suggestions for handling just this issue.

Perhaps the most serious objection to the S-R model of scientific explanation—as it was originally presented—is based upon the principle that *mere* statistical correlations explain nothing. A rapidly falling barometric reading is a sign of an imminent storm, and it is *highly correlated* with the onset of storms, but it certainly does not *explain* the occurrence of a storm. The S-R approach does, however, have a way of dealing with examples of this sort. A factor *C,* which is relevant to the occurrence of *B* in the presence of *A,* may be screened off in the presence of some additional factor *D;* the screening-off relation is defined by equations (3) and (4), which follow. To illustrate, given a series of days (*A*) in some particular locale, the probability of a storm occurring (*B*) is in general quite different from the probability of a storm if there has been a recent sharp drop in the barometric reading (*C*). Thus *C* is statistically relevant to *B* within *A.* If, however, we take into consideration the further fact that there is an actual drop in atmospheric pressure (*D*) in the region, then it is irrelevant whether that drop is registered on a barometer. In the presence of *D* and *A, C* becomes irrelevant to *B;* we say that *D* screens off *C* from *B*—in symbols,

$$P(B|A.C.D) = P(B|A.D). \tag{3}$$

However, *C* does not screen off *D* from *B,* that is,

$$P(B|A.C.D) \neq P(B|A.C), \tag{4}$$

for barometers sometimes malfunction, and it is the atmospheric pressure, not the reading on the barometer per se, that is directly relevant to the occurrence of the storm. A factor that has been screened off is irrelevant, and according to the definition of the S-R basis (condition 7), it is not to be included in the explanation. The falling barometer does not explain the storm.

Screening off is frequent enough and important enough to deserve further illustration. A study, reported in the news media a few years ago, revealed a positive correlation between coffee drinking and heart disease, but further investigation showed that this correlation results from a correlation between coffee drinking and cigarette smoking. It turned out that cigarette smoking screened off coffee drinking from heart disease, thus rendering coffee drinking statistically (as well as causally and explanatorily) irrelevant to heart disease. Returning to a previous example for another illustration, one could reasonably suppose that there is some correlation between low socioeconomic status and paresis, for there may be a higher degree of sexual promiscuity, a higher incidence of venereal disease, and less likelihood of adequate medical attention if the disease occurs. But the contraction of syphilis screens off such factors as degree of promiscuity, and the fact of syphilis going untreated screens off any tendency to fail to get adequate medical care. Thus when an individual has latent untreated syphilis, all other such circumstances as low socioeconomic status are screened off from the development of paresis.

As the foregoing examples show, there are situations in which one circumstance or occurrence is correlated with another because of an indirect causal relationship. In such cases, it often happens that the more proximate causal factors screen off those that are more remote. Thus 'mere correlations' are replaced in explanatory contexts with correlations that are intuitively recognized to have explanatory force. In *Statistical Explanation and Statistical Relevance*, where the S-R model of statistical explanation was first explicitly named and articulated, I held out some hope (but did not try to defend the thesis) that all of the causal factors that play any role in scientific explanation could be explicated in terms of statistical relevance relations—with the screening-off relation playing a crucial role. As I shall explain in chapter 6, I no longer believe this is possible. A large part of the material in the present book is devoted to an attempt to analyze the nature of the causal relations that enter into scientific explanations, and the manner in which they function in explanatory contexts. After characterizing the S-R model, I wrote:

> One might ask on what grounds we can claim to have characterized explanation. The answer is this. When an explanation (as herein explicated) has been provided, we know exactly how to regard any *A* with respect to the property *B*. We know which ones to bet on, which to bet against, and at what odds. We know precisely what degree of expectation is rational. We know how to face uncertainty about an *A*'s being a *B* in the most reasonable, practical, and efficient way. We know every factor that is relevant to an *A* having the property *B*. We know exactly what weight should have been attached to the prediction that this *A* will be a *B*. We know all of the regularities (universal and statistical) that are relevant to our original question. What more could one ask of an explanation? (Salmon et al., 1971, p. 78)

The answer, of course, is that we need to know something about the causal relationships as well.

In acknowledging this egregious shortcoming of the S-R model of scientific explanation, I am not abandoning it completely. The attempt, rather, is to supplement it in suitable ways. While recognizing its incompleteness, I still think it constitutes a sound basis upon which to erect a more adequate account. And at a fundamental level, I still think it provides important insights into the nature of scientific explanation.

In the introduction to *Statistical Explana-*

tion and Statistical Relevance, I offered the following succinct comparison between Hempel's I-S model and the S-R model:

I-S model: an explanation is an *argument* that renders the explanandum *highly probable.*
S-R model: an explanation is an *assembly of facts statistically relevant* to the explanandum, *regardless of the degree of probability* that results.

It was Richard Jeffrey (1969) who first explicitly formulated the thesis that (at least some) statistical explanations are not arguments; it is beautifully expressed in his brief paper, "Statistical Explanation vs. Statistical Inference," which was reprinted in *Statistical Explanation and Statistical Relevance.* In (Salmon, 1965, pp. 145–146), I had urged that positive relevance rather than high probability is the desideratum in statistical explanation. In (Salmon, 1970), I expressed the view, which many philosophers found weird and counter-intuitive (e.g., L. J. Cohen, 1975), that statistical explanations may even embody *negative* relevance—that is, an explanation of an event may, in some cases, show that the event to be explained is less probable than we had initially realized. I still do not regard that thesis as absurd. In an illuminating discussion of the explanatory force of positively and negatively relevant factors, Paul Humphreys (1981) has introduced some felicitous terminology for dealing with such cases, and he has pointed to an important constraint. Consider a simple example. Smith is stricken with a heart attack, and the doctor says, "*Despite* the fact that Smith exercised regularly and had given up smoking several years ago, he contracted heart disease *because* he was seriously overweight." The "because" clause mentions those factors that are positively relevant and the "despite" clause cites those that are negatively relevant. Humphreys refers to them as *contributing causes* and *counteracting causes,* respectively. When we discuss causal explanation in later chapters, we will want to say that a complete explanation of an event must make mention of the causal factors that tend to prevent its occurrence as well as those that tend to bring it about. Thus it is *not* inappropriate for the S-R basis to include factors that are negatively relevant to the explanandum-event. As Humphreys points out, however, we would hardly consider as appropriate a putative explanation that had only negative items in the "despite" clause and no positive items in the "because" category. "Despite the fact that Jones never smoked, exercised regularly, was not overweight, and did not have elevated levels of triglycerides and cholesterol, he died of a heart attack," would hardly be considered an acceptable *explanation* of his fatal illness.

Before concluding this chapter on models of statistical explanation, we should take a brief look at the deductive-homological-probabilistic (D-N-P) model of scientific explanation offered by Peter Railton (1978). By employing well-established statistical laws, such as that covering the spontaneous radioactive decay of unstable nuclei, it is possible to deduce the fact that a decay-event for a particular isotope has a certain probability of occurring within a given time interval. For an atom of carbon 14 (which is used in radiocarbon dating in archaeology, for example), the probability of a decay in 5,730 years is ½. The explanation of *the probability of the decay-event* conforms to Hempel's deductive-nomological pattern. Such an explanation does not, however, explain the actual occurrence of a decay, for, given the probability of such an event—however high or low—the event in question may not even occur. Thus the explanation does not qualify as an argument to the effect that the event-to-be-explained was to be expected with deductive certainty, given the explanans. Railton is, of course, clearly aware of the fact. He goes on to point out, nevertheless, that if we simply attach an "addendum" to the deductive argument stating that the event-to-be-explained did, in fact, occur in the case at hand, we can claim to have a probabilistic *account*—which is not a deductive or inductive argument—of the occurrence of the event. In this respect, Railton is in rather close agreement with Jeffrey (1969) that some explanations are not arguments. He also agrees with Jeffrey in emphasizing the importance of exhibiting the physical mechanisms that lead up to the probabilistic occurrence that is to be explained. Railton's theory—like that of Jeffrey—has some deep affinities to the S-R model. In including a reference to physical mechanisms as an essential part of his D-N-P model, however, Railton goes beyond the view that statistical relevance relations, in and of themselves, have explanatory import. His theory of scientific explanation can be more appropriately characterized as causal or mechanistic. It is closely related to the two-tiered causal-statistical account that I am attempting to elaborate as an improvement upon the S-R model.

Although, with Kyburg's help, I have offered what seem to be damaging counterexamples to the D-N model—for instance, the one about the man who explains his own avoidance of pregnancy on the basis of his regular consumption of his wife's birth control pills (Salmon et al., 1971, p. 34)—the major emphasis has been upon statistical explanation, and that continues to be the case in what follows. Aside from the fact that contemporary science obviously provides many statistical explanations of many types of phenomena, and that any philosophical theory of statistical explanation has only lately come forth, there is a further reason for focusing upon statistical explanation. As I maintained in chapter 1, we can identify three distinct approaches to scientific explanation that do not seem to differ from one another in any important way as long as we confine our attention to contexts in which all of the explanatory laws are universal generalizations. I shall argue in chapter 4, however, that these three general conceptions of scientific explanation can be seen to differ radically from one another when we move on to situations in which statistical explanations are in principle the best we can achieve. Close consideration of statistical explanations, with sufficient attention to their causal ingredients, provides important insight into the underlying philosophical questions relating to our scientific understanding of the world.

NOTES

1. Ilkka Niiniluoto (1981, p. 444) suggests that "Peirce should be regarded as the true founder of the theory of inductive-probabilistic explanation" on account of this statement, "The statistical syllogism may be conveniently termed the explanatory syllogism" (Peirce, 1932, 2:716). I am inclined to disagree, for one isolated and unelaborated statement of that sort can hardly be considered even the beginnings of any genuine theory.

2. As Carnap reports in the preface, the seminar proceedings were recorded and transcribed by his wife. Martin Gardner edited—it would probably be more accurate to say "wrote up"—the proceedings and submitted them to Carnap, who rewrote them extensively. There is little doubt that Carnap saw and approved the passages I have quoted.

3. Lucretius writes: "A woman makes conception more difficult by offering a mock resistance and accepting Venus with a wriggling body. She diverts the furrow from the straight course of the ploughshare and makes the seed fall wide of the plot. These tricks are employed by prostitutes for their own ends, so that they may not conceive *too frequently* and be laid up by pregnancy" (1951, p. 170, italics added).

4. See (Wessels, 1982), for an illuminating discussion of the history of the statistical interpretation of quantum mechanics.

5. These are spelled out in detail in (Salmon, 1965a). See (Salmon, 1967, pp. 109–111) for a discussion of the 'almost-deduction' conception of inductive inference.

6. We shall see in chapter 3 that the requirement of maximal specificity, as formulated by Hempel in his (1965) and revised in his (1968), does not actually do the job. Nevertheless, this was clearly its intent.

7. It should be mentioned in passing that Hempel (1965, pp. 380–381) offers still another model of scientific explanation that he characterizes as deductive-statistical (D-S). In an explanation of this type, a statistical regularity is explained by deducing it from other statistical laws. There is no real need, however, to treat

such explanations as a distinct type, for they fall under the D-N schema, just given, provided we allow that at least one of the laws may be statistical. In the present context, we are concerned only with statistical explanations of nondeductive sorts.

8. Many examples are presented and analyzed in (Salmon et al., 1971, pp. 33–40). Nancy Cartwright (1983, pp. 26–27) errs when she attributes to Hempel the requirement that a statistical explanation increase the probability of the explanandum; this thesis, which I first advanced in (Salmon, 1965), was never advocated by Hempel. Shortly thereafter (1983, 28–29), she provides a correct characterization of the relationships among the views of Hempel, Suppes, and me.

9. In Hempel's most recent discussion of statistical explanation, he appears to maintain the astonishing view that although such examples have *psychologically* misleading features, they do qualify as *logically* satisfactory explanations (1977, pp. 107–111).

10. I am extremely sympathetic to the thesis, expounded in (Humphreys, 1983), that probabilities—including those appearing in the S-R basis—are important tools in the construction of scientific explanations, but that they do not constitute any part of a scientific explanation per se. This thesis allows him to relax considerably the kinds of maximal specificity or homogeneity requirements that must be satisfied by statistical or probabilistic explanations. A factor that is statistically relevant may be causally irrelevant because, for example, it does not convert any contributing causes to counteracting causes or vice versa. This kind of relaxation is attractive in a theory of scientific explanation, for factors having small statistical relevance often seem otiose. Humphreys' approach does not show, however, that such relevance relations can be omitted from the S-R basis; on the contrary, the S-R basis must include such factors in order that we may ascertain whether they can be omitted from the causal explanation or not. I shall return to Humphreys' concept of aleatory explanation in chapter 9.

11. Suppose, for example, that we wish to compute $P(B_i|A.D_k)$, where $D_k = C_{j1} \vee \ldots \vee C_{j4}$, the cells C_{jr} being mutually exclusive. This can be done as follows. We are given $P(C_j|A)$ and $P(B_i|A.C_j)$. By the multiplication theorem,

$$P(D_k.B_i|A) = P(D_k|A) \times P(B_i|A.D_k)$$

Assuming $P(D_k \mid A) \neq 0$, we have,

$$P(B_i \mid A.D_k) = P(D_k.B_i \mid A)/P(D_k|A) \qquad (*)$$

By the addition theorem

$$P(D_k \mid A) = \sum_{r=1}^{q} P(C_{j_r} \mid A)$$

$$P(D_k.B_i \mid A) = \sum_{r=1}^{q} P(C_{j_r}B_i \mid A)$$

By the multiplication theorem

$$P(D_k.B_i \mid A) = \sum_{r=1}^{q} P(C_{j_r} \mid A) \times P(B_i \mid A.C_{j_r})$$

Substitution in (*) yields the desired relation:

$$P(B_i \mid A.D_k) = \frac{\sum_{r=1}^{q} P(C_{j_r} \mid A) \times P(B_i \mid A.C_{j_r})}{\sum_{r=1}^{q} P(C_{j_r} \mid A)}$$

12. Cartwright (1983, p. 27) asserts that on Hempel's account, "what counts as a good explanation is an objective, person-independent matter," and she applauds him for holding that view. I find it difficult to reconcile her characterization with Hempel's repeated emphatic assertion (prior to 1977) of the doctrine of essential epistemic relativization of inductive-statistical explanation. In addition, she complains that my way of dealing with problems concerning the proper formulation of the explanation-seeking why-question—that is, problems concerning the choice of an appropriate initial reference class—"makes explanation a subjective matter" (ibid., p. 29). "What explains what," she continues, "depends on the laws and facts true in our world, and cannot be adjusted by shifting our interest or our focus" (ibid.). This criticism seems to me to be mistaken. Clarification of the question is often required to determine what it is that is to be explained, and this may have pragmatic dimensions. However, once the explanandum has been unambiguously specified, on my view, the identification of the appropriate explanans is fully objective. I am in complete agreement with Cartwright concerning the desirability of such objectivity; moreover, my extensive concern with objective homogeneity is based directly upon the desire to eliminate from the theory of statistical explanation such subjective features as epistemic relativization.

Nancy Cartwright

THE TRUTH DOESN'T EXPLAIN MUCH

INTRODUCTION

Scientific theories must tell us both what is true in nature, and how we are to explain it. I shall argue that these are entirely different functions and should be kept distinct. Usually the two are conflated. The second is commonly seen as a by-product of the first. Scientific theories are thought to explain by dint of the descriptions they give of reality. Once the job of describing is done, science can shut down. That is all there is to do. To describe nature—to tell its laws, the values of its fundamental constants, its mass distributions . . .— is *ipso facto* to lay down how we are to explain it.

This is a mistake, I shall argue; a mistake which is fostered by the covering law model of explanation. The covering law model supposes that all we need to know are the laws of nature—and a little logic, perhaps a little probability theory—and then we know which factors can explain which others. For example, in the simplest deductive-nomological version,[1] the covering law model says that one factor explains another just in case the occurrence of the second can be deduced from the occurrence of the first given the laws of nature.

But the D-N model is just an example. In the sense which is relevant to my claims here, most models of explanation offered recently in the philosophy of science are covering law models. This includes not only Hempel's own inductive statistical model,[2] but also Patrick Suppes' probabilistic model of causation,[3] Wesley Salmon's statistical relevance model,[4] and even Bengt Hanson's contextualistic model.[5] All these accounts rely on the laws of nature, and just the laws of nature, to pick out which factors we can use in explanation.

A good deal of criticism has been aimed at Hempel's original covering law models. Much of the criticism objects that these models let in too much. On Hempel's account it seems we can explain Henry's failure to get pregnant by his taking birth control pills, and we can explain the storm by the falling barometer. My objection is quite the opposite. Covering law models let in too little. With a covering law model we can explain hardly anything, even the things of which we are most proud—like the role of DNA in the inheritance of genetic characteristics, or the formation of rainbows when sunlight is refracted through raindrops. We cannot explain these phenomena with a covering law model, I shall argue, because we don't have any laws which cover them. Covering laws are scarce.

Many phenomena which have perfectly good scientific explanations are not covered by any laws. No true laws, that is. They are at best covered by *ceteris paribus* generalizations—generalizations which hold only under special conditions, usually ideal conditions. The literal translation is "other things being equal"; but it would be more apt to read "ceteris paribus" as "other things being *right.*"

Sometimes we act as if this doesn't matter. We have in the back of our minds an "understudy" picture of *ceteris paribus* laws: *ceteris paribus* laws are real laws; they can stand in when the laws we would like to see aren't available and they can perform all the same functions, only not quite so well. But this won't do. For *ceteris paribus* generalizations read literally—without the "ceteris paribus" modifier—as laws, are false. They

Reprinted by permission from *American Philosophical Quarterly,* Vol. 17, no. 2, April 1980.

are not only false, but held by us to be false; and there is no ground in the covering law picture for false laws to explain anything. On the other hand, with the modifier the *ceteris paribus* generalizations may be true, but they cover only those few cases where the conditions are right. For most cases, either we have a law which purports to cover, but can't explain because it is acknowledged to be false, or we have a law which doesn't cover. Either way, it's bad for the covering law picture.

I. CETERIS PARIBUS LAWS

When I first started talking about the scarcity of covering laws, I tried to summarize my view by saying "There are no exceptionless generalizations." Then Merrilee Salmon asked, "How about 'All men are mortal'?" She was right. I had been focussing too much on the equations of physics. A more plausible claim would have been that there are no exceptionless quantitative laws in physics. Indeed, not only are there no exceptionless laws, but in fact our best candidates are known to fail. This is something like the Popperian thesis that *every theory is born refuted.* Every theory we have proposed in physics, even at the time when it was most firmly entrenched, was known to be deficient in specific and detailed ways. I think this is also true for every precise quantitative law within a physics theory.

But this is not the point I had wanted to make. For some laws are treated, at least for the time being, as if they were exceptionless, whereas others are not—even though they remain "on the books". Snell's law (about the angle of incidence and the angle of refraction for a ray of light) is a good example of this latter kind. In the optics text I use for reference (Miles V. Klein, *Optics*),[6] it first appears on page 21, and without qualification:

Snell's Law: At an interface between dielectric media, there is (also) *a refracted ray* in the second medium, lying in the plane of incidence, making an angle θ_t with the normal, and obeying Snell's law:

$$\sin \theta / \sin \theta_t = n_2 / n_1$$

where v_1 and v_2 are the velocities of propagation in the two media, and $n_1 = (c/v_1)$, $n_2 = (c/v_2)$ are the indices of refraction.
(θ is the angle of incidence. Italics added.)

It is only some 500 pages later, when the law is derived from the "full electromagnetic theory of light" that we learn that Snell's law as stated on page 21 is true only for media whose optical properties are *isotropic.* (In anisotropic media, "there will generally be *two* transmitted waves."[7]) So what is deemed true is not really Snell's law as stated on page 21, but rather a refinement of Snell's law:

Refined Snell's Law: For any two media which are optically isotropic, at an interface between dielectrics there is a refracted ray in the second medium, lying in the plane of incidence, making an angle θ_t with the normal, such that:

$$\sin \theta / \sin \theta_t = n_2 / n_1 .$$

The Snell's law of page 21 in Klein's book is an example of a *ceteris paribus* law, a law that holds only in special circumstances—in this case when the media are both isotropic. Klein's statement on page 21 is clearly not to be taken literally. Charitably, we are inclined to put the modifier "ceteris paribus" in front to hedge it. But what does this *ceteris paribus* modifier do? With an eye to statistical versions of the covering law model (Hempel's I-S picture, or Salmon's statistical relevance model, or Suppes' probabilistic model of causation) we may suppose that the unrefined Snell's law is not intended to be a universal law, as literally stated, but rather some kind of statistical law. The obvious candidate is a crude statistical law: *for the most part,* at an interface between dielectric media there is *a* refracted ray . . . But this won't do. For *most* media are optically anisotropic, and in an anisotropic medium there are *two* rays. I think there are no more satisfactory alternatives. If *ceteris paribus* laws are to be true laws, there are no statistical laws they can generally be identified with.

II. WHEN LAWS ARE SCARCE

Why do we keep Snell's law on the books when we both know it to be false and have a more accurate refinement available? There are obvious pedagogic reasons. But are there serious scientific ones? I think there are, and these reasons have to do with the task of explaining. I claim that specifying which factors are explanatorily relevant to which others is a job done by science over and above the job of laying out the laws of nature. Once the laws of nature are known, we still have to decide what kinds of factors can be cited in explanation.

One thing that *ceteris paribus* laws do is to express our explanatory commitments. They tell what kinds of explanations are permitted. We know from the refined Snell's law that in any isotropic medium, the angle of refraction can be explained by the angle of incidence, according to the equation $\sin \Theta/\sin \Theta_t = n_2/n_1$. To leave the unrefined Snell's law on the books is to signal that the same kind of explanation can be given even for some anisotropic media. The pattern of explanation derived from the ideal situation is employed even where the conditions are less than ideal; and we assume that we can understand what happens in *nearly* isotropic media by rehearsing how light rays behave in pure isotropic cases.

This assumption is a delicate one, and it obviously derives from certain metaphysical views we hold about the continuity of physical processes. I wish I had more to say about it. But for the moment I intend only to point out that it *is* an assumption, and an assumption which (prior to the "full electromagnetic theory") goes well beyond our knowledge of the facts of nature. We *know* that in isotropic media, the angle of refraction is due to the angle of incidence under the equation $\sin \theta/\sin \theta_t = n_2/n_1$. We *decide* to explain the angles for the two refracted rays in anisotropic media in the same manner. We may have good reasons for the decision—in this case if the media are nearly isotropic, the two rays will be very close together, and close to the angle predicted by Snell's law; we believe in continuity of physical processes; etc.—but still this decision is not forced by our knowledge of the laws of nature.

Obviously this decision could not be taken if we also had on the books a second refinement of Snell's law, implying that in any anisotropic media, the angles are quite different from those given by Snell's law. But, as I shall argue, laws are scarce, and often we have no law at all about what happens in conditions which are less than ideal.

Covering law theorists will tell a different story about the use of *ceteris paribus* laws in explanation. From their point of view, *ceteris paribus* explanations are elliptical for genuine covering law explanations from true laws which we don't yet know. When we use a *ceteris paribus* "law" which we know to be false, the covering law theorist supposes us to be making a bet about what form the true law takes. For example, to retain Snell's unqualified law would be to bet that the (at the time unknown) law for anisotropic media will entail values "close enough" to those derived from the original Snell law.

I have two difficulties with this story. The first arises from an extreme metaphysical possibility, which I in fact believe in. Covering law theorists tend to think that nature is well-regulated; in the extreme, that there is a law to cover every case. I do not. I imagine that natural objects are much like people in societies. Their behavior is constrained by some specific laws and by a handful of general principles, but it is not determined in detail, even statistically. What happens on most occasions is dictated by no law at all.

This is not a metaphysical picture that I urge. My claim is that this picture is as plausible as the alternative. God may have written just a few laws and grown tired. Determinists, or whoever, may contend that nature must be simple, tidy, an object of beauty and admiration. . . . But there is one outstanding empirical dictum in favor of untidiness: if we must make metaphysical models of reality, we had best make the model as much like our experience as possible. So I would model the Book of Nature on the best current Encyclopedia of Science; and current encyclopedias of science are a piecemeal hodgepodge of different theories for different kinds of phenomena, with only here and there the odd connecting law for overlapping domains.

The best policy is to remain agnostic, or at least not to let other important philosophical issues depend on the outcome. We don't know whether we are in a tidy universe or an untidy one. But whichever universe we are in, the ordinary commonplace activity of giving explanations ought to make sense. It may turn out that in the Last Judgment God allows us to look at the Book of Nature and we see that it is woefully incomplete. We ought not to have an analysis of explanation that tells us, then, that we never were explaining all along, that the activity didn't make sense most of the time we did it.

The second difficulty for the ellipsis version of the covering law account is more pedestrian. But it is based on the same fundamental point: we should adopt no account of explanation which dictates that most of the time we think we're explaining, we're not. The covering law account of *ceteris paribus* laws has just this consequence. For elliptical explanations aren't explanations: they are at best assurances that explanations are to be had. The law which is supposed to appear in the complete, correct D-N explanation is not a law we have in our theory, a law that we can state, let alone test. There may be covering law explanations in these cases. But those explanations are not our explanations; and those unknown laws cannot be our grounds for saying of a nearly isotropic medium, "$\sin \Theta_t \approx k(n_2/n_1)$ *because* $\sin \Theta = k$."

What then are our grounds? I claim only what they are not: they are not the laws of nature. The laws of nature that we know at any time are not enough to tell us what kinds of explanations can be given at that time. That requires a decision; and it is just this decision that covering law theorists make when they wager about the existence of unknown laws. We may believe in these unknown laws, but we do so on no ordinary grounds: they have not been tested, nor are they derived from a higher level theory. Our grounds for believing in them are only as good as our reasons for adopting the corresponding explanatory strategy, and no better.

III. WHEN LAWS CONFLICT

I have been maintaining that there aren't enough covering laws to go around. Why? The view depends on the picture of science that I mentioned earlier. Science is broken into various distinct domains: hydrodynamics, genetics, laser theory, . . . We have a lot of very detailed and sophisticated theories about what happens within the various domains. But we have little theory about what happens in the intersection of domains.

Diagrammatically, we have laws like

$$\textit{ceteris paribus}, (x)\,(S(x) \rightarrow I(x))$$

and

$$\textit{ceteris paribus}, (x)\,(A(x) \rightarrow -I(x)).$$

For example, (*ceteris paribus*) adding salt to water decreases the cooking time of potatoes; taking the water to higher altitudes increases it. Refining, if we spoke more carefully we might say instead, "Adding salt to water while keeping the altitude constant decreases the cooking time; whereas increasing the altitude while keeping the saline content fixed increases it;" or

$$(x)\,(S(x)\ \&\ -A(x) \rightarrow I(x))$$

and

$$(x)\,(A(x)\ \&\ -S(x) \rightarrow -I(x)).$$

But neither of these tells what happens when we both add salt to the water and move to higher altitudes.

Here we think that probably somewhere in the books there is a precise answer about what would happen, even though it is not part of our common folk wisdom. But this is not always the case. An example which I have discussed before[7] will illustrate. Flow processes like diffusion, heat transfer, or electric current are described by various well-known phenomenological laws—Fick's law for diffusion; Fourier's for heat flow; Newton's law for shearing force; and Ohm's law for electric current. But these are not true laws: each is a *ceteris paribus* law which describes what happens only so long as a single cause (e.g., a concentration gradient or a temperature gradient) is at work. Most real life cases involve some combination of causes; and general laws which describe what happens in these complex cases are not available. There is no general theory for how to combine the effects of the separate phenomenological laws.

The same is true for other disciplines as well. For example, although both quantum theory and relativity are highly developed, detailed, and sophisticated, there is no satisfactory theory of relativistic quantum mechanics. Where theories intersect, laws are usually hard to come by.

IV. WHEN EXPLANATIONS CAN BE GIVEN ANYWAY

So far, I have only argued half the case. I have argued that covering laws are scarce, and that *ceteris paribus* laws are no true laws. *Ceteris paribus* laws, read literally as descriptions or regularities in nature, are either false, if the *ceteris paribus* modifier is omitted, or irrelevant to much real life, if it is included. It remains to argue that, nevertheless, *ceteris paribus* laws have a fundamental explanatory role. But this is easy, for most of our explanations are explanations from *ceteris paribus* laws.

Let me illustrate with a humdrum example. Last year I planted camelias in my garden. I know that camelias like rich soil so I planted them in composted manure. On the other hand, the manure was still warm, and I also know that camelia roots can't take high temperatures. So I did not know what to expect. But when many of my camelias died, despite otherwise perfect care, I knew what went wrong. The camelias died because they were planted in hot soil.

This is surely the right explanation to give. Of course, I cannot be absolutely certain that this explanation is the correct one. Some other factor may have been responsible, nitrogen deficiency or some genetic defect in the plants, a factor which I didn't notice, or may not even have known to be relevant. But this uncertainty is not peculiar to cases of explanation. It is just the uncertainty that besets all of our judgments about matters of fact. We must allow for oversight; still, since I made a reasonable effort to eliminate other menaces to my camelias, we may have some confidence that this is the right explanation.

So, we have an explanation for the death of my camelias. But it is not an explanation from any true covering law. There is no law that says that camelias just like mine, planted in soil which is both hot and rich, die. To the contrary, they do not all die. Some thrive; and probably those that do, do so *because* of the richness of the soil they are planted in. We may insist that there must be some differentiating factor which brings the case under a covering law—in soil which is rich and hot, camelias of one kind die; those of another thrive. I will not deny that there may be such a covering law. I merely repeat that our ability to give this humdrum explanation precedes our knowledge of that law. In the Day of Judgment, when all laws are known, these may suffice to explain all phenomena. Nevertheless, in the meantime we do give explanations; and it is the job of science to tell us what kinds of explanations are admissible.[8]

In fact, I want to urge a stronger thesis.

If, as is possible, the world is not a tidy deterministic system, this job of telling how we are to explain will be a job which is still left when the descriptive task of science is complete. Imagine for example (what I suppose to actually be the case) that the facts about camelias are irreducibly statistical. Then it is possible to know all the general nomological facts about camelias which there are to know—for example, that 62% of all camelias in just the circumstances of my camelias die, and 38% survive.[9] Still, one would not thereby know how to explain what happened in my garden. You would still have to look to the *Sunset Garden Book* to learn that the *heat* of the soil explains the perishing, and the *richness* explains the plants which thrive.

V. CONCLUSION

I have said that in general scientific explanations use *ceteris paribus* laws, laws which read literally as descriptive statements of fact are false, not only false, but deemed false even in the context of use. This is no accident. Explanatory laws by their very nature have exceptions; only by unlikely circumstance will such a law be literally true. Our picture of the explanatory structure of nature requires this. We suppose that there are certain fundamental laws at work in nature. (At these meetings last year, Ernan McMullin called these "structural laws".[10]) What objectively happens is a consequence of the *interplay* of these fundamental laws. The fundamental laws themselves do not describe objectively occurring regularities; rather, the regularities which occur in nature are the result of the operation and interference of these fundamental laws. It is part of the nature of an explanatory law that it hold only *ceteris paribus*—that is, that it not really hold at all. The laws which explain are not laws in any literal sense. They do not tell what truly happens in nature; and conversely, a full knowledge of what truly happens in nature, even what happens regularly and of necessity, does not tell how to explain. The tasks of describing nature and of telling how to explain it are distinct.

NOTES

1. Hempel, C. G. "Scientific Explanation" in C. G. Hempel, *Aspects of Scientific Explanation.* (New York: Free Press, 1965).

2. Hempel, C. G., *ibid.*

3. Suppes, Patrick. *A Probabilistic Theory of Causality.* (Amsterdam, 1970).

4. Salmon, Wesley. "Statistical Explanation" in Wesley Salmon (ed.), *Statistical Explanation and Statistical Relevance.* (Pittsburgh, 1971).

5. Hanson, Bengt. "Explanations—Of What?" (Mimeographed: Stanford University, 1974).

6. Klein, Miles V., *Optics.* (New York, 1970).

7. Klein, *loc. cit.*, p. 602.

8. Cartwright, Nancy. "How Do We Apply Science?" in R. S. Cohen *et al.* (eds.), *PSA 1974.* (Dordrecht-Holland: Reidel, 1976).

9. Various writers, especially Suppes (note 3) and Salmon (note 4), have urged that knowledge of more sophisticated statistical facts will suffice to determine what factors can be used in explanation. I do not believe that this claim can be carried out. *Cf.* Cartwright, Nancy, "Causal Laws and Effective Strategies," in *Nous* (Nov., 1979). In that paper I argue that one already needs a full complement of causal knowledge before one can use statistical laws to fix explanatory relevance. This causal knowledge, however, is knowledge of no singular facts, nor even of any regularities; so full knowledge of the regularities of nature is not enough to determine how we are to explain.

10. McMullin, Ernan, "Structural Explanation," *American Philosophical Quarterly*, vol. 15, no. 2, (1979), pp. 139–147.

B. A. Brody

TOWARDS AN ARISTOTELEAN THEORY OF SCIENTIFIC EXPLANATION

In this paper, I consider a variety of objections against the covering-law model of scientific explanation, show that Aristotle was already aware of them and had solutions for them, and argue that these solutions are correct. These solutions involve the notions of nonHumean causality and of essential properties. There are a great many familiar objections, both methodological and epistemological, to introducing these concepts into the methodology of science, but I show that these objections are based upon misunderstandings of these concepts.

Let us begin by considering the following explanation of why it is that sodium normally combines with chlorine in a ratio of one-to-one[1]:

(A) (1) sodium normally combines with bromine in a ratio of one-to-one
(2) everything that normally combines with bromine in a ratio of one-to-one normally combines with chlorine in a ratio of one-to-one

---

(3) therefore, sodium normally combines with chlorine in a ratio of one-to-one.

This purported explanation meets all of the requirements laid down by Hempel's covering law model for scientific explanation ([5], pp. 248–249). After all, the law to be explained is deduced from two other general laws which are true and have empirical content. Nevertheless, this purported explanation seems to have absolutely no explanatory power. And even if one were to say, as I think it would be wrong to say. that it does have at least a little explanatory power, why is it that it is not as good an explanation of the law in question (that sodium normally combines with chlorine in a ratio of one-to-one) as the explanation of that law in terms of the atomic structure of sodium and chlorine and the theory of chemical bonding? The covering law model, as it stands, seems to offer us no answer to that question.

A defender of the covering law model would, presumably, offer the following reply: both of these explanations, each of which meets the requirements of the model, are explanations of the law in question, but the explanation in terms of atomic structure is to be preferred to the explanation in terms of the way that sodium combines with bromine because the former contains in its explanans more powerful laws than the latter. The laws about atomic structure and the theory of bonding are more powerful than the law about the ratio with which sodium combines with bromine because more phenomena can be explained by the former than by the latter.

I find this answer highly unsatisfactory, partially because I don't see that (A) has any explanatory power at all. But that is not the real problem. The real problem is that this answer leaves something very mysterious. I can see why, on the grounds just mentioned, one would prefer to have laws like the ones about atomic structure rather than laws like

---

Reprinted by permission from *Philosophy of Science*, Vol. 39, no. 1, March 1972, pp. 20–31.

the one about the ratio with which sodium and bromine combine. But why does that make explanations in terms of the latter type of laws less preferable? Or to put the question another way, why should laws that explain more explain better?

So much for my first problem for the covering-law model, a problem with its account of the way in which we explain scientific laws. I should now like to raise another problem for it, a problem with its account of the way in which we explain particular events. Consider the following three explanations:

(B) (1) If the temperature of a gas is constant, then its pressure is inversely proportional to its volume
(2) at time t_1, the volume of the container c was v_1 and the pressure of the gas in it was p_1
(3) the temperature of the gas in c did not change from t_1 to t_2
(4) the pressure of the gas in container c at t_2 is $2p_1$

---

(5) the volume of x at t_2 is $\frac{1}{2}v_1$.

(C) (1) if the temperature of a gas is constant, then its pressure is inversely proportional to its volume
(2) at time t_1, the volume of the container c was v_1 and the pressure of the gas in it was p_1
(3) the temperature of the gas in c did not change from t_1 to t_2
(4) the volume of the gas in container c at t_2 is $\frac{1}{2}v_1$

---

(5) the pressure of c at t_2 is $2p_1$.

(D) (1) if the temperature of a gas is constant, then its pressure is inversely proportional to its volume
(2) at time t_1, the volume of the container c was v_1 and the pressure of the gas in it was p_1
(3) the temperature of the gas in c did not change from t_1 to t_2
(4) by t_2, I had so compressed the container by pushing on it from all sides that its volume was $\frac{1}{2}v_1$

---

(5) the pressure of c at t_2 is $2p_1$.

All three of these purported explanations meet all of the requirements of Hempel's model. The explanandum, in each case, is deducible from the explanans which, in each case, contain at least one true general law with empirical content. And yet, there are important differences between the three. My intuitions are that (B) is no explanation at all (thereby providing us with a clear counter-example to Hempel's model), that (C) is a poor explanation, and that (D) is a much better one. But if your intuitions are that (B) is still an explanation, even if not a very good one, that makes no difference for now. The important point is that there is a clear difference between the explanatory power of these three explanations, and the covering law model provides us with no clue as to what it is.

These problems and counter-examples are not isolated cases. I shall give, later on in this paper after I offer my own analysis and solution of them, a recipe for constructing loads of additional problems and counter-examples. Now the existence of these troublesome cases led me to suspect that there is something fundamentally wrong with the whole covering law model and that a new approach to the understanding of scientific explanation is required. At the same time, however, I felt that this model, which fits so many cases and seems so reasonable, just couldn't be junked entirely. This left me in a serious dilemma, one that I only began to see my way out of after I realized that Aristotle, in the *Posterior Analytics,* had already seen these problems and had offered a solution to them, one that contained both elements of Hempel's model and some other elements entirely foreign to it. So let me begin my presentation of my solution to these problems by looking at some aspects of Aristotle's theory of scientific explanation.

Aristotle (*ibid.*, I, 13), wanted to draw a distinction between knowledge of the fact (knowledge that *p* is so) and knowledge of the reasoned fact (knowledge why *p* is so) and he did so by asking us to consider the following two arguments, the former of which only provides us with knowledge of the fact while the latter of which provides us with knowledge of the reasoned fact:

(E) (1) the planets do not twinkle
(2) all objects that do not twinkle are near the earth

---

(3) therefore, the planets are near the earth.

(F) (1) the planets are near the earth
(2) all objects that are near the earth do not twinkle

---

(3) therefore, the planets do not twinkle.

The interesting thing about this point, for our purposes, is that while both of these arguments fit Hempel's model,[2] only one of them, as Aristotle already saw, provides us with an explanation of its conclusion. Moreover, his account of why this is so seems just right:

> . . . of two reciprocally predicable terms the one which is not the cause may quite easily be the better known and so become the middle term of the demonstration. . . . This syllogism, then, proves not the reasoned fact but only the fact; since they are not near because they do not twinkle. The major and middle of the proof, however, may be reversed, and then the demonstration will be of the reasoned fact . . . since its middle term is the proximate cause. (78^a 28–78^b 3)

In other words, nearness is the cause of not twinkling, and not vice versa, so the nearness of the planets to the earth explains why they do not twinkle, but their not twinkling does not explain why they are near the earth.

It is important to note that such an account is incompatible with the logical empiricists' theory of causality as constant conjunction. After all, given the truth of the premises of (E) and (F), nearness and nontwinkling are each necessary and sufficient for each other, so, on the constant conjunction account each is equally the cause of the other.[3] And even if the constant conjunction account is supplemented in any of the usual ways, nearness and nontwinkling would still equally be the cause of each other. After all, both 'all near celestial objects twinkle' and 'all twinkling celestial objects are near' contain purely qualitative predicates, have a potentially infinite scope, are deducible from higher-order scientific generalizations and support counterfactuals. In other words, both of these generalizations are law-like generalizations, and not mere accidental ones, so each of the events in question is, on a sophisticated Humean account, the cause of the other. So Aristotle's account presupposes the falsity of the constant conjunction account of causality. But that is okay. After all, the very example that we are dealing with now, where nearness is clearly the cause of nontwinkling but not vice versa, shows us that the constant conjunction theory of causality (even in its normal more-sophisticated versions) is false.

Now if we apply Aristotle's account to our example with the gas, we get a satisfactory account of what is involved there. The decrease in volume (due, itself, to my pressing on the container from all sides) is the cause of the increase in pressure, but not vice versa, so the former explains the latter but not vice versa. And Aristotle's account also explains a phenomenon called to our attention by Bromberger [2], viz. that while we can deduce both the height of a flagpole from the length of the shadow it casts and the position of the sun in the sky and the length of the shadow it casts from the height of the flagpole and the position of the sun in the sky, only the latter deduction can be used in an explanation. It is easy to see why this is so; it is the sun striking at a given angle the flagpole of the given height that causes its shadow to have the length that it does, but the sun striking the flagpole when its shadow has the length of the shadow is surely not the cause of the height of the flagpole.

Generalizing this point, we can add a new requirement for explanation: a deductive-nomological explanation of a particular event is a satisfactory explanation of the event when (besides meeting all of Hempel's requirements) its explanans contains essentially a description of the event which is the cause of the event described in the explanandum. If they do not then, it may not be a satisfactory explanation. And similarly, a deductive-nomological explanation of a law is a satisfactory explanation of that law when (besides meeting all of Hempel's requirements) every event which is a case of the law to be explained is caused by an event which is a case of one (in each case, the same) of the laws contained essentially in the explanans.[4]

It might be thought that what we have said so far is sufficient to explain why it is that (A) is not an explanation and why it is that the explanation of sodium and chlorine's combining in a one-to-one ratio in terms of the atomic structure of sodium and chlorine is an explanation. After all, no event which is a case of sodium and chlorine combining in a one-to-one ratio is caused by any event which is a case of sodium and bromine combining in a one-to-one ratio. So, given our requirements, deduction (A), even though it meets all of Hempel's requirements, need not be (and indeed is not) an explanation. But every event which is a case of sodium and chlorine combining in a one-to-one ratio is caused by the sodium and chlorine in question having the atomic structure that they do (after all, if they had a different atomic structure, they would combine in a different ratio). So an explanation involving the atomic structure would meet our new requirement and would therefore be satisfactory.

The trouble with this account is that it incorrectly presupposes that it is the atomic structure of sodium and chlorine that cause them to combine in a one-to-one ratio. A whole essay would be required to show, in detail, what is wrong with this presupposition; I can, here, only briefly indicate the trouble and hope that this brief indication will be sufficient for now: a given case of sodium combining with chlorine is the same event as that sodium combining with that chlorine in a one-to-one ratio, and, like all other events, that event has only one cause.[5] It is, perhaps, that event which brings it about that the sodium and chlorine are in proximity to each other under the right conditions. That is the cause of the event in question, and not the atomic structure of the sodium and chlorine in question (which, after all, were present long before they combined). To be sure, these atomic structures help explain one aspect of the event in question, the ratio in which they combine, but that does not make them the cause of the event.[6]

To say that the atomic structure of the atoms in question is not the cause of their combining in a one-to-one ratio is *not* to say that a description of that structure is not an essential part of any causal explanation of their combining. It obviously is. But equally well, to say that a description of it is a necessary part of any causal explanation is *not* to say that it is (or is part of) the cause of their combining. There is a difference, after all, between causal explanations and causes and between parts of the former and parts of the latter. Similarly, to say that the atomic structure is not the cause of their combining is *not* to say that that event had no cause; indeed, we suggested one (the event which brought about the proximity of the atoms) and others can also be suggested (the event of the atoms acquiring certain specific electrical and quantum mechanical properties). It is only to say that the atomic structure is not the cause.

Keeping these two points in mind, we can see that all that we said before was that the perfectly satisfactory explanation, in terms of the atomic structure of the atoms, of their combining in a one-to-one ratio does not meet the condition just proposed because it contains no description of the event which caused the combining to take place. But since it obviously is a good explanation, some additional types of explanations must be allowed for.

So Aristotle's first suggestion, while quite helpful in solving some of our problems, does not solve all of them. There is, however, another important suggestion that he makes that will, I believe, solve the rest of them. Aristotle says:

> Demonstrative knowledge must rest on necessary basic truths; for the object of scientific knowledge cannot be other than it is. Now attributes attaching essentially to their subjects attach necessarily to them. . . . It follows from this that premises of the demonstrative syllogism must be connections essential in the sense explained: for all attributes must inhere essentially or else be accidental, and accidental attributes are not necessary to their subjects. (*Posterior Analytics* 74^b 5–12)

There are many aspects of this passage that I do not want to discuss now. But one part of it seems to me to suggest a solution to our problem. It is the suggestion that a demonstration can be used as an explanation (can provide us with "scientific knowledge") when at least one of the explanans essential to the derivation states, that a certain class of objects has a certain property, and (although the explanans need not state this) that property is possessed by those objects essentially.

Let us, following that suggestion, now look at our two proposed explanations as to why sodium combines with chlorine in a ratio of one-to-one. In one of them, we are supposed to explain this in terms of the fact that sodium combines with bromine in a one-to-one ratio. In the other explanation, we are supposed to explain this in terms of the atomic structure of sodium and chlorine. Now in both of these cases, we can demonstrate from the fact in question (and certain additional facts) that sodium does combine with chlorine in a one-to-one ratio. But there is an important difference between these two proposed explanations. The atomic structure of some chunk of sodium or mass of chlorine is an essential property of that object. Something with a different atomic number would be (numerically) a different object. But the fact that it combines with bromine in a one-to-one ratio is not an essential property of the sodium chunk, although it may be true of every chunk of sodium. One can, after all, imagine situations[7] in which it would not combine in that ratio but in which it would still be (numerically) the same object. Therefore, one of our explanans, the one describing the atomic structure of sodium and chlorine, contains a statement that attributes to the sodium and chlorine a property which is an essential property of that sodium and chlorine (even if the statement does not say that it is an essential property), while the other of our explanans, the one describing the way in which sodium combines with bromine, does not. And it is for just this reason that the former explanans, but not the latter, explains the phenomenon in question.

Generalizing this Aristotelean point, we can set down another requirement for explanations as follows: a deductive-nomological explanation of a particular event is a satisfactory explanation of that event when (besides meeting all of Hempel's requirements) its explanans contains essentially a statement attributing to a certain class of objects a property had essentially by that class of objects (even if the statement does not say that they have it essentially) and when at least one object involved in the event described in the explanandum is a member of that class of objects. If this requirement is unfulfilled, then it may not be a satisfactory explanation. And similarly, a deductive-nomological explanation of a law is a satisfactory explanation of that law when (besides meeting all of Hempel's requirements) each event which is a case of the law which is the explanandum, involves an entity which is a member of a class (in each case, the same class) such that the explanans contain a statement attributing to that class a property which each of its members have essentially (even if the statement does not say that they have it essentially).

It is important to note that such an account is incompatible with the logical empiricist conception of theoretical statements as instruments and not as statements describing the world. For after all, many of these essential attributions are going to be theoretical statements, and they can hardly be statements attributing to a class of objects an essential property if they aren't really statements at all. But that is okay, for it just gives us one more reason for rejecting an account, more notable for the audacity of its proponents in proposing it than for its plausability or for the illumination it casts.

There are two types of objections to essential explanations that we should deal with immediately. The first really has its origin in Duhem's critique of the idea that scientific theories explain the observable world ([3], Ch. 1). Duhem argued that if we view a theory as an explanation of an observable phenomenon, we would have to suppose that the theory gives us an account of the physical reality underlying what we observe. Such claims about the true nature of reality are, however, empirically unverifiable metaphysical hypotheses, which scientists should shun, and therefore we must not view a theory as an attempt to explain what we observe. Now contemporary theories of explanation, like the deductive-nomological model, avoid this problem, by not requiring of an explanation that its explanans describe the true reality underlying the observed explanandum. But if we now claim that a deductive-nomological explanation is a satisfactory explanation when (among other possibilities) its explanans describe essential properties of some objects involved in the explanandum event, aren't we introducing these disastrous, because empirically undecidable, issues about the true nature of the reality of these objects into science? After all, the scientist will now have to decide, presumably by nonempirical means, whether the explanans do describe the essence of the objects in question.

The trouble with this objection is that it just assumes, without any arguments, that claims about the essences of objects would have to be empirically undecidable claims, claims that could be decided only upon the basis of metaphysical assumptions. This presupposition, besides being unsupported, just seems false. After all, the claim that the essential property of sodium is its atomic number (and not its atomic weight, or its color, or its melting point) can be defended empirically, partially by showing that for this property, unlike the others just mentioned, there are no obvious cases of sodium which do not have it, and partially by showing how all objects that have this property behave alike in many important respects while objects which do not have this property in common do not behave alike in these important respects. Now a lot more has to be said about the way in which we determine empirically the essence of a given object (or of a given type of object), and we will return to this issue below, but enough has been said, I think, to justify the claim that the idea that scientific explanation is essential explanation does not mean that scientific explanation involves empirically undecidable claims.

It should be noted, by the way, that this idea of the discovery of essences by empirical means is not new to us. It was already involved in Aristotle's theory of *epagoge* (*op. cit.,* II, 19). I do not now want to enter into the question as to exactly what Aristotle had in mind, if he did have anything exact in mind, when he was describing that process. It is sufficient to note that he, like all other true adherents to the theory of essential explanation, saw our knowledge of essences as the result of reflection upon what we have observed and not as the result of some strange sort of metaphysical knowledge.

The second objection to essential explanations has been raised by Popper. He writes:

> The essentialist doctrine that I am contesting is solely the doctrine that science aims at ultimate explanation; that is to say, an explanation which (essentially, or, by its very nature) cannot be further explained, and which is in no need of any further explanation. Thus my criticism of essentialism does not aim at establishing the nonexistence of essences; it merely aims at showing the obscurantist character of the role played by the idea of essences in the Galilean philosophy of science. ([8], p. 105)

Popper's point really is very simple. If our explanans contain a statement describing essential properties (e.g., sodium has the following atomic structure . . .), then there is nothing more to be said by way of explaining these explanans themselves. After all, what could we say by way of answering the question "why does sodium contain the atomic structure that it does"? So the use of essential explanations leads us to unexplainable explanans, and therefore to no new insights gained in the search for explanations of these explanans, and therefore to scientific sterility. Therefore, science should reject essential explanations.

There are, I believe, two things wrong with this objection. To begin with, Popper assumes that essential explanations will involve unexplainable explanans, and this is usually only partially true. Consider, after all, our explanation of sodium's combining with chlorine in a one-to-one ratio in terms of the atomic structure of sodium and chlorine. The explanans of that explanation, besides containing statements attributing to sodium and chlorine their essences (viz. their atomic number), also contain the general principles of chemical bonding, and these are not unexplainable explanans since they are not claims about essences. In general, even essential explanations leave us with some part (usually the most interesting part) of their explanans to explain, and they do not therefore lead to sterility in future enquiry. But, in addition, even if we did have an essential explanation all of whose explanans were essential statements and therefore unexplainable explanans, what are we to do according to Popper? Should we reject the explanation? Should we keep it but believe that it is not an essential explanation? Neither of these strategies seem very plausible in those cases where we have good reasons both for supposing that the explanation is correct and for supposing that the explanans do describe the essential properties of the objects in question. It cannot after all, be a good scientific strategy to reject what we have good reasons to accept. So even if Popper's claim about their sterility for future scientific enquiry is true for some essential explanations, I cannot see that it gives us any reasons for rejecting these explanations, or for rejecting their claim to be essential explanations, when these explanations and claims are empirically well supported.

There is, of course, a certain point to Popper's objection, a point that I gladly concede. As is shown by his example from the history of gravitational theory, people may rush to treat a property as essential, without adequate empirical evidence for that claim, and then it may turn out that they were wrong. They may even have good evidence for the claim that the property is essential and still be wrong. In either case, enquiry has been blocked where it should not have been blocked. We should certainly therefore be cautious in making claims about essential properties and should, even when we make them on the basis of good evidence, realize that they may still be wrong. But these words of caution are equally applicable to all scientific claims; the havoc wreaked by false theories that lead enquiry along mistaken paths can be as bad as the havoc wreaked by false essential claims that block enquiry. And since they are only words of caution, they should not lead us to give up either theoretical explanations in general or essential explanations in particular.

Let us see where we now stand. We have, so far, rejected Hempel's requirements for an explanation on the grounds that they are not sufficient and we have suggested two alternative Aristotelean conditions such that, for the set of explanations meeting Hempel's requirements, any explanation meeting either of these requirements is an adequate explanation.[8] Doing this is sufficient to help us deal with the problem of self-explanation, another problem that the covering-law model has had difficulty with. As Hempel already recognized in his original presentation of the covering-law model, we need some additional requirement to rule out such obvious self-explanations as

$$\text{(G)}\quad \frac{\begin{array}{l}(x)Px\\ Qa\end{array}}{Qa}$$

and slightly less obvious self-explanations such as

$$\text{(H)}\quad \frac{\begin{array}{l}(x)(Px \cdot Qa)\\ Qa \vee \sim Qa\end{array}}{Qa}$$

He proposed a simple solution to that problem but Eberle, Kaplan, and Montague showed that it wouldn't do [4]. Consider, they said, the following example of a bad explanation, that meets all of Hempel's requirements, of an object's having a property H. Let us take any true law of the form $(x)Fx$ (where there is no connection between an object's having the property F and its having the property H). From that law it follows that (where G is any third unrelated property)

(1) $(x)(y)[Fx \vee (Gy \supset Hy)]$

It also follows from Ha, the fact to be explained, that

(2) $(Fb \vee \sim Ga) \supset Ha.$

But from these two true statements, we can derive Ha, and this derivation, a subtle form of self-explanation, meets all of Hempel's requirements, so Hempel still had not solved the problem of self-explanation. Now there exist several syntactic solutions to this problem, solutions that are partially ad hoc and partially intuitively understandable (see [6] and [7]). As such, they are not entirely satisfactory. Writing about one of them, Hempel admits that:

> . . . it would be desirable to ascertain more clearly to what extent the additional requirement is justifiable, not on the ad hoc grounds that it blocks those proofs, but in terms of the rationale of scientific explanation. ([5], p. 295)

Now our theory offers a simple, non-ad hoc, solution to this problem. The derivation used by Eberle, Kaplan, and Montague does not meet either of our two conditions. Neither (1) nor (2) describe the cause of Ha. And neither (1) nor (2) ascribe an essential property to the members of a certain class of objects of which a is a member. Therefore, although their derivation meets all of Hempel's requirements, it need not be (and indeed is not) an adequate explanation.

By now, the advantages of our theory are obvious. It provides intuitively satisfactory, non-ad hoc, solutions to problems that the covering-law model cannot handle. And at the same time, it incorporates (by keeping Hempel's requirements) the elements of truth in the covering-law model. It only remains, therefore, to consider the one serious objection to this whole Aristotelean theory, an objection that we have already touched upon when we dealt with Duhem. Given what we mean by 'causality' and 'essence', can we ever know that e_1 is the cause of e_2 or that P is an essential property of e_1, and if so, how can we know this?

This problem can be sharpened considerably. There is no problem, in principle, about our coming to know that events of type E_1 are constantly conjoined with events of type E_2. All that we have to do is to observe that this is so in enough varied cases. And if 'e_1 causes e_2' only means that 'e_1 is an event of type E_1 and e_2 is an event of type E_2 such that E_1 is constantly conjoined with E_2', we can easily see how we could come to know that e_1 is the cause of e_2. But if, as our Aristotelean account demands, 'e_1 causes e_2' means something more than that, can we know that it is true, and if we can, how can we know that it is true? Similarly, there is no problem, in principle, about our coming to know that objects of type O_1 have a certain property P_1 in common. All that we have to do is to observe that this is so in enough varied cases. And if 'P_1 is an essential property of o_1' only means that 'o_1 is an object of type O_1 and all objects of type O_1 have P_1', we can easily see how we could come to know that P_1 is an essential property of o_1. But if, as our Aristotelean theory demands, 'P_1 is an essential property of o_1' means something more than that, can we know that this is true, and if we can, how can we know that this is true?

There is an important difference between these questions. If we conclude that we cannot, or do not, know the truth of statements of the form 'e_1 causes e_2' or 'P_1 is an essential property of o_1' (where these statements are meant in the strong sense required by our theory), then our theory must be rejected. After all, knowledge of the truth of statements of that form is, according to our theory, a necessary condition for knowing that we have (although not for having) adequate explanations. And we obviously do know, in at least some cases, that a given explanation is adequate. So if we cannot, or do not, know statements of the above form, our theory is false. However, if we conclude that we can, and do, know the truth of statements of the above-mentioned form, but we don't know how we know their truth, then all that we have left is a research project, viz. to find out how we know their truth; what we don't have is an objection to our theory.

This is an extremely important point. I shall show, in a moment, that we do have, and a fortiori can have, knowledge of these statements. But, to be quite frank, I have no adequate account (only the vague indications mentioned above when talking about Duhem) of how we have this knowledge. So, on the basis of this last point, I conclude that the Aristotelean theory of explanation faces a research problem about knowledge (hence the title of this paper), but no objection about knowledge.

Now for the proof that we do, and a fortiori can, have knowledge of the above-mentioned type. Our examples will, I am afraid, be familiar ones. It seems obvious that we know that

(1) if the temperature of a gas is constant, then an increase in its pressure is invariably accompanied by an inversely proportional decrease in its volume
(2) if the temperature of a gas is constant, then a decrease in its volume is invariably accompanied by an inversely proportional increase in its pressure
(3) if the temperature of a gas is constant, then an increase in its pressure does not cause an inversely proportional decrease in its volume
(4) if the temperature of a gas is constant, then a decrease in its volume does cause an inversely proportional increase in its pressure.

Here we have causal knowledge of the type required, since the symmetry between (1) and (2) and the asymmetry between (3) and (4) show that the causal knowledge that we have when we know (3) and (4) is not mere knowledge about constant conjunctions. Similarly, it seems obvious that we know that

(1) all sodium has the property of normally combining with bromine in a one-to-one ratio

(2) all sodium has the property of having the atomic number 11
(3) the property of normally combining with bromine in a one-to-one ratio is not an essential property of sodium
(4) the property of having the atomic number 11 is an essential property of sodium.

Here we have essential knowledge of the type required, since the symmetry between (1) and (2) and the asymmetry between (3) and (4) show that the essential knowledge that we have when we know (3) and (4) is not mere knowledge about all members of a certain class having a certain property.

I conclude, therefore, that we have every good reason to accept, but none to reject, the Aristotelean theory of explanation sketched in this paper. And I also conclude that it therefore behooves us to find out how we have the type of knowledge mentioned above, the type of knowledge that lies behind our knowledge that certain explanations that we offer really are adequate explanations.

NOTES

1. I first called attention to the problems raised by this type of explanation in my [1].

2. Leaving aside the question, irrelevant for us now, about the truth of these premises, we shall throughout this discussion just assume that they are true.

3. Unless one adds the requirement that the cause must be before the effect, the normal way of drawing an asymmetry between causes and effects when each are necessary and sufficient for the other, in which case neither is the cause of the other and Aristotle's account still won't do.

4. We have made this condition sufficient, but not necessary, for reasons that will emerge below. It will also be seen there that Aristotle, who had a broader notion of cause, could have made it necessary as well.

5. To be sure, e_1 can have as causes both e_2 and e_3 (where $e_2 \neq e_3$) but only when either e_2 is the cause of e_3 or e_3 the cause of e_2. That exception is of no relevance here.

6. It might, at least, be maintained that they are still the cause of that aspect of the event. But that is just a confusion—it is events, and not their aspects, that have causes.

7. Even ones in which all currently believed scientific laws hold, but in which the initial conditions are quite different from the ones that now normally hold.

8. We have not, however, required as a necessary condition that any explanation must meet one of these two conditions. This is so, partially because of the problem of statistical explanations, but partially because of the possibility, raised by Aristotle, that there are additional types of explanations. After all, our two conditions let in explanations in terms of Aristotle's efficient and formal causes. We still have to consider, but will not in this paper, possible explanations in terms of what he would call material and final causes.

REFERENCES

1. Brody, B. A. "Natural Kinds and Real Essences." *Journal of Philosophy* (1967).

2. Bromberger, S. "Why Questions." *Introductory Readings in the Philosophy of Science.* Edited by B. A. Brody. Englewood Cliffs: Prentice Hall, 1969.

3. Duhem, P. *La Théorie Physique, Son Objet et sa Structure.* Paris: Chevalier et Rivière, 1914.

4. Eberle, R. A., Kaplan, D., and Montague, R. "Hempel and Oppenheim on Explanation." *Philosophy of Science* (1961).

5. Hempel, C. G. *Aspects of Scientific Explanation.* New York: Free Press, 1955.

6. Kaplan, D. "Explanation Revisited." *Philosophy of Science* (1961).

7. Kim, J. "Discussion: On the Logical Conditions of Deductive Explanation." *Philosophy of Science* (1963).

8. Popper, K. R. *Conjectures and Refutations.* London: 1963.

Peter Achinstein

CAN THERE BE A MODEL OF EXPLANATION?

1. INTRODUCTION

Since 1948, when Hempel and Oppenheim[1] published their pioneering article, various models of explanation have appeared. But each has had its counterexamples, and observers of the philosophical scene may wonder whether models of the kind sought are really possible. Are their proponents engaged in a fruitless task of inquiry?

Hempel and other modelists are particularly concerned with explanations that answer questions of the form

(1) Why is it the case that *p*?[2]

The sentence replacing '*p*' in (1) Hempel calls the *explanandum.* It describes the phenomenon, or event, or fact, to be explained. The answer to an explanation-seeking why-question of form (1) Hempel calls the *explanans.* It is a sentence, or set of sentences, that provides the explanation. We can speak of the explanans as explaining the explanandum. And we can say that an explanans *potentially* explains an explanandum when, if the sentences of the explanans were true, the explanans would correctly explain the explanandum.

Thus, if the explanation-seeking why-question is

Q: Why is it the case that this metal expanded,

then the explanandum is

(2) This metal expanded.

If, in reply to *Q*, an explainer claims that

(3) This metal was heated; and all metals that are heated expand,

then (3) is the explanans for the explanandum (2). And (3) potentially explains (2) if, given the truth of (3), (3) would correctly explain (2).

A *model* of explanation is a set of necessary and sufficient conditions that determine whether the explanans correctly explains the explanandum (where the explanation-seeking question is of form (1)). It can also be described as a set of conditions that determine whether the explanans potentially explains the explanandum. If the conditions are satisfied by a given explanans and explanandum, then the former correctly explains the latter, provided that the former is true.

In what follows, my concern will be with models as sets of sufficient (rather than necessary) conditions for correct explanations; and as providing such conditions for explanations of particular events or facts rather than of general laws. Most of the counterexamples in the literature have been raised against models construed in this way. I shall argue that one important reason for the failure of these models is that their proponents want to impose requirements which, in effect, destroy the efficacy of their models.

Reprinted by permission from *Theory and Decision,* Vol. 13, no. 3, September 1981.

2. TWO REQUIREMENTS FOR A MODEL

The first is that no singular sentence or conjunction of such sentences in the explanans can entail the explanandum.[3] I will call this the No-Entailment-By-Singular-Sentence requirement, or NES for short. What is the justification for it?

There are, I suggest, three reasons that modelists I have in mind support it. First, it precludes certain 'self-explanations' and 'partial self-explanations'. Suppose we want to explain why a particular metal expanded. Assume that the explanandum in this case is

(1) This metal expanded.

The NES requirement precludes (1) itself from being or, being part of, an explanans for (1). It also precludes from an explanans for (1) sentences such as "This metal was heated and expanded", and "This metal expanded, and all metals that are heated expand", which would be regarded as partial self-explanations of (1).

Secondly, modelists emphasize the importance of general laws in an explanans. Such laws provide an essential link between the singular sentences of the explanans and the singular sentence that constitutes the explanandum. Intuitively, to explain a particular event involves relating it to other particular events via a law; if the singular sentences of the explanans themselves entail the explanandum laws become unnecessary, on this view.

Thirdly, the NES requirement in effect removes from an explanans certain sentences which involve explanatory connectives such as 'explains', 'because', 'on account of', 'due to', 'reason', and 'causes'. Let me call sentences in which such terms connect phrases or other sentences *explanation-sentences*. Here are some examples:

(2) This metal's being heated explains why it expanded
This metal expanded because it was heated
This metal expanded on account of its being heated
The reason this metal expanded is that it was heated
This metal's expanding is due to its being heated
The fact that this metal was heated caused it to expand.

NES precludes any of the explanation-sentences in (2) from being, or being part of, an explanans for (1), since each of them is a singular sentence that entails (1).[4] Without a condition such as NES one could simply require, e.g., that an explanans for an explanandum *p* be a singular explanation-sentence of the form '*q* explains (why) *p*' or '*p* because *q*'. Any such explanans, if it were true, would correctly explain the explanandum *p*. Of course, any of the six sentences in (2), if true, could be cited in correctly explaining why this metal expanded. Modelists need not deny this. Their claim is that the sentences in (2) do not correctly explain (1) in the right sort of way. They would exclude sentences in (2) from an explanans for (1) because they think that an adequate explanans for (1) must reconstruct the sentences in (2) so that the explanatory connectives in the latter are, in effect, analyzed in non-explanatory terms. One of the purposes of a model of explanation is to define terms such as 'explains', 'because', 'reason', and 'cause', and not to allow them to be used as primitives within an explanans. By providing some analysis of explanation modelists want to show why it is that this metal's being heated explains why it expanded. There is little enlightenment in saying that this is so because (2) is true.

NES does not exclude all singular explanation-sentences (or all explanatory connectives) from an explanans.[5] But by precluding those that entail the explanandum it does eliminate ones that, from the viewpoint of the modelists, most seriously reduce the possibility of philosophical enlightenment from the resulting explanation. (Whether such modelists would advocate a broader requirement eschewing all explanation-sentences from an explanans is a possibility I shall not discuss.)

NES also excludes certain sentences from an explanans that do not explicitly invoke explanatory connectives but are importantly like those that do which are excluded. Suppose we want to explain why the motion of this particle was accelerated. Our explanandum is

(3) The motion of this particle was accelerated.

Consider the explanans

(4) An electrical force accelerated the motion of this particle.

Although (4) itself contains no explicit explanatory connective such as 'explains', 'because', or 'causes', it, nevertheless carries a causal implication concerning the event to be explained. It is roughly equivalent to the following explanation-sentences which do have such connectives:

(5) An electrical force caused the motion of this particle to be accelerated
The motion of this particle was accelerated because of the presence of an electrical force.

NES precludes (4) as well as (5) from an explanans for (3), since (4) is a singular sentence that entails (3). Those who support NES would emphasize that (4), no less than (5), invokes an essentially explanatory connection between an explanans-event and the explanandum-event which it is the task of a model of explanation to explicate.

The second requirement, which I shall call the *a priori* requirement, is that the only *empirical* consideration in determining whether the explanans correctly explains the explanandum is the truth of the explanans; all other considerations are *a priori*. Accordingly, whether an explanans potentially explains an explanandum is a matter that can be settled by *a priori* means (e.g., by appeal to the meanings of words, and to deductive relationships between sentences). A model must thus impose conditions on potential explanations the satisfaction of which can be determined non-empirically. A condition such as this would therefore be precluded:

> An explanans potentially explains an explanandum only if there is a (true) universal or statistical law relating the explanans and explanandum.

Whether there is such a law is not an *a priori* matter.

The idea is that a model of explanation should require that sufficient information be incorporated into the explanans that it becomes an *a priori* question whether the explanans, if true, would correctly explain the explanandum. There is an analogy between this and what various logicians and philosophers say about the concepts of *proof* and *evidence*.

Often a scientist will claim that a proposition q can be proved from a proposition p, or that e is evidence that h is true, even though the scientist is tacitly making additional empirical background assumptions which have a bearing on the validity of the proof or on the truth of the evidence claim. If all of these assumptions are made explicit as additional premises in the proof, or as additional conjuncts to the evidence, then whether such and such is a proof, or is evidence for a hypothesis, is settleable *a priori*. Similarly, often a scientist who claims that a certain explanans correctly explains an explanandum will be making relevant empirical background assumptions not incorporated into the explanans; if the latter are made explicit and added to the explanans, it becomes an *a priori* question whether the explanans, if true, would correctly explain the explanandum.

There is an additional similarity alleged

between these concepts. A scientist would not regard a proof as correct—i.e., as proving what it purports to prove—unless its premises are true. Nor would he regard *e* as evidence that *h* (or *e* as confirming or supporting *h*) unless *e* is true. (That John has those spots is not evidence that he has measles unless he does have those spots.)[6] And whether the premises of the proof, or the evidence report, is true is, in the empirical sciences at least, not an *a priori* question. Nevertheless, deductive logicians, as well as inductive logicians in the Carnapian tradition, believe that they can isolate an *a priori* aspect of proof and evidence such that the only empirical consideration in determining whether a proof or a statement of the form '*e* is evidence that *h*' is correct is the truth of the premises of the proof or of the *e*-statement in the evidence claim; all other considerations are *a priori*. What I have been calling the *a priori* requirement makes the corresponding claim about the concept of explanation: the only empirical consideration in determining whether an explanation is correct is the truth of the explanans.

3. MODELS PURPORTING TO SATISFY THESE REQUIREMENTS

a. The Basic D–N Model

Consider this model as providing a set of sufficient conditions for explanations of particular events, facts, etc. The explanation-seeking question is of the form 'why is it the case that *p*?', where *p*, the explanandum, is a sentence describing the event to be explained. The explanans is a set containing sentences of two sorts. One sort purports to describe particular conditions that obtained prior to, or at the same time as, the event to be explained. The other are lawlike sentences (sentences that if true are laws). The model requires that the explanans entail the explanandum and that the explanans be true.

No singular sentence (or conjunction of such sentences) that entails the explanandum will appear in the explanans.[7] Any such sentence which is an explanation-sentence that someone might utter in explaining something will itself be analyzed as a D–N explanation, i.e., as a deductive argument in which no premises that are singular sentences entail the conclusion. For example, if someone utters the singular explanation-sentence

(1) This metal's being heated explains why it expanded (caused it to expand, etc.)

in explaining why the metal expanded, a D–N theorist will restructure (1) as an argument such as this

(2) This metal was heated
Any metal that is heated expands
Therefore,
This metal expanded.

And he will identify the premises of (2) as the explanans of the explanation and the conclusion as the explanandum. The premise in (2) that is a singular sentence does not entail the conclusion.

The *a priori* requirement also seems to be satisfied by this model. The only empirical consideration in determining whether the explanans correctly explains the explanandum is the truth of the explanans; all other considerations are *a priori*. What are these other considerations? They are whether the explanans (but not the conjunction of singular sentences in it) deductively entails the explanandum and whether it contains at least some sentences that are lawlike. The former is not an empirical question, nor is the latter, as construed by Hempel, since whether a sentence is lawlike depends only on its syntactical form and the semantical interpretation of its terms.[8]

The D–N model as a set of sufficient conditions for particular events is very broad, and one might seek to add further restrictions. Three more limited versions will be noted.

b. The D–N Dispositional Model[9]

Here the explanandum is a sentence with a form such as

(3) *X* manifested *P* when conditions of type *C* obtained.

And the explanans contains sentences of the form

(4) *X* has *F*, and conditions of type *C* obtained

(5) Anything with *F* manifests *P* when conditions of type *C* obtain.

For an explanans consisting of (4) and (5) to provide a correct D–N dispositional explanation of (3) the model requires that *F* be a disposition-term, that (5) be lawlike, that (4) and (5) entail (3), and that (4) and (5) be true. The singular sentence in the explanans is not to entail the explanandum. And the satisfaction of all the conditions of the model, save for the truth of the explanans, is determined *a priori*. The only condition in addition to those of the basic *D–N* model is that *F* be a disposition-term, something settleable syntactically and/or semantically.

c. The D–N Motivational Model[10]

Here the explanandum is a sentence saying that some agent acted in a certain way. The explanans contains a singular sentence attributing a desire (motive, end) to that agent, a singular sentence attributing the belief to that agent that performing the act described in the explanandum is, in the circumstances, a (the best, the only) way to satisfy that desire, and a lawlike sentence relating desires, beliefs, and actions of the kind in question. For example, the explanandum might be a sentence of the form.

(6) Agent *X* peformed act *A*,

and the explanans might contain sentences of the form

(7) *X* desired *G*.

(8) *X* believed that doing *A* is, in the circumstances, a (the best, the only) way to obtain *G*

(9) Whenever an agent desires something *G* and believes that the performance of a certain act is, in the circumstances, a (etc.) way to obtain *G* he performs that act.[11]

For an explanans consisting of (7)–(9) to provide a correct D–N *motivational* explanation of (6) the model requires that (9) be lawlike, that (7)–(9) entail (6), and that (7)–(9) be true. In this model, like the others, the singular sentences in the explanans are not to entail the explanandum, and the satisfaction of all the requirements of the model, save for the truth of the explanans, is settleable *a priori*.

d. Woodward's Functional Interdependence Model[12]

This proposes adding to the basic D–N conditions the following additional necessary condition:

(10) *Condition of functional interdependence:* the law occurring in the explanans for the explanandum *p* must be stated in terms of variables or parameters, variations in the values of which will permit the derivation of other explananda which are appropriately different from *p* (p. 46).

Suppose that the explanation-seeking question is 'Why is it the case that this pendulum has a period of 2.03 seconds?', for which the explanandum is

(11) This pendulum has a period of 2.03 seconds.

Consider the following D–N argument:

(12) This pendulum is a simple pendulum
The length of this pendulum is 100 cm
The period T of a simple pendulum is related to the length L by the formula $T = 2\pi(L/g)^{1/2}$, where $g = 980$ cm/sec^2
Therefore,
This pendulum has a period of 2.03 seconds.

The third premise in (12) is a law satisfying Woodward's condition (10). Its variables are the period T and the length L. And variations in the values of these variables will permit the derivation of explananda which Woodward regards as appropriately different from (11). For example, if we change the explanandum (11) to

(13) This pendulum has a period of 3.14 seconds

the law in (12) allows the derivation of (13) if the value of L is changed to 245 cm. Woodward is impressed by the fact that explanations, particularly in science, permit a variety of possible phenomena to be explained. He writes:

> The laws in examples [of this sort] formulate a systematic relation between . . . variables. They show us how a range of different changes in certain of these variables will be linked to changes in others of these variables. In consequence, these generalizations are such that when the variables in them assume one set of values (when we make certain assumptions about boundary and initial conditions) the explananda in the . . . explanations are derivable, and when the variables in them assume other sets of values, a range of other explananda is derivable (p. 46).

The satisfaction of condition (10) is settleable *a priori*. If this condition is the only one to be added to those of the basic D–N model, then the resulting model satisfies the *a priori* requirement and NES.[13]

4. VIOLATION OF THE A PRIORI REQUIREMENT

Despite the claims of these models the *a priori* requirement is not really satisfied (or else we will have to call certain explanations correct which are clearly not so). In order to show this I shall make use of some of the many counterexamples that have been employed against the D–N model. In these examples, the explanans is true, and the other D–N conditions are satisfied. Yet the explanandum-event did not occur because of the explanans-event, but for some other reason; and this can only be known empirically.

Consider this example:

(a) Jones ate a pound of arsenic at time t
Anyone who eats a pound of arsenic dies within 24 hours
Therefore,
Jones died within 24 hours of t.

Assume that the premises of (a) are true. Then it is supposed to be settleable *a priori* whether these premises correctly explain the conclusion. According to the D–N model all we need to determine is whether the second premise is lawlike (let us assume that it is), and whether the conjunction of premises (but not the first premise alone) entails the conclusion (it does). Since these D–N conditions are satisfied, the explanans should correctly explain the explanandum; and assuming the truth of the explanans, this matter is settleable on *a priori* grounds alone, no matter what other empirical propositions are true. However, the matter is not settleable *a priori*, since Jones could have died within 24 hours of *t* for some unrelated reason. For example, he might have died in a car accident not brought on by his arsenic feast, which, given the information in the explanans, could only be determined empirically. Suppose he did die from being hit by a car. Then the explanans in (a) does not correctly explain the explanandum, even though all the conditions of the D–N model are satisfied. Assuming the truth of the premises in (a) it is not settleable *a priori* whether these premises correctly explain the explanandum.

A similar problem besets all the more specialized versions of the D–N model cited above. Thus consider these D–N arguments:

(b) Disposition example
That bar is magnetic, and a small piece of iron was placed near it
Any magnetic bar is such that when a small piece of iron is placed near it the iron moves toward the bar
Therefore,
This small piece of iron moved toward the bar.

Suppose, however, that a much more powerful contact force had been exerted on the small piece of iron, and that it moved toward the bar because of this force, not because the bar is magnetic. (Assume that the magnetic force is negligible by comparison with the mechanical force.)

(c) Motivational example
Smith desired to buy eggs and he believed that going to the store is the only way to buy eggs
Whenever, etc. (law relating beliefs and desires to actions)
Therefore,
Smith went to the store.

But suppose Smith went to the store because he wanted to see his girlfriend who works in the store, not because he wanted to buy eggs.

(d) Functional interdependence example
This pendulum is a simple pendulum
The period of this pendulum is 2.03 seconds
The period T of a simple pendulum is related to the length L by the formula $T = 2\pi(L/g)^{1/2}$, where $g = 980$ cm/sec^2
Therefore,
This pendulum has a length of 100 cm.

A pendulum has the period it has because of its length, but not *vice versa*. (This type of counterexample is like the others in so far as it invokes an explanans-fact that is inoperative with respect to the explanandum-fact; but the case is also different because there is no intervening cause here, although there is in the others.)

With each of the D–N models considered, whether a particular example satisfies the *requirements of the model* (with the exception of the truth-requirement for the explanans) is a settleable *a priori*. Yet in all of these cases, given the truth of the explanans, whether the latter correctly explains the explanandum is not settleable *a priori*. Thus in example (c), even if Smith desired to buy eggs, and he believed that going to the store is the only way to do so, and the lawlike sentence relating beliefs and desires to actions is true, it does not follow that

(1) Smith went to the store *because* he desired to buy eggs and believed that going to the store is the only way to buy eggs.

The explanans of (c) correctly explains the explanandum only if (1) is true. Yet given the truth of the explanans of (c) it is not settleable *a priori*, but only empirically, whether (1) is true.

We could, of course, see to it that the matter is settleable *a priori* by changing the motivational model so as to incorporate (1) into the explanans of (c). Assuming that this enlarged explanans is true, whether the latter correctly explains the explanandum is *a priori*—indeed trivially so. But now, of course, the NES requirement is violated, since (1) is a singular sentence that entails the explanandum of (c).

In example (d), even if the explanans is true and Woodward's condition of functional interdependence is satisfied, it does not follow that

(2) This pendulum has a length of 100 cm *because* it is a simple pendulum with a period of 2.03 seconds and the law of the simple pendulum holds.[14]

The explanans of (d) correctly explains the explanandum only if (2) is true. Yet assuming the truth of the explanans of (d), the truth-value of (2) is not settleable *a priori*, but only empirically. Assuming that there is a lawlike connection between the period and length of a simple pendulum, whether the pendulum has the period it does because of its length, or whether it has the length it does because of its period, or whether neither of these is true, is not knowable *a priori*.[15]

More generally, in the explanans in each of these models some factors are cited, together with a lawlike sentence relating these factors to the type of event to be explained. But given that the factors were present and that the lawlike sentence is true, there is no *a priori* guarantee that the event in question occurred because of those factors. Whether it did is an empirical question whose answer even the truth of the lawlike sentence does not completely determine. And if we include in the explanans a sentence to the effect that the event in question did occur because of those factors we violate the NES requirement.

It might be replied that we should tighten the conditions on the lawlike sentence in the explanans by requiring not simply that it relate the factors cited in the explanans to the type of event in the explanandum, but that it do so in an explicitly explanatory way. Thus in (a) we might require the lawlike sentence not to be simply 'Anyone who eats a pound of arsenic dies within 24 hours' but

(3) Anyone who eats a pound of arsenic dies within 24 hours because he had done so.

It is settleable *a priori* whether an explanans consisting of (3) together with 'Jones ate a pound of arsenic' is such that, if true, it would correctly explain the explanandum in (a).

This solution, however, would not be an attractive one for D–N theorists, since (3) is just a generalized explanation-sentence containing an explanatory connective that such theorists are trying to analyse by means of their model. Moreover, tightening the lawlike sentence in this way will produce many false explanations, since such tightened sentences will often be false even though their looser counterparts are true. For example, (3), construed as lawlike, is false since people who eat a pound of arsenic can die from unrelated causes. And if we weaken (3)—still keeping the explanatory clause—by writing

(4) Anyone who eats a pound of arsenic *can* die within 24 hours because he has done so,

we obtain a sentence that is true but not powerful enough for the job. It is not settleable *a priori* whether an explanans consisting of (4) together with "Jones ate a pound of arsenic at t," if true, would correctly explain why Jones died within $t + 24$, since he could have died for a different reason even though this explanans is true.

Another possible way of tightening the conditions on the lawlike sentence in the explanans is to require that it relate spatio-temporally contiguous events. (This would mean that the explanans would have to describe an event—or chain of events—that is spatio-temporally contiguous with the explanandum-event.) Jaegwon Kim has discussed laws of this sort, and he provides schemas for them which are roughly equivalent to the following:

(5) (x) $(t)(t')(x$ has P at t, and $\text{loc}(x,t)$ is spatially contiguous with $\text{loc}(x,t')$, and t is temporally contiguous with $t' \rightarrow x$ has Q at $t')$

(6) (x) $(y)(t)(t')(x$ has P at t, and $\text{loc}(x,t)$ is spatially contiguous with $\text{loc}(y,t')$, and t is temporally contiguous with $t' \rightarrow y$ has Q at $t')$.[16]

'$\text{loc}(x,t)$' means the location of x at time t. Kim does not specify a precise meaning for the arrow in (5) and (6), except that it is to convey the idea of 'causal or nomological implication' (p. 229, note 19). Under the present proposal, the arsenic explanation (a) would be precluded, since the only law invoked in (a) is not of forms (5) or (6). It does not express a relationship between types of events that are spatio-temporally contiguous. And, indeed, the explanans-event in (a) is not spatio-temporally contiguous with the explanandum-event.

This solution, like the previous one, may succeed in excluding intervening cause counterexamples such as (a). But it would not, I think, be welcomed by D–N theorists. If the arrow in (5) and (6) is to be construed causally as meaning (something like) 'causes it to be the case that', then, as with (3), the laws in D–N explanations will be generalized explanation-sentences containing an explanatory connective that D–N theorists seek to define by means of their model. Furthermore, requiring laws of forms (5) or (6) in an explanans will disallow explanations that D–N theorists, and many others, find perfectly acceptable. For example, it will not permit an explanation of a particle's acceleration due to the gravitational or electrical force of another body acting over a spatial distance. It will not permit explaining why a certain amount of a chemical compound was formed by appeal simply to (macro-) laws governing chemical reactions—where the formation of that amount of the compound takes time and is not temporally contiguous with the mixing of the reactants. Nor will the present proposal suffice to preclude all of the previous counterexamples. In particular, the pendulum example (d)—in which the pendulum's length is explained by reference to its period—is not disallowed. Assuming that the arrow in (5) and (6) represents nomological but not causal implication, we can express the following 'law':

(x) $(t)(t')(x$ is a simple pendulum with a period T at time t, and $\text{loc}(x,t)$ is spatially contiguous with $\text{loc}(x,t')$, and t is temporally contiguous with $t' \rightarrow x$ has a length L at t' which is related to T by the formula $T = 2\pi(L/g)^{1/2})$.

This, being of form (5), can be used in the explanans in (d), which, when suitably modified, will permit an explanation of the pendulum's length by reference to its period. For these reasons the present proposal does not seem promising.

Our observations regarding the various D–N models can be generalized. Assume that the explanans satisfies the NES requirement. In the explanans we can describe an event of a type always associated with an event of the sort described in the explanandum. We can include a law saying that such events are invariably and necessarily associated. The truth of the explanans event-description and of the law is no guarantee that the explanandum-event occurred because of the explanans-event. It could have occurred because some event unrelated to the one in the explanans was operative whereas the explanans-event was not. And this cannot be known *a priori* from the explanans. We can make it *a priori* by including in the explanans an appropriate singular sentence that entails the explanandum (e.g., an explanation-sentence that says in effect that the explanandum is true because the explanans is, or that the explanans-event caused the explanandum-event). But then the NES requirement would be violated. Or, we can make it *a priori* by using a generalized explanation-sentence. But since this is contrary to the philosophical spirit of such models and will, in any case, tend to produce false explanations, it will not be considered a viable solution. We can also make it *a priori* by requiring laws of forms (5) or (6) and construing the arrow causally. But this too does not satisfy the intent of such models, and, in addition, will not permit wanted explanations. On the other hand, if the arrow is understood nomologically but not causally, then whether the explanans, if true, correctly explains the explanandum is, in general, not knowable *a priori*.

To avoid the kind of problem in question we can say that it is an empirical, not an *a priori*, question whether an explanans describing events and containing laws relating these types of events to the explanandum-event correctly explains the latter. Or we can include in the explanans some singular sentence—either an explanation-sentence or something like it (e.g., (4) in Section 2)—that entails the explanandum. In the first case the *a priori* requirement is violated, in the second, NES. For this reason, I suggest, D–N models which attempt to provide sufficient conditions for correct explanations in such a way as to satisfy both these requirements will not be successful.

5. EMPIRICAL MODELS

The models I shall now mention satisfy NES but overtly violate the *a priori* requirement. Their proponents seem to recognize that if the former requirement is to be satisfied it is not an *a priori* but an empirical question whether the explanans if true would correctly explain the explanandum. However, I shall argue, the empirical considerations they introduce are not of the right sort to avoid the problem discussed above.

a. Salmon's Statistical-Relevance (S–R) model[17]

This is a model for the explanation of particular events. Salmon construes such an explanation as answering a question of the form

(1) Why is X, which is a member of class A, a member of class B?

Although Salmon does not do so, I shall say that the explanandum in such a case has the form

(2) X, which is a member of class A, is a member of class B,

which is presupposed by (1). The explanans consists of a set of empirical probability laws relating classes A and B, together with a class inclusion sentence for X, as follows:

(3) $p(B,A.C_1) = p_1$
$p(B,A.C_2) = p_2$
.
.
.
$p(B,A.C_n) = p_n$
$X \epsilon C_k \ (1 \leq k \leq n)$.

Salmon imposes two conditions on the explanans. One is that the probability values $p_1, \ldots, p_n$ all be different. (I shall not go into the reason for this.) The other is

(4) *The homogeneity condition:*
$A.C_1, A.C_2, \ldots, A.C_n$ is a partition of A, and each $A.C_i$ is homogeneous with respect to B.

$A.C_1, A.C_2, \ldots, A.C_n$ is a *partition* of A if and only if these sets comprise a set of mutually exclusive and exhaustive subsets of A. To say that a set A is *homogeneous* with respect to B is to say that there is no way, even in principle, to effect a partition of A that is statistically relevant to B without already knowing which members of A are also members of B. (C is statistically relevant to B within A if and only if $p(B,A.C) \neq p(B,A)$.) Intuitively, if A is homogeneous with respect to B then A is a random class with respect to B. Unlike the D–N models, Salmon's model does not require that the explanans show that the event in the explanandum was to be expected, but only with what probability it was to be expected.

I shall assume that for Salmon if the explanans (3) and the explanandum (2) are true, then the explanans correctly explains the explanandum, provided that Salmon's two conditions are satisfied.[18] Let's consider a simple example in which the probabilities have the values 0 and 1. There is a wire connected in a circuit to a live battery and a working bulb, and we want to explain why the bulb is lit, or more precisely, what, if anything, the wire does which contributes to the lighting of the bulb. Putting this in Salmon's form (1), the explanatory question becomes 'Why is this wire, which is a member of the class of things connected in a circuit to a live battery and working bulb, a member of the class of circuits containing a bulb that is lit?' The explanandum is

(5) This wire, which is a member of the class of things connected in a circuit to a live battery and a working bulb, is a member of the class of circuits containing a bulb that is lit.

Letting

A = The class of things connected in a circuit to a live battery and a working bulb
B = the class of circuits containing a bulb that is lit
C_1 = the class of things that conduct electricity
C_2 = the class of things that don't conduct electricity
X = this wire,

we can construct the following explanans for (5):

(6) $p(B,A.C_1) = 1$
$p(B,A.C_2) = 0$
$X \epsilon A.C_1$.

Salmon's two conditions are satisfied since the probability values are different, and since $A.C_1$ and $A.C_2$ is a partition of A and each $A.C_i$ is homogeneous with respect to B. Roughly, (6) explains why the bulb is lit by pointing out that the probability that the bulb in the circuit will be lit, given that the wire conducts electricity, is 1, that the probability that it will be lit, given that the wire does not conduct electricity, is 0, and that the wire does in fact conduct electricity.

Salmon's model satisfies the NES requirement since the only singular sentence in the explanans will not entail the explanandum. (Otherwise at least one of the probability laws in the explanans would be *a priori*, not empirical.) However, the homogeneity condition prevents his model from satisfying the *a priori* requirement.[19] Whether $A.C_1$, $A.C_2$, . . . ,$A.C_n$ is a partition of A, i.e., whether these classes have any members in common and every member of A belongs to one of them, is not in general an *a priori* question (though it happens to be in the above example). Nor is the question of whether each $A.C_i$ is homogeneous with respect to B. For example, it cannot be decided *a priori* whether there is some subclass of the class of electrical conductors such that the probability of the bulb being lit is different in this subclass from what it is in the class as a whole; this is an empirical issue. Accordingly, whether the explanans (6), if true, would correctly explain the explanandum (5) is not settleable *a priori*.

Does the inclusion of the empirical homogeneity condition avoid the kind of problem earlier discussed plaguing the D–N models? Unfortunately not. This can be seen if we change our circuit example a bit. Let A and B be the same classes as before. We now introduce

C_3 = the class of things that conduct heat
C_4 = the class of things that do not conduct heat.

I shall make the simplifying assumption that it is a law that something conducts heat if and only if it conducts electricity. Now consider this explanans for (5):

(7) $p(B,A.C_3) = 1$
$p(B,A.C_4) = 0$
$X \epsilon A.C_3$.

Although Salmon's two conditions are satisfied, (7) ought not to be regarded as a correct explanation of (5), even if (7) is true. Intuitively, if we took (7) to correctly explain (5) we would be saying that the bulb is lit because the wire conducts heat (where the probability that it is lit, given that the wire does (not) conduct heat, is 1 (0)). But this seems incorrect. The bulb is lit because the wire conducts electricity not heat, though to be sure it does both, and that it does one if and only if it does the other is a law of nature. Admittedly, by our assumption, the class $A.C_3$ = the class $A.C_1$. But it is not a class which explains for Salmon, but a sentence indicating that an item is a member of a class. If the class is described in one way the explanation may be correct, while not if described in another way. In sentences of the form

(8) *X*'s being an $A.C_i$ (a member of the class $A.C_i$)—together with such and such probability laws—correctly explains why *X*, which is a member of *A*, is a member of *B*,

the '*A.C*' position is referentially opaque. A sentence obtained from (8) by substituting an expression referring to the same class as 'the class $A.C_i$' will not always have the same truth-value.

The kind of example here used against Salmon's model[20] is similar to those raised earlier, in the following respect. In the explanans a certain fact about the wire is cited, viz., that it conducts heat, which (under the conditions of the set-up) is nomologically associated, albeit indirectly, with the fact to be explained, viz., the bulb's being lit. However, it is not the explanans-fact that is the operative one in this case but the fact that the wire conducts electricity. By invoking the homogeneity condition Salmon in effect recognizes that the question of the explanatory operativeness of the explanans is not an *a priori* matter. The problem is that his homogeneity condition is not sufficient to guarantee that the explanans-fact is operative with respect to the explanandum-fact.

b. Brody's Essential Property Model

Brody construes this model as providing a set of sufficient conditions for explanations of particular events.[21] These conditions are those of the basic D–N model together with the following:

Essential property condition: 'The explanans contains essentially a statement attributing to a certain class of objects a property had essentially by that class of objects (even if the statement does not say that they have it essentially) and . . . at least one object involved in the event described in the explanandum is a member of that class of objects' (p. 26).

For example, since Brody thinks that atomic numbers are essential properties of the elements he would regard the following explanation as correct, provided its premises are true, since the D–N conditions plus his essential property condition are satisfied:

(9) This substance is copper
Copper has the atomic number 29
Anything with the atomic number 29 conducts electricity
Therefore,
This substance conducts electricity

Brody proposes the essential property condition in order to preclude certain counterexamples to the basic D–N model. Moreover, he regards the satisfaction of this condition as an empirical matter.[22] Brody's model, in cases in which the explanans does not say explicitly that the property in question is essential, does not satisfy the *a priori* condition.[23] To know whether (9) is a correct explanation if its premises are true is not an *a priori* matter, since we must know whether having the atomic number 29 is an essential property of copper; and this knowledge is empirical, according to Brody. On the other hand, the NES requirement is satisfied since the singular premises in an explanans will not entail the explanandum.

One might object to Brody's model on grounds of obscurity in the notion of an essential property (which, by the way, he seems to distinguish from mere properties a thing has necessarily). However, the problem I want to raise is not this, and so I shall suppose the model is reasonably clear; indeed I shall stick to atomic number, which is the sort of property Brody claims to be essential to the element which has it.

Consider now the following argument which satisfies Brody's essential property condition plus the other requirements of the D–N model.

(10) Jones ate a pound of the substance in that jar
The substance in that jar is arsenic
Arsenic has the atomic number 33
Anyone who eats a pound of substance whose atomic number is 33 dies within 24 hours
Therefore,
Jones died within 24 hours of eating a pound of the substance in that jar.

Suppose, however, that Jones died in an unrelated car accident, and not because he ate arsenic. Although Brody's essential property condition is satisfied, as are the conditions of the D–N model, the explanans in (10) does not, even though true, correctly explain the explanandum. Brody may in effect recognize that it is not an *a priori* but an empirical question whether a D–N explanans if true correctly explains its explanandum. Nevertheless, the empirical requirement which his model invokes—the essential property condition—is not of the right sort to avoid the kind of problem plaguing this and previous models. Like Salmon's homogeneity condition, Brody's essential property condition is not sufficient to guarantee that the explanans-fact is explanatorily operative with respect to the explanandum-fact.

Both Salmon and Brody seem to recognize that if the NES requirement is to be satisfied it is an empirical, not an *a priori*, question whether an explanans if true correctly explains an explanandum. Yet the empirical considerations their models deploy are not sufficient to insure that if satisfied an explanation will be correct if its explanans is true. Can problems of the sort generated by these models be avoided in any way other than by abandoning NES?

6. TWO CAUSAL MODELS

The final two models I shall discuss seem to offer a solution. They satisfy NES, violate the *a priori* requirement, and yet avoid the previous problems. However, whether they provide accounts that would be welcome to most of those seeking models of explanation is quite dubious.

a. Brody's Causal Model

As in the case of his essential property model, Brody regards this as providing a set of sufficient conditions for explanations of particular events. These conditions are those of the basic D–N model together with the

Causal condition: The explanans "contains essentially a description of the event which is the cause of the event described in the explanandum."[24]

To see how this is supposed to work let us reconsider

(1) Jones ate a pound of arsenic at time *t*
Anyone who eats a pound of arsenic dies within 24 hours
Therefore,
Jones died within 24 hours of *t*

The problem we noted with the basic D–N model occurs, e.g., if both premises of (1) are true but Jones died within 24 hours of *t* for some unrelated reason. Brody's causal condition saves the day since in such a case the event described in the first premise of the explanans was not the cause of the event described in the explanandum. Hence, on this model we cannot conclude that the explanans if true correctly explains the explanandum.

The present model, like those in Section 5, violates the *a priori* requirement. Whether the explanans if true correctly explains the explanandum is not an *a priori* question, since the causal condition must be satisfied; and whether the explanans-event caused the explanandum-event is, in general, an empirical matter. It is not completely clear whether Brody wants to exclude from the explanans itself singular causal sentences that entail the explanandum, but I shall consider that version of his model which makes this exclusion. Like the models in Section 5, this model, I shall assume, is to satisfy the NES requirement.

Woodward is another modelist who proposes the need for a causal condition to supplement the basic D–N model and his own functional interdependence condition:

> These examples suggest that a fully acceptable model of scientific explanation will need to embody some characteristically causal notions (e.g., some notion of causal priority), or some more generalized analogue of these (e.g., some notion of explanatory priority) (*op. cit.*, p. 53).

However, unlike Brody, he leaves open the question of how such a causal condition should be formulated.[25]

b. The Causal-Motivational Model[26]

Here, as in the D–N motivational model (Section 3), the explanandum is a sentence saying that some agent acted in a certain way. The explanans contains a sentence attributing a desire (motive, etc.) to that agent, and a sentence attributing the belief to that agent that performing the act described in the explanandum is, in the circumstances, a (the best, the only) way to satisfy that desire. Thus the explanandum might be a sentence of the form

Agent *X* performed act *A*,

and the explanans might contain sentences of the form

X desired *G*
X believed that doing *A* is, in the circumstances, a (the best, the only) way to obtain *G*.

Unlike the D–N motivational model, however, a law in the explanans relating beliefs and desires to actions is not required. What is required is the satisfaction of a

Causal condition: *X*'s desire and his belief (described in the explanans) caused *X* to perform act *A*.

The counterexample cited earlier against the D–N motivational model is now avoided, since in that example it was not the agent's belief and desire mentioned in the explanans, but some other belief and desire, that caused him to act. As in the case of Brody's causal model, the *a priori* requirement is not satisfied, but NES is.

I shall not here try to defend or criticize these two models.[27] I shall assume for the sake of the argument that each avoids, or can be modified so as to avoid, the kind of problem I have been concerned with. However, each does so by violating NES in spirit, whereas the earlier models satisfy this requirement both in spirit and in letter. In order to apply the present causal models one must determine the truth of sentences such as these:

(2) Jones' eating a pound of arsenic at time *t* caused him to die within 24 hours of *t*

(3) Smith's desire to buy eggs and his belief that going to the store is the only way to do so caused him to go to the store.

But these are singular explanation-sentences that entail the explanandum. To be sure, neither model requires such sentences to be in the explanans. Still in each model to determine whether the explanans if true correctly explains the explanandum one has to determine the truth of such sentences. I am not criticizing the models on these grounds. But I believe that many of those who seek models of explanation will want to do so. They will say that in order to know whether the explanans in such a model correctly explains the explanandum one has to determine, independently of the truth of the explanans, the truth of sentences of a sort these modelists want to exclude from the explanans itself. Morever, they will point out that there is not much difference between determining the truth of (2) and (3), on the one hand, and that of

> Jones' eating a pound of arsenic at *t* explains why he died within 24 hours of *t* (or Jones died within 24 hours of *t* because he ate a pound of arsenic at *t*)

> Smith's desire to buy eggs and his belief that going to the store is the only way to do so explains why he went to the store (or, Smith went to the store because he had this desire and belief)

on the other. Models which impose the above causal conditions, they are likely to say, provide insufficient philosophical clarification for the concept of explanation, even though the explanation-sentence expressing the causal relationship is not itself a part of the explanans.[28] If a central aim of modelists is to define terms such as 'explain', 'because', and 'causes', this excludes their employment as primitive notions in the explanans, or, as in the present case, in the conditions of the model.

Regardless of whether we view such criticism as important, the present models are of interest because to avoid the sorts of problems raised in Section 4 these models, unlike their D–N ancestors, require establishing the truth of empirical sentences to determine whether the explanans if true correctly explains the explanandum. In this respect they are like Salmon's S–R model and Brody's essential property model. However, unlike the latter, the empirical sentences whose truth they require establishing are themselves singular sentences that entail the explanandum.

7. CONCLUSIONS

Our discussion suggests the following conclusions: (i) If the explanans is not to contain singular sentences that entail the explanandum, then it will be an empirical not an *a priori* question whether the explanans if true correctly explains the explanandum. (ii) This empirical question will involve determining the truth-values of certain singular sentences (either explanation-sentences or something akin to them) that do entail the explanandum; otherwise factors will be citable in the explanans which are not explanatorily operative with respect to the explanandum-event.

Can there be a model of explanation? Specifically, can there be a set of sufficient conditions which are such that if they are satisfied by the explanans and explanandum the former correctly explains the latter? Our discussion suggests that there can be no such model if, like D–N theorists, we insist that it satisfy both the *a priori* and the NES requirements. Moreover, it also suggests that we will not be successful in discovering a model in which (a) the NES requirement is satisfied, and in which (b) it is not an explicit condition of the model that some singular sentence be true that entails the explanandum.

It does not follow from this that an explanans which appeals to causal factors, laws, dispositions, desires and beliefs, statistically relevant factors, or essential properties cannot correctly explain an explanandum. However, models of explanation of the sort I am considering do not simply list kinds of factors that can be explanatory. Their proponents want to supply sufficient conditions for correct explanations. If it is demanded that these conditions satisfy the NES and *a priori* requirements, or NES plus (b) above, then I am suggesting that such models will not be forthcoming.[29]

NOTES

1. Carl G. Hempel and Paul Oppenheim, 'Studies in the Logic of Explanation', reprinted in Hempel, *Aspects of Scientific Explanation* (N.Y., 1965), 245–290.

2. See Hempel, *Aspects,* p. 334.

3. Or, in a tighter formulation, no sentence or conjunction of sentences in the explanans is equivalent to a conjunction some of whose members are singular sentences which individually or in conjunction entail the explanandum. The notion of a singular sentence which Hempel and others use in characterizing their models is often employed in a more or less intuitive way. And when there is an attempt to make this notion more precise difficulties emerge, as Hempel himself is aware (see his *Aspects,* p. 356). Hempel suggests that the concept can be adequately defined for a formalized language containing quantificational notation; but there are problems even here (see my *Law and Explanation* (Oxford, 1971), pp. 36–7). However, the kinds of cases I will be concerned with are quite simple and would, I think, be classified by modelists both as singular sentences and as ones to be excluded from the explanation in question.

4. To classify them as singular, of course, is not to deny that they have certain implicitly general features which further analysis might separate from the singular ones. Some might claim that NES should be applied only to completely singular sentences—ones with no generality present at all (cf., Hempel). The sentences in (2) must first be reduced. For example, in the case of the last the reduction might be: 'this metal was heated; this metal expanded; and whenever a metal is heated it expands'. If the latter is now taken to be the reduced form of the explanans, then the reduced explanans contains a singular sentence—'this metal expanded'—which entails the explanandum (1).

5. For example, it permits the following. Explanandum: 'An event of type *C* occurred'. Explanans: 'An event of type *A* caused one of type *B;* whenever an event of type *B* occurs so does one of type *C*.'

6. See my 'Concepts of Evidence', *Mind* 87 (1978), 22–45.

7. This is required in Hempel's informal and formal characterizations of his model. See his *Aspects,* pp. 248, 273, 277.

8. See Hempel's discussion of lawlikeness, *Aspects,* pp. 271–272, 292, 340. I am here giving a simplified account of the D–N model; the same considerations apply to the more complete account given by Hempel and Oppenheim in Section 7 of their classic paper.

9. See Hempel, *Aspects,* 462.

10. See Hempel, *Aspects,* p. 254.

11. I do not for a moment believe that (9) is true. But this is not the problem I want to deal with here.

12. James Woodward, 'Scientific Explanation', *British Journal for the Philosophy of Science* 30 (1979), 41–67.

13. Woodward does not claim that (10) plus the basic D–N model provides a set of sufficient conditions for correct explanations—a point to which I will return later. But for the moment I want to treat his model as if it did purport to provide such a set.

14. We might be willing to say that this pendulum *must* have a length of 100 cm because it is a simple pendulum with a period of 2.03 seconds, etc. But here the explanandum is different.

15. Woodward recognizes the pendulum example (d) as a genuine counter-example to the D–N model even when supplemented by his functional interdependence condition (p. 55). He believes that a causal condition, which he does

not formulate, will need to be added to the D–N model in addition to his condition. This type of proposal will be examined in section 6 when Brody's causal model is discussed.

16. Kim, 'Causation, Nomic Subsumption, and the Concept of an Event', *Journal of Philosophy* 70 (1973), 217–236. This is a modification of Kim, whose formulations need a slight repair.

17. Wesley C. Salmon, *Statistical Explanation and Statistical Relevance* (Pittsburgh, 1971).

18. Salmon in his book does not explicitly say this, but this seems to be the most reasonable interpretation of his position. See e.g., his remarks on pp. 79–80 in which he is distinguishing homogeneity from epistemic homogeneity, and in which he compares his model with Hempel's. Salmon might, of course, say that he is supplying conditions only for the concept of a well-confirmed or justified explanation. But I am construing his model in a stronger sense, and in private conversation he assures me that this is the correct interpretation.

19. It is possible to construe Salmon's model as requiring the satisfaction of the homogeneity condition to be stated in the explanans itself. If so the model would purport to satisfy the *a priori* requirement. However, in what follows I shall continue to assume that only sentences of the type in (3) comprise the explanans, and thus that the model is an empirical one. On either construction the same difficulty will emerge.

20. A variety of such examples, as well as other trenchant criticism, can be found in John B. Meixner's doctoral dissertation, *Salmon's Statistical Model of Explanation* (The Johns Hopkins University, Baltimore, 1976).

21. B. A. Brody, 'Towards an Aristotelian Theory of Scientific Explanation', *Philosophy of Science* 39 (1972), pp. 20–31. There is a corresponding model for the explanation of laws, which I won't discuss.

22. See *ibid.*, p. 27.

23. If the explanans explicitly says that the property is essential, then the model purports to satisfy the *a priori* condition. But I want here to consider a model that explicitly violates this condition. (In either case the model turns out to be unsatisfactory.)

24. Brody, *op. cit.*, p. 23.

25. In his most recent writings, Salmon too proposes adding a causal condition to his S–R model. Unlike Brody and Woodward, he attempts to define the concept of causation he utilizes, although, by his own admission, the definition is not entirely adequate. See his 'Why ask "Why?"', *Proceedings and Addresses of the American Philosophical Association* 51 (1977), 683–705.

26. See C. J. Ducasse, 'Explanation, Mechanism, and Teleology', *Journal of Philosophy* XXII (1925), pp. 150–55; Donald Davidson, 'Actions, Reasons, and Causes', *Journal of Philosophy* LX (1963), pp. 685–700; Alvin I. Goldman, *A Theory of Human Action* Englewood Cliffs, N.J.), p. 78.

27. A cogent attack on Brody's model can be found in Timothy McCarthy, 'On an Aristotelian Model of Scientific Explanation', *Philosophy of Science* 44 (1977), 159–166. See Davidson, 'Psychology as Philosophy', in J. Glover (ed.), *Philosophy of Mind* (Oxford, 1976), pp. 103–104, for criticism of the causal-motivational model (which Davidson himself once supported).

28. Indeed, some formulations of the causal-motivational model (e.g., Ducasse's and Goldman's) seem to allow as an explicit part of the explanans a singular explanation-sentence that entails the explanandum. In such cases the NES requirement is violated in letter as well as in spirit.

29. Material in this paper is from a book I am writing on the nature of explanation. I am indebted to the National Science Foundation for support of research.

Bas C. van Fraassen

THE PRAGMATICS OF EXPLANATION

There are two problems about scientific explanation. The first is to describe it: when is something explained? The second is to show why (or in what sense) explanation is a virtue. Presumably we have no explanation unless we have a good theory; one which is independently worthy of acceptance. But what virtue is there in explanation over and above this? I believe that philosophical concern with the first problem has been led thoroughly astray by mistaken views on the second.

I. FALSE IDEALS

To begin I wish to dispute three ideas about explanation that seem to have a subliminal influence on the discussion. The first is that explanation is a relation simply between a theory or hypothesis and the phenomena or facts, just like truth for example. The second is that explanatory power cannot be logically separated from certain other virtues of a theory, notably truth or acceptability. And the third is that explanation is the overriding virtue, the end of scientific inquiry.

When is something explained? As a foil to the above three ideas, let me propose the simple answer: *when we have a theory which explains.* Note first that "have" is not "have on the books"; I cannot claim to have such a theory without implying that this theory is acceptable all told. Note also that both "have" and "explains" are tensed; and that I have allowed that we can have a theory which does not explain, or "have on the books" an unacceptable one that does. Newton's theory explained the tides but not the advance in the perihelion of mercury; we used to have an acceptable theory, provided by Newton, which bore (or bears timelessly?) the explanation relationship to some facts but not to all. My answer also implies that we can intelligibly say that the theory explains, and not merely that people can explain by means of the theory. But this consequence is not very restrictive, because the former could be an ellipsis for the latter.

There are questions of usage here. I am happy to report that the history of science allows systematic use of both idioms. In Huygens and Young the typical phrasing seemed to be that phenomenon may be explained *by means of* principles, laws and hypotheses, or *according to* a view.[1] On the other hand, Fresnel writes to Arago in 1815 "Tous ces phénomènes . . . sont réunis et expliqués par la même théorie des vibrations," and Lavoisier says that the oxygen hypothesis he proposes *explains* the phenomena of combustion.[2] Darwin also speaks in the latter idiom: "In scientific investigations it is permitted to invent any hypothesis, and if it explains various large and independent classes of facts it rises to the rank of a well-grounded theory"; though elsewhere he says that the facts of geographical distribution are *explicable on* the theory of migration.[3]

My answer did separate acceptance of the theory from its explanatory power. Of course the second can be a reason for the first; but *that* requires their separation. Various philosophers have held that explanation logically requires true (or acceptable) theo-

Reprinted by permission from *American Philosophical Quarterly*, Vol. 14, no. 2, April 1977.

ries as premises. Otherwise, they hold, we can at most mistakenly believe that we have an explanation.

This is also a question of usage, and again usage is quite clear. Lavoisier said of the phlogiston hypothesis that it is too vague and consequently "s'adapte a toutes les explications dans lesquelles on veut le faire entrer."[4] Darwin explicitly allows explanations by false theories when he says "It can hardly be supposed that a false theory would explain, in so satisfactory a manner as does the theory of natural selection, the several large classes of facts above specified."[5] More recently, Gilbert Harman has argued similarly: that a theory explains certain phenomena is part of the evidence that leads us to accept it. But that means that the explanation-relation is visible beforehand. Finally, we criticize theories selectively: a discussion of celestial mechanics around the turn of the century would surely contain the assertion that Newton's theory does explain many planetary phenomena, though not the advance in the perihelion of Mercury.

There is a third false ideal, which I consider worst: that explanation is the *summum bonum* and exact aim of science. A virtue could be overriding in one of two ways. The first is that it is a minimal criterion of acceptability. Such is consistency with the facts in the domain of application (though not necessarily with all data, if these are dubitable!). Explanation is not like that, or else a theory would not be acceptable at all unless it explained all facts in its domain. The second way in which a virtue may be overriding is that of being required when it can be had. This would mean that if two theories pass other tests (empirical adequacy, simplicity) equally well, then the one which explains more must be accepted. As I have argued elsewhere,[6] and as we shall see in connection with Salmon's views below, a precise formulation of this demand requires hidden variables for indeterministic theories. But of course, hidden variables are rejected in scientific practice as so much "metaphysical baggage" when they make no difference in empirical predictions.

II. A BIASED HISTORY

I will outline the attempts to characterize explanation of the past three decades, with no pretense of objectivity. On the contrary, the selection is meant to illustrate the diagnosis, and point to the solution, of the next section.

1. Hempel

In 1966, Hempel summarized his views by listing two main criteria for explanation. The first is the criterion of *explanatory relevance:* "the explanatory information adduced affords good grounds for believing that the phenomenon to be explained did, or does, indeed occur."[7] That information has two components, one supplied by the scientific theory, the other consisting of auxiliary factual information. The relationship of providing good grounds is explicated as (a) implying (D–N case), or (b) conferring a high probability (I–S case), which is not lowered by the addition of other (available) evidence.

As Hempel points out, this criterion is not a sufficient condition for explanation: the red shift gives us good grounds for believing that distant galaxies are receding from us, but does not explain why they do. The classic case is the *barometer example:* the storm will come exactly if the barometers fall, which they do exactly if the atmospheric conditions are of the correct sort; yet only the last factor explains. Nor is the criterion a necessary condition; for this the classic case is the *paresis example.* We explain why the mayor, alone among the townsfolk, contracted paresis by his history of latent, contracted syphilis; yet such histories are followed by paresis in only a small percentage of cases.

The second criterion is the requirement

of *testability;* but since all serious candidates for the role of scientific theory meet this, it cannot help to remove the noted defects.

2. Beckner, Putnam, and Salmon

The criterion of explanatory relevance was revised in one direction, informally by Beckner and Putnam and precisely by Salmon. Morton Beckner, in his discussion of evolution theory, pointed out that this often explains a phenomenon only by showing how it could have happened, given certain possible conditions.[8] Evolutionists do this by constructing models of processes which utilize only genetic and natural selection mechanisms, in which the outcome agrees with the actual phenomenon. Parallel conclusions were drawn by Hilary Putnam about the way in which celestial phenomena are explained by Newton's theory of gravity: celestial motions could indeed be as they are, given a certain possible (though not, known) distribution of masses in the universe.[9]

We may take the paresis example to be explained similarly. Mere consistency with the theory is of course much too weak, since that is implied by logical irrelevance. Hence Wesley Salmon made this precise as follows: to explain is to exhibit (the) statistically relevant factors.[10] (I shall leave till later the qualifications about "screening off.") Since this sort of explication discards the talk about modelling and mechanisms of Beckner and Putnam, it may not capture enough. And indeed, I am not satisfied with Salmon's arguments that his criterion provides a sufficient condition. He gives the example of an equal mixture of Uranium 238 atoms and Polonium 214 atoms, which makes the Geiger counter click in interval $(t, t + m)$. This means that one of the atoms disintegrated. Why did it? The correct answer will be: because it was a Uranium 238 atom, if that is so—although the probability of its disintegration is much higher relative to the previous knowledge that the atom belonged to the described mixture.[11] The problem with this argument is that, on Salmon's criterion, we can explain not only why there was a disintegration, but also why *that* atom disintegrated *just then.* And surely that is exactly one of those facts which atomic physics leaves unexplained?

But there is a more serious general criticism. Whatever the phenomenon is, we can amass the statistically relevant factors, as long as the theory does not rule out the phenomenon altogether. "What more could one ask of an explanation?" Salmon inquires.[12] But in that case, as soon as we have an empirically adequate theory, we have an explanation of every fact in its domain. We may claim an explanation as soon as we have shown that the phenomenon can be embedded in some model allowed by the theory—that is, does not throw doubt on the theory's empirical adequacy.[13] But surely that is too sanguine?

3. Global Properties

Explanatory power cannot be identified with empirical adequacy; but it may still reside in the performance of the theory as a whole. This view is accompanied by the conviction that science does not explain individual facts but general regularities and was developed in different ways by Michael Friedman and James Greeno. Friedman says explicitly that in his view, "the kind of understanding provided by science is global rather than local" and consists in the simplification and unification imposed on our world picture.[14] That S_1 explains S_2 is a conjunction of two facts: S_1 implies S_2 relative to our background knowledge (and/or belief) K, *and* S_1 unifies and simplifies the set of its consequences relative to K. Friedman will no doubt wish to weaken the first condition in view of Salmon's work.

The precise explication Friedman gives of the second condition does not work, and is not likely to have a near variant that does.[15] But here we may look at Greeno's proposal.[16] His abstract and closing statement subscribe to the same general view as Friedman. But he takes as his model of a theory one which specifies a single probability space Q as the correct one, plus two partitions (or random variables) of which one is designated *explanandum* and the other *explanans*. An example: sociology cannot explain why Albert, who lives in San Francisco and whose father has a high income, steals a car. Nor is it meant to. But it does explain delinquency in terms of such other factors as residence and parental income. The degree of explanatory power is measured by an ingeniously devised quantity which measures the information I the theory provides of the explanandum variable M on the basis of explanans S. This measure takes its maximum value if all conditional probabilities $P(M_i/S_j)$ are zero or one (D–N case), and its minimum value zero if S and M are statistically independent.

Unfortunately, this way of measuring the unification imposed on our data abandons Friedman's insight that scientific understanding cannot be identified as a function of grounds for rational expectation. For if we let S and M describe the behavior of the barometer and coming storms, with P(barometer falls) = P(storm comes) = 0.2, P(storm comes/barometer falls) = 1, and P(storm comes/barometer does not fall) = 0, then the quantity I takes its maximum value. Indeed, it does so whether we designate M or S as explanans.

It would seem that such asymmetries as exhibited by the red shift and barometer examples must necessarily remain recalcitrant for any attempt to strengthen Hempel's or Salmon's criteria by global restraints on theories alone.

4. The Major Difficulties

There are two main difficulties, illustrated by the old paresis and barometer examples, which none of the examined positions can handle. The first is that there are cases, clearly in a theory's domain, where the request for explanation is nevertheless rejected. We can explain why John, rather than his brothers contracted paresis, for he had syphilis; but not why he, among all those syphilitics, got paresis. Medical science is incomplete, and hopes to find the answer some day. But the example of the uranium atom disintegrating just then rather than later, is formally similar and we believe the theory to be complete. We also reject such questions as the Aristotelians asked the Galileans: why does a body free of impressed forces retain its velocity? The importance of this sort of case, and its pervasive character, has been repeatedly discussed by Adolf Grünbaum.

The second difficulty is the asymmetry revealed by the barometer: even if the theory implies that one condition obtains when and only when another does, it may be that it explains the one in terms of the other and not vice versa. An example which combines both the first and second difficulty is this: according to atomic physics, each chemical element has a characteristic atomic structure and a characteristic spectrum (of light emitted upon excitation). Yet the spectrum is explained by the atomic structure, and the question why a substance has that structure does not arise at all (except in the trivial sense that the questioner may need to have the terms explained to him).

5. Causality

Why are there no longer any Tasmanian natives? Well, they were a nuisance, so the white settlers just kept shooting them till there were none left. The request was not for population statistics, but for the story; though in some truncated way, the statistics "tell" the story.

In a later paper Salmon gives a primary place to causal mechanisms in explanation.[17] Events are bound into causal chains by two relations: spatio-temporal continuity and statistical relevance. Explanation requires the exhibition of such chains. Salmon's point of departure is Reichenbach's *principle of the common cause:* every relation of statistical relevance ought to be explained by one of causal relevance. This means that a correlation of simultaneous values must be explained by a prior common cause. Salmon gives two statistical conditions that must be met by a common cause C of events A and B:

(a) $P(A \& B/C) = P(A/C)P(B/C)$
(b) $P(A/B \& C) = P(A/C)$ "*C screens off B from A.*"

If $P(B/C) \neq 0$ these are equivalent, and symmetric in A and B.

Suppose that explanation is typically the demand for a common cause. Then we still have the problem: when does this arise? Atmospheric conditions explain the correlation between barometer and storm, say; but are still prior causes required to explain the correlation between atmospheric conditions and falling barometers?

In the quantum domain, Salmon says, causality is violated because "causal influence is not transmitted with spatio-temporal continuity." But the situation is worse. To assume Reichenbach's principle to be satisfiable, continuity aside, is to rule out all genuinely indeterministic theories. As example, let a theory say that C is invariably followed by one of the incompatible events A, B, or D, each with probability ⅓. Let us suppose the theory complete, and its probabilities irreducible, with C the complete specification of state. Then we will find a correlation for which only C could be the common cause, but it is not. Assuming that A, B, D are always preceded by C and that they have low but equal prior probabilities, there is a statistical correlation between $\phi = (A \text{ or } D)$ and $\psi = (B \text{ or } D)$, for $P(\phi/\psi) = P(\psi/\phi) = \frac{1}{2} \neq P(\phi)$. But C, the only available candidate, does not screen off ϕ from ψ: $P(\phi/C \& \psi) = P(\phi/\psi) = \frac{1}{2} \neq P(\phi/C)$ which is ⅔. Although this may sound complicated, the construction is so general that almost any irreducibly probabilistic situation will give a similar example. Thus Reichenbach's *principle of the common cause* is in fact a demand for hidden variables.

Yet we retain the feeling that Salmon has given an essential clue to the asymmetries of explanation. For surely the crucial point about the barometer is that the atmospheric conditions screen off the barometer fall from the storm? The general point that the asymmetries are totally bound up with causality was argued in a provocative article by B. A. Brody.[18] Aristotle certainly discussed examples of asymmetries: the planets do not twinkle because they are near, yet they are near if and only if they do not twinkle (*Posterior Analytics*, I, 13). Not all explanations are causal, says Brody, but the others use a second Aristotelian notion, that of essence. The spectrum angle is a clear case: sodium has that spectrum because it has this atomic structure, which is its essence.

Brody's account has the further advantage that he can say when questions do not arise: other properties are explained in terms of essence, but the request for an explanation of the essence does not arise. However, I do not see how he would distinguish between the questions why the uranium atom disintegrated and why it disintegrated just then. In addition there is the problem that modern science is not formulated in terms of causes and essences, and it seems doubtful that these concepts can be redefined in terms which do occur there.

6. Why-Questions

A why-question is a request for explanation. Sylvain Bromberger called P the *presupposition* of the question *Why-P?* and restated the problem of explanation as that of giving the conditions under which proposition Q is a correct answer to a why-question with presupposition P.[19] However, Bengt Hannson has pointed out that "Why was it John who ate the apple?" and "Why was it the apple which John ate?" are different why-questions, although the comprised proposition is the same.[20] The difference can be indicated by such phrasing, or by emphasis ("Why did *John* . . . ?") or by an auxiliary clause ("Why did John rather than . . . ?"). Hannson says that an explanation is requested, not of a proposition or fact, but of an *aspect* of a proposition.

As is at least suggested by Hannson, we can cover all these cases by saying that we wish an explanation of why *P* is true in contrast to other members of a set *X* or propositions. This explains the tension in our reaction to the paresis-example. The question why the mayor, in contrast to other townfolk generally, contracted paresis *has* a true correct answer: because of his latent syphilis. But the question why he did in contrast to the other syphilitics in his country club, has no true correct answer. Intuitively we may say: *Q* is a correct answer to *Why P in contrast to X?* only if *Q* gives reasons to expect that *P*, in contrast to the other members of *X*. Hannson's proposal for a precise criterion is: the probability of *P* given *Q* is higher than the average of the probabilities of *R* given *Q*, for members *R* of *X*.

Hannson points out that the set *X* of alternatives is often left tacit; the two questions about paresis might well be expressed by the same sentence in different contexts. The important point is that explanations are not requested of propositions, and consequently a distinction can be drawn between answered and rejected requests in a clear way. However, Hannson makes *Q* a correct answer to *Why P in contrast to X?* when *Q* is statistically irrelevant, when *P* is already more likely than the rest; or when *Q* implies *P* but not the others. I do not see how he can handle the barometer (or red shift, or spectrum) asymmetries. On his precise criterion, that the barometer fell is a correct answer to why it will storm as opposed to be calm. The difficulty is very deep: if *P* and *R* are necessarily equivalent, according to our accepted theories, how can *Why P in contrast to X?* be distinguished from *Why R in contrast to X?*

III. THE SOLUTION

1. Prejudices

Two convictions have prejudiced the discussion of explanation, one methodological and one substantive.

The first is that a philosophical account must aim to produce necessary and sufficient conditions for theory *T* explaining phenomenon *E*. A similar prejudice plagued the discussion of counter-factuals for twenty years, requiring the exact conditions under which, if *A* were the case, *B* would be. Stalnaker's liberating insight was that these conditions are largely determined by context and speaker's interest. This brings the central question to light: what *form* can these conditions take?

The second conviction is that explanatory power is a virtue of theories by themselves, or of their relation to the world, like simplicity, predictive strength, truth, empirical adequacy. There is again an analogy with counterfactuals: it used to be thought that science contains, or directly implies, counterfactuals. In all but limiting cases, however, the proposition expressed is highly context-dependent, and the implication is there at most relative to the determining contextual factors, such as speakers' interest.

2. Diagnosis

The earlier accounts lead us to the format: *C* explains *E* relative to theory *T* exactly if (a) *T* has certain global virtues, and (b) *T* implies a certain proposition $\phi(C,E)$ expressible in the language of logic and probability theory. Different accounts directed themselves to the specification of what should go into (a) and (b). We may add, following Beckner and Putnam, that *T* explains *E* exactly if there is a proposition *C* consistent with *T* (and presumably, background beliefs) such that *C* explains *E* relative to *T*.

The significant modifications were proposed by Hannson and Brody. The former pointed out that the explanadum E cannot be reified as a proposition: we request the explanation of something F in contrast to its alternatives X (the latter generally tacitly specified by context). This modification is absolutely necessary to handle some of our puzzles. It requires that, in (b) above, we replace "$\phi(C,E)$" by the formula form "$\psi(C,F,X)$." But the problem of asymmetries remains recalcitrant, because if T implies the necessary equivalence of F and F' (say, atomic structure and characteristic spectrum), then T will also imply $\psi(C,F',X)$ if and only if it implies $\psi(C,F,X)$.

The only account we have seen which grapples at all successfully with this, is Brody's. For Brody points out that even properties which we believe to be constantly conjoined in all possible circumstances, can be divided into essences and accidents, or related as cause and effect. In this sense, the asymmetries were no problem for Aristotle.

3. The Logical Problem

We have now seen exactly what logical problem is posed by the asymmetries. To put it in current terms: how can we distinguish propositions which are true in exactly the same possible worlds?

There are several known approaches that use impossible worlds. David Lewis, in his discussion of causality, suggests that we should look not only to the worlds theory T allows as possible, but also to those it rules out as impossible, and speaks of counterfactuals which are counterlegal. Relevant logic and entailment draw distinctions between logically equivalent sentences and their semantics devised by Routley and Meyer use both inconsistent and incomplete worlds. I believe such approaches to be totally inappropriate for the problem of explanation, for when we look at actual explanations of phenomena by theories, we do not see any detours through circumstances or events ruled out as impossible by the theory.

A further approach, developed by Rolf Schock, Romane Clark, and myself distinguishes sentences by the facts that make them true. The idea is simple. That it rains, that it does not rain, that it snows, and that it does not snow, are four distinct facts. The disjunction that it rains or does not rain is made true equally by the first and second, and not by the third or fourth, which distinguishes it from the logically equivalent disjunction that it snows or does not snow.[21] The distinction remains even if there is also a fact of its raining or not raining, distinct or identical with that of its snowing or not snowing.

This approach can work for the asymmetries of explanation. Such asymmetries are possible because, for example, the distinct facts that light is emitted with wavelengths λ, μ, . . . conjointly make up the characteristic spectrum, while quite different facts conjoin to make up the atomic structure. So we have shown how such asymmetries *can* arise, in the way that Stalnaker showed how failures of transitivity in counterfactuals *can* arise. But while we have the distinct facts to classify asymmetrically, we still have the non-logical problem: whence comes the classification? The only suggestion so far is that it comes from Aristotle's concepts of cause and essence; but if so, modern science will not supply it.

4. The Aristotelian Sieve

I believe that we should return to Aristotle more thoroughly, and in two ways. To begin, I will state without argument how I understand Aristotle's theory of science. Scientific activity is divided into two parts, *demonstration* and *explanation,* the former treated mainly by the *Posterior Analytics* and the latter mainly by Book II of the *Physics.* Illustrations in the former are mainly examples of explanations in which the results of demonstration are *applied;* this is why the examples contain premises and conclusions which are not necessary and universal principles, although demonstration is only to and from such principles. Thus the division corresponds to our pure versus applied science. There is no reason to think that principles and demonstrations have such words as "cause" and "essence" in them, although looking at pure science from outside, Aristotle could say that its principles state causes and essences. In applications, the principles may be filtered through a conceptual sieve originating outside science.

The doctrine of the four "causes" (*aitiai*) allows for the systematic ambiguity or context-dependence of why-questions.[22] Aristotle's example (Physics II, 3; 195a) is of a lantern. In a modern example, the question why the porch light is on may be answered "because I flipped the switch" or "because we are expecting company," and the context determines which is appropriate. Probabilistic relations cannot distinguish these. Which factors are explanatory is decided not by features of the scientific theory but by concerns brought from outside. This is true even if we ask specifically for an "efficient cause," for how far back in the chain should we look, and which factors are merely auxiliary contributors?

Aristotle would not have agreed that essence is context-dependent. The essence is what the thing *is*, hence, its sum of classificatory properties. Realism has always asserted that ontological distinctions determine the "natural" classification. But which property is counted as explanatory and which as explained seems to me clearly context dependent. For consider Bromberger's flagpole example: the shadow is so long because the pole has this height, and not conversely. At first sight, no contextual factor could reverse this asymmetry, because the pole's height is a property it has in and by itself, and its shadow is a very accidental feature. The general principle linking the two is that its shadow is a function $f(x,t)$ of its height x and the time t (the latter determining the sun's elevation). But imagine the pole is the pointer on a giant sundial. Then the values of f have desired properties for each time t, and we appeal to these to explain why it is (had to be) such a tall pole.

We may again draw a parallel to counterfactuals. Professor Geach drew my attention to the following spurious argument: If John asked his father for money, then they would not have quarreled (because John is too proud to ask after a quarrel). Also if John asked and they hadn't quarreled, he would receive. By the usual logic of counterfactuals, it follows that if John asked his father for money, he would receive. But we know that he would not, because they have in fact quarreled. The fallacy is of equivocation, because "what was kept constant" changed in the middle of the monologue. (Or if you like, the aspects by which worlds are graded as more or less similar to this one.) Because science cannot dictate what speakers decide to "keep constant" it contains no counterfactuals. By exact parallel, *science contains no explanations.*

5. The Logic of Why-Questions

What remains of the problem of explanation is to study its logic, which is the logic of why-questions. This can be put to some extent, but not totally, in the general form developed by Harrah and Belnap and others.[23]

A question admits of three classes of response, *direct answers, corrections,* and *comments.* A *presupposition,* it has been held, is any proposition implied by all direct answers, or equivalently, denied by a correction. I believe we must add that the question "Why P, in contrast to X?" also presupposes that (a) P is a member of X, (b) P is true and the majority of X are not. This opens the door to the possibility that a question may not be uniquely determined by its set of direct answers. The question itself should decompose into factors which determine that set: the *topic* P, the *alternative* X, and a *request specification* (of which the doctrine of the four "causes" is perhaps the first description).

We have seen that the propositions involved in question and answer must be individuated by something more than the set of possible worlds. I propose that we use the facts that make them true (see footnote 21). The context will determine an asymmetric relation among these facts, of *explanatory relevance;* it will also determine the theory or beliefs which determine which worlds are *possible,* and what is *probable* relative to what.

We must now determine what direct answers are and how they are evaluated. They must be made true by facts (and only by facts forcing such) which are explanatorily relevant to those which make the topic true. Moreover, these facts must be statistically relevant, telling for the topic in contrast to the alternatives generally; this part I believe to be explicable by probabilities, combining Salmon's and Hannson's account. How strongly the answers count for the topic should be part of their evaluation as better or worse answers.

The main difference from such simple questions as "Which cat is on the mat?" lies in the relation of a why-question to its presuppositions. A why-question may fail to arise because it is ill-posed (*P* is false, or most of *X* is true), or because only question-begging answers tell probabilistically for *P* in contrast to *X* generally, or because none of the factors that do tell for *P* are explanatorily relevant in the question-context. Scientific theory enters mainly in the evaluation of possibilities and probabilities, which is only part of the process, and which it has in common with other applications such as prediction and control.

IV. SIMPLE PLEASURES

There are no explanations in science. How did philosophers come to mislocate explanation among semantic rather than pragmatic relations? This was certainly in part because the positivists tended to identify the pragmatic with subjective psychological features. They looked for measures by which to evaluate theories. Truth and empirical adequacy are such, but they are weak, being preserved when a theory is watered down. Some measure of "goodness of fit" was also needed, which did not reduce to a purely internal criterion such as simplicity, but concerned the theory's relation to the world. The studies of explanation have gone some way toward giving us such a measure, but it was a mistake to call this explanatory power. The fact that seemed to confirm this error was that we do not say that we *have* an explanation unless we have a theory which is acceptable, and victorious in its competition with alternatives, whereby we can explain. Theories are applied in explanation, but the peculiar and puzzling features of explanation are supplied by other factors involved. I shall now redescribe several familiar subjects from this point of view.

When a scientist campaigns on behalf of an advocated theory, he will point out how our situation will change if we accept it. Hitherto unsuspected factors become relevant, known relations are revealed to be strands of an intricate web, some terribly puzzling questions are laid to rest as not arising at all. We shall be in a much better position to explain. But equally, we shall be in a much better position to predict and control. The features of the theory that will make this possible are its empirical adequacy and logical strength, not special "explanatory power" and "control power." On the other hand, it is also a mistake to say explanatory power is nothing but those other features, for then we are defeated by asymmetries having no "objective" basis in science.

Why are *new* predictions so much more to the credit of a theory than agreement with the old? Because they tend to bring to light new phenomena which the older theories cannot explain. But of course, in doing so, they throw doubt on the empirical adequacy of the older theory: they show that a precondition for explanation is not met. As Boltzmann said of the radiometer, "the theories based on older hydrodynamic experience can never describe" these phenomena.[24] The failure in explanation is a by-product.

Scientific inference is inference to the best explanation. That does not rule at all for the supremacy of explanation among the virtues of theories. For we evaluate how good an explanation is given by how good a theory is used to give it, how close it fits to the empirical facts, how internally simple and coherent the explanation. There is a further evaluation in terms of a prior judgment of which kinds of factors are explanatorily relevant. If this further evaluation took precedence, overriding other considerations, explanation would be the peculiar virtue sought above all. But this is not so: instead, science schools our imagination so as to revise just those prior judgments of what satisfies and eliminates wonder.

Explanatory power is something we value and desire. But we are as ready, for the sake of scientific progress, to dismiss questions as not really arising at all. Explanation is indeed a virtue; but still, less a virtue than an anthropocentric pleasure.[25]

NOTES

1. I owe these and following references to my student Mr. Paul Thagard. For instance see C. Huygens, *Treatise on Light,* tr. by S. P. Thompson (New York, 1962), pp. 19, 20, 22, 63; Thomas Young, *Miscellaneous Works,* ed. by George Peacock (London, 1855), Vol. I, pp. 168, 170.

2. Augustin Fresnel, *Oeuvres Complètes* (Paris, 1866), Vol. I, p. 36 (see also pp. 254, 355); Antoine Lavoisier, *Oeuvres* (Paris, 1862), Vol. II, p. 233.

3. Charles Darwin, *The Variation of Animals and Plants* (London, 1868), Vol. I, p. 9; *On the Origin of the Species* (Facs. of first edition, Cambridge, Mass., 1964), p. 408.

4. Antoine Lavoisier, *op. cit.,* p. 640.

5. *Origin* (sixth ed., New York, 1962), p. 476.

6. "Wilfrid Sellars on Scientific Realism," *Dialogue,* Vol. 14 (1975), pp. 606–616.

7. C. G. Hempel, *Philosophy of Natural Science* (Englewood Cliffs, New Jersey, 1966), p. 48.

8. *The Biological Way of Thought* (Berkeley, 1968), p. 176; this was first published in 1959.

9. In a paper of which a summary is found in Frederick Suppe (ed.), *The Structure of Scientific Theories* (Urbana, Ill., 1974).

10. "Statistical Explanation," pp. 173–231 in R. G. Colodny (ed.), *The Nature and Function of Scientific Theories* (Pittsburgh, 1970); reprinted also in Salmon's book cited below.

11. *Ibid.*, pp. 207–209. Nancy Cartwright has further, unpublished, counter-examples to the necessity and sufficiency of Salmon's criterion.

12. *Ibid.*, p. 222.

13. These concepts are discussed in my "To Save the Phenomena," *The Journal of Philosophy,* Vol. 73 (1976), forthcoming.

14. "Explanation and Scientific Understanding," *The Journal of Philosophy,* Vol. 71 (1974), pp. 5–19.

15. See Philip Kitcher, "Explanation, Conjunction, and Unification," *The Journal of Philosophy,* Vol. 73 (1976), pp. 207–212.

16. "Explanation and Information," pp. 89–103 in Wesley Salmon (ed.), *Statistical Explanation and Statistical Relevance* (Pittsburgh, 1971). This paper was originally published with a different title in *Philosophy of Science,* Vol. 37 (1970), pp. 279–293.

17. "Theoretical Explanation," pp. 118–145 in Stephan Körner (ed.), *Explanation* (Oxford, 1975).

18. "Towards an Aristotelian Theory of Scientific Explanation," *Philosophy of Science,* Vol. 39 (1972), pp. 20–31.

19. "Why-Questions," pp. 86–108 in R. G. Colodny (ed.), *Mind and Cosmos* (Pittsburgh, 1966).

20. "Explanations—Of What?" (mimeographed: Stanford University, 1974).

21. Cf. my "Facts and Tautological Entailments," *The Journal of Philosophy,* Vol. 66 (1969), pp. 477–487 and in A. R. Anderson, *et al,* (ed.), *Entailment* (Princeton, 1975); and "Extension, Intension, and Comprehension" in Milton Munitz (ed.), *Logic and Ontology* (New York, 1973).

22. Cf. Julius Moravcik, "Aristotle on Adequate Explanations," *Synthese,* Vol. 28 (1974), pp. 3–18.

23. Cf. N. D. Belnap, Jr., "Questions: Their

Presuppositions, and How They Can Fail to Arise," *The Logical Way of Doing Things*, ed. by Karel Lambert (New Haven, 1969), pp. 23–39.

24. Ludwig Boltzmann, *Lectures on Gas Theory*, tr. by S. G. Brush (Berkeley, 1964), p. 25.

25. The author wishes to acknowledge helpful discussions and correspondence with Professors N. Cartwright, B. Hannson, K. Lambert, and W. Salmon, and the financial support of the Canada Council.

The Nature of Causality

Michael Scriven

CAUSATION AS EXPLANATION

0. INTRODUCTION

Some of the most widely used concepts in the logical framework of science—explanation, causation, and evaluation would seem to be good candidates for that honor—enjoy a curious love-hate relationship with philosophers of science. Though many philosophers have made serious and diverse efforts to analyze them, there have been others, equally eminent, who have completely denied their importance or their legitimacy. It has been argued that the task of science is to describe and not to explain; that reference to causation is a sign of an immature science and not to be found in the more advanced parts of the physical sciences today; and that value-free science should be regarded as the appropriate ideal.

I believe that in each case, the arguments for the rejection are fallacious and not even very subtly fallacious but I would agree that there is one great attraction about the nihilistic positions, which is that they avoid the exceedingly difficult task of providing a reasonable analysis of the concepts in question. It is even possible that the extraordinary recalcitrance of these concepts has driven or at least assisted philosophers to conclude that they are not respectable.

Social respectability is a matter of the directness with which one can be linked to indubitably respectable antecedents, and philosophical respectability is a matter of the directness of the linkage to philosophically respectable components. Given the rather limited group of respectable basic concepts with which positivist and neo-positivist philosophers of science began, there arose a rather snobbish view of a good many concepts in the scientist's working vocabulary. From my point of view, the time for this aprioristic approach has long gone, and I have spent a great many years arguing for a more plebeian approach to the philosophy of science, an approach which analyzes scientific language without reference to any social register from previous epistemological excursions. Sometimes this leads to its own, hopefully more natural, reductions, and I believe this is the case with the three concepts I have mentioned here.

A somewhat general way to put the point would be to say that these concepts are all pragmatic rather than syntactic or semantic concepts; that their function can only be understood or explicated by reference to a

Reprinted by permission of the author and the editor of *Nous*, Vol. 9 (1975):3–10.

number of contextual parameters; that these contextual parameters are of a kind that neo-positivists disdain as being subjective, or psychologistic, or non-logical; and that the specific function of these three notions is similar in that each serves as a focusing or encapsulating device in scientific discourse. Finally, it is suggested that causation can only be understood as a special case of explanation (and not as a specific case of the totally unrelated notion of correlation, which is the neo-positivist's line), just as evaluation can only be understood as a special case of description (and not as totally distinct from description). Parenthetically, one can append the thought that the connection between these notions is even closer than might appear, since causation *sometimes* involves evaluation.

Space will not permit us to explore all these concepts and their relationships here. A brief enumeration of some points that may be useful as a stimulus for discussion is all that will be attempted. But first, some illustrations of the use and potential use of causation in science today, illustrations that should be embarrassing for any remaining adherents of the causal nihilism position.

I. THE IMPORTANCE OF CAUSATION

I will give three contemporary examples to illustrate the centrality and hint at the indispensability of causation in contemporary science. The debate over the cause of economic events such as the present inflation/depression is one in which the eminent economists involved will not accept translations of the dispute into non-causal (e.g., correlation) terms. In some way, it seems clear that that refusal is connected with the close relationship between control and causation.

In the area where psychology, education, and policy studies overlap, the most important current methodological issue concerns the debated necessity for randomized control-group designs for social experiments (such as early childhood interventions). It is now clear that expert opinion has hardened and unified behind the view that *only* the full experimental approach can provide the clear causal conclusions necessary for policy decisions. The crucial design feature of this claim is the necessity for randomized allocation of subjects to the experimental and control groups, and the logically interesting point is the close connection between randomness and causation.

As a last example, one might as well pick one from the physical sciences instead of the medical or biosciences, just to emphasize the enormity of Russell's misconception of causation as a concept transcended by sciences as they mature. Particle theory and field theory provide many examples such as the dispute over "action at a distance", but let's focus on relativity theory. It's long been assumed that only those solutions of different equations of motion that lie within the so-called "light cone" are real. This turns out to be due, not to any mathematical necessity, but to the fact that solutions outside the light cone would correspond to events whose effects had preceded them. The view that such solutions are not real is thus based on a claim about the logic of cause. When it became clear that this claim was *not* analytic, then a series of interesting alternative solutions opened up for empirical investigation, as shown by Roger Newton. (Of course, the concept of retarded potential in the work of Feynman had already shown implicitly that the concept of action backwards in time did not lead to any obvious inconsistencies, but apparently the general implications of this had not been seen.) The logical moral of this example is the irrelevance of temporal sequencing to the essential nature of the cause-effect relationship. And it is an example where the philosophical analysis has had significant consequences for practicing science; there are many others today, for example in the interpretation of placebo studies, the analysis of variance, and cosmological theories.

The importance of causation is not limited to such particular instances of its application. Partly because of the attacks on it that have attempted either to impugn its scientific respectability or to reduce it to other notions, it has not received the attention it has deserved as a basis for *general* methodological procedures, at least since Mill. (On the other hand, its general importance has been considerably exaggerated at times, e.g., by supposing that the concept of universal determinism essentially involves causation. See [6].) I have argued that a good part of scientific (and, e.g., historiographical) procedure, both practical and theoretical, can be construed as an application of eliminative explanation or the trouble-shooting chart paradigm. (See [7].) One aspect of the philosophical significance of this is its apparent refutation of the *impact* (not the literal content) of the anti-inductivist position. Elaborations of this approach, (e.g., in [9] and J. R. Platt's theory of "Strong Inference") can provide extremely powerful procedures for the working scientist.

II. ASSUMPTIONS OF NON-PRAGMATIC AND PRAGMATIC ANALYSES

It is obvious that essentially pragmatic concepts will prove intransigent to more formalistic efforts at analysis. A fairly reliable symptom of this situation is the occurrence of heroic procrustean efforts to reduce the concept to a manageable shape, efforts which are notably unpersuasive to other analysts, some of whom will often propose the exact contrary of another's analysis as seminal. Some favorite oversimplifications include: causes must precede their effects, must not succeed them, must be separated from them in time, must be contiguous to them in time, be linked to them in time, be linked to them spatially, must not be linked to them logically, must be events, must be processes, must presuppose determinism, imply the absence of free will, be necessary conditions for their effects (or components of necessary conditions), must be sufficient conditions for their effects, be both necessary and sufficient conditions for their effects (or components thereof), etc.

I have provided counterexamples to the more interesting of these claims elsewhere (see [8]), but I want to pick out a couple for re-examination here since they are crucial for recent attempts to provide analyses of causation.

My opening procedural assumption is that we are trying to analyze the term "cause" as it is used in *scientific* assertions by trained scientists unfamiliar with the philosophical argumentation, and hence that it is a strike against an analysis that it has to reject or correct what a scientist would say, *unless* equally well-trained scientists would accept the correction as *scientifically* correct (the emphasis on the terms "scientific" and "scientifically" in the preceding is to exclude metascientific, i.e., philosophical, discussions by scientists, with regard to which they have no right to be considered any more reliable as authorities than the dissenting philosophers of science). One strike against an analysis does not disqualify it, but shifts the burden of proof.

My second procedural assumption is that any approach to causation which can cope with the concept without committing itself to errors such as those listed is prima facie preferable to any that cannot. It is almost obvious that any such analysis will be "less precise", more "vague", than most of its predecessors. But the quoted terms indicate demerits only where the quality indicated is present to an *inappropriate* degree. (It is not a demerit of the term "precise" that it is not precise: it would be a demerit of that term if it were used where a more precise term could be usefully employed.) Finally, I assume it to be an autonomous merit in an analysis of a scientific concept if it leads to improvements in scientific practice; that would also be an indicator of probable philosophical merit. These assumptions embody the cash value of the naturalist or non-reconstructionist approach to the philosophy of science, an approach that in some sense corresponds to the "ordinary language" approach to non-scientific language. It is no different from the scientific approach to any phenomenon and is only different in degree from the reconstructionist or aprioristic approach, which is here defined pejoratively as an approach which *over*weights systematization at the expense of misrepresentation.

It may be especially useful to try for a "naturalistic" reconceptualization of causation at this time, since it is clear that the alternatives are gradually being forced into extremely convoluted forms. The regularity or correlational accounts have been pushed into paroxysms over incompletely specifiable laws and over what is to count as a, or the, correct description of cause/effect; the counterfactual analyses face problems with indeterminism and overdetermination and with the explorations of possible and similar worlds. It may not now seem quite such a sell-out to define causation in terms of pragmatic concepts not previously admitted to the inner sanctums of logic.

III. ILLUSTRATIVE ERRORS

I shall only take up two claims, one commonly made by counterfactualists and one from the regularists. (Neither is essential for the position so labeled.)

"If the cause had not been, the effect never had existed"; thus David Lewis paraphrases Hume and sets out the axiom for his own analysis ([5]: 557). Jaegwon Kim rightly points out in his comments on Lewis's paper, a few pages later [4], that this includes many non-causal cases of dependency, i.e., is too inclusive. It is also too restrictive because of the overdetermination cases, of which Lewis only discusses one. I find it hard to understand his discussion, but let us try to apply it to an actual case. Suppose we hit an unstable trans-uranic atom with a hopped-up proton in an accelerator and kick an electron out of the outer ring. Suppose that the atom would soon have emitted that electron spontaneously in the natural decay process if we hadn't intervened. On the face of it, Hume looks wrong. Lewis's attempt at salvage ([5]: 567) relies on postulating an intervening causal chain, which doesn't seem to have any application in our example (deliberately so constructed). If we try to apply his solution to the example by extrapolating to a degenerate version of it required for a causal chain with no intervening links, then it appears we have to abandon the claim of causation altogether, which refutes the analysis. As I say, I may not properly understand the proposal. On the other hand, the counterexample has some awkward features for most analyses of causation,[1] and it has lethal cousins, e.g., the problem of identifying the cause of *over*determined deaths in the forensic area.

It seems to be absolutely inescapable that causes are not, in any reasonable sense, necessary conditions for their effects. And that would condemn Lewis's analysis even if it avoided the half-dozen other problems ably catalogued by Berofsky, the other commentator at the symposium [1]. Nevertheless, it seems to me that there is *something* to the general formulation of it in terms of possible worlds, which does not involve the necessary condition analysis. (It is perhaps better evidenced in Clendinnen [2].)

The regularists or correlationists or covering-law people run into another kind of trouble, that of "phantom laws". Since, despite Mill and others, causes are certainly not sufficient conditions, the best one can hope for is that they are components of sufficient conditions. The trouble is then that most causes are not components of conditions that are known to be sufficient in the sense that stateable general laws can be given that combine with the cause and other surrounding conditions to make the effect inferrable. Hence, one must either simply assert a primitive kind of sufficiency (Ennis

[3]), or assert that one has reason to suppose there exists such a law, even though one can't produce it. It's not clear that a regularity view, thus modified, is any more than a reaffirmation of faith in (at least statistical) determinism. The simple fact is that you can sometimes establish beyond reasonable doubt that *A* caused *X* in the circumstances (*C*) surrounding the experiment, but you cannot specify what elements of *C* are "essential"; nor can you provide a reliable exhaustive description of all the components in *C*. (For example, you may know that dropping your watch caused it to stop running). *Is* there a phantom law which would show *A* plus (part of) *C* to be a sufficient condition for *X*? At best a statistical one of course (because of quantum uncertainty); but that's not the real problem. The real problem is that such cases suggest the concept of cause is epistemologically prior to (certainly independent of) that of (universally quantified) law and, hence, that the regularity analysis is unsound. I realize that the criteria for a good analysis are debatable, but certainly the point emerges that the Hume-Mill criteria are not met, i.e., that dispensability is impossible in practice. Even "dispensability in principle" has a hollow ring, since even total knowledge of all laws would not enable one to do without the notion of cause in the watch or autopsy kind of example, because of the impossibility of determining all the necessary antecedent conditions. Of course, the regularity view is also open to a horrendous list of other difficulties (e.g., those listed by Lewis, [5]: 557), but its essential hollowness is perhaps the most damaging defect if one has any interest in pay-off for scientists. One can hardly expect much enthusiasm from a scientist who can establish particular causal claims quite easily but can't establish general laws, when one suggests that what he *means* by the former is something about the existence of an unknown combination of the latter, with unknown initial conditions. A bird in the hand, he may feel, is worth several hypothetical ones in the bush.

IV. A RADICAL ALTERNATIVE

It may be possible to give enough of a sketch of a very different approach to make it seem interesting. I'll say a little here and perhaps have a little more flesh to put on these bones by the time this paper is presented.

Let's begin by listing some of the questions that are either not asked or put aside as of secondary importance by the writers we have been discussing. Why doesn't something cause itself? Why is cause (usually) transitive?[2] Why isn't birth the cause of every particular death? Why is it that doctors will sometimes refuse to say that there was a cause of death in the usual sense, and instead say that the person died "of old age"? Why isn't it a good answer to the question, "What was the cause of the boat's sinking"? to say "Because it filled with water"? Is there any type of entity that can't be a cause or an effect? What makes "the cause" the cause instead of a cause? We won't get to all these questions, but they may keep our effort from drifting off into contemplating its own navel.

Now suppose we approach talk of causes with a model of highly pragmatic discourse in mind. One such model might be a request for directions. Thus, we might compare "What's the cause of death in this case"? with "How does one get to the party"? or "Where *is* the parish church"?. It's pretty clear that good answers to the latter questions are good only to the extent that they connect up with the resources of the inquirer, in particular, with what the inquirer already knows about the local geography; of course, the answers will also have to be consistent with the laws of nature and the laws of language and logic governing, for example, the relationship of directions such as North, South, East, and West. Beyond that, almost anything goes—in suitable circumstances. A good reply to the question about getting to the party might be "Come with us"; another might be "Go down the hill till you hear the noise"; or "It's at Karen's house". What makes one of these appropriate and others inappropriate? It is necessary to pause for a moment and reflect on the scenario that you fabricate to fit each of these as you read them. Within a fraction of a second you construct a set of circumstances which makes sense of the response and, satisfied, read on. For example, "Come with us" may have conjured up a picture of the inquirer standing at curbside, having addressed the inquiry to the driver of a car parked at that point. The driver's answer is appropriate because it involves an assessment of need (need that goes beyond the surface content of the inquiry), a consideration of resources, a plan that matches needs and resources, and a direct answer to the question that expresses the plan. Such complex feats of interpretation and reflection and reaction are commonplace parts of practical discourse, even though most of their components are hidden beneath the surface of conversation.

Now suppose we assert that requests for the cause of something are like requests for directions or guidance (What should I do now?). More specifically, they are requests for *single-factor explanations* (of a certain kind).

The correlative grammar of other causal concepts would then go something like this. *A* cause is *an* explanatory factor (of a particular kind). Causation is the relation between explanatory factors (of this kind) and what they explain.

We can amplify the concept of explanation on which this depends by saying, crudely speaking, that (i) explanations are symbolic vehicles for conveying understanding; (ii) understanding is acquired whenever the capacity for solving a certain appropriate range of problems is achieved without learning the solutions for each problem separately; (iii) understanding may thus be acquired by all sorts of inputs, scientific or not, depending simply on whether they produce this result; and (iv) the determination of the "appropriate range of problems" is a contextual requirement. Ingenious counterexamples about electro-shock "explanations" can easily be handled but will be set aside for our present crude purposes by saying that "explanation" refers to conventional communications. We should note, however, that the amount of understanding conveyed to a particular recipient by an explanation, including a causal explanation, varies enormously, from the point where all that's understood is the mere fact that strybocitosis caused the trouble (although the listener has no idea what strybocitocis *is*), to the point where he can answer a huge range of questions about the explained phenomenon. We tend to invest the scientific form of the explanation with the maximum benefits in comprehension that it *can* impart; thus, we may say of some jargonistic paragraph in an electronics text that *it* is the explanation of something or other, even though we cannot understand any of the non-logical terms in it. Thus, we make "explanation" into a dispositional property of the vehicles of understanding. "The explanation" is thus whatever (of the conventional types of communication) would impart full understanding to a suitably qualified listener/reader. This reification has misled formal analysts into thinking that some combination of formal properties of the scientific expression will encapsulate the notion of explanation; but the meaning of that concept springs entirely from the "psychologistic" notion of understanding for which it serves, in various circumstances, as a carrier. And so, mutatis mutandis, for causation.

Consider the request, made to his doctor, for the cause of some respiratory trouble experienced by a patient. The doctor looks for an answer that meets some of the following conditions. It must be a factor that explains or partyly explains the condition, and it must be of a certain kind, which we will examine in a moment. Only if there is just one such factor will an unequivocal answer to the request for *the* cause arise: otherwise, it may be possible to give a couple of factors that can be called causes, or causal factors. *The* cause, assuming for the moment that there is just one, will be whatever *does* explain the condition; explaining *it* to the patient is a separate task (i.e., explaining the explanation is not a redundancy).

We now must look at the considerations which narrow down the range of possible responses. These conditions are essentially contextual and communicational; that is, their nature is to be discovered by looking at the context of the communication and the nature of communication rather than at the events (etc.) themselves. There *is* an "objective correlative" of the cause—there are some facts about the situation to which the communication refers—which provide a necessary condition for the truth of the causal claim, but this condition is so weak that it includes most of the surrounding circumstances. So weak, in fact, that stating it cannot possibly be said to provide an analysis of cause, particularly since in any defensible form it cannot distinguish cause from effect except by invoking causal notions *and* phantom laws (see [8]). In fact, the only physical requirement that *can't* be derived from the communication analysis and that *can* be expressed without using causal terms is that the cause occurs. (It may occur before, after, or simultaneously with the effect, or any combination thereof.)

Consideration of the nature of a request for information, information that hopefully will produce understanding (i.e., consideration of this communication) yields at least the following results:

(i) It is obviously useless to provide information that the inquirer already has. One does not come to understand a phenomenon described by oneself as *X* by being told that it exists or that it is called *X* (hence explanation and a fortiori causation are irreflexive).

(ii) On the other hand, one must not suppose the whole content of a communication to lie in the event (etc.) it describes. *That* event may be familiar to the listener, but *that it* is the explanation may inform him. So, a better account than that of (i) would also appeal to the fact that the relation of an event to itself is not an explanatory one, if indeed it is a relation at all. It is not explanatory, because there is nothing one can be said to understand as a result of contemplating it, from the range of relevant problems associated with understanding an event.

(iii) If the phenomenon-to-be-explained (*X*) is explained by reference to another phenomenon, *P*, and that phenomenon is explained (in the same context of inquiry) by reference to phenomenon *A*, then of course *X* has to be explained in terms of *A* (hence causation is, generally speaking, transitive).

(iv) If *X* can be explained in terms of *P*, it's an open question whether the converse is true (hence causation is not antisymmetrical).

(v) The distinction between cause and effect, hard or impossible to make on the usual analysis, is the distinction between the phenomenon to be explained and the necessarily different (at least in description) phenomena that provide the explanation.

(vi) Birth isn't the cause of everyone's death (although it's both a necessary and a sufficient condition for it), because, one might say, everyone knows that everyone was born *and* knows the connection of birth with death and hence can't be asking for that when asking for the cause (explanation) of a particular death. Against this, one might use this counterexample: A child who asks how his BB gun got rusty while standing in the basement might properly be told that it was the result of water seepage, but could not properly be told that it was because of the oxygen in the earth's atmosphere, *even if* he didn't know the connection of oxygen with rust. Against the counterexample, we distinguish between the necessity and the sufficiency of the reason we gave against accepting birth as the cause of death. It is a sufficient reason for exclusion of an explanation that its occurrence and connections be known already. It's not the only such reason, i.e., a necessary reason, for such exclusions, however. Another reason would be irrelevance or lack of specificity to the problem with regard to which understanding is sought. In the oxygen case, the boy's question requests an explanation of why *his* gun *rusted,* his *specifically;* and an answer which refers to a factor common to *all* guns, including those which have *not* rusted, is not relevant, since it fails to explain the distinction implicit in the request.

An understanding of the extent to which a request for explanation involves an implicit contrast enables one to handle all the cases of what Lewis calls the "invidious" distinction between various 'potential causes' that results in identifying "the cause" in certain cases.

(vii) The astonishing variety of entities (and nonentities?) that can be called causes (including reasons, incidentally) is simply a reflection of the variety of things about which we may be ignorant. The doctor may identify the cause of the chest condition as "last week's fishing trip"; here the patient knows all about that, but did not know it could be connected with the bronchitis; or if he knew it could be, he did not know that in this case it was. Or the doctor may refer to infection by someone unknown in close physical contact with the patient—where the patient fills in the description from his own knowledge.

(viii) But what is this "connection" we are referring to in the last example? Isn't it *causal* connection? And if so, aren't we now completing a circularity in the analysis? That is, hasn't it turned out that the "particular kind" of explanation that causation was said to be, has turned out to be the *causal* kind? And how explanatory is *that?*

Well, one can reply by saying that the kind of conclusion that is relevant here is just a matter of the kind of explanation that is required to provide understanding to the patient, i.e., a matter of the ignorance of the patient. That is, the problem is still one of analyzing the context of the communication. But one might equally well say that we're looking here for factors over which we, or he, have or had control—for manipulable factors. And that's a causal notion. We use causal notions in talking of stellar or solar or micro events which cannot and probably could not be manipulated in a literal sense. But the extension of that sense to such cases is very simple, following exactly the extensions of concepts like heat, force, or agent. Looking for the factors that are responsible for a condition is looking for causal explanations. It isn't a non-causally defined alternative to looking for causal explanations.

What we've done is to approach the concept of a causal explanation from the explanation side, so to speak. Doing that may be somewhat enlightening. But is isn't reductionist; it doesn't eliminate the concept of cause, any more than approaching from the hard side or the counterfactual side does. One can pseudo-eliminate it by using the primitive-term recipe (Gasking), responsibility (Ennis), independence (Simon), manipulability (Flew), influence, or randomness; and even that may be more enlightening than it is deceitful. For the idea that circular definitions can't be useful is just another oversimplification from the same friendly folks who brought you the idea that causation can't be a basic concept. It *is* a basic concept, not reducible to others. But that doesn't mean that setting out its *relationship* to others isn't important. Analysis involves display as well as disappearance.

NOTES

1. I can see no way in which Davidson's analysis, for instance, can handle this case, except by adding the time to the event-description, a procedure which is both objectionable as reconstructivist and useless against co-terminous overdetermination examples.

2. Lewis, as most, assumes the exceptionless form here; "Causation must always be transitive . . ". ([5]: 563). I'm uneasy about cases like: This man's death was due to asphyxia; his death led many of his friends to reminisce about their days at Stanford together; (hence) asphyxia led many of his friends to reminisce. One feels that asphyxia is the kind of thing that could only be said to *stop* people reminiscing.

REFERENCES

1. Bernard Berofsky, "The Counterfactual Analysis of Causation", *Journal of Philosophy* 70(1973): 568–69.

2. F. John Clendinnen, "Causal Sequences and Explanation" (unpublished, 1973).

3. Robert Ennis, "On Causality" (unpublished, 1973).

4. Jaegwon Kim, "Causes and Counterfactuals", *Journal of Philosophy* 70(1973): 570–72.

5. David Lewis, "Causation", *Journal of Philosophy* 70(1973): 556–67.

6. Michael Scriven, "The Present State of Determinism in Physics", *Journal of Philosophy* 54(1957): 727–41.

7. —, "Truisms as the Grounds for Historical Explanations", in *Theories of History,* ed. by P. L. Gardiner (Glencoe, Ill.: Free Press, 1959): 443–75.

8. —, "The Logic of Cause", *Theory and Decision* 2(1971): 49–66.

9. —, "Maximizing the Power of Causal Investigations—The Modus Operandi Method", in *Evaluation in Education: Current Applications,* ed. by W. James Popham (Berkeley: McCutchan, 1974).

J. L. Mackie

CAUSES AND CONDITIONS

Asked what a cause is, we may be tempted to say that it is an event which precedes the event of which it is the cause, and is both necessary and sufficient for the latter's occurrence; briefly that a cause is a necessary and sufficient preceding condition. There are, however, many difficulties in this account. I shall try to show that what we often speak of as a cause is a condition not of this sort, but of a sort related to this. That is to say, this account needs modification, and can be modified, and when it is modified we can explain much more satisfactorily how we can arrive at much of what we ordinarily take to be causal knowledge; the claims implicit within our causal assertions can be related to the forms of the evidence on which we are often relying when we assert a causal connection.

1. SINGULAR CAUSAL STATEMENTS

Suppose that a fire has broken out in a certain house, but has been extinguished before the house has been completely destroyed. Experts investigate the cause of the fire, and they conclude that it was caused by an electrical short-circuit at a certain place. What is the exact force of their statement that this short-circuit caused this fire? Clearly the experts are not saying that the short-circuit was a necessary condition for this house's catching fire at this time; they know perfectly well that a short-circuit somewhere else, or the overturning of a lighted oil stove, or any one of a number of other things might, if it had occurred, have set the house on fire. Equally, they are not saying that the short-circuit was a sufficient condition for this house's catching fire; for if the short-circuit had occurred, but there had been no inflammable material near by, the fire would not have broken out, and even given both the short-circuit and the inflammable material, the fire would not have occurred if, say, there had been an efficient automatic sprinkler at just the right spot. Far from being a condition both necessary and sufficient for the fire, the short-circuit was, and is known to the experts to have been, neither necessary nor sufficient

Reprinted by permission from *American Philosophical Quarterly,* Vol. 17, no. 4, October 1965.

for it. In what sense, then, is it said to have caused the fire?

At least part of the answer is that there is a set of conditions (of which some are positive and some are negative), including the presence of inflammable material, the absence of a suitably placed sprinkler, and no doubt quite a number of others, which combined with the short-circuit constituted a complex condition that was sufficient for the house's catching fire—sufficient, but not necessary, for the fire could have started in other ways. Also, of *this* complex condition, the short-circuit was an indispensible part: the other parts of this condition, conjoined with one another in the absence of the short-circuit, would not have produced the fire. The short-circuit which is said to have caused the fire is thus an indispensable part of a complex sufficient (but not necessary) condition of the fire. In this case, then, the so-called cause is, and is known to be, an *insufficient* but *necessary* part of a condition which is itself *unnecessary* but *sufficient* for the result. The experts are saying, in effect, that the short-circuit is a condition of this sort, that it occurred, that the other conditions which conjoined with it form a sufficient condition were also present, and that no other sufficient condition of the house's catching fire was present on this occasion. I suggest that when we speak of the cause of some particular event, it is often a condition of this sort that we have in mind. In view of the importance of conditions of this sort in our knowledge of and talk about causation, it will be convenient to have a short name for them: let us call such a condition (from the initial letters of the words italicized above), an INUS condition.[1]

This account of the force of the experts' statement about the cause of the fire may be confirmed by reflecting on the way in which they will have reached this conclusion, and the way in which anyone who disagreed with it would have to challenge it. An important part of the investigation will have consisted in tracing the actual course of the fire; the experts will have ascertained that no other condition sufficient for a fire's breaking out and taking this course was present, but that the short-circuit did occur and that conditions were present which in conjunction with it were sufficient for the fire's breaking out and taking the course that it did. Provided that there is some necessary and sufficient condition of the fire—and this is an assumption that we commonly make in such contexts—anyone who wanted to deny the experts' conclusion would have to challenge one or another of these points.

We can give a more formal analysis of the statement that something is an INUS condition. Let 'A' stand for the INUS condition—in our example, the occurrence of a short-circuit at that place—and let 'B' and '$\bar{C}$' (that is, 'not-C', or the absence of C) stand for the other conditions, positive and negative, which were needed along with A to form a sufficient condition of the fire—in our example, B might be the presence of inflammable material, $\bar{C}$ the absence of a suitably placed sprinkler. Then the conjunction '$AB\bar{C}$' represents a sufficient condition of the fire, and one that contains no redundant factors; that is, $AB\bar{C}$ is a minimal sufficient condition for the fire.[2] Similarly, let $D\bar{E}F$, $\bar{G}\bar{H}I$, etc., be all the other minimal sufficient conditions of this result. Now provided that there is some necessary and sufficient condition for this result, the disjunction of all the minimal sufficient conditions for it constitutes a necessary and sufficient condition.[3] That is, the formula '$AB\bar{C}$ or $D\bar{E}F$ or $\bar{G}\bar{H}I$ or . . .' represents a necessary and sufficient condition for the fire, each of its disjuncts, such as '$AB\bar{C}$', represents a minimal sufficient condition, and each conjunct in each minimal sufficient condition, such as 'A', represents an INUS condition. To simplify and generalize this, we can replace the conjunction of terms conjoined with 'A' (here '$B\bar{C}$') by the single term 'X', and the formula representing the disjunction of all the other minimal sufficient conditions—here '$D\bar{E}F$ or $\bar{G}\bar{H}I$ or . . .'—by the single term '*Y*'. Then an INUS condition is defined as follows:

A is an INUS condition of a result *P* if and only if, for some *X* and for some *Y*, (*AX* or *Y*) is a necessary and sufficient condition of *P*, but *A* is not a sufficient condition of *P* and *X* is not a sufficient condition of *P*.

We can indicate this type of relation more briefly if we take the provisos for granted and replace the existentially quantified variables '*X*' and '*Y*' by dots. That is, we can say that *A* is an INUS condition of *P* when (*A* . . . or . . .) is a necessary and sufficient condition of *P*.

(To forestall possible misunderstandings, I would fill out this definition as follows.[4] First, there could be a set of minimal sufficient conditions of *P*, but no necessary conditions, not even a complex one; in such a case, *A* might be what Marc-Wogau calls a moment in a minimal sufficient condition, but I shall not call it an INUS condition. I shall speak of an INUS condition only where the disjunction of all the minimal sufficient conditions is also a necessary condition. Secondly, the definition leaves it open that the INUS condition *A* might be a conjunct in each of the minimal sufficient conditions. If so, *A* would be itself a necessary condition of the result. I shall still call *A* an INUS condition in these circumstances: it is not part of the definition of an INUS condition that it should *not* be necessary, although in the standard cases, such as that sketched above, it is not in fact necessary.[5] Thirdly, the requirement that *X* by itself should not be sufficient for *P* insures that *A* is a non-redundant part of the sufficient condition *AX;* but there is a sense in which it may not be strictly necessary or indispensable even as a part of *this* condition, for it may be replaceable: for example *KX* might be another minimal sufficient condition of *P*.[6] Fourthly, it *is* part of the definition that the minimal sufficient condition, *AX*, of which *A* is a non-redundant part, is not also a necessary condition, that there is another sufficient condition *Y* (which may itself be a disjunction of sufficient conditions). Fifthly, and similarly, it *is* part of the definition that *A* is not by itself sufficient for *P*. The fourth and fifth of these points amount to this: I shall call *A* an INUS condition only if there are terms which actually occupy the places occupied by '*X*' and '*Y*' in the formula for the necessary and sufficient condition. However, there may be cases where there is only one minimal sufficient condition, say *AX*. Again, there may be cases where *A* is itself a minimal sufficient condition, the disjunction of all minimal sufficient conditions being (*A* or *Y*); again, there may be cases where *A* itself is the only minimal sufficient condition, and is itself both necessary and sufficient for *P*. In any of these cases, as well as in cases where *A* is an INUS condition, I shall say that *A* is *at least an* INUS *condition.* As we shall see, we often have evidence which supports the conclusion that something is *at least* an INUS condition; we may or may not have other evidence which shows that it is *no more than* an INUS condition.)

I suggest that a statement which asserts a singular causal sequence, of such a form as '*A* caused *P*', often makes, implicitly, the following claims:

(i) *A* is at least an INUS condition of *P*—that is, there is a necessary and sufficient condition of *P* which has one of these forms: (*AX* or *Y*), (*A* or *Y*), *AX*, *A*.

(ii) *A* was present on the occasion in question.

(iii) The factors represented by the '*X*', if any, in the formula for the necessary and sufficient condition were present on the occasion in question.

(iv) Every disjunct in '*Y*' which does not contain '*A*' as a conjunct was absent on the occasion in question. (As a rule, this means that whatever '*Y*' represents was absent on this occasion. If '*Y*' represents a single conjunction of factors, then it was absent if at least one of its conjuncts was absent; if it represents a disjunction, then it was absent if each of its disjuncts was absent. But we do not wish to exclude the possibility that '*Y*' should be, or contain as a disjunct, a conjunction one of whose conjuncts is *A*, or to require that *this* conjunction should have been absent.)[7]

I do not suggest that this is the whole of what is meant by '*A* caused *P*' on any occasion, or even that it is a part of what is meant on every occasion: some additional and alternative parts of the meaning of such statements are indicated below.[8] But I am suggesting that this is an important part of the concept of causation; the proof of this suggestion would be that in many cases the falsifying of any one of the above-mentioned claims would rebut the assertion that *A* caused *P*.

This account is in fairly close agreement, in substance if not in terminology, with at least two accounts recently offered of the cause of a single event.

Konrad Marc-Wogau sums up his account thus: 'when historians in singular causal statements speak of a cause or the cause of a certain individual event β, then what they are referring to is another individual event α which is a moment in a minimal sufficient and at the same time necessary condition *post factum* β'.[9]

He explained his phrase 'necessary condition *post factum*' by saying that he will call an event a_1 a necessary condition *post factum* for *x* if the disjunction 'a_1 or a_2 or a_3 . . . or a_n' represents a necessary condition for *x*, and of these disjuncts only a_1 was present on the particular occasion when *x* occurred.

Similarly Michael Scriven has said:

> Causes are *not* necessary, even contingently so, they are not sufficient—but they are, to talk that language, *contingently sufficient*. . . . They are part of *a* set of conditions that does guarantee the outcome, and they are non-redundant in that the rest of *this* set (which does not include all the other conditions present) is not alone sufficient for the outcome. It is not even true that they are relatively necessary, i.e., necessary with regard to that set of conditions rather than the total circumstances of their occurrence, for there may be several possible replacements for them which happen not to be present. There remains a ghost of necessity; a cause is a factor from a set of possible factors the presence of one of which (*any* one) is necessary in order that a set of conditions actually present be sufficient for the effect.[10]

There are only slight differences between these two accounts, or between each of them and that offered above. Scriven seems to speak too strongly when he says that causes are not necessary: it is, indeed, not part of the definition of a cause of this sort that it should be necessary, but, as noted above, a cause, or an INUS condition, may be necessary, either because there is only one minimal sufficient condition or because the cause is a moment in each of the minimal sufficient conditions. On the other hand, Marc-Wogau's account of a minimal sufficient condition seems too strong. He says that a minimal sufficient condition contains 'only those moments relevant to the effect' and that a moment is relevant to an effect if 'it is a necessary condition for β: β would not have occurred if this moment had not been present'. This is less accurate than Scriven's statement that the cause only needs to be non-redundant.[11] Also, Marc-Wogau's requirement, in his account of a necessary condition *post factum,* that only one minimal sufficient condition (the one containing α) should be present on the particular occasion, seems a little too strong. If two or more minimal sufficient conditions (say a_1 and a_2) were present, but α was a moment in each of them, then though neither a_1 nor a_2 was necessary *post factum,* α would be so. I shall use this phrase 'necessary *post factum*' to include cases of this sort: that is, α is a necessary condition *post factum* if it is a moment in every minimal sufficient condition that was present. For example, in a cricket team the wicket-keeper is also a good batsman. He is injured during a match, and does not bat in the second inning, and the substitute wicket-keeper drops a vital catch that the original wicket-keeper would have taken. The team loses the match, but it would have won if the wicket-keeper had *both* batted *and* taken that catch. His injury was a moment in two minimal sufficient conditions for the loss of the match; either his not batting, or the catch's not being taken, would on its own have ensured the loss of the match. But we can certainly say that his injury caused the loss of the match, and that it was a necessary condition *post factum.*

This account may be summed up, briefly and approximately, by saying that the statement '*A* caused *P*' often claims that *A* was necessary and sufficient for *P* in the circumstances. This description applies in the standard cases, but we have already noted that a cause is non-redundant rather than necessary even in the circumstances, and we shall see that there are special cases in which it may be neither necessary or non-redundant.

2. DIFFICULTIES AND REFINEMENTS[12]

Both Scriven and Marc-Wogau are concerned not only with this basic account, but with certain difficulties and with the refinements and complications that are needed to overcome them. Before dealing with these I shall introduce, as a refinement of my own account, the notion of a causal field.[13]

This notion is most easily explained if we leave, for a time, singular causal statements and consider general ones. The question 'What causes influenza?' is incomplete and partially indeterminate. It may mean 'What causes influenza in human beings in general?' If so, the (full) cause that is being sought is a difference that will mark off cases in which human beings contract influenza from cases in which they do not; the causal field is then the region that is to be thus divided, *human beings in general.* But the question may mean, 'Given that influenza viruses are present, what makes some people contract the disease whereas others do not?' Here the causal field is *human beings in conditions where influenza viruses are present.* In all such cases, the cause is required to differentiate, within a wider region in which the effect sometimes occurs and sometimes does not, the sub-region in which it occurs: this wider region is the causal field. This notion can now be applied to singular causal questions and statements. 'What caused this man's skin cancer?'[14] may mean 'Why did this man develop skin cancer now when he did not develop it before?' Here the causal field is the career of this man: it is within this that we are seeking a difference between the time when skin cancer developed and times when it did not. But the same question may mean 'Why did this man develop skin cancer, whereas other men who were also exposed to radiation did not?' Here the causal field is the class of men thus exposed to radiation. And what is the cause in relation to one field may not be the cause in relation to another. Exposure to a certain dose of radiation may be the cause in relation to the former field: it cannot be the cause in relation to the latter field since it is part of the description of that field, and being present throughout that field it cannot differentiate one sub-region of it from another. In relation to the latter field, the cause may be, in Scriven's terms, 'some as-yet-unidentified constitutional factor.'

In our first example of the house which caught fire, the history of this house is the field in relation to which the experts were looking for the cause of the fire: their question was 'Why did this house catch fire on this occasion, and not on others?' However, there may still be some indeterminacy in this choice of a causal field. Does this house, considered as the causal field, include all its features, or all its relatively permanent features, or only some of these? If we take all its features, or even all of its relatively permanent ones, as constituting the field, then some of the things that we have treated as conditions—for example the presence of inflammable material near the place where the short-circuit occurred—would have to be regarded as parts of the field, and we could not then take them also as conditions which in relation to this field, as additions to it or intrusions into it, are necessary or sufficient for something else. We must, therefore, take the house, in so far as it constitutes the causal field, as determined only in a fairly general way, by only some of its relatively permanent features, and we shall then be free to treat its other features as conditions which do not constitute the field, and are not parts of it, but which may occur within it or be added to it. It is in general an arbitrary matter whether a particular feature is regarded as a condition (that is, as a possible causal factor) or as part of the field, but it cannot be treated in both ways at once. If we are to say that something happened to this house because of, or partly because of, a certain feature, we are implying that it would still have been *this* house, the house in relation to which we are seeking the cause of this happening, even if it had not had this particular feature.

I now propose to modify the account given above of the claims often made by singular causal statements. A statement of such a form as '*A* caused *P*' is usually elliptical, and is to be expanded into '*A* caused *P* in relation to the field *F*.' And then in place of the claim stated in (i) above, we require this:

(ia) *A* is at least an INUS condition of *P* in the field *F*—that is, there is a condition which, given the presence of whatever features characterize *F* throughout, is necessary and sufficient for *P*, and which is of one of these forms: (*AX* or *Y*), (*A* or *Y*), *AX*, *A*.

In analysing our ordinary causal statements, we must admit that the field is often taken for granted or only roughly indicated, rather than specified precisely. Nevertheless, the field in relation to which we are looking for a cause of this effect, or saying that such-and-such is a cause, may be definite enough for us to be able to say that certain facts or possibilities are irrelevant to the particular causal problem under consideration, because they would constitute a shift from the intended field to a different one. Thus if we are looking for the cause, or causes, of influenza, meaning its cause(s) in relation to the field *human beings*, we may dismiss, as not directly relevant, evidence which shows that some proposed cause fails to produce influenza in rats. If we are looking for the cause of the fire in *this house*, we may similarly dismiss as irrelevant the fact that a proposed cause would not have produced a fire if the house had been radically different, or had been set in a radically different environment.

This modification enables us to deal with the well-known difficulty that it is impossible, without including in the cause the whole environment, the whole prior state of the universe (and so excluding any likelihood of repetition), to find a genuinely sufficient condition, one which is 'by itself, adequate to secure the effect'.[15] It may be hard to find even a complex condition which was absolutely sufficient for this fire because we should have to include, as one of the negative conjuncts, such an item as the earth's not being destroyed by a nuclear explosion just after the occurrence of the suggested INUS condition; but it is easy and reasonable to say simply that such an explosion would, in more senses than one, take us outside the field in which we are considering this effect. That is to say, it may be not so difficult to find a condition which is sufficient in relation to the intended field. No doubt this means that causal statements may be vague, in so far as the specification of the field is vague, but this is not a serious obstacle to establishing or using them, either in science or in everyday contexts.[16]

It is a vital feature of the account I am suggesting that we can say that *A* caused *P*, in the sense described, without being able to specify exactly the terms represented by '*X*' and '*Y*' in our formula. In saying that *A* is at least an INUS condition for *P* in *F*, one is *not* saying what other factors, along with *A*, were both present and non-redundant, and one is *not* saying what other minimal sufficient conditions there may be for *P* in *F*. One is not even claiming to be able to say what they are. This is in no way a difficulty:

it is a readily recognizable fact about our ordinary causal statements, and one which this account explicitly and correctly reflects.[17] It will be shown (in § 5 below) that this elliptical or indeterminate character of our causal statements is closely connected with some of our characteristic ways of discovering and confirming causal relationships: it is precisely for statements that are thus 'gappy' or indeterminate that we can obtain fairly direct evidence from quite modest ranges of observation. On this analysis, causal statements implicitly contain existential quantifications; one can assert an existentially quantified statement without asserting any instantiation of it, and one can also have good reason for asserting an existentially quantified statement without having the information needed to support any precise instantiation of it. I can know that there is someone at the door even if the question 'Who is he?' would floor me.

Marc-Wogau is concerned especially with cases where 'there are two events, each of which independently of the other is a sufficient condition for another event'. There are, that is to say, two minimal sufficient conditions, both of which actually occurred. For example, lightning strikes a barn in which straw is stored, and a tramp throws a burning cigarette butt into the straw at the same place and at the same time. Likewise for a historical event there may be more than one 'cause', and each of them may, on its own, be sufficient.[18] Similarly Scriven considers a case where '. . . conditions (perhaps unusual excitement plus constitutional inadequacies) [are] present at 4.0 p.m. that guarantee a stroke at 4.55 p.m. and consequent death at 5.0 p.m.; but an entirely unrelated heart attack at 4.50 p.m. is still correctly called the cause of death, which, as it happens, does occur at 5.0 p.m.'[19]

Before we try to resolve these difficulties let us consider another of Marc-Wogau's problems: Smith and Jones commit a crime, but if they had not done so the head of the criminal organization would have sent other members to perform it in their stead, and so it would have been committed anyway.[20] Now in this case, if 'A' stands for the actions of Smith and Jones, what we have is that AX is one minimal sufficient condition of the result (the crime), but $\bar{A}Z$ is another, and both X and Z are present. A combines with one set of the standing conditions to produce the result by one route; but the absence of A would have combined with another set of the standing conditions to produce the same result by another route. In this case we *can* say that A was a necessary condition *post factum*. This sample satisfies the requirements of Marc-Wogau's analysis, and of mine, of the statement that A caused this result; and this agrees with what we would ordinarily say in such a case. (We might indeed add that there was *also* a deeper cause—the existence of the criminal organization, perhaps—but this does not matter; our formal analyses do not insure that a particular result will have a unique cause, nor does our ordinary causal talk require this.) It is true that in this case we cannot say what will usually serve as an informal substitute for the formal account, that the cause, here A, was necessary (as well as sufficient) in the circumstances; for $\bar{A}$ would have done just as well. We cannot even say that A was non-redundant. But this shows merely that a formal analysis may be superior to its less formal counterparts.

Now in Scriven's example, we might take it that the heart attack prevented the stroke from occurring. If so, then the heart attack *is* a necessary condition *post factum*: it is a moment in the only minimal sufficient condition that was present in full, for the heart attack itself removed some factor that was a necessary part of the minimal sufficient condition which has the excitement as one of its moments. This is strictly parallel to the Smith and Jones case. Again it is odd to say that the heart attack was in any way necessary, since the absence of the heart attack would have done just as well: this absence would have been a moment in that other minimal sufficient condition, one of those other moments was the excitement. Nevertheless, the heart attack was necessary *post factum*, and the excitement was not. Scriven draws the distinction, quite correctly, in terms of continuity and discontinuity of causal chains: 'the heart attack was, and the excitement was not the cause of death because the "causal chain" between the latter and death was interrupted, while the former's "went to completion".' But it is worth noting that a break in the causal chain corresponds to a failure to satisfy the logical requirements of a moment in a minimal sufficient condition that is also necessary *post factum*.

Alternatively, if the heart attack did not prevent the stroke, then we have a case parallel to that of the straw in the barn, or of the man who is shot by a firing squad, and two bullets go through his heart simultaneously. In such cases the requirements of my analysis, or Marc-Wogau's, or of Scriven's, are not met: each proposed cause *is* redundant and not even necessary *post factum*, though the disjunction of them is necessary *post factum* and non-redundant. But this agrees very well with the fact that we *would* ordinarily hesitate to say, of either bullet, that it caused the man's death, or of either the lightning or the cigarette butt that it caused the fire, or of either the excitement or the heart attack that it was the cause of death. As Marc-Wogau says, 'in such a situation as this we are unsure also how to use the word 'cause'.' Our ordinary concept of cause does not deal clearly with cases of this sort, and we are free to decide whether or not to add to our ordinary use, and to the various more or less formal descriptions of it, rules which allow us to say that where more than one at-least-INUS-condition, and its conjunct conditions, are present, each of them caused the result.[21]

The account thus far developed of singular causal statements has been expressed in terms of statements about necessity and sufficiency: it is therefore incomplete until we have added an account of necessity and sufficiency themselves. This question is considered in § 4 below. But the present account is independent of any particular analysis of necessity and sufficiency. Whatever analysis of these we finally adopt, we shall use it to complete the account of what it is to be an INUS condition, or to be at least an INUS condition. But in whatever way this account is completed, we can retain the general principle that at least part of what is often done by a singular causal statement is to pick out, as the cause, something that is claimed to be at least an INUS condition.

3. GENERAL CAUSAL STATEMENTS

Many general causal statements are to be understood in a corresponding way. Suppose, for example, that an economist says that the restriction of credit causes (or produces) unemployment. Again, he will no doubt be speaking with reference to some causal field; this is now not an individual object, but a class, presumably economies of a certain general kind; perhaps their specification will include the feature that each economy of the kind in question contains a large private enterprise sector with free wage-earning employees. The result, unemployment, is something which sometimes occurs and sometimes does not occur within this field, and the same is true of the alleged cause, the restriction of credit. But the economist is not saying that (even in relation to this field) credit restriction is either necessary or sufficient for unemployment, let alone both necessary and sufficient. There may well be other circumstances which must be present along with credit restriction, in an economy of the kind referred to, if unemployment is to result; these other circumstances will no doubt include various negative ones, the absence of various counteracting causal factors which, if they were present, would prevent this result. Also, the economist will probably be quite prepared to admit that in an economy of this kind unemployment would be brought about by other combinations of circumstances in which the restriction of credit plays no part. So once again the claim that he is making is merely that the restriction of credit is, in economies of this kind, a non-redundant part of one sufficient condition for unemployment: that is, an INUS condition. The economist is probably assuming that there is some condition, no doubt a complex one, which is both necessary and sufficient for unemployment in this field. This being assumed, what he is asserting is that, for some *X* and for some *Y* (*AX* or *Y*) is a necessary and sufficient condition for *P* in *F*, but neither *A* nor *X* is sufficient on its own, where '*A*' stands for the restriction of credit, '*P*' for unemployment, and '*F*' for the field, economies of such-and-such a sort. In a developed economic theory the field *F* may be specified quite exactly, and so may the relevant combinations of factors represented here by '*X*' and '*Y*'. (Indeed, the theory may go beyond statements in terms of necessity and sufficiency to ones of functional dependence, but this is a complication which I am leaving aside for the present.) In a preliminary or popular statement, on the other hand, the combinations of factors may either be only roughly indicated or be left quite undetermined. At one extreme we have the statement that (*AX* or *Y*) is a necessary and sufficient condition, where '*X*' and '*Y*' are given definite meanings; at the other extreme we have the merely existentially quantified statement that this holds for *some* pair *X* and *Y*. Our knowledge in such cases ordinarily falls somewhere between these two extremes. We can use the same convention as before, deliberately allowing it to be ambiguous between these different interpretations, and say that in any of these cases, where *A* is an INUS condition of *P* in *F* (*A* . . . or . . .) is a necessary and sufficient condition of *P* in *F*.

A great deal of our ordinary causal knowledge is of this form. We know that the eating of sweets causes dental decay. Here the field is human beings who have some of their own teeth. We do not know, indeed it is not true, that the eating of sweets by any such person is a sufficient condition for dental decay: some people have peculiarly resistant teeth, and there are probably measures which, if taken along with the eating of sweets, would protect the eater's teeth from decay. All we know is that sweet-eating combined with a set of positive and negative factors which we can specify, if at all, only roughly and incompletely, constitutes a minimal sufficient condition for dental decay—but not a necessary one, for there are other combinations of factors, which do not include sweet-eating, which would also make teeth decay, but which we can specify, if at all, only roughly and incompletely. That is, if '*A*' now represents sweet-eating, '*P*' dental decay, and '*F*' the class of human beings with some of their own teeth, we can say that, for some *X* and *Y* (*AX* or *Y*) is necessary and sufficient for *P* in *F*, and we *may* be able to go beyond this merely existentially quantified statement to at least a partial specification of the *X* and *Y* in question. That is, we can say that (*A* . . . or . . .) is a necessary and sufficient condition, but that *A* itself is only an INUS condition. And the same holds for many general causal statements of the form '*A* causes (or produces) *P*'. It is in this sense that the application of a potential difference to the ends of a copper wire produces an electric current in the wire; that a rise in the temperature of a piece of metal makes it expand; that moisture rusts steel; that exposure to various kinds of radiation causes cancer, and so on.

However, it is true that not all ordinary general causal statements are of this sort. Some of them are implicit statements of functional dependence. Functional dependence is a more complicated relationship of which necessity and sufficiency can be regarded as special cases. Here too what we commonly single out as causing some result is only one of a number of factors which jointly affect the result. Again, some causal statements pick out something that is not only an INUS condition, but also a necessary condition. Thus we may say that the yellow fever virus is the cause of yellow fever. (This statement is not, as it might appear to be, tautologous, for the yellow fever virus and the disease itself can be independently specified.) In the field in question—human beings—the injection of this virus is not by itself a sufficient condition for this disease, for persons who have once recovered from yellow fever are thereafter immune to it, and other persons can be immunized against it. The injection of the virus, combined with the absence of immunity (natural or artificial), and perhaps combined with some other factors, constitutes a sufficient condition for the disease. Beside this, the injection of the virus is a necessary condition of the disease. If there is more than one complex sufficient condition for yellow fever, the injection of the virus into the patient's bloodstream (either by a mosquito or in some other way) is a factor included in every such sufficient condition. If '*A*' stands for this factor, the necessary and sufficient condition has the form (*A* . . . or *A* . . . etc.), where *A* occurs in every disjunct. We sometimes note the difference between this and the standard case by using the phrase 'the cause'. We may say not merely that this virus *causes* yellow fever, but that it is *the cause* of yellow fever; but we would say only that sweet-eating *causes* dental decay, not that it is *the cause* of dental decay. But about an individual case we could say that sweet-eating was *the cause* of the decay of this person's teeth, meaning (as in § 1 above) that the only sufficient condition present here was the one of which sweet-eating is a non-redundant part. Nevertheless, there will not in general be any one item which has a unique claim to be regarded as *the cause* even of an individual event, and even after the causal field has been determined. Each of the moments in the minimal sufficient condition, or in each minimal sufficient condition, that was present can equally be regarded as the cause. They may be distinguished as predisposing causes, triggering causes, and so on, but it is quite arbitrary to pick out as 'main' and 'secondary', different moments which are equally non-redundant items in a minimal sufficient condition, or which are moments in two minimal sufficient conditions each of which makes the other redundant.[22]

4. NECESSITY AND SUFFICIENCY

One possible account of general statements of the forms '*S* is a necessary condition of *T*' and '*S* is a sufficient condition of *T*'—where '*S*' and '*T*' are general terms—is that they are equivalent to simple universal propositions. That is, the former is equivalent to 'All *T* are *S*' and the latter to 'All *S* are *T*'. Similarly, '*S* is necessary for *T* in the field *F*' would be equivalent to 'All *FT* are *S*', and '*S* is sufficient for *T* in the field *F*' to 'All *FS* are *T*'. Whether an account of this sort is adequate is, of course, a matter of dispute; but it is not disputed that these statements about necessary and sufficient conditions at least *entail* the corresponding universals. I shall work on the assumption that this account is adequate, that general statements of necessity and sufficiency are equivalent to universals: it will be worth while to see how far this account will take us, how far we are able, in terms of it, to understand how we use, support, and criticize these statements of necessity and sufficiency.

A directly analogous account of the corresponding singular statements is not satisfactory. Thus it will not do to say that 'A short-circuit here was a necessary condition of a fire in this house' is equivalent to 'All cases of this house's catching fire are cases of a short-circuit occurring here', because the latter is automatically true if this house has caught fire only once and a short-circuit has occurred on that occasion, but this is not enough to establish the statement that the short-circuit was a necessary condition of the fire; and there would be an exactly parallel objection to a similar statement about a sufficient condition.

It is much more plausible to relate singular statements about necessity and sufficiency to certain kinds of non-material conditionals. Thus 'A short-circuit here was a necessary condition of a fire in this house' is closely related to the counterfactual conditional 'If a short-circuit had not occurred here this house would not have caught fire', and 'A short-circuit here was a sufficient condition of a fire in this house' is closely related to what Goodman has called the factual conditional, 'Since a short-circuit occurred here, this house caught fire'.

However, a further account would still have to be given of these non-material conditionals themselves. I have argued elsewhere[23] that they are best considered as condensed or telescoped *arguments*, but that the statements used as premisses in these arguments are no more than simple factual universals. To use the above-quoted counterfactual conditional is, in effect, to run through an incomplete argument: 'Suppose that a short-circuit did not occur here, then the house did not catch fire'. To use the factual conditional is, in effect, to run through a similar incomplete argument, 'A short-circuit occurred here; therefore the house caught fire'. In each case the argument might in principle be completed by the insertion of other premisses which, together with the stated premiss, would entail the stated conclusion. Such additional premisses may be said to *sustain* the non-material conditional. It is an important point that someone can use a non-material conditional without completing or being able to complete the argument, without being prepared explicitly to assert premisses that would sustain it, and similarly that we can understand such a conditional without knowing exactly how the argument would or could be completed. But to say that a short-circuit here was a necessary condition of a fire in this house is to say that there is some set of true propositions which would sustain the above-stated counterfactual, and to say that it was a sufficient condition is to say that there is some set of true propositions which would sustain the above-stated factual conditional. If this is conceded, then the relating of singular statements about necessity and sufficiency to non-material conditionals leads back to the view that they refer indirectly to certain simple universal propositions. Thus, if we said that a short-circuit here was a necessary condition for a fire in this house, we should be saying that there are true universal propositions from which, together with true statements about the characteristics of this house, and together with the supposition that a short-circuit did not occur here, it would follow that the house did not catch fire. From this we could infer the universal proposition which is the more obvious, but unsatisfactory, candidate for the analysis of this statement of necessity, 'All cases of this house's catching fire are cases of a short-circuit occurring here', or, in our symbols, 'All *FP* are *A*'. We can use this to represent approximately the statement of necessity, on the understanding that it is to be a consequence of some set of wider universal propositions, and is not to be automatically true merely because there is only this one case of an *FP*, of this house's catching fire.[24] A statement that *A* was a sufficient condition may be similarly represented by 'All *FA* are *P*'. Correspondingly, if all that we want to say is that (*A* . . . or . . .) was necessary and sufficient for *P* in *F*, this will be represented approximately by the pair of universals 'All *FP* are (*A* . . . or . . .) and all *F* (*A* . . . or . . .) are *P*', and more accurately by the statement that there is some set of wider universal propositions from which, together with true statements about the features of *F*, this pair of universals follows. This, therefore, is the fuller analysis of the claim that in a particular case *A* is an INUS condition of *P* in *F*, and hence of the singular statement that *A* caused *P*. (The statement that *A* is *at least* an INUS condition includes other alternatives, corresponding to cases where the necessary and sufficient condition is (*A* or . . .), *A* . . . , or *A*).

Let us go back now to general statements of necessity and sufficiency and take *F* as a class, not as an individual. On the view that I am adopting, at least provisionally, the statement that Z is a necessary and sufficient condition for P in F is equivalent to 'All FP are Z and all FZ are P'. Similarly, if we cannot completely specify a necessary and sufficient condition for P in F, but can only say that the formula '(A . . . or . . .)' represents such a condition, this is equivalent to the pair of incomplete universals, 'All FP are (A . . . or . . .) and all F (A . . . or . . .) are P'. In saying that our general causal statements often do no more than specify an INUS condition, I am therefore saying that much of our ordinary causal knowledge is knowledge of such pairs of incomplete universals, of what we may call elliptical or *gappy* causal laws.

* * *

NOTES

1. This term was suggested by D. C. Stove who has also given me a great deal of help by criticizing earlier versions of this article.

2. The phrase 'minimal sufficient condition' is borrowed from Konrad Marc-Wogau, 'On Historical Explanation', *Theoria*, 28 (1962), 213–33. This article gives an analysis of singular causal statements, with special reference to their use by historians, which is substantially equivalent to the account I am suggesting. Many further references are made to this article, especially in n. 9 below.

3. Cf. n. 8 on p. 227 of Marc-Wogau's article, where it is pointed out that in order to infer that the disjunction of all the minimal sufficient conditions will be a necessary condition, 'it is necessary to presuppose that an arbitrary event C, if it occurs, must have sufficient reason to occur'. This presupposition is equivalent to the presupposition that there is some (possibly complex) condition that is both necessary and sufficient for C.

It is of some interest that some common turns of speech embody this presupposition. To say 'Nothing but X will do', or 'Either X or Y will do, but nothing else will', is a natural way of saying that X, or the disjunction (X or Y), is a *necessary* condition for whatever result we have in mind. But taken literally these remarks say only that there is no sufficient condition for this result other than X, or other than (X or Y). That is, we use to mean 'a necessary condition' phrases whose literal meanings would be 'the only sufficient condition', or 'the disjunction of all sufficient conditions'. Similarly, to say that Z is 'all that's needed' is a natural way of saying that Z is a sufficient condition, but taken literally this remark says that Z is the only necessary condition. But, once again, that the only necessary condition will also be a sufficient one follows only if we presuppose that some condition is both necessary and sufficient.

4. I am indebted to the referees for the suggestion that these points should be clarified.

5. Special cases where an INUS condition is also a necessary one are mentioned at the end of § 3.

6. This point, and the term 'non-redundant', are taken from Michael Scriven's review of Nagel's *The Structure of Science*, in *Review of Metaphysics*, 1964. See especially the passage on p. 408 quoted below.

7. See example of the wicket-keeper discussed below.

8. See §§7, 8.

9. See pp. 226–7 of the article referred to in n. 2 above. Marc-Wogau's full formulation is as follows:

"Let "msc" stand for minimal sufficient condition and "nc" for necessary condition. Then, suppose we have a class K of individual events a_1, $a_2 \ldots a_n$. (It seems reasonable to assume that K is finite; however, even if K were infinite the reasoning below would not be affected.) My analysis of the singular causal statement: α is the cause of β, where α and β stand for individual events, can be summarily expressed in the following statements:

(1) (EK) ($K = a_1, a_2, \ldots a_n$);

(2) (x) (x ;tv $K \equiv x$ msc β);

(3) (a_1 ;fn a_2 ;fn . . . a_n) nc β;

(4) (x) (x ;tv K x ;ue a_1) ;ve x is not fulfilled when α occurs);

(5) α is a moment in a_1.

(3) and (4) say that a_1 is a necessary condition *post factum* for β. If a_1 is a necessary condition *post factum* for β, then every moment in a_1 is a necessary condition *post factum* for β, and therefore also α. As has been mentioned before (note 6) there is assumed to be a temporal sequence between α and β; β is not itself an element in K."

10. Op. cit., p. 408.

11. However, in n. 7 on pp. 222–33, Marc-Wogau draws attention to the difficulty of giving an accurate definition of "a moment in a sufficient condition." Further complications are involved in the account given in §5 below of "clusters" of factors and the progressive localization of a cause. A condition which is minimally sufficient in relation to one degree of analysis of factors may not be so in relation to another degree of analysis.

12. This section is something of an aside: the main argument is resumed in §3.

13. This notion of a causal field was introduced by John Anderson. He used it, e.g., in "The Problem of Causality," first published in the *Australasian Journal of Psychology and Philosophy,* 16 (1938), and reprinted in *Studies in Empirical Philosophy* (Sydney, 1962), pp. 126–36, to overcome certain difficulties and paradoxes in Mill's account of causation. I have also used this notion to deal with problems of legal and moral responsibility, in "Responsibility and Language," *Australasian Journal of Philosophy,* 33 (1955), 143–59.

14. These examples are borrowed from Scriven, op. cit., pp. 409–10. Scriven discusses them with reference to what he calls a "contrast class," the class of cases where the effect did not occur with which the case where it did occur is being contrasted. What I call the causal field is the logical sum of the case (or cases) in which the effect is being said to be caused with what Scriven calls the contrast class.

15. Cf. Bertrand Russell, 'On the Notion of Cause', *Mysticism and Logic* (London, 1917), p. 187. Cf. also Scriven's first difficulty, op. cit., p. 409: 'First, there are virtually no known sufficient conditions, literally speaking, since human or accidental interference is almost inexhaustibly possible, and hard to exclude by specific qualification without tautology'. The introduction of the causal field also automatically covers Scriven's third difficulty and third refinement, that of the contrast class and the relativity of causal statements to contexts.

16. J. R. Lucas, 'Causation', *Analytical Philosophy,* ed. R. J. Butler (Oxford, 1962), pp. 57–9, resolves this kind of difficulty by an informal appeal to what amounts to this notion of a causal field: '. . . these circumstances [cosmic cataclysms, etc.] . . . destroy the whole causal situation in which we had been looking for *Z* to appear . . . predictions are not expected to come true when quite unforeseen emergencies arise'.

17. This is related to Scriven's second difficulty, op. cit., p. 409: 'There still remains the problem of saying what the other factors are which, with the cause, make up the sufficient condition. If they can be stated, causal explanation is then simply a special case of subsumption under a law. If they cannot, the analysis is surely mythological'. Scriven correctly replies that 'a combination of the thesis of macro-determinism . . . and observation-plus-theory frequently gives us the very best of reasons for saying that a certain factor combines with an unknown sub-set of the conditions present into a sufficient condition for a particular effect'. He gives a statistical example of such evidence, but the whole of my account of typical sorts of evidence for causal relationships in §§5 and 7 below [omitted from this volume] is an expanded defence of a reply of this sort.

18. Op. cit., pp. 228–33.

19. Op. cit., pp. 410–11: this is Scriven's fourth difficulty and refinement.

20. Op. cit., p. 232: the example is taken from P. Gardiner, *The Nature of Historical Explanation* (Oxford, 1952), p. 101.

21. Scriven's fifth difficulty and refinement are concerned with the direction of causation. This is considered briefly in §8 below.

22. Cf. Marc-Wogau's concluding remarks, op. cit., pp. 232–3.

23. 'Counterfactuals and Causal Laws', *Analytical Philosophy,* ed. R. J. Butler (Oxford, 1962), pp. 66–80.

24. This restriction may be compared with one which Nagel imposes on laws of nature: 'the vacuous truth of an unrestricted universal is not sufficient for counting it a law; it counts as a law only if there is a set of other assumed laws from which the universal is logically derivable' (*The Structure of Science* (New York, 1961), p. 60). It might have been better if he had added 'or if there is some other way in which it is supported (ultimately) by empirical evidence'. Cf. my remarks in 'Counterfactuals and Causal Laws', pp. 72–4, 78–80.

Hans Reichenbach

THE LAWS OF NATURE

The idea of causality has stood in the foreground of every theory of knowledge of modern times. The fact that nature lends itself to a description in terms of causal laws suggests the conception that reason controls the happenings of nature; and the foregoing presentation of the influence which Newton's mechanics had on philosophical systems (chap. 6) makes it evident that the concept of a synthetic a priori has its roots in a deterministic interpretation of the physical world. Since the physics of an era deeply influences its theory of knowledge, it will be necessary to study the development which the concept of causality underwent in the physics of the nineteenth and twentieth centuries—a development which led to a revision of the idea of laws of nature and terminated in a new philosophy of causality.

The exposition of this historical process will be greatly facilitated if it is preceded by an analysis of the meaning of causality. These considerations may be attached to the inquiry into the meaning of explanation (given above in chap. 2), according to which explanation is generalization. Since explanation is reduction to causes, the causal relation is to be given the same interpretation. In fact, by a causal law the scientist understands a relation of the form *if-then,* with the addition that the same relation holds at all times. To say that the electric current causes a deflection of the magnetic needle means that whenever there is an electric current there is always a deflection of the magnetic needle. The addition in terms of *always* distinguishes the causal law from a chance coincidence. It once happened that while the screen of a motion picture theater showed the blasting of lumber, a slight earthquake shook the theater. The spectators had a momentary feeling that the explosion on the screen caused the shaking of the theater. When we refuse to accept this interpretation, we refer to the fact that the observed coincidence was not repeatable.

Since repetition is all that distinguishes the causal law from a mere coincidence, the meaning of causal relation consists in the statement of an exceptionless repetition—it is unnecessary to assume that it means more. The idea that a cause is connected with its effect by a sort of hidden string, that the effect is forced to follow the cause, is anthropomorphic in its origin and is dispensable; *if-then always* is all that is meant by a causal relation. If the theater would always shake when an explosion is visible on the screen, then there would be a causal relationship. We do not mean anything else when we speak of causality.

True, we sometimes do not stop with the assertion of an exceptionless coincidence, but look for further explanation. Pressing a certain button always is accompanied by a ringing of a bell—this regular coincidence is explained by the laws of electricity, which reveal the ringing of the bell to be a consequence of the relations between electric current and magnetism. But if we proceed to a formulation of these laws, we find that they, in turn, consist in the statement of an *if-then always* relation. The superiority of the laws of nature over simple regularities of the push-button type consists merely in their greater generality. They formulate relations which are manifested in various individual applications of very different kinds. The laws of electricity, for instance, state relations of permanent coincidences

---

Reprinted by permission from Hans Reichenbach, *The Rise of Scientific Philosophy,* University of California Press, 1962.

observable in push-button bells, electric motors, radios, and cyclotrons.

The interpretation of causality in terms of generality, clearly formulated in the writings of David Hume, is now generally accepted by the scientist. Laws of nature are for him statements of an exceptionless repetition—not more. This analysis not only clarifies the meaning of causality; it also opens the path for an extension of causality which has turned out to be indispensable for the understanding of modern science.

The laws of statistics, originally observed for the results of games of chance, were soon discovered also to apply to many other domains. The first social statistics were compiled in the seventeenth century; the nineteenth century brought the introduction of statistical considerations into physics. The kinetic theory of gases, according to which a gas consists of a great many little particles, called molecules, which swarm in all directions, collide with each other, and describe zigzag paths at an enormous speed, was constructed by the help of statistical computations. The statistical method arrived at its greatest triumph when it succeeded in explaining the phenomena of *irreversibility,* which characterize all thermic processes and which are so closely connected with the direction of time.

Everybody knows that heat flows from the hotter body to the colder one, and not vice versa. When we throw an ice cube into a glass of water, the water becomes colder, its heat wandering into the ice and dissolving it. This fact cannot be derived from the law of the conservation of energy. The ice cube is not so very cold and it still contains a great amount of heat; so it might very well give off part of its heat to the surrounding water and make it warmer, the ice itself becoming colder. Such a process would be in agreement with the law of the conservation of energy, if the amount of heat given off by the ice equals the amount received by the water. The fact that a process of this kind does not happen, that heat energy moves only in one direction, must be formulated as an independent law; it is this law which we call the law of irreversibility. The physicist often calls it the second principle of thermodynamics, reserving the name of the first principle to the law of the conservation of energy.

The wording of the principle of irreversibility must be very carefully given. It is not true that heat always flows from the higher temperature to the lower one. Every refrigerator is an example to the contrary. The machine pumps heat from the interior of the ice box to the outside, thus making the interior cooler and the surroundings warmer. But it can do so only because it uses up a certain amount of mechanical energy supplied by the electric motor; this energy is transformed into heat of the average temperature of the room. The physicist has shown that the amount of mechanical energy transformed into heat is greater than the amount of heat energy withdrawn from the interior of the refrigerator. If we regard heat of a higher temperature, or mechanical or electric energy, as an energy of a higher level, there is more energy going down than going up in the refrigerator. The principle of irreversibility is to be formulated as a statement that if all processes involved are included in the consideration, the total energy goes down, so that on the whole there is a tendency to compensation.

It was the discovery of the Vienna physicist Boltzmann that the principle of irreversibility is explainable through statistical considerations. The amount of heat in a body is given by the motion of its molecules; the greater the average speed of the molecule, the higher the temperature. It must be realized that this statement refers only to the average speed of the molecule; the individual molecules may have very different speeds. If a hot body comes into contact with a cold body, their molecules will collide. It may occasionally happen that a slow molecule hitting a fast one loses all its speed and makes the fast molecule even faster. But that is the exception; on the average there will be an equalization of the speeds through the collisions. The irreversibility of heat processes is thus explained as a phenomenon of mixture, comparable to the shuffling of cards, or the mixing of gases and liquids.

Though this explanation makes the law of irreversibility appear plausible, it also leads to an unexpected and serious consequence. It deprives the law of its strictness and makes it a law of probability. When we shuffle cards, we cannot call it impossible that our shuffling will eventually lead to an arrangement in which the first half of the deck contains all the red cards and the second half all the black ones; to arrive at such an arrangement must merely be called very improbable. All statistical laws are of this type. They supply a high probability for unordered arrangements, and leave only a low probability for ordered arrangements. The larger the number involved, the smaller the probability of the ordered arrangements; but this probability will never become zero. The phenomena of thermodynamics refer to very large numbers of individual occurrences, since the number of molecules is very large, and therefore involve extremely high probabilities for processes going in the direction of a compensation. But a process going in the opposite direction cannot strictly be called impossible. For instance, we cannot exclude the possibility that some day the molecules of the air in our room, by pure chance, arrive at an ordered state such that the molecules of oxygen are assembled on one side of the room and those of nitrogen on the other. Unpleasant as the prospect of sitting on the nitrogen side of the room may be, the possibility of such an occurrence cannot be absolutely excluded. Similarly, the physicist cannot exclude the possibility that, when you put an ice cube into a glass of water, the water starts boiling and the ice cube gets as cold as the interior of a deep-freezing cabinet. It may be a consolation to know that this probability is much lower than the probability of a fire breaking out at the same time in each house of a city by independent causes.

Whereas the practical consequences of the statistical interpretation of the law of irreversibility are insignificant because of the low probabilities for processes in the contrary direction, the theoretical consequences are of greatest significance. What was before a strict law of nature has been revealed as being merely a statistical law; the certainty of the law of nature has been replaced by a high probability. With this result the theory of causality entered into a new stage. The question arose whether the same fate might befall other laws of nature, and whether there would remain any strict causal laws.

The discussion of the problem led to two opposite conceptions. According to the first conception the use of statistical laws merely represents an expression of ignorance: if the physicist were able to observe and calculate the individual motion of every molecule, he would not have to resort to statistical laws and would give a strictly causal account of thermodynamic processes. Laplace's superman could do so; for him the path of every molecule would be foreseeable like the path of the stars, and he would not need any statistical laws. This conception does not abandon the idea of a strict causality; it merely regards causality as inaccessible to human knowledge, which by reason of its imperfection has to resort to probability laws.

The second conception represents the opposite point of view. It does not adhere to the belief in a strict causality of the motion of the individual molecule. It advances the opinion that what we observe as a causal law of nature is always the product of a great number of atomic occurrences; the idea of a strict causality may therefore be conceived as an idealization of the regularities of the macroscopic environment in which we live, as a simplification into which we are led because the great number of elementary processes involved makes us regard as a strict law what actually is a statistical law. According to this conception we are not entitled to transfer the idea of strict causality to the microscopic domain. We have no reason to assume that molecules are controlled by strict laws; equal initial situations of a molecule might be followed by different future situations, and even Laplace's superman could not predict the path of a molecule.

The issue is whether causality is an ultimate principle or merely a substitute for statistical regularity, applicable to the macroscopic domain but inadmissible for the realm of the atoms. On the basis of the physics of the nineteenth century the question could not be answered. It was the physics of the twentieth century, with its analysis of atomic occurrences in terms of Planck's concept of the quantum, that gave the answer. From the investigations of modern quantum mechanics we know that the individual atomic occurrences do not lend themselves to a causal interpretation and are merely controlled by probability laws. This result, formulated in Heisenberg's famous principle of indeterminacy, constitutes the proof that the second conception is the correct one, that the idea of a strict causality is to be abandoned, and that the laws of probability take over the place once occupied by the law of causality.

If the logical analysis of causality, as set forth at the beginning of this chapter, is kept in mind, this result will appear as a natural extension of the older views. Causality was to be formulated as a law of exceptionless generality, as an *if-then always* relation. Probability laws are laws that have exceptions, but exceptions that occur in a regular percentage of instances. The probability law is an *if-then in a certain percentage* relation. Modern logic offers the means of dealing with such a relation, which in contradistinction to the *implication* of usual logic is called a *probability implication.* The causal structure of the physical world is replaced by a probability structure, and the understanding of the physical world presupposes the elaboration of a theory of probability.

It should be realized that even without the results of quantum mechanics, the analysis of causality shows that probability notions are indispensable. In classical physics the causal law is an idealization, and the actual occurrences are more complex than is assumed for the causal description. When a physicist calculates the trajectory of a bullet fired by a gun, he figures it out in terms of some major factors, such as the powder charge and the inclination of the barrel; but because he cannot take into account all the minor factors, like the direction of the wind and the moisture of the air, his calculation is limited in its exactness. That means he can predict the point where the bullet will hit only with a certain probability. Or if an engineer constructs a bridge, he can predict its capacity only with a certain probability; circumstances may occur which he did not anticipate and which make the bridge break down under a smaller load. The law of causality, even if true, holds only for ideal objects; the actual objects we deal with are controllable only within the limits of a certain high probability because we cannot exhaustively describe their causal structure. The significance of the probability concept was seen for such reasons before the discoveries of quantum mechanics. After these discoveries it is even more obvious that no philosopher can evade the concept of probability, if he wants to understand the structure of knowledge.

The philosophy of rationalism has at all times referred to causality for a demonstration of the rational character of this world. Spinoza's conception of a predetermined universe is unthinkable without a belief in causality. Leibniz' idea of a logical necessity, acting behind physical occurrences, is dependent on the assumption of a causal connection of all phenomena. Kant's theory of a synthetic a priori knowledge of nature quotes, in addition to the laws of space and time, the principle of causality as the foremost instance of such knowledge. Like the development of the problems of space and time, that of the principle of causality has led, ever since the death of Kant, to a disintegration of the synthetic a priori. The foundations of rationalism were shaken by the very discipline that had supplied—with its mathematical interpretation of nature—the rationalist's major support. The empiricist of modern times derives his most conclusive arguments from mathematical physics.

PART III Confirmation of Scientific Hypotheses

Introduction

We have assumed so far that scientists have at their disposal a collection of laws and theories used to explain and/or predict what they observe. It is necessary now to analyze the process by which the scientist acquires the collection. Considering that, at any given moment, there may be many alternative hypotheses that have been proposed, how does the scientist know which, if any, of these hypotheses are true? What justification can the scientist give for claiming that certain proposed laws and theories are correct? This is the problem of the confirmation of scientific hypotheses.

Many of the classical philosophers of science proposed the following model (often called the *hypothetico-deductive model*) of the confirmation of scientific hypotheses: The scientists deduce from a hypothesis that they want to test an observable consequence of that hypothesis. Then, they run an experiment to see if this hypothesis holds. If it does not, they know the hypothesis is false, but if it does, then the scientists have a confirming instance for the hypothesis. As the scientist finds more confirming instances for the hypothesis, the probability or degree of confirmation of that hypothesis rises, and when they accumulate a sufficient number and variety of confirming instances, they are justified in accepting the hypothesis as being true.

This model for the confirmation of scientific hypotheses is clearly preferable to the simpler model of *induction by enumeration*. According to this latter model, hypotheses of the form "All A's are B's" are acceptable when one has examined a sufficient number of A's and seen that they are all B's. The trouble with this simpler model is that it is only applicable to hypotheses of a particular form which are such that one can determine by observation whether or not objects have properties A and B. The hypothetico-deductive model, on the other hand, seems applicable to all hypotheses. For this reason, many philosophers of science have adopted some version of the hypothetico-deductive model of confirmation, and much work in the theory of confirmation has been devoted to an elaboration and formalization of this intuitive model.

Professor Hempel's "Studies in the Logic of Confirmation" offers an account of the nature of confirming instances. In particular, he defines a function that indicates whether some evidence confirms, disconfirms, or is irrelevant to a given hypothesis. Hempel offers several conditions that must be satisfied by an acceptable confirmation function, argues that various functions that have been proposed are incorrect precisely because they fail to satisfy these conditions, and then offers a new function, the *satisfaction function,* which is intuitively plausible and which does satisfy his conditions of adequacy.

Hempel raises in his paper a related problem about the notion of a confirming instance which has been widely discussed in recent years. This is the "paradox of the ravens." Hempel shows that the adoption of a function which (a) satisfies his equivalence condition and (b) satisfies our intuitive feeling that the observation of an A that is a B confirms the hypothesis that all A's are B's, leads to the seemingly paradoxical result that the observation of a white table confirms the hypothesis that all ravens are black. Hempel argues that this result should be accepted and that its paradoxical nature should be ascribed to a psychological illusion.

It is clear by now that even the most basic notion of the hypothetico-deductive model, the notion of a confirming instance, is difficult to define precisely. But as mentioned above, the hypothetico-deductive model also employs the notion of the degree of confirmation of a hypothesis. It claims that the degree of confirmation of a hypothesis rises as we get additional confirming evidence for that hypothesis. Is it possible to offer a formal precise account of the notion of the degree of confirmation of a hypothesis, given a certain body of evidence? There have been many attempts to do this, but Professor Carnap's account has attracted the most interest. The two selections from his writings reprinted in this section present both a formal and an informal account of his approach.

Carnap defines a function that states, for any given hypothesis and body of evidence, the degree of confirmation of that hypothesis on the basis of that body of evidence. He argues that such functions should be both regular and symmetrical (these two requirements are carefully explained in his second selection). There are, however, many functions that satisfy both of these requirements, and in the selections following, Carnap discusses two of them, c^* and cw. He argues that c^* is preferable to cw because the former, but not the latter, allows us to learn from experience, i.e., to modify our expectations in light of the observations we make. Carnap points out, however, that c^* is not the only regular and symmetrical function that enables us to learn from experience. Therefore, it remains an open question for the Carnapian school to find some justification for adopting one of these many possible functions as a definition of "degree of confirmation."

At the end of Carnap's second article, he shows how we can understand a variety of inductive inferences in light of his account of degrees of confirmation. One type of inductive inference presents a special problem for Carnap, however: This is the inference from an observed sample to a hypothesis of universal form, an inference that seems to be fundamental to the confirmation of scientific laws and theories. Unfortunately, the larger the universe in question, the smaller the value, in any of Carnap's systems, of the degree of confirmation of a universal hypothesis.

In an infinite universe, all universal statements, no matter how much evidence we collect, have a zero degree of confirmation. It would seem, therefore, that if we adopt a Carnapian system of inductive logic, we must conclude that scientific laws and theories are never well-confirmed. Because this result has seemed so counter-intuitive to many philosophers, they have questioned Carnap's whole approach. In attempting to resolve this difficulty, Carnap claims that the scientist is really concerned with the predictions he can derive from the law, but not with the law itself, and these predictions do not have the low (or zero) degree of confirmation that the law itself has.

There is one fundamental problem that Goodman has raised which needs to be elaborated upon here. Consider the hypothesis "all emeralds are green." It seems natural to suppose that the observation of green emeralds confirms that hypothesis and that, given the fact that people have observed emeralds under many diverse conditions and found them all green, the hypothesis in question has a high degree of confirmation. But now, consider the hypothesis that "all emeralds are grue," where *grue* means "examined before *t* and green or not examined before *t* and blue" (where *t* is some time after now). Since all the emeralds that we have examined are grue as well as green, it looks as though we are forced to conclude against our intuitions that our observations of emeralds confirm the grue hypothesis as well as the green hypothesis, and that the grue hypothesis has the same high degree of confirmation as the green hypothesis. Goodman points out, however, that this highly counter-intuitive result leads to the disastrous consequence that the observations we make offer an equal degree of confirmation to any prediction we care to make about any object or event. Any theory of confirming evidence or of degrees of confirmation must find, therefore, a way to prevent hypotheses like the grue hypothesis from being confirmed in the same way that ordinary hypotheses are confirmed. Goodman calls this problem the *new riddle of induction*.

Most philosophers considering this problem have attempted to solve it by finding some significant distinction between the predicate "green" and "grue" which can serve as the basis for ruling out the confirmation of hypotheses containing predicates like "grue" in the distinguishing respect. For example, Carnap has argued that hypotheses containing grue-like predicates involve a reference to specific temporal positions, and this enables us to distinguish them from legitimate hypotheses that can be confirmed. Goodman shows, however, that this distinction does not exist.

An apparently insuperable difficulty in carrying out Carnap's program for inductive logic was the need to provide a rational justification for choosing one probability distribution from among the many consistent distributions. Professor Savage's approach is to accept the impossibility of justifying a unique choice and to urge that the appropriate subject is the study of the family of acceptable probability assignments and the ways in which these are modified in the face of subsequent experience.

He prefers to call this approach "personal probability" as opposed to the term "subjective probability," which is often used by others for the same ideas. He tries to show that as the personal probability measure is modified by experience, it becomes objective because of its responsiveness to new evidence. He also argues briefly that the approach incorporates all that is reasonable of frequency approaches to probability and that it gives a better treatment of the probability of universal statements than the approach of Carnap and others.

Professor Harman takes a more radically different approach to induction still; according to his view, induction leads us to choose that hypothesis which provides the best explanation for the phenomena in question. This inverts the order which is normally assumed between induction and explanation, in which we decide by induction what is likely to be true and then choose explanations from among the likely hypotheses. For Harman the plausibility of hypotheses depends on their explanatory power. Of course a great deal of the consequences of this approach will depend on exactly what view we take of explanation, for example, whether we think of it as absolute and unique, or relative and context-dependent. Thus, all of the issues about explanation discussed in Section II would be relevant to questions about induction if Harman is correct.

Professor Glymour addresses a slightly different question than the other essays in this section. He asks how we determine which parts of a theory are tested by a particular experiment or observation. He describes three possible attitudes toward his question:

a. The hypothetico-deductive account gives a satisfactory answer.
b. An inductive method gives a satisfactory answer.
c. The question is ill-founded because all experiments always test the entire theory.

He argues briefly that all three of these attitudes are incorrect and then proceeds to develop his own approach. In somewhat simplified terms, his position is that an experiment or observation tests those parts of a theory which are used in computing a non-trivial quantity in a significant way. A non-trivial quantity is one which could have had values different from the determined one, and it is computed in a significant way if the value assigned can be checked either by other computations or by direct measurement. This statement is applicable only to theories with numerical functions, but he gives a more general statement for other theories. In the end, he allows that his approach could be seen as a particular refinement of the hypothetico-deductive account, but emphasizes the importance of the refinement to rule out trivial "confirmations."

Professor Lakatos considers the other side of the question of experiment: What do we learn from an experimental result that contradicts what is predicted by a theory? The falsificationist account of scientific methodology, developed most notably by Sir Karl Popper, tells us that the only rational course of action is to abandon the unsuccessful theory. However, the history of science is replete with examples of failed predictions where scientists persisted in pursuing the theory and ultimately found ways of reconciling the theory with the evidence. A notable example would be that in the nineteenth century Newtonian theory gave incorrect predictions for the orbit of Uranus. Instead of abandoning Newtonian theory Adams and Leverrier calculated (independently) what it would take to preserve the theory and his work led to the discovery of Neptune.

Still, we cannot, as Lakatos notes, perpetually disregard contrary evidence, and thus, there is a serious methodological problem concerning disconfirmation. His solution is to propose that the proper unit of appraisal is the research program. It is rational to continue to tinker with a theory and the auxiliary assumptions associated with it (e.g., assumptions about the number of planets), as long as the research program is generally progressive. Lakatos notes that a metamethodological criterion of the adequacy of his account is that according to it most, although not necessarily all, actual science should turn out to be rational.

In his book *Structure of Scientific Revolutions,* Professor Kuhn argued that no system of inductive logic could be developed for the evaluation of theories and that the only reasonable method for evaluating scientific hypotheses was to consider the judgment of the appropriate scientific community. This suggestion was criticized by many as making theory choice a matter of psychology or politics. In this essay Kuhn attempts to clarify his position and to avoid the serious misunderstanding embodied in the criticisms. He lists five traditional criteria for good theories—accuracy, consistency, simplicity, scope and fruitfulness—and indicates his complete agreement with these as the main relevant criteria. However, he argues, the evaluation of a given theory on each of these criteria is not a straightforward or routine matter. To have an idea that simplicity is an important objective virtue of a theory is not to have a method of measuring exactly how simple a theory is. Furthermore, in addition to the issues of evaluating each of the virtues, in comparing two theories the typical situation is one in which one theory may excel in some of the virtues (e.g., simplicity and scope) but be inferior in accuracy and fruitfulness. Thus, he concludes, the reasons for choosing theories can be quite objective without the choice being one that is determined by any rules that produce unequivocal results that all well-informed individuals would agree on.

Professor Laudan develops a new methodology for theory evaluation based on two principles. The first is that theory evaluation is always a matter of comparison: one cannot evaluate a theory in isolation but can only compare two contemporaneous theories. His second principle is that we must distinguish between the term "theory" construed in a specific way, i.e., Bohr's theory of the atom, and a broader use of "theory," as in referring to the atomic theory of matter. Bohr's theory of the atom is one attempt to give a specific form to the atomic theory of matter. When Bohr's theory was shown to be incorrect it was given up but another form of atomic theory took its place. For the latter, larger kind of theory, Laudan introduces the term "research tradition."

Each research tradition will be exemplified by a number of specific theories and will go through various modifications and developments. At any one time, some of the specific theories under the general heading of the research tradition may be mutually contradictory. For example, the research tradition of evolution includes as specific theories incompatible accounts of the exact rates and mechanisms of evolution. What distinguishes one research tradition from another is partly the theories that exemplify it but also the metaphysical and methodological commitments that constitute it.

He acknowledges that Kuhn and Lakatos have, in effect, developed accounts of research traditions (under the labels "disciplinary matrices" and "research programmes") but he is unsatisfied by their efforts for several reasons. He believes that neither Kuhn nor Lakatos take sufficiently into account the importance of conceptual, as opposed to empirical, problems in evaluating the progress of research traditions. He also disagrees with Kuhn's claims about the extent to which a research tradition is implicit in the practices of the community, and with Lakatos' total separation of questions of rational acceptability from issues of progress. In his evaluation of progress he stresses that there are at least two factors to be considered—the amount of progress a tradition has made and the rate at which progress is currently being made. And the latter proves, in his analysis, to be the more fundamental, for he argues that it is always rational to pursue the research tradition that has the highest rate of progress.

Professor Giere, like Kuhn, is critical of attempts to provide an over-arching methodology for theory evaluation. He responds to criticisms that Kuhn's approach leads to relativism by arguing that the alternatives provide no more absolute objectivity than does the approach Kuhn suggests and that he is elaborating. His own development of the Kuhnian idea is that the process of theory evaluation should be considered an evolutionary one; theories can only be judged in competition with one another in the particular environment constituted by the scientific setting of the time. A theory is to be preferred if it provides more or better solutions to the problems of present concern using whatever other information is available. An important consequence of this approach is that the history of science plays an indispensable role in the philosophy of science and is not simply a convenient source of examples.

Our previous selections have all dealt with questions about the rational evaluation of theories, whether it be acceptance or confirmation. There is another process that has historically been of interest to philosophers but was ignored for most of this century: the process of scientific discovery. According to the standard account, this process can be analyzed psychologically, sociologically and historically, but there is no possible logical account of this process. In other words, the standard account maintains that whereas we can offer a prescriptive account of the conditions under which a hypothesis once proposed is really confirmed, the most that we can offer is a descriptive account of the conditions under which scientists actually propose new hypotheses.

Opposing this point of view, Professor Hanson argues that the process of scientific discovery is amenable to logical analysis. He claims that there is an important difference between good reasons and bad reasons for suggesting a hypothesis, and that the logic of scientific discovery is a prescriptive analysis of the conditions in which it would be reasonable to suggest a given hypothesis.

Hanson is quite aware of the following objection that might be raised: Any considerations that could make the suggestion of a given hypothesis plausible would also help confirm that hypothesis. Consequently, there is no special logic of scientific discovery; the only special analyses of the process are the descriptive historical, psychological and sociological analyses.

Hanson's reply to this objection is that there are special considerations that make the suggestion of hypotheses plausible but which do not confirm those hypotheses. What are these special considerations? According to Hanson, they are analogical considerations which suggest that a certain type of hypothesis is likely to be correct. Although such considerations do not confirm any particular hypothesis, they do make plausible the suggestion of any hypothesis of that type. Hanson illustrates the type of analogical considerations he has in mind by an analysis of Kepler's work on planetary motion.

Professor Laudan explores the historical reasons why the quest for a logic of discovery was abandoned until Hanson's attempt at reviving the subject. He notes that the original motivation for a logic of discovery was for a method that would not only produce theories but also would produce theories that could be guaranteed in advance to be correct. As it became recognized that the process of producing novel hypotheses was separable from the process of evaluating hypotheses, the main efforts of philosophers was directed toward the latter. Evaluation seemed more appropriately the domain of rational inquiry, and the fact that the task required choosing among extant hypotheses made the task of finding a logic of evaluation more promising than a logic of invention. Having given an historical account of why the logic of discovery was dropped as a subject of investigation, he offers no conclusions as to whether it should be taken up again.

Professor Curd grants that there is probably no interesting study of the initial generation of hypotheses that would be properly called logic rather than psychology. However, unlike many writers who have criticized the conception of logic of discovery because discovery is an instantaneous individual process, Curd argues that there is an extended period during which a theory is being discovered and that processes that take place can be both rational and public. Moreover, he claims that there is an early stage of investigation, after a hypothesis is conceived but before it is tested, at which rational considerations can enter and which might be properly called a logic of discovery because it is the process by which new theories are developed and established. He distinguishes between the evaluation of a theory (a decision about its plausibility made after testing) from the initial decision as to which theory to pursue. To pursue a theory is to perform experiments to test it, or, at an even earlier stage, working out some of the experimental consequences. Thus, his suggestion is, in effect, that a logic of pursuit is a logic of discovery.

He considers objections from a hypothetico-deductive approach intended to show the impossibility of a logic of discovery and finds those objections unconvincing. He turns next to the claims of inductivists that probabilistic considerations would provide all that is needed of a logic of pursuit. Against this he argues that far too little information is available to assign probabilities to hypotheses at early stages. Finally, he considers Peirce's suggestion of abduction—arguments of the form:

> Event A occurred and is surprising.
> If hypothesis B were true, A would not be surprising.
> Therefore, B is likely to be true.

He points out that the conclusion does not follow, and in fact there will be many other hypotheses C, D and so on which would also make A unsurprising. Thus, abduction cannot provide even a logic of pursuit, since one cannot pursue all of the hypotheses that would make A unsurprising.

The Classical Approach

Carl G. Hempel

STUDIES IN THE LOGIC OF CONFIRMATION

To the memory of my wife,
Eva Ahrends Hempel

1. OBJECTIVE OF THE STUDY[1]

The defining characteristic of an empirical statement is its capability of being tested by a confrontation with experimental finding, i.e., with the results of suitable experiments or "focussed" observations. This feature distinguishes statements which have empirical content both from the statements of the formal sciences, logic and mathematics, which require no experimental test for their validation, and from the formulations of trans-empirical metaphysics, which do not admit of any.

The testability here referred to has to be understood in the comprehensive sense of "testability in principle"; there are many empirical statements which, for practical reasons, cannot be actually tested at present. To call a statement of this kind testable in principle means that it is possible to state just what experiential findings, if they were actually obtained, would constitute favourable evidence for it, and what findings or "data," as we shall say for brevity, would constitute unfavourable evidence; in other words, a statement is called testable in principle, if it is possible to describe the kind of data which would confirm or disconfirm it.

The concepts of confirmation and of disconfirmation as here understood are clearly more comprehensive than those of conclusive verification and falsification. Thus, e.g., no finite amount of experiential evidence can conclusively verify a hypothesis expressing a general law such as the law of gravitation, which covers an infinity of potential instances, many of which belong either to the as yet inaccessible future, or to the irretrievable past; but a finite set of relevant data may well be "in accord with" the hypothesis and thus constitute confirming evidence for it. Similarly, an existential hypothesis, asserting, say, the existence of an as yet unknown chemical element with certain specified characteristics, cannot be conclusively proved false by a finite amount of evidence which fails to "bear out" the hypothesis; but such unfavourable data may, under certain conditions, be considered as weakening the hypothesis in question, or as constituting disconfirming evidence for it.[2]

While, in the practice of scientific research, judgments as to the confirming or disconfirming character of experiential data obtained in the test of a hypothesis are often made without hesitation and with a wide consensus of opinion, it can hardly be said that these judgments are based on an explicit theory providing general criteria of confirmation and of disconfirmation. In this respect, the situation is comparable to the

Reprinted from *Mind* (1945) by permission of the editor of *Mind*.

manner in which deductive inferences are carried out in the practice of scientific research: This, too, is often done without reference to an explicitly stated system of rules of logical inference. But while criteria of valid deduction can be and have been supplied by formal logic, no satisfactory theory providing general criteria of confirmation and disconfirmation appears to be available so far.

In the present essay, an attempt will be made to provide the elements of a theory of this kind. After a brief survey of the significance and the present status of the problem, I propose to present a detailed critical analysis of some common conceptions of confirmation and disconfirmation and then to construct explicit definitions for these concepts and to formulate some basic principles of what might be called the logic of confirmation.

2. SIGNIFICANCE AND PRESENT STATUS OF THE PROBLEM

The establishment of a general theory of confirmation may well be regarded as one of the most urgent desiderata of the present methodology of empirical science.[3] Indeed, it seems that a precise analysis of the concept of confirmation is a necessary condition for an adequate solution of various fundamental problems concerning the logical structure of scientific procedure. Let us briefly survey the most outstanding of these problems.

(a) In the discussion of scientific method, the concept of relevant evidence plays an important part. And while certain "inductivist" accounts of scientific procedure seem to assume that relevant evidence, or relevant data, can be collected in the context of an inquiry prior to the formulation of any hypothesis, it should be clear upon brief reflection that relevance is a relative concept; experiential data can be said to be relevant or irrelevant only with respect to a given hypothesis; and it is the hypothesis which determines what kind of data or evidence are relevant for it. Indeed, an empirical finding is relevant for a hypothesis if and only if it constitutes either favourable or unfavourable evidence for it; in other words, if it either confirms or disconfirms the hypothesis. Thus, a precise definition of relevance presupposes an analysis of confirmation and disconfirmation.

(b) A closely related concept is that of instance of a hypothesis. The so-called method of inductive inference is usually presented as proceeding from specific cases to a general hypothesis of which each of the special cases is an "instance" in the sense that it "conforms to" the general hypothesis in question, and thus constitutes confirming evidence for it.

Thus, any discussion of induction which refers to the establishment of general hypotheses on the strength of particular instances is fraught with all those logical difficulties—soon to be expounded—which beset the concept of confirmation. A precise analysis of this concept is, therefore, a necessary condition for a clear statement of the issues involved in the complex problem of induction and of the ideas suggested for their solution—no matter what their theoretical merits or demerits may be.

(c) Another issue customarily connected with the study of scientific method is the quest for "rules of induction." Generally speaking, such rules would enable us to "infer," from a given set of data, that hypothesis or generalization which accounts best for all the particular data in the given set. Recent logical analyses have made it increasingly clear that this way of conceiving the problem involves a misconception: While the process of invention by which scientific discoveries are made is as a rule *psychologically guided and stimulated* by antecedent knowledge of specific facts, its results are *not logically determined* by them; the way in which scientific hypotheses or theories are discovered cannot be mirrored in a set of general rules of inductive inference.[4] One of the crucial considerations which lead to this conclusion is the following: Take a scientic theory such as the atomic theory of matter. The evidence on which it rests may be described in terms referring to directly observable phenomena, namely to certain "macroscopic" aspects of the various experimental and observational data which are relevant to the theory. On the other hand, the theory itself contains a large number of highly abstract, nonobservational terms such as "atom," "electron," "nucleus," "dissociation," "valence," and others, none of which figures in the description of the observational data. An adequate rule of induction would therefore have to provide, for this and for every conceivable other case, mechanically applicable criteria determining unambiguously, and without any reliance on the inventiveness or additional scientific knowledge of its user, all those new abstract concepts which need to be created for the formulation of the theory that will account for the given evidence. Clearly, this requirement cannot be satisfied by any set of rules, however ingeniously devised; there can be no general rules of induction in the above sense; the demand for them rests on a confusion of logical and psychological issues. What determines the soundness of a hypothesis is not the way it is arrived at (it may even have been suggested by a dream or a hallucination), but the way it stands up when tested, i.e., when confronted with relevant observational data. Accordingly, the quest for rules of induction in the original sense of canons of scientific discovery has to be replaced, in the logic of science, by the quest for general objective criteria determining (a) whether, and—if possible—even (b) to what degree, a hypothesis H may be said to be corroborated by a given body of evidence E. This approach differs essentially from the inductivist conception of the problem in that it presupposes not only E, but also H, as given and then seeks to determine a certain logical relationship between them. The two parts of this latter problem can be restated in somewhat more precise terms as follows:

(A) To give precise definitions of the two nonquantitative relational concepts of confirmation and of disconfirmation, i.e., to define the meaning of the phrases "E confirms H" and "E disconfirms H." (When E neither confirms nor disconfirms H, we shall say that E is neutral, or irrelevant, with respect to H.)

(B) (1) To lay down criteria defining a metrical concept "degree of confirmation of H with respect to E," whose values are real numbers; or, failing this,

(2) To lay down criteria defining two relational concepts, "more highly confirmed than" and "equally well confirmed with," which make possible a nonmetrical comparison of hypotheses (each with a body of evidence assigned to it) with respect to the extent of their confirmation.

Interestingly, problem (B) has received much more attention in methodological research than problem (A); in particular, the various theories of the "probability of hypotheses" may be regarded as concerning this problem complex; we have here adopted[5] the more neutral term "degree of confirmation" instead of "probability" because the latter is used in science in a definite technical sense involving reference to the relative frequency of the occurrence of a given event in a sequence, and it is at least an open question whether the degree of confirmation of a hypothesis can generally be defined as a probability in this statistical sense.

The theories dealing with the probability of hypotheses fall into two main groups: The "logical" theories construe probability as a logical relation between sentences (or propositions; it is not always clear which is meant)[6]; the "statistical" theories interpret the probability of a hypothesis in substance as the limit of the relative frequency of its confirming instances among all relevant cases.[7] Now it is a remarkable fact that none of the theories of the first type which have been developed so far provides an explicit general definition of the probability (or degree of confirmation) of a hypothesis H with respect to a body of evidence E; they all limit themselves essentially to the construction of an uninterpreted postulational system of logical probability. For this reason, these theories fail to provide a complete solution of problem (B). The statistical approach, on the other hand, would, if successful, provide an explicit numerical definition of the degree of confirmation of a hypothesis; this definition would be formulated in terms of the numbers of confirming and disconfirming instances for *H* which constitute the body of evidence *E*. Thus, a necessary condition for an adequate interpretation of degrees of confirmation as statistical probabilities is the establishment of precise criteria of confirmation and disconfirmation, in other words, the solution of problem (A).

However, despite their great ingenuity and suggestiveness, the attempts which have been made so far to formulate a precise statistical definition of the degree of confirmation of a hypothesis seem open to certain objections,[8] and several authors[9] have expressed doubts as to the possibility of defining the degree of confirmation of a hypothesis as a metrical magnitude, though some of them consider it as possible, under certain conditions, to solve at least the less exacting problem (B) (2), i.e., to establish standards of nonmetrical comparison between hypotheses with respect to the extent of their confirmation. An adequate comparison of this kind might have to take into account a variety of different factors;[10] but again the numbers of the confirming and of the disconfirming instances which the given evidence includes will be among the most important of those factors.

Thus, of the two problems, (A) and (B), the former appears to be the more basic one, first, because it does not presuppose the possibility of defining numerical degrees of confirmation or of comparing different hypotheses as to the extent of their confirmation; and second because our considerations indicate that any attempt to solve problem (B)—unless it is to remain in the stage of an axiomatized system without interpretation—is likely to require a precise definition of the concepts of confirming and disconfirming instance of a hypothesis before it can proceed to define numerical degrees of confirmation or to lay down nonmetrical standards of comparison.

(d) It is now clear that an analysis of confirmation is of fundamental importance also for the study of the central problem of what is customarily called epistemology; this problem may be characterized as the elaboration of "standards of rational belief" or of criteria of warranted assertibility. In the methodology of empirical science this problem is usually phrased as concerning the rules governing the test and the subsequent acceptance or rejection of empirical hypotheses on the basis of experimental or observational findings, while in its "epistemological" version the issue is often formulated as concerning the validation of beliefs by reference to perceptions, sense data, or the like. But no matter how the final empirical evidence is construed and in what terms it is accordingly expressed, the theoretical problem remains the same: to characterize, in precise and general terms, the conditions under which a body of evidence can be said to confirm, or to disconfirm, a hypothesis of empirical character; and that is again our problem (A).

(e) The same problem arises when one attempts to give a precise statement of the empiricist and operationalist criteria for the empirical meaningfulness of a sentence; these criteria, as is well known, are formulated by reference to the theoretical testability of the sentence by means of experimental evidence;[11] and the concept of theoretical testability, as was pointed out earlier, is closely related to the concepts of confirmation and disconfirmation.[12]

Considering the great importance of the concept of confirmation, it is surprising that no systematic theory of the non-quantitative relation of confirmation seems to have been developed so far. Perhaps this fact reflects the tacit assumption that the concepts of confirmation and of disconfirmation have a sufficiently clear meaning to make explicit definitions unnecessary or at least comparatively trivial. And indeed, as will be shown below, there are certain features which are rather generally associated with the intuitive notion of confirming evidence, and which, at first, seem well-suited to serve as defining characteristics of confirmation. Closer examination will reveal the definitions thus obtainable to be seriously deficient and will make it clear that an adequate definition of confirmation involves considerable difficulties.

Now the very existence of such difficulties suggests the question whether the problem we are considering does not rest on a false assumption: Perhaps there are no objective criteria of confirmation; perhaps the decision as to whether a given hypothesis is acceptable in the light of a given body of evidence is no more subject to rational, objective rules than is the process of inventing a scientific hypothesis or theory; perhaps, in the last analysis, it is a "sense of evidence," or a feeling of plausibility in view of the relevant data, which ultimately decides whether a hypothesis is scientifically acceptable.[13] This view is comparable to the opinion that the validity of a mathematical proof or of a logical argument has to be judged ultimately by reference to a feeling of soundness or convincingness; and both theses have to be rejected on analogous grounds: They involve a confusion of logical and psychological considerations. Clearly, the occurrence or nonoccurrence of a feeling of conviction upon the presentation of grounds for an assertion is a subjective matter which varies from person to person, and with the same person in the course of time; it is often deceptive, and can certainly serve neither as a necessary nor as a sufficient condition for the soundness of the given assertion.[14] A rational reconstruction of the standards of scientific validation cannot, therefore, involve reference to a sense of evidence; it has to be based on objective criteria. In fact, it seems reasonable to require that the criteria of empirical confirmation, besides being objective in character, should contain no reference to the specific subject matter of the hypothesis or of the evidence in question; it ought to be possible, one feels, to set up purely formal criteria of confirmation in a manner similar to that in which deductive logic provides purely for malcriteria for the validity of deductive inferences.

With this goal in mind, we now turn to a study of the non-quantitative concept of confirmation. We shall begin by examining some current conceptions of confirmation and exhibiting their logical and methodological inadequacies; in the course of this analysis, we shall develop a set of conditions for the adequacy of any proposed definition of confirmation; and finally, we shall construct a definition of confirmation which satisfies those general standards of adequacy.

3. NICOD'S CRITERION OF CONFIRMATION AND ITS SHORTCOMINGS

We consider first a conception of confirmation which underlies many recent studies of induction and of scientific method. A very explicit statement of this conception has been given by Jean Nicod in the following passage: "Consider the formula or the law: *A entails B*. How can a particular proposition, or more briefly, a fact, affect its probability? If this fact consists of the presence of *B* in a case of *A*, it is favourable to the law '*A entails B*'; on the contrary, if it consists of the absence of *B* in a case of *A*, it is unfavourable to this law. It is conceivable that we have here the only two direct modes in which a fact can influence the probability of a law. . . . Thus, the entire influence of particular truths or facts on the probability of universal propositions or laws would operate by means of these two elementary relations which we shall call *confirmation* and *invalidation*."[15] Note that the applicability of the criterion is restricted to hypotheses of the form "*A entails B*." Any hypothesis *H* of this kind may be expressed in the notation of symbolic logic[16] by means of a universal conditional sentence, such as, in the simplest case,

$$(x)(P(x) \supset Q(x)),$$

i.e., "For any object *x:* If *x* is a *P*, then *x* is a *Q*," or also, "Occurrences of the quality *P* entails occurrence of the quality *Q*." According to the above criterion this hypothesis is confirmed by an object *a*, if *a* is *P* and *Q;* and the hypothesis is disconfirmed by *a* if *a* is *P*, but not *Q*. In other words, an object confirms a universal conditional hypothesis if and only if it satisfies both the antecedent (here: "$P(x)$") and the consequent (here: "$Q(x)$") of the conditional; it disconfirms the hypothesis if and only if it satisfies the antecedent, but not the consequent of the conditional; and (we add to this Nicod's statement) it is neutral, or irrelevant, with respect to the hypothesis if it does not satisfy the antecedent.

This criterion can readily be extended so as to be applicable also to universal conditionals containing more than one quantifier, such as "Twins always resemble each other," or, in symbolic notation, "$(x)(y)(\text{Twins}(x, y) \supset \text{Rsbl}(x, y))$." In these cases, a confirming instance consists of an ordered couple, or triple, etc., of objects satisfying the antecedent and the consequent of the conditional. (In the case of the last illustration, any two persons who are twins and who resemble each other would confirm the hypothesis; twins who do not resemble each other would disconfirm it; and any two persons not twins—no matter whether they resemble each other or not—would constitute irrelevant evidence.)

We shall refer to this criterion as Nicod's criterion.[17] It states explicitly what is perhaps the most common tacit interpretation of the concept of confirmation. While seemingly quite adequate, it suffers from serious shortcomings, as will now be shown.

(a) First, the applicability of this criterion is restricted to hypotheses of universal conditional form; it provides no standards for existential hypotheses (such as "There exists organic life on other stars," or "Poliomyelitis is caused by some virus") or for hypotheses whose explicit formulation calls for the use of both universal and existential quantifiers (such as "Every human being dies some finite number of years after his birth," or the psychological hypothesis, "You can fool all of the people some of the time and some of the people all of the time, but you cannot fool all of the people all of the time," which may be symbolized by "$(x)(Et)\text{Fl}(x, t) \cdot (Ex)(t)\text{Fl}(x, t) \cdot \sim (x)(t)\text{Fl}(x, t)$," (where "$\text{Fl}(x, t)$" stands for "You can fool (person) *x* at time *t*"). We note, therefore, the desideratum of establishing a criterion of confirmation which is applicable to hypotheses of any form.[18]

(b) We now turn to a second shortcoming of Nicod's criterion. Consider the two sentences:

$$S_1: (x)(\text{Raven}(x) \supset \text{Black}(x));$$

$$S_2: (x)(\sim\text{Black}(x) \supset \sim\text{Raven}(x))$$

(i.e. "All ravens are black" and "Whatever is not black is not a raven"), and let *a, b, c, d* be four objects such that *a* is a raven and black, *b* a raven but not black, *c* not a raven but black, and *d* neither a raven nor black. Then, according to Nicod's criterion, *a* would confirm S_1, but be neutral with respect to S_2; *b* would disconfirm both S_1 and S_2; *c* would be neutral with respect to both S_1 and S_2, and *d* would confirm S_2, but be neutral with respect to S_1.

But S_1 and S_2 are logically equivalent; they have the same content, they are different formulations of the same hypothesis. And yet, by Nicod's criterion, either of the objects *a* and *d* would be confirming for one of the two sentences, but neutral with respect to the other. This means that Nicod's criterion makes confirmation depend not only on the content of the hypothesis, but also on its formulation.[19]

One remarkable consequence of this situation is that every hypothesis to which the criterion is applicable—i.e. every universal conditional—can be stated in a form for which there cannot possibly exist any confirming instances. Thus, e.g., the sentence:

$(x)[(\text{Raven}(x) \cdot \sim\text{Black}(x)) \supset (\text{Raven}(x) \cdot \sim\text{Raven}(x)]$

is readily recognized as equivalent to both S_1 and S_2 above; yet no object whatever can confirm this sentence, i.e., satisfy both its antecedent and its consequent; for the consequent is contradictory. An analogous transformation is, of course, applicable to any other sentence of universal conditional form.

4. THE EQUIVALENCE CONDITION

The results just obtained call attention to a condition which an adequately defined concept of confirmation should satisfy, and in the light of which Nicod's criterion has to be rejected as inadequate: *Equivalence condition:* Whatever confirms (disconfirms) one of two equivalent sentences, also confirms (disconfirms) the other.

Fulfilment of this condition makes the confirmation of a hypothesis independent of the way in which it is formulated; and no doubt it will be conceded that this is a necessary condition for the adequacy of any proposed criterion of confirmation. Otherwise, the question as to whether certain data confirm a given hypothesis would have to be answered by saying: "That depends on which of the different equivalent formulations of the hypothesis is considered"—which appears absurd. Furthermore—and this is a more important point than an appeal to a feeling of absurdity—an adequate definition of confirmation will have to do justice to the way in which empirical hypotheses function in theoretical scientific contexts such as explanations and predictions; but when hypotheses are used for purposes of explanation or prediction,[20] they serve as premisses in a deductive argument whose conclusion is a description of the event to be explained or predicted. The deduction is governed by the principles of formal logic, and according to the latter, a deduction which is valid will remain so if some or all of the premisses are replaced by different, but equivalent statements; and indeed, a scientist will feel free, in any theoretical reasoning involving certain hypothesis to use the latter in whichever of their equivalent formulations is most convenient for the development of his conclusions. But if we adopted a concept of confirmation which did not satisfy the equivalence condition, then it would be possible, and indeed necessary, to argue in certain cases that it was sound scientific procedure to base a prediction on a given hypothesis if formulated in a sentence S_1, because a good deal of confirming evidence had been found for S_1; but that it was altogether inadmissible to base the prediction (say, for convenience of deduction) on an equivalent formulation S_2, because no confirming evidence for S_2 was available. Thus, the equivalence condition has to be regarded as a necessary condition for the adequacy of any definition of confirmation.

5. THE "PARADOXES" OF CONFIRMATION

Perhaps we seem to have been labouring the obvious in stressing the necessity of satisfying the equivalence condition. This impression is likely to vanish upon consideration of certain consequences which derive from a combination of the equivalence condition with a most natural and plausible assumption concerning a sufficient condition of confirmation.

The essence of the criticism we have levelled so far against Nicod's criterion is that it certainly cannot serve as a necessary condition of confirmation; thus, in the illustration given in the beginning of section 3, the object a confirms S_1 and should therefore also be considered as confirming S_2, while according to Nicod's criterion it is not. Satisfaction of the latter is therefore not a necessary condition for confirming evidence.

On the other hand, Nicod's criterion

might still be considered as stating a particularly obvious and important sufficient condition of confirmation. And indeed, if we restrict ourselves to universal conditional hypotheses in one variable[21]—such as S_1 and S_2 in the above illustration—then it seems perfectly reasonable to qualify an object as confirming such a hypothesis if it satisfies both its antecedent and its consequent. The plausibility of this view will be further corroborated in the course of our subsequent analyses.

Thus, we shall agree that if *a* is both a raven and black, then *a* certainly confirms S_1: "$(x)(\text{Raven}(x) \supset \text{Black}(x))$," and if *d* is neither black nor a raven, *d* certainly confirms S_2:

$$(x)(\sim\text{Black}(x) \supset \sim\text{Raven}(x)).$$

Let us now combine this simple stipulation with the equivalence condition: Since S_1 and S_2 are equivalent, *d* is confirming also for S_1; and thus, we have to recognize as confirming for S_1 any object which is neither black nor a raven. Consequently, any red pencil, any green leaf, any yellow cow, etc., becomes confirming evidence for the hypothesis that all ravens are black. This surprising consequence of two very adequate assumptions (the equivalence condition and the above sufficient condition of confirmation) can be further expanded: The following sentence can readily be shown to be equivalent to S_1: S_3: $(x)[(\text{Raven}(x) \vee \sim \text{Raven}(x)) \supset (\sim\text{Raven}(x) \vee \text{Black}(x))]$, i.e., "Anything which is or is not a raven is either no raven or black." According to the above sufficient condition, S_3 is certainly confirmed by any object, say *e*, such that (1) *e* is or is not a raven and, in addition, (2) *e* is not a raven or also black. Since (1) is analytic, these conditions reduce to (2). By virtue of the equivalence condition, we have therefore to consider as confirming for S_1 any object which is either no raven or also black (in other words: any object which is no raven at all, or a black raven).

Of the four objects characterized in section 3, *a, c* and *d* would therefore constitute confirming evidence for S_1, while *b* would be disconfirming for S_1. This implies that any non-raven represents confirming evidence for the hypothesis that all ravens are black.

We shall refer to these implications of the equivalence criterion and of the above sufficient condition of confirmation as the *paradoxes of confirmation.*

How are these paradoxes to be dealt with? Renouncing the equivalence condition would not represent an acceptable solution, as is shown by the consideration presented in section 4. Nor does it seem possible to dispense with the stipulation that an object satisfying two conditions, C_1 and C_2, should be considered as confirming a general hypothesis to the effect that any object which satisfies C_1, also satisfies C_2.

But the deduction of the above paradoxical results rests on one other assumption which is usually taken for granted, namely, that the meaning of general empirical hypotheses, such as that all ravens are black, or that all sodium salts burn yellow, can be adequately expressed by means of sentences of universal conditional form, such as "$(x)(\text{Raven}(x) \supset \text{Black}(x))$" and "$(x)(\text{Sod. Salt}(x) \supset \text{Burn Yellow}(x))$," etc. Perhaps this customary mode of presentation has to be modified; and perhaps such a modification would automatically remove the paradoxes of confirmation? If this is not so, there seems to be only one alternative left, namely to show that the impression of the paradoxical character of those consequences is due to misunderstanding and can be dispelled, so that no theoretical difficulty remains. We shall now consider these two possibilities in turn: The subsections 5.11 and 5.12 are devoted to a discussion of two different proposals for a modified representation of general hypotheses; in subsection 5.2, we shall discuss the second alternative, i.e. the possibility of tracing the impression of paradoxicality back to a misunderstanding.

5.11. It has often been pointed out that

while Aristotelian logic, in agreement with prevalent every day usage, confers "existential import" upon sentences of the form "All *P*'s are *Q*'s," a universal conditional sentence, in the sense of modern logic, has no existential import; thus, the sentence:

$$(x)(\text{Mermaid}(x) \supset \text{Green}(x))$$

does not imply the existence of mermaids; it merely asserts that any object either is not a mermaid at all, or a green mermaid; and it is true simply because of the fact that there are no mermaids. General laws and hypotheses in science, however—so it might be argued—are meant to have existential import; and one might attempt to express the latter by supplementing the customary universal conditional by an existential clause. Thus, the hypothesis that all ravens are black would be expressed by means of the sentence S_1: "$(x)(\text{Raven}(x) \supset \text{Black}(x)) \cdot (Ex)\text{Raven}(x)$"; and the hypothesis that no non-black things are ravens by S_2: "$(x)(\sim\text{Black}(x) \supset \sim\text{Raven}(x)) \cdot (Ex) \sim\text{Black}(x)$)." Clearly, these sentences are not equivalent, and of the four objects *a, b, c, d* characterized in section 3, part (*b*), only *a* might reasonably be said to confirm S_1, and only *d* to confirm S_2. Yet this method of avoiding the paradoxes of confirmation is open to serious objections:

(a) First of all, the representation of every general hypothesis by a conjunction of a universal conditional and an existential sentence would invalidate many logical inferences which are generally accepted as permissible in a theoretical argument. Thus, for example, the assertions that all sodium salts burn yellow, and that whatever does not burn yellow is no sodium salt are logically equivalent according to customary understanding and usage; and their representation by universal conditionals preserves this equivalence; but if existential clauses are added, the two assertions are no longer equivalent, as is illustrated above by the analogous case of S_1 and S_2.

(b) Second, the customary formulation of general hypotheses in empirical science clearly does not contain an existential clause, nor does it, as a rule, even indirectly determine such a clause unambiguously. Thus, consider the hypothesis that if a person after receiving an injection of a certain test substance has a positive skin reaction, he has diphtheria. Should we construe the existential clause here as referring to persons, to persons receiving the injection, or to persons who, upon receiving the injection, show a positive skin reaction? A more or less arbitrary decision has to be made; each of the possible decisions gives a different interpretation to the hypothesis, and none of them seems to be really implied by the latter.

(c) Finally, many universal hypotheses cannot be said to imply an existential clause at all. Thus, it may happen that from a certain astrophysical theory a universal hypothesis is deduced concerning the character of the phenomena which would take place under certain specified extreme conditions. A hypothesis of this kind need not (and, as a rule, does not) imply that such extreme contions ever were or will be realized; it has no existential import. Or consider a biological hypothesis to the effect that whenever man and ape are crossed, the offspring will have such and such characteristics. This is a general hypothesis; it might be contemplated as a mere conjecture, or as a consequence of a broader genetic theory, other implications of which may already have been tested with positive results; but unquestionably the hypothesis does not imply an existential clause asserting that the contemplated kind of crossbreeding referred to will, at some time, actually take place.

While, therefore, the adjunction of an existential clause to the customary symbolization of a general hypothesis cannot be considered as an adequate *general* method of coping with the paradoxes of confirmation, there is a purpose which the use of an existential clause may serve very well, as was pointed out to me by Dr. Paul Oppenheim:[22] If somebody feels that objects of the types *c* and *d* mentioned above are irrelevant rather than confirming for the hypothesis in question, and that qualifying them as confirming evidence does violence to the meaning of the hypothesis, then this may indicate that he is consciously or unconsciously construing the latter as having existential import; and this kind of understanding of general hypotheses is in fact very common. In this case, the "paradox" may be removed by pointing out that an adequate symbolization of the intended meaning requires the adjunction of an existential clause. The formulation thus obtained is more restrictive than the universal conditional alone; and while we have as yet set up no criteria of confirmation applicable to hypotheses of this more complex form, it is clear that according to every acceptable definition of confirmation objects of the types *c* and *d* will fail to qualify as confirming cases. In this manner, the use of an existential clause may prove helpful in distinguishing and rendering explicit different possible interpretations of a given general hypothesis which is stated in non-symbolic terms.

5.12. Perhaps the impression of the paradoxical character of the cases discussed in the beginning of section 5 may be said to grow out of the feeling that the hypothesis that all ravens are black is about ravens, and not about non-black things, nor about all things. The use of an existential clause was one attempt at expressing this presumed peculiarity of the hypothesis. The attempt has failed, and if we wish to reflect the point in question, we shall have to look for a stronger device. The idea suggests itself of representing a general hypothesis by the customary universal conditional, supplemented by the indication of the specific "field of application" of the hypothesis; thus, we might represent the hypothesis that all ravens are black by the sentence "$(x)(\text{Raven}(x) \supset \text{Black}(x))$" (or any one of its equivalents), plus the indication "Class of ravens" characterizing the field of application; and we might then require that every confirming instance should belong to the field of application. This procedure would exclude the objects *c* and *d* from those constituting confirming evidence and would thus avoid those undesirable consequences of the existential-clause device which were pointed out in 5.11 (c). But apart from this advantage, the second method is open to objections similar to those which apply to the first: (a) The way in which general hypotheses are used in science never involves the statement of a field of application; and the choice of the latter in a symbolic formulation of a given hypothesis thus introduces again a considerable measure of arbitrariness. In particular, for a scientific hypothesis to the effect that all *P*'s are *Q*'s, the field of application cannot simply be said to be the class of all *P*'s; for a hypothesis such as that all sodium salts burn yellow finds important applications in tests with negative results, i.e., it may be applied to a substance of which it is not known whether it contains sodium salts, nor whether it burns yellow; and if the flame does not turn yellow, the hypothesis serves to establish the absence of sodium salts. The same is true of all other hypotheses used for tests of this type. (b) Again, the consistent use of a domain of application in the formulation of general hypotheses would involve considerable logical complications, and yet would have no counterpart in the theoretical procedure of science, where hypotheses are subjected to various kinds of logical transformation and inference without any consideration that might be regarded as referring to changes in the fields of application. This method of meeting the paradoxes would therefore amount to dodging the problem by means of an *ad hoc* device which cannot be justified by reference to actual scientific procedure.

5.2. We have examined two alternatives to the customary method of representing general hypotheses by means of universal conditionals; neither of them proved an adequate means of precluding the paradoxes of confirmation. We shall now try to show that what is wrong does not lie in the customary way of construing and representing general hypotheses, but rather in our reliance on a misleading intuition in the matter: The impression of a paradoxical situation is not objectively founded; it is a psychological illusion.

(a) One source of misunderstanding is the view, referred to before, that a hypothesis of the simple form "Every *P* is a *Q*" such as "All sodium salts burn yellow," asserts something about a certain limited class of objects only, namely, the class of all *P*'s. This idea involves a confusion of logical and practical considerations: Our interest in the hypothesis may be focussed upon its applicability to that particular class of objects, but the hypothesis nevertheless asserts something about, and indeed imposes restrictions upon, *all* objects (within the logical type of the variable occurring in the hypothesis, which in the case of our last illustration might be the class of all physical objects). Indeed, a hypothesis of the form "Every *P* is a *Q*" forbids the occurrence of any objects having the property *P* but lacking the property *Q*, i.e., it restricts all objects whatsoever to the class of those which either lack the property *P* or also have the property *Q*. Now, every object either belongs to this class or falls outside it, and thus, every object—and not only the *P*'s—either conforms to the hypothesis or violates it; there is no object which is not implicitly "referred to" by a hypothesis of this type. In particular, every object which either is no sodium salt or burns yellow conforms to, and thus "bears out" the hypothesis that all sodium salts burn yellow; every other object violates that hypothesis.

The weakness of the idea under consideration is evidenced also by the observation that the class of objects about which a hypothesis is supposed to assert something is in no way clearly determined, and that it changes with the context, as was shown in 5.12 (a).

(b) A second important source of the appearance of paradoxicality in certain cases of confirmation is exhibited by the following consideration.

Suppose that in support of the assertion "All sodium salts burn yellow" somebody were to adduce an experiment in which a piece of pure ice was held into a colourless flame and did not turn the flame yellow. This result would confirm the assertion, "Whatever does not burn yellow is no sodium salt," and consequently, by virtue of the equivalence condition, it would confirm the original formulation. Why does this impress us as paradoxical? The reason becomes clear when we compare the previous situation with the case of an experiment where an object whose chemical constitution is as yet unknown to us is held into a flame and fails to turn it yellow, and where subsequent analysis reveals it to contain no sodium salt. This outcome, we should no doubt agree, is what was to be expected on the basis of the hypothesis that all sodium salts burn yellow—no matter in which of its various equivalent formulations it may be expressed; thus, the data here obtained constitute confirming evidence for the hypothesis. Now the only difference between the two situations here considered is that in the first case we are told beforehand the test substance is ice, and we happen to "know anyhow" that ice contains no sodium salt; this has the consequence that the outcome of the flame-colour test becomes entirely irrelevant for the confirmation of the hypothesis and thus can yield no new evidence for us. Indeed, if the flame should not turn yellow, the hypothesis requires that the substance contain no sodium salt—and we know beforehand that ice does not—and if the flame should turn yellow, the hypothesis would impose no further restrictions on the substance; hence, either of the possible outcomes of the experiment would be in accord with the hypothesis.

The analysis of this example illustrates a general point: In the seemingly paradoxical cases of confirmation, we are often not actually judging the relation of the given evidence, *E*, alone to the hypothesis *H* (we fail to observe the "methodological fiction," characteristic of every case of confirmation, that we have no relevant evidence for *H* other than that included in *E*); instead, we tacitly introduce a comparison of *H* with a body of evidence which consists of *E* in conjunction with an additional amount of information which we happen to have at our disposal; in our illustration, this information includes the knowledge (1) that the substance used in the experiment is ice, and (2) that ice contains no sodium salt. If we assume this additional information as given, then, of course, the outcome of the experiment can add no strength to the hypothesis under consideration. But if we are careful to avoid this tacit reference to additional knowledge (which entirely changes the character of the problem), and if we formulate the question as to the confirming character of the evidence in a manner adequate to the concept of confirmation as used in this paper, we have to ask: Given some object *a* (it happens to be a piece of ice, but this fact is not included in the evidence), and given the fact that *a* does not turn the flame yellow and is no sodium salt—does *a* then constitute confirming evidence for the hypothesis? And now—no matter whether *a* is ice or some other substance—it is clear that the answer has to be in the affirmative; and the paradoxes vanish.

So far, in section (b), we have considered mainly that type of paradoxical case which is illustrated by the assertion that any non-black non-raven constitutes confirming evidence for the hypothesis, "All ravens are black." However, the general idea just outlined applies as well to the even more extreme cases exemplified by the assertion that any non-raven as well as any black object confirms the hypothesis in question. Let us illustrate this by reference to the latter case. If the given evidence *E* (i.e., in the sense of the required methodological fiction, all our data relevant for the hypothesis) consists only of one object which, in addition, is black, then *E* may reasonably be said to support even the hypothesis that all objects are black, and *a fortiori E* supports the weaker assertion that all ravens are black. In this case, again, our factual knowledge that not all objects are black tends to create an impression of paradoxicality which is not justified on logical grounds. Other "paradoxical" cases of confirmation may be dealt with analogously, and it thus turns out that the "paradoxes of confirmation," as formulated above, are due to a misguided intuition in the matter rather than to a logical flaw in the two stipulations from which the "paradoxes" were derived.[23,24]

* * *

8. CONDITIONS OF ADEQUACY FOR ANY DEFINITION OF CONFIRMATION

The two most customary conceptions of confirmation, which were rendered explicit in Nicod's criterion and in the prediction criterion, have thus been found unsuitable for a general definition of confirmation. Besides this negative result, the preceding analysis has also exhibited certain logical characteristics of scientific prediction, explanation, and testing, and it has led to the establishment of certain standards which an adequate definition of confirmation has to satisfy. These standards include the equivalence condition and the requirement that the definition of confirmation be applicable to hypotheses of any degree of logical complexity, rather than to the simplest type of universal conditional only. An adequate definition of confirmation, however, has to satisfy several further logical requirements, to which we now turn.

First of all, it will be agreed that any sentence which is entailed by, i.e., a logical consequence of, a given observation report has to be considered as confirmed by that report: Entailment is a special case of confirmation. Thus, e.g., we want to say that the observation report "*a* is black" confirms the sentence (hypothesis) "*a* is black or grey"; and—to refer to one of the illustrations given in the preceding section—the observation sentence $R_2(a, b)$ should certainly be confirming evidence for the sentence $(Ez)R_2(a, z)$. We are therefore led to the stipulation that any adequate definition of confirmation must insure the fulfilment of the

(8.1) *Entailment condition.* Any sentence which is entailed by an observation report is confirmed by it.[25]

This condition is suggested by the preceding consideration, but of course not proved by it. To make it a standard of adequacy for the definition of confirmation means to lay down the stipulation that a proposed definition of confirmation will be rejected as logically inadequate if it is not constructed in such a way that (8.1) is unconditionally satisfied. An analogous remark applies to the subsequently proposed further standards of adequacy.

Second, an observation report which confirms certain hypotheses would invariably be qualified as confirming any consequence of those hypotheses. Indeed, any such consequence is but an assertion of all or part of the combined content of the original hypotheses and has therefore to be regarded as confirmed by any evidence which confirms the original hypotheses. This suggests the following condition of adequacy:

(8.2) *Consequence Condition.* If an observation report confirms every one of a class *K* of sentences, then it also confirms any sentence which is a logical consequence of *K*.

If (8.2) is satisfied, then the same is true of the following two more special conditions:

(8.21) *Special Consequence Condition.* If an observation report confirms a hypothesis *H*, then it also confirms every consequence of *H*.

(8.22) *Equivalence Condition.* If an observation report confirms a hypothesis *H*, then it also confirms every hypothesis which is logically equivalent with *H*.

(This follows from (8.21) in view of the fact that equivalent hypotheses are mutual consequences of each other.) Thus, the satisfaction of the consequence condition entails that of our earlier equivalence condition, and the latter loses its status of an independent requirement.

In view of the apparent obviousness of these conditions, it is interesting to note that the definition of confirmation in terms of successful prediction, while satisfying the equivalence condition, would violate the consequence condition. Consider, for example, the formulation of the prediction-criterion given in the earlier part of the preceding section. Clearly, if the observational findings B_2 can be predicted on the basis of the findings B_1 by means of the hypothesis *H*, the same prediction is obtainable by means of any equivalent hypothesis, but not generally by means of a weaker one.

On the other hand, any prediction obtainable by means of *H* can obviously also be established by means of any hypothesis which is stronger than *H*, i.e., which logically entails *H*. Thus, while the consequence condition stipulates in effect that whatever confirms a given hypothesis also confirms any weaker hypothesis, the relation of confirmation defined in terms of successful prediction would satisfy the condition that whatever confirms a given hypothesis also confirms every stronger one.

But is this "converse consequence condition," as it might be called, not reasonable enough, and should it not even be included among our standards of adequacy for the definition of confirmation? The second of these two suggestions can be readily disposed of: The adoption of the new condition, in addition to (8.1) and (8.2), would have the consequence that any observation report *B* would confirm any hypothesis *H* whatsoever. Thus, e.g., if *B* is the report "*a* is a raven" and *H* is Hooke's law, then, according to (8.1), *B* confirms the sentence "*a* is a raven," hence *B* would, according to the converse consequence condition, confirm the stronger sentence "*a* is a raven, and Hooke's law holds"; and finally, by virtue of (8.2), *B* would confirm *H*, which is a consequence of the last sentence. Obviously, the same type of argument can be applied in all other cases.

But is it not true, after all, that very often observational data which confirm a hypothesis *H* are considered also as confirming a stronger hypothesis? Is it not true, for example, that those experimental findings which confirm Galileo's law, or Kepler's laws, are considered also as confirming Newton's law of gravitation?[26] This is indeed the case, but this does not justify the acceptance of the converse entailment condition as a general rule of the logic of confirmation; for in the cases just mentioned, the weaker hypothesis is connected with the stronger one by a logical bond of a particular kind: It is essentially a substitution instance of the stronger one; thus, e.g., while the law of gravitation refers to the force obtaining between any two bodies, Galileo's law is a specialization referring to the case where one of the bodies is the earth, the other an object near its surface. In the preceding case, however, where Hooke's law was shown to be confirmed by the observation report that *a* is a raven, this situation does not prevail; and here, the rule that whatever confirms a given hypothesis also confirms any stronger one becomes an entirely absurd principle. Thus, the converse consequence condition does not provide a sound general condition of adequacy.[27]

A third condition remains to be stated:[28]

(8.3) *Consistency Condition.* Every logically consistent observation report is logically compatible with the class of all the hypotheses which it confirms.

The two most important implications of this requirement are the following:

(8.31) Unless an observation report is self-contradictory,[29] it does not confirm any hypothesis with which it is not logically compatible.

(8.32) Unless an observation report is self-contradictory, it does not confirm any hypotheses which contradict each other.

The first of these corollaries will readily be accepted; the second, however,—and consequently (8.3) itself—will perhaps be felt to embody a too severe restriction. It might be pointed out, for example, that a finite set of measurements concerning the variation of one physical magnitude, *x*, with another, *y*, may conform to, and thus be said to confirm, several different hypotheses as to the particular mathematical function in terms of which the relationship of *x* and *y* can be expressed; but such hypotheses are incompatible because to at least one value of *x*, they will assign different values of *y*.

No doubt it is possible to liberalize the formal standards of adequacy in line with these considerations. This would amount to dropping (8.3) and (8.32) and retaining only (8.31). One of the effects of this measure would be that when a logically consistent observation report *B* confirms each of two hypotheses, it does not necessarily confirm their conjunction; for the hypotheses might be mutually incompatible, hence their conjunction self-contradictory; consequently, by (8.31), *B* could not confirm it. This consequence is intuitively rather awkward, and one might therefore feel inclined to suggest that while (8.3) should be dropped and (8.31) retained, (8.32) should be replaced by the requirement (8.33): If an observation sentence confirms each of two hypotheses, then it also confirms their conjunction. But it can readily be shown that by virtue of (8.2) this set of conditions entails the fulfilment of (8.32).

If, therefore, the condition (8.3) appears to be too rigorous, the most obvious alternative would seem to lie in replacing (8.3) and its corollaries by the much weaker condition (8.31) alone; and it is an important problem whether an intuitively adequate definition of confirmation can be constructed which satisfies (8.1), (8.2), and (8.31), but not (8.3). One of the great advantages of a definition which satisfies (8.3) is that it sets a limit, so to speak, to the strength of the hypotheses which can be confirmed by given evidence.[30]

The remainder of the present study, therefore, will be concerned exclusively with the problem of establishing a definition of confirmation which satisfies the more severe formal conditions represented by (8.1), (8.2), and (8.3) together.

The fulfilment of these requirements, which may be regarded as general laws of the logic of confirmation, is of course only a necessary, not a sufficient, condition for the adequacy of any proposed definition of confirmation. Thus, e.g., if "*B* confirms *H*" were defined as meaning "*B* logically entails *H*," then the above three conditions would clearly be satisfied; but the definition would not be adequate because confirmation has to be a more comprehensive relation than entailment (the latter might be referred to as the special case of *conclusive* confirmation). Thus, a definition of confirmation, to be acceptable, also has to be materially adequate: It has to provide a reasonably close approximation to that conception of confirmation which is implicit in scientific procedure and methodological discussion. That conception is vague and to some extent quite unclear, as I have tried to show in earlier parts of this paper; therefore, it would be too much to expect full agreement as to the material adequacy of a proposed definition of confirmation; on the other hand, there will be rather general agreement on certain points; thus, e.g., the identification of confirmation with entailment, or the Nicod criterion of confirmation as analyzed above, or any definition of confirmation by reference to a "sense of evidence" will probably now be admitted not to be adequate approximations to that concept of confirmation which is relevant for the logic of science.

On the other hand, the soundness of the logical analysis (which, in a clear sense, always involves a logical reconstruction) of a theoretical concept cannot be gauged simply by our feelings of satisfaction at a certain proposed analysis; and if there are, say, two alternative proposals for defining a term on the basis of a logical analysis, and if both appear to come fairly close to the intended meaning, then the choice has to be made largely by reference to such features as the logical properties of the two reconstructions, and the comprehensiveness and simplicity of the theories to which they lead.

9. THE SATISFACTION CRITERION OF CONFIRMATION

As has been mentioned before, a precise definition of confirmation requires reference to some definite "language of science," in which all observation reports and all hypotheses under consideration are assumed to be formulated, and whose logical structure is supposed to be precisely determined. The more complex this language, and the richer its logical means of expression, the more difficult it will be, as a rule, to establish an adequate definition of confirmation for it. However, the problem has been solved at least for certain cases: With respect to languages of a comparatively simple logical structure, it has been possible to construct an explicit definition of confirmation which satisfies all of the above logical requirements, and which appears to be intuitively rather adequate. An exposition of the technical details of this definition has been published elsewhere;[31] in the present study, which is concerned with the general logical and methodological aspects of the problem of confirmation rather than with technical detail, it will be attempted to characterize the definition of confirmation thus obtained as clearly as possible with a minimum of technicalities.

Consider the simple case of the hypothesis *H:* $(x)(\text{Raven}(x) \supset \text{Black}(x))$, where "Raven" and "Black" are supposed to be terms of our observational vocabulary. Let *B* be an observation report to the effect that Raven (a) · Black(*a*) · ~ Raven(*c*) · Black(*c*) · ~ Raven(*d*) · ~ Black(*d*). Then *B* may be said to confirm *H* in the following sense: There are three objects altogether mentioned in *B*, namely *a, c,* and *d;* and as far as these are concerned, *B* informs us that all those which are ravens (i.e., just the object *a*) are also black.[32] In other words, from the information contained in *B* we can infer that the hypothesis *H* does hold true within the finite class of those objects which are mentioned in *B*.

Let us apply the same consideration to a hypothesis of a logically more complex structure. Let *H* be the hypothesis "Everybody likes somebody"; in symbols: $(x)(Ey)\text{Likes}(x, y)$, i.e. for every (person) *x*, there exists at least one (not necessarily different person) *y* such that *x* likes *y*. (Here again, "Likes" is supposed to be a relation-term which occurs in our observational vocabulary.) Suppose now that we are given an observation report *B* in which the names of two persons, say *e* and *f*, occur. Under what conditions shall we say that *B* confirms *H*? The previous illustration suggests the answer: If from *B* we can infer that *H* is satisfied within the finite class $\{e, f\}$; i.e. that within $\{e, f\}$ everybody likes somebody. This in turn means that *e* likes *e* or *f*, and *f* likes *e* or *f*. Thus, *B* would be said to confirm *H* if *B* entailed the statement "*e* likes *e* or *f*, and *f* likes *e* or *f*." This latter statement will be called the development of *H* for the finite class $\{e, f\}$.

The concept of *development of a hypothesis, H, for a finite class of individuals, C,* can be defined in a general fashion; the development of *H* for *C* states what *H* would assert if there existed exclusively those objects which are elements of *C*. Thus, e.g., the development of the hypothesis $H_1 = (x)(P(x) \vee Q(x))$ (i.e., "Every object has the property *P* or the property *Q*") for the class $\{a, b\}$ is $(P(a) \vee Q(a)) \cdot (P(b) \vee Q(b))$ (i.e., "*a* has the property *P* or the property *Q*, and *b* has the property *P* or the property *Q*"); the development of the existential hypothesis H_2 that at least one object has the property *P*, i.e., $(Ex)P(x)$, for $\{a, b\}$ is $P(a) \vee P(b)$; the development of a hypothesis which contains no quantifiers, such as $H_3{:}P(c) \vee Q(c)$ is defined as that hypothesis itself, no matter what the reference class of individuals is.

A more detailed formal analysis based on considerations of this type leads to the introduction of a general relation of confirmation in two steps; the first consists in defining a special relation of direct confirmation along the lines just indicated; the second step then defines the general relation of confirmation by reference to direct confirmation.

Omitting minor details, we may summarize the two definitions as follows:

(9.1 Df.) An observation report *B* directly confirms a hypothesis *H* if *B* entails the development of *H* for the class of those objects which are mentioned in *B*.

(9.2 Df.) An observation report *B* confirms a hypothesis *H* if *H* is entailed by a class of sentences each of which is directly confirmed by *B*.

The criterion expressed in these definitions might be called the satisfaction criterion of confirmation because its basic idea consists in construing a hypothesis as confirmed by a given observation report if the hypothesis is satisfied in the finite class of those individuals which are mentioned in the report. Let us now apply the two definitions to our last examples: The observation report B_1: $P(a) \cdot Q(b)$ directly confirms (and therefore also confirms) the hypothesis H_1, because it entails the development of H_1 for the class $\{a, b\}$, which was given above. The hypothesis H_3 is not directly confirmed by B, because its development, i.e., H_3 itself, obviously is not entailed by B_1. However, H_3 is entailed by H_1, which is directly confirmed by B_1; hence, by virtue of (9.2), B_1 confirms H_3.

Similarly, it can readily be seen that B_1 directly confirms H_2.

Finally, to refer to the first illustration given in this section: The observation report Raven(a) · Black(a) · ~ Raven(c) · ~ Black(c) · ~ Raven(d) · ~ Black(d) confirms (even directly) the hypothesis (x)(Raven(x) ⊃ Black(x)), for it entails the development of the latter for the class $\{a, c, d\}$, which can be written as follows: (Raven(a) ⊃ Black(a)) · (Raven(c) ⊃ Black(c)) · (Raven(d) ⊃ Black(d)).

It is now easy to define disconfirmation and neutrality:

(9.3 Df.) An observation report B disconfirms a hypothesis H if it confirms the denial of H.

(9.4 Df.) An observation report B is neutral wth respect to a hypothesis H if B neither confirms nor disconfirms H.

By virtue of the criteria laid down in (9.2), (9.3), (9.4), every consistent observation report, B, divides all possible hypotheses into three mutually exclusive classes: Those confirmed by B, those disconfirmed by B, and those with respect to which B is neutral.

The definition of confirmation here proposed can be shown to satisfy all the formal conditions of adequacy embodies in (8.1), (8.2), and (8.3) and their consequences; for the condition (8.2) this is easy to see; for the other conditions the proof of more complicated.[33]

Furthermore, the application of the above definition of confirmation is not restricted to hypotheses of universal conditional form (as Nicod's criterion is, for example), nor to universal hypotheses in general; it applies, in fact, to any hypothesis which can be expressed by means of property and relation terms of the observational vocabulary of the given language, individual names, the customary connective symbols for "not," "and," "or," "if-then," and any number of universal and existential quantifiers.

Finally, as is suggested by the preceding illustrations as well as by the general considerations which underlie the establishment of the above definition, it seems that we have obtained a definition of confirmation which also is materially adequate in the sense of being a reasonable approximation to the intended meaning of confirmation.

* * *

NOTES

1. The present analysis of confirmation was to a large extent suggested and stimulated by a cooperative study of certain more general problems which were raised by Dr. Paul Oppenheim, and which I have been investigating with him for several years. These problems concern the form and the function of scientific laws and the comparative methodology of the different branches of empirical science. The discussion with Mr. Oppenheim of these issues suggested to me the central problem of the present essay. The more comprehensive problems just referred to will be dealt with by Mr. Oppenheim in a publication which he is now preparing.

In my occupation with the logical aspects of confirmation, I have benefited greatly by discussions with several students of logic, including Professor R. Carnap, Professor A. Tarski, and particularly Dr. Nelson Goodman, to whom I am indebted for several valuable suggestions which will be indicated subsequently.

A detailed exposition of the more technical aspects of the analysis of confirmation presented in this article is included in my article "A Purely Syntactical Definition of Confirmation," *The Journal of Symbolic Logic,* Vol. VIII (1943).

2. This point as well as the possibility of conclusive verification and conclusive falsification will be discussed in some detail in Sec. 10 of the present paper.

3. Or of the "logic of science," as understood by R. Carnap; cf. *The Logical Syntax of Language* (New York and London, 1937), Sec. 72, and the supplementary remarks in *Introduction to Semantics* (Cambridge, Mass., 1942), p. 250.

4. See the lucid presentation of this point in Karl Popper's *Logik der Forschung* (Wien, 1935), esp. Secs. 1, 2, 3, and 25, 26, 27; cf. also Albert Einstein's remarks in his lecture *On the Method of Theoretical Physics* (Oxford, 1933), pp. 11–12. Also of interest in this context is the critical discussion of induction by H. Feigl in "The Logical Character of the Principle of Induction," *Philosophy of Science,* Vol. I (1934).

5. Following R. Carnap's usage in "Testability and Meaning," *Philosophy of Science,* Vols. III (1936) and IV (1937); esp. Sec. 3 (in Vol. III).

6. This group includes the work of such writers as Janina Hosiasson-Lindenbaum (cf. for instance, her article "Induction et analogie: Comparaison de leur fondement," *Mind,* Vol. L (1941); (also see n. 24), H. Jeffreys, J. M. Keynes, B. O. Koopman, J. Nicod (see n. 15), St. Mazurkiewicz, F. Waismann. For a brief discussion of this conception of probability, see Ernest Nagel, *Principles of the Theory of Probability* (Internat. Encyclopedia of Unified Science, Vol. I, No. 6, Chicago, 1939), esp. Secs. 6 and 8.

7. The chief proponent of this view is Hans Reichenbach; cf. especially "Ueber Induktion und Wahrscheinlichkeit," *Erkenntnis,* Vol. V (1935), and *Experience and Prediction* (Chicago, 1938), Chap. v.

8. Cf. Karl Popper, *Logik der Forschung* (Wien, 1935), Sec. 80; Ernest Nagel, *loc. cit.,* Sec. 8, and "Probability and the Theory of Knowledge," *Philosophy of Science,* Vol. VI (1939); C. G. Hempel, "Le problème de la vérité," *Theoria* (Göteborg), Vol. III (1937), Sec. 5, and "On the Logical Form of Probability Statements," *Erkenntnis,* Vol. VII (1937–38), esp. Sec. 5. Cf. also Morton White, "Probability and Confirmation," *The Journal of Philosophy,* Vol. XXXVI (1939).

9. See, for example, J. M. Keynes, *A Treatise on Probability* (London, 1929), esp. Chap. iii; Ernest Nagel, *Principles of the Theory of Probability* (cf. n. 6 above), esp. p. 70. Compare also the somewhat less definitely sceptical statement by Carnap, *loc. cit.* (see n. 5), Sec. 3, p. 427.

10. See especially the survey of such factors given by Ernest Nagel in *Principles of the Theory of Probability* (cf. n. 6), pp. 66–73.

11. Cf. for example, A. J. Ayer, *Language, Truth and Logic* (London and New York, 1936), Chap. i; R. Carnap, "Testability and Meaning" (cf. n. 5), Secs. 1, 2, 3; H. Feigl, "Logical Empiricism" in *Twentieth Century Philosophy,* ed. Dagobert D. Runes (New York, 1943); P. W. Bridgman, *The Logic of Modern Physics* (New York, 1928).

12. It should be noted, however, that in his essay "Testability and Meaning" (cf. n. 5), R. Carnap has constructed definitions of testability and confirmability which avoid reference to the concept of confirming and of disconfirming evidence; in fact, no proposal for the definition of these latter concepts is made in that study.

13. A view of this kind has been expressed, for example, by M. Mandelbaum in "Causal Analyses in History," *Journal of the History of Ideas,* Vol. III (1942); cf. esp. pp. 46–47.

14. See Karl Popper's pertinent statement, *loc. cit.,* Sec. 8.

15. Jean Nicod, *Foundations of Geometry and Induction,* trans. P. P. Wiener (London, 1930), p. 219; cf. also R. M. Eaton's discussion of "Confirmation and Infirmation," which is based on Nicod's views; it is included in Chap. iii of his *General Logic* (New York, 1931).

16. In this paper, only the most elementary devices of this notation are used; the symbolism is essentially that of *Principia Mathematica,* except that parentheses are used instead of dots, and that existential quantification is symbolized by "(*E*)" instead of by the inverted "*E*."

17. This term is chosen for convenience, and in view of the above explicit formulation given by Nicod; it is not, of course, intended to imply that this conception of confirmation originated with Nicod.

18. For a rigorous formulation of the problem, it is necessary first to lay down assumptions as to the means of expression and the logical structure of the language in which the hypotheses are supposed to be formulated; the desideratum then calls for a definition of confirmation applicable to any hypotheses which can be expressed in the given language. Generally speaking, the problem becomes increasingly difficult with increasing richness and complexity of the assumed "language of science."

19. This difficulty was pointed out, in substance, in my article "Le problème de la vértité," *Taeoria* (Göteborg), Vol. III (1937), esp. p. 222.

20. For a more detailed account of the logical structure of scientific explanation and prediction, cf. C. G. Hempel, "The Function of General Laws in History," *The Journal of Philosophy,* Vol. XXXIX (1942), esp. Secs. 2, 3, 4. The characterization, given in that paper as well as in the above text, of explanations and predictions as arguments of a deductive logical structure, embodies an oversimplification: as will be shown in Sec. 7 of the present essay, explanations and predictions often involve "quasi-inductive" steps besides deductive ones. This point, however, does not affect the validity of the above argument.

21. This restriction is essential: In its general form, which applies to universal conditionals in any number of variables, Nicod's criterion cannot even be construed as expressing a sufficient condition of confirmation. This is shown by the following rather surprising example: Consider the hypothesis S_1:

$$(x)(y)[\sim(R(x, y)\cdot R(y, x)) \supset (R(x, y)\cdot\sim R(y, x))].$$

Let a, b be two objects such that $R(a, b)$ and $\sim R(b, a)$. Then clearly, the couple (a, b) satisfies both the antecedent and the consequent of the universal conditional S_1; hence, if Nicod's criterion in its general form is accepted as stating a sufficient condition of confirmation, (a, b) constitutes confirming evidence for S_1. However, S_1 can be shown to be equivalent to

$$S_2\text{: } (x)(y)R(x, y)$$

Now, by hypothesis, we have $\sim R(b,a)$; and this flatly contradicts S_2 and thus S_1. Thus, the couple (a, b), although satisfying both the antecedent and the consequent of the universal conditional S_1 actually constitutes disconfirming evidence of the strongest kind (conclusively disconfirming evidence, as we shall say later) for that sentence. This illustration reveals a striking and—as far as I am aware—hitherto unnoticed weakness of that conception of confirmation which underlies Nicod's criterion. In order to realize the bearing of our illustration upon Nicod's original formulation, let A and B be $\sim(R(x,y) \cdot R(y,x))$ and $R(x, y) \cdot \sim R(y, x)$ respectively. Then S_1 asserts that A entails B, and the couple (a, b) is a case of the presence of B in the presence of A; this should, according to Nicod, be favourable to S_1.

22. This observation is related to Mr. Oppenheim's methodological studies referred to in n. 1.

23. The basic idea of sect. (b) in the above analysis of the "paradoxes of confirmation" is due to Dr. Nelson Goodman, to whom I wish to reiterate my thanks for the help he rendered me, through many discussions, in clarifying my ideas on this point.

24. The considerations presented in section (b) above are also influenced by, though not identical in content with, the very illuminating discussion of the "paradoxes" by the Polish methodologist and logician Janina Hosiasson-Lindenbaum; cf. her article "On Confirmation," *The Journal of Symbolic Logic,* Vol. V (1940), especially Sec. 4. Dr. Hosiasson's attention had been called to the paradoxes by the article referred to in n. 2, and by discussions with the author. To my knowledge, hers has so far been the only publication which presents an explicit attempt to solve the problem. Her solution is based on a theory of degrees of confirmation, which is developed in the form of an uninterpreted axiomatic system (cf. n. 6 and part (b) in Sec. 1 of the present article), and most of her arguments presuppose that theoretical framework. I have profited, however, by some of Miss Hosiasson's more general observations which proved relevant for the analysis of the paradoxes of the non-gradated relation of confirmation which forms the object of the present study.

One point in those of Miss Hosiasson's comments which rest on her theory of degrees of confirmation is of particular interest, and I should like to discuss it briefly. Stated in reference to the raven-hypothesis, it consists in the suggestion that the finding of one non-black object which is no raven, while constituting confirming evidence for the hypothesis, would increase the degree of confirmation of the hypothesis by a smaller amount than the finding of one raven which is black. This is said to be so because the class of all ravens is much less numerous than that of all non-black objects, so that—to put the idea in suggestive though somewhat misleading terms—the finding of one black raven confirms a larger portion of the total content of the hypothesis than the finding of one non-black non-raven. In fact, from the basic assumptions of her theory, Miss Hosiasson is able to derive a theorem according to which the above statement about the relative increase in degree of confirmation will hold provided that actually the number of all ravens is small compared with the number of all non-black objects. But is this last numerical assumption actually warranted in the present case and analogously in all other "paradoxical" cases? The answer depends in part upon the logical structure of the language of science. If a "coordinate language" is used, in which, say, finite space-time regions figure as individuals, then the raven-hypothesis assumes some such form as "Every space-time region which contains a raven, contains something black"; and even if the total number of ravens ever to exist is finite, the class of space-time regions containing a raven has the power of the continuum, and so does the class of space-time regions containing something non-black; thus, for a coordinate language of the type under consideration, the above numerical assumption is not warranted. Now the use of a coordinate language may appear quite artificial in this particular illustration; but it will seem very appropriate in many other contexts, such as, e.g., that of physical field theories. On the other hand, Miss Hosiasson's numerical assumption may well be justified on the basis of a "thing language," in which physical objects of finite size function as individuals. Of course, even on this basis, it remains an empirical question, for every hypothesis of the form "All *P*'s are *Q*'s," whether actually the class of non-*Q*'s is much more numerous than the class of *P*'s; and in many cases this question will be very difficult to decide.

25. As a consequence of this stipulation, a contradictory observation report, such as {Black(a), ~Black(a)} confirms every sentence, because it has every sentence as a consequence. Of course, it is possible to exclude the possibility of contradictory observation reports altogether by a slight restriction of the definition of "observation report." There is, however, no important reason to do so.

26. Strictly speaking, Galileo's law and Kepler's laws can be deduced from the law of gravitation only if certain additional hypotheses—including the laws of motion—are presupposed; but this does not affect the point under discussion.

27. William Barrett, in a paper entitled "Discussion on Dewey's Logic" (*The Philosophical Review,* Vol. L (1941), pp. 305 ff., esp. p. 312) raises some questions closely related to what we have called above the consequence condition and the converse consequence condition. In fact, he invokes the latter (without stating it explicitly) in an argument which is designed to show that "not every observation which confirms a sentence need also confirm all its consequences," in other words, that the special consequence condition (8.21) need not always be satisfied. He supports his point by reference to "the simplest case: the sentence *C* is an abbreviation of $A \cdot B$, and the observation 0 confirms *A, and so C,* but is irrelevant to *B*, which is a consequence of *C*." (Italics mine.)

For reasons contained in the above discussion of the consequence condition and the converse consequence condition, the application of the latter in the case under consideration seems to us unjustifiable, so that the illustration does not prove the author's point; and indeed, there seems to be every reason to preserve the unrestricted validity of the consequence condition. As a matter of fact, Mr. Barrett himself argues that "the degree of confirmation for the consequence of a sentence cannot be less than that of the sentence itself"; this is indeed quite sound; but it is hard to see how the recognition of this principle can be reconciled with a renunciation of the special consequence condition, since the latter may be considered simply as the correlate, for the non-gradated relation of confirmation, of the former principle which is adapted to the concept of degree of confirmation.

28. For a fourth condition, see n. 33.

29. A contradictory observation report confirms every hypothesis (cf. n. 8) and is, of course, incompatible with every one of the hypotheses it confirms.

30. This was pointed out to me by Dr. Nelson Goodman. The definition later to be outlined in this essay, which satisfies conditions (8.1), (8.2), and (8.3), lends itself, however, to certain generalizations which satisfy only the more liberal conditions of adequacy just considered.

31. In my article referred to in n. 1. The logical structure of the languages to which the definition in question is applicable is that of the lower functional calculus with individual constants, and with predicate constants of any degree. All sentences of the language are assumed to be formed exclusively by means of predicate constants, individual constants, individual variables, universal

and existential quantifiers for individual variables, and the connective symbols of denial, conjunction, alternation, and implication. The use of predicate variables or of the identity sign is not permitted.

As to the predicate constants, they are all assumed to belong to the observational vocabulary, i.e. to denote a property or a relation observable by means of the accepted techniques. ("Abstract" predicate terms are supposed to be defined in terms of those of the observational vocabulary and then actually to be replaced by their *definientia,* so that they never occur explicitly.)

As a consequence of these stipulations, an observation report can be characterized simply as a conjunction of sentences of the kind illustrated by $P(a)$, $\sim P(b)$, $R(c, d)$, $\sim R(e, f)$, etc., where P, R, etc., belong to the observational vocabulary, and a, b, c, d, e, f, etc., are individual names, denoting specific objects. It is also possible to define an observation report more liberally as any sentence containing no quantifiers, which means that besides conjunctions also alternations and implication sentences formed out of the above kind of components are included among the observation reports.

32. I am indebted to Dr. Nelson Goodman for having suggested this idea; it initiated all those considerations which finally led to the definition to be outlined below.

33. For these proofs, see the article referred to in Part I, n. 1. I should like to take this opportunity to point out and to remedy a certain defect of the definition of confirmation which was developed in that article, and which has been outlined above: This defect was brought to my attention by a discussion with Dr. Olaf Helmer.

It will be agreed that an acceptable definition of confirmation should satisfy the following further condition which might well have been included among the logical standards of adequacy set up in Sec. 8 above: (8.4). If B_1 and B_2 are logically equivalent observation reports and B_1 confirms (disconfirms, is neutral with respect to) a hypothesis H, then B_2, too, confirms (disconfirms, is neutral with respect to) H. This condition is indeed satisfied if observation reports are construed, as they have been in this article, as classes or conjunctions of observation sentences. As was indicated at the end of n. 14, however, this restriction of observation reports to a conjunctive form is not essential; in fact, it has been adopted here only for greater convenience of exposition, and all the preceding results, including especially the definitions and theorems of the present section, remain applicable without change if observation reports are given the more liberal interpretation characterized at the end of n. 14. (In this case, if P and Q belong to the observational vocabulary, such sentences as $P(a) \vee Q(a)$, $P(a) \vee Q(b)$, etc., would qualify as observation reports.) This broader conception of observation reports was therefore adopted in the article referred to in Part I, n. 1; but it has turned out that in this case, the definition of confirmation summarized above does not generally satisfy the requirement (8.4). Thus, e.g., the observation reports, $B_1 = P(a)$ and $B_2 = P(a) \cdot (Q(b) \vee \sim Q(b))$ are logically equivalent, but while B_1 confirms (and even directly confirms) the hypothesis $H_1 = (x)P(x)$, the second report does not do so, essentially because it does not entail $P(a) \cdot P(b)$, which is the development of H_1 for the class of those objects mentioned in B_2. This deficiency can be remedied as follows: The fact that B_2 fails to confirm H_1 is obviously due to the circumstance that B_2 contains the individual constant b, without asserting anything about b: The object b is mentioned only in an analytic component of B_2. The atomic constituent $Q(b)$ will therefore be said to occur (twice) inessentially in B_2. Generally, an atomic constituent A of a molecular sentence S will be said to occur inessentially in S if by virtue of the rules of the sentential calculus S is equivalent to a molecular sentence in which A does not occur at all. Now an object will be said to be mentioned inessentially in an observation report if it is mentioned only in such components of that report as occur inessentially in it. The sentential calculus clearly provides mechanical procedures for deciding whether a given observation report mentions any object inessentially, and for establishing equivalent formulations of the same report in which no object is mentioned inessentially. Finally, let us say that an object is mentioned essentially in an observation report if it is mentioned, but not only mentioned inessentially, in that report. Now we replace (9.1) by the following definition:

(9.1*a*) An observation report B directly confirms a hypothesis H if B entails the development of H for the class of those objects which are mentioned essentially in B.

The concept of confirmation as defined by (9.1*a*) and (9.2) now satisfies (8.4) in addition to (8.1), (8.2), (8.3) even if observation reports are construed in the broader fashion characterized earlier in this footnote.

Rudolf Carnap

STATISTICAL AND INDUCTIVE PROBABILITY

If you ask a scientist whether the term "probability" as used in science has always the same meaning, you will find a curious situation. Practically everyone will say that there is only one scientific meaning; but when you ask that it be stated, two different answers will come forth. The majority will refer to the concept of probability used in mathematical statistics and its scientific applications. However, there is a minority of those who regard a certain nonstatistical concept as the only scientific concept of probability. Since either side holds that its concept is the only correct one, neither seems willing to relinquish the term "probability." Finally, there are a few people—and among them this author—who believe that an unbiased examination must come to the conclusion that both concepts are necessary for science, though in different contexts.

I will now explain both concepts—distinguishing them as "statistical probability" and "inductive probability"—and indicate their different functions in science. We shall see, incidentally, that the inductive concept, now advocated by a heretic minority, is not a new invention of the twentieth century, but was the prevailing one in an earlier period and only forgotten later on.

Published by The Galois Institute of Mathematics and Art, 1955. Reprinted by permission of the author and the publisher.

The *statistical concept of probability* is well known to all those who apply in their scientific work the customary methods of mathematical statistics. In this field, exact methods for calculations employing statistical probability are developed and rules for its application are given. In the simplest cases, probability in this sense means the relative frequency with which a certain kind of event occurs within a given reference class, customarily called the "population." Thus, the statement "The probability that an inhabitant of the United States belongs to blood group A is p" means that a fraction p of the inhabitants belongs to this group. Sometimes a statement of statistical probability refers, not to an actually existing or observed frequency, but to a potential one, i.e., to a frequency that would occur under certain specifiable circumstances. Suppose, for example, a physicist carefully examines a newly made die and finds it is a geometrically perfect and materially homogeneous cube. He may then assert that the probability of obtaining an ace by a throw of this die is 1/6. This means that *if* a sufficiently long series of throws with this die were made, the relative frequency of aces would be 1/6. Thus, the probability statement here refers to a potential frequency rather than to an actual one. Indeed, if the die were destroyed before any throws were made, the assertion would still be valid.

Exactly speaking, the statement refers to the physical microstate of the die; without specifying its details (which presumably are not known), it is characterized as being such that certain results would be obtained if the die were subjected to certain experimental procedures. Thus the statistical concept of probability is not essentially different from other disposition concepts which characterize the objective state of a thing by describing reactions to experimental conditions, as, for example, the I.Q. of a person, the elasticity of a material object, etc.

Inductive probability occurs in contexts of another kind; it is ascribed to a hypothesis with respect to a body of evidence. The hypothesis may be any statement concerning unknown facts, say, a prediction of a future event, e.g., tomorrow's weather or the outcome of a planned experiment or of a presidential election, or a presumption concerning the unobserved cause of an observed event. Any set of known or assumed facts may serve as evidence; it consists usually in results of observations which have been made. To say that the hypothesis h has the probability p (say, 3/5) with respect to the evidence e, means that for anyone to whom this evidence but no other relevant knowledge is available, it would be reasonable to believe in h to the degree p or, more exactly, it would be unreasonable for him to bet on h at odds higher than p:(1 − p) (in the example, 3:2). Thus inductive probability measures the strength of support given to h by e or the *degree of confirmation* of h on the basis of e. In most cases in ordinary discourse, even among scientists, inductive probability is not specified by a numerical value but merely as being high or low or, in a comparative judgment, as being higher than another probability. It is important to recognize that every inductive probability judgment is relative to some evidence. In many cases no explicit reference to evidence is made; it is then to be understood that the totality of relevant information available to the speaker is meant as evidence. If a member of a jury says that the defendant is very probably innocent or that, of two witnesses A and B who have made contradictory statements, it is more probable that A lied than that B did, he means it with respect to the evidence that was presented in the trial plus any psychological or other relevant knowledge of a general nature he may possess. Probability as understood in contexts of this kind is not frequency. Thus, in our example, the evidence concerning the defendant, which was presented in the trial, may be such that it cannot be ascribed to any other person; and if it could be ascribed to several people, the juror would not know the relative frequency of innocent persons among them. Thus the probability concept used here cannot be the statistical one. While a statement of statistical probability asserts a matter of fact, a statement of inductive probability is of a purely logical nature. If hypothesis and evidence are given, the probability can be determined by logical analysis and mathematical calculation.

One of the basic principles of the theory of inductive probability is the *principle of indifference*. It says that, if the evidence does not contain anything that would favor either of two or more possible events, in other words, if our knowledge situation is symmetrical with respect to these events, then they have equal probabilities relative to the evidence. For example, if the evidence e_1 available to an observer X_1 contains nothing else about a given die than the information that it is a regular cube, then the symmetry condition is fulfilled and therefore each of the six faces has the same probability 1/6 to appear uppermost at the next throw. This means that it would be unreasonable for X_1 to bet more than one to five on any one face. If X_2 is in possession of the evidence e_2 which, in addition to e_1, contains the knowledge that the die is heavily loaded in favor of one of the faces without specifying which one, the probabilities for X_2 are the same as for X_1. If, on the other hand, X_3 knows e_3 to the effect that the load favors the ace, then the probability of the ace on the basis of e_3 is higher than 1/6. Thus, inductive probability, in contradistinction to statistical probability, cannot be ascribed to a material object by itself, irrespective of an observer. This is obvious in our example; the die is the same for all three observers and hence cannot have different properties for them. Inductive probability characterizes a hypothesis relative to available information; this information may differ from person to person and vary for any person in the course of time.

A brief look at the historical development of the concept of probability will give us a better understanding of the present controversy. The mathematical study of problems of probability began when some mathematicians of the sixteenth and seventeenth centuries were asked by their gambler friends about the odds in various games of chance. They wished to learn about probabilities as a guidance for their betting decisions. In the beginning of its scientific career, the concept of probability appeared in the form of inductive probability. This is clearly reflected in the title of the first major treatise on probability, written by Jacob Bernoulli and published posthumously in 1713; it was called *Ars Conjectandi,* the art of conjecture, in other words, the art of judging hypotheses on the basis of evidence. This book may be regarded as marking the beginning of the so-called classical period of the theory of probability. This period culminated in the great systematic work by Laplace, *Theorie analytique des probabilités* (1812). According to Laplace, the purpose of the theory of probability is to guide our judgments and to protect us from illusions. His explanations show clearly that he is mostly concerned, not with actual frequencies, but with methods for judging the acceptability of assumptions, in other words, with inductive probability.

In the second half of the last century and still more in our century, the application of statistical methods gained more and more ground in science. Thus attention was increasingly focussed on the statistical concept of probability. However, there was no clear awareness of the fact that this development constituted a transition to a fundamentally different meaning of the word "probability." In the 1920's the first probability theories based on the frequency interpretation were proposed by men like the statistician R. A. Fisher, the mathematician R. von Mises, and the physicist-philosopher H. Reichenbach. These authors and their followers did not explicitly suggest to abandon that concept of probability which had prevailed since the classical period and to replace it by a new one. They rather believed that their concept was essentially the same as that of all earlier authors. They merely claimed that they had given a more exact definition for it and had developed more comprehensive theories on this improved foundation. Thus, they interpreted Laplace's word "probability" not in his inductive sense, but in their own statistical sense. Since there is a strong, though by far not complete analogy between the two concepts, many mathematical theorems hold in both interpretations, but others do not. Therefore these authors could accept many of the classical theorems but had to reject others. In particular, they objected strongly to the principle of indifference. In the frequency interpretation, this principle is indeed absurd. In our earlier example with the observer X_1, who knows merely that the die has the form of a cube, it would be rather incautious for him to assert that the six faces will appear with equal frequency. And if the same assertion were made by X_2, who has information that the die is biased, although he does not know the direction of the bias, he would contradict his own knowledge. In the inductive interpretation, on the other hand, the principle is valid even in the case of X_2, since in this sense it does not predict frequencies but merely says, in effect, that it would be arbitrary for X_2 to have more confidence in the appearance of one face than in that of any other face and therefore it would be unreasonable for him to let his betting decisions be guided by such arbitrary expectations. Therefore it seems much more plausible to assume that Laplace meant the principle of indifference in the inductive sense rather than to assume that one of the greatest minds of the eighteenth century in mathematics, theoretical physics, astronomy, and philosophy chose an obvious absurdity as a basic principle.

The great economist John Maynard Keynes made the first attempt in our century to revive the old but almost forgotten inductive concept of probability. In his *Treatise on Probability* (1921) he made clear that the inductive concept is implicitly used in all our thinking on unknown events both in everyday life and in science. He showed that the classical theory of probability in its application to concrete problems was understandable only if it was interpreted in the inductive sense. However, he modified and restricted the classical theory in several important points. He rejected the principle of indifference in its classical form. And he did not share the view of the classical authors that it should be possible in principle to assign a numerical value to the probability of any hypothesis whatsoever. He believed that this could be done only under very special, rarely fulfilled conditions, as in games of chance where there is a well determined number of possible cases, all of them alike in their basic features, e.g., the six possible results of a throw of a die, the possible distributions of cards among the players, the possible final positions of the ball on a roulette table, and the like. He thought that in all other cases at best only comparative judgments of probability could be made, and even these only for hypotheses which belong, so to speak, to the same dimension. Thus one might come to the result that, on the basis of available knowledge, it is more probable that the next child of a specified couple will be male rather than female; but no comparison could be made between the probability of the birth of a male child and the probability of the stocks of General Electric going up tomorrow.

A much more comprehensive theory of inductive probability was constructed by the geophysicist Harold Jeffreys (*Theory of Probability,* 1939). He agreed with the classical view that probability can be expressed numerically in all cases. Furthermore, in view of the fact that science replaces statements in qualitative terms (e.g., "the child to be born will be very heavy") more and more by those in terms of measurable quantities ("the weight of the child will be more than eight pounds"), Jeffreys wished to apply probability also to hypotheses of quantitative form. For this reason, he set up an axiom system for probability much stronger than that of Keynes. In spite of Keynes's warning, he accepted the principle of indifference in a form quite similar to the classical one: "If there is no reason to believe one hypothesis rather than another, the probabilities are equal." However, it can easily be seen that the principle in this strong form leads to contradictions. Suppose, for example, that it is known that every ball in an urn is either blue or red or yellow but that nothing is known either of the color of any particular ball or of the numbers of blue, red, or yellow balls in the urn. Let B be the hypothesis that the first ball to be drawn from the urn will be blue, R, that it will be red, and Y, that it will be yellow. Now consider the hypotheses B and non-B. According to the principle of indifference as used by Laplace and again by Jeffreys, since nothing is known concerning B and non-B, these two hypotheses have equal probabilities, i.e., one half. Non-B means that the first ball is not blue, hence either red or yellow. Thus "R or Y" has probability one half. Since nothing is known concerning R and Y, their probabilities are equal and hence must be one fourth each. On the other hand, if we start with the consideration of R and non-R, we obtain the result that the probability of R is one half and that of B one fourth, which is incompatible with the previous result. Thus Jeffreys's system as it stands is inconsistent. This defect can-

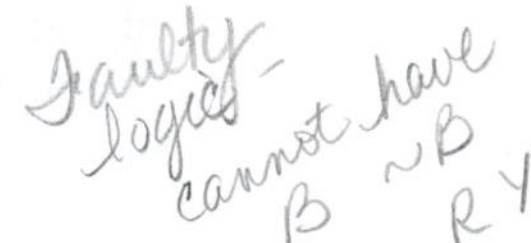

not be eliminated by simply omitting the principle of indifference. It plays an essential role in the system; without it, many important results can no longer be derived. In spite of this defect, Jeffreys's book remains valuable for the new light it throws on many statistical problems by discussing them for the first time in terms of inductive probability.

Both Keynes and Jeffreys discussed also the statistical concept of probability, and both rejected it. They believed that all probability statements could be formulated in terms of inductive probability and that therefore there was no need for any probability concept interpreted in terms of frequency. I think that in this point they went too far. Today an increasing number of those who study both sides of the controversy which has been going on for thirty years are coming to the conclusion that here, as often before in the history of scientific thinking, both sides are right in their positive theses, but wrong in their polemic remarks about the other side. The statistical concept, for which a very elaborate mathematical theory exists, and which has been fruitfully applied in many fields in science and industry, need not at all be abandoned in order to make room for the inductive concept. Both concepts are needed for science, but they fulfill quite different functions. Statistical probability characterizes an objective situation, e.g., a state of a physical, biological, or social system. Therefore it is this concept which is used in statements concerning concrete situations or in laws expressing general regularities of such situations. On the other hand, inductive probability, as I see it, does not occur *in* scientific statements, concrete or general, but only in judgments *about* such statements; in particular, in judgments about the strength of support given by one statement, the evidence, to another, the hypothesis, and hence about the acceptability of the latter on the basis of the former. Thus, strictly speaking, inductive probability belongs not to science itself but to the methodology of science, i.e., the analysis of concepts, statements, theories, and methods of science.

The theories of both probability concepts must be further developed. Although a great deal of work has been done on statistical probability, even here some problems of its exact interpretation and its application, e.g., in methods of estimation, are still controversial. On inductive probability, on the other hand, most of the work remains still to be done. Utilizing results of Keynes and Jeffreys and employing the exact tools of modern symbolic logic, I have constructed the fundamental parts of a mathematical theory of inductive probability or inductive logic (*Logical Foundations of Probability,* 1950). The methods developed make it possible to calculate numerical values of inductive probability ("degree of confirmation") for hypotheses concerning either single events or frequencies of properties and to determine estimates of frequencies in a population on the basis of evidence about a sample of the population. A few steps have been made towards extending the theory to hypotheses involving measurable quantities such as mass, temperature, etc.

It is not possible to outline here the mathematical system itself. But I will explain some of the general problems that had to be solved before the system could be constructed and some of the basic conceptions underlying the construction. One of the fundamental questions to be decided by any theory of induction is whether to accept a principle of indifference and, if so, in what form. It should be strong enough to allow the derivation of the desired theorems, but at the same time sufficiently restricted to avoid the contradictions resulting from the classical form.

The problem will become clearer if we use a few elementary concepts of inductive logic. They will now be explained with the help of the first two columns of the accompanying diagram. We consider a set of four individuals, say four balls drawn from an urn. The individuals are described with respect to a given division of mutually exclusive properties; in our example, the two properties black (B) and white (W). An *individual distribution* is specified by ascribing to each individual one property. In our example, there are sixteen individual distributions; they are pictured in the second column (e.g., in the individual distribution No. 3, the first, second, and fourth ball are black, the third is white). A *statistical distribution,* on the other hand, is characterized by merely stating the number of individuals for each property. In the example, we have five statistical distributions, listed in the first column (e.g., the statistical distribution No. 2 is described by saying that there are three B and one W, without specifying *which* individuals are B and which W).

By the *initial probability* of a hypothesis ("probability a priori" in traditional terminology) we understand its probability before any factual knowledge concerning the individuals is available. Now we shall see that, if any initial probabilities which sum up to one are assigned to the individual distributions, all other probability values are thereby fixed. To see how the procedure works, put a slip of paper on the diagram alongside the list of individual distributions and write down opposite each distribution a fraction as its initial probability; the sum of the sixteen fractions must be one, but otherwise you may choose them just as you like. We shall soon consider the question whether some choices might be preferable to others. But for the moment we are only concerned with the fact that any arbitrary choice constitutes one and only one *inductive method* in the sense that it leads to one and only one system of probability values which contain an initial probability for any hypothesis (concerning the given individuals and the given properties) and a relative probability for any hypothesis with respect to any evidence. The procedure is as follows. For any given statement we can, by perusing the list of individual distributions, determine those in which it holds (e.g., the statement "among the first three balls there is exactly one W" holds in distributions Nos. 3, 4, 5, 6, 7, 9). Then we assign to it as initial probability the sum of the initial probabilities of the individual distributions in which it holds. Suppose that an evidence statement e (e.g., "The first ball is B, the second W, the third B") and a hypothesis h (e.g., "The fourth ball is B") are given. We ascertain first the individual distributions in which e holds (in the example, Nos. 4 and 7), and then those among them in which also h holds (only No. 4). The former ones determine the initial probability of e; the latter ones determine that of e and h together. Since the latter are among the former, the latter initial probability is a part (or the whole) of the former. We now divide the latter initial probability by the former and assign the resulting fraction to h as its relative probability with respect to e. (In our example, let us take the values of the initial probabilities of individual distributions given in the diagram for methods I and II, which will soon be explained. In method I the values for Nos. 4 and 7—as for all other individual distributions—are 1/16; hence the initial probability of e is 2/16. That of e and h together is the value of No. 4 alone, hence 1/16. Dividing this by 2/16, we obtain 1/2 as the probability of h with respect to e. In method II, we find for Nos. 4 and 7 in the last column the values 3/60 and 2/60 respectively. Therefore the initial probability of e is here 5/60, that of e and h together 3/60; hence the probability of h with respect to e is 3/5.)

The problem of choosing an inductive method is closely connected with the problem of the principle of indifference. Most authors since the classical period have accepted some form of the principle and have thereby avoided the otherwise unlimited arbitrariness in the choice of a

	STATISTICAL DISTRIBUTIONS		INDIVIDUAL DISTRIBUTIONS	METOHD I	METHOD II	
	Number of Blue	Number of White		Initial Probability of Individual Distributions	Initial Probability of Statistical Distributions	Initial Probability of Individual Distributions
1.	4	0	1. ● ● ● ●	1/16	1/5	1/5 = 12/60
2.	3	1	2. ● ● ● ○	1/16	1/5	1/20 = 3/60
			3. ● ● ○ ●	1/16		1/20 = 3/60
			4. ● ○ ● ●	1/16		1/20 = 3/60
			5. ○ ● ● ●	1/16		1/20 = 3/60
3.	2	2	6. ● ● ○ ○	1/16	1/5	1/30 = 2/60
			7. ● ○ ● ○	1/16		1/30 = 2/60
			8. ● ○ ○ ●	1/16		1/30 = 2/60
			9. ○ ● ● ○	1/16		1/30 = 2/60
			10. ○ ● ○ ●	1/16		1/30 = 2/60
			11. ○ ○ ● ●	1/16		1/30 = 2/60
4.	1	3	12. ● ○ ○ ○	1/16	1/5	1/20 = 3/60
			13. ○ ● ○ ○	1/16		1/20 = 3/60
			14. ○ ○ ● ○	1/16		1/20 = 3/60
			15. ○ ○ ○ ●	1/16		1/20 = 3/60
5.	0	4	16. ○ ○ ○ ○	1/16	1/5	1/5 = 12/60

Inductive Probability Methods. (From Rudolf Carnap, "What is Probability?" *Scientific American,* September, 1953.)

method. On the other hand, practically all authors in our century agree that the principle should be restricted to some well-defined class of hypotheses. But there is no agreement as to the class to be chosen. Many authors advocate either method I or method II, which are exemplified in our diagram. Method I consists in applying the principle of indifference to individual distributions, in other words, in assigning equal initial probabilities to individual distributions. In method II the principle is first applied to the statistical distributions and then, for each statistical distribution, to the corresponding individual distributions. Thus, in our example, equal initial probabilities are assigned in method II to the five statistical distributions, hence 1/5 to each; then this value 1/5 or 12/60 is distributed in equal parts among the corresponding individual distributions, as indicated in the last column.

If we examine more carefully the two ways of using the principle of indifference, we find that either of them leads to contradictions if applied without restriction to all divisions of properties. (The reader can easily check the following results by himself. We consider, as in the diagram, four individuals and a division D_2 into two properties; blue (instead of black) and white. Let h be the statement that all four individuals are white. We consider, on the other hand, a division D_3 into three properties: dark blue, light blue, and white. For division D_2, as used in the diagram, we see that h is an individual distribution (No. 16) and also a statistical distribution (No. 5). The same holds for division D_3. By setting up the complete diagram for the latter division, one finds that there are fifteen statistical distributions, of which h is one, and 81 individual distributions (viz., 3 × 3 × 3 × 3), of which h is also one. Applying method I to division D_2, we found as the initial probability of h 1/16; if we apply it to D_3, we find 1/81; these two results are incompatible. Method II applied to D_2 led to the value 1/5; but applied to D_3 it yields 1/15. Thus this method likewise furnishes incompatible results.) We, therefore, restrict the use of either method to one division, viz. The one consisting of all properties which can be distinguished in the given universe of discourse (or which we wish to distinguish within a given context of investigation). If modified in this way, either method is consistent. We may still regard the examples in the diagram as representing the modified methods I and II, if we assume that the difference between black and white is the only difference among the given individuals, or the only difference relevant to a certain investigation.

How shall we decide which of the two methods to choose? Each of them is regarded as *the* reasonable method by prominent scholars. However, in my view, the chief mistake of the earlier authors was their failure to specify explicitly the main characteristic of a reasonable inductive method. It is due to this failure that some of them chose the wrong method. This characteristic is not difficult to find. Inductive thinking is a way of judging hypotheses concerning unknown events. In order to be reasonable, this judging must be guided by our knowledge of observed events. More specifically, other things being equal, a future event is to be regarded as the more probable, the greater the relative frequency of similar events observed so far under similar circumstances. This *principle of learning from experience* guides, or rather ought to guide, all inductive thinking in everyday affairs and in science. Our confidence that a certain drug will help in a present case of a certain disease is the higher the more frequently it has helped in past cases. We would regard a man's behavior as unreasonable if his expectation of a future event were the higher the less frequently he saw it happen in the past, and also if he formed his expectations for the future without any regard to what he had observed in the past. The principle of learning from experience seems indeed so obvious that it might appear superfluous to emphasize it explicitly. In fact, however, even some authors of high rank have advocated an inductive method that violates the principle.

Let us now examine the methods I and II from the point of view of the principle of learning from experience. In our earlier example we considered the evidence e saying that of the four balls drawn the first was B, the second W, the third B; in other words, that two B and one W were so far observed. According to the principle, the prediction h that the fourth ball will be black should be taken as more probable than its negation, non-h. We found, however, that method I assigns probability 1/2 to h, and therefore likewise 1/2 to non-h. And we see easily that it assigns to h this value 1/2 also on any other evidence concerning the first three balls. Thus method I violates the principle. A man following this method sticks to the initial probability value for a prediction, irrespective of all observations he makes. In spite of this character of method I, it was proposed as the valid method of induction by prominent philosophers, among them Charles Sanders Peirce (in 1883) and Ludwig Wittgenstein (in 1921), and even by Keynes in one chapter of his book, although in other chapters he emphasizes eloquently the necessity of learning from experience.

We saw earlier that method II assigns, on the evidence specified, to h the probability 3/5, hence to non-h 2/5. Thus the principle of learning from experience is satisfied in this case, and it can be shown that the same holds in any other case. (The reader can easily verify, for example, that with respect to the evidence that the first three balls are black, the probability of h is 4/5 and therefore that of non-h 1/5.) Method II in its modified, consistent form was proposed by the author in 1945. Although it was often emphasized throughout the historical development that induction must be based on experience, nobody as far as I am aware, succeeded in specifying a consistent inductive method satisfying the principle of learning from experience. (The method proposed by Thomas Bayes (1763) and developed by Laplace—sometimes called "Bayes's rule" or "Laplace's rule of succession"—fulfills the principle. It is essentially method II, but in its unrestricted form; therefore it is inconsistent.) I found later that there are infinitely many consistent inductive methods which satisfy the principle (*The Continuum of Inductive Methods,* 1952). None of them seems to be as simple in its definition as method II, but some of them have other advantages.

Once a consistent and suitable inductive method is developed, it supplies the basis for a *general method of estimation,* i.e., a method for calculating, on the basis of given evidence, an estimate of an unknown value of any magnitude. Suppose that, on the basis of the evidence, there are n possibilities for the value of a certain magnitude at a given time, e.g., the amount of rain tomorrow, the number of persons coming to a meeting, the price of wheat after the next harvest. Let the possible values be x_1, $x_2, \ldots, x_n$, and their inductive probabilities with respect to the given evidence p_1, p_2, $\ldots$, p_n, respectively. Then we take the product p_1x_1 as the expectation value of the first case at the present moment. Thus, if the occurrence of the first case is certain and hence $p_1 = 1$, its expectation value is the full value x_1; if it is just as probable that it will occur as that it will not, and hence $p_1 = 1/2$, its expectation value is half its full value ($p_1x_1 = x_1/2$), etc. We proceed similarly with the other possible values. As estimated or total expectation value of the magnitude on the given evidence we take the sum of the expectation values for the possible cases, that is, $p_1x_1 + p_2x_2 + \ldots + p_nx_n$. (For example, suppose someone considers buying a ticket for a lottery and, on the basis of his knowledge of the lottery procedure, there is a probability of 0.01 that the ticket will win the first prize of $200 and a probability of 0.03 that it will win $50; since there are no other prizes, the probability that it will win nothing is 0.96. Hence the estimate of the gain in dollars is $0.01 \times 200 + 0.03 \times 50 + 0.96 \times 0 = 3.50$. This is the value of the ticket for him and it would be irrational for him to pay more for it.) The same method may be used in order to make a rational decision in a situation where one among various possible actions is to be chosen. For example, a man considers several possible ways for investing a certain amount of money. Then he can—in principle, at least—calculate the estimate of his gain for each possible way. To act rationally, he should then choose that way for which the estimated gain is highest.

Bernoulli and Laplace and many of their followers envisaged the idea of a theory of inductive probability which, when fully developed, would supply the means for evaluating the acceptability of hypothetical assumptions in any field of theoretical research and at the same time methods for determining a rational decision in the affairs of practical life. In the more sober cultural atmosphere of the late nineteenth century and still more in the first half of the twentieth, this idea was usually regarded as a utopian dream. It is certainly true that those audacious thinkers were not as near to their aim as they believed. But a few men dare to think today that the pioneers were not mere dreamers and that it will be possible in the future to make far-reaching progress in essentially that direction in which they saw their vision.

Rudolf Carnap

ON INDUCTIVE LOGIC

§1. INDUCTIVE LOGIC

Among the various meanings in which the word "probability" is used in everyday language, in the discussion of scientists, and in the theories of probability, there are especially two which must be clearly distinguished. We shall use for them the terms "probability$_1$" and "probability$_2$." Probability$_1$ is a logical concept, a certain logical relation between two sentences (or, alternatively, between two propositions); it is the same as the concept of degree of confirmation. I shall write briefly "c" for "degree of confirmation," and "$c(h, e)$" for "the degree of confirmation of the hypothesis h on the evidence e"; the evidence is usually a report on the results of our observations. On the other hand, probability$_2$ is an empirical concept; it is the relative frequency in the long run of one property with respect to another. The controversy between the so-called logical conception of probability, as represented, e.g., by Keynes,[1] and Jeffreys,[2] and others, and the frequency conception, maintained, e.g., by Von Mises[3] and Reichenbach,[4] seems to me futile. These two theories deal with two different probability concepts which are both of great importance for science. Therefore, the theories are not incompatible, but rather supplement each other.[5]

In a certain sense we might regard deductive logic as the theory of L-implication (logical implication, entailment). And inductive logic may be construed as the theory of degree of confirmation, which is, so to speak, partial L-implication. e L-implies h says that h is implicitly given with e, in other words, that the whole logical content of h is contained in e. On the other hand, $c(h, e) = 3/4$ says that h is not entirely given with e but that the assumption of h is supported to the degree 3/4 by the observational evidence expressed in e.

In the course of the last years, I have constructed a new system of inductive logic by laying down a definition for degree of confirmation and developing a theory based on this definition. A book containing this theory is in preparation.[6] The purpose of the present paper is to indicate briefly and informally the definition and a few of the results found; for lack of space, the reasons for the choice of this definition and the proofs for the results cannot be given here. The book will, of course, provide a better basis than the present informal summary for a critical evaluation of the theory and of the fundamental conception on which it is based.[7]

§2. SOME SEMANTICAL CONCEPTS

Inductive logic is, like deductive logic, in my conception a branch of semantics. However, I shall try to formulate the present outline in such a way that it does not presuppose knowledge of semantics.

Let us begin with explanations of some semantical concepts which are important both for deductive logic and for inductive logic.[8]

The system of inductive logic to be outlined applies to an infinite sequence of finite language systems L_N ($N = 1, 2, 3$, etc.) and an infinite language system L_∞. L_∞ refers to an infinite universe of individuals, designated by the individual constants a_1, a_2, etc.

(or a, b, etc.), while L_N refers to a finite universe containing only N individuals designated by $a_1, a_2, \ldots a_N$. Individual variables x_1, x_2, etc. (or x, y, etc.), are the only variables occurring in these languages. The languages contain a finite number of predicates of any degree (number of arguments), designating properties of the individuals or relations between them. There are, furthermore, the customary connectives of negation (~, corresponding to not), disjunction (V, or), conjunction (·, and); universal and existential quantifiers (for every x, there is an x); the sign of identity between individuals =, and t as an abbreviation for an arbitrarily chosen tautological sentence. (Thus the languages are certain forms of what is technically known as the lower functional logic with identity.) (The connectives will be used in this paper in three ways, as is customary: (1) between sentences, (2) between predicates (§8), (3) between names (or variables) of sentences (so that, if i and j refer to two sentences, iVj is meant to refer to their disjunction).)

A sentence consisting of a predicate of degree n with n individual constants is called an *atomic sentence* (e.g., Pa_1, i.e., a_1 has the property P, or Ra_3a_5, i.e., the relation R holds between a_3 and a_5). The conjunction of all atomic sentences in a finite language L_N describes one of the possible states of the domain of the N individuals with respect to the properties and relations expressible in the language L_N. If we replace in this conjunction some of the atomic sentences by their negations, we obtain the description of another possible state. All the conjunctions which we can form in this way, including the original one, are called *state-descriptions* in L_N. Analogously, a state-description in $L\infty$ is a class containing some atomic sentences and the negations of the remaining atomic sentences; since this class is infinite, it cannot be transformed into a conjunction.

In the actual construction of the language systems, which cannot be given here, semantical rules are laid down determining for any given sentence j and any state-description i whether j holds in i, that is to say whether j would be true if i described the actual state among all possible states. The class of those state-descriptions in a language system L (either one of the systems L_N or $L\infty$) in which j holds is called the *range* of j in L.

The concept of range is fundamental both for deductive and for inductive logic; this has already been pointed out by Wittgenstein. If the range of a sentence j in the language system L is universal, i.e., if j holds in every state-description (in L), j must necessarily be true independently of the facts; therefore we call j (in L) in this case *L-true* (logically true, analytic). (The prefix L- stands for "logical"; it is not meant to refer to the system L.) Analogously, if the range of j is null, we call j *L-false* (logically false, self-contradictory). If j is neither L-true nor L-false, we call it *factual* (synthetic, contingent). Suppose that the range of e is included in that of h. Then in every possible case in which e would be true, h would likewise be true. Therefore we say in this case that e *L-implies* (logically implies, entails) h. If two sentences have the same range, we call them *L-equivalent;* in this case, they are merely different formulations for the same content.

The L-concepts just explained are fundamental for deductive logic and therefore also for inductive logic. Inductive logic is constructed out of deductive logic by the introduction of the concept of degree of confirmation. This introduction will here be carried out in three steps: (1) the definition of regular c-functions (§3), (2) the definition of symmetrical c-functions (§5), (3) the definition of the degree of confirmation c^* (§6).

§3. REGULAR *C*-FUNCTIONS

A numerical function m ascribing real numbers of the interval 0 to 1 to the sentences of a finite language L_N is called a regular m-function if it is constructed according to the following rules:

(1) We assign to the state-descriptions

in L_N as values of m any positive real numbers whose sum is 1.

(2) For every other sentence j in L_N, the value $m(j)$ is determined as follows:

(a) If j is not L-false, $m(j)$ is the sum of the m-values of those state-descriptions which belong to the range of j.

(b) If j is L-false and hence its range is null, $m(j) = 0$.

(The choice of the rule (2)(*a*) is motivated by the fact that j is L-equivalent to the disjunction of those state-descriptions which belong to the range of j and that these state-descriptions logically exclude each other.)

If any regular m-function m is given, we define a corresponding function c as follows:

(3) For any pair of sentences e, h in L_N, where e is not L-false, $c(h, e) = \frac{m(e \cdot h)}{m(e)}$.

$m(j)$ may be regarded as a measure ascribed to the range of j; thus the function m constitutes a metric for the ranges. Since the range of the conjunction $e \cdot h$ is the common part of the ranges of e and of h, the quotient in (3) indicates, so to speak, how large a part of the range of e is included in the range of h. The numerical value of this ratio, however, depends on what particular m-function has been chosen. We saw earlier that a statement in deductive logic of the form *e L-implies h* says that the range of e is entirely included in that of h. Now we see that a statement in inductive logic of the form $c(h, e) = 3/4$ says that a certain part—in the example, three fourths—of the range of e is included in the range of h.[9] Here, in order to express the partial inclusion numerically, it is necessary to choose a regular m-function for measuring the ranges. Any m chosen leads to a particular c as defined above. All functions c obtained in this way are called *regular c-functions.*

One might perhaps have the feeling that the metric m should not be chosen once for all but should rather be changed according to the accumulating experiences.[10] This feeling is correct in a certain sense. However, it is to be satisfied not by the function m used in the definition (3) but by another function m dependent upon e and leading to an alternative definition (5) for the corresponding c. If a regular m is chosen according to (1) and (2), then a corresponding function m_e is defined for the state-descriptions in L_N as follows:

(4) Let i be a state-descriptin in L_N, and e a non-L-false sentence in L_N.

(a) If e does not hold in i, $m_e(i) = 0$.

(b) If e holds in i, $m_e(i) = \frac{m(i)}{m(e)}$.

Thus m_e represents a metric for the state-descriptions which changes with the changing evidence e. Now $m_e(j)$ for any other sentence j in L_N is defined in analogy to (2)(a) and (b). Then we define the function c corresponding to m as follows:

(5) For any pair of sentences e, h in L_N, where e is not L-false, $c(h, e) = m_e(h)$.

It can easily be shown that this alternative definition (5) yields the same values as the original definition (3).

Suppose that a sequence of regular m-functions is given, one for each of the finite languages L_N ($N = 1, 2$, etc.). Then we define a corresponding m-function for the infinite language as follows:

(6) $m(j)$ in L_∞ is the limit of the values $m(j)$ in L_N for $N \to \infty$.

c-functions for the finite languages are based on the given m-functions according to (3). We define a corresponding c-function for the infinite language as follows:

(7) $c(h, e)$ in L_∞ is the limit of the values $c(h, e)$ in L_N for $N \to \infty$.

The definitions (6) and (7) are applicable

only in those cases where the specified limits exist.

We shall later see how to select a particular sub-class of regular *c*-functions (§5) and finally one particular *c*-function c^* as the basis of a complete system of inductive logic (§6). For the moment, let us pause at our first step, the definition of regular *c*-functions just given, in order to see what results this definition alone can yield, before we add further definitions. The theory of regular *c*-functions, i.e., the totality of those theorems which are founded on the definition stated, is the first and fundamental part of inductive logic. It turns out that we find here many of the fundamental theorems of the classical theory of probability, e.g., those known as the theorem (or principle) of multiplication, the general and the special theorems of addition, the theorem of division, and, based upon it, Bayes's theorem.

One of the cornerstones of the classical theory of probability is the principle of indifference (or principle of insufficient reason). It says that, if our evidence *e* does not give us any sufficient reason for regarding one of two hypotheses *h* and *h′* as more probable than the other, then we must take their probabilities$_1$ as equal: $c(h, e) = c(h', e)$. Modern authors, especially Keynes, have correctly pointed out that this principle has often been used beyond the limits of its original meaning and has then led to quite absurd results. Moreover, it can easily be shown that, even in its original meaning, the principle is by far too general and leads to contradictions. Therefore the principle must be abandoned. If it is and we consider only those theorems of the classical theory which are provable without the help of this principle, then we find that these theorems hold for all regular *c*-functions. The same is true for those modern theories of probability$_1$ (e.g., that by Jeffreys, *op. cit.*) which make use of the principle of indifference. Most authors of modern axiom systems of probability$_1$ (e.g., Keynes, *op. cit.*, Waismann, *op. cit.*, Mazurkiewicz[11], Hosiasson[12], von Wright[13]) are cautious enough not to accept that principle. An examination of these systems shows that their axioms and hence their theorems hold for all regular *c*-functions. Thus these systems restrict themselves to the first part of inductive logic, which, although fundamental and important, constitutes only a very small and weak section of the whole of inductive logic. The weakness of this part shows itself in the fact that it does not determine the value of *c* for any pair *h, e* except in some special cases where the value is 0 or 1. The theorems of this part tell us merely how to calculate further values of *c* if some values are given. Thus it is clear that this part alone is quite useless for application and must be supplemented by additional rules. (It may be remarked incidentally, that this point marks a fundamental difference between the theories of probability$_1$ and of probability$_2$ which otherwise are analogous in many respects. The theorems concerning probability$_2$ which are analogous to the theorems concerning regular *c*-functions constitute not only the first part but the whole of the logico-mathematical theory of probability$_2$. The task of determining the value of probability$_2$ for a given case is—in contradistinction to the corresponding task for probability$_1$—an empirical one and hence lies outside the scope of the logical theory of probability$_2$.)

§4. THE COMPARATIVE CONCEPT OF CONFIRMATION

Some authors believe that a metrical (or quantitative) concept of degree of confirmation, that is, one with numerical values, can be applied, if at all, only in certain cases of a special kind and that in general we can make only a comparison in terms of higher or lower confirmation without ascribing numerical values. Whether these authors are right or not, the introduction of a merely comparative (or topological) concept of confirmation not presupposing a metrical concept is, in any case, of interest. We shall now discuss a way of defining a concept of this kind.

For technical reasons, we do not take the concept "more confirmed" but "more or equally confirmed." The following discussion refers to the sentences of any finite language L_N. We write, for brevity, $MC(h, e, h', e')$ for h is confirmed on the evidence e more highly or just as highly as h' on the evidence e'.

Although the definition of the comparative concept MC at which we aim will not make use of any metrical concept of degree of confirmation, let us now consider, for heuristic purposes, the relation between MC and the metrical concepts, i.e., the regular c-functions. Suppose we have chosen some concept of degree of confirmation, in other words, a regular c-function c, and further a comparative relation MC; then we shall say that MC is in accord with c if the following holds:

(1) For any sentences h, e, h', e', if $MC(h, e, h', e')$ then $c(h, e) \geqq c(h', e')$.

However, we shall not proceed by selecting one c-function and then choosing a relation MC which is in accord with it. This would not fulfill our intention. Our aim is to find a comparative relation MC which grasps those logical relations between sentences which are, so to speak, prior to the introduction of any particular m-metric for the ranges and of any particular c-function; in other words, those logical relations with respect to which all the various regular c-functions agree. Therefore we lay down the following requirement:

(2) The relation MC is to be defined in such a way that it is in accord with *all* regular c-functions; in other words, if $MC(h, e, h', e')$, then for every regular c, $c(h, e) \geqq c(h', e')$.

It is not difficult to find relations which fulfill this requirement (2). First let us see whether we can find quadruples of sentences h, e, h', e' which satisfy the following condition occurring in (2):

(3) For every regular c, $c(h, e) \geqq c(h', e')$.

It is easy to find various kinds of such quadruples. (For instance, if e and e' are any non-L-false sentences, then the condition (3) is satisfied in all cases where e L-implies h, because here $c(h, e) = 1$; further in all cases where e' L-implies $\sim h'$, because here $c(h', e') = 0$; and in many other cases.) We could, of course, define a relation MC by taking some cases where we know that the condition (3) is satisfied and restricting the relation to these cases. Then the relation would fulfill the requirement (2); however, as long as there are cases which satisfy the condition (3) but which we have not included in the relation, the relation is unnecessarily restricted. Therefore we lay down the following as a second requirement for MC:

(4) MC is to be defined in such a way that it holds in all cases which satisfy the condition (3); in such a way, in other words, that it is the most comprehensive relation which fulfills the first requirement (2).

These two requirements (2) and (4) together stipulate that $MC(h, e, h', e')$ is to hold if and only if the condition (3) is satisfied; thus the requirements determine uniquely one relation MC. However, because they refer to the c-functions, we do not take these requirements as a definition for MC, for we intend to give a purely comparative definition for MC, a definition which does not make use of any metrical concepts but which leads nevertheless to a relation MC which fulfills the requirements (2) and (4) referring to c-functions. This aim is reached by the following definition (where $=_{Df}$ is used as sign of definition).

(5) $MC(h, e, h', e') =_{Df}$ the sentences h, e, h', e' (in L_N) are such that e and e' are not L-false and at least one of

the following three conditions is fulfilled:

(a) e L-implies h,

(b) e' L-implies $\sim h'$,

(c) $e' \cdot h'$ L-implies $e \cdot h$ and simultaneously e L-implies $h \vee e'$.

((a) and (b) are the two kinds of rather trivial cases earlier mentioned; (c) comprehends the interesting cases; an explanation and discussion of them cannot be given here.)

The following theorem can then be proved concerning the relation *MC* defined by (5). It shows that this relation fulfills the two requirements (2) and (4).

(6) For any sentences h, e, h', e' in L_N the following holds:

(a) If $MC(h, e, h' e')$, then, for every regular c, $c(h, e) \geqq c(h', e')$.

(b) If, for every regular c, $c(h, e) \geqq c(h', e')$, then $MC(h, e, h', e')$.

(With respect to $L\infty$, the analogue of (6)(a) holds for all sentences, and that of (6)(b) for all sentences without variables.)

§5. SYMMETRICAL *C*-FUNCTIONS

The next step in the construction of our system of inductive logic consists in selecting a narrow sub-class of the comprehensive class of all regular c-functions. The guiding idea for this step will be the principle that inductive logic should treat all individuals on a par. The same principle holds for deductive logic; for instance, if . .*a*. .*b*. . L-implies - -*b*- -*c*- - (where the first expression is meant to indicate some sentence containing *a* and *b*, and the second another sentence containing *b* and *c*), then L-implication holds likewise between corresponding sentences with other individual constants, e.g., between . .*d*. .*c*. . and - -*c*- -*a*- -. Now we require that this should hold also for inductive logic, e.g., that *c*(- -*b*- -*c*- -, . .*a*. .*b*. .) = *c*(- -*c*- -*a*- -, . .*d*. .*c*. .). It seems that all authors on probability$_1$ have assumed this principle—although it has seldom, if ever, been stated explicitly—by formulating theorems in the following or similar terms: "On the basis of observations of s things of which s_1 were found to have the property M and s_2 not to have this property, the probability that another thing has this property is such and such." The fact that these theorems refer only to the number of things observed and do not mention particular things shows implicitly that it does not matter which things are involved; thus it is assumed, e.g., that $c(Pd, Pa \cdot Pb \cdot \sim Pc) = c(Pc, Pa \cdot Pd \cdot \sim Pb)$.

The principle could also be formulated as follows. Inductive logic should, like deductive logic, make no discrimination among individuals. In other words, the value of c should be influenced only by those differences between individuals which are expressed in the two sentences involved; no differences between particular individuals should be stipulated by the rules of either deductive or inductive logic.

It can be shown that this principle of non-discrimination is fulfilled if c belongs to the class of symmetrical c-functions which will now be defined. Two state-descriptions in a language L_N are said to be *isomorphic* or to have the same structure if one is formed from the other by replacements of the following kind: We take any one-one relation R such that both its domain and its converse domain is the class of all individual constants in L_N, and then replace every individual constant in the given state-description by the one correlated with it by R. If a regular m-function (for L_N) assigns to any two isomorphic state-descriptions (in L_N) equal values, it is called a symmetrical m-function; and a c-function based upon such an m-function in the way explained earlier (see (3) in §3) is then called a *symmetrical c-function.*

§6. THE DEGREE OF CONFIRMATION *C**

Let i be a state-description in L_N. Suppose there are n_i state-descriptions in L_N isomorphic to i (including i itself), say i, i', i'', etc. These n_i state-descriptions exhibit one and the same structure of the universe of L_N with respect to all the properties and relations designated by the primitive predicates in L_N. This concept of structure is an extension of the concept of structure or relation-number (Russell) usually applied to one dyadic relation. The common structure of the isomorphic state-descriptions i, i', i'', etc., can be described by their disjunction $i \vee i' \vee i'' \vee \ldots$. Therefore we call this disjunction, say j, a *structure-description* in L_N. It can be shown that the range of j contains only the isomorphic state-descriptions i, i', i'', etc. Therefore (see (2)(a) in §3) $m(j)$ is the sum of the m-values for these state-descriptions. If m is symmetrical, then these values are equal, and hence

(1) $$m(j) = n_i \times m(i).$$

And, conversely, if $m(j)$ is known to be q, then

(2) $$m(i) = m(i') = m(i'') = \ldots = q/n_i.$$

This shows that what remains to be decided is merely the distribution of m-values among the structure-descriptions in L_N. We decide to give them equal m-values. This decision constitutes the third step in the construction of our inductive logic. This step leads to one particular m-function m^* and to the c-function c^* based upon m^*. According to the preceding discussion, m^* is characterized by the following two stipulations:

(3) (a) m^* is a symmetrical m-function;
(b) m^* has the same value for all structure-descriptions (in L_N).

We shall see that these two stipulations characterize just one function. Every state-description (in L_N) belongs to the range of just one structure-description. Therefore, the sum of the m^*-values for all structure-descriptions in L_N must be the same as for all state-descriptions, hence 1 (according to (1) in §3). Thus, if the number of structure-descriptions in L_N is m, then, according to (3)(b),

(4) for every structure-description j in L_N,
$$m^*(j) = \frac{1}{m}.$$

Therefore, if i is any state-description in L_N and n_i is the number of state descriptions isomorphic to i, then, according to (3)(a) and (2),

(5) $$m^*(i) = \frac{1}{mn_i}.$$

(5) constitutes a definition of m^* as applied to the state-descriptions in L_N. On this basis, further definitions are laid down as explained above (see (2) and (3) in §3): first a definition of m^* as applied to all sentences in L_N, and then a definition of c^* on the basis of m^*. Our inductive logic is the theory of this particular function c^* as our concept of degree of confirmation.

It seems to me that there are good and even compelling reasons for the stipulation (3)(a), i.e., the choice of a symmetrical function. The proposal of any non-symmetrical c-function as degree of confirmation could hardly be regarded as acceptable. The same can not be said, however, for the stipulation (3)(b). No doubt, to the way of thinking which was customary in the classical period of the theory of probability, (3)(b) would appear as validated, like (3)(a), by the principle of indifference. However, to modern, more critical thought, this mode of reasoning appears as invalid because the structure-descriptions (in contradistinction to the individual constants) are by no means alike in their logical features but show very conspicuous differences. The definition of c^* shows a great simplicity in comparison with other concepts which may be taken into consideration. Although this fact may influence our decision to choose c^*, it cannot, of course, be regarded as a sufficient reason for this choice. It seems to me that the choice of c^* cannot be justified by any features of the definition which are immediately recognizable, but only by the consequences to which the definition leads.

There is another c-function c_W which at the first glance appears not less plausible than c^*. The choice of this function may be suggested by the following consideration. Prior to experience, there seems to be no reason to regard one state-description as less probable than another. Accordingly, it might seem natural to assign equal m-values to all state-descriptions. Hence, if the number of the state-descriptions in L_N is n, we define for any state-description i

$$(6) \qquad m_w(i) = \frac{1}{n}.$$

This definition (6) for m_W is even simpler than the definition (5) for m^*. The measure ascribed to the ranges is here simply taken as proportional to the cardinal numbers of the ranges. On the basis of the m_w-values for the state-descriptions defined by (6), the values for the sentences are determined as before (see (2) in §3), and then c_W is defined on the basis of m_W (see (3) in §3).[14]

In spite of its apparent plausibility, the function c_W can easily be seen to be entirely inadequate as a concept of degree of confirmation. As an example, consider the language L_{101} with P as the only primitive predicate. Let the number of state-descriptions in this language be n (it is 2^{101}). Then for any state-description, $m_W = 1/n$. Let e be the conjunction $Pa_1 \cdot Pa_2 \cdot Pa_3 \ldots Pa_{100}$ and let h be Pa_{101}. Then $e \cdot h$ is a state-description and hence $m_W(e \cdot h) = 1/n$. e holds only in the two state-descriptions $e \cdot h$ and $e \cdot \sim h$; hence $m_W(e) = 2/n$. Therefore $c_W(h, e) = 1/2$. If e' is formed from e by replacing some or even all of the atomic sentences with their negations, we obtain likewise $c_W(h, e') = 1/2$. Thus the c_W-value for the prediction that a_{101} is P is always the same, no matter whether among the hundred observed individuals the number of those which we have found to be P is 100 or 50 or 0 or any other number. Thus the choice of c_W as the degree of confirmation would be tantamount to the principle never to let our past experiences influence our expectations for the future. This would obviously be in striking contradiction to the basic principle of all inductive reasoning.

§7. LANGUAGES WITH ONE-PLACE PREDICATES ONLY

The discussions in the rest of this paper concern only those language systems whose primitive predicates are one-place predicates and hence designate properties, not relations. It seems that all theories of probability constructed so far have restricted themselves, or at least all of their important theorems, to properties. Although the definition of c^* in the preceding section has been stated in a general way so as to apply also to languages with relations, the greater part of our inductive logic will be restricted to properties. An extension of this part of inductive logic to relations would require certain results in the deductive logic of relations, results which this discipline, although widely developed in other respects, has not yet reached (e.g., an answer to the apparently simple question as to the number of structures in a given finite language system).

Let $L_N{}^p$ be a language containing N individual constants $a_1, \ldots a_N$, and p one-place primitive predicates $P_1, \ldots P_p$. Let us consider the following expressions (sentential matrices). We start with $P_1x \cdot P_2x \ldots P_px$; from this expression we form others by negating some of the conjunctive components, until we come to $\sim P_1x \cdot \sim P_2x \ldots \sim P_px$, where all components are negated. The number of these expressions is $k = 2^p$; we abbreviate them by $Q_1x, \ldots Q_kx$. We call the k properties expressed by those k expressions in conjunctive form and now designated by the k new Q-predicates the *Q-properties* with respect to the given language $L_N{}^p$. We see easily that these Q-properties are the strongest properties expressible in this language (except for the L-empty, i.e., logically self-contradictory, property); and further, that they constitute an exhaustive and non-overlapping classification, that is to say, every individual has one and only one of the Q-properties. Thus, if we state for each individual which of the Q-properties it has, then we have described the individuals completely. Every state-description can be brought into the form of such a statement, i.e., a conjunction of N Q-sentences, one for each of the N individuals. Suppose that in a given state-description i the number of individuals having the property Q_1 is N_1, the number for Q_2 is N_2, . . . that for Q_k is N_k. Then we call the numbers N_1, N_2, . . . N_k the *Q-numbers* of the state-description i; their sum is N. Two state-descriptions are isomorphic if and only if they have the same Q-numbers. Thus here a structure-description is a statistical description giving the Q-numbers N_1, N_2, etc., without specifying which individuals have the properties Q_1, Q_2, etc.

Here—in contradistinction to languages with relations—it is easy to find an explicit function for the number m of structure-descriptions and, for any given state-description i with the Q-numbers N_1, . . . N_k, an explicit function for the number n_i of state-descriptions isomorphic to i, and hence also a function for $m^*(i)$.[15]

Let j be a non-general sentence (i.e., one without variables) in $L_N{}^p$. Since there are effective procedures (that is, sets of fixed rules furnishing results in a finite number of steps) for constructing all state-descriptions in which j holds and for computing m^* for any given state-description, these procedures together yield an effective procedure for computing $m^*(j)$ (according to (2) in §3). However, the number of state-descriptions becomes very large even for small language systems (it is k^N, hence, e.g., in $L_7{}^3$ it is more than two million). Therefore, while the procedure indicated for the computation of $m^*(j)$ is effective, nevertheless in most ordinary cases it is impracticable; that is to say, the number of steps to be taken, although finite, is so large that nobody will have the time to carry them out to the end. I have developed another procedure for the computation of $m^*(j)$ which is not only effective but also practicable if the number of individual constants occurring in j is not too large.

The value of m^* for a sentence j in the infinite language has been defined (see (6) in §3) as the limit of its values for the same sentence j in the finite languages. The question arises whether and under what conditions this limit exists. Here we have to distinguish two cases. (i) Suppose that j contains no variable. Here the situation is simple; it can be shown that in this case $m^*(j)$ is the same in all finite languages in which j occurs; hence it has the same value also in the infinite language. (ii) Let j be general, i.e., contain variables. Here the situation is quite different. For a given finite language with N individuals, j can of course easily be transformed into an L-equivalent sentence j'_N without variables, because in this language a universal sentence is L-equivalent to a conjunction of N components. The values of $m^*(j'_N)$ are in general different for each N; and although the simplified procedure mentioned above is available for the computation of these values, this procedure becomes impracticable even for moderate N. Thus for general sentences the problem of the existence and the practical computability of the limit becomes serious. It can be shown that for every general sentence the limit exists; hence m^* has a value for all sentences in the infinite language. Moreover, an effective procedure for the computation of $m^*(j)$ for any sentence j in the infinite language has been constructed. This is based on a procedure for transforming any given general sentence j into a non-general sentence j' such that j and j', although not necessarily L-equivalent, have the same m^*-value in the infinite language and j' does not contain more individual constants than j; this procedure is not only effective but also practicable for sentences of customary length. Thus, the computation of $m^*(j)$ for a general sentence j is in fact much simpler for the infinite language than for a finite language with a large N.

With the help of the procedure mentioned, the following theorem is obtained:

> If j is a purely general sentence (i.e., one without individual constants) in the infinite language, then $m^*(j)$ is either 0 or 1.

§8. INDUCTIVE INFERENCES

One of the chief tasks of inductive logic is to furnish general theorems concerning inductive inferences. We keep the traditional term "inference"; however, we do not mean by it merely a transition from one sentence to another (viz., from the evidence or premiss e to the hypothesis or conclusion h) but the determination of the degree of confirmation $c(h, e)$. In deductive logic it is sufficient to state that h follows with necessity from e; in inductive logic, on the other hand, it would not be sufficient to state that h follows—not with necessity but to some degree or other—from e. It must be specified to what degree h follows from e; in other words, the value of $c(h, e)$ must be given. We shall now indicate some results with respect to the most important kinds of inductive inference. These inferences are of special importance when the evidence or the hypothesis or both give statistical information, e.g., concerning the absolute or relative frequencies of given properties.

If a property can be expressed by primitive predicates together with the ordinary connectives of negation, disjunction, and conjunction (without the use of individual constants, quantifiers, or the identity sign), it is called an *elementary property*. We shall use M, M', M_1, M_2, etc., for elementary properties. If a property is empty by logical necessity (e.g., the property designated by $P \cdot \S P$), we call it L-empty; if it is universal by logical necessity (e.g., $P \vee \S P$), we call it L-universal. If it is neither L-empty nor L-universal (e.g., P_1, $P_1 \cdot \S P_2$), we call it a *factual property;* in this case it may still happen to be universal or empty, but if so, then contingently, not necessarily. It can be shown that every elementary property which is not L-empty is uniquely analyzable into a disjunction (i.e., or-connection) of Q-properties. If M is a disjunction of n Q-properties ($n \geqq 1$), we say that the (logical) *width* of M is n; to an L-empty property we ascribe the width 0. If the width of M is w ($\geqq 0$), we call w/k its *relative width* (k is the number of Q-properties).

The concepts of width and relative width are very important for inductive logic. Their neglect seems to me one of the decisive defects in the classical theory of probability which formulates its theorems "for any property" without qualification. For instance, Laplace takes the probability a priori that a given thing has a given property, no matter of what kind, to be 1/2. However, it seems clear that this probability cannot be the same for a very strong property (e.g., $P_1 \cdot P_2 \cdot P_3$) and for a very weak property (e.g., $P_1 \vee P_2 \vee P_3$). According to our definition, the first of the two properties just mentioned has the relative width 1/8, and the second 7/8. In this and in many other cases the probability or degree of confirmation must depend upon the widths of the properties involved. This will be seen in some of the theorems to be mentioned later.

§9. THE DIRECT INFERENCE

Inductive inferences often concern a situation where we investigate a whole population (of persons, things, atoms, or whatever else) and one or several samples picked out of the population. An inductive inference from the whole population to a sample is called a direct inductive inference. For the sake of simplicity, we shall discuss here and in most of the subsequent sections only the case of one property M, hence a classification of all individuals into M and $\sim M$. The theorems for classifications with more properties are analogous but more complicated.

In the present case, the evidence e says that in a whole population of n individuals there are n_1 with the property M and $n_2 = n - n_1$ with $\sim M$; hence the relative frequency of M is $r = n_1/n$. The hypothesis h says that a sample of s individuals taken from the whole population will contain s_1 individuals with the property M and $s_2 = s - s_1$ with $\sim M$. Our theory yields in this case the same values as the classical theory.[16]

If we vary s_1, then c^* has its maximum in the case where the relative frequency s_1/s in the sample is equal or close to that in the whole population.

If the sample consists of only one individual c, and h says that c is M, then $c^*(h, e) = r$.

As an approximation in the case that n is very large in relation to s, Newton's theorem holds.[17] If furthermore the sample is sufficiently large, we obtain as an approximation Bernoulli's theorem in its various forms.

It is worthwhile to note two characteristics which distinguish the direct inductive inference from the other inductive inferences and make it, in a sense, more closely related to deductive inferences:

(i) The results just mentioned hold not only for c^* but likewise for all symmetrical c-functions; in other words, the results are independent of the particular m-metric chosen provided only that it takes all individuals on a par.

(ii) The results are independent of the width of M. This is the reason for the agreement between our theory and the classical theory at this point.

§10. THE PREDICTIVE INFERENCE

We call the inference from one sample to another the predictive inference. In this case, the evidence e says that in a first sample of s individuals, there are s_1 with the property M, and $s_2 = s - s_1$ with $\sim M$. The hypothesis h says that in a second sample of s' other individuals, there will be s'_1 with M, and $s'_2 = s' - s'_1$ with $\sim M$. Let the width of M be w_1; hence the width of $\sim M$ is $w_2 = k - w_1$.[18]

The most important special case is that where h refers to one individual e only and says that c is M. In this case,

$$(1) \qquad c^*(h, e) = \frac{s_1 + u_1}{s + k}.$$

Laplace's much debated rule of succession gives in this case simply the value $(s_1 + 1)/(s + 2)$ for any property whatever; this, however, if applied to different properties, leads to contradictions. Other authors state the value s_1/s, that is, they take simply the observed relative frequency as the probability for the prediction that an unobserved individual has the property in question. This rule, however, leads to quite implausible results. If $s_1 = s$, e.g., if three individuals have been observed and all of them have been found to be M, the last-mentioned rule gives the probability for the next individual being M as 1, which seems hardly acceptable. According to (1), c^* is influenced by the following two factors (though not uniquely determined by them):

(i) w_1/k, the relative width of M;
(ii) s_1/s, the relative frequency of M in the observed sample.

The factor (i) is purely logical; it is determined by the semantical rules. (ii) is empirical; it is determined by observing and counting the individuals in the sample. The value of c^* always lies between those of (i) and (ii). Before any individual has been observed, c^* is equal to the logical factor (i). As we first begin to observe a sample, c^* is influenced more by this factor than by (ii). As the sample is increased by observing more and more individuals (but not including the one mentioned in h), the empirical factor (ii) gains more and more influence upon c^* which approaches closer and closer to (ii); and when the sample is sufficiently large, c^* is practically equal to the relative frequency (ii). These results seem quite plausible.[19]

The predictive inference is the most important inductive inference. The kinds of inference discussed in the subsequent sections may be construed as special cases of the predictive inference.

§11. THE INFERENCE BY ANALOGY

The inference by analogy applies to the following situation. The evidence known to us is the fact that individuals b and c agree in certain properties and, in addition, that b has a further property; thereupon we consider the hypothesis that c too has this property. Logicians have always felt that a peculiar difficulty is here involved. It seems plausible to assume that the probability of the hypothesis is the higher the more properties b and c are known to have in common; on the other hand, it is felt that these common properties should not simply be counted but weighed in some way. This becomes possible with the help of the concept of width. Let M_1 be the conjunction of all properties which b and c are known to have in common. The known similarity between b and c is the greater the stronger the property M_1, hence the smaller its width. Let M_2 be the conjunction of all properties which b is known to have. Let the width of M_1 be w_1, and that of M_2, w_2. According to the above description of the situation, we presuppose that M_2 L-implies M_1 but is not L-equivalent to M_1; hence $w_1 > w_2$. Now, we take as evidence the conjunction $e \cdot j$; e says that b is M_2, and j says that c is M_1. The hypothesis h says that c has not only the properties ascribed to it in the evidence but also the one (or several) ascribed in the evidence to b only, in other words, that c has all known properties of b, or briefly that c is M_2. Then

$$(1) \qquad c^*(h, e \cdot j) = \frac{w_2 + 1}{w_1 + 1}.$$

j and h speak only about c; e introduces the other individual b which serves to connect the known properties of c expressed by j with its unknown properties expressed by h. The chief question is whether the degree of confirmation of h is increased by the analogy between c and b, in other words, by the addition of e to our knowledge. A theorem[20] is found which gives an affirmative answer to this question. However, the increase of c^* is under ordinary conditions rather small; this is in agreement with the general conception according to which reasoning by analogy, although admissible, can usually yield only rather weak results.

Hosiasson[21] has raised the question mentioned above and discussed it in detail. She says that an affirmative answer, a proof for the increase of the degree of confirmation in the situation described, would justify the universally accepted reasoning by analogy. However, she finally admits that she does not find such a proof on the basis of her axioms. I think it is not astonishing that neither the classical theory nor modern theories of probability have been able to give a satisfactory account of and justification for the inference by analogy. For, as the theorems mentioned show, the degree of confirmation and its increase depend here not on relative frequencies but entirely on the logical widths of the properties involved, thus on magnitudes neglected by both classical and modern theories.

The case discussed above is that of simple analogy. For the case of multiple analogy, based on the similarity of c not only with one other individual but with a number n of them, similar theorems hold. They show that c^* increases with increasing n and approaches 1 asymptotically. Thus, multiple analogy is shown to be much more effective than simple analogy, as seems plausible.

§12. THE INVERSE INFERENCE

The inference from a sample to the whole population is called the inverse inductive inference. This inference can be regarded as a special case of the predictive inference with the second sample covering the whole remainder of the population. This inference is of much greater importance for practical statistical work than the direct inference, because we usually have statistical information only for some samples and not for the whole population.

Let the evidence e say that in an observed sample of s individuals there are s_1 individuals with the property M and $s_2 = s - s_1$ with $\sim M$. The hypothesis h says that in the whole population of n individuals, of which the sample is a part, there are n_1 individuals with M and n_2 with $\sim M$ ($n_1 \geqq s_1$, $n_2 \geqq s_2$). Let the width of M be w_1, and that of $\sim M$ be $w_2 = k - w_1$. Here, in distinction to the direct inference, $c^*(h, e)$ is dependent not only upon the frequencies but also upon the widths of the two properties.[22]

§13. THE UNIVERSAL INFERENCE

The universal inductive inference is the inference from a report on an observed sample to a hypothesis of universal form. Sometimes the term "induction" has been applied to this kind of inference alone, while we use it in a much wider sense for all non-deductive kinds of inference. The universal inference is not even the most important one; it seems to me now that the role of universal sentences in the inductive procedures of science has generally been overestimated. This will be explained in the next section.

Let us consider a simple law l, i.e., a factual universal sentence of the form "all M are M'" or, more exactly, "for every x, if x is M, then x is M'," where M and M' are elementary properties. As an example, take "all swans are white." Let us abbreviate $M \cdot \sim M'$ ("non-white swan") by M_1 and let the width of M_1 be w_1. Then l can be formulated thus: "M_1 is empty," i.e. "there is no individual (in the domain of individuals of the language in question) with the property M_1" ("there are no non-white swans"). Since l is a factual sentence, M_1 is a factual property; hence $w_1 > 0$. To take an example, let w_1 be 3; hence M_1 is a disjunction of three Q-properties, say $Q \vee Q' \vee Q''$. Therefore, l can be transformed into: "Q is empty, and Q' is empty, and Q'' is empty." The weakest factual laws in a language are those which say that a certain Q-property is empty; we call them Q-laws. Thus we see that l can be transformed into a conjunction of w_1 Q-laws. Obviously l asserts more if w_1 is larger; therefore we say that the law l has the strength w_1.

Let the evidence e be a report about an observed sample of s individuals such that we see from e that none of these s individuals violates the law l; that is to say, e ascribes to each of the s individuals either simply the property $\sim M_1$ or some other property L-implying $\sim M_1$. Let l, as above, be a simple law which says that M_1 is empty, and w_1 be the width of M_1; hence the width of $\sim M_1$ is $w_2 = k - w_1$. For finite languages with N individuals, $c^*(l, e)$ is found to decrease with increasing N, as seems plausible.[23] If N is very large, c^* becomes very small; and for an infinite universe it becomes 0. The latter result may seem astonishing at first sight; it seems not in accordance with the fact that scientists often speak of "well-confirmed" laws. The problem involved here will be discussed later.

So far we have considered the case in which only positive instances of the law l have been observed. Inductive logic must, however, deal also with the case of negative instances. Therefore let us now examine another evidence e' which says that in the observed sample of s individuals there are s_1 which have the property M_1 (non-white swans) and hence violate the law l, and that $s_2 = s - s_1$ have $\sim M_1$ and hence satisfy the law l. Obviously, in this case there is no point in taking as hypothesis the law l in its original forms, because l is logically incompatible with the present evidence e', and hence $c^*(l, e') = 0$. That all individuals satisfy l is excluded by e'; the question remains whether at least all unobserved individuals satisfy l. Therefore we take here as hypothesis the restricted law l' corresponding to the original unrestricted law l; l' says that all individuals not belonging to the sample of s individuals described in e' have the property $\sim M_1$. w_1 and w_2 are, as previously, the widths of M_1 and $\sim M_1$ respectively. It is found that $c^*(l', e')$ decreases with an increase of N and even more with an increase in the number s_1 of violating cases.[24] It can be shown that, under ordinary circumstances with large N, c^* increases moderately when a new individual is observed which satisfies the original law l. On the other hand, if the new individual violates l, c^* decreases very much, its value becoming a small fraction of its previous value. This seems in good agreement with the general conception.

For the infinite universe, c^* is again 0, as in the previous case. This result will be discussed in the next section.

§14. THE INSTANCE CONFIRMATION OF A LAW

Suppose we ask an engineer who is building a bridge why he has chosen the building materials he is using, the arrangement and dimensions of the supports, etc. He will refer to certain physical laws, among them some general laws of mechanics and some specific laws concerning the strength of the materials. On further inquiry as to his confidence in these laws he may apply to them phrases like "very reliable," "well founded," "amply confirmed by numerous experiences." What do these phrases mean? It is clear that they are intended to say something about probability$_1$ or degree of confirmation. Hence, what is meant could be formulated more explicitly in a statement of the form "$c(h, e)$ is high" or the like. Here the evidence e is obviously the relevant observational knowledge of the engineer or of all physicists together at the present time. But what is to serve as the hypothesis h? One might perhaps think at first that h is the law in question, hence a universal sentence l of the form: "For every space-time point x, if such and such conditions are fulfilled at x, then such and such is the case at x." I think, however, that the engineer is chiefly interested not in this sentence l, which speaks about an immense number, perhaps an infinite number, of instances dispersed through all time and space, but rather in one instance of l or a relatively small number of instances. When he says that the law is very reliable, he does not mean to say that he is willing to bet that among the billion of billions, or an infinite number, of instances to which the law applies there is not one counter-instance, but merely that this bridge will not be a counter-instance, or that among all bridges which he will construct during his lifetime, or among those which all engineers will construct during the next one thousand years, there will be no counter-instance. Thus h is not the law l itself but only a prediction concerning one instance or a relatively small number of instances. Therefore, what is vaguely called the reliability of a law is measured not by the degree of confirmation of the law itself but by that of one or several instances. This suggests the subsequent definitions. They refer, for the sake of simplicity, to just one instance; the case of several, say one hundred, instances can then easily be judged likewise. Let e be any non-L-false sentence without variables. Let l be a simple law of the form earlier described (§13). Then we understand by the *instance confirmation* of l on the evidence e, in symbols $c^*_i(l, e)$, the degree of confirmation, on the evidence e, of the hypothesis that a new individual not mentioned in e fulfills the law l.[25]

The second concept, now to be defined, seems in many cases to represent still more accurately what is vaguely meant by the reliability of a law l. We suppose here that l has the frequently used conditional form mentioned earlier: "For every x, if x is M, then x is M'" (e.g., "all swans are white"). By the *qualified-instance confirmation* of the law that all swans are white we mean the degree of confirmation for the hypothesis h' that the next swan to be observed will likewise be white. The difference between the hypothesis h used previously for the instance confirmation and the hypothesis h' just described consists in the fact that the latter concerns an individual which is already qualified as fulfilling the condition M. That is the reason why we speak here of the qualified-instance confirmation, in symbols c^*_{qi}.[26] The results obtained concerning instance confirmation and qualified-instance confirmation[27] show that the values of these two functions are independent of N and hence hold for all finite and infinite universes. It has been found that, if the number s_1 of observed counter-instances is a fixed small number, then, with the increase of the sample s, both c^*_i and c^*_{qi} grow close to 1, in contradistinction to c^* for the law itself. This justifies the customary manner of speaking of "very reliable" or "well-founded" or "well-confirmed" laws, provided we interpret these phrases as referring to a high value of either of our two concepts just introduced. Understood in this sense, the phrases are not in contradiction to our previous results that the degree of confirmation of a law is very small in a large domain of individuals and 0 in the infinite domain (§13).

These concepts will also be of help in situations of the following kind. Suppose a scientist has observed certain events, which are not sufficiently explained by the known physical laws. Therefore he looks for a new law as an explanation. Suppose he finds two incompatible laws l and l', each of which would explain the observed events satisfactorily. Which of them should he prefer? If the domain of individuals in question is finite, he may take the law with the higher degree of confirmation. In the infinite domain, however, this method of comparison fails, because the degree of confirmation is 0 for either law. Here the concept of instance confirmation (or that of qualified-instance confirmation) will help. If it has a higher value for one of the two laws, then this law will be preferable, if no reasons of another nature are against it.

It is clear that for any deliberate activity predictions are needed, and that these predictions must be "founded upon" or "(inductively) inferred from" past experiences, in some sense of those phrases. Let us examine the situation with the help of the following simplified schema. Suppose a man X wants to make a plan for his actions and, therefore, is interested in the prediction h that c is M'. Suppose further, X has observed (1) that many other things were M and that all of them were also M', let this be formulated in the sentence e; (2) that c is M, let this be j. Thus he knows e and j by observation. The problem is, how does he go from these premisses to the desired conclusion h? It is clear that this cannot be done by deduction; an inductive procedure must be applied. What is this inductive procedure? It is usually explained in the following way. From the evidence e, X infers inductively the law l which says that all M are M'; this inference is supposed to be inductively valid because e contains many positive and no negative instances of the law l; then he infers h ("c is white") from l ("all swans are white") and j ("c is a swan") deductively. Now let us see what the procedure looks like from the point of view of our inductive logic. One might perhaps be tempted to transcribe the usual description of the procedure just given into technical terms as follows. X infers l from e inductively because $c^*(l, e)$ is high; since $l \cdot j$ L-implies h, $c^*(h, e \cdot j)$ is likewise high; thus h may be inferred inductively from $e \cdot j$. However, this way of reasoning would not be correct, because, under ordinary conditions, $c^*(l, e)$ is not high but very low, and even 0 if the domain of individuals is infinite. The difficulty disappears when we realize on the basis of our previous discussions that X does not need a high c^* for l in order to obtain the desired high c^* for h; all he needs is a high c^*_{qi} for l; and this he has by knowing e and j. To put it in another way, X need not take the roundabout way through the law l at all, as is usually believed; he can instead go from his observational knowledge $e \cdot j$ directly to the prediction h. That is to say, our inductive logic makes it possible to determine $c^*(h, e \cdot j)$ directly and to find that it has a high value, without making use of any law. Customary thinking in every-day life likewise often takes this short-cut, which is now justified by inductive logic. For instance, suppose somebody asks Mr. X what color he expects the next swan he will see to have. Then X may reason like this: He has seen many white swans and no non-white swans; therefore he presumes, admittedly not with certainty, that the next swan will likewise be white; and he is willing to bet on it. He does perhaps not even consider the question whether all swans in the universe without a single exception are white; and if he did, he would not be willing to bet on the affirmative answer.

We see that the use of laws is not indispensable for making predictions. Nevertheless it is expedient of course to state universal laws in books on physics, biology, psychology, etc. Although these laws stated by scientists do not have a high degree of confirmation, they have a high qualified-instance confirmation and thus serve us as efficient instruments for finding those highly confirmed singular predictions which we need for guiding our actions.

§15. THE VARIETY OF INSTANCES

A generally accepted and applied rule of scientific method says that for testing a given law we should choose a variety of specimens as great as possible. For instance, in order to test the law that all metals expand by heat, we should examine not only specimens of iron, but of many different metals. It seems clear that a greater variety of instances allows a more effective examination of the law. Suppose three physicists examine the law mentioned; each of them makes one hundred experiments by heating one hundred metal pieces and observing their expansion; the first physicist neglects the rule of variety and takes only pieces of iron; the second follows the rule to a small extent by examining iron and copper pieces; the third satisfies the rule more thoroughly by taking his one hundred specimens from six different metals. Then we should say that the third physicist has confirmed the law by a more thoroughgoing examination than the two other physicists; therefore he has better reasons to declare the law well founded and to expect that future instances will likewise be found to be in accordance with the law; and in the same way the second physicist has more reasons than the first. Accordingly, if there is at all an adequate concept of degree of confirmation with numerical values, then its value for the law, or for the prediction that a certain number of future instances will fulfill the law, should be higher on the evidence of the report of the third physicist about the positive results of his experiments than for the second physicist, and higher for the second than for the first. Generally speaking, the degree of confirmation of a law on the evidence of a number of confirming experiments should depend not only on the total number of (positive) instances found but also on their variety, i.e., on the way they are distributed among various kinds.

Ernest Nagel[28] has discussed this problem in detail. He explains the difficulties involved in finding a quantitative concept of degree of confirmation that would satisfy the requirement we have just discussed, and he therefore expresses his doubt whether such a concept can be found at all. He says (pp. 69ff.): "It follows, however, that the degree of confirmation for a theory seems to be a function not only of the absolute number of positive instances but also of the kinds of instances and of the relative number in each kind. It is not in general possible, therefore, to order degrees of confirmation in a linear order, because the evidence for theories may not be comparable in accordance with a simple linear schema;

and a fortiori degrees of confirmation cannot, in general, be quantized." He illustrates his point by a numerical example. A theory T is examined by a number E of experiments all of which yield positive instances; the specimens tested are taken from two non-overlapping kinds K_1 and K_2. Nine possibilities $P_1, \ldots P_9$ are discussed with different numbers of instances in K_1 and in K_2. The total number E increases from 50 in P_1 to 200 in P_9. In P_1, 50 instances are taken from K_1 and none from K_2; in P_9, 198 from K_1 and 2 from K_2. It does indeed seem difficult to find a concept of degree of confirmation that takes into account in an adequate way not only the absolute number E of instances but also their distribution among the two kinds in the different cases. And I agree with Nagel that this requirement is important. However, I do not think it impossible to satisfy the requirement; in fact, it is satisfied by our concept c^*.

This is shown by a theorem in our system of inductive logic, which states the ratio in which the c^* of a law l is increased if s new positive instances of one or several different kinds are added by new observations to some former positive instances. The theorem, which is too complicated to be given here, shows that c^* is greater under the following conditions: (1) if the total number s of the new instances is greater, *ceteris paribus;* (2) if, with equal numbers s, the number of different kinds from which the instances are taken is greater; (3) if the instances are distributed more evenly among the kinds. Suppose a physicist has made experiments for testing the law l with specimens of various kinds and he wishes to make one more experiment with a new specimen. Then it follows from (2), that the new specimen is best taken from one of those kinds from which so far no specimen has been examined; if there are no such kinds, then we see from (3) that the new specimen should best be taken from one of those kinds which contain the minimum number of instances tested so far. This seems in good agreement with scientific practice. (The above formulations of (2) and (3) hold in the case where all the kinds considered have equal width; in the general and more exact formulation, the increase of c^* is shown to be dependent also upon the various widths of the kinds of instances.) The theorem shows further that c^* is much more influenced by (2) and (3) than by (1); that is to say, it is much more important to improve the variety of instances than to increase merely their number.

The situation is best illustrated by a numerical example. The computation of the increase of c^*, for the nine possible cases discussed by Nagel, under certain plausible assumptions concerning the form of the law l and the widths of the properties involved, leads to the following results. If we arrange the nine possibilities in the order of ascending values of c^*, we obtain this: P_1, P_3, P_7, P_9; P_2, P_4, P_5, P_6, P_8. In this order we find first the four possibilities with a bad distribution among the two kinds, i.e., those where none or only very few (two) of the instances are taken from one of the two kinds, and these four possibilities occur in the order in which they are listed by Nagel; then the five possibilities with a good or fairly good distribution follow, again in the same order as Nagel's. Even for the smallest sample with a good distribution (viz., P_2, with 100 instances, 50 from each of the two kinds) c^* is considerably higher—under the assumptions made, more than four times as high—than for the largest sample with a bad distribution (viz., P_9, with 200 instances, divided into 198 and 2). This shows that a good distribution of the instances is much more important than a mere increase in the total number of instances. This is in accordance with Nagel's remark (p. 69): "A large increase in the number of positive instances of one kind may therefore count for less, in the judgment of skilled experimenters, than a small increase in the number of positive instances of another kind."

Thus we see that the concept c^* is in satisfactory accordance with the principle of the variety of instances.

§16. THE PROBLEM OF THE JUSTIFICATION OF INDUCTION

Suppose that a theory is offered as a more exact formulation—sometimes called a "rational reconstruction"—of a body of generally accepted but more or less vague beliefs. Then the demand for a justification of this theory may be understood in two different ways. (1) The first, more modest task is to validate the claim that the new theory is a satisfactory reconstruction of the beliefs in question. It must be shown that the statements of the theory are in sufficient agreement with those beliefs; this comparison is possible only on those points where the beliefs are sufficiently precise. The question whether the given beliefs are true or false is here not even raised. (2) The second task is to show the validity of the new theory and thereby of the given beliefs. This is a much deeper going and often much more difficult problem.

For example, Euclid's axiom system of geometry was a rational reconstruction of the beliefs concerning spatial relations which were generally held, based on experience and intuition, and applied in the practices of measuring, surveying, building, etc. Euclid's axiom system was accepted because it was in sufficient agreement with those beliefs and gave a more exact and consistent formulation for them. A critical investigation of the validity, the factual truth, of the axioms and the beliefs was only made more than two thousand years later by Gauss.

Our system of inductive logic, that is, the theory of c^* based on the definition of this concept, is intended as a rational reconstruction, restricted to a simple language form, of inductive thinking as customarily applied in everyday life and in science. Since the implicit rules of customary inductive thinking are rather vague, any rational reconstruction contains statements which are neither supported nor rejected by the ways of customary thinking. Therefore, a comparison is possible only on those points where the procedures of customary inductive thinking are precise enough. It seems to me that on these points sufficient agreement is found to show that our theory is an adequate reconstruction; this agreement is seen in many theorems, of which a few have been mentioned in this paper.

An entirely different question is the problem of the validity of our or any other proposed system of inductive logic, and thereby of the customary methods of inductive thinking. This is the genuinely philosophical problem of induction. The construction of a systematic inductive logic is an important step towards the solution of the problem, but still only a preliminary step. It is important because without an exact formulation of rules of induction, i.e., theorems on degree of confirmation, it is not clear what exactly is meant by "inductive procedures," and therefore the problem of the validity of these procedures cannot even be raised in precise terms. On the other hand, a construction of inductive logic, although it prepares the way towards a solution of the problem of induction, still does not by itself give a solution.

Older attempts at a justification of induction tried to transform it into a kind of deduction, by adding to the premisses a general assumption of universal form, e.g., the principle of the uniformity of nature. I think there is fairly general agreement today among scientists and philosophers that neither this nor any other way of reducing induction to deduction with the help of a general principle is possible. It is generally acknowledged that induction is fundamentally different from deduction, and that any prediction of a future event reached inductively on the basis of observed events can never have the certainty of a deductive conclusion; and, conversely, the fact that a prediction reached by certain inductive procedures turns out to be false does not show that those inductive procedures were incorrect.

The situation just described has sometimes been characterized by saying that a theoretical justification of induction is not possible, and, hence, that there is no problem of induction. However, it would be better to say merely that a justification in the old sense is not possible. Reichenbach[29] was the first to raise the problem of the justification of induction in a new sense and to take the first step towards a positive solution. Although I do not agree with certain other features of Reichenbach's theory of induction, I think it has the merit of having first emphasized these important points with respect to the problem of justification: (1) The decisive justification of an inductive procedure does not consist in its plausibility, i.e., its accordance with customary ways of inductive reasoning, but must refer to its success in some sense; (2) the fact that the truth of the predictions reached by induction cannot be guaranteed does not preclude a justification in a weaker sense; (3) it can be proved (as a purely logical result) that induction leads in the long run to success in a certain sense, provided the world is "predictable" at all, i.e., such that success in that respect is possible. Reichenbach shows that his rule of induction R leads to success in the following sense: R yields in the long run an approximate estimate of the relative frequency in the whole of any given property. Thus suppose that we observe the relative frequencies of a property M in an increasing series of samples, and that we determine on the basis of each sample with the help of the rule R the probability q that an unobserved thing has the property M, then the values q thus found approach in the long run the relative frequency of M in the whole. (This is, of course, merely a logical consequence of Reichenbach's definition or rule of induction, not a factual feature of the world.)

I think that the way in which Reichenbach examines and justifies his rule of induction is an important step in the right direction, but only a first step. What remains to be done is to find a procedure for the examination of any given rule of induction in a more thoroughgoing way. To be more specific, Reichenbach is right in the assertion that any procedure which does not possess the characteristic described above (viz., approximation to the relative frequency in the whole) is inferior to his rule of induction. However, his rule, which he calls "the" rule of induction, is far from being the only one possessing that characteristic. The same holds for an infinite number of other rules of induction, e.g., for Laplace's rule of succession (see above, §10; here restricted in a suitable way so as to avoid contradictions), and likewise for the corresponding rule of our theory of c^* (as formulated in theorem (1), §10). Thus our inductive logic is justified to the same extent as Reichenbach's rule of induction, as far as the only criterion of justification so far developed goes. (In other respects, our inductive logic covers a much more extensive field than Reichenbach's rule; this can be seen by the theorems on various kinds of inductive inference mentioned in this paper.) However, Reichenbach's rule and the other two rules mentioned yield different numerical values for the probability under discussion, although these values converge for an increasing sample towards the same limit. Therefore we need a more general and stronger method for examining and comparing any two given rules of induction in order to find out which of them has more chance of success. I think we have to measure the success of any given rule of induction by the total balance with respect to a comprehensive system of wagers made according to the given rule. For this task, here formulated in vague terms, there is so far not even an exact formulation; and much further investigation will be needed before a solution can be found.

NOTES

1. J. M. Keynes, *A Treatise on Probability,* 1921.
2. H. Jeffreys, *Theory of Probability,* 1939.
3. R. von. Mises, *Probability, Statistics, and Truth* (orig. 1928), 1939.

4. H. Reichenbach, *Wahrscheinlichkeitslehre,* 1935.

5. The distinction briefly indicated here is discussed more in detail in my paper "The Two Concepts of Probability," which appears in *Philos. and Phenom. Research,* V, No. 4, 1945.

6. The reader is referred to the later works of Carnap, especially *Logical Foundations of Probability* (University of Chicago Press, 1950) (2nd ed., 1962), and "The Aim of Inductive Logic" in *Logic, Methodology and Philosophy of Science,* eds. E. Nagel, P. Suppes, and A. Tarski (Stanford University Press, 1962.)

7. In an article by C. G. Hempel and Paul Oppenheim in the present issue of this journal, a new concept of degree of confirmation is proposed, which was developed by the two authors and Olaf Helmer in research independent of my own.

8. For more detailed explanations of some of these concepts see my *Introduction to Semantics,* 1942.

9. See F. Waismann, "Logische Analyse des Wahrscheinlichkeitsbegriffs," *Erkenntnis,* I (1930), 228–48.

10. See Waismann, op. cit., p. 242.

11. St. Mazurkeiwicz, "Zur Axiomatik der Wahrscheinlichkeitsrechnung," *C. R. Soc. Science Varsovie,* Cl. III, Vol. 25, 1932, 1–4.

12. Janina Hosiasson-Lindenbaum, "On Confirmation," *Journal of Symbolic Logic,* V, (1940), 133–48.

13. G. H. von Wright, *The Logical Problem of Induction* (Acta Phil. Fennica, 1941, Fasc. III). See also C. D. Broad, *Mind,* LIII, 1944.

14. It seems that Wittgenstein meant this function c_w in his definition of probability, which he indicates briefly without examining its consequences. In his *Tractatus Logico-Philosophicus,* he says: "A proposition is the expression of agreement and disagreement with the truth-possibilities of the elementary [i.e., atomic] propositions" (*4.4); "The world is completely described by the specification of all elementary propositions plus the specification, which of them are true and which false" (*4.26). The truth-possibilities specified in this way correspond to our state-descriptions. Those truth-possibilities which verify a given proposition (in our terminology, those state-descriptions in which a given sentence holds) are called the truth-grounds of that proposition (*5.101). "If T_r is the number of the truth-grounds of the proposition r, T_{rs} the number of those truth-grounds of the proposition s which are at the same time truth-grounds of r, then we call the ratio T_{rs}: T_r the measure of the *probability* which the proposition r gives to the proposition s" (*5.15). It seems that the concept of probability thus defined coincides with the function c_w.

15. The results are as follows.

$$(1)\qquad m = \frac{(N + k - 1)!}{N!(k - 1)!}$$

$$(2)\qquad n_i = \frac{N!}{N_1!N_2! \cdots N_k!}$$

Therefore (according to (5) in §6):

$$(3)\qquad m^*(i) = \frac{N_1!N_2! \cdots N_k!(k - 1)!}{(N + k - 1)!}.$$

16. The general theorem is as follows:

$$c^*(h, e) = \frac{\binom{n_1}{s_1}\binom{n_2}{s_2}}{\binom{n}{s}}$$

17. $(c)^*h, e) = \binom{s}{s_1} r^{s1}(1 - r)^{s2}.$

18. The general theorem is as follows:

$$c^*(h, e) = \frac{\binom{s_1+s_1'+w_1-1}{s_1'}\binom{s_2+s_2'+w_2-1}{s_2'}}{\binom{s+s'+k-1}{s'}}.$$

19. Another theorem may be mentioned which deals with the case where, in distinction to the case just discussed, the evidence already gives some information about the individual *c* mentioned in *h*. Let M_1 be a factual elementary property with the width w_1 ($w_1 \geqq 2$); thus M_1 is a disjunction of w_1 *Q*-properties. Let M_2 be the disjunction of w_2 among those w_1 *Q*-properties ($1 \leqq w_2 < w_1$); hence M_2 L-implies M_1 and has the width w_2. *e* specifies first how the *s* individuals of an observed sample are distributed among cer-

tain properties, and, in particular, it says that s_1 of them have the property M_1 and s_2 of these s_1 individuals have also the property M_2; in addition, e says that c is M_1; and h says that c is also M_2. Then,

$$c^*(h, e) = \frac{s_2 + w_2}{s_1 + w_1}.$$

This is analogous to (1); but in the place of the whole sample we have here that part of it which shows the property M_1.

20. $$\frac{c^*(h, e \cdot j)}{c^*(h, j)} = 1 + \frac{w_1 - w_2}{w_2(w_1 + 1)}$$

This theorem shows that the ratio of the increase of c^* is greater than 1, since $w_1 > w_2$.

21. Janina Lindenbaum-Hosiasson, "Induction et analogie: Comparaison de leur fondement," *Mind,* L, (1941), 351–65; see especially pp. 361–65.

22. The general theorem is as follows:

$$c^*(h, e) = \frac{\binom{n_1+w_1-1}{s_1+w_1-1}\binom{n_2+w_2-1}{s_2+w_2-1}}{\binom{n+k-1}{n-s}}.$$

Other theorems, which cannot be stated here, concern the case where more than two properties are involved, or give approximations for the frequent case where the whole population is very large in relation to the sample.

23. The general theorem is as follows:

(1) $$c^*(l, e) = \frac{\binom{s+k-1}{w_1}}{\binom{N+k-1}{w_1}}.$$

In the special case of a language containing M_1 as the only primitive predicate, we have $w_1 = 1$ and $k = 2$, and hence $c^*(l, e) =$

$$\frac{s + 1}{N + 1}.$$

The latter value is given by some author as holding generally (see Jeffreys, op. cit., p. 106 (16)). However, it seems plausible that the degree of confirmation must be smaller for a stronger law and hence depend upon w_1.

If s, and hence N, too, is very large in relation to k, the following holds as an approximation:

(2) $$c^*(l, e) = \left(\frac{s}{N}\right)^{\frac{w_1}{k}}.$$

For the infinite language $L \infty$ we obtain, according to definition (7) in §3:

(3) $$c^*(l, e) = 0.$$

24. The theorem is as follows:

$$c^*(l', e') = \frac{\binom{s+k-1}{s_1+w_1}}{\binom{N+k-1}{s_1+w_1}}.$$

25. In technical terms, the definition is as follows: $c^*_i(l, e) =_{\text{Df}} c^*(h, e)$, where h is an instance of l formed by the substitution of an individual constant not occurring in e.

26. The technical definition will be given here. Let l be "for every x, if x is M, then x is M'." Let l be non-L-false and without variables. Let c be any individual constant not occurring in e; let j say that c is M, and h' that c is M'. Then the qualified-instance confirmation of l with respect to M and M' on the evidence e is defined as follows: $c^*_{qi}(M, M', e) =_{\text{Df}} c^*(h', e \cdot j)$.

27. Some of the theorems may here be given. Let the law l say, as above, that all M are M'. Let M_1 be defined, as earlier, by $M \cdot \sim M'$ ("non-white swan") and M_2 by $M \cdot M'$ ("white swan"). Let the widths of M_1 and M_2 be w_1 and w_2 respectively. Let e be a report about s observed individuals saying that s_1 of them are M_1 and s_2 are M_2, while the remaining ones are $\sim M$ and hence neither M_1 nor M_2. Then the following holds:

(1) $$c^*_i(l, e) = 1 - \frac{s_1 + w_1}{s + k}.$$

(2) $$c^*_{qi}(M, M', e) = 1 - \frac{s_1 + w_1}{s_1 + w_1 + s_2 + w_2}.$$

The values of c^*_i and c^*_{qi} for the case that the observed sample does not contain any individuals violating the law l can easily be obtained from the values stated in (1) and (2) by taking $s_1 = 0$.

28. E. Nagel, *Principles of the Theory of Probability*. Int. Encycl. of Unified Science, I, No. 6, 1939; see pp. 68–71.

29. Hans Reichenbach, *Experience and Prediction*, 1938, §§38 ff., and earlier publications.

Nelson Goodman

THE NEW RIDDLE OF INDUCTION

Confirmation of a hypothesis by an instance depends rather heavily upon features of the hypothesis other than its syntactical form. That a given piece of copper conducts electricity increases the credibility of statements asserting that other pieces of copper conduct electricity, and thus confirms the hypothesis that all copper conducts electricity. But the fact that a given man now in this room is a third son does not increase the credibility of statements asserting that other men now in this room are third sons, and so does not confirm the hypothesis that all men now in this room are third sons. Yet in both cases our hypothesis is a generalization of the evidence statement. The difference is that in the former case the hypothesis is a *lawlike* statement; while in the latter case, the hypothesis is a merely contingent or accidental generality. Only a statement that is *lawlike*—regardless of its truth or falsity or its scientific importance—is capable of receiving confirmation from an instance of it; accidental statements are not. Plainly, then, we must look for a way of distinguishing lawlike from accidental statements.

So long as what seems to be needed is merely a way of excluding a few odd and unwanted cases that are inadvertently admitted by our definition of confirmation, the problem may not seem very hard or very pressing. We fully expect that minor defects will be found in our definition and that the necessary refinements will have to be worked out patiently one after another. But some further examples will show that our present difficulty is of a much graver kind.

Suppose that all emeralds examined before a certain time t are green. At time t, then, our observations support the hypothesis that all emeralds are green; and this is in accord with our definition of confirmation. Our evidence statements assert that emerald a is green, that emerald b is green, and so on; and each confirms the general hypothesis that all emeralds are green. So far, so good.

Now let me introduce another predicate less familiar than "green." It is the predicate "grue" and it applies to all things examined before t just in case they are green but to other things just in case they are blue. Then at time t we have, for each evidence statement asserting that a given emerald is green, a parallel evidence statement asserting that that emerald is grue. And the statements that emerald a is grue, that emerald b is grue, and so on, will each confirm the general hypothesis that all emeralds are grue. Thus according to our definition, the prediction that all emeralds subsequently examined will be green and the prediction that all will be grue are alike confirmed by evidence

---

statements describing the same observations. But if an emerald subsequently examined is grue, it is blue and hence not green. Thus although we are well aware which of the two incompatible predictions is genuinely confirmed, they are equally well confirmed according to our present definition. Moreover, it is clear that if we simply choose an appropriate predicate, then on the basis of these same observations we shall have equal confirmation, by our definition, for any prediction whatever about other emeralds—or indeed about anything else. As in our earlier example, only the predictions subsumed under lawlike hypotheses are genuinely confirmed; but we have no criterion as yet for determining lawlikeness. And now we see that without some such criterion, our definition not merely includes a few unwanted cases, but is so completely ineffectual that it virtually excludes nothing. We are left once again with the intolerable result that anything confirms anything. This difficulty cannot be set aside as an annoying detail to be taken care of in due course. It has to be met before our definition will work at all.

Nevertheless, the difficulty is often slighted because on the surface there seem to be easy ways of dealing with it. Sometimes, for example, the problem is thought to be much like the paradox of the ravens. We are here again, it is pointed out, making tacit and illegitimate use of information outside the stated evidence: the information, for example, that different samples of one material are usually alike in conductivity, and the information that different men in a lecture audience are usually not alike in the number of their older brothers. But while it is true that such information is being smuggled in, this does not by itself settle the matter as it settles the matter of the ravens. There the point was that when the smuggled information is forthrightly declared, its effect upon the confirmation of the hypothesis in question is immediately and properly registered by the definition we are using. On the other hand, if to our initial evidence we add statements concerning the conductivity of pieces of other materials or concerning the number of older brothers of members of other lecture audiences, this will not in the least affect the confirmation, according to our definition, of the hypothesis concerning copper or of that concerning other lecture audiences. Since our definition is insensitive to the bearing upon hypotheses of evidence so related to them, even when the evidence is fully declared, the difficulty about accidental hypotheses cannot be explained away on the ground that such evidence is being surreptitiously taken into account.

A more promising suggestion is to explain the matter in terms of the effect of this other evidence not directly upon the hypothesis in question but *in*directly through other hypotheses that *are* confirmed, according to our definition, by such evidence. Our information about other materials does by our definition confirm such hypotheses as that all pieces of iron conduct electricity, that no pieces of rubber do, and so on; and these hypotheses, the explanation runs, impart to the hypothesis that all pieces of copper conduct electricity (and also to the hypothesis that none do) the character of lawlikeness—that is, amenability to confirmation by direct positive instances when found. On the other hand, our information about other lecture audiences *dis*confirms many hypotheses to the effect that all the men in one audience are third sons, or that none are; and this strips any character of lawlikeness from the hypothesis that all (or the hypothesis that none) of the men in *this* audience are third sons. But clearly if this course is to be followed, the circumstances under which hypotheses are thus related to one another will have to be precisely articulated.

The problem, then, is to define the relevant way in which such hypotheses must be alike. Evidence for the hypothesis that all iron conducts electricity enhances the lawlikeness of the hypothesis that all zirconium conducts electricity, but does not similarly affect the hypothesis that all the objects on my desk conduct electricity. Wherein lies the difference? The first two hypotheses fall under the broader hypothesis—call it *H*—that every class of things of the same material is uniform in conductivity; the first and third fall only under some such hypothesis as—call it *K*—that every class of things that are either all of the same material or all on a desk is uniform in conductivity. Clearly the important difference here is that evidence for a statement affirming that one of the classes covered by *H* has the property in question increases the credibility of any statement affirming that another such class has this property; while nothing of the sort holds true with respect to *K*. But this is only to say that *H* is lawlike and *K* is not. We are faced anew with the very problem we are trying to solve: the problem of distinguishing between lawlike and accidental hypotheses.

The most popular way of attacking the problem takes its cue from the fact that accidental hypotheses seem typically to involve some spatial or temporal restriction, or reference to some particular individual. They seem to concern the people in some particular room, or the objects on some particular person's desk; while lawlike hypotheses characteristically concern all ravens or all pieces of copper whatsoever. Complete generality is thus very often supposed to be a sufficient condition of lawlikeness; but to define this complete generality is by no means easy. Merely to require that the hypothesis contain no term naming, describing, or indicating a particular thing or location will obviously not be enough. The troublesome hypothesis that all emeralds are grue contains no such term; and where such a term does occur, as in hypotheses about men in *this room,* it can be suppressed in favor of some predicate (short or long, new or old) that contains no such term but applies only to exactly the same things. One might think, then, of excluding not only hypotheses that actually contain terms for specific individuals but also all hypotheses that are equivalent to others that do contain such terms. But, as we have just seen, to exclude only hypotheses of which *all* equivalents are free of such terms is to exclude nothing. On the other hand, to exclude all hypotheses that have *some* equivalent containing such a term is to exclude everything; for even the hypothesis

All grass is green

has as an equivalent

All grass in London or elsewhere is green.

The next step, therefore, has been to consider ruling out predicates of certain kinds. A syntactically universal hypothesis is lawlike, the proposal runs, if its predicates are "purely qualitative" or "nonpositional." This will obviously accomplish nothing if a purely qualitative predicate is then conceived either as one that is equivalent to some expression free of terms for specific individuals, or as one that is equivalent to no expression that contains such a term; for this only raises again the difficulties just pointed out. The claim appears to be rather that at least in the case of a simple enough predicate we can readily determine by direct inspection of its meaning whether or not it is purely qualitative. But even aside from obscurities in the notion of "the meaning" of a predicate, this claim seems to me wrong. I simply do not know how to tell whether a predicate is qualitative or positional, except perhaps by completely begging the question at issue and asking whether the predicate is "well-behaved"—that is, whether simple syntactically universal hypotheses applying it are lawlike.

This statement will not go unprotested. "Consider," it will be argued, "the predicates 'blue' and 'green' and the predicate 'grue' introduced earlier, and also the predicate 'bleen' that applies to emeralds examined before time *t* just in case they are blue and to other emeralds just in case they are green. Surely it is clear," the argument runs, "that the first two are purely qualitative and the second two are not; for the meaning of each of the latter two plainly involves reference to a specific temporal position." To this I reply that indeed I do recognize the first two as well-behaved predicates admissible in lawlike hypotheses, and the second two as ill-behaved predicates. But the argument that the former but not the latter are purely qualitative seems to me quite unsound. True enough, if we start with "blue" and "green," then "grue" and "bleen" will be explained in terms of "blue" and "green" and a temporal term. But equally truly, if we start with "grue" and "bleen," then "blue" and "green" will be explained in terms of "grue" and "bleen" and a temporal term; "green," for example, applies to emeralds examined before time *t* just in case they are grue, and to other emeralds just in case they are bleen. Thus qualitativeness is an entirely relative matter and does not by itself establish any dichotomy of predicates. This relativity seems to be completely overlooked by those who contend that the qualitative character of a predicate is a criterion for its good behavior.

Of course, one may ask why we need worry about such unfamiliar predicates as "grue" or about accidental hypotheses in general, since we are unlikely to use them in making predictions. If our definition works for such hypotheses as are normally employed, isn't that all we need? In a sense, yes; but only in the sense that we need no definition, no theory of induction, and no philosophy of knowledge at all. We get along well enough without them in daily life and in scientific research. But if we seek a theory at all, we cannot excuse gross anomalies resulting from a proposed theory by pleading that we can avoid them in practice. The odd cases we have been considering are the clinically pure cases that, though seldom encountered in practice, nevertheless display to best advantage the symptoms of a widespread and destructive malady.

We have so far neither any answer nor any promising clue to an answer to the question what distinguishes lawlike or confirmable hypotheses from accidental or nonconfirmable ones; and what may at first have seemed a minor technical difficulty has taken on the stature of a major obstacle to the development of a satisfactory theory of confirmation. It is this problem that I call the new riddle of induction.

Other Approaches

Leonard J. Savage

IMPLICATIONS OF PERSONAL PROBABILITY FOR INDUCTION*

INTRODUCTION

Statistical inference and philosophy evidently bear on each other. Exploration of their connections and common ground is accelerating but is not easy. Philosophers find the statistical literature dilute, discordant, philosophically unrigorous, and technical, and therefore hard to winnow. As a statistician driven toward philosophy by interest in the foundations of statistics, I find myself impeded by corresponding difficulties, unable to cover even the most pertinent chapters of philosophy or to determine which are pertinent.

Notwithstanding the reference to implication in my title, I shall attempt no demonstrations here. Rather, I shall grope to share with you some possible insights into induction inspired by study of personal probability and statistics. Genuine demonstrations in philosophy seem rare or nonexistent, though philosophical discussion is often couched in pithy little logical-sounding arguments and sometimes in even more treacherous long ones. How often is such seeming logic advanced with a conviction of rigor and how often as a sort of figure of speech hinting at something vague and insecure?

Three-line arguments utterly demolishing the concept of personal probability are widespread. Each has an even shorter and more devastating refutation, and so on. Such repartee can bear fruit, but only by slow growth on the soil of humility.

Some of the most untrustworthy of philosophical demonstrations have been among the most valuable. Warriors can overtake tortoises; yet Zeno convinces us that there is more to motion than meets the eye. And the importance of Hume's argument against induction—the keynote of this symposium—is undoubted; though perhaps most philosophers view it, like the arguments of Zeno, only as a challenge to search out manifest fallacy. Some of us, however, find Hume's conclusion not paradoxical but close to the mark.

The next section introduces personal probability, necessarily briefly. Though thorough discussion of personal probability cannot be an objective of this paper, a critical section will intensify its introduction, forestall unnecessary misunderstanding, and provide a natural setting for some remarks on induction. We can then look

---

*To be presented in an APA Symposium on Subjective Probability, December 27, 1967. Commentators will be J. Sayer Minas (University of Waterloo, Ontario) and Ernest Nagel (Columbia University).

The research for this paper was supported by the Army, Navy, Air Force, and NASA under a contract administered by the Office of Naval Research.

Reprinted by permission of the estate of the author and *The Journal of Philosophy,* Vol. LXIV, no. 19 (October 5, 1967), pp. 593–607.

directly at the riddle of induction through the eyes of personal probability. Finally, several questions commonly associated with induction will be touched upon in a section on universal propositions.

PERSONAL PROBABILITY

The concept of personal probability was discovered several times between 1921 and 1940 and has older roots. Since about 1950, it has been known to statisticians and is having an increasing influence on them. For history, readings, and bibliography, see the anthology of Kyburg and Smokler.[1] Not all theories of personal probability are quite the same, and in presenting the concept to you, I shall attempt to portray scarcely any view but my own.

Personal probability can be regarded as part of a certain theory of coherent preference in the face of uncertainty. This preference theory is normative; its goal is to help us make better decisions by exposing to us possible incoherencies in our attitudes toward real and hypothetical alternatives.

If as a daily beverage I prefer water to wine and vinegar to water, you may disagree with me and even pity me, but my bizarre tastes are no ground for taxing me with incoherency. If, however, I go on to express a preference for wine as opposed to vinegar, this preference (however normal in itself) is absurd in the presence of my other preferences. If it is called to my attention, I would do well to review my expressed preferences and alter at least one of them.

Various systems of postulates, such as the postulate of transitivity of preference, though qualitative in approach, lead to an arithmetization of the value judgments and opinions of an ideally coherent person. In technical terms, such a person acts in the face of uncertainty so as to maximize the expected *utility* of his experiences with respect to his *personal probability measure* of the events that might affect those experiences.

The utility function of a person is a certain behaviorally defined expression of the value for him of the experiences to which it applies. The theory makes no attempt to brand some utility functions as more appropriate than others; *de gustibus non disputandum est.* This is not to say that drinking vinegar in preference to wine is normal but simply that there cannot be an objective right and wrong about such matters, as there is about an expression of intransitivity of preference.

As utilities express values or tastes, so a person's system of personal probabilities expresses his opinion in an arithmetic way. Nothing in the theory of personal probability precludes his believing that Elizabeth I wrote *Hamlet.* Though bizarre to you and me, in the light of what we and the person all know, this opinion need not be incoherent. Yet, being subject to a personal probability does impose much objective discipline on a person's opinions. He can believe that Elizabeth I probably wrote *Hamlet;* also that it will probably snow in Rio tomorrow. But then coherency will require him to consider that it will snow in Rio tomorrow more probable than that *Hamlet* was written by a commoner.

Though utility is no less fundamental to the preference theory than is personal probability, the latter is much more important for this paper and must therefore be more fully described.

The ideally coherent person, frequently called by the apt technical term 'you', is said to regard the event A as more probable than B under this condition: If you could receive a particular prize if and only if A obtains or else if and only if B obtains, then you would prefer the alternative that associates the prize with A.

Coherency seems to demand that, if C is incompatible with A and B, then the union of A and C will be more probable for you than the union of B and C if and only if A is more probable for you than B. If also there exist for you partitions of the universe into arbitrarily many equally probable events, then there is necessarily a unique (finitely additive) probability measure so defined on all events that A is more probable for you than B if and only if the numerical probability of A exceeds the numerical probability of B. The partition assumption, which can be somewhat weakened[2] is not really an assumption of coherency but rather an assumption of a sufficient richness of contemplated events.

All currently active versions of the preference theory, explicitly or implicitly, exclude dependence of your personal probabilities on what the prize is. The personal probability measure P does vary with the person and with his initial body of knowledge, or data, but we need not here complicate the notation with an explicit indication of this dependence. It is, however, important to describe how opinion changes under the impact of new bits of knowledge such as that the event D obtains; so *conditional probability*, or *probability given* D is introduced.

You are said to hold A to be more probable than B given D under this condition: You would rather have a prize contingent on the intersection of A and D than the same prize contingent on the intersection of B and D. The situation is almost verbatim the same as before and, therefore, leads generally to a new probability measure on events A, dependent on the conditioning event D. If the initial probability $P(D)$ is not 0, then the conditional probability of A given D is $P(A|D) = P(A \cap D)/P(D)$. This is not a mere convention, but an immediate deduction from the qualitative definition of conditional probability. Nor is the qualitative definition itself unmotivated, as will now be explained.

Suppose that you are to be allowed to associate the prize with A or B but may defer your choice until learning which element D_i of a partition (that is, a disjoint and exhaustive finite sequence of events) actually obtains. This amounts to making several simultaneous decisions, one for each i. According to one of the criteria of coherency, you must definitely prefer to associate the prize with A in case D_i does obtain if and only if A is more probable than B given D_i in the qualitative sense. This not only clarifies the definition but illustrates how there is within the preference theory a natural interpretation (of at least one important sense) of the phrase 'learning by experience'.

If you are to choose one or another act in the light of which of the possible outcomes D_i occur in some experiment, then, planning now, you would agree to have your behavior after the experiment governed by your conditional probabilities, given whichever D_i actually obtains. Therefore, since you are coherent, these conditional probabilities will indeed be your effective probabilities when you have seen the outcome of the experiment. (This, incidentally, shows why, at any given moment, you must use the probabilities conditional on all that you have thus far learned, a point which has sometimes seemed puzzling.[3]) By elaborating, any sort of contingency planning can be represented in the preference theory. This is noteworthy; for the theory itself is atemporal and makes no scientific or philosophical commitments about time.

CRITICAL DISCUSSION OF PERSONAL PROBABILITY

My central claim for personal probability is that the preference theory, of which personal probability is an aspect, is a valuable framework for disciplining our behavior and attitudes in the face of uncertainty. How well the claim can be defended is of course open to debate and experience, and only after some discussion of that can we turn to the natural, but secondary, question of what personal probability has to do with probability.

Save through the criterion of coherency, the preference theory makes no distinction between right and wrong opinion. It does not censure the neighbor whom we find superstitious or paranoid nor recognize any notion of the correct inference from data beyond what is implied by the definition and analysis of conditional probability. Some find in this open-mindedness a deadly objection against the theory. Two lines of reply suggest themselves. First, a theory that does some things well is not to be discarded merely for not doing everything. Second, a coherent person strongly but not absolutely rigidly convinced, for example, that 13 is a lucky number for him at roulette would reach a different opinion if he failed to win with exceptionally high frequency in a trial of many bets on that number. The theory does thus require holders of extremely diverse systems of opinion to agree closely with one another when presented with suitable common evidence. This alone seems to me an adequate model of the ostensible objectivity of scientific knowledge.

Sometimes theories of statistics based on a frequency concept of probability are called "objective" and the theory of statistics based on personal probability is called "subjective." This is natural, because the probabilities of a successful frequency theory would be objective, and personal probabilities are clearly subjective. But the employment of frequentistic theories of statistics also involves subjective judgments, as is usually recognized by their proponents. In such theories, the subjective judgments are not fully under that orderly discipline, coherency, which is demanded by the preference theory. Thus arises a paradox: Some frequency enthusiasts disparage personalistic statistics for dealing in opinions rather than facts, though their own theory of statistics actually proves to be more subjective than the personalistic one and in fact virtually becomes the personalistic theory when certain criteria of coherency are recognized.[4]

Holders of what I have called *necessary* views of probability hope, in effect, to improve upon the concept of personal probability by finding such strong rules governing the probability of one event (or proposition) in the light of another that there will be no room for personal differences, given common knowledge. Should this program be possible, it could not but be welcome, but all of its proponents admit to being very short of their goal. No purported steps toward it seem valid to me, and it might even be "demonstrated" that none are possible. Attempts to construct necessary probability seem generally to be affected, explicitly or implicitly, by the dubious notion, promulgated in Wittgenstein's *Tractatus,* of atomic propositions as the natural irreducible propositions of which all others are disjunctions. Necessary theories are, apparently inevitably, based on notions of symmetry, such as that knowledge of each of two or more things is exactly the same in every relevant respect. However, the judgment that those attributes which distinguish the similar objects or events are irrelevant is really a subjective one, for which there has not been and, in my judgment, cannot be any valid objective prescription. Successful construction of necessary probability would, it seems to me, negate just what is most convincing in Hume's skepticism. Modern necessary theories, descended as they are from naive old notions of equally likely cases, are designed to escape certain well-known disasters, but succeed only in postponing them.

According to a frequent criticism, a person's probability for an event will be high if he desires that event and low if he does not, or just the opposite, depending on his temperament; so the theory of personal probability is thought to encourage the errors of optimism and pessimism. These are indeed errors, and it is an important psychological truth that we cannot protect ourselves against them merely by logical care, as in principle we can against outright fallacies. However, the preference theory by its very structure exhorts us to appraise the probabilities of events, apart from any actual consequences they may have for us, by considering only certain hypothetical consequences. The counsel of dispassionate comparison is built in, though of course men of flesh and blood will not always be able to follow it.

Revolving as it does around pleasure and pain, profit and loss, the preference theory is sometimes thought to be too mundane to guide pure science or idle curiosity. Should there indeed be a world of action and a separate world of the intellect and should the preference theory be a valid guide for the one, yet utterly inferior to some other guide for the other, then even its limited range of applicability would be vast in interest and importance; but this dualistic possibility is for me implausible on the face of it and not supported by the theories advanced in its name.

Of course, the goods and ills to which the theory refers need not be mundane, but may reflect the most heroic aspirations—or the most vile. The philosophical puzzle is how the theory can bear on situations in which any notion of motive seems inapplicable. For my part, though perhaps without justification, I can hardly imagine betting at even odds against *A* rather than for *A* (should such a choice be forced upon me) if from the point of view of pure science or idle curiosity I felt quite sure of *A*.

To illustrate with a sufficiently idle question: Was Caesar wearing a new toga when he was assassinated? I guess not; perhaps you do too, and reasons are easy to adduce. Correspondingly, an enforceable document entitling the bearer to ten dollars in case the toga was new would be worth about one dollar to me. The strong doubt and the preference to sell are for me inextricable. The notion of this hypothetical preference to sell does involve contrafactual propositions, which are philosophically puzzling. Such contrafactuals seem essential to the whole theory, not only to its applications to idle curiosity. Whether that is bad and what to do about it are questions beyond my present depth.

Some personalists and necessarians are altogether immune to the criticism of worldliness because, for them, that *A* is more probable than *B* for Mr. Smith is an intuitive and unanalyzable notion. They have been justifiably suspected, however, of not knowing what they are talking about. What use can their unanalyzable concept be for either the world of action or that of the intellect? This divergence between two kinds of personalists, though sharp when focused upon, has not been intensely disruptive or even prevented both kinds from residing in one head.

Interest without material interest is not without meaning, but, no matter how pure an investigator's science may be, he must come down to earth if he reflects how he should next spend his time, not to mention his or the government's money.

Does personal probability have any claim on the name 'probability'? Does it alone have such a claim? These questions are, as I have said, relatively secondary, but we are bound to ask them sooner or later, and thinking about them increases familiarity with personal probability.

Personal probability does have many of the attributes suggested by 'probability'; it is a probability measure (in the mathematical sense) that helps guide action. A natural and, in the presence of controversy, a generous supposition is that there are several kinds of probability, but I am unable to share that view.

In a trivial sense, there are indeed many kinds of probability because the mathematical properties of probability apply to many things having no connection, and only a formal parallelism, with any extra-mathematical notion of probability. For example, the distribution of the total mass of the furniture in this room would be a mathematical probability. But, speaking seriously of more than one valid interpretation of probability, we mean interpretations concerning uncertainty or indeterminable behavior. Carnap, for example, proposes to recognize both a necessary probability and a frequency probability. Others, while respecting the notion of personal probability, feel that a frequency probability is also meaningful and important. Though those who put forward pluralistic views normally seem to take for granted that the different kinds of probabilities have something to do with one another, they seem not to delve into the relationships. Otherwise, they would, I suspect, find that one kind of probability subsumes the others as special cases. At any rate, that is close to the conclusion of radical personalists like me.

Each attempt to define probability is of course based on something genuine which any complete analysis must take into account. For example, probability calculations often are based on the judgment that certain events are equally probable; these situations evoke symmetry and, more recently, necessary definitions of probability. Certainly too, observation of a long part of an even longer sequence of events judged to be similar (in a sense to be made clear soon) leads us to evaluate the probability of each as yet untried event in the sequence as practically equal to the frequency of success, the success ratio, in the events that have been tried. Such situations evoke frequentistic definitions of probability. The personalist understands both of these phenomena in terms of personal probability.

If, for example, a person judges in a certain situation that every card in the deck has the same probability of being drawn, then this probability must be 1/52; the probability of a red card must be 1/2, and so on. But an objective final criterion for such judgments of symmetry seems to be a will-o'-the-wisp. Each person must stand on his own two feet, judging for himself in the light of his other opinions and experiences.

Turning to frequentistic views, they seem to us to involve serious circularities or lacunae. An excellent analysis of the situation that evokes ideas of frequency has been given by de Finetti (*op. cit.*), and though it is too long and mathematical to be fully repeated here, anyone seriously contesting that personal probability has within itself the tenable part of the notion of frequency probability ought to reply to de Finetti's analysis. Even certain fragments of it, which are brief and easy to state, are enlightening. A person accepts a sequence of events as similar in the spirit of the frequentistic notion of probability, or exchangeable, exactly when the probability for him that all of any given k of the events obtain does not depend on which k they are. In the presence of this symmetry judgment and certain other judgments, which can loosely be described as a moderately open mind about what the success ratio in any subset of events will be, the probability of any event of the sequence, given the outcomes of many others, must be nearly equal to the success ratio for the observed events. Once again, familiar calculations are based on a judgment of symmetry, which is a special instance of the judgments called personal probabilities.

This analysis of frequentistic notions can be carried further. After a large number of events in the exchangeable sequence has been observed, the coherent person who began with a moderately open mind will necessarily be rather sure that the success ratio already observed is close to all the success ratios to be observed in that sequence in the indefinite future.

To summarize my opinion, the foundational difficulties in the definition of personal probability are less than those of other attempts to define probability, and the truth behind other attempts to define probability is correctly expressible through the theory of personal probability.

IS INDUCTION RATIONAL?

In a learned, thorough, and plainspoken article, which has been invaluable to me in preparing this paper, Wesley Salmon[5] has stated the riddle of induction, given its history, and explored the strengths and weaknesses of attempts to answer or escape it. He concludes that the riddle has not been answered, but refuses to despair and makes a stirring comparison with the riddle of the infinitesimal calculus, which was solved only after more than a century of resistance, with enormous benefit to mathematics and to the human mind.

The riddle of induction can be put thus: What rational basis is there for any of our beliefs about the unobserved?

The theory of personal probability touches on the domain of the riddle and can even be construed as giving a partial answer. The theory prescribes, presumably compellingly, exactly how a set of beliefs should change in the light of what is observed. It can help you say, "My opinions today are the rational consequence of what they were yesterday and of what I have seen since yesterday." In principle, yesterday's opinions can be traced to the day before, but even given a coherent demigod able to trace his present opinions back to those with which he was born and to what he has experienced since, the theory of personal probability does not pretend to say with what system of opinions he ought to have been born. It leaves him, just as Hume would say, without rational foundation for his beliefs of today.

Can there be any such foundation? The theory as such is silent, but I am led by study of it to doubt that there is a rational basis for what we believe about the unobserved. In fact, Hume's arguments, and modern variants of them such as Goodman's discussion of 'bleen' and 'grue', appeal to me as correct and realistic. That all my beliefs are but my personal opinions, no matter how well some of them may coincide with opinions of others, seems to me not a paradox but a truism. The grandiose image of a demigod tracing his beliefs back to the cradle only to find an impasse there seems a valid metaphor. If there is rational basis for beliefs going beyond mere coherency, then there are some specific opinions that a rational baby demigod must have. Put that way, the notion of any such basis seems to me quite counterintuitive.

We may understand better if we explore why the skeptical answer repels so many. One philosopher will ask, "Can you be sincere in saying that you do not know that you have two hands?" Indeed, I believe firmly enough in my two hands to stake my whole life on them against a trifling gain, as when I climb a ladder for a look around, and almost too firmly to imagine what evidence would convince me that they are not there. Does not so perfect a degree of belief deserve to be called knowledge? No, not in the spirit of the riddle; for the question is not how firm or widespread the belief is but whether it is rational in such a sense that I would be irrational to believe otherwise.

Turn now to a different example. If the first twenty balls drawn from a box are black, is it not rational to believe that most of the balls in the box are black? Not at all, as housewives and other statisticians know. Often the only sound cherries in a box are on top; why should it not be so with the black balls? Perhaps that is unfair, because the original statement was elliptic and should be amplified. How? Perhaps by some reference to shuffling. A good counsel, but with too much practical and physical experience behind it to find any place in a fundamental definition of the rational. Perhaps we were supposed to understand that in each drawing every ball still in the box had an equal chance of being drawn. That does go part way toward justifying the conclusion (via exchangeable events), but the only interpretation I know for 'equal chance' is symmetry of opinion.

Even if each ball does have an equal chance, the conclusion still does not follow. For if, before the twenty drawings, you were strongly of the opinion that white balls were slightly in the majority, you might still quite plausibly think so; for example, even had the first twenty babies born in Shanghai last year been girls, I would remain rather firmly of the opinion that most babies born in Shanghai last year were boys. But suppose, as a counterskeptic might insist, that there is no information bearing on the contents of the box. What can it mean to be devoid of relevant information? The correct opinion of the completely uninformed mind brings us to a less grandiose but no more tractable version of the demigod in his crib and the dubious program of defining necessary probability.

To proceed not only without initial information but without any initial opinion, personal or public, is the slogan of most frequentistic statisticians. Among the efforts under this slogan, at least two important directions are distinguishable: fiducial probability, as in Fisher's volume of 1956[6] and earlier; and inductive behavior as opposed to inductive inference, as initiated by Neyman in 1938.[7] Neither direction seems successful to me. The first does apparently purport to answer the riddle in some cases, but the claim to rationality as opposed to mere objectivity eludes me and perhaps would not actually be pressed by its proponents. The second direction seeks to escape the riddle by emphasizing behavior as opposed to opinion, but the riddle can be asked as provocatively about behavior as it can about opinion. Actually, it is the followers of this direction who are even more subjective than we personalists.

If there is no rational basis for induction, why does induction work? As I have been trying to bring out by examples, no sharply defined method that works has actually been put forward. Rather there is a vaguely defined method, a general and indefinitely ramified art, that we all learn more or less well. To be sure, logic cannot be fully written down either; whether there is any analogy here, I can only ask, but surely it is far from perfect and does not make induction more rational.

If all coherent opinion must be equally respected, why is astronomy better than astrology? Should your whole outlook and the facts (which latter we may pretend to be common to us both) lead you to believe astrology to be effective, then reason alone cannot assail your position. Actually, practically all of us who disbelieve strongly in astrology do so for diffuse and subtle reasons; few have directly tried astrology and found it wanting or even reflected carefully on it. St. Augustine gave astrology some thought, perhaps overworking the pertinent fact that he and the son of his mother's slave were born at the same time. Jung[8] too was open-minded about astrology and made an empirical test of what he regarded as one of its predictions. The test proved difficult to appraise, but, in principle, though Jung and I differed greatly in outlook, we had enough in common so that a better test might well have brought us to a common opinion about the aspect of astrology under test—negative, I would wager.

UNIVERSAL PROPOSITIONS

How do we know universals? Some philosophers seem to regard this as the first and simplest question about induction.[9] But it seems to be an advanced and complicated one. For without it we have already been discussing induction, and universals differ greatly from one another in analysis and status.

Since I see no objective grounds for any specific belief beyond immediate experience, I see none for believing a universal other than one that is tautological, given what has been observed, as it is when it is a purely mathematical conclusion or when every possible instance has been observed. Reflection on the meaning and role of universals, especially through examples, will bring out some interesting points more or less pertinent to induction, though they can hardly be new to some of you.

Ironically, the most traditional universal, "All men are mortal", is by no means the least complicated, and one complicating feature makes the proposition particularly compelling. The counterinstance, "Smith is not mortal", is itself a universal and (in its ordinary mundane sense) in conflict not merely with what we believe about people but with what we believe about the solar system and the galaxy.

Consider more modest generalizations about the fragility of man. Has any man ever lived more than n years? For $n = 100$, surely; for $n = 150$, possibly; for $n = 200$, probably not; for $n = 500$, surely not. These responses are crude expressions of my own personal probability. Yours need not be the same. The example illustrates that what we would ordinarily call knowledge of a universal is acceptance with a high probability of a universal with finite domain or of many such, vaguely specified. My opinion that no man has ever lived 500 years is of course justified in that arguments adducible for it would convince many, notwithstanding the reputation of Methuselah. But, as in any other induction, justification in the sense that one who rejects it is guilty of an error comparable to a logical fallacy is not available.

Will some man ever live 1,000 years? The respondent can hardly imagine all that he is being asked to contemplate. Try substituting for 'ever' the phrase 'within the next *n* years'. For modest values of *n*, speculation does not seem meaningless. Conceivably, though for me not probably, even the next few decades will bring such scientific control of aging that lifespans of 1,000 years for some men then alive will become plausible. The probability of such a technological advance within the next 100, 1,000, or 10,000 years is of course somewhat more probable. But even with rather high tolerance for hypothetical questions, one loses track of meanings if asked to speculate about humanity hundreds of thousands, let alone hundreds of millions, of years into the future.

Are all emeralds green? If by 'emerald' we mean a certain kind of green gem, then the answer is "yes," by convention. But perhaps geologists mean such a thing by 'emerald' that, for them, there are already nongreen emeralds, or at least the possibility of nongreen ones is not closed by convention.

What has just been said about the greenness of emeralds might be paraphrased for the blackness of crows, which suggests still other points. That it would be cheating to paint a crow nonblack helps to bring out once more what a vast amount is vaguely implicit in the simplest-sounding propositions of ordinary language. When we say that all crows are black, do we mean to imply that there are no albino crows or only that the classic downfall of the generalization about white swans is not about to repeat itself in some new Australia where whole species of crows are white? For me, there may well be an occasional albino crow, and there might already, not to mention the distant future, be some island where all the crows are albino. Or, conceivably, some present or future biologist could show convincingly that albinism is not possible for any crow. The observation, by me and my neighbors, of millions of crows, all of them black, does not really go far toward establishing that all crows are black in any very wide sense, even granting that we would know a brown crow with a red breast if we saw one. Yet we may well be convinced of the blackness of the next thousand crows to be encountered by us in our usual haunts. (Stop press: *The Encyclopedia Britannica* mentions a species of gray crow in England.)

You can know that all the golf balls in a cigar box are white, in the practical everyday sense of knowledge, by looking into the open box and examining the half dozen balls that are there. Even this is not knowledge with a capital K. You cannot altogether escape the dangers of hallucination, fatigue, and tricks of light—a point to which I must return. The presence of several balls rather than one only is not really an interesting feature here; the 'all' in the example is trivial.

About 51 per cent of the babies born alive in Boston are boys. Such frequency generalizations seem closely related to universals, and many of the phenomena illustrated in connection with universals apply again. Thus, interpreted narrowly, the proposition might be merely a statement of a fact already observed; interpreted too broadly, it could involve science fiction-like predictions that none of us would really venture. An idealized situation in which the proposition could be personally justifiable is this. The statement is taken to refer to some such finite class of live births regarded as exchangeable as first deliveries of young women in Boston in 1967, all live births in Boston in 1967, all live births in Boston in the past, all live births in Boston from 1,000 A.D. to 2,000 A.D., etc. The larger the class, the cruder will be the approximate exchangeability. If, before making observations, you are not strongly opinionated against the immediate neighborhood of 51 per cent and if you find 51 per cent males in a sample of ten thousand or more, then coherency will require you to be pretty sure that the success ratio in the whole set (the population) is within less than 1 per cent of what it is in the observed set (the sample). In practice, exchangeability may only crudely approximate your opinion, though well enough to justify the conclusion.

A frequency generalization, like the universal propositions and theories of science, is supported not by just one line of evidence but by a network of evidence. For example, sex ratios vary more than can be reasonably accounted for by chance from one community and one circumstance to another, but they have always been close to 51 per cent. Therefore, far from having an initial prejudice against this frequency's applying to Boston, I happen to be prejudiced in favor of its vicinity.

Evidently, my attitude toward universals tends to be reductionist. I would analyze them away as elliptical and, often more or less deliberately, ambiguous statements of a variety of finite conjunctions. But I must confess a serious difficulty in this reductionist program. Any ordinary proposition of ostensibly particular form, such as "This ball is red", is intended and understood to imply many things not yet observed and indeed many propositions that would ordinarily be regarded as universals. For example, "This object will look red to me and others whenever examined by daylight for some ill-specified time to come", "It has about the same diameter in every direction", "It will not soon change shape if left undisturbed", etc. We can attempt more cautious particular propositions, such as "I see white in the upper-left quadrant", hoping thus to avoid being deceived by appearances inasmuch as we report only appearances. But universals lurk even in such reports of sense data. The notions of "I", "upper", "right", and "white" all seem to take their meanings from orderly experience. Indeed, I cannot imagine communication in the absence of expectation of continued order in domains as yet unperceived. To be sure, each universal implicit in an ostensible particular can itself be subjected to reductionist analysis like other universals, but the ideal of eliminating universals altogether seems impossible to me. We have come once more, but along a different path, to the place where personalists disagree with necessarians in expecting no solution to the problem of the *tabula rasa*.

NOTES

1. Henry E. Kyburg, Jr., and Howard E. Smokler, *Studies in Subjective Probability* (New York: Wiley, 1964). Includes an English translation of Bruno de Finetti, "La prévision: Ses lois logiques, ses sources subjectives," *Annales de l'Institut Henri Poincaré,* VII (1937): 1–68; and also a reprinting of my "The Foundations of Statistics Reconsidered," pp. 575–586 in *Proceedings of the Fourth [1960] Berkeley Symposium on Mathematical Statistics and Probability,* vol. I, Jerzy Neyman, ed. (Berkeley: University of California Press, 1961).

A few other works pertinent to personal probability are listed in my "Difficulties in the Theory of Personal Probability," *Philosophy of Science,* XXXIV (1967), to be published.

2. Leonard J. Savage, *The Foundations of Statistics* (New York: Wiley, 1954), chap. 4.

3. Alfred J. Ayer, "The Conception of Probability as a Logical Relation," in *The Problem of Knowledge* (New York: Penguin, 1956), pp. 67–73.

4. Leonard J. Savage, "The Foundations of Statistics Reconsidered," pp. 575–586 in *Proceedings of the Fourth* [1960] *Berkeley Symposium on Mathematical Statistics and Probability,* vol. I, ed. Jerzy Neyman (Berkeley, University of California Press), 1961.

5. Wesley C. Salmon, "The Foundations of Scientific Inference," in *Mind and Cosmos: Essays in Contemporary Science and Philosophy,* vol. III, University of Pittsburgh Series in the Philosophy of Science (Pittsburgh, Pa.: University Press, 1966).

6. Sir Ronald A. Fisher, *Statistical Methods and Scientific Inference* (New York: Hafner, 1956).

7. Jerzy Neyman, "L'estimation statistique, traitée comme un problème classique de probabilité," pp. 25–57 of *Actualités Scientifiques et Industrielles,* 739 (Paris: Hermann et Cie., 1938).

8. Carl G. Jung, "Synchronizität als ein Prinzip akausaler Zusammenhänge," *Naturerklärung und Psyche* (Studien aus dem C. G. Jung-Institut, IV; Zurich: Rascher, 1952). English translation: "Synchronicity: An Acausal Connecting Principle," pp. 417–552 in *Structure and Dynamics of the Psyche,* vol. VIII of *Collected Works,* ed. by H. Read, *et al.,* translated by R.F.C. Hull (New York: Pantheon, 1960).

9. Jean Nicod, "The Logical Problem of Induction" (orig. Paris, 1923), in *Foundations of Geometry and Induction* (New York: Harcourt, Brace & World, 1930).

Gilbert H. Harman

THE INFERENCE TO THE BEST EXPLANATION[1]

I wish to argue that enumerative induction should not be considered a warranted form of nondeductive inference in its own right.[2] I claim that, in cases where it appears that a warranted inference is an instance of enumerative induction, the inference should be described as a special case of another sort of inference, which I shall call "the inference to the best explanation."

The form of my argument in the first part of this paper is as follows: I argue that even if one accepts enumerative induction as one form of nondeductive inference, one will have to allow for the existence of "the inference to the best explanation." Then I argue that all warranted inferences which may be described as instances of enumerative induction must also be described as instances of the inference to the best explanation.

So, on my view, either (a) enumerative induction is not always warranted or (b) enumerative induction is always warranted but is an uninteresting special case of the more general inference to the best explanation. Whether my view should be expressed as (a) or (b) will depend upon a particular interpretation of "enumerative induction."

In the second part of this paper, I attempt to show how taking the inference to the best explanation (rather than enumerative induction) to be the basic form of nondeductive inference enables one to account for an interesting feature of our use of the word "know." This provides an additional reason for describing our inferences as instances of the inference to the best explanation rather than as instances of enumerative induction.

I

"The inference to the best explanation" corresponds approximately to what others have called "abduction," "the method of hypothesis," "hypothetic inference," "the method of elimination," "eliminative induction," and "theoretical inference." I prefer my own terminology because I believe that it

Reprinted by permission from the author and from *Philosophical Review,* Vol. 74, no. 1, January 1965.

avoids most of the misleading suggestions of the alternative terminologies.

In making this inference one infers, from the fact that a certain hypothesis would explain the evidence, to the truth of that hypothesis. In general, there will be several hypotheses which might explain the evidence, so one must be able to reject all such alternative hypotheses before one is warranted in making the inference. Thus one infers, from the premise that a given hypothesis would provide a "better" explanation for the evidence than would any other hypothesis, to the conclusion that the given hypothesis is true.

There is, of course, a problem about how one is to judge that one hypothesis is sufficiently better than another hypothesis. Presumably such a judgment will be based on considerations such as which hypothesis is simpler, which is more plausible, which explains more, which is less *ad hoc*, and so forth. I do not wish to deny that there is a problem about explaining the exact nature of these considerations; I will not, however, say anything more about this problem.

Uses of the inference to the best explanation are manifold. When a detective puts the evidence together and decides that it *must* have been the butler, he is reasoning that no other explanation which accounts for all the facts is plausible enough or simple enough to be accepted. When a scientist infers the existence of atoms and subatomic particles, he is inferring the truth of an explanation for various data which he wishes to account for. These seem the obvious cases; but there are many others. When we infer that a witness is telling the truth, our inference goes as follows: (i) we infer that he says what he does because he believes it; (ii) we infer that he believes what he does because he actually did witness the situation which he describes. That is, our confidence in his testimony is based on our conclusion about the most plausible explanation for that testimony. Our confidence fails if we come to think there is some other possible explanation for his testimony (if, for example, he stands to gain a great deal from our believing him). Or, to take a different sort of example, when we infer from a person's behavior to some fact about his mental experience, we are inferring that the latter fact explains better than some other explanation what he does.

It seems to me that these examples of inference (and, of course, many other similar examples) are easily described as instances of the inference to the best explanation. I do not see, however, how such examples may be described as instances of enumerative induction. It may seem plausible (at least prima facie) that the inference from scattered evidence to the proposition that the butler did it may be described as a complicated use of enumerative induction; but it is difficult to see just how one would go about filling in the details of such an inference. Similar remarks hold for the inference from testimony to the truth of that testimony. But whatever one thinks about these two cases, the inference from experimental data to the theory of subatomic particles certainly does not seem to be describable as an instance of enumerative induction. The same seems to be true for most inferences about other people's mental experiences.

I do not pretend to have a conclusive proof that such inferences cannot be made out to be complicated uses of enumerative induction. But I do think that the burden of proof here shifts to the shoulders of those who would defend induction in this matter, and I am confident that any attempt to account for these inferences as inductions will fail. Therefore, I assert that even if one permits himself the use of enumerative induction, he will still need to avail himself of at least one other form of nondeductive inference.

As I shall now try to show, however, the opposite does not hold. If one permits himself the use of the inference to the best explanation, one will not still need to use enumerative induction (as a separate form of inference). Enumerative induction, as a separate form of nondeductive inference, is superfluous. All cases in which one appears to be using it may also be seen as cases in which one is making an inference to the best explanation.

Enumerative induction is supposed to be a kind of inference that exemplifies the following form. From the fact that all observed *A*'s are *B*'s we may infer that all *A*'s are *B*'s (or we may infer that at least the next *A* will probably be a *B*). Now, in practice we always know more about a situation than that all observed *A*'s are *B*'s, and before we make the inference, it is good inductive practice for us to consider the total evidence. Sometimes, in the light of the total evidence, we are warranted in making our induction, at other times not. So we must ask ourselves the following question: under what conditions is one permitted to make an inductive inference?

I think it is fair to say that, if we turn to inductive logic and its logicians for an answer to this question, we shall be disappointed. If, however, we think of the inference as an inference to the best explanation, we can explain when a person is and when he is not warranted in making the inference from "All observed *A*'s are *B*'s" to "All *A*'s are *B*'s." The answer is that one is warranted in making this inference whenever the hypothesis that all *A*'s are *B*'s is (in the light of all the evidence) a better, simpler, more plausible (and so forth) hypothesis than is the hypothesis, say, that someone is biasing the observed sample in order to make us think that all *A*'s are *B*'s. On the other hand, as soon as the total evidence makes some other, competing hypothesis plausible, one may not infer from the past correlation in the observed sample to a complete correlation in the total population.

The inference from "All observed *A*'s are *B*'s" to "The next observed *A* will be *B*" may be handled in the same way. Here, one must compare the hypothesis that the next *A* will be different from the preceding *A*'s with the hypothesis that the next *A* will be similar to preceding *A*'s. As long as the hypothesis that the next *A* will be similar is a better hypothesis in the light of all the evidence, the supposed induction is warranted. But if there is no reason to rule out a change, then the induction is unwarranted.

I conclude that inferences which appear to be applications of enumerative induction are better described as instances of the inference to the best explanation. My argument has been (1) that there are many inferences which cannot be made out to be applications of enumerative induction but (2) that we can account for when it is proper to make inferences which appear to be applications of enumerative induction, if we describe these inferences as instances of the inference to the best explanation.

II

I now wish to give a further reason for describing our inferences as instances of the inference to the best explanation rather than enumerative induction.[3] Describing our inference as enumerative induction disguises the fact that our inference makes use of certain lemmas, whereas, as I show below, describing the inference as one to the best explanation exposes these lemmas. These intermediate lemmas play a part in the analysis of knowledge based on inference. Therefore, if we are to understand such knowledge, we must describe our inference as inference to the best explanation.

Let me begin by mentioning a fact about the analysis of "know" which is often overlooked.[4] It is now generally acknowledged by epistemologists that, if a person is to know, his belief must be both true and warranted. We shall assume that we are now speaking of a belief which is based on a (warranted) inference.[5] In this case, it is not sufficient for knowledge that the person's final belief be true. If these intermediate propositions are warranted but false, then the person cannot be correctly described as *knowing* the conclusion. I will refer to this necessary condition of knowledge as "the condition that the lemmas be true."

To illustrate this condition, suppose I read on the philosophy department bulletin board that Stuart Hampshire is to read a paper at Princeton tonight. Suppose further that this warrants my believing that Hampshire will read a paper at Princeton tonight. From this belief, we may suppose I infer that Hampshire will read a paper (somewhere) tonight. This belief is also warranted. Now suppose that, unknown to me, tonight's meeting was called off several weeks ago, although no one has thought to remove the announcement from the bulletin board. My belief that Hampshire will read a paper at Princeton tonight is false. It follows that I do not know whether or not Hampshire will read a paper (somewhere) tonight, even if I am right in believing that he will. Even if I am accidentally right (because Hampshire has accepted an invitation to read a paper at N.Y.U.), I do not know that Hampshire will read a paper tonight. The condition that the lemmas be true has not been met in this case.

I will now make use of the condition that the lemmas be true in order to give a new reason for describing the inferences on which belief is based as instances of the inference to the best explanation rather than of enumerative induction. I will take two different sorts of knowledge (knowledge from authority and knowledge of mental experiences of other people) and show how our ordinary judgment of when there is and when there is not knowledge is to be accounted for in terms of our belief that the inference involved must make use of certain lemmas. Then I will argue that the use of these lemmas can be understood only if the inference is in each case described as the inference to the best explanation.

First, consider what lemmas are used in obtaining knowledge from an authority. Let us imagine that the authority in question either is a person who is an expert in his field or is an authoritative reference book. It is obvious that much of our knowledge is based on authority in this sense. When an expert tells us something about a certain subject, or when we read about the subject, we are often warranted in believing that what we are told or what we read is correct. Now one condition that must be satisfied if our belief is to count as knowledge is that our belief must be true. A second condition is this: what we are told or what we read cannot be there by mistake. That is, the speaker must not have made a slip of the tongue which affects the sense. Our belief must not be based on reading a misprint. Even if the slip of the tongue or the misprint has changed a falsehood into truth, by accident, we still cannot get knowledge from it. This indicates that the inference which we make from testimony to truth must contain as a lemma the proposition that the utterance is there because it is believed and not because of a slip of the tongue or typewriter. Thus our account of this inference must show the role played by such a lemma.

My other example involves knowledge of mental experience gained from observing behavior. Suppose we come to know that another person's hand hurts by seeing how he jerks it away from a hot stove which he has accidentally touched. It is easy to see that our inference here (from behavior to pain) involves as lemma the proposition that the pain is responsible for the sudden withdrawal of the hand. (We do not know the hand hurts, even if we are right about the pain being there, if in fact there is some alternative explanation for the withdrawal.) Therefore, in accounting for the inference here, we will want to explain the role of this lemma in the inference.

My claim is this: if we describe the inferences in the examples as instances of the inference to the best explanation, then we easily see how lemmas such as those described above are an essential part of the inference. On the other hand, if we describe the inferences as instances of enumerative induction,[6] then we obscure the role of such lemmas. When the inferences are described as basically inductive, we are led to think that the lemmas are, in principle, eliminable. They are not so eliminable. If we are to account properly for our use of the word "know," we must remember that these inferences are instances of the inference to the best explanation.

In both examples, the role of the lemmas

in our inference is explained only if we remember that we must infer an explanation of the data. In the first example we infer that the best explanation for our reading or hearing what we do is given by the hypothesis that the testimony is the result of expert belief expressed without slip of tongue or typewriter. From this intermediate lemma we infer the truth of the testimony. Again, in making the inference from behavior to pain, we infer the intermediate lemma that the best explanation for the observed behavior is given by the hypothesis that this behavior results from the agent's suddenly being in pain.

If in the first example we think of ourselves as using enumerative induction, then it seems in principle possible to state all the relevant evidence in statements about the correlation between (on the one hand) testimony of a certain type of person about a certain subject matter, where this testimony is given in a certain manner, and (on the other hand) the truth of that testimony. Our inference appears to be completely described by saying that we infer from the correlation between testimony and truth in the past to the correlation in the present case. But, as we have seen, this is not a satisfactory account of the inference which actually does back up our knowledge, since this account cannot explain the essential relevance of whether or not there is a slip of the tongue or a misprint. Similarly, if the inference used in going from behavior to pain is thought of as enumerative induction, it would again seem that getting evidence is in principle just a matter of finding correlations between behavior and pain. But this description leaves out the essential part played by the lemma whereby the inferred mental experience must figure in the explanation for the observed behavior.

If we think of the inferences which back up our knowledge as inferences to the best explanation, then we shall easily understand the role of lemmas in these inferences. If we think of our knowledge as based on enumerative induction (and we forget that induction is a special case of the inference to the best explanation), then we will think that inference is solely a matter of finding correlations which we may project into the future, and we will be at a loss to explain the relevance of the intermediate lemmas. If we are adequately to describe the inferences on which our knowledge rests, we must think of them as instances of the inference to the best explanation.

I have argued that enumerative induction should not be considered a warranted form of inference in its own right. I have used two arguments: (a) we can best account for when it is proper to make inferences which appear to be applications of enumerative induction by describing these inferences as instances of the inference to the best explanation; and (b) we can best account for certain necessary conditions of one's having knowledge (for example, which is knowledge from authority or which is knowledge of another's mental experience gained through observing his behavior) if we explain these conditions in terms of the condition that the lemmas be true and if we think of the inference on which knowledge is based as the inference to the best explanation rather than as enumerative induction.

NOTES

1. This paper is based on one read at the December 1963 meetings in Washington of the Eastern Division of the American Philosophical Association. I wish to thank J. J. Katz, R. P. Wolff, and a reader for the *Philosophical Review* for their helpful comments.

2. Enumerative induction infers from observed regularity to universal regularity or at least to regularity in the next instance.

3. In what follows, when I speak of "describing an inference as an instance of enumerative induction," I understand this phrase to rule out thought of the inference as an instance of the inference to the best explanation. I have no objection to talking of enumerative induction where one recognizes the inference as a special case of the inference to the best explanation.

4. But see Edmund L. Gettier, "Is Justified

True Belief Knowledge?," *Analysis,* 23 (1963), 121–123 and Clark, "Knowledge and Grounds: A Comment on Mr. Gettier's Paper," *Analysis,* 24 (1963), 46–48.

5. Cf. "How Belief Is Based on Inference," *The Journal of Philosophy,* LXI (1964), 353–360.

6. See note 3.

Clark Glymour

RELEVANT EVIDENCE

Scientists often claim that an experiment or observation tests certain hypotheses within a complex theory but not others. Relativity theorists, for example, are unanimous in the judgment that measurements of the gravitational red shift do not test the field equations of general relativity; psychoanalysts sometimes complain that experimental tests of Freudian theory are at best tests of rather peripheral hypotheses; astronomers do not regard observations of the positions of a single planet as a test of Kepler's third law, even though those observations may test Kepler's first and second laws. Observations are regarded as relevant to some hypotheses in a theory but not relevant to others in that same theory. There is another kind of scientific judgment that may or may not be related to such judgments of relevance: determinations of the accuracy of the predictions of some theories are not held to provide tests of those theories, or, at least, positive results are not held to support or confirm the theories in question. There are, for example, special relativistic theories of gravity that predict the same phenomena as does general relativity, yet the theories are regarded as mere curiosities.[1]

Prima facie, such judgments either may be conventional and properly explained entirely by sociological factors, or else they may have an underlying rationale and so may be explained as applications of general principles of scientific inference. At least with regard to the first kind of judgments, that is, those which are explicitly judgments of relevance, three different philosophical views are common: (1) the hypothetico-deductive method provides an obvious and well-understood rationale for such discriminations; (2) one or another system of inductive logic provides a rationale for such discriminations; and (3) there is no rationale for the judgments in question, and they must really be entirely the result of convention.[2] All three opinions are, I believe, quite wrong; there are principles that explain and provide a rationale for scientific judgments of relevance, but they are not exactly hypothetico-deductive principles nor are they principles of a probabilistic kind. The principles that provide a rationale for judgments of relevance also provide a partial rationale for other central features of scientific method; notably, they also explain why some theories are not supported by determinations of the accuracy of predictions derived from them. One consequence is that, although theories may be underdetermined by all possible evidence of a specified kind, they need not be so radically or so easily underdetermined as some writers, including myself,[3] have thought.

Consider the first of the above positions: One might suppose that some hypotheses in a theory are, in conjunction with initial con-

Reprinted by permission from the author and *The Journal of Philosophy,* Vol. LXXII, no. 14, (August 14, 1975), pp. 403–426.

ditions, *essential* to the deduction of a sentence that is decidable by experiment or observation. Such hypotheses would then be tested by the appropriate experiments or observations whereas other hypotheses in the theory—those not essential to the deduction—would not be so tested. An account of this kind is satisfactory only if the notion of an "essential" hypothesis can be made precise; and there are good reasons to believe that such a clarification is not trivial and perhaps not even possible, for the difficulties in making precise the notion of essential hypotheses are exactly those which meet any attempt to provide a criterion of cognitive significance of the kind long sought by the positivists. The positivists proposed to divide the predicates of a theory into two disjoint classes, one of which would comprise the "observation terms" of the theory. A sentence in the language of the theory was to be deemed significant if it was testable, and testability was to be defined solely in terms of the consequence relation holding between, on the one hand, sentences, or classes of sentences, in the language of the theory, and, on the other hand, sentences whose only nonlogical terms were observational. Every attempt to provide such a criterion has failed, and the catalogue of failures is familiar.[4] But if we could specify in precise logical terms what it is for a hypothesis, in conjunction with initial conditions, to be essential to the deduction of an experimentally decidable sentence, then taking the observation terms to be those nonlogical terms occurring in the experimentally decidable sentence or in the statement of initial conditions, we would have an account of testability of the kind the positivists required. We must expect that all the technical sorts of objections that told against empiricist criteria of cognitive significance would tell against any attempt to give a hypothetico-deductive account of epistemic relevance. Some of those who were themselves once part of the positivist tradition saw this connection fairly clearly and drew very strong holist conclusions from the failure of significance criteria. David Kaplan[5] reports that when Carnap was presented with a class of counter-examples (devised by Kaplan) to his last attempt at a significance criterion, "he reflected that he had been quite wrong for about 30 years, and that his critics who had been arguing that theories must be accepted or rejected as a whole (he mentioned at least Quine and Hempel) were very likely correct." And Hempel, at the end of his negative review of attempts at empiricist significance criteria, proposed that theories be evaluated in terms of their clarity and precision, and by such holist canons as simplicity, explanatory and predictive power, and the extent to which they have, as a whole, been confirmed by experience.

Which brings us to the second position. Hempel's own qualitative theory of confirmation[6] has the property that, if e is an evidence statement and p any sentence, consistent with e, that is not a logical consequence of a sentence all of whose nonlogical terms occur in e, then e confirms neither p nor the negation of p. But most of the evidence for complex theories is stated in terms that use only fragments of the vocabularies of the theories. For example, the positions of the planets on the celestial sphere supports Kepler's laws, but this evidence is stated in terms of times, ascensions, and declinations: the notions of a period of an orbit, a mean distance from the sun, and so on, do not occur in the statement of such evidence. Accordingly, despite the fact that his intent was to give an account of epistemic relevance,[7] Hempel's theory cannot explain why such evidence provides support for the theory as a whole or for particular hypotheses within the theory. Quantitative theories of confirmation using logical measure functions—Carnap's m^* for example—do better, but they share some of the limitations of Hempel's system; for example, if a hypothesis and an evidence statement share no nonlogical vocabulary, then the second generally cannot confirm or disconfirm the first.

Several contemporary accounts of scientific inference suppose it to proceed by the formation of conditional probabilities by means of Bayes's theorem in the theory of probability. That is, it is assumed that there are prior probabilities assigned to all hypotheses in question, and the new or posterior probability of (or degree of belief in) a hypothesis *h* on new evidence *e* is just the conditional probability of *h* on *e* (and whatever old evidence there may be). Richard Jeffrey has generalized this strategy so that it need not be assumed that the evidence statement, *e*, is certain.[8] A test that results in evidence *e* is taken to be relevant to hypothesis *h* if and only if the posterior probability of *h*, that is, its conditional probability on *e*, is different from the prior probability of *h*. Analyses of this sort may perhaps be made consistent with the sorts of judgments of relevance described at the outset, but I think we should doubt that they explain such judgments or provide a rationale for them. In order to determine the conditional probability of *h* on *e* by Bayes's rule we must know the prior probabilities of *h* and of *e*, and we must know the conditional probability of *e* on *h*. Frequentists maintain that such prior probabilities are objective frequencies; more particularly, Reichenbach proposed that the prior probability of a theory or hypothesis be taken as the frequency of success in a suitable reference class of theories of the same kind as the theory in question. He gave, unfortunately, no account of how the success of past theories might, without circularity, be determined, nor did he indicate with any concrete examples just how the required groupings might be effected. Reichenbach himself seems to have understood his account as a proposal for future practice: "Should we some day reach a stage in which we have as many statistics on theories as we have today on cases of disease and subsequent death . . . the choice of the reference class for the probability of theories would seem as natural as that of the reference class for the probability of death."[9] Whatever the merits or difficulties with the proposal, one thing is clear: it cannot provide a rationale for those detailed judgments of relevance which scientists now make and have long been making, nor can it explain the great agreement scientists in the same field show about such matters. For we simply do not have statistics of the kind Reichenbach envisioned, nor do we have any idea of what their values would be or even of how to collect them.

Subjective probability theorists, who regard the probabilities of hypotheses as measures of our degrees of belief in them, are not affected by such criticism. But of course, on a strict subjectivist view, the assignments of prior probabilities are quite arbitrary so long as they accord with the requirements of the theory of probability. If, then, judgments of relevance are to be explained ultimately in terms of prior probability distributions, and those distributions are without rationale, the judgments of relevance will also be without rationale.[10] The bare subjectivist account seems to be a version of the third position above: judgments of relevance are conventional.

The conventionalist view would presumably attribute the agreement about relevance to such factors as the education of graduate students: young scientists are told by old scientists what is relevant to what. All relativity texts say that certain experiments do not test certain hypotheses because that was what all relativity textbook writers were taught. There are two difficulties: these suppositions do not explain how judgments of relevance came to be established in the first place, and they do not explain how it is that, with very little controversy, judgments of relevance are made in new cases. The latter fact, especially, suggests that, if scientific education determines scientific judgments about the relevance of evidence to theory, it must do so by teaching, explicitly or tacitly, principles and not merely cases. On the other hand, the conventionalist view has for its support the fundamental consideration that no plausible principles are known that would warrant the discrimination in question. I shall try to remove that support.

II

It is widely thought that, save in exceptional circumstances, universal hypotheses are supported or confirmed by their positive instances. If the hypothesis contains anomalous predicates—"grue," for example—then it will fail to be confirmed by positive instances, and, likewise, if the hypothesis is entailed by some well-confirmed theory, and a positive instance of the hypothesis is inconsistent with that theory, then the instance may serve to reduce our reasons to believe the hypothesis. But, barring circumstances such as these, we expect that universal hypotheses will be confirmed by their positive instances, and, in particular, we expect that a quantitative hypothesis stated as an equation will be confirmed by a set of values for the magnitudes[11] occurring in the hypothesis if the set is a solution to the equation. Now the trouble is that our experiments, observations, and measurements do not appear to provide us with positive instances of the hypotheses of our theories; in the quantitative case, for example, the magnitudes we determine by experiment or observation are generally not those, or not all of those, which occur in our theories concerning the phenomena observed.

Scientists seem to know very well how to get values of magnitudes occurring in their theories from values of magnitudes determined experimentally. Their strategy is to use hypotheses of the very theory to be tested to compute values of other magnitudes from experimentally determined magnitudes. To take a very simple example, suppose our theory consists of the single hypothesis that, for any sample of gas, so long as no gas is added to or removed from the sample, the product of the pressure and volume of the gas is proportional to the temperature of the gas. In other terms, under the given conditions

$$PV = kT$$

where k is an undetermined constant. Suppose further that we have means for measuring P, V, and T, but no means for measuring k. Then the hypothesis may be tested by obtaining two sets of values for P, V, and T, using the first set of values together with the very hypothesis to be tested to determine a value for k

$$k = \frac{pv}{t}$$

and using the value k thus obtained together with the second set of values for P, V, and T either to instantiate or to contradict the hypothesis.

In the example the very hypothesis to be tested was used to determine, from experiment, a value for a quantity occurring in it, and the determination was very simple. Cases of this kind abound in scientific literature,[12] but in general the situation is considerably more complicated. Typically, the theory in question will contain a great many hypotheses, and a given experiment or collection of experiments may fail to measure values of more than one quantity in the theory. To determine a value for one of the latter quantities the use of several hypotheses in the theory may be required, and the determination may proceed through the computation of values for intermediate quantities, or combinations of such. Such a determination or computation may be represented by a finite graph. The initial, or zero-level, nodes of the graph will be experimentally determined quantities; n-level nodes will be quantities or combinations of quantities such that, for each n-level node, some hypothesis of the theory determines a unique value of that node from suitable values of all the $(n-1)$-level nodes with which it is connected. The graph will have a single maximal element, and that element will be a single quantity. We permit that two connected or unconnected nodes may correspond to the same quantity or combination of quantities. I will call such a graph a *computation.*

The graph associated with the computation of the constant in the ideal-gas law is obvious, but it may not be clear what happens in a more complicated case. Let us consider a theory developed in a recent psychological paper;[13] since our considerations are almost entirely structural, we need not concern ourselves with much of the detail regarding the interpretation—which happens to be complicated—of the quantities occurring in the theory. The theory consists of the following set of linear equations, together with their consequences (with respect to real algebra):

$$
\begin{aligned}
&(1)\quad A_1 = E_1\\
&(2)\quad B_1 = G_1 + G_2 + E_2\\
&(3)\quad A_2 = E_1 + E_2\\
&(4)\quad B_2 = G_1 + G_2\\
&(5)\quad A_3 = G_1 + E_1\\
&(6)\quad B_3 = G_2 + E_2
\end{aligned}
$$

The *A*s and *B*s are supposed to be quantities that we know how to estimate experimentally. Suppose then that we do an experiment that gives us values for the quantities A_1, B_1, A_3, and B_3. Naturally we could use equation (1) to compute a value for E_1 immediately from the experimental value of A_1. But it is also possible to compute a value for E_1 from the values of B_1, A_3, and B_3 in the following way:

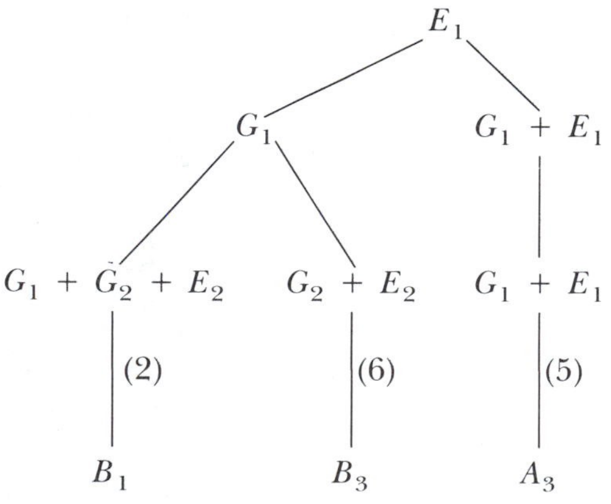

As we have seen, a given set of data may permit the computation of a value for a quantity in more than one way. If the data are consistent with the theory, then these different computations must agree in the value they determine for the computed quantity, but, if the data are inconsistent with the theory, then different computations of the same quantity may give different results. Further, and most important, what quantities in a theory may be computed from a given set of initial data depends both on the initial data and on the structure of the theory. In the example above we supposed given values for A_1, B_1, A_3, and B_3. These permit us to compute values for E_1 and for G_1, but, as the authors of the paper from which we have taken the equations put it, "two of the parameters, G_2 and E_2, occur only together in the expectations with the same coefficients, and are therefore inseparable. We can therefore estimate only G_1, E_1, and $(G_2 + E_2)$" (*ibid.*, 317). That is, we cannot, with this theory, get values of G_2 and of E_2 with these data. Similar things happen with other sets of possible initial values. If we have values of A_1, B_1, A_2, B_2 only, then we cannot compute values for G_1 or for G_2. If, initially, we have values for A_2, B_2, A_3, B_3 only, then we cannot compute values for any of the quantities that appear on the right-hand side of the preceding equations.

It is clear, then, I hope, how scientists may use hypotheses in their theories for the determination of values of quantities that are not in fact measured or estimated by standard statistical methods. The examples already given suffice, I believe, to show that the strategy is in fact used explicitly in some cases. The question is, to what end is this strategy used? More particularly, if experiment permits the computation of values for all quantities occurring in a hypothesis, and these values accord with the hypothesis, does the positive instance thus obtained support or confirm the hypothesis? The answer cannot always be affirmative. Consider the example just discussed; suppose we determine A_1 by experiment and use the hypothesis:

(1) $$A_1 = E_1$$

to compute a value for E_1. We then have values for both A_1 and E_1, and these values are in accord with hypothesis (1) and provide a positive instance of that hypothesis. But clearly it would be wrong to think that this instance provides any support for the hypothesis. Intuitively, the difficulty is that the value of E_1 has been determined in such a way that, no matter what the value of A_1, it could not possibly fail to provide a positive instance of the hypothesis. To test a hypothesis we must do something that could result in presumptive evidence against the hypothesis. So a plausible necessary condition for a set I of values of quantities to test hypothesis h with respect to theory T is that there exist computations (using hypotheses in T & h) from I of values for the quantities occurring in h, and there exist a set J of possible values for the same initial quantities such that the same computations from J result in a negative instance of h—that is, the values of the quantities occurring in h which are computed from J must contradict h. Actually, it is not necessary that all the quantities occurring in h be computable from the initial data, for some of them may occur vacuously. For example, to test an equation of the form

$$a(x^2 + y) + bx - ay = 0$$

we do not require a value for y. The quantity y is vacuous in the equation because, given any value v of x for which there exists a value u of y such that (v, u) is a solution to the equation, then (v, z) is also a solution for all possible values, z, of y. The generalization to cases with more quantities is obvious.

There is another useful condition which, for many theories, is equivalent to that just given. Suppose a hypothesis is equivalent to an equation of the form:

$$X(Q_1 \cdots Q_j) = 0$$

where X is some functional form, and where it is understood that two hypotheses are equivalent if every set of values which is a solution of one is a solution of the other and vice versa. Suppose further that a value for every quantity occurring in the hypothesis can be computed (by using hypotheses of a given theory T) from a set of values for experimentally determined quantities $E_1 \ldots E_k$. Now, for any quantity Q_i occurring in the hypothesis, the computation for Q_i specifies Q_i as a single-valued function of the quantities whose nodes are immediately connected to the Q_i node in the graph of the computation. Similarly, the quantities at the nth-level nodes are, each of them, specified as single-valued functions of the quantities at the $(n - 1)$-level nodes with which they are connected. Thus, ultimately, by composing all these functions, Q_i itself is specified as a single-valued function $f_i(E_1 \ldots E_k)$ of the experimentally determined quantities $E_1 \ldots E_k$. Replacing each Q_i in the hypothesis by $f_1(E_1 \ldots E_k)$ we obtain the equation

$$X(f_1(E_1 \ldots E_k), \ldots, f_i(E_1 \ldots E_k))$$

in which the only quantities are those experimentally determined. We shall say that this equation *represents* the hypothesis for this set of computations. For example, if the hypothesis is (1) above, that is,

$$A_1 = E_1$$

and the only computation is that of E_1 illustrated previously, then the representative of the hypothesis for this computation is

$$A_1 = (A_3 + B_3 - B_1)$$

Now the following is obvious: If the representative of a hypothesis for a set of computations holds identically, that is, if every set of possible values for the quantities occurring in the representative is a solution of the representative, then the computations cannot test the hypothesis, because the necessary condition given before will not obtain. Something more is true. If the functional form X of the hypothesis, and the functions f_i, are composed of operators that determine unique values for all possible sets of values of the quantities they operate on, then the hypothesis will be tested by a set of computations from initial data if the equation representing the hypothesis is not an identity.

We have, in effect, an account of theory testing, and one that naturally evolves from a few elementary observations: *ceteris paribus,* hypotheses are supported by positive instances, disconfirmed by negative; instances, whether positive or negative, of a hypothesis in a theory are got by using the hypotheses of that theory itself (or, conceivably, some other) to make computations from values got from experiment, observation, or independent theoretical considerations; the computations must be carried out in such a way as to admit the possibility that the resulting instance of the hypothesis tested will be negative. Hypotheses, on this account, are not generally tested or supported or confirmed absolutely, but only *relative to a theory*. The general idea is certainly not new. Herman Weyl,[14] for example, seems to have had it:

> The requirements which emerge from our discussion for a correct theory of the course of the world may be formulated as follows:
>
> 1. *Concordance.* The definite value which a quantity occurring in the theory assumes in a certain individual case will be determined from the empirical data on the basis of the theoretically posited connections. *Every such determination has to yield the same result* . . . Not infrequently a (relatively) direct observation of the quantity in question . . . is compared with a computation on the basis of other observations. . . .
>
> 2. It must in principle always be possible to determine on the basis of observational data the definite value which a quantity occurring in the theory will have in a given individual case. This expresses the postulate that the theory in its explanation of the phenomena, must not contain redundant parts (121/2).

Again, in "Testability and Meaning"[15] Carnap proposed to regard hypotheses as confirmed by observation statements if the hypotheses, or instances of them, could be deduced from premises consisting of the observation statements and certain special hypotheses. The special hypotheses—bilateral reduction sentences—were in effect allegedly privileged hypotheses of a theory; privileged in being immune from disconfirmation and in being analytic. But the appeal to analytic truth is quite independent of the main idea, namely, to confirm hypotheses by deducing instances of them by means of other hypotheses in the same theory.

III

Before turning to the questions with which we began, some objections to this account of theory testing need to be considered.

One objection is that the foregoing account is an account of testing for quantitative theories only; it does not seem to apply to qualitative theories or to theories construed as deductively closed, axiomatizable sets of first-order sentences. But the account is straightforwardly extended to first-order theories, and thereby to qualitative theories if the logical form of their hypotheses is known.

By a "quantity" we will mean an open atomic formula. By a "value" for a quantity we will mean an atomic sentence or its negation containing the same predicate constant as the quantity. It certainly must be allowed that, if initial data I (that is, a set of values for quantities) and theory T are consistent, then I disconfirms h with respect to T if T and I together entail $\sim h$ but T alone does not. Conversely, if T and I are consistent and T and I entail h but T alone does not, then I must count as confirming h with respect to T. The more typical and more complicated cases arise when T and I together neither entail nor refute h unless T does so alone. For these cases we may give a quasi-Hempelian analysis:

I confirms h with respect to T if

(i) T and I are consistent with each other and with h.

(ii) There exists a set, call it S, of values

for quantities such that there are computations from I of the values in S and, further, such that S entails the development (in Hempel's sense[16]) of h for the individual constants occurring in members of S.

(iii) There exists a set J of possible values for the initial quantities such that the same computations (as in ii) from J given values of the quantities in S that entail the development of the negation of h.

I disconfirms h if I confirms the negation of h.

I should like briefly to note some features of this account. If I is inconsistent with T, then I neither confirms nor disconfirms any hypothesis with respect to T; but in that case I may nonetheless confirm or disconfirm various hypotheses with respect to sub-theories of T. Hempel's consistency and equivalence conditions are satisfied so long as the theory is kept fixed. The same initial data may, however, confirm inconsistent hypotheses with respect to different theories. Because of condition iii, Hempel's special consequence condition is not satisfied, and neither, of course, is the converse consequence condition.

On Hempel's theory, $\sim R(a)$ confirms both $\forall x \sim Rx$ and $\forall x(Rx \supset Bx)$, but, on the account just given, it does not, because no value of $R(x)$ will, by itself, entail the development of the negation of the second hypothesis, and so condition iii is not met. The "paradox" of the ravens arises in the new account just as in Hempel's, but it is at least confined: if initial data Ra,Ba confirm a hypothesis of universal conditional form with respect to theory T, it is not always the case that $\sim Ra, \sim Ba$ also confirm that hypothesis with respect to T. For example, if the hypothesis is $\forall x(Cx \supset Dx)$ and the theory is $\forall x(Rx \supset Cx)$ & $\forall x(Dx \equiv Bx)$, then the first set of initial data, Ra,Ba confirms the hypothesis, but $\sim Ra, \sim Ba$ does not confirm the hypothesis.

Although I think that most of the features of the foregoing account for first-order theories are plausible enough, I shall not defend them now. There are a variety of ways in which the general strategy I have outlined in the previous section might be extended to formalized theories, and the quasi-Hempelian account just given is only one of them. One can, for example, try to preserve the consequence condition by replacing iii with a radically weaker condition, e.g.,

(iii*) If h has a representative for the set of computations in ii, the representative is not a valid formula.

but then one will have to allow that $\sim Ra$ confirms $\forall x(Rx \supset Bx)$. Again, it is straightforward to adapt the general strategy to a Popperian viewpoint, so that hypotheses of universal form may be tested but hypotheses of existential or mixed form never are. The point is that the account *can be* extended to formalized theories, and the extension need not be much less plausible—I think not any less plausible at all—than accounts of confirmation that are confined to "observation" statements.

A serious difficulty, urged by Professor Hempel, is this: typically, the hypotheses of a theory of themselves determine nothing about experimental or observational data; something definite about experimental outcomes can be inferred from the theory—or values of theoretical quantities can be inferred from the data—only if special, empirically untested, assumptions are made. Hempel calls such assumptions "qualifying clauses" or "provisos." One example, alleged by several writers, is that no observable consequences about the motions of heavenly bodies follow from Newton's three laws and the law of universal gravitation unless one makes some assumption about what forces are acting, e.g., that only gravitational forces act between the bodies of the solar system.

There may indeed be many cases in which a theory can be applied to a system only if it is assumed that the system has some property of a kind that is not determined experimentally; even when that is so, however, one must still be able to say what hypotheses in the theory are tested by the experimental results on the supposition that the qualifying clause is met, and our account proposes an answer to that question. Of course, one wants to know something more about when it is reasonable to assume that qualifying clauses are satisfied, and what role they may play in the assessment of a whole theory, but that is beyond our scope at present.

It is not clear to me how often such qualifying clauses are really essential. Consider Newton again. In book III of the *Principia* Newton uses his first two laws to deduce from Kepler's laws that there is a centripetal force acting on the planets in inverse proportion to the square of their distances from the sun. He further shows, using terrestrial experiments and the third law, that this centripetal force between two bodies must be proportional to the product of their masses. Now, as deductivists like Duhem[17] have insisted, these deductions do not result in an instance of the gravitational-force law because that law requires that the gravitational force acting between *any two* bodies be proportional to the product of the masses and inversely proportional to the square of the distance between them; but the total gravitational force acting on any planet must be the sum of the forces due to the sun and to the other bodies in the solar system, and hence the total gravitational force acting on a planet ought not to be inversely proportional to the square of its distance from the sun. Newton's conclusions are inconsistent with his law. Duhem's objection fails entirely, however, if we recognize that Kepler's laws need not be taken as strictly correct initial data, but rather as very good approximations subject to whatever errors there may be in the observations of planetary positions and times. The question then becomes whether the planetary perturbations are sufficiently small that the deviation in the total force acting on a planet from that calculated by Newton using Kepler's laws is less than the error of the computed result due to error in the initial data. Such a determination in turn, requires, besides some idea of the error of the observations, an estimate of the relative masses of the planets to the sun. For any planet with a satellite, the ratio of the planet's mass to the sun's can be estimated from data independent of those used to compute the circumsolar force; Newton is thus able to argue without circularity that the gravitational interaction of the planets is very small in comparison with the solar force.[18]

In effect, the method of testing described in this paper is Newton's method, save that in Newton's case the matter is complicated by the use of empirical laws as initial data and the use of approximations. Not only Newton, but Newtonian scientists of the eighteenth and nineteenth centuries claimed to deduce their laws from the phenomena. Perhaps they overstated their case, but they had, nonetheless, a case to state. The scorn heaped on their method by Duhem is undeserved.

Another objection is that the account is, after all, just the old hypothetico-deductive account. For, if a set of initial data confirms a hypothesis with respect to a theory according to the preceding account, then surely there is a valid deduction of some of the propositions in the initial data set from premises consisting of the rest of the propositions in the initial data set, the hypothesis tested, and the theorems of the theory that are used in the computations. Further, if the data disconfirm the hypothesis, the negation of some proposition in the initial data set must be deducible in an analogous way. And surely H-D theorists would agree that in some contexts only some particular hypothesis or hypotheses from among all those which might appear in such deductions are in fact tested.

It is true, I think, that any test can be converted into a deductive argument in the way suggested; but the converse is not true. Not all deductions of singular statements from putative laws and initial conditions can be transformed into tests. For example, suppose hypothesis h is tested by data I with respect to theory T. For each predicate occurring in h or in T but not occurring in I, choose two new, distinct predicates, and replace each occurrence of each predicate, P say, by the disjunction of the two new predicates associated with P. Then h is changed into a new hypothesis h^*, and T is changed into a new theory T^*, and, further, if there is a valid deduction of a proposition in I from the rest of I, h, and theorems of T, then, by the substitution theorem, there is also a valid deduction of that proposition in I from the rest of I, h^*, and T^*. But, in general, I will not test h^* with respect to T^*. That is exactly as it should be, for no scientist would take evidence to support a theory like T^* when another like T was available. The H-D method has us deduce singular statements from laws; the new procedure, in effect, has us deduce *instances* of laws from singular statements and other laws. The two are not the same. I have no doubt that H-D advocates agree that sometimes data test certain hypotheses and not others; what I doubt is that their principles afford any explanation of those judgments.

IV

We still have to consider what the account of theory testing can contribute to the questions with which we began. What grounds can there be for claims to the effect that one or another experiment has no bearing on one or another hypothesis within a theory? In general terms our answer is clear enough: depending on the nature of the experiment or observation and the structure of the theory in question, a given hypothesis may or may not be tested according to the scheme outlined in previous sections. In particular cases, detailing the application of the scheme may be very complex, and the psychoanalytic and relativity examples mentioned at the outset are certainly too complex to discuss here.[19] It is, however, fairly easy to see how the account of theory testing can explain the claim that observations of a single planet do not, of themselves, provide a test of Kepler's third law.

Kepler's first and second laws specify features of the motion of any planetary body moving about the sun. The third law, however, relates features of the orbits of any two bodies; specifically it claims that the ratio of the periods of any two planets equals the 3/2 power of the ratio of their mean distances from the sun. The parameters that uniquely determine the Keplerian orbit at any time can be estimated from several observations of the planet on the celestial sphere; in fact, three suitably chosen observations suffice for the computations,[20] and a fourth observation of a single planet permits a test of Kepler's first and second laws. But, however many observations we may have of the location of a single planet on the celestial sphere, those are not, by assumption, observations of the location of any *other* planet on the celestial sphere. To test Kepler's third law, we need estimates of the periods and mean distances from the sun of at least two planets. But from the observations of one planet alone we cannot compute, using Kepler's laws and their consequences, the parameters of the orbit of any other planet. We can, of course, compute under those circumstances the *ratio* of the square of the period to the cube of the mean distance from the sun for any planet whatsoever, but only by *using* Kepler's third law itself. So, even if we count such a ratio as one quantity, the representative of Kepler's third law (see p. 412 above) for the requisite computations will be a trivial identity, and hence the third law will not be tested.

The account of theory testing helps to account for a good deal more about scientific methodology. A standard methodological principle is that a theory is better supported by a variety of evidence than by a narrow spectrum of evidence. The substance of the principle is, however, unclear so long as we lack some account of what constitutes relevant variety. One view, which I believe is incorrect, is that what constitutes a relevant variety of evidence for a theory is entirely determined by what other theories happen to be in competition with the first.[21] On the contrary, if, as I have argued, a given piece of evidence may be evidence for some hypothesis in a theory even while it is irrelevant to other hypotheses in that theory, then we surely want our pieces of evidence to be various enough to provide tests of as many different hypotheses in that theory as possible, regardless of what, in historical context, the competing theories may be. There is a further complication. In assessing a theory we are judging how well it is supported with respect to itself, and this reflexive feature of theory testing makes for certain difficulties. If a hypothesis is confirmed by observations and computations using another hypothesis in the theory, then it is always possible that the agreement between hypothesis and evidence is spurious: both the hypothesis tested and some hypothesis used in the computations of the test may be in error, but the errors in one hypothesis may be exactly (or exactly enough) compensated for by the errors in the other. Conversely, a true hypothesis may be disconfirmed by observations and computations using other hypotheses in the theory if one or more of the hypotheses used in the computations are incorrect. The only means available for guarding against such errors is to have a variety of evidence, so that as many hypotheses as possible are tested in as many different ways as possible. What makes one way of testing relevantly different from another is that the hypotheses used in the one computation are different from the hypotheses used in the other computation. Part of what makes one piece of evidence relevantly different from another piece of evidence is that some test is possible from the first that is not possible from the second, or that in the two cases there is some difference in the precision of computed values of theoretical quantities.

Kepler's laws again provide a simple example. Kepler did not determine elliptical orbits for planets as simply the best fit for the data; on the contrary, he gave a physical argument for the area rule—his second law—and used the area rule together with the data to infer that the planetary orbits are ellipses. Seventeenth-century astronomers were able to confirm Kepler's first law only by using his second, and they were able to confirm his second only by using his first. Understandably, there remained considerable disagreement and uncertainty as to whether the two laws were correct, or whether the errors in one were compensated for by the errors in the other. Not until the invention of the micrometer and Flamsteed's observations of Jupiter and its satellites, late in the seventeenth century, was a confirmation of Kepler's second law obtained without any assumption concerning the planet's orbit.[22] I doubt that this example is singular; quite the reverse: it seems unlikely to me that the development and testing of any complex modern theory in physics or in chemistry can be understood without some appreciation of the way a variety of evidence serves to separate hypotheses.

At the outset it was observed that some theories are regarded chiefly as curiosities and rarely taken seriously, despite the fact that they account for all the evidence accounted for by some theory taken very seriously and are not known to be irreconcilable with any other phenomena. In many cases this kind of scientific discrimination can plausibly be explained as the result of applying the principles of evidential relevance that we are concerned to describe.

Some years ago Walter Thirring[22] published a special relativistic theory of gravitation. Thirring's theory supposes that spacetime has a flat metric η_{uv} like that of special relativity and that gravitation is due to a tensor field, ψ_{uv}, that has no effect on the metric. Writing down equations for these quantities, Thirring was able to show that his theory accounts for many of the phenomena that are usually taken to confirm general relativity. His theory is almost universally regarded as a curiosity; such an assessment might of course result from mere prejudice or from any of a variety of obscurely motivated methodological opinions, e.g., the view that a phenomenon confirms a theory only if the theory literally *predicts* the phenomenon. But I think the account of relevant evidence developed in the preceding sections best explains this assessment, and also best explicates what physicists typically say in justifying that assessment. What they say is that Thirring's theory is defective because his metric, η_{uv}, is not "observable."[24] A better word would be 'determinable', and, if we understand the authors in that way, then the complaint makes perfect sense. Free-falling particles do not follow geodesics of Thirring's metric, η_{uv}, nor do clocks measure time according to it, nor rods distance. What, according to the theory, such systems measure are geodesics, time, distances, as determined by the quantity:

$$\eta_{uv} - f\psi_{uv}$$

where f is a suitable function. By making compensatory changes in ψ_{uv}, an infinite variety of different flat metrics η_{uv} can be made compatible with all data about rods, clocks, test particles, etc. This is not just experimental uncertainty, or a failure to obtain perfect accuracy in our measurements. We noted earlier that, if in a theory a quantity A is replaced throughout by an algebraic combination of new quantities B, C, D, then hypotheses formerly tested by various initial data may be turned into hypotheses not tested by those data, because values for B, C, D cannot be computed even approximately. That is in effect what happens in Thirring's theory: the general relativistic metric, g_{uv}, which is determinable in principle from the behavior of material objects, is replaced by an algebraic combination—$(\eta_{uv} - f\psi_{uv})$—of new quantities. The result is that values for the new quantities cannot be computed from the relevant initial data, and so, although it might be possible to determine evidence *against* Thirring's theory, it is not possible to determine evidence *for* its central hypotheses because they cannot be instantiated. The physicists' principle is that we should prefer theories whose hypotheses are positively tested by our evidence to theories that, even though consistent with our data and affording an explanation of it, are not positively tested by it. The principle is a good one.

Theories with quantities whose values cannot be determined by the evidence are, in an intuitive way, less simple than theories without undeterminable quantities or with fewer of them. Still, it is a mistake to see this discrimination as no more than a manifestation of our preference for simple theories; I think we do better to try to understand whatever rational preference there may be for simpler scientific hypotheses as derivative from our preference for better tested theories, and the account presented here provides at least a partial rationale for our attachment to simplicity. Quine, for one, seems to think differently:

> Yet another principle that may be said to figure as a tacit guide of science is that of sufficient reason. A lingering trace of this venerable principle seems recognizable, at any rate, in the scientist's shunning of gratuitous singularities. If he arrives at laws of dynamics that favor no one frame of reference over others that are in motion with respect to it, he forthwith regards the notion of absolute rest and hence of absolute position as untenable. This rejection is not, as one is tempted to suppose, a rejection of the empirically undefinable; empirically unexceptionable definitions of rest are ready to hand, in the arbitrary adoption of any of various specifiable frames of reference. It is a rejection of the gratuitous. This principle may, however, plausibly be subsumed under the demand for simplicity, thanks to the looseness of the latter idea.[25]

Though it is perfectly correct that we can always make determinable an undeterminable quantity in a theory merely by adding a further hypothesis, that is not enough. For it is not in the least obvious that we can always add a hypothesis which will be tested by the evidence available or which will not be tested negatively either by the evidence available or by evidence easily produced. In Newtonian theory there is no way to compute which unaccelerated trajectories through space-time are truly at rest with respect to absolute space. One can easily add to the theory untestable hypotheses about the rest frame—e.g., that the center of mass of the universe is at rest; and one can easily add hypotheses one has every reason to believe false or at best contingently true—e.g., that inertially moving cabbages are at rest. Doing better is hard. Identification of the rest frame with the reference system in which particular physical systems—whether cabbages or the Sun—are at rest is unsatisfactory, for such correlations cannot be even approximate laws because the physical systems can be accelerated. The aether, were there one, would perhaps have done the job, but there is not one, and the importance of that fact in the history of physics underscores the point: what theoretical magnitudes we can determine depends on what lawlike hypotheses are available to us, and that, in turn, depends on what kinds of things there are.

There is another kind of case where judgments often attributed to a taste for simple things can at least partially be attributed instead to a taste for well-tested things. First a word about error. Suppose the measurements that comprise some body of evidence are subject to error and, though the exact error of any measurement is unknown, an upper bound to the error is known. Then each measurement may be regarded as determining an interval of possible values of the measured quantity, within which the true value must lie. This is, I believe, a typical circumstance in scientific measurement. Computations of theoretical quantities may proceed as before, but what is determined from the data is a *set* of values of any computed quantity. Again, a test of a theoretical relation is understood as before, but with the following complication: what is required for an instance of a hypothesis is, for each quantity in the hypothesis, a set of values for that quantity such that there can be drawn from the respective sets a collection of precise values—one for each quantity—satisfying the hypothesis. This is the obvious generalization of our account when error is present.

Suppose a theorist is entertaining hypotheses about the functional form of the relation between two quantities, X and Y, which he can determine experimentally. We assume that he has no well-established theory to guide him, and we suppose his measurements of X to be subject to some error of known bound. If getting values for X and for Y is difficult, costly, and tedious, our theorist will doubtless wish to draw his conclusions from but a few data points if that is possible. Suppose he has six points and, to within the tolerable error, they lie on a line: our theorist claims that the relation between X and Y is linear. Why does he think the linear hypothesis better than some other polynomial relation? In particular, the six data points are perfectly consistent with the hypothesis that

$$Y = a_0 + a_1X + a_2X^2 + a_3X^3 + a_4X^4 + a_5X^5$$

and, because of the error, the coefficients of quadratic and higher powers of X need not be zero. Why the linear hypothesis rather than the fifth-degree hypothesis? Can there be any more to it than a taste for simple things?

A Popperian answer is that the simpler, linear hypothesis can be falsified by fewer data points than can the fifth-degree hypothesis. This cannot be exactly the right reason, for, *given* the six data points, the linear and fifth-degree hypothesis each require the same number of *additional* data points for a possible falsification, namely, one. The reason for the preference, I suggest, is straightforward: two data points permit a computation of intervals of values for the undetermined coefficients of the linear hypothesis, and four more data points permit four tests of that hypothesis; but the six values of X and Y permit only a computation of intervals of values of the constant coefficients in the fifth-degree hypothesis; they do not permit any test of it. The theorist should prefer the linear hypothesis for the straightforward reason that he has more positive evidence for it than for any other polynomial relation.

V

There are two theses which have recently gained such wide assent among empiricist philosophers that they deserve to be regarded as new dogmas of empiricism. I have in mind the claim that our theories may be underdetermined by all possible evidence, and the further claim that each theory is tested as a whole. Dogmas may of course be true, and, with suitable qualifications, these dogmas are. I should like to conclude by saying something about the qualifications.

For some theories, at some stages of their development, a set of quantities can plausibly be demarcated such that the evidence for or against the theory in question consists of values for these quantities for various systems. When such a demarcation can plausibly be made, it not only makes sense to ask whether the theory is uniquely determined by all possible evidence of the relevant kind, but, further, we can sometimes hope to get an answer to this question. Of course an answer, whether affirmative or negative, says nothing about what sorts of underdetermination may occur if novel kinds of evidence are discovered. For example, the state of absolute rest is undeterminable in Newtonian gravitational theory, but, had the combination of Newtonian theory with Maxwell's electrodynamics proved correct, optical experiments would have permitted a determination of the rest frame.[26] Again, for certain models of general relativity, it can be shown that no measurements of the quantities peculiar to that theory suffice to determine the global topology of space-time,[27] but, even if our universe is in fact one of these topologically underdetermined universes, it is still possible that other branches of physics—plasma physics for example—might provide evidence and theory sufficient to determine a unique topology.

If we confine consideration to a given kind of evidence, we can inquire whether evidence of that kind uniquely determines a best theory that explains it. Conceivably, all possible such evidence might fail to determine a unique theory for either of two kinds of reasons. First, there might occur two or more theories that are not intertranslatable but all of whose hypotheses are tested positively by the evidence so that every methodological demand met by one theory is met by the other. I know of no plausible candidates for this kind of case, but I see no reason why they should not exist. Second, there might occur two or more theories that differ only in hypotheses that cannot be tested, and, for some reason or other, every plausible theory accounting for the evidence also contains such a hypothesis. There are a great many examples of this kind of case, and analyzing when this sort of underdetermination arises is a standard problem in the social sciences.[28]

Demonstrating underdetermination is sometimes possible, but it is not as easy as some writers have supposed. Reichenbach,[29] for example, argued that, even in the context of classical physics, the theory of the geometry of space is underdetermined; for, given any geometry, we can suppose it to be the true one and explain the coincidence behavior of material bodies in terms of this geometry and the action of a "universal force." But, if one sets out actually to write down such a theory, one quickly discovers that it is obtainable only by dividing the Euclidean metric of Newtonian theory into two new quantities, just as Thirring divided the metric field of general relativity into two new quantities. The result is a theory which, on the same evidence, is less well tested than Newtonian theory. We cannot demonstrate underdetermination by substituting for one or more predicates of a theory a combination of new predicates, since the result of the substitution is a theory less well tested than the original.

Early in this century both Duhem and Frege urged that a theory must be tested as a whole. Reductive programs, like Carnap's *Aufbau,* would have avoided holism had they succeeded, but they did not succeed. Later, a number of philosophers, notably Carnap and C. I. Lewis, tried to avoid holism by putting analytic truth to work. They kept in common some version of the claim that, given a collection of analytic truths, or truths by convention, each hypothesis in a theory has its own, independent connections with experience. It is understandable that a new romance with holism should be the concomitant of estrangement from the distinction between analytic truths and synthetic truths.

Part of what has been said or suggested on behalf of holism is false, and part of it is true. It is true that a great part of a theory may be involved in the confirmation of any of its hypotheses, and it is further true that the assessment of any hypothesis in a theory in the face of negative evidence requires the assessment of all hypotheses in that theory. It is false that a piece of evidence is evidence indiscriminately for all hypotheses in a theory or for none of them, and it is false as well that theories must be accepted or rejected as a whole. For positive evidence may fail to provide any support for some hypotheses in a theory—support, that is, with respect to the theory itself—even while confirming other hypotheses. And, if the total evidence is of sufficient variety, evidence inconsistent with a theory may still leave us with a fragment that is best confirmed with respect to itself. If we are lucky, in some axiomatizations of the theory we may even be able to single out a particular axiom that deserves the blame. A naive holism that supposes theory to confront experience as an unstructured, blockish whole will inevitably be perplexed by the power of scientific argument to distribute praise and to distribute blame among our beliefs.

NOTES

1. See, for example, Ya. B. Zeldovich and I. D. Novikov, *Relativistic Astrophysics* (Chicago: University Press, 1971), pp. 66–71.

2. For the third position, see Kaplan, *infra;* the view is certainly suggested by many of Quine's remarks, but I find it nowhere explicitly in his writings. The second position is perhaps the most popular: cf. C. I. Lewis, *An Analysis of Knowledge and Valuation* (La Salle, Ill.: Open Court, 1946); H. Reichenbach, *Experience and Prediction* (Chicago: University Press, 1938); R. C. Jeffrey, "Probability and Falsification," unpublished; I can cite no texts for the first view, but philosophers at the University of Chicago and at Indiana University, where earlier versions of this paper were read, urged it. I am indebted to them for their criticism, and to the National Science Foundation for support of research. I owe special thanks to Richard Jeffrey and to Carl Hempel for reading and criticizing drafts of this essay.

3. "Theoretical Realism and Theoretical Equivalence," in R. Buck and R. Cohen, eds., *Boston Studies in the Philosophy of Science,* vol. VIII (Boston: Reidel, 1971).

4. Cf. Hempel, "Empiricist Criteria of Cognitive Significance: Problems and Changes," in *Aspects of Scientific Explanation* (New York: Free Press, 1965).

5. "Homage to Carnap," in Buck and Cohen, *op. cit.,* pp. xlvi–xlvii.

6. "Studies in the Logic of Confirmation," in *Aspects of Scientific Explanation, op. cit.*

7. See *ibid,* p. 5/6. Hempel was, of course, aware of the difficulty and entertained remedies. One remedy, the converse consequence condition, he rightly rejected, and subsequent attempts to revive it [cf. B. Brody, "Confirmation and Explanation," this JOURNAL, LXV, 10 (May 16, 1968): 282–299] have not proved fruitful.

8. *The Logic of Decision* (New York: McGraw-Hill, 1965), ch. 11.

9. *Theory of Probability* (Berkeley: Univ. of California Press, 1949). This passage is taken from S. Luckenbach, *Probabilities, Problems and Paradoxes* (Encino, Calif.: Dickenson, 1972), p. 44.

10. The standard Bayesian response to criticisms that turn on the arbitrariness of prior probabilities is by appeal to stable estimation theorems; i.e., to proofs that, under certain conditions whatever the prior distributions may be, the posterior distributions will be nearly the same given sufficient evidence. [Cf. Edwards, Lindman, and Savage, "Bayesian Statistical Inference for Psychological Research," *Psycholog-*

ical Review, LXX (1963).] But I know of no such theorems for the kind of case under consideration, that is, when the evidence statements are confined to a proper sublanguage of the language in which the hypotheses may be formulated.

11. I shall use the terms 'magnitude' or 'quantity' either to signify abstract objects, e.g., the type of the token 'kinetic energy', or else to signify properties under a description. The important point is that for my purposes "mean kinetic energy" . . . and "temperature" must count as different quantities even though temperature is mean kinetic energy.

12. For example, some of Jean Perrin's tests of equations of the kinetic theory are exactly of the kind illustrated. Perrin had, for instance, to use one of the equations to be tested to determine a value for a constant (Avogadro's number) it contained.

13. J. Jinks and D. Fulker, "Comparison of the Biometrical, Genetical MAVA and Classical Approaches to the Analysis of Human Behavior," *Psychological Bulletin,* LXXIII, 5 (May 1970). The equations given are taken from p. 316.

14. *Philosophy of Mathematics and Natural Science* (New York: Atheneum, 1963).

15. *Philosophy of Science,* III, 4 (October 1936): 419–471; IV, 1 (January 1937): 1–40; reprinted in H. Feigl and M. Brodbeck, *Readings in the Philosophy of Science* (New York: Appleton-Century-Crofts, 1953).

16. Cf. "Studies in the Logic of Confirmation," *op. cit.*

17. Cf. *The Aim and Structure of Physical Theory* (Princeton, N.J.: University Press, 1954), *passim.*

18. This discussion ignores many historical niceties. Newton assumes, for example, that the center of gravity of the solar system moves inertially, and this assumption, having no experimental support, is presumably just the sort of thing Hempel would call a "proviso." But Newton's argument does not in fact require the assumption. A more careful account of Newton's argument is given in my "Physics and Evidence," to appear in *Pittsburgh Studies in the Philosophy of Science.*

19. For a very qualitative application of the strategy to Freudian theory, see my "Freud, Kepler and the Clinical Evidence," in R. Wollheim, ed., *Freud* (New York: Doubleday, 1975). The explanation I should offer of why the field equations of general relativity are not tested by measurements of the gravitational red shift turns on the imprecision of these measurements and closely follows the account given by John Anderson in his *Principles of Relativity Physics* (New York: Academic Press, 1967), ch. 12.

20. The classic treatment is Gauss, *Theory of the Motion of Heavenly Bodies Moving about the Sun in Conic Sections.* A translation from the Latin is published by Dover, New York, 1963.

21. For this view see, for example, Peter Achinstein, "Inference to Scientific Laws," in R. Stuewer, ed., *Minnesota Studies in the Philosophy of Science,* vol. V (Minneapolis: Univ. of Minnesota Press, 1970), p. 95.

22. Cf. Curtis Wilson, "From Kepler's Laws, So-called, to Universal Gravitation: Empirical Factors," *Archive for the History of Exact Sciences,* VI (1969): 89–170.

23. "An Alternative Approach to the Theory of Gravitation," *Annals of Physics,* XVI (1961): 96–117.

24. Zeldovich and Novikov, *loc. cit.* Thirring makes essentially the same criticism of his own theory. More recent analyses have shown that the theory is in fact inconsistent.

25. *Word and Object* (Cambridge, Mass.: M.I.T. Press, 1960), p. 21.

26. A discussion of this case is given in M. Friedman, *Foundations of Space-Time Theories,* unpublished Ph.D. thesis, Princeton, 1972.

27. Cf. my "Topology, Cosmology and Convention," *Synthese,* XXIV, 2 (August 1972): 195–218.

28. Cf. Franklin Fisher, *The Identification Problem in Econometrics* (New York: McGraw-Hill, 1966).

29. *The Philosophy of Space and Time* (New York: Dover, 1957).

Imre Lakatos

THE ROLE OF CRUCIAL EXPERIMENTS IN SCIENCE*

Exactly how and what do we learn about scientific theories from experiment? The term 'crucial experiment' indicates that from some experiments we learn more than from others. Immense numbers of experiments are never recorded, and of those recorded and published most are forgotten and buried in the annals of science on dusty bookshelves possibly never to be looked at again.

From what class of experiments then do we learn? There are several competing theories of learning which define this class very differently.[1]

According to strict inductivists an experiment is basic (rather than 'crucial') if one can induce some important law of nature from it. Those logicians who recognized the invalidity of inductive generalizations saw the force of 'crucial' experiment in securing the truth of a scientific theory in a different way. Some thought it was possible to enumerate *a priori* all the possible rival theories, and regarded those experiments as 'crucial' which refuted $n-1$ rival theories and thereby proved the n-th. It is *reason* which conjectures and it is *experiment* which disproves and proves. But, as many skeptics pointed out, rival theories are always indefinitely many and therefore the *proving* power of experiment vanishes. One cannot learn from experience about the truth of any scientific theory, only at best about its falsehood: *confirming instances have no epistemic value whatsoever.*

But are all unrefuted theories equally conjectural? According to some they are. According to others, some are more probable than others. Such probability values are calculated by modern inductive logicians by way of definitions of possible states of affairs which, of course, are bound to rely on strong assumptions. Once we agree to these assumptions, each confirming instance may assume some tiny epistemic relevance for the probability of theories.

Another suggested possibility was to withhold the status of proof from any theory, but to assess its verisimilitude by counting the number of its defeated serious rivals: one does not know how much we learn in this way about human imagination (in inventing new alternatives and designing crucial experiments) and how much about nature, for the most counter-intuitive characteristic of this learning theory is that it puts a tremendous premium on the invention of new false alternatives: degree of corroboration may be the hallmark of the perverse inventiveness of the human mind rather than of the theory's nature-dependent verisimilitude.

Can we then at least learn from experiment that some theories are false? I have shown elsewhere and shall argue again later that we cannot. Fries pointed out that no proposition can be proved from facts, but even if we were to *accept as true,* by methodological decision, certain factual propositions, the conventionalists' case for the

*This paper was delivered as a talk at the International Colloquium on the Meaning and Role of Philosophy and Science in Contemporary Society, at the Pennsylvania State University in September 1971. It was prepared on the basis of my [1968*a*], [1968*b*], [1970], [1971*a*] and [1971*b*]. Concerning the points which could here be discussed only cursorily, or in a simplified fashion, the reader might consult the detailed exposition in these papers.

Reprinted by permission from *Studies in History and Philosophy of Science, Vol. 4, no. 4, 1974.*

never-excludable possibility of rescue operations shows disproofs of specific theories to be impossible. We cannot learn from experience the falsehood of any theory.

Two philosophers, Grünbaum and Popper, tried to save falsificationism (and hence falsificationist learning theory) from this *impasse*. Grünbaum ([1969] and [1971]) withdrew considerably in the face of the conventionalists' arguments that theories cannot be conclusively refuted. His only remaining claim seems to be that we can learn about the high probability of the falsehood of *some* scientific theories. And, after enumerating some interesting examples where falsifications were ignored or discussed and revised, he leaves the question of the empirical appraisal of *most* scientific theories *completely open*. Learning from experience is then in disarray. Even, however, to sustain his restricted claim—that *some* scientific theories are falsifiable—he needs to show: (1) that some factual ('basic') propositions are reliable; and (2) that some background knowledge can be so highly probable as to be true 'beyond reasonable doubt'. I do not see how he can do either. Popper, on the other hand, offers a different, and, indeed, *general* solution. He accepts that all scientific propositions, basic or universal, are equally conjectural; he then specifies a 'game of science' by which we 'accept' some of them, 'reject' others. Popper's game of science is governed primarily by the moral maxim that one must not stick to one's theories forever in the light of unfavourable evidence; conventionalism is morally wrong: we *must* learn from experience. But at the end of his classic *Logik der Forschung* he offers a methodology without an epistemology or learning theory, and confesses explicitly that his methodology may lead us epistemologically astray, and implicitly, that *ad hoc* stratagems might lead us to Truth.[2]

Popper's 'game of science' (or 'logic of scientific discovery', or 'methodology', or 'system of appraisals', or 'demarcation criterion', or 'definition of science'[3]) is a set of *standards* for scientific theories.[4]

Popper's logic of discovery contains 'proposals', 'conventions' about when a theory should be taken seriously (when a crucial experiment could, and indeed has been, devised against it) and when a theory should be rejected (when it has failed a test). In Popper's logic of discovery—as in Pascal's, Bernard's, or Grünbaum's—scientific theories are not based on, established, or 'probabilified', by 'facts', but rather eliminated by them. Progress consists of an incessant, ruthless, revolutionary confrontation of bold, speculative theories and repeatable observations, and of the subsequent speedy elimination of the defeated theories: 'The method of trial and error is a *method of eliminating false theories* by observation statements'.[5] 'Conjectures [are] boldly put forward for trial, to be eliminated if they clash with observations'.[6] Thus, the history of science is seen as a series of duels between theory and experiment, duels in which only experiments can score decisive victories. The theoretician proposes a scientific theory; some basic statements contradict it; if one of these becomes 'accepted',[7] the theory is 'refuted' and must be rejected, and a new one has to take its place. 'What ultimately decides the fate of a theory is the result of a test, *i.e.*, an agreement about basic statements'.[8] Popper realizes, of course, that we always test large systems of theories rather than isolated ones. But he does not regard this as an insurmountable difficulty: he suggests that we should *agree* which part of such a system is responsible for the refutation (that is, which part is to be regarded as false), perhaps helped by independent tests of some portions of the system. Within Popper's philosophy this kind of agreement is absolutely indispensable: if one were allowed to blame refutations upon the initial conditions *all the time*, no major theory need ever be rejected.[9] He is not content with tests which are designed to test large systems: he calls on the scientist to specify, beforehand, those experiments which will, if their outcome is negative, lead to the falsification of the very heart of the system.[10] He demands of the scientist that he specify in advance under what experimental conditions he would give up his *most basic* assumptions.[11] This moral demand, indeed, is the gist of Popper's 'demarcation criterion' or, to use another term, of his definition of science.[12]

Popper's definition of science can best be put in terms of 'conventions' or 'rules' governing the '*game of science*'.[13]

The opening move must be a *consistent, falsifiable hypothesis:* that is, a consistent hypothesis which has agreed-on potential falsifiers. A potential falsifier is a 'basic statement' whose truth-value is decidable with the help of the experimental techniques of the time. The scientific jury must agree unanimously that there is an experimental technique which will enable them to assign a truth-value to the 'basic statement'. (Unanimity can, of course, be reached by expelling the minority as pseudo-scientists or cranks.[14])

The next move is the repeated performance of the test in a controlled experiment,[15] and the second decision of the jury on what actual truth-value (truth or falsehood) to attribute to the potential falsifier. (If this second decision is not unanimous, there are two possible moves: either the potential falsifier must be withdrawn and, unless a replacement is found, the opening move cancelled; or, alternatively, the dissenting minority of the jury must be declared cranks and excluded from the jury.[16])

If the second verdict is *negative,* and the potential falsifier is rejected, then the hypothesis is declared 'corroborated', which only means that it invites further challenges. If the second verdict is *positive,* and the potential falsifier accepted, then the hypothesis is declared 'falsified', which means that it is *rejected,* 'overthrown', 'dropped', buried with military honours.[17] (In 1960 Popper introduced a new rule: military pomp can only be awarded to an eliminated hypothesis if, before it was falsified, it was at least once—in a different experiment—corroborated.[18])

After the burial a new hypothesis is invited. This new hypothesis must, however, explain the partial success, if any, of its predecessor, and also something more. A hypothesis, however novel in its intuitive aspects, will not be allowed to be proposed, unless it has novel empirical content in excess of its predecessor. If it has no such excess content, the referee will declare it '*ad hoc*' and make the proposer withdraw it. If the new hypothesis is not *ad hoc,* the standard procedure for falsifiable hypotheses, as described above, is followed for the new hypothesis.[19]

This 'scientific game', if properly played, will 'progress' in the sense that the theories subsequently proposed will have increasing generality (or 'empirical content'); they will pose ever deeper *questions* about the universe.[20]

Just as the rules of chess do not explain why some people should play the game and, indeed, devote their lives to it, the rules of science do not explain why some people should play the game of science and, indeed, devote their lives to it. The rules decide whether a particular *move* is 'proper' (or 'scientific') or not, but they remain silent about whether the *game as a whole* is 'proper' (or 'rational') or not. The rules say nothing either about the (psychological) motives of the players or about the (rational) purpose of the game. One can, of course, play the game as a genuine game and enjoy it for itself, without caring for its purpose or being aware of one's motives.

The rules of the game are *conventions,* and can be formulated in terms of a *definition.*[21] How can one criticize a definition, in particular, if one interprets it nominalistically?[22] A definition is then a mere abbreviation, a tautology. What can one criticize about a tautology? Popper claims that his definition of science is 'fruitful': 'that a great many points can be clarified and explained with its help'. He quotes Menger: 'Definitions are dogmas; only the conclusions drawn from them can afford us any new insight'.[23] But how can a definition have explanatory power or afford new insights? Popper's answer is this: 'It is only from the consequences of my definition of empirical science, and from the methodological decisions which depend upon this definition, that the scientist will be able to see how far it conforms to his intuitive idea of the goal of his endeavours'.[24]

This answer complies with Popper's general position that conventions can be criticized by discussing their 'suitability' relative to some purpose: 'As to the suitability of any convention opinions may differ; and a reasonable discussion of these questions is only possible between parties having some purpose in common. The choice of that purpose . . . goes beyond rational argument'.[25] Popper, in his *Logik der Forschung* never specifies a *purpose* of the game of science that would go beyond what is contained in its rules. The idea that the *aim* of science is *truth,* occurs in his writings for the first time in 1957.[26] In his *Logik der Forschung* the quest for truth may be a psychological *motive* of scientists—it is not a rational *purpose* of science.[27]

Even in Popper's later writings we find no suggestion of how to appraise one consistent set of rules (or demarcation criterion) as leading more successfully towards truth than another. Indeed, the thesis that any such argument connecting method and success is impossible has been a cornerstone of Popper's philosophy from 1920 to 1971. Thus I conclude that Popper never offered a theory of rational criticism of consistent conventions. He does not answer the question: '*Under what conditions would you give up your demarcation criterion*'?[28]

But the question can be answered. I shall give my answer in two stages: first a naive and then a more sophisticated answer. I start by recalling how Popper, according to his own account, had arrived at his criterion. He thought, like the best scientists of his time, that Newton's theory, although refuted, was a wonderful scientific achievement; that Einstein's theory was still better; and that astrology, Freudianism and twentieth-century Marxism were pseudo-scientific. His problem was to find a definition of science from which these '*basic judgements*' concerning each of these theories followed; and he offered a novel solution. Now let us agree *provisionally* on the meta-criterion that a *a rationality theory—or demarcation criterion—is to be rejected if it is inconsistent with accepted 'basic value judgements' of the scientific community.*[29] Indeed, this metamethodological rule would seem to correspond to the falsificationist methodological rule that a scientific theory is to be rejected if it is inconsistent with an ('empirical') basic statement unanimously accepted by the scientific community. Popper's whole methodology rests on the contention that there exist (relatively) singular statements on whose truth-value scientists can reach unanimous agreement; without such agreement there would be a 'new Babel' and 'the soaring edifice of science would soon lie in ruins'.[30] Now even if there is agreement about 'basic' statements, but on the other hand no agreement whatsoever about how to appraise scientific achievement relative to this 'empirical basis', would not the soaring edifice of science equally soon lie in ruins? No doubt it would, Surprisingly, while there has been little agreement concerning a *universal* criterion of the scientific character of theories, there has been considerable agreement over the last two centuries concerning single achievements. While there has been no *general* agreement concerning a theory of scientific rationality, there has been considerable agreement concerning the rationality of a particular step in the game—was it scientific or crankish? A general definition of science thus must reconstruct the acknowledgedly best games and the most esteemed gambits as 'scientific'; if it fails to do so, it has to be rejected.[31]

However, if we apply this meta-criterion (*which I am going to reject later*), the falsificationist demarcation criterion has to be rejected.

The falsificationist demarcation criterion can indeed be easily 'falsified' by showing that in the light of this meta-criterion the best scientific achievements were unscientific and that the best scientists, in their greatest moments, broke the falsificationist rules of science.

In Popper's version of falsificationism the basic rule is that *the scientist must specify in advance under what experimental conditions he will give up even his most basic assumptions:* Criteria of refutation have to be laid down beforehand: it must be agreed which observable situations, if actually observed, mean that the theory is refuted. But what kind of clinical responses would refute to the satisfaction of the analyst *not merely a particular clinical diagnosis but psychoanalysis itself*? And have such criteria even been discussed or agreed upon by analysts?[32] In the case of psychoanalysis Popper was right: no answer has been forthcoming. Freudians have been nonplussed by Popper's basic challenge concerning scientific honesty. They have refused to specify experimental conditions under which they would give up their basic assumptions. For Popper this is the hallmark of their intellectual dishonesty. But what if we put Popper's question to the Newtonian scientist: 'What kind of observation would refute to the satisfaction of the Newtonian, not merely a particular Newtonian explanation, but Newtonian dynamics and gravitational theory itself? And have such criteria even been discussed or agreed upon by Newtonians?' The Newtonian will, alas, scarcely be able to give a positive answer.[33] But then if psychoanalysts are to be condemned as dishonest by Popper's standards, must not Newtonians be similarly condemned?

Popper may certainly withdraw his celebrated challenge and demand falsifiability—and rejection on falsification—only for *systems* of theories, including initial conditions and all sorts of auxiliary and observational theories. This is a very considerable withdrawal, for it allows the imaginative scientist to save his pet theory by suitable lucky alterations in some odd corner of the theoretical maze. But even Popper's mitigated rule will make life impossible for the most brilliant scientist. For in large research programmes there are always known anomalies: normally the researcher puts them aside and follows the positive heuristic of the programme.[34] In general he rivets his attention on the positive heuristic rather than on the distracting anomalies, and hopes that the 'recalcitrant instances' will be turned into confirming instances as the programme progresses. On Popper's terms, even great scientists use forbidden gambits, *ad hoc* stratagems: instead of regarding Mercury's anomalous perihelion as a falsification of the Newtonian theory of our planetary system and thus as a reason for its rejection, most of them shelved it as a problematic instance to be solved at some later stage—or offered *ad hoc* solutions. This methodological attitude of treating as *anomalies* what Popper would regard as *counter examples* is commonly accepted by the best scientists. Some of the research programmes now held in highest esteem by the scientific community progressed in an ocean of anomalies.[35] Rejection of such work by Popper as irrational ('uncritical') implies—at least on our quasi-Polanyiite meta-criterion—a falsification of his definition.

Moreover, for Popper, an inconsistent system does not forbid any observable state of affairs and working on it must invariably be regarded as irrational: 'A self-contradictory system must be rejected . . . [because it] is uninformative . . . No statement is singled out . . . since all are derivable'.[36] But in such cases the best scientists' rule is frequently: '*Allez en avant et la foi vous viendra*.'[37] This anti-Popperian rule secured a sanctuary for the infinitesimal calculus hounded by Bishop Berkeley, and for naïve set theory in the period of the first paradoxes. Indeed, if the game of science had been played according to Popper's rule book, Bohr's 1913 paper would never have been published since it was inconsistently grafted on to Maxwell's theory, and Dirac's delta functions would have been suppressed until Schwartz.

In general, both Popper and Grünbaum stubbornly overestimate the immediate striking force of purely negative criticism, whether empirical or logical. 'Once a mistake, or a contradiction, is pinpointed, there can be no verbal evasion: it can be proved, and that is that'.[38] Grünbaum seems to think that the 'negative result' embodied in the Michelson-Morley experiment played a crucial logical role in the genesis of relativity theory.[39] But I have shown that prior to the emergence of relativity theory the Michelson-Morley experiment was in no 'logical' sense a 'negative result' for classical physics.[40]

This is how some of the 'basic' appraisals of the scientific élite 'falsify' the falsificationist definition of science and falsificationist morality.

I have tried to amend the falsificationist definition of science so that it no longer rules out essential gambits of actual science. I tried to bring about such an amendment, *primarily by shifting the problem of appraising theories to the problem of appraising historical series of theories, or, rather, of 'research programmes', and by changing the falsificationist rules of theory rejection.*

First, *one may 'accept' not only basic but also universal statements as conventions: indeed, this is the most important clue to the continuity of scientific growth.*[41] The basic unit of appraisal must be not an isolated theory or conjunction of theories but rather a *research programme,* with a conventionally accepted (and thus, by provisional decision, 'irrefutable') *hard core* and with a *positive heuristic* which defines problems, foresees anomalies and turns them victoriously into examples according to a preconceived plan. The scientist lists anomalies, but as long as his research programme sustains its momentum, he ignores them. *It is primarily the positive heuristic of his programme, not the anomalies, which dictate the choice of his problems.* Only when the driving force of the positive heuristic weakens, may more attention be given to anomalies. (The methodology of research programmes can explain in this way *the relative autonomy of theoretical science;* disconnected chains of conjectures and refutations cannot.)

(In my approach, then, we learn from experience primarily through a few verifying instances, but learning is a very complicated and theoretical process. For falsificationists learning comes only from negative instances. As Agassi put it in 1964: 'Learning from experience is learning from a refuting instance. The refuting instance then becomes a problematic instance' (p. 201). In 1969 Agassi again emphasizes that 'we learn from experience by refutations' (p. 169), and adds that one can learn *only* from refutation but not from corroboration (p. 167). But this is a very poor theory of learning from experience.[42] Feyerabend [1969] says that '*negative instances suffice in science*'.)

The appraisal of large units like research programmes is in one sense much more liberal and in another much more strict than Popper's appraisal of theories. This new appraisal is *more tolerant* in the sense that it allows a research programme to outgrow infantile diseases, such as inconsistent foundations and occasional *ad hoc* moves. Anomalies, inconsistencies, *ad hoc* strategems, even alleged negative 'crucial' experiments, can be consistent with the overall progress of a research programme. The old rationalist dream of a mechanical, semi-mechanical or at least fast-acting method for showing up falsehood, unprovenness, meaningless rubbish or even irrational choice has to be given up. But this new appraisal is also *more strict* in that it demands not only that a research programme should successfully predict novel facts, but also that the protective belt of its auxiliary hypotheses should be largely built according to a preconceived unifying idea, laid down in advance in the positive heuristic of the research programme.[43]

It is very difficult to decide, especially if one does not demand progress at each single step, when a research programme has degenerated hopelessly; or when one of two rival programmes has achieved a decisive advantage over the other. In this sense there can be no 'instant rationality'. *Neither the logician's proof of inconsistency nor the experimental scientists's verdict of anomaly can defeat a research programme at one blow.* The falsificationist can be 'wise' only after the event if he wants to apply falsificationism to research programmes rather than to isolated theories. Nature may shout 'No' but human ingenuity—contrary to Weyl and Popper[44]—may always be able to shout louder. With sufficient brilliance, and some luck, any theory, even if it is false, can be defended 'progressively' for a long time. Grünbaum admits this now: but then what remains of his falsificationism?[45]

But when should a particular theory, or a whole research programme, be rejected? I claim, only if there is a better one to replace it.[46] Thus I separate Popperian 'falsification' and 'rejection', the conflation of which turned out to be the main weakness of his 'naive falsificationism'.[47] *One learns not by accepting or rejecting one single theory but by comparing one research programme with another for theoretical, empirical and heuristic progress.*[48]

My modification then presents a very different picture of the game of science from Popper's. The best opening gambit is not a falsifiable (and therefore consistent) hypothesis, but a research programme. Mere 'falsifications' (that is, anomalies) are recorded but need not be acted upon. *'Crucial experiments' in the falsificationist sense do not exist:* at best they are honorific titles conferred on certain anomalies *long after the event* when one programme has been defeated by another one. For the falsificationist a crucial experiment is described by an accepted basic statement which is inconsistent with a theory.[49] I, for one, hold that no accepted basic statement alone entitles us to reject a theory. Such a clash may present a problem (major or minor), but in no circumstance a 'victory'. No experiment is crucial at the time it is performed (except perhaps psychologically). The falsificationist pattern of 'conjectures and refutations', that is, the pattern of trial-by-hypothesis followed by error-shown-by-experiment breaks down. A theory can only be eliminated by a *better* theory, that is, by one which has excess empirical content over the corroborated content of its predecessors, some of which is subsequently confirmed. And for this replacement of one theory by a better one, the first theory does not even have to be 'falsified' in the orthodox sense of the term.[50] Thus progress and learning are marked by instances verifying excess content rather than by falsifying instances,[51] and 'falsification' and 'rejection' become logically independent.[52] Popper says explicitly that 'before a theory has been refuted we can never know in what way it may have to be modified'.[53] In my view it is rather the opposite way round: before a theory has been modified we can never know in what way it has been 'refuted', and some of the most interesting modifications are motivated by the 'positive heuristic' of the research programme rather than by anomalies.[54]

Thus I offered a falsification of the falsificationist theory of 'crucial experiments'. But an opponent could claim that the falsification of my own new criterion is not much more difficult than Grünbaum's and Popper's. What about the immediate impact of great crucial experiments, like that of the falsification of the parity principle? Or the long, pedestrian, trial-and-error procedures which occasionally precede the announcement of a major research programme? Will not the judgment of the scientific élite go against my—or, indeed, against *any*—universal rules?

I should like to present my answer in two stages. First, I should like to amend slightly my previously announced provisional metacriterion,[55] and then replace it altogether with a better one.

First, the slight amendment. If a universal rule clashes with a particular 'normative basic judgment', one should allow some time for the scientific community to ponder about the clash: they may give up their particular judgment and submit to the general rule.[56] These 'second-order' falsifications must not be rushed.

Secondly, if we abandon negative crucial experiments in *method,* why stick to it in *metamethod?* We can easily have a second-order methodology of *methodological* (as opposed to scientific) research programmes: the methodology of research programmes self-applied.

While maintaining that a theory of rationality has to try to organize basic value judgments in universal, coherent frameworks, we do not have to reject such a framework immediately, merely because of some anomalies or other inconsistencies. On the other hand, a good rationality theory must anticipate further basic value judgments unexpected in the light of their predecessors or even lead to the revision of previously held basic value judgments. We reject a rationality theory only for a better one, for one which, in this quasi-empirical sense, represents a *progressive shift.* Thus this new—more lenient—metacriterion enables us to compare rival logics of discovery and discern growth in 'metascientific' knowledge.

For instance, the falsificationist theory of scientific rationality need not be seen as 'falsified' simply because it clashes with some actual basic judgments of leading scientists. On the contrary, on our new criterion it represents progress over its justificationist predecessors. For, contrary to these predecessors, it rehabilitated the scientific status of falsified theories like the phlogiston theory, thus reversing a *value judgement* (of inductivist historians) which expelled the latter from the history of science proper into the history of irrational beliefs. Likewise, it reversed the appraisal of the falling star of the 1920s: of the Bohr-Kramers-Slater theory.[57] In the light of most justificationist theories of rationality, the history of science is, at its best, a history of *pre*scientific preludes to some *future* history of science.[58] Falsificationist methodology enabled the historian to interpret more of the actual value judgments (as seen at the time) in the history of science as rational: falsificationism constituted progress compared with inductivism.

On the other hand, I hope that my methodology will be seen, in turn—on the criterion I specified—as a further step forward. For it seems to offer a coherent account of *more* old, isolated basic value judgments as rational; indeed, it has led to new and, at least for the justificationist or naive falsificationist, *surprising* basic value judgments. For instance, for the falsificationist, it becomes *irrational* to approve of (and therefore retain and further elaborate) Newton's gravitational theory after the discovery of Mercury's anomalous perihelion; or it becomes *irrational* to approve of (and therefore boldly develop) Bohr's old quantum theory based on inconsistent foundations: it may even have been irrational to approve of Einstein's early relativity theory, at least without the shock of the Michelson-Morley experiment. From my point of view these were perfectly *rational* developments. According to my theory, unlike that of the falsificationists, Newtonians, Bohr and Einstein were right. Also, as seen from the point of view of my methodology, some rearguard skirmishes for defeated programmes were perfectly rational, and not signs of dogmatic behaviour; and thus it enables us to reverse those standard judgments of later historiography which led to the disappearance of many of these skirmishes from history of science textbooks.[59] Such rearguard skirmishes were previously deleted both by the inductivist and by the falsificationist party histories.

Progress in the theory of rationality happens to be marked by historical discoveries or rediscoveries: by the reconstruction of a growing bulk of value-impregnated history as rational.[60]

I, of course, can easily answer the question when I would give up my criterion of demarcation: when another one is proposed which is better on my metacriterion.[61] (I have not yet answered the question under what circumstances I shall give up my metacriterion; but one must always stop somewhere.)

CONCLUSION

The problem of appraisal of scientific theories (of which the problem of demarcation is a zero-case) is one of the basic problems of the philosophy of science. Its solution determines the normative content of science-learning theory; the outline of our code of intellectual honesty; and also our historiographical outlook. (It also, by the way, determines a specific formulation of the problem of induction.)

There are three major approaches to the solution of this generalized demarcation problem:

(1) One may try to offer a universal demarcation criterion like the ones proposed by probabilists or falsificationists or by the methodology of scientific research programmes. This is Leibnitz's, Carnap's, Popper's, Grünbaum's (and my own) approach.

(2) One may agree that one anomaly may be more conclusive than another; one theory may be better than another; but there is and can be no universal demarcation criterion to decide. Each case has to be dealt with on its own merit and the judgment of authority (of the great scientists) adhered to. This is Polanyi's and Kuhn's approach.[62]

(3) One may deny that any theory is epistemically superior to any other approach; therefore, there are only competing beliefs, some of them called 'scientific'. This cultural relativism originating with ancient skepticism is widely spread now in contemporary anti-science movements; its most articulate expression is to be found in Feyerabend's recent 'epistemological anarchism'.

I view the third approach with horror: I view the second as abject philosophical surrender to authority. Unless we achieve progress in the solution of the generalized problem of demarcation, many branches of science may well degenerate into tribal specializations with standards uncheckable from the outside. This is where I see the most important challenge to the philosophy of science.

NOTES

1. 'Learning from experience' is a normative idea; all the different theories I am going to discuss have normative character. Moreover, all purely empirical learning theories miss the heart of the problem. Also *cf.* my [1970], 123, text to footnote 2.

2. These difficulties may only be 'solved' through a superimposition on this game of some—merely posited—'inductive principle' as I argued in [1968*a*] and [1971*b*]. On the epistemological level there has been no progress in the skeptic-dogmatist controversy since Pyrrho and Hume. In particular, Popper's contribution to the solution of the problem of induction, contrary to his own claim, is nil.

3. This profusion of synonyms has proved to be rather confusing.

4. Incidentally, this problem of standards is altogether alien to 'hermeneutics', so vigorously represented at this conference by Professor Apel.

5. Popper [1963], 56; his own italics.

6. Popper [1963], 46.

7. For the conditions of acceptance of basic statements, *cf.* Popper [1935], section 22, and my [1970], 107–8.

8. Popper [1935], section 30.

9. Grünbaum, I am sure, abhors this Popperian conventionalism. This is why he wishes—to my mind, unsuccessfully—to assign high epistemic value to both basic statements *and* to background knowledge.

10. For references *cf.* footnotes 33 and 47.

11. *Cf.* text to footnote 32. Also *cf.* my [1970], 107.

12. *Cf.* my [1970], 109. For an interesting discussion *cf.* also Musgrave [1968].

13. Popper [1935], section 11 and also 85. The first paragraph in section 11 explains why he gave the title *The Logic of Scientific Discovery* to his book and is worth quoting:

Methodological rules are here regarded as *conventions.* They might be described as the rules of the game of empirical science. They differ from the rules of pure logic rather as do the rules of chess, which few would regard as part of *pure* logic; seeing that the rules of pure logic govern transformations of linguistic formulae, the result of an inquiry into the rules of chess could perhaps be entitled *The Logic of Chess,* but hardly *Logic* pure and simple. (Similarly, the result of an inquiry into the rules of the game of science—that is, of scientific discovery—may be entitled *The Logic of Scientific Discovery.*)

14. I am afraid Popper did not spell out this implication; although he mentions, as if it were a matter of fact, that cranks do not 'seriously disturb the working of various social institutions which have been designed to further scientific objectivity . . .' (Popper [1945], II, 218). Then he goes on: 'Only political power . . . can impair (their) functioning. . . '. (Also *cf.* his [1957*a*], 32.) I wonder.

15. For the concept of 'controlled experiment', *cf.* my [1970], III, footnote 6.

16. *Cf.* footnote 14.

17. Popper [1935], sections 3 and 4.

18. *Cf.* Popper [1963], 242–5.

19. Following Popper's new rule referred to in the previous footnote, the anti-adh cness rules may also be tightened; and we have to distinguish between *ad hoc*$_1$ and *ad hoc*$_2$; *cf.* my [1968*a*], 375–90, especially 389, footnote 1.

20. Popper [1935], section 85, last sentence.

21. *Cf.* Popper [1935], sections 4 and 11.

22. For an excellent discussion of the distinction between nominalism and realism (or, as Popper prefers to call it, 'essentialism') in the theory of definitions, *cf.* Popper [1945], Chapter 11, and [1963], 20.

23. Popper [1935], section 11.

24. *Ibid.*

25. Popper [1935], section 4.

26. Popper [1957*b*].

27. Popper, in 1935, called the search for truth 'the strongest (unscientific) motive' ([1935], section 85).

28. This flaw is the more serious since Popper himself has expressed qualifications about his criterion. For instance, in his [1963] he describes 'dogmatism', that is, treating anomalies as a kind of 'background noise', as something that is 'to some extent necessary' (p. 49). But on the next page he identifies this 'dogmatism' with 'pseudoscience'. Is then pseudoscience 'to some extent necessary?' Also, *cf.* my [1970], 177, footnote 3.

29. 'Basic value judgments' sounds better in German: '*normative Basissätze*'.

30. Popper [1935], section 29.

31. This approach, of course, does not mean that we *believe* that the scientists 'basic judgments' are unfailingly rational; it only means that we *accept* them in order to criticize universal definitions of science. (If we add that no such *universal* criterion has been found and no such *universal* criterion will ever be found, the stage is set for Polanyi's conception of the lawless closed autocracy of science.)

The idea of this meta-criterion may be seen as a 'quasi-empirical' self-application of Popperian falsificationism. I had introduced this 'quasi-empiricalness' earlier in the context of mathematical philosophy. We may abstract from *what* flows in the logical channels of a deductive system, whether it is something certain or something fallible, whether it is truth and falsehood or probability and improbability, or even moral or scientific desirability and undesirability: it is the *how* of the flow which decides whether the system is negativist, 'quasi-empirical', dominated by *modus tollens* or whether it is justificationist, 'quasi-Euclidean', dominated by *modus ponens.* (*Cf.* my [1967]). This 'quasi-empirical' approach may be applied to *any* kind of normative knowledge like ethical or aesthetic, as has already been done by Watkins in his [1963] and [1967]. But now I prefer another approach.

32. Popper [1963], 38, footnote 3; my italics. This, of course, is equivalent to his celebrated 'demarcation criterion' between science and pseudo-science—or, as he put it, 'metaphysics'. (For this point, also *cf.* Agassi [1964], section VI.)

33. *Cf.* my [1970], 100–1.

34. *Cf.* my [1970], especially 135ff.

35. *Ibid.*, 138ff.

36. *Cf.* Popper [1935], section 24.

37. *Cf.* my [1970], especially 140ff.

38. Popper [1959], 394. He adds: 'Frege did

not try evasive manoeuvres when he received Russell's criticism'. But, of course, he did. (*Cf.* Frege's *Postscript* to the second edition of his *Grundgesetze.*) This historiographical mistake may also be related to Popper's earlier overconfidence in the unambiguity of mathematical reasoning. Also *cf.* my [1968*a*], 357, footnote 2.

39. E.g. Grünbaum [1963], ch. 12.

40. *Cf.* my [1970], 159–65. [*Added in press*]: *Cf.* also Zahar [1973], and the discussion that followed it in the *British Journal for the Philosophy of Science* in 1974.

41. Popper does not permit this:

There is a vast difference between my views and conventionalism. I hold that what characterizes the empirical method is just this: our conventions determine the acceptance of the *singular,* not of *universal* statements. (Popper [1935], section 30.)

Grünbaum, too, rejects the idea of treating theories as conventions.

42. *Cf.* my [1970], 121, footnote 1, and 123.

43. In my [1970] I called patched-up developments which did not meet such criteria *ad* hoc_3 strategems. Planck's first correction of the Lummer-Pringsheim formula was *ad hoc* in *this* sense. A particularly good example is Meehl's anomaly (*cf.* my [1970], 175, footnote 3, and 176, footnote 1). This conceptions of '*ad* hoc_3' is partly anticipated by Grünbaum [1964], 1411.

44. Popper [1935], section 85.

45. Grunbaum [1971], 126–7.

46. *Cf.* my [1968*a*], 383–6, my [1968*b*], 162–7, and my [1970], 116ff. and 155ff.

47. *One important consequence is the difference between Popper's and Grünbaum's discussions of the 'Duhem-Quine argument' and mine; cf.* on the one hand Popper [1935], last paragraph of section 18 and section 19, footnote 1; Popper [1957*a*], 131–3; Popper [1963], 112, footnote 26, 238–9 and 243; and Grünbaum, [1960], [1969], and [1971]; and on the other hand, *cf.* my [1970], 184–9.

48. *Cf.* my [1970], 132–8. [*Added in press*]: Also *cf.* Zahar [1973], 99–104.

49. As a consequence of my criticism, Popper withdrew from this position. *Now* he says that only *important 'real'* falsifiers should make us reject a theory. As he recently put it: 'The first *real* discrepancy can refute [a theory]'. But when is an accepted basic statement, inconsistent with a theory, a *real* falsifier? Obviously this is a matter for the scientific *élite* to decide. For instance, according to Popper, Mercury's anomalous perihelion was *not* a 'real' discrepancy. A planet moving in a square would be a 'real' one.

By 1970 Popper had to choose: will he go on searching for a better universal demarcation criterion and accept the methodology of scientific research programmes, or will he become a Polanyiite. He chose the latter. (*Cf.* Popper [1971], 9.) Also *cf.* footnote 62.

50. Popper occasionally—and Feyerabend systematically—stressed the *catalytic* role of alternative theories in devising so-called 'crucial experiments'. But alternatives are not merely catalysts, which can be later removed in the rationed reconstruction, but are *necessary* parts of the falsifying process. (*Cf.* my [1970], 121, footnote 4.)

51. *Cf.* especially my [1970], 120–1.

52. *Cf.* especially my [1968*a*], 385 and my [1970], 121.

53. Popper [1963], 51.

54. *Cf.* especially my [1970], 135–8.

55. *Cf.* above, 364.

56. There is a certain analogy between this pattern and the occasional appeals procedure of the theoretical scientist against the verdict of the experimental jury; *cf.* my [1970], 127–31.

57. Van der Waerden thought that the Bohr–Kramers–Slater theory was bad: Popper's theory showed it to be good. *Cf.* Van der Waerden [1967], 13 and Popper [1963], 242ff.; for a critical discussion, *cf.* my [1970], 168, footnote 4 and 169, footnote 1.

58. The attitude of some modern logicians to the history of mathematics is a typical example; *cf.* my [1963–4], 3.

59. *Cf.* my [1970], section 3(c).

60. There is nothing necessary about this process. I need not say that no rationality theory can or should explain *all* history of science as rational: even the greatest scientists make wrong steps and fail in their judgment.

61. [*Added in press*]: Since this paper was prepared, such a methodology has indeed been proposed: *cf.* Zahar [1973], 99–104.

62. For Popper's recent conversion to Polanyi's position above, *cf.* footnote 49.

REFERENCES

J. Agassi, [1964]: 'Scientific Problems and Their Roots in Metaphysics', in M. Bunge (*ed.*), *The Critical Approach to Science and Philosophy. The Free Press of Glencoe,* New York, pp. 189–211.

J. Agassi, [1969]: 'Popper on Learning from Experience', in Rescher (*ed.*), *Studies in the Philosophy of Science, American Philosophical Quarterly Monograph Series,* pp. 162–172.

P. K. Feyerabend, [1969]: 'A Note on Two "Problems" of Induction', *British Journal for the Philosophy of Science,* **13,** pp. 319–323.

A. Grünbaum, [1960]: 'The Duhemian Argument', *Philosophy of Science,* **27,** pp. 75–87.

A. Grünbaum, [1963]: *Philosophical Problems of Space and Time, Routledge and Kegan Paul,* London.

A. Grünbaum, [1964]: 'The Bearing of Philosophy on History of Science', *Science,* **143,** pp. 1406–1412.

A. Grünbaum, [1969]: 'Can we Ascertain the Falsity of a Scientific Hypothesis?', *Studium Generale,* **22,** pp. 1061–1093.

A. Grünbaum, [1971]: 'Can we Ascertain the Falsity of a Scientific Hypothesis?', in M. Mandelbaum (*ed.*), *Observation and Theory in Science, The Johns Hopkins Press,* Baltimore and London pp. 69–129.

I. Lakatos, [1963–1964]: 'Proofs and Refutations', *British Journal for the Philosophy of Science,* **14,** pp. 1–25, 120–139, 221–243, 296–342.

I. Lakatos, [1967]: 'A Renaissance of Empiricism in the Recent Philosophy of Mathematics', in Lakatos (*ed.*) *Problems in the Philosophy of Mathematics, North Holland,* Amsterdam, pp. 199–202.

I. Lakatos, [1968*a*]: 'Changes in the Problem of Inductive Logic', in Lakatos (*ed.*) *The Problem of Inductive Logic, North Holland, Amsterdam,* pp. 315–417.

I. Lakatos, [1968*b*]: 'Criticism and the Methodology of Scientific Research Programmes', *Proceedings of the Aristotelian Society,* **69,** pp. 149–186.

I. Lakatos, [1970]: 'Falsification and the Methodology of Scientific Research Programmes', in Lakatos and Musgrave (*eds.*) *Criticism and the Growth of Knowledge, Cambridge University Press,* London and New York, pp. 91–195.

I. Lakatos, [1971*a*]: 'History of Science and its Rational Reconstructions', in R.C. Buck and R.S. Cohen (*eds.*), *Boston Studies in the Philosophy of Science,* **8,** *Reidel, Dordrecht,* pp. 91–136.

I. Lakatos, [1971*b*]: 'Popper zum Abgrenzungs–und Induktionsproblem', in H. Lenk (*ed.*), *Neue Aspekte der Wissenschaftstheorie, Vieweg,* Braunschweig, German version of Lakatos [1974].

I. Lakatos, [1974]: 'Popper on Demarcation and Induction', in Schlipp (*ed.*), *The Philosophy of Sir Karl Popper, Open Court,* Lasalle.

A. Musgrave, [1968]: 'On a Demarcation Dispute', in Lakatos and Musgrave (*eds.*) *Problems in the Philosophy of Science, North Holland,* Amsterdam, pp. 78–85.

K. Popper, [1935]: *Logik der Forschung, Julius Springer,* Vienna, expanded English edition, Popper [1959].

K. Popper, [1945]: *The Open Society and Its Enemies, Routledge and Kegan Paul,* London Vol. 2.

K. Popper, [1957*a*]: *The Poverty of Historicism, Routledge and Kegan Paul,* London.

K. Popper, [1957*b*]: 'The Aim of Science', *Ratio* **1,** pp. 24–35.

K. Popper, [1959]: *The Logic of Scientific Discovery, Hutchinson,* London.

K. Popper, [1963]: *Conjectures and Refutations, Routledge and Kegan Paul,* London.

K. Popper, [1971]: 'Conversations with Philosophers—Sir Karl Popper talks about some of his basic ideas with Brian Magee', in *The Listener,* London, 7 January 1971, p. 8–12.

B.L. Van der Waerden, [1967]: *Sources of Quantum Mechanics, North Holland,* Amsterdam.

J.W.N. Watkins, [1963]: 'Negative Utilitarianism', *Aristotelian Society Supplementary Volume,* **37,** pp. 95–114.

J. W. N. Watkins, [1967]: 'Decision and Belief', in R. Hughes (*ed.*), *Decision Making, British Broadcasting Corporation,* London.

E.G. Zahar, [1973]: 'Why did Einstein's Programme Supersede Lorentz's', *The British Journal for the Philosophy of Science,* **24,** pp. 95–123, pp. 223–262.

Thomas S. Kuhn

OBJECTIVITY, VALUE JUDGMENT, AND THEORY CHOICE

Previously unpublished Machette Lecture, delivered at Furman University, 30 November 1973.

In the penultimate chapter of a controversial book first published fifteen years ago, I considered the ways scientists are brought to abandon one time-honored theory or paradigm in favor of another. Such decision problems, I wrote, "cannot be resolved by proof." To discuss their mechanism is, therefore, to talk "about techniques of persuasion, or about argument and counterargument in a situation in which there can be no proof." Under these circumstances, I continued, "lifelong resistance [to a new theory] . . . is not a violation of scientific standards. . . . Though the historian can always find men—Priestley, for instance—who were unreasonable to resist for as long as they did, he will not find a point at which resistance becomes illogical or unscientific."[1] Statements of that sort obviously raise the question of why, in the absence of binding criteria for scientific choice, both the number of solved scientific problems and the precision of individual problem solutions should increase so markedly with the passage of time. Confronting that issue, I sketched in my closing chapter a number of characteristics that scientists share by virtue of the training which licenses their membership in one or another community of specialists. In the absence of criteria able to dictate the choice of each individual, I argued, we do well to trust the collective judgment of scientists trained in this way. "What better criterion could there be," I asked rhetorically, "than the decision of the scientific group?"[2]

A number of philosophers have greeted remarks like these in a way that continues to surprise me. My views, it is said, make of theory choice "a matter of mob psychology."[3] Kuhn believes, I am told, that "the decision of a scientific group to adopt a new paradigm cannot be based on good reasons of any kind, factual or otherwise."[4] The debates surrounding such choices must, my critics claim, be for me "mere persuasive displays without deliberative substance."[5] Reports of this sort manifest total misunderstanding, and I have occasionally said as much in papers directed primarily to other ends. But those passing protestations have had negligible effect, and the misunderstandings continue to be important. I conclude that it is past time for me to describe, at greater length and with greater precision, what has been on my mind when I have uttered statements like the ones with which I just began. If I have been reluctant to do so in the past, that is largely because I have

Reprinted by permission from Thomas S. Kuhn *The Essential Tension: Selected Studies in the Scientific Tradition and Change,* 1974.

preferred to devote attention to areas in which my views diverge more sharply from those currently received than they do with respect to theory choice.

What, I ask to begin with, are the characteristics of a good scientific theory? Among a number of quite usual answers I select five, not because they are exhaustive, but because they are individually important and collectively sufficiently varied to indicate what is at stake. First, a theory should be accurate: within its domain, that is, consequences deducible from a theory should be in demonstrated agreement with the results of existing experiments and observations. Second, a theory should be consistent, not only internally or with itself, but also with other currently accepted theories applicable to related aspects of nature. Third, it should have broad scope: in particular, a theory's consequences should extend far beyond the particular observations, laws, or subtheories it was initially designed to explain. Fourth, and closely related, it should be simple, bringing order to phenomena that in its absence would be individually isolated and, as a set, confused. Fifth—a somewhat less standard item, but one of special importance to actual scientific decisions—a theory should be fruitful of new research findings: it should, that is, disclose new phenomena or previously unnoted relationships among those already known.[6] These five characteristics—accuracy, consistency, scope, simplicity, and fruitfulness—are all standard criteria for evaluating the adequacy of a theory. If they had not been, I would have devoted far more space to them in my book, for I agree entirely with the traditional view that they play a vital role when scientists must choose between an established theory and an upstart competitor. Together with others of much the same sort, they provide *the* shared basis for theory choice.

Nevertheless, two sorts of difficulties are regularly encountered by the men who must use these criteria in choosing, say, between Ptolemy's astronomical theory and Copernicus's, between the oxygen and phlogiston theories of combustion, or between Newtonian mechanics and the quantum theory. Individually the criteria are imprecise: individuals may legitimately differ about their application to concrete cases. In addition, when deployed together, they repeatedly prove to conflict with one another; accuracy may, for example, dictate the choice of one theory, scope the choice of its competitor. Since these difficulties, especially the first, are also relatively familiar, I shall devote little time to their elaboration. Though my argument does demand that I illustrate them briefly, my views will begin to depart from those long current only after I have done so.

Begin with accuracy, which for present purposes I take to include not only quantitative agreement but qualitative as well. Ultimately it proves the most nearly decisive of all the criteria, partly because it is less equivocal than the others but especially because predictive and explanatory powers, which depend on it, are characteristics that scientists are particularly unwilling to give up. Unfortunately, however, theories cannot always be discriminated in terms of accuracy. Copernicus's system, for example, was not more accurate than Ptolemy's until drastically revised by Kepler more than sixty years after Copernicus's death. If Kepler or someone else had not found other reasons to choose heliocentric astronomy, those improvements in accuracy would never have been made, and Copernicus's work might have been forgotten. More typically, of course, accuracy does permit discriminations, but not the sort that lead regularly to unequivocal choice. The oxygen theory, for example, was universally acknowledged to account for observed weight relations in chemical reactions, something the phlogiston theory had previously scarcely attempted to do. But the phlogiston theory, unlike its rival, could account for the metals' being much more alike than the ores from which they were formed. One theory thus matched experience better in one area, the other in another. To choose between them on the basis of accuracy, a scientist would need to decide the area in which accuracy was more significant. About that matter chemists could and did differ without violating any of the criteria outlined above, or any others yet to be suggested.

However important it may be, therefore, accuracy by itself is seldom or never a sufficient criterion for theory choice. Other criteria must function as well, but they do not eliminate problems. To illustrate I select just two—consistency and simplicity—asking how they functioned in the choice between the heliocentric and geocentric systems. As astronomical theories both Ptolemy's and Copernicus's were internally consistent, but their relation to related theories in other fields was very different. The stationary central earth was an essential ingredient of received physical theory, a tight-knit body of doctrine which explained, among other things, how stones fall, how water pumps function, and why the clouds move slowly across the skies. Heliocentric astronomy, which required the earth's motion, was inconsistent with the existing scientific explanation of these and other terrestrial phenomena. The consistency criterion, by itself, therefore, spoke unequivocally for the geocentric tradition.

Simplicity, however, favored Copernicus, but only when evaluated in a quite special way. If, on the one hand, the two systems were compared in terms of the actual computational labor required to predict the position of a planet at a particular time, then they proved substantially equivalent. Such computations were what astronomers did, and Copernicus's system offered them no labor-saving techniques; in that sense it was not simpler than Ptolemy's. If, on the other hand, one asked about the amount of mathematical apparatus required to explain, not the detailed quantitative motions of the planets, but merely their gross qualitative features—limited elongation, retrograde motion and the like—then, as every schoolchild knows, Copernicus required only one circle per planet, Ptolemy two. In that sense the Copernican theory was the simpler, a fact vitally important to the choices made by both Kepler and Galileo and thus essential to the ultimate triumph of Copernicanism. But that sense of simplicity was not the only one available, nor even the one most natural to professional astronomers, men whose task was the actual computation of planetary position.

Because time is short and I have multiplied examples elsewhere, I shall here simply assert that these difficulties in applying standard criteria of choice are typical and that they arise no less forcefully in twentieth-century situations than in the earlier and better-known examples I have just sketched. When scientists must choose between competing theories, two men fully committed to the same list of criteria for choice may nevertheless reach different conclusions. Perhaps they interpret simplicity differently or have different convictions about the range of fields within which the consistency criterion must be met. Or perhaps they agree about these matters but differ about the relative weights to be accorded to these or to other criteria when several are deployed together. With respect to divergences of this sort, no set of choice criteria yet proposed is of any use. One can explain, as the historian characteristically does, why particular men made particular choices at particular times. But for that purpose one must go beyond the list of shared criteria to characteristics of the individuals who make the choice. One must, that is, deal with characteristics which vary from one scientist to another without thereby in the least jeopardizing their adherence to the canons that make science scientific. Though such canons do exist and should be discoverable (doubtless the criteria of choice with which I began are among them), they are not by themselves sufficient to determine the decisions of individual scientists. For that purpose the shared canons must be fleshed out in ways that differ from one individual to another.

Some of the differences I have in mind

result from the individual's previous experience as a scientist. In what part of the field was he at work when confronted by the need to choose? How long had he worked there; how successful had he been; and how much of his work depended on concepts and techniques challenged by the new theory? Other factors relevant to choice lie outside the sciences. Kepler's early election of Copernicanism was due in part to his immersion in the Neoplatonic and Hermetic movements of his day; German Romanticism predisposed those it affected toward both recognition and acceptance of energy conservation; nineteenth-century British social thought had a similar influence on the availability and acceptability of Darwin's concept of the struggle for existence. Still other significant differences are functions of personality. Some scientists place more premium than others on originality and are correspondingly more willing to take risks; some scientists prefer comprehensive, unified theories to precise and detailed problem solutions of apparently narrower scope. Differentiating factors like these are described by my critics as subjective and are contrasted with the shared or objective criteria from which I began. Though I shall later question that use of terms, let me for the moment accept it. My point is, then, that every individual choice between competing theories depends on a mixture of objective and subjective factors, or of shared and individual criteria. Since the latter have not ordinarily figured in the philosophy of science, my emphasis upon them has made my belief in the former hard for my critics to see.

What I have said so far is primarily simply descriptive of what goes on in the sciences at times of theory choice. As description, furthermore, it has not been challenged by my critics, who reject instead my claim that these facts of scientific life have philosophic import. Taking up that issue, I shall begin to isolate some, though I think not vast, differences of opinion. Let me begin by asking how philosophers of science can for so long have neglected the subjective elements which, they freely grant, enter regularly into the actual theory choices made by individual scientists? Why have these elements seemed to them an index only of human weakness, not at all of the nature of scientific knowledge?

One answer to that question is, of course, that few philosophers, if any, have claimed to possess either a complete or an entirely well-articulated list of criteria. For some time, therefore, they could reasonably expect that further research would eliminate residual imperfections and produce an algorithm able to dictate rational, unanimous choice. Pending that achievement, scientists would have no alternative but to supply subjectively what the best current list of objective criteria still lacked. That some of them might still do so even with a perfected list at hand would then be an index only of the inevitable imperfection of human nature.

That sort of answer may still prove to be correct, but I think no philosopher still expects that it will. The search for algorithmic decision procedures has continued for some time and produced both powerful and illuminating results. But those results all presuppose that individual criteria of choice can be unambiguously stated and also that, if more than one proves relevant, an appropriate weight function is at hand for their joint application. Unfortunately, where the choice at issue is between scientific theories, little progress has been made toward the first of these desiderata and none toward the second. Most philosophers of science would, therefore, I think, now regard the sort of algorithm which has traditionally been sought as a not quite attainable ideal. I entirely agree and shall henceforth take that much for granted.

Even an ideal, however, if it is to remain credible, requires some demonstrated relevance to the situations in which it is supposed to apply. Claiming that such demonstration requires no recourse to subjective factors, my critics seem to appeal, implicitly or explicitly, to the well-known distinction between the contexts of discovery and of justification.[7] They concede, that is, that the subjective factors I invoke play a significant role in the discovery or invention of new theories, but they also insist that that inevitably intuitive process lies outside of the bounds of philosophy of science and is irrelevant to the question of scientific objectivity. Objectivity enters science, they continue, through the processes by which theories are tested, justified, or judged. Those processes do not, or at least need not, involve subjective factors at all. They can be governed by a set of (objective) criteria shared by the entire group competent to judge.

I have already argued that that position does not fit observations of scientific life and shall now assume that that much has been conceded. What is now at issue is a different point: whether or not this invocation of the distinction between contexts of discovery and of justification provides even a plausible and useful idealization. I think it does not and can best make my point by suggesting first a likely source of its apparent cogency. I suspect that my critics have been misled by science pedagogy or what I have elsewhere called textbook science. In science teaching, theories are presented together with exemplary applications, and those applications may be viewed as evidence. But that is not their primary pedagogic function (science students are distressingly willing to receive the word from professors and texts). Doubtless *some* of them were *part* of the evidence at the time actual decisions were being made, but they represent only a fraction of the considerations relevant to the decision process. The context of pedagogy differs almost as much from the context of justification as it does from that of discovery.

Full documentation of that point would require longer argument than is appropriate here, but two aspects of the way in which philosophers ordinarily demonstrate the relevance of choice criteria are worth noting. Like the science textbooks on which they are often modelled, books and articles on the philosophy of science refer again and again to the famous crucial experiments: Foucault's pendulum, which demonstrates the motion of the earth; Cavendish's demonstration of gravitational attraction; or Fizeau's measurement of the relative speed of sound in water and air. These experiments are paradigms of good reason for scientific choice; they illustrate the most effective of all the sorts of argument which could be available to a scientist uncertain which of two theories to follow; they are vehicles for the transmission of criteria of choice. But they also have another characteristic in common. By the time they were performed no scientist still needed to be convinced of the validity of the theory their outcome is now used to demonstrate. Those decisions had long since been made on the basis of significantly more equivocal evidence. The exemplary crucial experiments to which philosophers again and again refer would have been historically relevant to theory choice only if they had yielded unexpected results. Their use as illustrations provides needed economy to science pedagogy, but they scarcely illuminate the character of the choices that scientists are called upon to make.

Standard philosophical illustrations of scientific choice have another troublesome characteristic. The only arguments discussed are, as I have previously indicated, the ones favorable to the theory that, in fact, ultimately triumphed. Oxygen, we read, could explain weight relations, phlogiston could not; but nothing is said about the phlogiston theory's power or about the oxygen theory's limitations. Comparisons of Ptolemy's theory with Copernicus's proceed in the same way. Perhaps these examples should not be given since they contrast a developed theory with one still in its infancy. But philosophers regularly use them nonetheless. If the only result of their doing so were to simplify the decision situation, one could not object. Even historians do not claim to deal with the full factual complexity of the situations they describe. But these simplifications emasculate by making choice totally unproblematic. They eliminate, that is, one essential element of the decision situations that scientists must resolve if their field is to move ahead. In those situations there are always at least some good reasons for each possible choice. Considerations relevant to the context of discovery are then relevant to justification as well; scientists who share the concerns and sensibilities of the individual who discovers a new theory are ipso facto likely to appear disproportionately frequently among that theory's first supporters. That is why it has been difficult to construct algorithms for theory choice, and also why such difficulties have seemed so thoroughly worth resolving. Choices that present problems are the ones philosophers of science need to understand. Philosophically interesting decision procedures must function where, in their absence, the decision might still be in doubt.

That much I have said before, if only briefly. Recently, however, I have recognized another, subtler source for the apparent plausibility of my critics' position. To present it, I shall briefly describe a hypothetical dialogue with one of them. Both of us agree that each scientist chooses between competing theories by deploying some Bayesian algorithm which permits him to compute a value for $p(T,E)$, i.e., for the probability of a theory T on the evidence E available both to him and to the other members of his professional group at a particular period of time. "Evidence," furthermore, we both interpret broadly to include such considerations as simplicity and fruitfulness. My critic asserts, however, that there is only one such value of p, that corresponding to objective choice, and he believes that all rational members of the group must arrive at it. I assert, on the other hand, for reasons previously given, that the factors he calls objective are insufficient to determine in full any algorithm at all. For the sake of the discussion I have conceded that each individual has an algorithm and that all their algorithms have much in common. Nevertheless, I continue to hold that the algorithms of individuals are all ultimately different by virtue of the subjective considerations with which each must complete the objective criteria before any computations can be done. If my hypothetical critic is liberal, he may now grant that these subjective differences do play a role in determining the hypothetical algorithm on which each individual relies during the early stages of the competition between rival theories. But he is also likely to claim that, as evidence increases with the passage of time, the algorithms of different individuals converge to the algorithm of objective choice with which his presentation began. For him the increasing unanimity of individual choices is evidence for their increasing objectivity and thus for the elimination of subjective elements from the decision process.

So much for the dialogue, which I have, of course, contrived to disclose the non sequitur underlying an apparently plausible position. What converges as the evidence changes over time need only be the values of p that individuals compute from their individual algorithms. Conceivably those algorithms themselves also become more alike with time, but the ultimate unanimity of theory choice provides no evidence whatsoever that they do so. If subjective factors are required to account for the decisions that initially divide the profession, they may still be present later when the profession agrees. Though I shall not here argue the point, consideration of the occasions on which a scientific community divides suggests that they actually do so.

My argument has so far been directed to two points. It first provided evidence that the choices scientists make between competing theories depend not only on shared criteria—those my critics call objective—but also on idiosyncratic factors dependent on individual biography and personality. The latter are, in my critics' vocabulary, subjective, and the second part of my argument has attempted to bar some likely ways of denying their philosophic import. Let me now shift to a more positive approach, returning briefly to the list of shared criteria—accuracy, simplicity, and the like—with which I began. The considerable effectiveness of such criteria does not, I now wish to suggest, depend on their being sufficiently articulated to dictate the choice of each individual who subscribes to them. Indeed, if they were articulated to that extent, a behavior mechanism fundamental to scientific advance would cease to function. What the tradition sees as eliminable imperfections in its rules of choice I take to be in part responses to the essential nature of science.

As so often, I begin with the obvious. Criteria that influence decisions without specifying what those decisions must be are familiar in many aspects of human life. Ordinarily, however, they are called, not criteria or rules, but maxims, norms, or values. Consider maxims first. The individual who invokes them when choice is urgent usually finds them frustratingly vague and often also in conflict one with another. Contrast "He who hesitates is lost" with "Look before you leap," or compare "Many hands make light work" with "Too many cooks spoil the broth." Individually maxims dictate different choices, collectively none at all. Yet no one suggests that supplying children with contradictory tags like these is irrelevant to their education. Opposing maxims alter the nature of the decision to be made, highlight the essential issues it presents, and point to those remaining aspects of the decision for which each individual must take responsibility himself. Once invoked, maxims like these alter the nature of the decision process and can thus change its outcome.

Values and norms provide even clearer examples of effective guidance in the presence of conflict and equivocation. Improving the quality of life is a value, and a car in every garage once followed from it as a norm. But quality of life has other aspects, and the old norm has become problematic. Or again, freedom of speech is a value, but so is preservation of life and property. In application, the two often conflict, so that judicial soul-searching, which still continues, has been required to prohibit such behavior as inciting to riot or shouting fire in a crowded theater. Difficulties like these are an appropriate source for frustration, but they rarely result in charges that values have no function or in calls for their abandonment. That response is barred to most of us by an acute consciousness that there are societies with other values and that these value differences result in other ways of life, other decisions about what may and what may not be done.

I am suggesting, of course, that the criteria of choice with which I began function not as rules, which determine choice, but as values, which influence it. Two men deeply committed to the same values may nevertheless, in particular situations, make different choices as, in fact, they do. But that difference in outcome ought not to suggest that the values scientists share are less than critically important either to their decisions or to the development of the enterprise in which they participate. Values like accuracy, consistency, and scope may prove ambiguous in application, both individually and collectively; they may, that is, be an insufficient basis for a *shared* algorithm of choice. But they do specify a great deal: what each scientist must consider in reaching a decision, what he may and may not consider relevant, and what he can legitimately be required to report as the basis for the choice he has made. Change the list, for example by adding social utility as a criterion, and some particular choices will be different, more like those one expects from an engineer. Subtract accuracy of fit to nature from the list, and the enterprise that results may not resemble science at all, but perhaps philosophy instead. Different creative disciplines are characterized, among other things, by different sets of shared values. If philosophy and engineering lie too close to the sciences, think of literature or the plastic arts. Milton's failure to set *Paradise Lost* in a Copernican universe does not indicate that he agreed with Ptolemy but that he had things other than science to do.

Recognizing that criteria of choice can function as values when incomplete as rules has, I think, a number of striking advantages. First, as I have already argued at length, it accounts in detail for aspects of scientific behavior which the tradition has seen as anomalous or even irrational. More important, it allows the standard criteria to function fully in the earliest stages of theory choice, the period when they are most needed but when, on the traditional view, they function badly or not at all. Copernicus was responding to them during the years required to convert heliocentric astronomy from a global conceptual scheme to mathematical machinery for predicting planetary position. Such predictions were what astronomers valued; in their absence, Copernicus would scarcely have been heard, something which had happened to the idea of a moving earth before. That his own version convinced very few is less important than his acknowledgment of the basis on which judgments would have to be reached if heliocentricism were to survive. Though idiosyncrasy must be invoked to explain why Kepler and Galileo were early converts to Copernicus's system, the gaps filled by their efforts to perfect it were specified by shared values alone.

That point has a corollary which may be more important still. Most newly suggested theories do not survive. Usually the difficulties that evoked them are accounted for by more traditional means. Even when this does not occur, much work, both theoretical and experimental, is ordinarily required before the new theory can display sufficient accuracy and scope to generate widespread conviction. In short, before the group accepts it, a new theory has been tested over time by the research of a number of men, some working within it, others within its traditional rival. Such a mode of development, however, *requires* a decision process which permits rational men to disagree, and such disagreement would be barred by the shared algorithm which philosophers have generally sought. If it were at hand, all conforming scientists would make the same decision at the same time. With standards for acceptance set too low, they would move from one attractive global viewpoint to another, never giving traditional theory an opportunity to supply equivalent attractions. With standards set higher, no one satisfying the criterion of rationality would be inclined to try out the new theory, to articulate it in ways which showed its fruitfulness or displayed its accuracy and scope. I doubt that science would survive the change. What from one viewpoint may seem the looseness and imperfection of choice criteria conceived as rules may, when the same criteria are seen as values, appear an indispensable means of spreading the risk which the introduction or support of novelty always entails.

Even those who have followed me this far will want to know how a value-based enterprise of the sort I have described can develop as a science does, repeatedly producing powerful new techniques for prediction and control. To that question, unfortunately, I have no answer at all, but that is only another way of saying that I make no claim to have solved the problem of induction. If science did progress by virtue of some shared and binding algorithm of choice, I would be equally at a loss to explain its success. The lacuna is one I feel acutely, but its presence does not differentiate my position from the tradition.

It is, after all, no accident that my list of the values guiding scientific choice is, as nearly as makes any difference, identical with the tradition's list of rules dictating choice. Given any concrete situation to which the philosopher's rules could be applied, my values would function like his rules, producing the same choice. Any justification of induction, any explanation of why the rules worked, would apply equally to my values. Now, consider a situation in which choice by shared rules proves impossible, not because the rules are wrong but because they are, as rules, intrinsically incomplete. Individuals must then still choose and be guided by the rules (now values) when they do so. For that purpose, however, each must first flesh out the rules, and each will do so in a somewhat different way even though the decision dictated by the variously completed rules may prove unanimous. If I now assume, in addition, that the group is large enough so that individual differences distribute on some normal curve, then any argument that justifies the philosopher's choice by rule should be immediately adaptable to my choice by value. A group too small, or a distribution excessively skewed by external historical pressures, would, of course, prevent the argument's transfer.[8] But those are just the circumstances under which scientific progress is itself problematic. The transfer is not then to be expected.

I shall be glad if these references to a normal distribution of individual differences and to the problem of induction make my position appear very close to more traditional views. With respect to theory choice, I have never thought my departures large and have been correspondingly startled by such charges as "mob psychology," quoted at the start. It is worth noting, however, that the positions are not quite identical, and for that purpose an analogy may be helpful. Many properties of liquids and gases can be accounted for on the kinetic theory by supposing that all molecules travel at the same speed. Among such properties are the regularities known as Boyle's and Charles's law. Other characteristics, most obviously evaporation, cannot be explained in so simple a way. To deal with them one must assume that molecular speeds differ, that they are distributed at random, governed by the laws of chance. What I have been suggesting here is that theory choice, too, can be explained only in part by a theory which attributes the same properties to all the scientists who must do the choosing. Essential aspects of the process generally known as verification will be understood only by recourse to the features with respect to which men may differ while still remaining scientists. The tradition takes it for granted that such features are vital to the process of discovery, which it at once and for that reason rules out of philosophical bounds. That they may have significant functions also in the philosophically central problem of justifying theory choice is what philosophers of science have to date categorically denied.

What remains to be said can be grouped in a somewhat miscellaneous epilogue. For the sake of clarity and to avoid writing a book, I have throughout this paper utilized some traditional concepts and locutions about the viability of which I have elsewhere expressed serious doubts. For those who know the work in which I have done so, I close by indicating three aspects of what I have said which would better represent my views if cast in other terms, simultaneously indicating the main directions in which such recasting should proceed. The areas I have in mind are: value invariance, subjectivity, and partial communication. If my views of scientific development are novel—a matter about which there is legitimate room for doubt—it is in areas such as these, rather than theory choice, that my main departures from tradition should be sought.

Throughout this paper I have implicitly assumed that, whatever their initial source, the criteria or values deployed in theory choice are fixed once and for all, unaffected by their participation in transitions from one theory to another. Roughly speaking, but only very roughly, I take that to be the case. If the list of relevant values is kept short (I have mentioned five, not all independent) and if their specification is left vague, then such values as accuracy, scope, and fruitfulness are permanent attributes of science. But little knowledge of history is required to suggest that both the application of these values and, more obviously, the relative weights attached to them have varied markedly with time and also with the field of application. Furthermore, many of these variations in value have been associated with particular changes in scientific theory. Though the experience of scientists provides no philosophical justification for the values they deploy (such justification would solve the problem of induction), those values are in part learned from that experience, and they evolve with it.

The whole subject needs more study (historians have usually taken scientific values, though not scientific methods, for granted), but a few remarks will illustrate the sort of variations I have in mind. Accuracy, as a value, has with time increasingly denoted quantitative or numerical agreement, sometimes at the expense of qualitative. Before early modern times, however, accuracy in that sense was a criterion only for astronomy, the science of the celestial region. Elsewhere it was neither expected nor sought. During the seventeenth century, however, the criterion of numerical agreement was extended to mechanics, during the late eighteenth and early nineteenth centuries to chemistry and such other subjects as electricity and heat, and in this century to many parts of biology. Or think of utility, an item of value not on my initial list. It too has figured significantly in scientific development, but far more strongly and steadily for chemists than for, say, mathematicians and physicists. Or consider scope. It is still an important scientific value, but important scientific advances have repeatedly been achieved at its expense, and the weight attributed to it at times of choice has diminished correspondingly.

What may seem particularly troublesome about changes like these is, of course, that they ordinarily occur in the aftermath of a theory change. One of the objections to Lavoisier's new chemistry was the roadblocks with which it confronted the achievement of what had previously been one of chemistry's traditional goals: the explanation of qualities, such as color and texture, as well as of their changes. With the acceptance of Lavoisier's theory such explanations ceased for some time to be a value for chemists; the ability to explain qualitative variation was no longer a criterion relevant to the evaluation of chemical theory. Clearly, if such value changes had occurred as rapidly or been as complete as the theory changes to which they related, then theory choice would be value choice, and neither could provide justification for the other. But, historically, value change is ordinarily a belated and largely unconscious concomitant of theory choice, and the former's magnitude is regularly smaller than the latter's. For the functions I have here ascribed to values, such relative stability provides a sufficient basis. The existence of a feedback loop through which theory change affects the values which led to that change does not make the decision process circular in any damaging sense.

About a second respect in which my resort to tradition may be misleading, I must be far more tentative. It demands the skills of an ordinary language philosopher, which I do not possess. Still, no very acute ear for language is required to generate discomfort with the ways in which the terms "objectivity" and, more especially, "subjectivity" have functioned in this paper. Let me briefly suggest the respects in which I believe language has gone astray. "Subjective" is a term with several established uses: in one of these it is opposed to "objective," in another to "judgmental." When my critics describe the idiosyncratic features to which I appeal as subjective, they resort, erroneously I think, to the second of these senses. When they complain that I deprive science of objectivity, they conflate that second sense of subjective with the first.

A standard application of the term "subjective" is to matters of taste, and my critics appear to suppose that that is what I have made of theory choice. But they are missing a distinction standard since Kant when they do so. Like sensation reports, which are also subjective in the sense now at issue, matters of taste are undiscussable. Suppose that, leaving a movie theater with a friend after seeing a western, I exclaim: "How I liked that terrible potboiler!" My friend, if he disliked the film, may tell me I have low tastes, a matter about which, in these circumstances, I would readily agree. But, short of saying that I lied, he cannot disagree with my report that I liked the film or try to persuade me that what I said about my reaction was wrong. What is discussable in my remark is not my characterization of my internal state, my exemplification of taste, but rather my *judgment* that the film was a potboiler. Should my friend disagree on that point, we may argue most of the night, each comparing the film with good or great ones we have seen, each revealing, implicitly or explicitly, something about how he *judges* cinematic merit, about his aesthetic. Though one of us may, before retiring, have persuaded the other, he need not have done so to demonstrate that our difference is one of judgment, not taste.

Evaluations or choices of theory have, I think, exactly this character. Not that scientists never say merely, I like such and such a theory, or I do not. After 1926 Einstein said little more than that about his opposition to the quantum theory. But scientists may always be asked to explain their choices, to exhibit the bases for their judgments. Such judgments are eminently discussable, and the man who refuses to discuss his own cannot expect to be taken seriously. Though there are, very occasionally, leaders of scientific taste, their existence tends to prove the rule. Einstein was one of the few, and his increasing isolation from the scientific community in later life shows how very limited a role taste alone can play in theory choice. Bohr, unlike Einstein, did discuss the bases for his judgment, and he carried the day. If my critics introduce the term "subjective" in a sense that opposes it to judgmental—thus suggesting that I make theory choice undiscussable, a matter of taste—they have seriously mistaken my position.

Turn now to the sense in which "subjectivity" is opposed to "objectivity," and note first that it raises issues quite separate from those just discussed. Whether my taste is low or refined, my report that I liked the film is objective unless I have lied. To my judgment that the film was a potboiler, however, the objective-subjective distinction does not apply at all, at least not obviously and directly. When my critics say I deprive theory choice of objectivity, they must, therefore, have recourse to some very different sense of subjective, presumably the one in which bias and personal likes or dislikes function instead of, or in the face of, the actual facts. But that sense of subjective does not fit the process I have been describing any better than the first. Where factors dependent on individual biography or personality must be introduced to make values applicable, no standards of factuality or actuality are being set aside. Conceivably my discussion of theory choice indicates some limitations of objectivity, but not by isolating elements properly called subjective. Nor am I even quite content with the notion that what I have been displaying are limitations. Objectivity ought to be analyzable in terms of criteria like accuracy and consistency. If these criteria do not supply all the guidance that we have customarily expected of them, then it may be the meaning rather than the limits of objectivity that my argument shows.

Turn, in conclusion, to a third respect, or set of respects, in which this paper needs to be recast. I have assumed throughout that the discussions surrounding theory choice are unproblematic, that the facts appealed to in such discussions are independent of theory, and that the discussions' outcome is appropriately called a choice. Elsewhere I have challenged all three of these assumptions, arguing that communication between proponents of different theories is inevitably partial, that what each takes to be facts depends in part on the theory he espouses, and that an individual's transfer of allegiance from theory to theory is often better described as conversion than as choice.

Though all these theses are problematic as well as controversial, my commitment to them is undiminished. I shall not now defend them, but must at least attempt to indicate how what I have said here can be adjusted to conform with these more central aspects of my view of scientific development.

For that purpose I resort to an analogy I have developed in other places. Proponents of different theories are, I have claimed, like native speakers of different languages. Communication between them goes on by translation, and it raises all translation's familiar difficulties. That analogy is, of course, incomplete, for the vocabulary of the two theories may be identical, and most words function in the same ways in both. But some words in the basic, as well as in the theoretical vocabularies of the two theories—words like "star" and "planet," "mixture" and "compound," or "force" and "matter"—do function differently. Those differences are unexpected and will be discovered and localized, if at all, only by repeated experience of communication breakdown. Without pursuing the matter further, I simply assert the existence of significant limits to what the proponents of different theories can communicate to one another. The same limits make it difficult or, more likely, impossible for an individual to hold both theories in mind together and compare them point by point with each other and with nature. That sort of comparison is, however, the process on which the appropriateness of any word like "choice" depends.

Nevertheless, despite the incompleteness of their communication, proponents of different theories can exhibit to each other, not always easily, the concrete technical results achievable by those who practice within each theory. Little or no translation is required to apply at least some value criteria to those results. (Accuracy and fruitfulness are most immediately applicable, perhaps followed by scope. Consistency and simplicity are far more problematic.) However incomprehensible the new theory may be to the proponents of tradition, the exhibit of impressive concrete results will persuade at least a few of them that they must discover how such results are achieved. For that purpose they must learn to translate, perhaps by treating already published papers as a Rosetta stone or, often more effective, by visiting the innovator, talking with him, watching him and his students at work. Those exposures may not result in the adoption of the theory; some advocates of the tradition may return home and attempt to adjust the old theory to produce equivalent results. But others, if the new theory is to survive, will find that at some point in the language-learning process they have ceased to translate and begun instead to speak the language like a native. No process quite like choice has occurred, but they are practicing the new theory nonetheless. Furthermore, the factors that have led them to risk the conversion they have undergone are just the ones this paper has underscored in discussing a somewhat different process, one which, following the philosophical tradition, it has labelled theory choice.

NOTES

1. *The Structure of Scientific Revolutions,* 2d ed. (Chicago, 1970), pp. 148, 151–52, 159. All the passages from which these fragments are taken appeared in the same form in the first edition, published in 1962.

2. Ibid., p. 170.

3. Imre Lakatos, "Falsification and the Methodology of Scientific Research Programmes," in I. Lakatos and A. Musgrave, eds., *Criticism and the Growth of Knowledge* (Cambridge, 1970), pp. 91–195. The quoted phrase, which appears on p. 178, is italicized in the original.

4. Dudley Shapere, "Meaning and Scientific Change," in R. G. Colodny, ed., *Mind and Cosmos: Essays in Contemporary Science and Philosophy.* University of Pittsburgh Series in the Philosophy of Science, vol. 3 (Pittsburgh, 1966), pp. 41–85. The quotation will be found on p. 67.

5. Israel Scheffler, *Science and Subjectivity* (Indianapolis, 1967), p. 81.

6. The last criterion, fruitfulness, deserves

more emphasis than it has yet received. A scientist choosing between two theories ordinarily knows that his decision will have a bearing on his subsequent research career. Of course he is especially attracted by a theory that promises the concrete successes for which scientists are ordinarily rewarded.

7. The least equivocal example of this position is probably the one developed in Scheffler, *Science and Subjectivity,* chap. 4.

8. If the group is small, it is more likely that random fluctuations will result in its members' sharing an atypical set of values and therefore making choices different from those that would be made by a larger and more representative group. External environment—intellectual, ideological, or economic—must systematically affect the value system of much larger groups, and the consequences can include difficulties in introducing the scientific enterprise to societies with inimical values or perhaps even the end of that enterprise within societies where it had once flourished. In this area, however, great caution is required. Changes in the environment where science is practiced can also have fruitful effects on research. Historians often resort, for example, to differences between national environments to explain why particular innovations were initiated and at first disproportionately pursued in particular countries, e.g., Darwinism in Britain, energy conservation in Germany. At present we know substantially nothing about the minimum requisites of the social milieux within which a sciencelike enterprise might flourish.

Larry Laudan

FROM THEORIES TO RESEARCH TRADITIONS

The intellectual function of an established conceptual scheme is to determine the patterns of theory, the meaningful questions, the legitimate interpretations . . .
S. TOULMIN (1970), p. 40

Theories are inevitably involved in the solution of problems; the very aim of theorizing is to provide coherent and adequate solutions to the empirical problems which stimulate inquiry. Theories, moreover, are designed to avoid (or to resolve) the various conceptual and anomalous problems which their predecessors generate. If one looks at inquiry in this way, if one views theories from this perspective, it becomes clear that *the central cognitive test of any theory involves assessing its adequacy as a solution of certain empirical and conceptual problems.* Having developed in earlier chapters a taxonomy for describing the kinds of problems which confront theories, we must now lay down adequacy conditions for determining when a theory provides an acceptable solution to the problems which confront it.

But before we embark on that task, we must clarify what theories are and how they function, for a failure to make some rudimentary distinctions here has brought grief to more than one major philosophy of science. Entire books have been devoted to the structure of scientific theory; I am attempting nothing that ambitious. Rather, I shall want to insist on only two major points with respect to an analysis of theories.

Reprinted by permission from Larry Laudan, *Progress and Its Problems.* University of California Press, 1977.

In the first place, to make explicit what has been implicit all along, *the evaluation of theories is a comparative matter.* What is crucial in any cognitive assessment of a theory is how it fares with respect to its competitors. Absolute measures of the empirical or conceptual credentials of a theory are of no significance; decisive is the judgment as to how a theory stacks up against its known contenders. Much of the literature in the philosophy of science has been based upon the assumption that theoretical evaluation occurs in a competitive vacuum. By contrast, I shall be assuming that assessments of theories always involve comparative modalities. We ask: is this theory better than that one? Is this doctrine the best among the available options?

The second major claim of this chapter is that *it is necessary to distinguish, within the class of what are usually called "scientific theories," between two different sorts of propositional networks.*

In the standard literature on scientific inference, as well as in common scientific practice, the term "theory" refers to (at least) two very types of things. We often use the term "theory" to denote a very specific set of related doctrines (commonly called "hypotheses" or "axioms" or "principles") which can be utilized for making specific experimental predictions and for giving detailed explanations of natural phenomena. Examples of this type of theory would include Maxwell's theory of electromagnetism, the Bohr-Kramers-Slater theory of atomic structure, Einstein's theory of the photoelectric effect, Marx's labor theory of value, Wegener's theory of continental drift, and the Freudian theory of the Oedipal complex.

By contrast, the term "theory" is also used to refer to much more general, much less easily testable, sets of doctrines or assumptions. For instance, one speaks about "the atomic theory," or "the theory of evolution," or "the kinetic theory of gases." In each of these cases, we are referring not to a single theory, but to a whole spectrum of individual theories. The term "evolutionary theory" for instance, does not refer to any single theory but to an entire family of doctrines, historically and conceptually related, all of which work from the assumption that organic species have common lines of descent. Similarly, the term "atomic theory" generally refers to a large set of doctrines, all of which are predicated on the assumption that matter is discontinuous. A particularly vivid instance of one theory which includes a wide variety of specific instantiations is offered by recent "quantum theory." Since 1930, that term has included (among other things) quantum field theories, group theories, so-called S-matrix theories, and renormalized field theories—between any two of which there are huge conceptual divergences.

The differences between the two types of theories outlined above are vast: not only are there contrasts of generality and specificity between them, but the modes of appraisal and evaluation appropriate to each are radically different. It will be the central claim of this chapter that *until we become mindful of the cognitive and evaluational differences between these two types of theories, it will be impossible to have a theory of scientific progress which is historically sound or philosophically adequate.*

But it is not only fidelity to scientific practice and usage which requires us to take these larger theoretical units seriously. Much of the research done by historians and philosophers of science in the last decade suggests that these more general units of analysis exhibit many of the epistemic features which, although most characteristic of science, elude the analyst who limits his range to theories in the narrower sense. Specifically, it has been suggested by Kuhn and Lakatos that *the more general theories,* rather than the more specific ones, *are the primary tool for understanding and appraising scientific progress.*

I share this conviction in principle, but find that the accounts hitherto given of what these larger theories are, and how they evolve, are not fully satisfactory. Because the bulk of this chapter will be devoted to outlining a new account of the more global theories (which I shall be calling *research traditions*), it is appropriate that I should indicate what I find chiefly wanting in the best known efforts to grapple with this problem. Of the many theories of scientific evolution that have been developed, two specifically address themselves to the question of the nature of these more general theories.

KUHN'S THEORY OF SCIENTIFIC "PARADIGMS"

In his influential *Structure of Scientific Revolutions,* Thomas Kuhn offers a model of scientific progress whose primary element is the "paradigm." Although Kuhn's notion of paradigms has been shown to be systematically ambiguous[1] (and thus difficult to characterize accurately), they do have certain identifiable characteristics. They are, to begin with, "ways of looking at the world"; broad quasi-metaphysical insights or hunches about how the phenomena in some domain should be explained. Included under the umbrella of any well-developed paradigm will be a number of specific theories, each of which presupposes one or more elements of the paradigm. Once a paradigm is accepted by scientists (and one of Kuhn's more extreme claims is that in any "mature" science,[2] *every* scientist will accept the *same* paradigm most of the time), they can proceed with the process of "paradigm articulation," also known as "normal science." In periods of normal science, the dominant paradigm will itself be regarded as unalterable and immune from criticism. Individual, specific theories (which represent efforts "to articulate the paradigm," i.e., to apply it to an ever wider range of cases) may well be criticized, falsified and abandoned; but the paradigm itself is unchallenged. It remains so until enough "anomalies"[3] accumulate (Kuhn never indicates how this point is determined) that scientists begin to ask whether the dominant paradigm is really appropriate. Kuhn calls this time a period of "crisis." During a crisis, scientists begin for the first time to consider seriously alternative paradigms. If one of those alternatives proves to be more *empirically successful* than the former paradigm, a scientific revolution occurs, a new paradigm is enthroned, and another period of normal science ensues.

There is much that is valuable in Kuhn's approach. He recognizes clearly that maxi-theories have different cognitive and heuristic functions than mini-theories. He has probably been the first thinker to stress the tenacity and persevering qualities of global theories—even when confronted with serious anomalies.[4] He has correctly rejected the (widely assumed) cumulative character of science.[5] But for all its many strengths, Kuhn's model of scientific progress suffers from some acute conceptual and empirical difficulties. For instance, Kuhn's account of paradigms and their careers has been extensively criticized by Shapere, who has highlighted the obscure and opaque character of the paradigm itself by pointing out many inconsistencies in Kuhn's use of the notion.[6] Feyerabend[7] and others have stressed the historical incorrectness of Kuhn's stipulation that "normal science" is in any way typical or normal. Virtually every major period in the history of science is characterized both by the co-existence of numerous competing paradigms, with none exerting hegemony over the field, and by the persistent and continuous manner in which the foundational assumptions of every paradigm are debated within the scientific community. Numerous critics have noted the arbitrariness of Kuhn's theory of crisis: if (as Kuhn says) a few anomalies do not produce a crisis, but "many" do, how does the scientist determine the "crisis point?" There are other serious flaws as well. In my view, the most significant of these are:

1. Kuhn's failure to see *the role of conceptual problems* in scientific debate and in paradigm evaluation. Insofar as Kuhn grants that there are any rational criteria for

paradigm choice, or for assessing the "progressiveness" of a paradigm, those criteria are the traditional positivist ones such as: Does the theory explain more facts than its predecessor? Can it solve some empirical anomalies exhibited by its predecessor? The whole notion of conceptual problems and their connection with progress finds no serious exemplification in Kuhn's analysis.

2. Kuhn never really resolves the crucial question of *the relationship between a paradigm and its constituent theories*. Does the paradigm entail or merely inspire its constituent theories? Do these theories, once developed, justify the paradigm, or does the paradigm justify them? It is not even clear, in Kuhn's case, whether a paradigm precedes its theories or arises *nolens volens* after their formulation. Although this issue is extremely complex, any adequate theory of science is going to have to come to grips with it more directly than Kuhn has.

3. Kuhn's paradigms have a rigidity of structure which precludes them from evolving through the course of time in response to the weaknesses and anomalies which they generate. Moreover, because he makes the core assumptions of the paradigm immune from criticism, *there can be no corrective relationship between the paradigm and the data.* Accordingly, it is very difficult to square the inflexibility of Kuhnian paradigms with the historical fact that many maxi-theories have evolved through time.

4. Kuhn's paradigms, or "disciplinary matrices," are always implicit, never fully articulated.[8] As a result, it is difficult to understand how he can account for the many theoretical controversies which have occurred in the development of science, since scientists can presumably only debate about assumptions which have been made reasonably explicit. When, for instance, a Kuhnian maintains that the ontological and methodological frameworks for Cartesian or Newtonian physics, for Darwinian biology, or for behavioristic psychology were only implicit and never received overt formulation, he is running squarely in the face of the historical fact that the core assumptions of all these paradigms were explicit even from their inception.

5. Because paradigms are so implicit and can only be identified by pointing to their "exemplars" (basically an archetypal application of a mathematical formulation to an experimental problem), it follows that whenever two scientists utilize the same exemplars, they are, for Kuhn, *ipso facto* committed to the same paradigm. Such an approach ignores the persistent fact that different scientists often utilize the same laws or exemplars, yet subscribe to radically divergent views about the most basic questions of scientific ontology and methodology. (For instance, both mechanists and energeticists accepted identical conservation laws.) To this extent, analysing science in terms of paradigms is unlikely to reveal that "strong network of commitments—conceptual, theoretical, instrumental, and metaphysical"[9] which Kuhn hoped to localize with his theory of paradigms.

LAKATOS' THEORY OF "RESEARCH PROGRAMMES"

Largely in response to Kuhn's assault on some of the cherished assumptions of traditional philosophy of science, Imre Lakatos has developed an alternative theory about the role of these "super-theories" in the evolution of science. Calling such general theories "research programmes," Lakatos argues that research programmes have three elements: (1) a "hard-core" (or "negative heuristic") of fundamental assumptions which cannot be abandoned or modified without repudiation of the research programme;[10] (2) the "positive heuristic," which contains "a partially articulated set of suggestions or hints on how to change, . . . modify, sophisticate [*sic*]"[11] our specific theories whenever we wish to improve them, and (3) "a series of theories, T_1, T_2, T_3, . . ." where each subsequent theory "results from adding auxiliary clauses to . . . the previous theory."[12] Such theories are the specific instantiations of the general research programme. Research programmes can be progressive or regressive in a variety of ways: but progress, for Lakatos even more than for Kuhn, is a function exclusively of the *empirical* growth of a tradition. It is the possession of greater "empirical content," or of a higher "degree of empirical corroboration" which makes one theory superior to, and more progressive than, another.

Lakatos' model is, in many respects, a decided improvement on Kuhn's. Unlike Kuhn, Lakatos allows for, and stresses, the historical importance of the co-existence of several alternative research programmes at the same time, within the same domain. Unlike Kuhn, who often takes the view that paradigms are incommensurable[13] and thus not open to rational comparison, Lakatos insists that we can objectively compare the relative progress of competing research traditions. More than Kuhn, Lakatos tries to grapple with the thorny question of the relation of the super-theory to its constituent mini-theories.

But against that, Lakatos' model of research programs shares many of the flaws of Kuhn's paradigms, and introduces some new ones as well:

1. As with Kuhn, Lakatos' conception of progress is exclusively empirical; the only progressive modifications in a theory are those which increase the scope of its empirical claims.

2. The sorts of changes which Lakatos allows within the mini-theories which constitute his research programme are extremely restricted. In essence, Lakatos only permits, as the relation between any theory and its successor within a research programme, the addition of a new assumption or a semantic re-interpretation of terms in the predecessor theory. On this remarkable view of things, *two theories can only be in the same research programme if one of the two entails the other.* As we shall see shortly, in the vast majority of cases, the succession of specific theories within a maxi-theory involves the *elimination* as well as the addition of assumptions, and there are rarely successive theories which entail their predecessors.

3. A fatal flaw in the Lakatosian notion of research programmes is its dependence upon the Tarski-Popper notions of "empirical and logical content." *All* Lakatos' measures of progress require a comparison of the empirical content of every member of the series of theories which constitutes any research programme.[14] As Grünbaum and others have shown convincingly, the attempt to specify content measures for scientific theories is extremely problematic if not literally impossible.[15] Because comparisons of content are generally impossible, neither Lakatos nor his followers have been able to identify *any* historical case to which the Lakatosian definition of progress can be shown strictly to apply.[16]

4. Because of Lakatos' idiosyncratic view that the acceptance of theories can scarcely if ever be rational, he cannot translate his assessments of progress (assuming he could make them!) into recommendations about cognitive action.[17] Although one research programme may be more progressive than another, we can, on Lakatos' account, deduce nothing from that about which research programme should be preferred or accepted. As a result, there can never be a connection between a theory of progress and a theory of rational acceptability (or, to use Lakatos' language, between methodological "appraisal" and "advice").

5. Lakatos' claim that the accumulation of anomalies has no bearing on the appraisal of a research programme is massively refuted by the history of science.

6. Lakatos' research programmes, like Kuhn's paradigms, are rigid in their hard-core structure and admit of no fundamental changes.[18]

What should be clear, even from this very brief survey of two of the major theories of scientific change, is that there are a number of analytical and historical difficulties confronting existing attempts to understand the nature and role of maxi-theories. With some of those difficulties in mind, we can turn now to explore an alternative model of scientific progress, built upon elements outlined in the previous chapters. One crucial test of that model will be whether it can avoid some of the problems which handicap its predecessors. Although there are numerous common elements between my model and those of Kuhn and Lakatos (and I readily concede a great debt to their pioneering work), there are a sufficiently large number of differences that I shall try to develop the notion of a research tradition more or less from scratch.

THE NATURE OF RESEARCH TRADITIONS

We have already referred to a few classic research traditions: Darwinism, quantum theory, the electromagnetic theory of light. Every intellectual discipline, scientific as well as nonscientific, has a history replete with research traditions: empiricism and nominalism in philosophy, voluntarism and necessitarianism in theology, behaviorism and Freudianism in psychology, utilitarianism and intuitionism in ethics, Marxism and capitalism in economics, mechanism and vitalism in physiology, to name only a few. Such research traditions have a number of common traits:

1. Every research tradition has a number of specific theories which exemplify and partially constitute it; some of these theories will be contemporaneous, others will be temporal successors of earlier ones;

2. Every research tradition exhibits certain *metaphysical* and *methodological* commitments which, as an ensemble, individuate the research tradition and distinguish it from others;

3. Each research tradition (unlike a specific theory) goes through a number of different, detailed (and often mutually contradictory) formulations and generally has a long history extending through a significant period of time. (By contrast, theories are frequently short-lived.)

These are by no means the only important characteristics of research traditions, but they should serve, for the time being, to identify the kinds of objects whose properties I would like to explore.

In brief, a research tradition provides a set of guidelines for the development of specific theories. Part of those guidelines constitute an ontology which specifies, in a general way, the types of fundamental entities which exist in the domain or domains within which the research tradition is embedded. The function of specific theories within the research tradition is to explain all the empirical problems in the domain by "reducing" them to the ontology of the research tradition. If the research tradition is behaviorism, for instance, it tells us that the only legitimate entities which behavioristic theories can postulate are directly and publicly observable physical and physiological signs. If the research tradition is that of Cartesian physics, it specifies that only matter and minds exist, and that theories which talk of other types of substances (or of "mixed" mind and matter) are unacceptable. Moreover, the research tradition *outlines the different modes by which these entities can interact.* Thus, Cartesian particles can only interact by contact, not by action-at-a-distance. Entities, within a Marxist research tradition, can only interact by virtue of the economic forces influencing them.

Very often, the research tradition will also specify certain modes of procedure which constitute the legitimate *methods of inquiry* open to a researcher within that tradition. These methodological principles will be wide-ranging in scope, addressing themselves to experimental techniques, modes of theoretical testing and evaluation, and the like. For instance, the methodological posture of the scientist in a strict Newtonian research tradition is inevitably inductivist, allowing for the espousal of only those theories which have been "inductively inferred" from the data. The methods of procedure outlined for a behavioristic psychologist are what is usually called "operationalist." Put simplistically, *a research tradition is thus a set of ontological and methodological "do's" and "don'ts."* To attempt what is forbidden by the metaphysics and methodology of a research tradition is to put oneself outside that tradition and to repudiate it. If, for instance, a Cartesian physicist starts talking about forces acting-at-a-distance, if a behaviorist starts talking about subconscious drives, if a Marxist begins speculating about ideas which do not arise in response to the economic substructure; in each of these cases, the activity indicated puts the scientist in question beyond the pale. By breaking with the ontology or the methodology of the research tradition within which he has worked, he has violated the strictures of that research tradition and divorced himself from it. Needless to say, that is not necessarily a bad thing. Some of the most important revolutions in scientific thought have come from thinkers who had the ingenuity to break with the research traditions of their day and to inaugurate new ones. But what we must preserve, if we are to understand either the logic or the history of the natural sciences, is the notion of the *integrity* of a research tradition, for it is precisely that integrity which stimulates, defines and delimits what can count as a solution to many of the most important scientific problems.[19]

Although it is vital to distinguish between the ontological and the methodological components of a research tradition, the two are often intimately related, and for a very natural reason: namely, that one's views about the appropriate *methods* of inquiry are generally compatible with one's views about the *objects* of inquiry. When, for instance, Charles Lyell defined the "uniformitarian" research tradition in geology, his ontology was restricted to presently acting causes and his methodology insisted that we should "explain past effects in terms of presently acting causes." Without a "presentist" ontology, his uniformitarian methodology would have been inappropriate; and without the latter, the presentist ontology would not have allowed Lyell to explain the geological past. Similarly, the mathematical ontology of the Cartesian research tradition (an ontology which argued that *all* physical changes were entirely changes of *quantity*) was very closely connected with the (mathematically inspired) deductivist and axiomatic methodology of Cartesianism. As we shall see later, it does not always happen that the ontology and methodology of a research tradition are so closely intertwined (for instance, the inductivist methodology of the Newtonian research tradition had only the weakest of connections with that tradition's ontology), but such cases are the exception rather than the rule.

So a preliminary, working definition of a research tradition could be put as follows: *a research tradition is a set of general assumptions about the entities and processes in a domain of study, and about the appropriate methods to be used for investigating the problems and constructing the theories in that domain.*

* * *

THE EVALUATION OF RESEARCH TRADITIONS

Our focus thus far has been on the temporal dynamics of research traditions. We have learned something about how such traditions evolve, how they interact with their constituent theories and with wider elements of the worldview and the problem situation.

However, I have said nothing yet about how, if at all, it is possible for scientists to make sensible choices between alternative research traditions, nor about how a single tradition can be appraised relative to its acceptability. This is a crucial issue, for until and unless we can articulate workable criteria for choice between the larger units I am calling research traditions, then we have neither a theory of scientific rationality, nor a theory of progressive, cognitive growth.

In the next few pages, I shall be defining some criteria for the evaluation of research traditions, and discussing some of the different contexts in which cognitive evaluations can be made.

Adequacy and Progress

Even though research traditions in themselves entail *no* observable consequences, there are several different ways in which they can be rationally evaluated and thus compared. Two chief modes of appraisal, however, are the most common and the most decisive. One of these modes is synchronic, the other is diachronic and developmental.

We may, to begin with, ask about the (momentary) *adequacy* of a research tradition. We are essentially asking here how effective the *latest* theories within the research tradition are at solving problems. This, in turn, requires us to determine the problem-solving effectiveness of those theories which presently constitute the research tradition (ignoring their predecessors). Since we already discussed how to evaluate the problem-solving effectiveness of individual theories,[37] we need only combine those appraisals to find the adequacy of the broader research tradition.

Alternatively, we may ask about the *progressiveness* of a research tradition. Here our chief concern is to determine whether the research tradition has, in the course of time, increased or decreased the problem-solving effectiveness of its components, and thus its own (momentary) adequacy. This matter is, of course, unavoidably *temporal;* without a knowledge of the history of the research tradition, we can say nothing whatever about its progressiveness. Under this general rubric, there are two subordinate measures which are particularly important:

1. *the general progress of a research tradition*—this is determined by comparing the adequacy of the sets of theories which constitute the oldest and those which constitute the most recent versions of the research tradition;

2. *the rate of progress of a research tradition*—here, the changes in the momentary adequacy of the research tradition during any specified time span are identified.

It is important to note that the general progress and the rate of progress of a research tradition may be widely at odds. For instance, a research tradition may show a high degree of general progress, and yet show a low *rate* of progress, especially in its recent past. Alternatively, a research tradition may have a high rate of progress during its recent past while exhibiting limited general progress.

Likewise, and even more importantly, the appraisals of a research tradition based upon its progressiveness (either general or time-dependent) may be very different from those based on its momentary adequacy. One can conceive of cases, for example, where the adequacy of a research tradition is relatively high and yet it shows no general progress or even a negative rate of progress. (In fact, many actual research traditions have this character.) Alternatively, there are cases (e.g., behavioristic psychology and early quantum theory) where the general progress and the rate of progress of a research tradition are high, but where the momentary adequacy of the tradition is still quite low.

Needless to say, the appraisals will not always point in contrary directions, but the very fact that they can (and sometimes have) emphasizes the need to attend very carefully to the various *contexts* in which cognitive appraisals of research traditions are made. It is that issue which must occupy us next.

The Modalities of Appraisal: Acceptance and Pursuit

Almost all the standard writings on scientific appraisal, whether we look to philosophical or historical discussions of science, have two common features: they assume that there is only *one* cognitively legitimate context in which theories can be appraised; and they assume that this context has to do with determinations of the empirical well-foundedness of scientific theories. Both these assumptions probably need to be abandoned: the first because it is false, the second because it is too limited.

I shall be arguing that a careful examination of scientific practice reveals that there are generally *two* quite different contexts within which theories and research traditions are evaluated.[38] I shall suggest that, within each of these contexts of inquiry, very different sorts of questions are raised about the cognitive credentials of a theory, and that much scientific activity which appears irrational—if we insist on a uni-contextual analysis—can be perceived as highly rational if we allow for the divergent goals of the following two contexts:

The context of acceptance. Beginning with the more familiar of the two, it is clear that scientists often choose *to accept* one among a group of competing theories and research traditions, i.e., *to treat it as if it were true.* Particularly in cases where certain experiments or practical actions are contemplated, this is the operative modality. When, for instance, a research immunologist must prescribe medication for a volunteer in an experiment, when a physicist decides what measuring instrument to use for studying a problem, when a chemist is seeking to synthesize a compound with certain properties; in all these cases, the scientist must commit himself, however tentatively, to the acceptance of one group of theories and research traditions and to the rejection of others.

How can he make a coherent decision? There are a wide range of possible answers here: inductivists will say "choose the theory with the highest degree of confirmation"; or "choose the theory with the highest utility"; falsificationists—if they give any advice at all—will say "choose the theory with the greatest degree of falsifiability." Still others, such as Kuhn, would insist that *no* rational choice can be made.[39] I have already indicated why none of these answers are satisfactory. My own reply to the question, of course, would be, "*choose the theory (or research tradition) with the highest problem-solving adequacy.*"

On this view, the rationale for accepting or rejecting any theory is thus fundamentally based on the idea of problem-solving *progress.* If one research tradition has solved more important problems than its rivals, then accepting that tradition is rational precisely to the degree that we are aiming to "progress," i.e., to maximize the scope of solved problems. In other words, *the choice of one tradition over its rivals is a progressive (and thus a rational) choice precisely to the extent that the chosen tradition is a better problem solver than its rivals.*

This way of appraising research traditions has three distinct advantages over previous models of evaluation: (1) it is *workable:* unlike both inductivist and falsificationist models, the basic evaluation measures seem (at least in principle) to pose fewer difficulties; (2) it simultaneously offers an account of rational *acceptance* and of scientific *progress* which shows the two to be linked together in ways not explained by previous models; and (3) it comes closer to being widely applicable to the actual history of science than alternative models have been.

The context of pursuit. Even if we had an adequate account of theory choice within the context of acceptance, however, we would still be very far from possessing a full account of rational appraisal. The reason for this is that there are many important situations where scientists evaluate competing theories by criteria which have nothing directly to do with the acceptability or "warranted assertibility" of the theories in question.

The actual occurrence of such situations has often been observed. Paul Feyerabend in particular, has identified many historical cases where scientists have investigated and pursued theories or research traditions which were patently less acceptable, less worthy of belief, than their rivals. Indeed, *the emergence of virtually every new research tradition occurs under just such circumstances.* Whether we look to Copernicanism, the early stages of the mechanical philosophy, the atomic theory in the first half of the nineteenth century, early psychoanalytic theory, the preliminary efforts at the quantum mechanical approach to molecular structure, we see the same pattern: scientists often begin to pursue and to explore a new research tradition long before its problem-solving success (or its inductive support, or its degree of falsifiability, or its novel predictions) qualifies it to be accepted over its older, more successful rivals.

Another side to the same coin is the historical fact that *a scientist can often be working alternately in two different, and even mutually inconsistent, research traditions.* Particularly during periods of "scientific revolution," it is commonly the case that a scientist will spend part of his time working on the dominant research tradition and a part of his time working on one or more of its less successful, less fully developed rivals. If we take the view that it is rational to work with and explore only the theories one accepts (and its corollary that one ought not accept or believe mutually inconsistent theories) then there can be no way of making sense of this common phenomenon.

Hence neither the use of mutually inconsistent theories nor the investigation of less successful theories—both well-attested historical phenomena—can be explained if we insist that the context of acceptance exhausts scientific rationality. Confronted by such cases, we would have to conclude, with Feyerabend and Kuhn,[40] that the history of science is largely irrational. But if, on the other hand, we realize that *scientists can have good reasons for working on theories that they would not accept,* then this frequent phenomenon may be more comprehensible.

To see what could count as "good reasons" here, we must return to some earlier discussions. It has often been suggested in this essay that the solution of a maximum number of empirical problems, and the generation of a minimum number of conceptual problems and anomalies is the central aim of science. We have seen that such a view entails that we should accept at any time those theories or research traditions which have shown themselves to be the most successful problem solvers. But need the *acceptance* of a given research tradition preclude us from exploring and investigating alternatives which are inconsistent with it? Under certain circumstances, the answer to this question is decidedly negative. To see why, we need only consider the following general kind of case: suppose we have two competing research traditions, *RT* and *RT′;* suppose further that the momentary adequacy of *RT* is much higher than that of *RT′*, but that the *rate* of progress of *RT′* is greater than the related value for *RT. So far as acceptance is* concerned, *RT* is clearly the only acceptable one of the pair. We may nonetheless decide to work on, further articulate, and explore the problem-solving merits of *RT′*, precisely on the grounds that it has recently shown itself to be capable of generating new solutions to problems at an impressive rate. This is particularly appropriate if *RT′* is a relatively new research tradition. It is common knowledge that most new research traditions bring new analytic and conceptual techniques to bear on the solution of problems. These new techniques constitute (in the cliché) "fresh approaches" which, particularly over the short run, are likely to pay problem-solving dividends. To *accept* a budding research tradition merely because it has had a high rate of progress would, of course, be a mistake; but it would be equally mistaken to refuse to pursue it if it has exhibited a capacity to solve some problems (empirical *or* conceptual) which its older, and generally more acceptable, rivals have failed to solve.

Putting the point generally, we can say that *it is always rational to pursue any research tradition which has a higher rate of progress than its rivals* (even if the former has a lower problem-solving effectiveness). Our specific motives for pursuing such a research tradition could be one of many: we might have a hunch that, with further development, *RT′* could become more successful than *RT;* we might have grave doubts about *RT′* ever becoming generally successful, but feel that some of its more progressive elements could eventually be incorporated within *RT*. Whatever the vagaries of the individual case, if our general aim is increasing the number of problems we can solve, we cannot be accused of inconsistency or irrationality if we pursue (without accepting) some highly progressive research tradition, regardless of its momentary inadequacy (in the sense defined above).

In arguing that the rationality of pursuit is based on relative progress rather than overall success, I am making explicit what has been implicitly described in scientific usage as "promise" or "fecundity." There are numerous cases in the history of science which illustrate the role which an appraisal of promise or progressiveness can have in earning respectability for a research tradition.

The Galilean research tradition, for instance, could not in its early years begin to stack up against its primary competitor, Aristotelianism. Aristotle's research tradition could solve a great many more important empirical problems than Galileo's. Equally, for all the conceptual difficulties of Aristotelianism, it really posed fewer crucial conceptual problems than Galileo's early brand of physical Copernicanism—a fact that tends to be lost sight of in the general euphoria about the scientific revolution. But what Galilean astronomy and physics did have going for it was its impressive ability to explain successfully some well-known phenomena which constituted empirical anomalies for the cosmological tradition of Aristotle and Ptolemy. Galileo could explain, for example, why heavier bodies fell no faster than lighter ones. He could explain the irregularities on the surface of the moon, the moons of Jupiter, the phases of Venus, and the spots on the sun. Although Aristotelian scientists ultimately were able to find solutions for these phenomena (after Galileo drew their attention to them), the explanations proferred by them smacked of the artificial and the contrived. Galileo was taken so seriously by later scientists of the seventeenth century, not because his system as a whole could explain more than its medieval and renaissance predecessors (for it palpably could *not*), but rather because it showed promise by being able, in a short span of time, to offer solutions to problems which constituted anomalies for the other research traditions in the field.

Similarly, Daltonian atomism generated so much interest in the early years of the nineteenth century largely because of its scientific promise, rather than its concrete achievements. At Dalton's time, the dominant chemical research tradition was concerned with elective affinities. Eschewing any attempt to theorize about the microconstituents of matter, elective affinity chemists sought to explain chemical change in terms of the differential tendencies of certain chemical elements to unite with others. That chemical tradition had been enormously successful in correlating and predicting how different chemical substances combine. Dalton's early atomic doctrine could claim nothing like the overall problem-solving success of elective affinity chemistry (this is hardly surprising, for the affinity tradition was a century old by the time of Dalton's *New System of Chemical Philosophy*); still worse, Dalton's system was confronted by numerous serious anomalies.[41] What Dalton was able to do, however, was to predict—as no other chemical system had done before—that chemical substances would combine in certain definite ratios and multiples thereof, no matter how much of the various reagents was present. This phenomenon, summarized by what we now call the laws of definite and multiple proportions, created an immediate stir throughout European science in the decade after Dalton's atomic program was promulgated. Although most scientists refused to accept the Daltonian approach, many nonetheless were prepared to take it seriously, claiming that the serendipity of the Daltonian system made it at least sufficiently promising to be worthy of further development and refinement.

Whether the approach taken here to the problem of "rational pursuit" will eventually prevail is doubtful, for we have only begun to explore some of the complex problems in this area; what I would claim is that the linkage between progress and pursuit outlined above offers us a healthy middle ground between (on the one side) the insistence of Kuhn and the inductivists that the pursuit of alternatives to the dominant paradigm is *never rational* (except in times of crisis) and the anarchistic claim of Feyerabend and Lakatos that the pursuit of *any* research tradition—no matter how regressive it is—*can always be rational.*

Ronald N. Giere*

PHILOSOPHY OF SCIENCE NATURALIZED

In arguing a "role for history," Kuhn was proposing a naturalized philosophy of science. That, I argue, is the only viable approach to the philosophy of science. I begin by exhibiting the main general objections to a naturalistic approach. These objections, I suggest, are equally powerful against nonnaturalistic accounts. I review the failure of two nonnaturalistic approaches, methodological foundationism (Carnap, Reichenbach, and Popper) and metamethodology (Lakatos and Laudan). The correct response, I suggest, is to adopt an "evolutionary perspective." This perspective is defended against one recent critic (Putnam). To argue the plausibility of a naturalistic approach, I next sketch a naturalistic account of theories and of theory choice. This account is then illustrated by the recent revolution in geology. In conclusion I return to Kuhn's question about the role of history in developing a naturalistic theory of science.

*The support of the National Science Foundation is hereby gratefully acknowledged. My colleagues at Indiana and a reviewer supplied many helpful suggestions.

Reprinted by permission from the author and *Philosophy of Science,* Vol. 52, September 1985.

1. KUHN'S NATURALISM

In the very first chapter of *The Structure of Scientific Revolutions,* Kuhn sought to establish "a role for history." Part of that role, he implied, is as data for "a theory of scientific inquiry." And by "theory" he meant something comparable to theories in the sciences themselves. Thus, referring to standard philosophical distinctions, such as that between "discovery" and "justification," he wrote:

> Rather than being elementary logical or methodological distinctions, which would thus be prior to the analysis of scientific knowledge, they now seem integral parts of a traditional set of substantive answers to the very questions upon which they have been deployed. That circularity does not at all invalidate them. But it does make them parts of a theory and, by doing so, subjects them to the same scrutiny regularly applied to theories in other fields. If they are to have more than pure abstraction as their content, then that content must be discovered by observing them in application to the data they are meant to eluci-

date. How could history of science fail to be a source of phenomena to which theories about knowledge may legitimately be asked to apply? ([1962] 1970, p. 9)

Although he did not use exactly these words, Kuhn was advocating a *naturalized* philosophy of science.

The many philosophical criticisms of Kuhn's work focused mainly on the details of his naturalistic account. His account of revolutionary theory change, which invokes only naturalistic notions like gestalt switches and persuasion, was a frequent target. Few critics, however, raised the general question whether *any* purely naturalistic theory of science might be correct. No doubt this was due to the unquestioned presumption that no such account could be correct. It is precisely this presumption I wish now explicitly to challenge.

For some, I admit, no challenge is necessary. Some philosophers regard the philosophy of science as merely a branch of epistemology. And some of these philosophers follow Quine (1969) in the project of naturalizing epistemology. Others embrace a version of evolutionary epistemology. But these are still minorities. Naturalism in the philosophy of science is generally rejected not only by the successors to logical empiricism but also by most of those who agree with Kuhn in adopting a historical methodology. Thus Lakatos (1970), Toulmin (1972), Laudan (1977), and Shapere (1984) have each sought to show that the process of scientific inquiry is not only historical but *rational* as well. Rationality is not a concept that can appear in a naturalistic theory of science—unless reduced to naturalistic terms.[1]

My argument will be both negative and positive. I will first exhibit what seem to be the main general objections to a naturalistic approach. These objections, I suggest, are too strong. There seems no viable non-naturalistic way around them either. I review the failure of two such non-naturalistic approaches, methodological foundationism (Carnap, Reichenbach, and Popper) and metamethodology (Lakatos and Laudan). The correct response, I suggest, is to adopt an "evolutionary perspective." This perspective is defended against one recent critic (Putnam). To argue the plausibility of a naturalistic approach, I next sketch a naturalistic account of theories and of theory choice. This account is illustrated by the recent revolution in geology. In conclusion I return to Kuhn's question about the role of history in developing a naturalistic theory of science.

2. SOME ARGUMENTS AGAINST NATURALISM

The following are some general forms of argument that one would expect to be raised against any proposal to naturalize the philosophy of science.

The circle argument. The general idea behind the circle argument is that the use of scientific methods to investigate scientific methods must be circular, beg the question, or lead to a regress. A more explicit version of the argument might go something like this: One of the things any study of science must investigate is the methods (criteria, canons, etc.) scientists use in evaluating evidence. To pursue such an investigation *scientifically* requires using data about scientific practice to reach conclusions about scientific methods. Thus, any empirical investigation aimed at discovering the criteria that scientists use for evaluating evidence would necessarily presuppose at least some of the criteria it was supposedly setting out to discover. So not *all* the methods of science could be discovered by scientific investigation. At least *some* must be discoverable by other means.[2]

The circle argument is a version of classic arguments concerning the justification of induction. This relationship may partly explain the power of the argument within the philosophical community.

The argument from norms. This argument appeals to the distinction between facts and norms. A naturalistic study of science, it is claimed, could at most *describe* the methods scientists use in coming to adopt hypotheses or theories. The goal of the philosophy of science, however, is not merely to *describe* the methods scientists employ, but to *prescribe* what methods they *should* employ. We want to know not merely what criteria scientists in fact use in adopting theories; we want to know which are the *right* criteria. A naturalistic philosophy of science would be powerless to answer such questions.

The argument from relativism. This argument may be viewed as a corollary to the argument from norms, but relativism has been so much discussed of late that this form of argument deserves independent billing. The argument has the form of a *reductio*. A naturalistic philosophy of science, it is claimed, would be powerless to distinguish good from bad science. It would, for example, have to treat "creation theory" on a par with evolutionary theory. Such a philosophy of science would be at best worthless, at worst pernicious.

These, in brief, are some of the main arguments against a naturalized philosophy of science. A reply takes a bit longer.

3. METHODOLOGICAL FOUNDATIONISM.

Circle arguments have always been among the most powerful in the philosopher's arsenal. Their use, however, commits the user to constructing a defense against a similar attack. Regarding our particular circle argument, the traditional defense has been some form of *methodological foundationism*—the construction of a method whose correctness can be certified a priori.

One connection between naturalism and foundationism has been well charted by Quine (1969). For Quine it was the foundationist inability to reduce mathematics to logic (or semantics to behavior) that left us no alternative but to naturalize epistemology. It requires only a little elaboration to see that a similar connection exists within the philosophy of science.

When Kuhn's book first appeared, the methodologically foundationist programs of Carnap, Reichenbach, and Popper were among the most active areas of research in the philosophy of science. Carnap and Reichenbach, though not Popper, were also tempted by foundationism with regard to the data furnished by experience. Since this latter aspect of foundationist programs is not at issue here, I shall say no more about it.

Carnap was originally attracted to Russell's foundationism which utilized the method of logical construction developed in the context of the foundations of mathematics. Here the regress stops at a foundation of *logic*. Logic also provides a normative component for scientific reasoning and a bulwark against relativism.

Carnap and the early Logical Empiricists gave up Russell's strict form of methodological foundationism not so much for technical reasons as for broadly empirical reasons. They concluded that the methods of logical constructions could not yield the laws of physics as they understood them, and they were unwilling to reject the laws of physics as philosophically unsound. But few were willing to give up the idea that logic provides the foundation for scientific method.[3]

Even Popper, who otherwise was quite critical of many Logical Empiricist doctrines, rested his methodology on the simple logical rule of *modus tollens*. Here again, some of the most severe criticisms of Popper's methodology have been broadly empirical. Adopting Kuhn's claim that all theories have faced anomalies throughout their careers, Lakatos (1971), for example, argued that if we follow Popper's rules, we should have to regard all theories as falsified. Assuming falsified theories should be rejected, we reach the empirically unacceptable conclusion that all theories should be rejected.

Among the major Logical Empiricist figures, Reichenbach was unusual in seeking a methodological foundation not in logic but in a pragmatic rule of action. Assuming that the ultimate goal of science is to determine limiting relative frequencies in infinite sequences, he offered an *a priori* argument that his inductive rule of inference, the "straight rule," guaranteed success in reaching this goal so long as the goal was obtainable at all. Unfortunately, there is a continuum of "crooked rules" for which the same justification can be given. Hacking (1968) delivered the *coup de grâce* to the program by showing that any long-run justification sufficient to justify only the straight rule required the *empirical* assumption that the sequences in question be *random.* This exposed the program to the regress argument it was explicitly designed to elude. Reichenbach's program could also have been criticized on the *empirical* ground that science in fact has stronger goals than the long-run discovery of limiting relative frequencies. But this was not the argument that led to its demise.

During the 1940s, after moving to the United States, Carnap took up Keynes's program of developing an *inductive logic* that would be a formal generalization of deductive logic. His own semantic theories of the previous decade provided the formal background for this attempt. Being a logic, this inductive logic would stop the regress and provide the norms to defeat relativism. Carnap's inductive logic has been criticized on the empirical ground that it is too simple to tell us anything about the evaluation of actual scientific theories. The main reason most people gave up the program, however, was more technical. The logic requires a measure of the initial probability of all hypotheses. But the space of possible measures, like the space of Reichenbachian rules of inference, is so large that there seems no *a priori* and nonarbitrary way to justify a unique measure.

Recalling that the first chapter of Carnap's *Logical Foundations of Probability* (1950) was entitled "On Explication," we are reminded of another program for grounding a particular inductive logic, namely, as an explication of our prereflective concept of evidential support. But how do we know when we have correctly captured this concept, or even that "we" have a univocal concept of this sort? Carnap says only that the "explicatum" must be "similar to" the "explicandum." If this similarity is to be determined empirically, and there seems no other way, then Carnap too was caught in the circle argument.

Richard Jeffrey (1973) once argued that the correct inductive logic would be the one that eventually agrees with our inductive intuitions when the Carnapian program is sufficiently developed for more complex languages. To avoid the circle, this view must assume that our intuitions are given directly without empirical investigation. Moreover, this interpretation leaves it open whether the program ever will be sufficiently developed—which makes it impossible for us currently to say whether science ever has been or is now a rational enterprise. This is a rather weak foundation. And like any explication, it provides no protection from relativism. The logic is at best *descriptive* of *our* intuitions. It does not insure us that our intuitions themselves are correct.

The major remaining stronghold of methodological foundationism is "Bayesian inference" or one of its near relatives. Here the problem of picking out a unique initial probability measure is avoided by relativizing to an individual agent. Individuals supply their own initial measures. Rationality then consists in how one *revises* probability assignments in light of new evidence. It has often been objected that Bayesian inference goes too far in the direction of relativism. Moreover, the same type of problem that plagued both Carnap and Reichenbach arises here too because there are many different logically possible ways of "conditionalizing" on the evidence, and no *a priori* way of singling out one way as uniquely rational. One is reduced either to appealing to something like "explication," or to investigating actual reasoning, which reintroduces the circle.[4]

Adopting Quine's form of inference, I should like now to conclude that methodological foundationism is a hopeless program and thus that naturalism, in spite of the circle argument, is our only alternative. There is, however, a further line of inquiry that must be considered, if only because it has been so prominent in the post-Kuhnian literature.

4. METAMETHODOLOGY

Imre Lakatos (1971) introduced the term "metamethodology" to describe his method for investigating the relative superiority of any proposed theory of scientific method. Laudan (1977) adopted a similar strategy—though differing in detail. For brevity of exposition, I will concentrate on Laudan's approach.[5]

The connection between metamethodology and the circle argument arises as follows. If Lakatos and Laudan really had been taking a naturalistic approach to methodology, they would have adopted the reflexive strategy of applying their methodology to itself. This, however, is not their official doctrine. That they deliberately avoided a reflexive strategy *because* of its obvious circularity, I cannot say. Their metamethodologies, however, are not reflexive and thus not blatantly circular. Whether they can achieve their ends while still avoiding circularity is another question.

In Laudan's theory of scientific rationality, the measure of progress, and therefore of rationality, is problem-solving effectiveness. Roughly speaking, the problem-solving effectiveness of a research tradition is the weighted number of empirical problems solved by its latest theory minus the weighted number of outstanding anomalies and conceptual problems. Problems are weighted by their importance to the research tradition. The relative acceptability of one research tradition over another is determined by its relative problem-solving effectiveness. This measure, Laudan claims, provides a *rational* way of deciding the relative acceptability of two research traditions.

Laudan's metamethodological strategy is to seek first a set of "preferred pre-analytic intuitions about scientific rationality" (PIs). That is, looking at the history of science, we find

> a subclass of cases of theory-acceptance and theory rejection about which most scientifically educated persons have strong (and similar) intuitions. This class would include within it many (perhaps even all) of the following: (1) it was rational to accept Newtonian mechanics and to reject Aristotelian mechanics by, say, 1800; . . . ; (4) it was irrational after 1920 to believe that the chemical atom had no parts;. . . . (1977, p. 160)

The next step is to apply the methodology (Laudan's theory) to the PIs in order to determine the relative problem-solving effectiveness of the traditions in question. This assumes, of course, that we can indeed identify, count, and weigh the relevant problems. Comparison of computed problem-solving effectiveness will tell us which tradition was in fact most progressive and thus which should have been accepted according to Laudan's methodology. "The degree of adequacy of any theory of scientific appraisal is proportional to how many of the PIs it can do justice to" (1977, p. 161).

Assume, for the sake of argument, that Laudan's methodology agrees with *all* the PIs. Could we then be confident that it is "a sound explication of what we mean by rationality" (1977, p. 161)? I think not. The most one could conclude is that Laudan has identified a highly reliable *symptom* of the basis for our pre-analytic judgments of theory-acceptance and theory-rejection.

Suppose, contrary to Laudan, that our pre-analytic judgments are really based on an assessment of the approximate *truth* of the theories in question, and that we take problem-solving effectiveness as our best *evidence* for approximate truth. Laudan's method of assessment would then yield the same judgments of acceptance and rejection, but fail to capture the real basis of our judgments. The trouble is that comparison with our gross judgments of acceptance and rejection does not test the fine structure of the methodological theory. To test the fine structure, however, would require a more detailed empirical inquiry, and this would immediately raise the problem of circularity.

It may be, however, that Laudan would not be all that bothered by learning that he has not avoided circularity. His main concern, like that of Lakatos before him, is to have a *normative* theory of rationality. Let us, therefore, move on to the argument from norms.

Does the type of metamethodology advocated by Lakatos and Laudan yield methodological principles which are genuinely normative? Not really. At its most successful, the metamethodology would tell us only that we had discovered a general *description* of situations which we intuitively regard as clear cases of rational acceptance or pursuit. We might have correctly identified the descriptive component of the methodology, without capturing its normative force. To claim we had captured the normative component would require that we make the judgments we do *because* of considerations based on problem-solving effectiveness. In Kantian terms, Laudan's metamethodology could at most show only that we are acting in accord with his methodology, not that we are acting out of regard for that methodology. It cannot show that his methodology is actually embodied as a norm in our judgments.

This point is all the more pronounced if we consider not merely our own current preferred intuitions, but those of the historical actors in the episodes considered. Laudan does not attempt to show that actual scientists in historical contexts made the judgments they did because of considerations of problem-solving effectiveness. He is content to point out the correlation between their judgments and our calculations of actual problem-solving effectiveness. That is scant evidence that such considerations were normatively operative at the time.

Being at bottom a strategy for explication, not justification, Laudan's metamethodology also fails to provide a strong defense against relativism. Questions about the rationality of the whole western scientific tradition are ruled out because the metamethodology begins with the assumption that some judgments (the PIs) are rational. It is these we use to test the theory of rationality. This Laudan freely admits (p. 161). He fails to point out, however, that this leaves us defenseless against the cultural anthropologist who claims that the belief systems of non-Western cultures cannot rationally be judged by the standards of Western science.

5. AN EVOLUTIONARY PERSPECTIVE

I conclude that neither methodological foundationism nor metamethodology can break the circle or provide the norms needed to defeat relativism. This hardly proves that there is no way to achieve these ends. It does, however, provide some motivation for seeking to *understand* how a naturalized philosophy of science might fruitfully be pursued. I would suggest that evolutionary theory, together with recent work in cognitive science and the neurosciences, provides a basis for such an understanding. The following is the barest sketch of how the story might go.[6]

Human perceptual and other cognitive capacities have evolved along with human bodies. We share many of these capacities with other primates and even lower mammals. Indeed, those parts of our brains responsible for our more advanced linguistic abilities are built upon and linked to those parts that we share with other mammals. There can be no denying that these capacities are fairly well adapted to the environment in which they evolved. Without considerable adaptation, we would very likely not be here. Nor are these capacities trivial. The amount of perceptual and neural processing required just for a human to walk without falling or bumping into things is fantastically large and very complex.

The capacities evolution favors, of course, are just those that confer biological fitness, that is, the ability to survive and leave offspring. The ability to do modern science had nothing to do with the evolution of our perceptual and cognitive capacities—indeed, doing science may very well be detrimental to our survival as a species. The general problem faced by a naturalistic philosophy of science, then, is to explain how creatures with our natural endowments manage to learn so much about the detailed structure of the world—about atoms, stars and nebulae, entropy and genes. This problem calls for a *scientific* explanation.

Empiricist philosophers emphasized the role of immediate perceptual experience in their analyses of knowledge because of the high degree of subjective certainty attached to such experience. From an evolutionary perspective, the subjective certainty is indeed causally connected with the more direct source of the reliability of such judgments, which lies in our evolved capacities for interacting with our world. But the operation of these capacities is largely unrecorded in our conscious experience. Rationalist philosophers, on the other hand, focused on our more general subjective institutions, such as, that space has three dimensions and that time exhibits a linear structure. These judgments seem to be built into the way we think. And indeed they are, for the aspects of the world relevant to our biological fitness have roughly that structure.

Neither empiricists nor rationalists could see how to get beyond their subjective experience or intuitions. This led to the familiar philosophical views that the world is nothing more than the sum total of our sense experience or that it is totally unknowable. In fact, we possess built-in mechanisms for quite direct interaction with aspects of our environment. The operations of these mechanisms largely bypass our conscious experience and linguistic or conceptual abilities.[7]

Thinkers struggling to understand the nature of their own knowledge in the seventeenth and eighteenth centuries may be forgiven for not appreciating evolutionary theory or contemporary neurobiology. A century after Darwin a similar lack of appreciation is less forgivable.

The traditional philosophical skeptic would of course seek to reintroduce the circle argument. To invoke evolutionary theory to understand how we know about the world, he would say, simply begs the question. Evolutionary theory is a fairly advanced, and therefore problematic, form of scientific knowledge. Our problem is to *justify* that knowledge using something less problematic, such as, what one can "directly" experience or intuit.

At this point, however, the skeptic's reply is equally question-begging. Three hundred years of modern science and over a hundred years of biological investigation have led us to the firm conclusion that no humans have ever faced the world guided only by their own subjectively accessible experience and intuitions. Rather, we now know that our capacities for operating in the world are highly adapted to that world. The skeptic asks us to set all this aside in favor of a project that denies our conclusion. And he does so on the basis of what we claim to be an outmoded and mistaken theory about how knowledge is, in fact, acquired.

It should be noted that the above appeal to evolutionary theory is far more modest than that of numerous advocates of "evolutionary epistemology." It is limited to explaining why we need not worry about our failure to break the circle argument. Others have advanced the more extensive claim that evolutionary theory itself provides a good model for the overall development of scientific knowledge. I doubt that it is a very good model for this more ambitious purpose, and I shall suggest a far different account. We agree, however, that the issue is an empirical one, to be settled by scientific procedures, and not by philosophical arguments.[8]

Finally, an evolutionary perspective provides a program for dealing with norms and the problem of relativism. At some stage in the evolutionary process, the evolution of human organisms and human societies became coextensive. Even modestly complex societies require some social organization. Norms make it possible to maintain the requisite degree of social organization. Nor need the naturalist regard these as mere regularities in social behavior. Norms are taught and enforced by various means of social control. The regularity is a product of these social actions. What the naturalist denies is that there is any basis for the norms that transcends the society in its actual physical context. But does this view not leave us open to a radical form of relativism?

An evolutionary perspective places definite limits on how different a human society on earth could be. It is not physically possible that there should exist on earth a culture totally alien to us. Humans walk, talk, eat, sleep, and procreate. Correspondingly, they must acquire food and shelter. We could not fail to understand these activities. How a society goes about doing these things, on the other hand, is not uniquely determined by our biological nature, even if we include the physical circumstances of that society. There is always more than one way to skin a cat. Moreover, there is no supracultural basis for the norm that cats are to be skinned one particular way (or perhaps not at all). At this level, cultural relativism is correct. Does this imply that "creation theory" is as good as evolutionary theory? No more than it implies that prayer is as effective as penicillin for curing infections. Vindicating this reply, however, requires a positive theory of science.

6. REALISM, REFERENCE, AND RATIONALITY

Hilary Putnam (1982) has recently presented several arguments against the possibility of naturalistic or evolutionary epistemologies. One argument is that evolutionary epistemology presupposes metaphysical realism which, he claims to have shown, is incoherent. A second, more general argument is that naturalistic epistemologies attempt to eliminate normative reason. But reason, being both "immanent" and "transcendent," cannot be eliminated without committing "mental suicide" (1982, p. 22). Just explicating these arguments, let alone refuting them, would be a major undertaking. Here I can only attempt to locate some main points of disagreement and suggest where Putnam goes wrong.

Let us adopt Putnam's simple characterization of metaphysical realism as the view that "there is exactly one true and complete description of 'the way the world is'" (1981, p. 49). Must the evolutionary epistemologist or naturalistic philosopher of science make any such supposition? I don't see why. The naturalistic position is that our cognitive capacities are an evolutionary development of those possessed by lower primates and other animals. It is these same capacities the naturalistic philosopher of science employs in attempting to study the scientific activities of his fellow humans. Surely our primate ancestors could not be accused of being metaphysical realists. In so far as our cognitive abilities are continuous with theirs, why should we be any different? Perhaps some evolutionary epistemologists have indeed espoused metaphysical realism, maybe even claiming evolutionary support for such a position. But this is surely no necessary feature of an evolutionary perspective in epistemology.

In Putnam's terms, my naturalistic philosopher of science might be called an "internal realist." But naturalistic philosophers of science holding internal realism are, Putnam claims, no better off. He sees such a view as an attempt of *define* "rationality" in terms of the use of evolved capacities. The suggested formula is: Rational beliefs are those arrived at using evolved capacities for forming beliefs. But this formula is either obviously false or vacuous depending on whether we include all beliefs or only the rational ones. Thus, Putnam concludes, "The evolutionary epistemologist must either presuppose a 'realist' (i.e., metaphysical) notion of truth or see his formula collapse into vacuity" (1982, p. 5).

Here Putnam assumes that one of the tasks of a naturalistic epistemology would be to provide a *definition* of rationality. But one of the main points of an evolutionary perspective is that there is no sharp boundary between animals and humans, and thus between irrational and rational. From an evolutionary perspective, different organisms deal with aspects of their environments in more or less effective ways. Doing science is one of the ways we humans deal with aspects of our environment. Turning our attention to that process itself, we should expect to find that, in various respects, some people are more effective than others. And we would seek to explain why and how this comes about. Attempting to draw a fundamental distinction between rational and irrational activities is itself not an effective way to understand science, or any other human activity.

Of course I do not deny that providing a characterization of rationality is a well-entrenched feature of epistemology. By defining man as the rational animal, Aristotle bequeathed to philosophy the task of discovering the essence of rationality. We have given up essentialism in biology. It is about time we gave it up in epistemology, and for similar reasons.

As noted above, Putnam also has more general arguments purporting to show why reason (and by implication epistemology and the philosophy of science) cannot be naturalized. One line of argument is that reason requires language, which requires reference, which cannot be naturalized. Moreover, reason and language necessarily involve *values,* which also cannot be naturalized. I could not begin to untangle these arguments here. The most I can do is point out that if Putnam is correct, then there are genuinely *emergent* properties, for example, the property of being rational.[9] Somewhere along the line from fishes to philosophers there emerged fundamentally irreducible properties that science alone cannot explain.

Arguments against emergentism have been given by many philosophers, including, a generation ago, Putnam himself (Putnam and Oppenheim 1958). I shall not attempt to review them here. I only marvel that anyone could think these arguments refuted by an analysis of the possible reference of 'cat' and 'cherries' (Putnam 1981, chap. 2).

From a naturalistic perspective, the urge to find some essential difference between animals and humans is itself something to be explained. The evolutionary process produced a species of creatures that has spent much of its history denying its evolutionary origins. Why do humans keep insisting on their special (if not outright superior) place in nature? Psychologists, sociologists, and historians of religion have, in various guises, attempted to answer this question. What strikes me is how self-serving the emergentist program can be. Humans arguing that humans are a breed apart. One wonders if the rejection of a naturalistic approach to the philosophy of science (and philosophy generally) does not serve a much narrower self-interest. If the philosophy of science is naturalized, philosophers of science are on the same footing with historians, psychologists, sociologists, and others for whom the study of science is itself a scientific enterprise. The most philosophers of science could claim is to be the *theoreticians* of a developing science of science on the model of theoretical, as opposed to experimental, physics. Would that not be status enough?

7. MODELS AND THEORIES

As is clear from the form of the circle argument, a crucial test for any naturalistic theory of science is its account of *theory choice.* Since it would be impossible adequately to develop and defend a naturalistic account of theory choice in a short space, I will only present enough to show that such an account is both possible and at least somewhat plausible. Before one can discuss theory choice, however, it is necessary to say something about the objects of choice, namely, theories.

Since Euclid there has existed a more or less continuous tradition of representing theoretical knowledge in the form of an axiomatic system. Newton was part of this tradition, and so were the founders of modern logic. For most of this century, philosophers who have drawn their inspiration from logic and the foundations of mathematics have assumed that a theory is some type of formal, axiomatic system. The fact that scientists in the twentieth century rarely present theories in axiomatic form has not been very troubling because the philosopher's task has been seen as one of reconstruction, conceptual analysis, or justification—not description. If, however, one takes the descriptive task as fundamental, the axiomatic account clearly is not adequate. For the most part it is simply not true that theoretical scientists are engaged in developing axiomatic systems. This point is obvious for the major recent theoretical developments in sciences such as biology and geology, but it holds even for physics. Where are we to find a better account of scientific theories?

If we restrict ourselves to recent science (since 1900 or 1945), the task is easier because the transmission of theoretical knowledge has become quite uniform. It relies heavily on the advanced *textbook.* Until beginning dissertation research, most scientists in most fields learn what theory they know from textbooks (in conjunction with lectures, which also follow a textbook format). Thus, if we wish to learn what a theory is from the standpoint of scientists who use that theory, a good way to proceed is by examining the textbooks from which they learned much of what they know about that theory.

Classical mechanics provides a good example. Many sciences were modeled on mechanics and borrowed heavily from its mathematical techniques. And for many scientists and engineers today, classical mechanics provides their first experience with a real theory. In addition, classical mechanics has been a standard example for philosophers advocating an axiomatic account of theories. It thus allows a direct comparison of the merits of any rival account.

Looking at typical upper-division or graduate-level texts, what do we find? Often there is a chapter of mathematical preliminaries. The first substantive chapter, however, almost invariably presents Newton's three laws of motion. One needs a force function. The following chapters, therefore, are typically devoted to the use of Newton's laws of motion with various force-functions. A not too advanced text might devote a chapter to uniform forces—Galileo's problem of falling bodies. A typical next chapter takes up the case of a linear restoring force in one dimension, Hooke's Law—which yields a linear harmonic oscillator. Later one meets the Law of Universal Gravitation, the inverse-square force that yields orbital motion in two dimensions. And so on.

Within each chapter one finds, among other things, the following: (i) mathematical solutions to the equations of motion incorporating the specific force-function at issue, and (ii) examples of kinds of real systems to which these particular equations of motion might be applied. One learns, for example, that a linear restoring force yields a sinusoidal motion, and that the horizontal motion of a pendulum is approximately sinusoidal.

One of the most significant other things one learns is that none of the systems cited as examples *exactly* fit the equations. The horizontal restoring force of the pendulum, for example, is only linear in the limit as the angle of swing approaches zero. Regarding the equations as straight-forward statements which are then either true or false is, therefore, bound to misrepresent the situation. How, then, should we represent it?

I suggest we take the equations as characterizing an abstract, idealized system, for example, the simple harmonic oscillator. Calling such a system a "model" (or theoretical model) agrees pretty well with both scientific and philosophical usage. Claims about real systems, then, have the form: the real system *is similar to* the model. A pendulum with small amplitude, for example, is similar to a simple harmonic oscillator. I will call such claims "theoretical hypotheses." Implicit in any theoretical hypothesis is a specification of the respects and degrees in and to which the similarity is claimed to hold. At this point one could introduce truth and falsity for theoretical hypotheses, but a claim of truth here would be redundant, serving only to facilitate semantic assent.

The typical advanced text, then, presents the student with a *cluster of models* (really a cluster of clusters) together with a number of hypotheses about real things claimed to be similar to one or another of the models. For the purpose of developing a naturalistic theory of science, I suggest we understand the word 'theory' as including both the cluster or models and a broad range of hypotheses utilizing these models. Restricting 'theory' either to the models or to hypotheses produces too great a variance with how scientists use the term. For all sorts of reasons, it is best to stick as closely to scientific usage as is compatible with developing an overall, adequate theory of science.

In working through a standard text, students learn many things that are best not regarded as explicitly part of the theory, but that are very important nonetheless. They learn the accepted *interpretation* of general terms such as 'position', 'mass', and 'force'. They also learn how to *identify* particular positions, masses, and forces. Any theory of science must assume that scientists have the ability to make these sorts of interpretations and identifications. Securing a better understanding of how this is done, however, can safely be left to linguistics or, more generally, to the cognitive sciences.

It is evident that the above account of theories is realistic without going to the extreme of "metaphysical realism." Indeed, it is compatible with some recent forms of anti-realism. I would call it "constructive realism." It is "constructive" because models are humanly constructed abstract entities. It is realistic because it understands hypotheses as asserting a genuine similarity of structure between models and real systems without imposing any distinction between "theoretical" and "observational" aspects of reality. It is not "metaphysical" in that it makes no claim that there is one true and complete description of any real system. A constructive realist need not claim, for example, that there is a uniquely correct classical model for describing any actual pendulum. Nor must one claim similarity with the real world for *every* aspect of a model. One can be selective in choosing those respects in which the similarity is claimed to hold.[10]

From a naturalistic perspective, then, the theory of classical mechanics appears not to have the structure of an axiomatic system. At best an axiomatic structure could be imposed on one particular type of model, for example, systems of particles subject only to inverse-square forces. Nor, contrary to Popper's philosophy, do *universal* statements play a major role. No longer does one find sweeping Laplacian generalizations about all bodies in the universe. The typical hypothesis only asserts a similarity between a model and a more or less restricted class of

real systems such as pendulums. There are many more lessons to be learned from a serious study of science textbooks, but these are sufficient to proceed to a sketch of naturalistic theory choice.

8. NATURALISTIC THEORY CHOICE

In the philosophical literature, the problem of theory choice has almost universally been understood as one of characterizing *rational* choice. Most philosophers have been willing to grant that it would be rational to choose theories that are true (or at least approximately true). The trouble is, of course, that we do not have an independent check on which ones are true.

Philosophical treatments of theory choice, therefore, have generally proceeded by focusing on properties other than truth, and then attempted to establish a general principle saying it is rational to choose theories with the specified properties. Among the many properties of theories suggested for this role have been: simplicity, falsifiability, high degree of logical probability, high degree of corroboration, predictive power, explanatory power, fruitfulness, and so on. The preferred way of establishing the required general principle is by demonstrating a connection between the specified properties and truth. Despairing of establishing any such connection with truth, however, many philosophers have argued for the rationality of theory choice in terms of these other properties themselves.

The post-Positivist switch to larger units of analysis (paradigms, research programmes, or research traditions) has not significantly changed the general strategy. The difference is that now one focuses on properties of the larger unit, such as progressiveness, and then argues that it is rational to choose a tradition with these properties. The choice of theories is subordinated to the choice of the corresponding larger unit.

All of these approaches assume the more general principle of rationality that scientists generally strive to make a *rational* choice, however this is defined. Other than this general principle, philosophical accounts of theory choice make scant reference to the actual flesh-and-blood scientists who do the choosing. The approach is almost totally "top down." A naturalistic approach to theory choice is explicitly "bottom up." It begins with real agents facing various choices in the course of their actual scientific lives. It assumes that choosing theories is not too dissimilar from choosing anything else, and then looks at how humans in fact make choices.

If our naturalistic theory of science is not to be *merely* historical, we need a *theory* of theory choice. I would suggest that decision theory includes some models of choice that can provide at least a start. Decision theory, however, has a split personality. Sometimes it operates as an account of *rational* choice; other times it is more descriptive. Here we want the descriptive mode, which may be viewed as a specialized part of ordinary belief-desire psychology.

Taken descriptively, decision-theoretic models begin with a *decision problem* that may be represented as a matrix defined by a set of possible options and a set of possible states of the world. The agent's *desires* (or values) are represented by a ranking or utility measure over the option-state pairs, the *outcomes* of the decision process. The result of adding the agent's values is a completed value (or "payoff") matrix. The role of the agent's beliefs in decision making is more complicated, as will be illustrated below.

The focus of rational decision-theory has always been on the *decision rule* (or decision strategy) that defines *the* rational choice as a function of the payoff matrix. The problem of rational decision-theory has been to establish a uniquely rational decision strategy. Descriptive decision-theory looks instead at the characteristics of the decision strategies that are actually used.

Among the most promising descriptive strategies is *satisficing.* Agents following a satisficing strategy must have a good idea of their minimum satisfactory payoff—their satisfaction level. They then survey their options to see whether any have at least a satisfactory payoff for each possible state of the world. If such an option exists, that is the one chosen. If there is no satisfactory option, agents must either lower their satisfaction level or otherwise change the decision problem. Following a satisficing strategy thus guarantees at least a satisfactory payoff—unless, perhaps, no decision is made.[11]

One could, of course, go on to argue that satisficing is rational. But there is no need to do so. Rather, we can take the satisficing strategy as part of our theoretical model of human decision making. We can then investigate the characteristics of the model and inquire of the circumstances, if any, in which humans fit this model. The fit need not be perfect. Like many theoretical models, this one is highly idealized. My hypothesis is that scientists typically follow something approximating a satisficing strategy when faced with the problem of choosing among scientific theories. If this is correct, we have a good scientific explanation of theory choice in science.

9. THE REVOLUTION IN GEOLOGY

A naturalistic approach to theories and theory choice can be nicely illustrated by the recent revolution in the earth sciences.[12]

In the early decades of this century, earth scientists, that is, geologists, geophysicists, climatologists, etc., described the earth as having originated as a much warmer body that since cooled and contracted. In the process, it was thought, the heavier material tended to collect at the core, leaving the lighter material in the mantle and crust. I would say that earth scientists had constructed a *cluster of models* built around the idea of a slowly rotating molten sphere suspended in space. The *theoretical hypotheses* implicit in the texts of the time asserted that the earth is similar in many respects to such models. Much scientific work consisted in working out the consequences of these hypotheses using lots and lots of further information such as the relative abundances of various elements in the earth's crust.

"Contractionist" models were the product of many individual decisions by earth scientists over a long period of time. Many scientists, of course, never made any explicit decisions about any features of these models. They just learned what they were taught. For them, these models formed the background of their practice as earth scientists. The models were part of what is now called a "paradigm" or "research tradition." Here we are not concerned with how this tradition came to be, but how it came to be abandoned in the 1960s.

It is a consequence of contractionist hypotheses that the oceans and continents are relatively fixed. There may be some vertical motion as the planet continues to contract, but little horizontal motion. Geographers, of course, had long noted the remarkable fit in the coastlines of Africa and South America. This immediately suggests that the two were once joined and later drifted apart. There was, however, no known way to reconcile such horizontal motions with hypotheses based on contractionist models. There were no known forces that could operate horizontally in the crust and, to make matters even worse, the ocean floors are made of harder material than the continents. These seem to have been among the main factors in the decision, by prominent earth scientists, not to take seriously hypotheses based on drift models.

Between 1910 and 1915, drift models of the earth were revived by the German meteorologist, Alfred Wegener. Rather than attempting to review the development of Wegener's thought, I will examine briefly the distinction (recently emphasized by Laudan [1977]) between *pursuit* and *acceptance*. In my view, pursuit focuses on models, acceptance on hypotheses.

In 1910, "under the direct impression produced by the coastlines on either side of the Atlantic," Wegener (1966, p. 1) thought about drift models, but did not *pursue* their elaboration. In 1911 he did pursue such models in connection with reports of evidence for the prior existence of "land bridges" between Brazil and Africa. By the beginning of 1912, he had pretty much *accepted* some drift hypotheses. That is, he had concluded that drift models pretty well fitted the actual history of the earth's development. The 1915 German edition of *The Origin of Continents and Oceans* elaborated one particular model, but was primarily designed to convince others that, contrary to the prevailing view, some corresponding drift hypotheses were correct. The book did not succeed in getting many other scientists to accept any of his drift hypotheses, but it did convince a few others that drift models were worth pursuing.

The distinction between pursuit and acceptance is a naturalistic counterpart to the Logical Empiricists' distinction between discovery and justification. The Logical Empiricists held that there is a "logic" of justification, but not of discovery. Their critics argued that there is also a logic of discovery. From a naturalistic viewpoint, there is neither a logic of discovery nor of justification. Both pursuit and acceptance are alike simply as examples of human decision making. Instead of asking, "Is there a logic of discovery?" the critics should have asked, "Is there is a logic of justification?"

There was, however, a germ of truth in the Logical Empiricists' position. Decisions to pursue a type of model (a "program"?) seem to be much more complex, and thus more difficult to study, than decisions to accept corresponding hypotheses. What leads people like Wegener to spend time developing models that neither they (at least initially) nor the vast majority of their professional colleagues believe to fit the real world? What, in decision-theoretic terms, is their payoff matrix? I have considered this question not only for Wegener, but also for later figures in the story such as Arthur Holmes, J. Tuzo Wilson, Harry Hess, and Fred Vine. It is difficult to find any general patterns. The decisions seem about as varied as the individuals.

Not that acceptance decisions are always transparent. Wegener's decision to accept drift hypotheses had to have been idiosyncratic since few of the readers of his book made similar decisions. By contrast, the decision to accept a modified drift hypothesis in the 1960s was widely shared. This, I think, was because the scientific context in the 1960s so strongly structured everyone's payoff matrix that individual differences tended not to matter. The decision was robust under exchange of professionally competent individuals.

Most commentators agree that the crucial episode in the 1960s revolution was the verification of the Vine-Matthews-Morley hypothesis (VMM) and of Wilson's related hypothesis regarding transform faults. For simplicity I will concentrate on VMM.[13]

In retrospect, it is possible to identify two relatively independent lines of development during the 1950s that made possible the revolution in the 1960s. One, in oceanography, was the discovery of large systems of ocean ridges running roughly in a north-south direction. The mid-Atlantic ridge and the eastern Pacific ridges were among the first to be explored. The second line of development, in paleomagnetism, was the discovery of several apparently global reversals in the earth's magnetic field.

In 1960, Harry Hess, reviving an idea of Arthur Holmes, suggested that ocean ridges were formed by currents of molten material rising from the core and spreading laterally, east and west, from the center of the ridge. This hypothesis, named "sea floor spreading" by Dietz a year later, immediately suggested a mechanism for continental drift. The continents are carried along on top of the spreading sea floor material.

In 1963, Fred Vine and Drummond Matthews (and, independently, Lawrence Morley) put together Hess's model of sea floor spreading with the possibility of magnetic-field reversals. The resulting hypothesis implies that there should be stripes of oppositely directed remnant magnetism in the material of the ocean floor parallel to the ridges. The pattern of magnetic reversals found on land should be duplicated on the sea floor in the pattern of "stripes" parallel to and symmetrical with the ridges. Since the reversals seemed not to occur at regular intervals, the pattern carries a very distinctive "signature."

It is not clear exactly when people like Hess and Vine switched from pursuit to acceptance of VMM and a generalized drift hypothesis. It is clear that few people in the larger earth sciences community took up pursuit of drift models, let alone acceptance of drift hypotheses. This changed dramatically in 1966–67.

In 1966, a research vessel taking magnetic soundings across the Pacific-Antarctic Ridge brought back clear evidence of the magnetic pattern predicted by VMM. About the same time, a similar pattern of reversals was observed in cores of sea floor sediment. The impact on the earth sciences community was swift and complete. Within a year just about everyone with professionally competent knowledge of the situation accepted some sort of drift hypothesis.

Why did the community of earth scientists rush to accept drift hypotheses? Part of the explanation, I suggest, is that the payoff matrix for the decision to accept drift was clear and simple to just about everyone. First, the options were clear. One option was to accept the hypothesis that some drift model fits the earth better than a static model. That is, accept the hypothesis that there are large-scale horizontal movements involving both the sea floor and the continents. The single alternative was to retain the hypothesis that a static, contractionist model is basically correct. This implies rejecting the hypothesis of large-scale horizontal movement.

Among the relevant states of the world are: (i) that the tectonic structure of the world is more similar (in the relevant respects) to drift models than to static models; and (ii) the reverse. Are these the only relevant states? In general, clearly not. It would be relevant to almost anyone whether or not a majority of their professional peers made the same decision. Few people place great value on being proven correct, as Wegener was, only after they are dead. This factor can be neutralized if we restrict our attention to scientists whose professional work was directly affected by which type of model they chose. Such people cannot comfortably wait to see which way the wind is blowing. They have a strong interest in being right at the right time.

For our suitably restricted class of earth scientists, then, the basic structure of the decision problem is as shown in figure 1. The value ranking depends on only two fairly weak, and very plausible, assumptions. One is that being objectively right is regarded as satisfactory even if one would prefer that the world were otherwise. This does not mean assuming that scientists place an *intrinsic* positive value on being objectively right. They may in fact believe that in this case being right will yield valuable short-term professional payoffs. The second assumption is that most earth scientists who were not directly involved in the research on oceanography or paleomagnetism would have preferred that contractionist hypotheses had been correct. This is simply because their training and skills were developed in the context of such models. Switching carries the cost of acquiring new knowledge and new skills. These assumptions yield the ranking of outcomes exhibited in figure 1.

	DRIFT MODELS APPROXIMATELY CORRECT	STATIC MODELS APPROXIMATELY CORRECT
ADOPT DRIFT MODELS	SATISFACTORY	TERRIBLE
RETAIN STATIC MODELS	BAD	EXCELLENT

Figure 1. Decision Problem for Geologists in 1966.

As it stands, the matrix does not make obvious which choice one would prefer. But there is some vital information that has not yet been factored into the decision problem. This is in two parts. The first is that finding the magnetic profiles predicted by VMM was thought to be quite likely if Hess's model were roughly correct. The second is that finding such profiles on any remotely plausible static model was thought to be quite unlikely. Static models had few known resources for accommodating the existence of such a pattern on so large a scale. These judgments were widely shared by people pursuing, or otherwise developing, either type of model. And they support a clearly satisfactory decision rule: If VMM is verified, accept drift models; if not, continue accepting static models. What makes this rule satisfactory is that, given the above judgments, a satisfactory or excellent outcome was very likely while a bad or terrible outcome was very unlikely. A satisficer would ask for no more. If earth scientists are satisficers, we have a plausible explanation of their choice.

Notice how far from methodological foundationism this account is. It assumes agreement that the technology for measuring magnetic profiles is reliable. The Duhem-Quine problem is set aside by the fact that one can build, or often purchase commercially, the relevant measuring technology. The background knowledge (or auxiliary hypotheses) are embodied in proven technology.

The above account also assumes agreement on what is likely or not depending on which type of hypothesis is correct. The fact that there were *logically* possible contractionist models that could yield VMM was irrelevant. What mattered was whether anyone in the opposing camp thought they could come up with rival hypotheses that fitted the data. In this case, most of those who had been developing static models simply gave up. It was only in the context of this vast background of shared judgments that the data was able decisively to force the decision on a typical satisficer.

The above sketch must suffice. It is hardly enough to convince anyone to accept my particular naturalistic hypotheses regarding even just this one case. I hope it is enough to convince some that such models are worth pursuing.

10. A ROLE FOR HISTORY

Kuhn was of course correct in thinking that a naturalized philosophy of science would provide a role for history. The role, he suggested, was as *evidence* for theoretical claims about science. Yet the use of history by philosophers of science (recall the metamethodology of Lakatos and Laudan) suggests that this evidential relationship is more complex than it might seem.

It is useful to consider how some other sciences use the historical record as evidence for their theories. Evolutionary biology and economics provide appropriate models because they seem nicely to bracket a proposed theoretical science of science.

Turning first to evolutionary biology, it is generally thought that the fossil record provides historical evidence for evolutionary theory. I am far from convinced that this record, by itself, provides a satisfactory basis for deciding that any evolutionary theory is correct. Here, however, I am concerned with a narrower issue. Those who have used the fossil record as evidence for evolutionary theory have generally assumed that the underlying mechanisms of evolution, whatever they might be, are relatively stable. Few biologists have ever argued that we might need different models of evolution for different epochs. The major recent controversies over punctuated equilibria or mass extinctions concern the nature and rate of changes in the environment—not our models of the underlying evolutionary mechanisms.

In contrast to evolutionary biology, the most successful theoretical models in economics, whether macro or micro, are *equilibrium* models. The data for such models are, therefore, not historical in the sense that they follow economic developments over time. For models that do use genuinely historical data, one must turn to theories of economic *development.* These, however, are generally thought to provide the least successful models in all of economics, Marxism being the most obvious example. The generally accepted reason for the poor record of models of development is that the economic mechanisms themselves change over time. It is not simply a matter of looking at the same mechanisms operating in a different environment. A rural, agrarian society, for example, seems to embody different economic mechanisms than an urban, industrial society.

Following Kuhn, historically minded philosophers of science have argued, using historical examples, that not only the content of science changes with time. Its aims and methods change as well. This seems to imply that the relation between theories of science and the history of science follows the economic rather than the biological pattern. Indeed, the Kuhnian model of development—normal science, crisis, revolution, new normal science—seems to have as much, or as little, theoretical content as the Marxian stages of economic development. One wonders whether any theory of scientific development that includes changes in aims and methods could do much better. Yet most historically oriented philosophers of science since Kuhn seem to be aiming at a similarly grand theory of development. Few would describe their own aims in these terms, of course. But illustrating the same point using historical cases ranging from Newton, through Lavoisier, to Einstein, and even J. D. Watson, betrays the intent.

The options for a naturalistic theory of science, then, are these. The first is an ambitious strategy that seeks mechanisms of scientific development that can explain not only changes in content but also changes in aims and methods. One could then claim to have similar mechanisms operating over long periods, say from the seventeenth century to the present. The danger in this strategy is ending up with only vaguely defined models of science. A second, much less ambitious strategy would be to restrict attention to shorter epochs such as science in the seventeenth century or since World War II. Here the danger is ending up with models of only very restricted applicability.[14]

Following my own theory of science, I would suggest a third, hybrid strategy. Begin with the less ambitious strategy, and then try to link up the various models so as to obtain a cluster of partially overlapping models covering several epochs, perhaps, even, most of science since Newton. That, I suspect, is the most that can be done.

The suggested model of science provides some hope for thinking that the third strategy can be successful. The activities of model construction and model choice abstract from the scientific context in much the same way as models of mutation and selection abstract from the biological environment. These activities may take place in many different social and economic settings. Different aims, or values, may be reflected in the structure of decisions concerning specific hypotheses. So may the information yielded by new methodologies. Whether this is enough to provide informative similarities among widely separated epochs remains to be seen.

My aim in this paper, however, has not been to argue for a particular strategy or a particular model of science. These have been noted only to illustrate the possibilities opened up by a naturalistic approach. The main thesis is that the study of science must itself be a science. The only viable philosophy of science is a naturalized philosophy of science.

NOTES

1. Among prominent evolutionary or naturalistic epistemologists, I count Donald Campbell (1974) and Abner Shimony (1971, 1981). Toulmin's (1972) view is evolutionary but perhaps not naturalistic. Popper (1972) appropriates the title "evolutionary" without adequate justification. Friedman (1979) reaches conclusions that are naturalistic but not evolutionary. Arthur Fine's (1984) recently espoused "natural ontological attitude" encompasses a natural epistemological attitude as well. Outside the philosophy of science, advocates of naturalistic or evolutionary epistemologies are too numerous even to begin mentioning.

2. I first formulated this argument (in Giere 1973) as an expression of what I then took to be a majority view among philosophers of science. I did not intend to argue that it was impossible to establish a connection between the philosophy of science and the historical practice of science; only that the authors under review had failed adequately to address the most serious difficulties. Indeed, my own solution at the time (Giere 1975) was basically naturalistic, though not evolutionary.

3. The importance of broadly empirical considerations in the Logical Empiricists' rejection of Russellian foundationism has been emphasized recently by Hempel (1983). In this paper, Hempel distinguishes "normative" from "descriptive-naturalistic" methodologies, and argues for a mixed approach.

4. For a summary of the relevant literature, and many references, see Giere (1979). Advocates of "Bayesian inference" seem to assume that their reconstruction of scientific inference, while not strictly reducible to deductive logic, nevertheless somehow carries the normative force associated with deductive logic. I do not understand the basis for this assumption. I am not even convinced that deductive logic possesses the normative powers commonly ascribed to it.

5. The following discussion applies only to the Laudan of *Progress and Its Problems*. I understand he no longer subscribes to this methodology. Lakatos is somewhat ambiguous as to whether his metamethodology is just his ordinary methodology applied at the meta-level or a different methodology altogether.

6. What follows owes at least part of its inspiration to Paul Churchland's (1979) notion of an "epistemic engine." See also Patricia Smith Churchland and Paul M. Churchland (1983).

7. For some recent neurobiological findings relevant to the mechanisms underlying spatial coordination among mammals, see O'Keefe and Nadel (1978), and Pellionisz and Llinas (1982).

8. Here I am thinking particularly of Donald Campbell (1974) and his followers.

9. From informal comments at a conference in May, 1984, I infer that Putnam himself might agree that his view is a variety of emergentism.

10. The label "constructive realism" was originally intended as a direct contrast to van Fraassen's (1980) "constructive empiricism." See Giere (1984 and forthcoming). My view of theories is a liberal version of his "semantic" conception of theories, and similar to Frederick Suppe's (1973) conception. Van Fraassen's distinction between "observable" and "theoretical" seems to me a philosophical imposition. It is very difficult to interpret the actual practice of scientists as honoring such a distinction. I find Nancy Cartwright's (1983) anti-realism much more congenial, perhaps even compatible with a constructive realism. I could also agree with much of what Putnam (1978) says about "internal realism." There are some general similarities between my view and the "structuralist" approach of Sneed (1971) or Stegmüller (1979). This school, however, seems primarily interested in reconstruction and philosophical vindication, and little concerned with description.

11. Satisficing has been developed primarily by Herbert Simon. See his 1972 for further details and references.

12. Historians and philosophers of science have recently begun to give the revolution in geology the attention it deserves. See, for example, Frankel (1982), Rachel Laudan (1981), or Ruse (1981), and the references cited therein.

13. For technical details, see Frankel (1982). The reader is invited to compare Frankel's Laudanistic interpretation of this episode with my decision-theoretic account.

14. For an example of the ambitious strategy, see L. Laudan (1984).

REFERENCES

Campbell, D. R. (1974), "Evolutionary Epistemology", in *The Philosophy of Karl Popper,* P. A. Schilpp (ed.). La Salle: Open Court, pp. 413–63.

Carnap, R. [1950] (1962), *Logical Foundations of Probability.* 2nd edition. Chicago: University of Chicago Press.

Cartwright, N. D. (1983), *How the Laws of Physics Lie.* Cambridge: Cambridge University Press.

Churchland, P. M. (1979), *Scientific Realism and the Plasticity of Mind.* Cambridge: Cambridge University Press.

Churchland, P. S., and Churchland, P. M. (1983), "Stalking the Wild Epistemic Engine", *Noûs 17:* 5–18.

Fine, A. (1984), "And Not Anti-Realism Either", *Noûs 18:* 51–65.

Frankel, H. (1982), "The Development, Reception and Acceptance of the Vine-Matthews-Morley Hypothesis", *Historical Studies in the Physical Sciences 13:* 1–39.

Friedman, M. (1979), "Truth and Confirmation", *The Journal of Philosophy 76:*361–82.

Giere, R. N. (1973), "History and Philosophy of Science: Intimate Relationship or Marriage of Convenience?", *British Journal for the Philosophy of Science 24:* 282–97.

______. (1975), "The Epistemological Roots of Scientific Knowledge", in *Induction, Probability, and Confirmation,* G. Maxwell and R. M. Anderson, Jr. (eds.). Minnesota Studies in the Philosophy of Science, vol. 6. Minneapolis: University of Minnesota Press, pp. 212–61.

______. (1979), "Foundations of Probability and Statistical Inference", in *Current Research in Philosophy of Science,* P. D. Asquith and Henry E. Kyburg, Jr. (eds.). East Lansing: Philosophy of Science Association, pp. 503–33.

______. (1984), "Toward a Unified Theory of Science", in *Science and Reality,* J. T. Cushing, C. F. Delaney, and G. Gutting (eds.). Notre Dame: University of Notre Dame Press.

______. (forthcoming), "Constructive Realism", in *Images of Science,* P. M. Churchland and C. Hooker (eds.). Chicago: University of Chicago Press.

Hacking, I. (1968), "One Problem about Induction", in *The Problem of Inductive Logic,* I. Lakatos (ed.). Amsterdam: North-Holland, pp. 44–58.

Hempel, C. G. (1983), "Valuation and Objectivity in Science", in *Physics, Philosophy and Psychoanalysis,* R. S. Cohen and L. Laudan (eds.). Dordrecht: D. Reidel, pp. 73–100.

Kuhn, T. S. [1962] (1970), *The Structure of Scientific Revolutions.* 2nd edition. Chicago: University of Chicago Press.

Jeffrey, R. (1973), "Carnap's Inductive Logic", *Synthese 25:* 299–306.

Lakatos, I. (1970), "Falsification and the Methodology of Scientific Research Programmes", in *Criticism and the Growth of Knowledge,* I. Lakatos and A. Musgrave (eds.). Cambridge: Cambridge University Press.

______. (1971), "History of Science and Its Rational Reconstructions", in *PSA 1970,* R. S. Cohen and R. C. Buck (eds.). Boston Studies in the Philosophy of Science, vol. 8. Dordrecht: D. Reidel, pp. 91–135.

Laudan, L. (1977), *Progress and Its Problems.* Berkeley and Los Angeles: University of California Press.

______. (1984), *Science and Values: The Aims of Science and Their Role in Scientific Debate.* Berkeley and Los Angeles: University of California Press.

Laudan, R. (1981), "The Recent Revolution in Geology and Kuhn's Theory of Scientific Change", in *PSA 1978,* P. D. Asquith and I. Hacking (eds.). Vol. 2. East Lansing: Philosophy of Science Association, pp. 227–39.

O'Keefe, J., and Nadel, L. (1978), *The Hippocampus as a Cognitive Map.* Oxford: Clarendon Press.

Pellionisz, A., and Llinas, R. (1982), "Space-Time Representation in the Brain: The Cerebellum as a Predictive Space-Time Metric Tensor", *Neuroscience 7:* 2949–70.

Popper, K. R. (1972), *Objective Knowledge.* Oxford: Clarendon Press.

Putnam, H. (1978), *Meaning and the Moral Sciences.* London: Routledge and Kegan Paul.

______. (1981), *Reason, Truth and History.* Cambridge: Cambridge University Press.

______. (1982), "Why Reason Can't Be Naturalized", *Synthese 52:* 3–23.

Putnam, H., and Oppenheim, R. (1958), "Unity of Science as a Working Hypothesis", in *Concepts, Theories and the Mind-Body Problem,* H. Feigl, M. Scriven, and G. Maxwell (eds.). Minnesota Studies in the Philosophy of Science, vol.

2. Minneapolis: University of Minnesota Press, pp. 3–36.

Quine, W. V. O. (1969), "Epistemology Naturalized", in *Ontological Relativity and Other Essays.* New York: Columbia University Press.

Ruse, M. (1981), "What Kind of a Revolution Occurred in Geology?", in *PSA 1978,* P. D. Asquith and I. Hacking (eds.). Vol. 2. East Lansing: Philosophy of Science Association, pp. 240–73.

Shapere, D. (1984), *Reason and the Search for Knowledge.* Dordrecht: D. Reidel.

Shimony, A. (1971), "Perception from an Evolutionary Point of View", *The Journal of Philosophy 67:* 571–83.

______. (1981), "Integral Epistemology", in *Scientific Inquiry and the Social Sciences,* M. B. Brewer and B. E. Collins (eds.). San Francisco: Jossey-Bass, pp. 98–123.

Simon, H. A. (1972), "Theories of Bounded Rationality", in *Decision and Organization,* R. Radner and C. B. McGuire (eds.). Amsterdam: North-Holland, pp. 161–76.

Sneed, J. D. (1971), *The Logical Structure of Mathematical Physics.* Dordrecht: D. Reidel.

Stegmüller, W. (1979), *The Structuralist View of Theories.* New York: Springer.

Suppe, F. (1973), "Theories, Their Formulations, and the Operational Imperative", *Synthese 25:* 129–64.

Toulmin, S. (1972), *Human Knowledge.* Princeton: Princeton University Press.

van Fraassen, B. C. (1980), *The Scientific Image.* Oxford: Oxford University Press.

Wegener, A. (1966), *The Origin of Continents and Oceans.* New York: Dover.

The Logic of Discovery

Norwood Russell Hanson

IS THERE A LOGIC OF SCIENTIFIC DISCOVERY?

Is there a logic of scientific discovery? The approved answer to this is "No." Thus Popper argues:[1] "The initial stage, the act of conceiving or inventing a theory, seems to me neither to call for logical analysis nor to be susceptible of it." Again, "There is no such thing as a logical method of having new ideas, or a logical reconstruction of this process." Reichenbach writes that philosophy of science "cannot be concerned with [reasons for suggesting hypotheses], but only with [reasons for accepting hypotheses]."[2] Braithwaite elaborates: "The solution of these historical problems involves the individual psychology of thinking and the sociology of thought. None of these questions are our business here."[3]

Against this negative chorus, the "Ayes" have *not* had it. Aristotle (*Prior Analytics* II, 25) and Peirce[4] hinted that in science there may be more problems for the logician than just analyzing the arguments supporting already invented hypotheses. But contemporary philosophers are unreceptive to this. Let us try once again to discuss the distinction F. C. S. Schiller made between the Logic of Proof and the Logic of Discovery.[5] We may be forced, with the majority, to conclude "Nay." But only after giving Aristotle and Peirce a sympathetic hearing. Is there *anything* in the idea of a "logic of discovery" which merits the attention of a tough-minded, analytic logician?

It is unclear what a logic of discovery is a logic of. Schiller intended nothing more than "a logic of inductive inference." Doubtless his colleagues were so busy sectioning syllogisms that they ignored inference which mattered in science. All the attention philosophers now give to inductive reasoning, probability, and the principles of theory construction would have pleased Schiller. But, for Peirce, the work of Popper, Reichenbach, and Braithwaite would read less like a *Logic of Discovery* than like a *Logic of the Finished Research Report.* Contemporary logicians of science have described how one sets out reasons in support of a hypothesis once proposed. They have said nothing about the conceptual context within which such a hypothesis is initially proposed. Both Aristotle and Peirce insisted that the proposal of a hypothesis can be a reasonable affair. One can have good reasons, or bad, for suggesting one kind of hypothesis initially, rather than some other kind. These reasons may differ in type from those which lead one to accept a hypothesis once suggested. This is not to deny that one's reasons for proposing a hypothesis initially may be identical with his reasons for later accepting it.

One thing must be stressed. When Popper, Reichenbach, and Braithwaite urge that there is no logical analysis appropriate to the psychological complex which attends the conceiving of a new idea, they are saying nothing which Aristotle or Peirce would reject. The latter did not think themselves to be writing manuals to help scientists make discoveries. There could be no such manual.[6] Apparently they felt that there is a *conceptual* inquiry, one properly called "a logic of discovery," which is *not* to be confounded with the psychology and sociology appropriate to understanding how some investigator stumbled on to an improbable idea in unusual circumstances. There are factual discussions such as these latter. Historians like Sarton and Clagett have undertaken such circumstantial inquiries. Others—for example, Hadamard and Poincaré—have dealt with the psychology of discovery. But these are not logical discussions. They do not even turn on conceptual distinctions. Aristotle and Peirce thought they were doing something other than psychology, sociology, or history of discovery; they purported to be concerned with a *logic* of discovery.

This suggests caution for those who reject wholesale any notion of a logic of discovery on the grounds that such an inquiry can *only* be psychology, sociology, or history. That Aristotle and Peirce deny just this has made no impression. Perhaps Aristotle and Peirce were wrong. Perhaps there is no room for logic between the psychological dawning of a discovery and the justification of that discovery via successful predictions. But this should come as the conclusion of a discussion, not as its preamble. If Peirce is correct, nothing written by Popper, Reichenbach, or Braithwaite cuts against him. Indeed, these authors do not discuss what Peirce wishes to discuss.

Let us begin this uphill argument by distinguishing

(1) reasons for accepting a hypothesis *H,* from
(2) reasons for suggesting *H* in the first place.

This distinction is in the spirit of Peirce's thesis. Despite his arguments, most philosophers deny any *logical* difference between these two. This must be faced. But let us shape the distinction before denting it with criticism.

What would be our reasons for accepting *H*? These will be those we might have for thinking *H* true. But the reasons for suggesting *H* originally, or for formulating *H* in one way rather than another, may not be those one requires before thinking *H* true. They are, rather, those reasons which make *H* a *plausible type of conjecture*. Now, no one will deny *some* differences between what is required to show *H* true and what is required for deciding *H* constitutes a plausible kind of conjecture. The question is: Are these logical in nature, or should they more properly be called "psychological" or "sociological"?

Or one might urge, as does Professor Feigl, that the difference is just one of refinement, degree, and intensity. Feigl argues that considerations which settle whether *H* constitutes a plausible conjecture are of the *same type* as those which settle whether *H* is true. But since the initial proposal of a hypothesis is a groping affair, involving guesswork amongst sparse data, there *is* a distinction to be drawn; but this, Feigl urges, concerns two ends of a spectrum, ranging all the way from inadequate and badly selected data to that which is abundant, well diversified, and buttressed by a battery of established theories. The issue therefore remains: Is the difference between reasons for accepting *H* and reasons for suggesting it originally one of logical type, or one of degree, or of psychology, or of sociology?

Already a refinement is necessary if our original distinction is to survive. The distinction just drawn must be reset in the following, more guarded, language. Distinguish now

(1′) reasons for accepting a particular, minutely specified hypothesis *H*, from

(2′) reasons for suggesting that, whatever specific claim the successful *H* will make, it will, nonetheless, be a hypothesis of one *kind* rather than another.

Neither Aristotle, nor Peirce, nor (if you will excuse the conjunction) myself in earlier writings,[7] sought this distinction on these grounds. The earlier notion was that it was some particular, minutely specified *H* which was being looked at in two ways: (1) What would count for the acceptance of that *H*, and (2) what would count in favor of suggesting that same *H* initially.

This latter way of putting it is objectionable. The issue is whether, *before* having hit a hypothesis which succeeds in its predictions, one can have good reasons for anticipating that the hypothesis will be one of some particular *kind*. Could Kepler, for example, have had good reasons, *before* his elliptical-orbit hypothesis was established, for supposing that the successful hypothesis concerning Mars's orbit would be of the noncircular kind?[8] He *could* have argued that, whatever path the planet *did* describe, it would be a closed, smoothly curving, plane geometrical figure. Only this *kind* of hypothesis could entail such observation-statements as that Mars's apparent velocities at 90 degrees and at 270 degrees of eccentric anomaly were greater than any circular-type *H* could explain. Other *kinds* of hypotheses were available to Kepler: for example, that Mars's *color* is responsible for its high velocities, or that the dispositions of Jupiter's moons are responsible. But these would not have struck Kepler as capable of explaining such surprising phenomena. Indeed, he would have thought it *un*-reasonable to develop such hypotheses at all, and would have argued thus [Braithwaite counters: "But exactly which hypothesis was to be rejected was a matter for the 'hunch' of the physicists."[9] However, which *type* of hypothesis Kepler chose to reject was not just a matter of "hunch."]

I may still be challenged. Some will continue to berate my distinction between reasons for suggesting which type of hypothesis *H* will be, and reasons for accepting *H* ultimately.[10] There may indeed be "psychological" factors, the opposition concedes, which make certain types of hypotheses "look" as if they might explain phenomena. Ptolemy knew, as well as did Aristarchus before him and Copernicus after him, that a kind of astronomy which displaced the earth would be theoretically simpler, and easier to manage, than the hypothesis of a geocentric, geostatic universe. *But,* philosophers challenge, for psychological, sociological, or historical reasons, alternatives to geocentricism did not "look" as if they could explain the absence of stellar parallax. This cannot be a matter of logic, since for Copernicus one such alternative *did* "look" as if it could explain this. Insofar as scientists have *reasons* for formulating types of hypotheses (as opposed to hunches and intuitions), there are just the kinds of reasons which later show a particular *H* to be true. Thus, if the absence of stellar parallax constitutes more than a psychological reason for Ptolemy's resistance to alternatives to geocentricism, then it *is* his reason for rejecting such alternatives as *false.* Conversely, his reason for developing a geostatic type of hypothesis (again, absence of parallax) was his reason for taking some such hypothesis as *true.* And Kepler's reasons for rejecting Mars's color or Jupiter's moons as indicating the kinds of hypotheses responsible for Mars's accelerations were reasons which also served later in establishing some hypothesis of the noncircularity type.

So the objection to my distinction is: The only *logical* reason for proposing *H* will be of a certain type is that *data* incline us to think some *particular H* true. What Hanson advocates is psychological, sociological, or historical in nature; it has no logical import for the differences between proposing and establishing hypotheses.

Kepler again illustrates the objection. Every historian of science knows how the idea of uniform circular motion affected astronomers before 1600. Indeed, in 1591 Kepler abandoned a hypothesis because it entailed other-than-uniform circular orbits—something simply inconceivable for him. So psychological pressure against forming alternative types of hypotheses was great. But *logically* Kepler's reasons for entertaining a type of Martian motion other than uniformly circular were his reasons for accepting that as astronomical truth. He first encountered this type of hypothesis on perceiving that no simple adjustment of epicycle, deferent, and eccentric could square Mars's observed distances, velocities, and apsidal positions. These were also reasons which led him to assert that the planet's orbit is not the effect of circular motions, but of an elliptical path. Even after other inductive reasons confirmed the truth of the latter hypothesis, these early reasons were *still* reasons for accepting *H* as true. So they cannot have been reasons merely for proposing which type of hypothesis *H* would be, and nothing more.

This objection has been made strong. If the following cannot weaken it, then we shall have to accept it; we shall have to grant that there is *no* aspect of discovery which has to do with logical, or conceptual considerations.

When Kepler published *De Motibus Stellae Martis,* he had established that Mars's orbit was an ellipse, inclined to the ecliptic, and had the sun in one of the foci. Later (in the *Harmonices Mundi*) he generalized this for other planets. Consider the hypothesis *H′*: *Jupiter's* orbit is of the noncircular type.

The reasons which led Kepler to formulate *H′* were many. But they included this: that *H* (the hypothesis that *Mars's* orbit is elliptical) is true. Since Eudoxos, Mars had been the typical planet. (*We* know why. Mars's retrogradations and its movement around the empty focus—all this we observe with clarity from earth because of earth's spatial relations with Mars.) Now, Mars's dynamical properties are usually found in the other planets. If its orbit is ellipsodial, then it is reasonable to expect that, whatever the exact shape of the other orbits (for example, Jupiter's), they will all be of the noncircular type.

But such reasons would not *establish H′*. Because what makes it reasonable to anticipate that *H′* will be of a certain type is *analogical* in character. (Mars does *x;* Mars is a typical planet; so perhaps all planets do the same kind of thing as *x*.) Analogies cannot establish hypotheses, not even *kinds* of hypotheses. Only observations can do that. In this the hypothetico-deductive account (or Popper, Reichenbach, and Braithwaite) is correct. To establish H' requires plotting its successive positions on a smooth curve whose equations can be determined. It may then be possible to assert that Jupiter's orbit is an ellipse, an oviform, an epicycloid, or whatever. But it would not be reasonable to expect this when discussing only what type of hypothesis is likely to describe Jupiter's orbit. Nor is it right to characterize this difference between "*H*-as-illustrative-of-a-type-of hypothesis" and "*H*-as-empirically established" as a difference of psychology only. *Logically,* Kepler's analogical reasons for proposing that H' would be of a certain type were good reasons. But, logically, they would not then have been good reasons for asserting the truth of a specific value for H'—something which could be done only years later.

What are and are not good reasons for reaching a certain conclusion is a logical matter. No further observations are required to settle such issues, any more than we require experiments to decide, on the basis of one's bank statements, whether one is bankrupt. Similarly, whether or not Kepler's reasons for anticipating that H' will be of a certain kind are *good* reasons is a matter for logical inquiry.

Thus, the differences between reasons for expecting that some as yet undiscovered *H* will be of a certain type and those that establish this *H* are greater than is conveyed by calling them "psychological," "sociological," or "historical."

Kepler reasoned initially by analogy. Other kinds of reasons which make it plausible to propose that an *H*, once discovered, will be of a certain type, might include, for example, the detection of a formal symmetry in sets of equations or arguments. At important junctures Clerk Maxwell and Einstein detected such structural symmetries. This allowed them to argue, before getting their final answers, that those answers would be of a clearly describable type.

In the late 1920's, before anyone had explained the "negative-energy" solutions in Dirac's electron theory, good analogical reasons could have been advanced for the claim that, whatever specific assertion the ultimately successful *H* assumed, it would be of the Lorentz-invariant type. It could have been conjectured that the as yet undiscovered *H* would be compatible with the Dirac explanation of Compton scattering and doublet atoms, and would fail to confirm Schrödinger's hunch that the phase waves within configuration space actually described observable physical phenomena. All this could have been said before Weyl, Oppenheimer, and Dirac formulated the "hole theory of the positive electron." Good analogical reasons for supposing that this *type* of *H* would succeed could have been and, as a matter of fact, were advanced. Indeed, Schrödinger's attempt to rewrite the Dirac theory so that the negative-energy solutions disappeared was *rejected* for failing to preserve the Lorentz invariance.

Thus reasoning from observations of *A*'s as *B*'s to the proposal "All *A*'s are *B*'s" is different in type from reasoning analogically from the fact that *C*'s are *D*'s to the proposal "The hypothesis relating *A*'s and *B*'s will be of the same type as that relating *C*'s and *D*'s." (Here it is the *way* *C*'s are *D*'s which seems analogous to the way *A*'s are *B*'s.) And both of these are typically different from reasoning involving the detection of symmetries in equations describing *A*'s and *B*'s.

Indeed, put this way, what *could* an objection to the foregoing consist of? Establishing a hypothesis and proposing by analogy that a hypothesis is likely to be of a particular type surely follow reasoning which is different in type. Moreover, both procedures have a fundamentally logical or conceptual interest.

An objection: "Analogical arguments, and those based on the recognition of formal symmetries, are used because of inductively established beliefs in the reliability of arguments of that type. So the cash value of such appeals ultimately collapses into just those accounts given by *H-D* theorists."

Agreed. But we are not discussing the *genesis* of our faith in these types of arguments, only the *logic* of the arguments themselves. *Given* an analogical premise, or one based on symmetry considerations—or even on enumeration of particulars—one argues *from* these in logically different ways. Consider what further moves are necessary to convince one who doubted such arguments. A challenge to "All *A*'s are *B*'s" when this is based on induction by enumeration, could only be a challenge to justify induction, or at least to show that the particulars are being correctly described. This is inappropriate when the arguments rest on analogies or on the recognition of formal symmetries.

Another objection: "Analogical reasons, and those based on symmetry are *still* reasons for *H* even after it is (inductively) established. They are reasons *both* for proposing that *H* will be of a certain type and for accepting *H*."

Agreed, again. But, analogical and symmetry arguments could never *by themselves* establish particulars *H*'s. They can only make it plausible to suggest that *H* (when discovered) will be of a certain type. However, inductive arguments can, by themselves, establish particular hypotheses. So they must differ from arguments of the analogical or symmetrical sort.

H-D philosophers have been most articulate on these matters. So, let us draw out a related issue on which Popper, Reichenbach, and Braithwaite seem to me not to have said the last word.

J.S. Mill was wrong about Kepler (*A System of Logic,* III, 2–3). It is impossible to reconcile the delicate adjustment between theory, hypothesis, and observation recorded in *De Motibus Stellae Martis* with Mill's statement that Kepler's first law is but "a compendius expression for the one set of directly observed facts." Mill did not understand Kepler (as Peirce notes [*Collected Papers,* I, p. 31]). (It is equally questionable whether Reichenbach understood him: "Kepler's laws of the elliptic motion of celestial bodies were inductive generalizations of observed fact . . . [he] observed a series of . . . positions of the planet Mars and found that they may be connected by a mathematical relation . . .")[11] Mill's *Logic* is as misleading about scientific discovery as any account proceeding via what Bacon calls *"inductio per enumerationem simplicem ubi non reperitur instantia contridictoria."* (Indeed, Reichenbach observes: "It is the great merit of John Stuart Mill to have pointed out that all empirical inferences are reducible to the *inductio per enumerationem simplicem. . . .*")[12] The accounts of *H-D* theorists are equally misleading.

An *H-D* account of Kepler's first law would treat it as a high-level hypothesis in an *H-D* system. (This is Braithwaite's language.) It is regarded as a quasi-axiom, from whose assumption observation-statements follow. If these are true—if, for example, they imply that Uranus's orbit is an ellipse and that its apparent velocity at 90 degrees is greater than at aphelion—then the first law is confirmed. (Thus Braithwaite writes: "A scientific system consists of a set of hypotheses which form a deductive system . . . arranged in such a way that from some of the hypotheses as premises all the other hypotheses logically follow . . . the establishment of a system as a set of true propositions depends upon the establishment of its lowest level hypotheses . . .")[13]

This describes physical theory more adequately than did pre-Baconian accounts in terms of simple enumeration, or even post-Millian accounts in terms of ostensibly not-so-simple enumerations. It tells us about the logic of laws, and what they do in finished arguments and explanations. *H-D* accounts do not, however, tell us anything about the context in which laws are proposed in the first place; nor, perhaps, were they even intended to do so.

The induction-by-enumeration story *did*

intend to do this. *It* sought to describe good reasons for initially proposing *H*. The *H-D* account must be silent on this point. Indeed, the two accounts are not strict alternatives. (As Braithwaite suggests they are when he remarks of a certain higher-level hypothesis that it "will not have been established by induction by simple enumeration; it will have been obtained by the hypothetico-deductive method. . . .")[14] They are thoroughly compatible. Acceptance of the second is no reason for rejecting the first. A law *might* have been inferred from just an enumeration of particulars (for example, Boyle's law in the seventeenth century, Bode's in the eighteenth, the laws of Ampere and Faraday in the nineteenth, and much of meson theory now). It could *then* be built into an *H-D* system as a higher order proposition. If there is anything wrong with the older view, *H-D* accounts do not reveal this.

There *is* something wrong. It is false. Scientists do not always discover every feature of a law by enumerating and summarizing observables. (Thus, even Braithwaite[15] says: "Sophisticated generalizations (such as that about the proton-electron constitution of the hydrogen atom) . . . [were] certainly not derived by simple enumeration of instances . . .") But *this* does not strengthen the *H-D* account as against the inductive view. There is *no H-D* account of how "sophisticated generalizations" are *derived.* On his own principles, the *H-D* theorist's lips are sealed on this matter. But there are conceptual considerations which help us understand the *reasoning* that is sometimes successful in determining the type of an as-yet-undiscovered hypothesis.

Were the *H-D* account construed as a description of scientific practice, it would be misleading. (Braithwaite's use of "derived" is thus misleading. So is his announcement [p. 11] that he is going to explain "how we *come to make* use of sophisticated generalizations.") Natural scientists do not "start from" hypotheses. They start from data. And even then not from commonplace data, but from surprising anomalies. (Thus Aristotle remarks[16] that knowledge begins in astonishment. Peirce makes perplexity the trigger of scientific inquiry.[17] And James and Dewey treat intelligence as the result of mastering problem situations.)[18]

By the time a law gets fixed into a *H-D* system, the *original* scientific thinking is over. The pedestrian process of deducing observation-statements begins only after the physicist is convinced that the proposed hypothesis is at least of the right type to explain the initially perplexing data. Kepler's assistant could work out the consequences of *H′* and check its validity by seeing whether Jupiter behaved as *H′* predicts. This was possible because of Kepler's argument that what *H* had done for Mars, *H′* might do for Jupiter. The *H-D* account is helpful here; it analyzes *the argument of a completed research report.* It helps us see how experimentalists elaborate a theoretician's hypotheses. And the *H-D* account illuminates yet another aspect of science, but its proponents have not stressed it. Scientists often dismiss explanations alternative to that which has won their provisional assent along lines that typify the *H-D* method. Examples are in Ptolemy's *Almagest,* when (on observational grounds) he rules out a moving earth, in Copernicus's *De Revolutionibus . . . ,* when he rejects Ptolemy's lunar theory, in Kepler's *De Motibus Stellae Martis,* when he denies that the planes of the planetary orbits intersect in the center of the ecliptic, and in Newton's *Principia,* when he discounts the idea that the gravitational force law might be of an inverse cube nature. These mirror formal parts of Mill's *System of Logic* or Braithwaite's *Scientific Explanation.*

Still, the *H-D* analysis remains silent on reasoning which often conditions the discovery of laws—reasoning that determines which type of hypothesis is likely to be most fruitful to propose.

The induction-by-enumeration story views scientific inference as being from observations to the law, from particulars to the general. There is something true about this which the *H-D* account must ignore. Thus Newton wrote: "The main business of natural philosophy is to argue from phenomena. . . ."[19]

This inductive view, however, ignores what Newton never ignored: The inference is also from *explicanda* to an *explicans*. Why a beveled mirror shows spectra in sunlight is not explained by saying that all beveled mirrors do this. Why Mars moves more rapidly at 270 degrees and 90 degrees than could be expected of circular-uniform motions is not explained by saying that Mars (or even all planets) always move thus. On the induction view, these latter might count as laws. But only when it is explained why beveled mirrors show spectra and why planets apparently accelerate at 90 degrees will we have laws of the type suggested: Newton's laws of refraction and Kepler's first law. And even before such discoveries were made, arguments in favor of those *types* of laws were possible.

So the inductive view rightly suggests that laws are somehow related to inference *from* data. It wrongly suggests that the resultant law is but a summary of these data, instead of being an explanation of these data. A logic of discovery, then, might consider the structure of arguments in favor of one *type* of possible explanation in a given context as opposed to other *types*.

H-D accounts all agree that laws explain data. (Thus Braithwaite says: "A hypothesis to be regarded as a natural law must be a general proposition which can be thought to explain its instances; if the reason for believing the general proposition is solely direct knowledge of the truth of its instances, it will be felt to be a poor sort of explanation of these instances . . ." [op. cit., p. 302].) *H-D* theorists, however, obscure the initial connection between thinking about data and thinking about what kind of hypothesis will most likely lead to a law. They suggest that the fundamental inference in science is from higher-order hypotheses to observation-statements. This may characterize the setting out of one's reasons for making a prediction after *H* is formulated and provisionally established. It need not be a way of setting out reasons in favor of proposing originally of what type *H* is likely to be.

Yet the original suggestion of a hypothesis type is often a reasonable affair. It is not as dependent on intuition, hunches, and other imponderables as historians and philosophers suppose when they make it the province of genius but not of logic. If the establishment of *H* through its predictions has a logic, so has the initial suggestion that *H* is likely to be of one kind rather than another. To form the first specific idea of an elliptical planetary orbit, or of constant acceleration, or of universal gravitational attraction does indeed require genius—nothing less than a Kepler, a Galileo, or a Newton. But this does not entail that reflections leading to these ideas are nonrational. Perhaps *only* Kepler, Galileo, and Newton had intellects mighty enough to fashion these notions initially; but to concede this is not to concede that their reasons for first entertaining concepts of such a type surpass rational inquiry.

H-D accounts begin with the hypothesis as given, as cooking recipes begin with the trout. Recipes, however, sometimes suggest, "First catch your trout." The *H-D* account is a recipe physicists often use after catching hypotheses. However, the conceptual boldness which marks the history of physics shows more in the ways in which scientists *caught* their hypotheses than in the ways in which they elaborated these once caught.

To study only the verification of hypotheses leaves a vital part of the story untold—namely, the reasons Kepler, Galileo, and Newton had for thinking their hypotheses would be of one kind rather than another. In a letter to Fabricus, Kepler underlines this:

Prague, July 4, 1603

Dear Fabricius,

. . . You believe that I start with imagining some pleasant hypothesis and please myself in embellishing it, examining it only later by observations. In this you are very much mistaken. The truth is that after having built up an hypothesis on the ground of observations and given it proper foundations, I feel a peculiar desire to investigate whether I might discover some natural, satisfying combination between the two . . .

Had any *H-D* theorist ever sought to give an account of the way in which hypotheses in science *are discovered,* Kepler's words are for him. Doubtless *H-D* philosophers have tried to give just such an account. Thus, Braithwaite[20] writes: "Every science *proceeds* . . . by thinking of general hypotheses . . . from which particular consequences are deduced which can be tested by observation . . . ," and again, "Galileo's deductive system was . . . presented as deducible from . . . Newton's laws of motion and . . . his law of universal gravitation . . ."

How would an *H-D* theorist analyze the law of gravitation?

(1) First, the hypothesis *H:* that between any two particles in the universe exists an attracting force varying inversely as the square of the distance between them ($F = \lambda Mm/r^2$).
(2) Deduce from this (in accordance with the *Principia*)
   (a) *Kepler's* Laws, and
   (b) *Galileo's* Laws.
(3) But particular instances of (a) and (b) square with what is observed.
(4) Therefore *H* is, to this extent, confirmed.

The *H-D* account says nothing about how *H* was first puzzled out. But now consider why, here, the *H-D* account is prima facie plausible.

Historians remark that Newton's reflections on this problem began in 1680 when Halley asked: "If between a planet and the sun there exists an attraction varying inversely as the square of their distance, what then would be the path of the planet?" Halley was astonished by the immediate answer: "An ellipse." The astonishment arose not because Newton *knew* the path of a planet, but because he had apparently deduced this from the hypothesis of universal gravitation. Halley begged for the proof, but it was lost in the chaos of Newton's room. Sir Isaac's promise to work it out anew terminated in the writing of the *Principia* itself. Thus the story unfolds as an *H-D* plot: (1) from the suggestion of a hypothesis (whose genesis is a matter of logical indifference—that is, psychology, sociology, or history) to (2) the deduction of observation statements (the laws of Kepler and Gallieo), which turn out true, thus (3) establishing the hypothesis.

Indeed, the entire *Principia* unfolds as the plot requires—from propositions of high generality through those of restricted generality, terminating in observation-statements. Thus, Braithwaite[21] observes: "Newton's *Principia* [was] modelled on the Euclidean analogy and professed to prove [its] later propositions—those which were confirmed by confrontation with experience—by deducing them from original first principles . . ."

Despite this, the orthodox account is suspicious. The answer Newton gave Halley is not unique. He could have said "a circle" or "a parabola," and have been equally correct. The general answer is: "A conic section." The greatest mathematician of his time is not likely to have dealt with so mathematical a question as that concerning the possibility of a formal demonstration with an answer which is but a single value of the correct answer.

Yet the reverse inference, the retroduction, *is* unique. Given that the planetary orbits are ellipses, and allowing Huygen's law of centripetal force and Kepler's rule (that the square of a planet's period of revolution is proportional to the cube of its distance from the sun), the *type* of the law of gravitation can be inferred. Thus the question, "If the planetary orbits are ellipses, what form will the gravitational force law take?" invites the unique answer, "An inverse square type of law."

Given the datum that Mars moves in an ellipse, one can (by way of Huygen's law and Kepler's third law) explain this uniquely by suggesting how it might follow from a law of the inverse square type, such as the law of universal gravitation was later discovered to be.

The rough idea behind all this: Given an ellipsodial eggshell, imagine a tiny pearl moving inside it along the maximum elliptical orbit. What *kind* of force must the eggshell exert on the pearl to keep the latter in this path? Huygen's weights, when whirled on strings, required a force in the string, and in Huygen's arm, of $F_{(k)} \propto r/T^2$ (where r signifies distance, T time, and k is a constant of proportionality). This restraining force kept the weights from flying away like stones from David's sling. And something like this force would be expected in the eggshell. Keplers' third law gives $T^2 \propto r^3$. Hence, $F_{(k)} \propto r/r^3 \propto 1/r^2$. The force the shell exerts on the pearl will be of a kind which varies inversely as the square of the distance of the pearl from that focus of the ellipsoidal eggshell where the force may be supported to be centered. This is not yet the law of gravitation. But it certainly is an argument which suggests that the law is likely to be of an inverse square type. This follows by what Peirce called "retroductive reasoning." But what *is* this retroductive reasoning whose superiority over the *H-D* account has been so darkly hinted at?

Schematically, it can be set out thus:

(1) Some surprising, astonishing phenomena p_1, p_2, p_3 . . . are not encountered.[22]
(2) But p_1, p_2, p_3 . . . would not be surprising were a hypothesis of H's type to obtain. They would follow as a matter of course from something like H and would be explained by it.
(3) Therefore there is good reason for elaborating a hypothesis of the type of H; for proposing it as a possible hypothesis from whose assumption p_1, p_2, p_3 . . . might be explained.[23]

How, then, could the discovery of universal gravitation fit this account?

(1) The astonishing discovery that all planetary orbits are elliptical was made by Kepler.
(2) But such an orbit would not be surprising if, in addition to other familiar laws, a law of "gravitation," of the inverse square type obtained. Kepler's first law would follow as a matter of course; indeed that kind of hypothesis might even explain why (since the sun is in but one of the foci) the orbits are ellipses on which the planets travel with nonuniform velocity.
(3) Therefore there is good reason for further elaborating hypotheses of this kind.

This says something about the rational context within which a hypothesis of H's type might come to be "caught" in the first place. It begins where all physics begins—with problematic phenomena requiring explanation. It suggests what might be done to particular hypotheses once proposed—namely, the *H-D* elaboration. And it points up how much philosophers have yet to learn about the kinds of reasons scientists might have for thinking that one kind of hypothesis may explain initial perplexities; why, for example, an inverse square type of hypothesis may be preferred over others, *if* it throws the initially perplexing data into patterns within which determinate modes of connection can be perceived. At least it appears that the ways in which scientists sometimes reason their way *towards* hypotheses, by eliminating those which are certifiably of the wrong type, may be as legitimate an area for conceptual inquiry as are the ways in which they reason their way *from* hypotheses.

Recently, in the Lord Portsmouth collection in the Cambridge University Library, a document was discovered which bears on our discussion. There, in "Additional manuscripts 3968. No. 41, bundle 2," is the following draft in Newton's own hand:

And in the same year [1665, twenty years before the *Principia*] I began to think of gravity extending to ye orb of the Moon, and (having found out how to estimate the force with which a globe revolving within a sphere presses the surface of the sphere), from Kepler's rule . . . I deduced that the forces which keep the planets in their Orbs must be reciprocally as the squares of their distances from the centres about which they revolve . . .

This manuscript corroborates our argument. ("Deduce," in this passage, is used as when Newton speaks of deducing laws from phenomena—which is just what Aristotle and Peirce would call "retroduce.") Newton *knew* how to estimate the force of a small globe on the inner surface of a sphere. (To compare this with Halley's question and our pearl-within-eggshell reconstruction, note that a sphere can be regarded as a degenerate ellipsoid—that is, where the foci superimpose.) From this and from Kepler's rule, $T^2 \propto r^3$, Newton determined that, whatever the final form of the law of gravitation, it would very probably be of an inverse-square type. These were the reasons which led Newton to think further about the details of universal gravitation. The reasons for accepting one such hypothesis of this type *as a law* are powerfully set out later in the *Principia* itself; and they are much more comprehensive than anything which occurred to him at this early age. But without such preliminary reasoning Newton might have had no more grounds than Hooke or Wren for thinking the gravitation law to be an inverse-square type.

The morals of all this for our understanding of contemporary science are clear. With such a rich profusion of data and techniques as we have, the arguments necessary for *eliminating* hypotheses of the wrong type become a central research inquiry. Such arguments are not always of the *H-D* type; but if for that reason alone we refuse to scrutinize the conceptual content of the reasoning which precedes the actual proposal of definite hypotheses, we will have a poorer understanding of scientific thought in our time. For our own sakes, we must attend as much to how scientific hypotheses are caught, as to how they are cooked.

NOTES

1. Karl Popper, *The Logic of Scientific Discovery* (New York: Basic Books, 1959), pp. 31–32.
2. Hans Reichenbach, *Experience and Prediction* (Chicago: University of Chicago Press, 1938), p. 382.
3. R. B. Braithwaite, *Scientific Explanation* (Cambridge: Cambridge University Press, 1955), pp. 21–22.
4. C. S. Peirce, *Collected Papers* (Cambridge, Mass.: Harvard University Press, 1931), Vol. I, Sec. 188.
5. F. C. S. Schiller, "Scientific Discovery and Logical Proof," Charles Singer, ed., *Studies in the History and the Methods of the Sciences* (Oxford: Clarendon Press, 1917), Vol. I.
6. "There is no science which will enable a man to bethink himself of that which will suit his purpose," J. S. Mill, *A System of Logic*, III, Chap. I.
7. Cf. *Patterns of Discovery* (Cambridge, Mass.: Harvard University Press, 1958), pp. 85–92; "The Logic of Discovery," in *Journal of Philosophy*, LV, 25, 1073–89, 1958; More on "The Logic of Discovery," *op. cit.*, LVII, 6, 182–88, 1960.
8. Cf. *De Motibus Stellae Martis* (Munich), pp. 250ff.
9. *Op. cit.*, p. 20.
10. Reichenbach writes that philosophy "cannot be concerned with the first, but only with the latter" (*op. cit.*, p. 382).
11. Reichenbach, *op. cit.*, p. 371.
12. *Ibid.*, p. 389.
13. Braithwaite, *op. cit.*, pp. 12–13.
14. *Ibid.*, p. 303.
15. *Ibid.*, p. 11.
16. Aristotle, *Metaphysics* 982b, 11ff.
17. Peirce, *op. cit.*, Vol. II, Book III, Chap. ii, Part III.
18. Cf. John Dewey, *How We Think*, (London: Heath & Co., 1909), pp. 12f.
19. Newton, *Principia*, Preface.
20. *Op. cit.*, pp. xv, xi, 18.
21. *Op. cit.*, p. 352.
22. The astonishment may consist in the fact that p is at variance with accepted *theories*—for example, the discovery of discontinuous emission of radiation by hot black bodies, or the photoelectric effect, the Compton effect, and the continuous β-ray spectrum, or the orbital aberra-

tions of Mercury, the refrangibility of white light, and the high velocities of Mars at 90 degrees. What is important here is *that* the phenomena are encountered as anomalous, not *why* they are so regarded.

23. This is a free development of remarks in Aristotle (*Prior Analytics,* II, 25) and Peirce (*op. cit.*). Peirce amplifies: "It must be remembered that retroduction, although it is very little hampered by logical rules, nevertheless, is logical inference, asserting its conclusion only problematically, or conjecturally, it is true, but nevertheless having a perfectly definite logical form" (*op. cit.,* I, 188).

Larry Laudan

WHY WAS THE LOGIC OF DISCOVERY ABANDONED?

I

It is difficult to find a problem area in the philosophy of science about which more nonsense has been talked and in which more confusion reigns than 'the philosophy of discovery'. It is even hard to keep the characters straight. Russ Hanson, who thought the logic of discovery was a good thing, advocated the method of abduction, which was a method for the evaluation, not the discovery, of hypotheses. Hans Reichenbach, who was notorious for insisting that the 'context of discovery' is of no philosophical significance, was a proponent of the straight rule of induction, a technique for the discovery of natural regularities if ever there was one. Not to be slighted here is Karl Popper who wrote a book called the *Logic of Scientific Discovery,* which denies the existence of any referent for its title.

In the circumstances, it will perhaps not be taken amiss if I suggest that some historical explorations may be appropriate. With any luck, an analysis of the evolution of ideas about the logic of discovery may put us in a position to see what significance, if any, should be attached to recent efforts to get clear about the philosophical problems of discovery.

Not the least of my concerns will be that of utilizing history to ascertain what problems, what important questions, a logic of discovery was thought to resolve. If the recently revived program for finding a logic of discovery is worth taking seriously, it must be because such a logic, once propounded, might illuminate or clarify some important philosophical problems about the nature of science. Perhaps by examining the historical development of that tradition, we can tell what problems lie behind the quest for a logic of discovery.

Before we get to the history itself, a clarificatory preamble is in order. The term 'logic of discovery', like 'discovery' itself, is notoriously ambiguous. If one views the logic of justification as concerned exclusively with a study of the evidence relevant to the proverbial 'finished research report', then the logic of discovery—construed as dealing with the development and articulation of an idea at every stage in its history prior to its ultimate ratification—has a very wide scope indeed. It would include

---

Reprinted by permission from *Scientific Discovery, Logic and Rationality,* T. Nickles, ed., 1980. Copyright by D. Reidel Publishing Co.

an account of how a theory was first invented, how it was preliminarily evaluated and tested, how it was modified, and the like. I do not intend to interpret discovery so broadly. Between the context of discovery and the context of ultimate justification, there is a nether region, which I have called the *context of pursuit.*[1]

In the context of pursuit, the constraints are (presumably) tighter than the constraints (if any there be) on discovery and significantly looser than those we insist upon where belief or acceptance is concerned. This 'three-fold way' has several virtues lacking in the usual dichotomy. For one thing, these three contexts mark the temporal, if not the logical, history of a concept. It is first discovered; if found worthy of pursuit, it is entertained; if further evaluation shows it to be worthy of belief, it is accepted. More importantly, this trichotomy prevents us from lumping together activities and modalities of appraisal which have frequently but erroneously been confused with one another. For instance, both Peirce and Hanson construed the method of 'abduction' as a logic of scientific discovery. But it is nothing of the kind. As Wesley Salmon and others have pointed out, abduction does *not* tell us how to invent or discover an hypothesis. It leaves that (possibly creative) process unanalysed and tells us instead when an idea is worthy of pursuit (namely, when it explains something we are curious about). By calling the abductive method a partial logic of discovery, Hanson and to a lesser degree Peirce are culpable of having obfuscated the real nature of a logic of discovery. Equally, many of Hanson's critics have erroneously concluded that because abduction is no method of discovery, it must necessarily belong to the context of justification. Neither party to the dispute has seen that the abductive method belongs most naturally to neither discovery nor justification, but rather to pursuit.

However, my purpose in the early part of this paper is not to engage in a contemporary philosophical debate. The aim, rather, is to look at the historical development of views about the logic of discovery in order to ascertain why the optimism and the urgency among our philosophical forebears about understanding the process of discovery has generally given way to pessimism about the very possibility of a philosophical account of discovery.

These contentious preliminaries will have served their purpose, however, if they permit me to say that my concerns here will be with views about scientific discovery, not scientific pursuit. Accordingly, I shall construe discovery rather narrowly as concerned with 'the *eureka* moment', *i.e.*, the time when a new idea or conception first dawns, and I shall view the logic of discovery as a set of rules or principles according to which new discoveries can be *generated.* The historical justification for such a construal is that this latter sense is precisely what was traditionally meant by the logic or philosophy of discovery. The philosophical rationale for interpreting the logic of discovery in its narrow, rather than its broad, sense is that *only* on this construal can any sense be given to the current debate about the existence of a logic of discovery; since I take it that no one would deny that there clearly are rules or general principles governing discovery, if that term is construed so broadly as to include pursuit.

II

An event of major significance occurred in the course of 19th-century philosophy of science. The task of articulating a logic of scientific discovery and concept formulation—a task which had been at the core of epistemology since Aristotle's *Posterior Analytics*—was abandoned. In its place was put the very different job of formulating a logic of *post hoc* theory evaluation, a logic which did not concern itself with how concepts were generated or how theories were first formulated. This transformation marks one of the central watersheds in the history of philosophical thought, a fundamental cleavage between two very different perspectives on how knowledge is to be legitimated.

Throughout the 17th and 18th centuries (as in antiquity), the enterprise of articulating a logic of discovery flourished. Bacon, Descartes, Boyle, Locke, Leibniz and Newton—to name only the more prominent—all believed it was possible to formulate rules which would lead to the discovery of 'useful' facts and theories about nature. By the last half of the nineteenth century, this enterprise was dead, unambiguously repudiated by such philosophers of science as Peirce, Jevons, Mach, and Duhem. This essay is a speculative attempt to explain the changing historical fortunes of the logic of discovery; speculative because the space available to me does not allow for the sort of detailed historical documentation which would be required to make the story convincing. I shall aim, more modestly, at making the story at least plausible.

The desire to develop a logic of discovery was generally based upon two quite different motives. On the one hand, there was the *heuristic* and *pragmatic* problem of how to accelerate the pace of scientific advances, of how to increase the rate at which new discoveries were made by articulating fruitful rules for invention and innovation. (This side of the issue was especially prominent in Bacon, although there are clear signs of it in Descartes and Leibniz as well.) On the other hand, and more importantly, there was the *epistemological* problem of how to provide a sound warrant for our claims about the world. If a fool-proof logic of discovery could be devised, it would solve both problems simultaneously. It would both be an *instrumentarium* for generating new theories and, because infallible, it would automatically guarantee that any theories produced by its use were epistemically well grounded.

This second function is the crucial one to stress here. As conceived by most 17th- and 18th-century authors, *a logic of discovery would function epistemologically as a logic of justification.* Unlike now, there was then no distinction drawn between the contexts of discovery and justification. It is not that our forebears could not recognize the difference between discovery and justification; they could, and often did. But they were convinced that an appropriate (*i.e.*, infallible) logic of discovery would automatically authenticate its products and that a separate logic of justification was therefore redundant and unnecessary. They were preoccupied with developing logics of discovery, not because they were indifferent to the epistemological problem of justifying knowledge claims, but precisely because they took the justificational problem to be central.

Let me unpack this a bit more. Simply put, thinkers since antiquity had explored three rival views about how to justify the claim of natural philosophy to be genuine *scientia.* On one possible account—which corresponds to a familiar 20th-century caricature of Descartes—scientific claims were self-authenticating. They had only to be understood to see that they were true and could not be otherwise. There is scarcely any major figure in the history of philosophy, including Descartes, who maintained that most scientific theories could be authenticated in this way. Virtually all writers agreed that scientific truths about the world could only be 'educed' from contingent empirical data about the world; they agreed that there was no warranting process for typical scientific theories which did not involve *a posteriori* evidence. But beyond this point of agreement, there was a significant parting of the ways. One group—whom I shall call the *consequentialists*—believed that theories or claims could be justified by comparing (a subset of) their consequences with observation. If an appropriately selected range of consequences proved to be true, this was thought to provide an epistemic justification for asserting the truth of the theory. A second, and more predominate, group, whom I shall call the *generators,* believed that theories could be established only by showing that they followed logically (using certain allegedly truth-preserving algorithms) from statements which were directly gleaned from observation.

Much of this paper will be about the debate between the consequentialists (who tended to stress hypothetico-deduction and *post hoc* confirmation) and the generators (who believed that generational algorithms were the ideal device for authentication). What needs to be stressed at the outset is that *both groups*—including the latter, who argued for a logic of discovery—*were primarily concerned with the epistemic problem of theory justification.* The historical vicissitudes of the generators' program for establishing a logic of discovery are utterly unintelligible, I submit, unless one realizes that the *raison d'être* for seeking a logic of discovery was to provide a legitimate logic of justification.

As Dewey and Popper have both stressed, much of the history of epistemology is characterised by an infallibilist bent. Aristotle and Plato, Locke and Leibniz, Descartes and Kant: all subscribed to the view that legitimate science consists of statements which are both true and known to be true. To provide an epistemic justification for an assertion, *i.e.*, to show that it is 'scientific', was to point to evidence and rules of inference which, collectively, guaranteed its truth. We shall shortly come to see that this infallibilist orientation was closely linked with generators and the logic of discovery.

Among both ancient astronomers and ancient physicians, consequentialism was a prevalent doctrine. The Hippocratic tradition and the saving-the-phenomena tradition advocated the view that beliefs were to be judged in terms of an analysis of the truth of their consequences.[2] Eschewing the belief that theories could somehow be derived from observation, these thinkers stressed the necessity for the *post hoc* evaluation of a proposed theory in light of how well it stood up to observational and experiential tests.

But Plato and Aristotle, thorough going infallibilists where *episteme* was concerned, quickly saw the flaw in consequentialism. To argue from the truth of a consequence of a theory to the truth of the theory itself is, as they perceived, a logical fallacy (the so-called fallacy of affirming the consequent). They quite correctly argued that an infallibilist view of science was strictly incompatible with any form of consequentialism, *i.e.*, with any form of *post hoc* empirical testing, since the latter was logically inconclusive. If theories were to be demonstrably true, such demonstration could not come from any (nonexhaustive) survey of the truth of their consequences.

Because consequentialism was generally acknowledged to be epistemically inconclusive, most of those writers who believed science was infallible knowledge opted for some form of generationism. After all, if truth-preserving rules could be found for moving from particulars to universals, *and* if reliable particulars could be got at (which few seriously doubted), then one would have all that was required to give a justification of science as infallible knowledge. The point is this: if one seeks infallible knowledge and if one grants the fallaciousness of affirming the consequent, then the only viable hope for a logic of justification will reside in the quest for a truth-preserving logic of discovery. Except in very special circumstances,[3] infallibilism leads ineluctably to generationism (although *not* vice versa) and to the obliteration of any significant distinction between the logics of discovery and justification.

III

A part, then, of the attractiveness of a logic of discovery was linked to the prevalence of an infallibilist view of science. So long as the latter was in fashion, the former would remain in vogue. But this linkage is only one part of a very complex story. Another piece of the puzzle is closely related to views about the *objects* of science.

I can motivate this linkage by beginning with an observation on our own time. If there is general scepticism today about the viability of a logic of discovery, it is in part because most of us cannot conceive that there might be rules that would lead us from laboratory data to theories as complex as quantum theory, general relativity, and the structure of DNA. Our shared archetypes of significant science virtually all involve theoretical entities and processes which are inferentially far removed from the data which they explain. That there might be rules to lead one from tracks on a photographic plate to claims about the fine structure of subatomic particles is, to say the least, implausible. But suppose our archetypes were rather different; that they were such lawlike statements as 'All crows are black'; 'All gases expand when heated'; 'All planets move in ellipses'. If we viewed these as the primary objects of scientific inquiry, the notion that there might be some algorithm for going from evidence to discovered theory would not seem so bizarre. (Indeed, induction by simple enumeration, Peirce's 'qualitative induction' and Reichenbach's version of the straight rule are precisely such algorithms.)

The point is this: if what we expect to discover are general statements concerning observable regularities, then mechanical rules for generating a universal from one of more of its singular instances are not out of the question. By contrast, if we expect to discover 'deep-structure', explanatory theories (some of whose central concepts have no observational analogues), then the existence of plausible rules of discovery seems much more doubtful.

The historical significance of this contrast is that we should expect to find logics of discovery much more fashionable in epochs when the object of science is seen as discovering empirical laws than in epochs when the stress is upon discovering explanatory, deep-structural theories. This conjecture is confirmed by the historical record. Thus, during much of the 18th century, when Bacon and Newton had persuaded most philosophers that speculative theories and unobservable entities were anathema, inductive logics of discovery were ubiquitous among empiricists. Hume, Reid, D'Alembert, Priestley, and a multitude of other enlightenment philosophers took for granted the existence of an inductive logic of discovery. Similarly in the 1920s and 1930s, when again philosophy was dominated by the Spartan view that science consists chiefly of inductive generalizations, thinkers like Schiller, Reichenbach, Cohen and Nagel could devote much energy to discussing various inductive rules of discovery.

By contrast, in epochs when empirical generalizations are viewed as mundane and where theories are conceived chiefly as grandiose ontological frameworks, replete with unobservable entities, inductive logics of discovery have been ignored or, in some cases, their very existence denied. This was true of many 19th-century thinkers (such as Whewell and Boltzmann) who were struck by the nonobservational character of many of the chief theories of their time (especially important here were the wave theory of light and the atomic theory). It also seems to be true of our own time when virtually no philosopher shows any interest in the discovery of empirical regularities. Then, as now, inductive logics of discovery were given short shrift.

But the historical record also establishes that our *a priori* analysis was overhasty in one respect. While it is true that logics of discovery that involve some form of enumerative induction are taken seriously only by philosophers who believe that 'observational laws' typify scientific inquiry, there are other, *non*inductive logics of discovery which have been invoked by those concerned with the analysis of full-blown theories. Indeed, if we look carefully at the writings of many 18th- and 19th-century philosophers of science (including Hartley, LeSage, Priestley and Peirce[4]), we frequently find a concern with modes or methods of discovery which are quite unlike enumerative induction. I shall call these 'self-corrective logics of discovery'. Such 'logics' involve the application of an algorithm to a complex conjunction which consists of a predecessor theory and a relevant observation (usually one that refutes the prior theory). The algorithm is designed to produce a new theory which is 'truer' than the old. Such logics were thought to be analogous to various self-corrective methods of approximation in mathematics, where an initial posit or hypothesis was successively modified so as to produce revised posits which were demonstrably closer to the true value.

The appeal of this form of the logic of discovery was that, unlike inductive logics of discovery, it did not restrict itself to sentences about observables. The avowed aim, in exploring such a logic of discovery, was to have a logic rich enough that it could deal with the genesis of deep-structural theories. Unfortunately, a century and a half of exploration by a succession of major thinkers failed to bring the self-corrective program to fruition. (Peirce, for instance, grappled with this program for forty years before abandoning it.) No one was able to suggest plausible rules for modifying earlier theories in the face of new evidence so as to produce clearly superior replacements.

Thus far I have not been offering history, but rather some conceptual preliminaries with which I hope to make the history intelligible. The historical story itself is easy to tell. Most 17th- and 18th-century writers were epistemic infallibilists; accordingly, they looked chiefly to a logic of discovery rather than a logic of testing to provide the indubitable warrant for genuine science. Among the empiricists of the period (especially Locke, Newton, and Hume), there was the added conviction that laws linking observables—rather than 'transductive' theories explaining observables—were the hallmark of genuine science. This, too, conduced to make a justificatory logic of discovery plausible.

By the 1750s, the picture becomes murkier. Explanatory theories became fashionable again (subtle fluids, atoms, aethers, *etc*.). Accordingly, there was a distinct shift away from 'inductive' logics of discovery towards self-corrective logics of theoretical discovery (these are prominent in the writings of Hartley, LeSage, and Priestley). Infallibilism, however, remained the epistemic orthodoxy.

By the 1820s and 1830s, infallibilism itself was crumbling. Herschel, Whewell and Comte all acknowledged that there is no formula for producing true theories. As fallibilism emerged, there was an unmistakable shift away from the analysis of genesis towards the *post hoc* evaluation of theories.

It was argued that theories could not be proven to be true and that the most we can expect is that they can be shown to be likely or probable. The task of justification within such an orientation becomes the more modest (if still troublesome) one of showing that theories are likely. As soon as epistemic fallibilism replaces infallibilism, the possibility of justifying theories by examining their consequences becomes viable again; for all the familiar arguments about the impotence of *post hoc* confirmation to prove theories cease to be relevant once justification is no longer perceived as a matter of proof.

Thus, nongenerational logics of justification and epistemic fallibilism emerge simultaneously in the 19th century and jointly render redundant the logic of discovery. The confluence of these ideas can be shown clearly in the works of Herschel and Whewell, who were among the first philosophers of science to stress that theories could be judged independently of a knowledge of their mode of generation.

Herschel puts the point succinctly:

> In the study of nature, we must not, therefore, be scrupulous as to *how* we reach to a knowledge of such general facts, i.e., laws and theories: provided only we verify them carefully when once dictated, we must be content to seize them wherever they are to be found. (1830, p. 164)

Herschel's point involves no denial of the existence of a logic of discovery; he is rather asserting the *irrelevance* of the manner of generation of a hypothesis to its evaluation or justification. Whewell goes a step farther and denies the very existence of a logic of scientific discovery:

> Scientific discovery must ever depend upon some happy thought, of which we cannot trace the origin; some fortunate cast of intellect, rising above all rules. No maxims can be given which inevitably lead to discovery. (1847, Vol. II, pp. 20–21)

What Herschel and Whewell are pointing out (and their contemporary Comte makes a similar argument) is that (1) theories can be appraised ('verified') independently of the circumstances of their generation, and (2) such modes of appraisal, even if fallible, are more germane to the process of justification than any fallible rules of discovery would be. Where earlier philosophers had believed that it was only *via* a logic of discovery that theories could be justified, Herschel and Whewell sever that link, insisting that justification need not be parasitic on discovery.

Their proposed divorce of justification from discovery quickly became the philosophical orthodoxy. The lack of serious resistance to that separation is understandable, *provided* one realizes that justification had been the central problem all along. As soon as justification was separated from proof and as soon as plausible modes of justification were suggested which circumvented the context of discovery, most philosophers of science willingly replaced the programme for finding a logic of discovery—which was foundering anyway for the reasons mentioned above—by a programme for defining post-discovery empirical support.

IV

This latter programme, of course, has characterised most of the philosophy of science since the middle of the 19th century. Some recent writers who would revive an interest in the logic of discovery see it as something very different from the logic of justification. In this sense, they are radically at odds with the traditional aims of the logic of discovery. The older programme for a logic of discovery at least had a clear philosophic rationale: it was addressed to the unquestionably important philosophical problem of providing an epistemic warrant for accepting scientific theories. The newer programme for the logic of discovery, by contrast, has yet to make clear what philosophical problems about science it is addressing.

I have sought to show in this essay that the primary motivation for seeking what we call a logic of discovery was to address and resolve an *epistemological* problem about the well-foundedness of knowledge claims about the world. The program for articulating an infallible logic of discovery never came to fruition; but that failure only partially explains its abandonment. Equally crucial here was the joint emergence of epistemic fallibilism and of *post hoc* logics of theory testing; developments which rendered redundant and gratuitous the logic of discovery so far as the epistemological issue is concerned. It remains redundant now.

At the same time, it should be stressed that there is also a *heuristic* problem about science: how can we maximize the rate at which new and promising theories and laws are generated? *Post hoc* logics of justification have nothing to contribute here and can in nowise fulfill the heuristic tasks conceived for a logic of discovery. But before one concludes that the logic of discovery still has a philosophical rationale, one must ask what is specifically *philosophical* about studying the genesis of theories. Simply put, a theory is an artifact, fashioned perhaps by certain tools (*e.g.*, implicit rules of 'search'). The investigation of the mode of manufacture of artifacts (whether clay pots, surgical scalpels, or vitamin pills) is not normally viewed as a philosophical activity. And quite rightly, for the techniques appropriate to such investigations are those of the empirical sciences, such as psychology, anthropology, and physiology. The philosopher of art is not concerned *qua* philosopher with how a sculpture is chiseled out of a piece of granite; nor is the philosopher of law concerned with the mechanics of drafting a piece of legislation. Similarly, the case has yet to be made that the rules governing the techniques whereby theories are invented (if such rules there be) are the sorts of things that philosophers should claim any interest in or competence at. If this essay provides a partial answer to the question "Why was the logic of discovery abandoned?", it poses afresh the challenge: "Why should the logic of discovery be revived?"

NOTES

1. For a discussion of this problem, see my *Progress and Its Problems* (1977*a*, pp. 108–114).

2. See especially Duhem (1973).

3. The circumstances to which I refer are of two kinds:

(a) if one believed that *all* the consequences of a theory could be examined, then consequentialism and infallibilism would be compatible;

(b) if one believed that all possible theories could be enumerated and rejected seriatim by a method of exhaustion, then consequentialism and infallibilism are also compatible. For those many writers who rejected both of these assumptions, infallibilism ruled out any form of consequentialism.

4. For lengthy discussions of the views of these writers on the philosophy of science, see my (1970), (1973), (1977*b*), and (forthcoming).

BIBLIOGRAPHY

Duhem, P.: 1973, *To Save the Phenomena,* Univ. of Chicago Press, Chicago.

Herschel, J. F. W.: 1830, *A Preliminary Discourse on the Study of Natural Philosophy,* London.

Laudan, L.: 1970, "Thomas Ried and the Newtonian turn of British methodological thought," in R. Butts and J. Davis (eds.), *The Methodological Heritage of Newton,* Univ. of Toronto Press, Toronto, pp. 103–131.

Laudan, L.: 1973, "Charles Sanders Peirce and the trivialization of the self-corrective thesis," in R. Giere and R. Westfall (eds.), *Foundations of Scientific Method in the 19th Century,* Indiana Univ. Press, Bloomington, pp. 275–306.

Laudan, L.: 1977*a*, *Progress and Its Problems,* Univ. of California Press, Berkeley.

Laudan, L.: 1977*b*, "The Sources of Modern Methodology," in J. Hintikka and R. Butts (eds.), *Historical and Philosophical Dimensions of Logic, Methodology and Philosophy of Science,* D. Reidel, Dordrecht, pp. 3–20.

Laudan, L.: forthcoming, "The Medium and Its Message: A Study of Some Philosophical Controversies About the Ether," in J. Hodge and G. Cantor (eds.), *The Subtler Forms of Matter,* Cambridge Univ. Press, Cambridge.

Whewell, W.: 1847, *Philosophy of the Inductive Sciences, Founded Upon Their History,* 2nd ed., London.

Martin V. Curd

THE LOGIC OF DISCOVERY: AN ANALYSIS OF THREE APPROACHES

Is there anything about the discovery of scientific theories that is of legitimate concern to the philosopher of science? The answer to this question is still controversial. Popper (1959), Hempel (1966), and Braithwaite (1955), for example, have categorically denied that anything like a logic of discovery is possible. On the other hand, the notion has been vigorously defended by C. S. Peirce and, more recently, by Hanson (1961), Achinstein (1970), Simon (1973), Shaffner (1974) and others.

The aim of this paper is to examine arguments for and against the possibility of a logic of discovery and to suggest a resolution of the debate. Three major positions will be analyzed:

(I) The hypothetico-deductive account (Popper, Hempel);
(II) The inductive-probability account (Reichenbach, Salmon);
(III) The abductive inference account (Peirce, Hanson).

Before proceeding to the discussion of these three positions, I shall draw some preliminary distinctions. Failure to attend to these distinctions has, I believe, been responsible for needless confusion in discussions of whether or not a logic of discovery is possible.

First, it should be noted that in everyday discourse 'discovery', like 'seeing' and 'recognizing', is a 'success' word. To say that a person *P* has discovered *X* (where *X* is a physical object) carries with it the strong implication that (1) what has been discovered actually exists, and (2) in reporting the discovery, *P* has correctly described *X*. The first condition is clearly essential to any legitimate claim to the discovery of a physical object. The second condition can be controversial in everyday contexts and especially in scientific priority disputes. (What, for example, constitutes giving the 'correct' description of a physical object?[1]) Since I am concerned in this paper with the discovery of theories, not the discovery of things, I shall not discuss the second condition further except to note that in everyday contexts, to say that *P* has discovered that *Y* (where *Y* is a proposition) implies that *Y* is true. It is a violation of ordinary language to say that *P* has discovered that *Y* if *Y* is false. If, however, we impose this condition on the discovery of scientific theories, it has the unacceptable consequence that a theory, *T*, has been discovered only if *T* is true. In order to avoid this unwarranted intrusion of considerations based on ordinary language, I recommend giving a neutral account of the discovery of *T* in terms of 'the period of theory generation'. On this account it remains an open question whether *T* is true or false after it has been discovered or 'generated'.[2]

When one speaks of the discovery of a scientific theory, one usually refers not to a specific moment but to an extended period of time which I shall call *the period of theory generation*. Pierre Duhem satirized the

Reprinted by permission from *Scientific Discovery, Logic and Rationality*, T. Nickles, ed., 1980.

'instant creation' view of the discovery of theories when he wrote:

> The ordinary layman judges the birth of physical theories as the child the appearance of the chick [from an egg]. He believes that this fairy whom he calls by the name of science has touched with his magic wand the forehead of a man of genius and that the theory immediately appeared alive and complete, like Pallas Athena emerging fully armed from the forehead of Zeus. (1955, p. 221)

Theories, no less than chickens' eggs, require a period of incubation before the final product emerges.

For the purposes of this paper, I shall define the period of theory generation as beginning at the moment when a scientist (or research group) first begins thinking seriously about a problem and ending when the theory—what Hanson called the 'finished research report'—is first written down in a form suitable, say, for publication in a scientific journal. I shall assume that up to this time the theory has not been experimentally tested and that the end of the period of theory generation marks, so to speak, the first public appearance of the theory in a fully articulated form. Thus, for example, in the case of the discovery of the double helix model of DNA, the period of theory generation began with the first collaboration of Watson and Crick on the problem in 1951 and ended with the appearance of their famous paper in the journal *Nature* in 1953.

The stipulation that during the period of theory generation the theory has not yet been tested is not as artificial as it might appear at first sight. What is crucial is what one means by a test. I understand by a test the derivation from a theory of a result *not previously known* and the investigation of whether this novel prediction obtains in a controlled experiment. Most theories are generated so that they are not inconsistent with any result which is accepted as true and so that they permit the derivation of results which are regarded as relevant to the domain of inquiry in question. One might express this by saying that theories are usually deliberately constructed so that they conform to a set of constraints.[3] Since these constraints are based on antecedently known results (data, facts, laws, well-established theories), satisfaction of them does not qualify as a test in the sense adopted here.[4]

In talking about a logic of discovery, we might be interested in two very different aspects of the period of theory generation. I shall call these aspects:

(A) a logic of prior assessment; and
(B) a logic of theory generation.[5]

(A) A logic of prior assessment concerns the methodological appraisal of hypotheses after they have been generated but before they have been tested. That there must be some logic of this sort is suggested by the fact noted by Peirce that "proposals for hypotheses inundate us in an overwhelming flood" (1931–1958, 5.602). Given a finite body of empirical data, there is always a potentially infinite number of alternative hypotheses from which the data can be deduced. We must have some way of deciding which hypotheses to take seriously and which to ignore since our resources, both intellectual and financial, are finite. Apparatus is expensive to build; experiments are costly to conduct; we have only a limited amount of scientific manpower at our disposal.

Now I wish to draw a clear distinction between two kinds of prior assessment:

(A1) a logic of probability;
(A2) a logic of pursuit.

A logic of probability is concerned with judgments of the following type: Hypothesis H is probable or likely to be true; H_1 is more probable or more likely to be true than H_2. A logic of pursuit, on the other hand, is concerned with much more down-to-earth questions. Which hypothesis should we work on? Which hypothesis should we bother taking seriously enough to test? A logic of pursuit issues in such judgments as: Hypothesis H is worthy of pursuit; H_1 is more worthy of pursuit than H_2.[6] Both kinds of appraisal, whether of probability or pursuit-worthiness, presuppose that the hypotheses to be judged or compared have already been articulated or generated. That the logic of probability differs from the logic of pursuit is indicated by the fact that even if there were an objective measure of the probability of hypotheses, we might still have good reasons for testing some improbable theories first or for pursuing hypotheses which, on the basis of presently available evidence, are less likely to be true.

I shall argue that not only is the logic of pursuit of more immediate practical relevance to scientific inquiry than the logic of probability but also that it is the only workable notion of a logic of discovery in the sense of a logic of prior assessment that one can formulate. This stands the usual account on its head. We are normally told that the decision to pursue a theory should be based on our judgment of its probability or the likelihood that it is true, *i.e.*, the logic of pursuit is deemed to be parasitic upon the logic of probability. But I maintain that we seldom have any way of telling which of our newly generated hypotheses are likely to be true or probable, whereas we do have the means for deciding which hypotheses to pursue, which decisions do not depend on our inability to make these probability judgments.[7]

My principal objection to the claim that we can know of an hypothesis before it is tested that it is probable or likely to be true stems from a consideration of the concept of truth. It follows from a definition of truth such as Tarski's that if the hypothesis H is true then the following conditions will be satisfied:

(1) H is free from internal contradiction;
(2) none of the consequences of H are false;
(3) H is consistent with other true statements.

Conditions (1), (2) and (3) are necessary but jointly insufficient to ensure the truth of H. Even if we could know that they are satisfied, at best they establish that H might be true.[8] But if the satisfaction of these conditions does not guarantee truth, does it not at least entitle us to claim that H is probable or likely to be true? I think not. Consider a hypothesis H_1 which satisfies the above conditions and which, in the terms introduced earlier, conforms to a set of empirical constraints. For the sake of brevity let us say that H_1 is empirically adequate. Now, given H_1, it will usually be possible to devise another hypothesis, H_2, by a fairly trivial modification of the original hypothesis so that H_2 is incompatible with H_1 and yet has the same degree of empirical adequacy. For example, H_1 might be an inverse square law hypothesis and H_2, the hypothesis that the exponent of $1/r$) differs from 2 in the tenth decimal place. Of course, H_1 will generally be simpler than H_2, embody more relevant analogies and be thought to provide a better explanation of the phenomena than H_2, but on what grounds are we justified in asserting that H_1 is more probable than H_2 or more likely to be true? Criteria such as simplicity and explanatoriness have no obvious connection with truth. In fact, many theories which in the past have best satisfied these criteria have turned out to be false.

The defender of probability judgments clearly stands in need of an adequate inductive logic which would assign $P(H_1/e)$ a higher value than $P(H_2/e)$. But none of the attempts thus far to formulate such a logic that can be applied to scientific theories has met with success. While it would be unsound to conclude from the failures of Reichenbach's frequency approach and Carnap's program to provide an adequate analysis of the probability of hypotheses that the concept is illegitimate ('having an adequate analysis of *X*' is not identical with 'having the concept of *X*'), these failures do militate against the use of probability or likelihood to be true in a logic of prior assessment.

(B) What a logic of theory generation might be is far less clear. Three possibilities that I shall consider are:

(B1) The specification of a procedure, possibly algorithmic, that will generate non-trivial hypotheses.

(B2) An historical narrative of the sequence of steps followed by an individual in his path to the hypothesis.

(B3) A classification and analysis of the inferences scientists make in reasoning to their hypotheses plus a philosophical justification of why these inferences are reasonable ones.

My claim is that when version (B3) of a logic of theory generation is correctly understood as a *rational reconstruction,* the justification for the inferences to theories depends on the categories of appraisal in the logic of prior assessment. And, since I maintain that the only viable notion of a logic of prior assessment is a logic of pursuit, it turns out that the logic of pursuit is the real key to the logic of discovery.

I: THE HYPOTHETICO-DEDUCTIVE ACCOUNT

The strongest opposition to the possibility of a logic of discovery has been voiced by such philosophers of science as Popper, Hempel and Braithwaite. Popper, for example, has written:

> . . . the work of the scientist consists in putting forward and testing theories.
>
> The initial stage, the act of conceiving or inventing a theory, seems to me neither to call for logical analysis nor to be susceptible of it. The question how it happens that a new idea occurs to a man—whether it is a musical theme, a dramatic conflict, or a scientific theory—may be of great interest to empirical psychology; but it is irrelevant to the logical analysis of scientific knowledge . . . My view of the matter . . . is that there is no such thing as a logical method of having new ideas, or a logical reconstruction of this process. My view may be expressed by saying that every discovery contains 'an irrational element', or 'a creative intuition' in Bergson's sense. (1959, pp. 31–32)

The view that there is not, and cannot be, a logic of discovery is commonly represented as a corollary to the hypothetico-deductive account of scientific method. On this account there are *deductive* inferences *from* theories to predictions and possibly (for non-Popperians, at least), *inductive* inferences from successful predictions back *to* the theory. But these inferences take place when the theory is tested. There are no inferences to the theory in the first place.

Theories and hypotheses are simply guesses, *happy* guesses one hopes, but guesses nonetheless. As Einstein (1933) put it, "a theory can be tested by experience, but there is no way from experience to the setting up of a theory," theories are "free creations of the human mind." Thus, according to the hypothetico-deductive account, the manner in which a theory was first suggested is purely a psychological matter—whether it be from a dream, a vision or through a long and arduous struggle with recalcitrant data—this has no bearing on the theory's initial plausibility, no relation to the logical analysis of scientific knowledge, and hence no relevance to the concerns of the philosopher of science. Any talk of a logic of discovery is therefore tantamount to an invitation to commit the genetic fallacy.

In the light of our earlier distinction between two conceptions of a logic of discovery, we see that the hypothetico-deductive account is directed not so much against the existence of a logic of prior assessment as against the possibility of a logic of theory generation. One could consistently defend the former conception while opposing the latter. And in fact Popper, Hempel and others do discuss the general requirements that a hypothesis must satisfy before we are warranted, in their view, in taking it seriously enough to bother testing it. The factors considered by Popper, for example, in the prior assessment of a hypothesis, *H*, are:

(1) The internal consistency of *H*. Is *H* self-contradictory?
(2) The empirical character of *H*. Is *H* nontautological?
(3) The implications of *H* for the growth of scientific knowledge. Would *H* "constitute a scientific advance should it survive our various tests?" (Popper, 1959, p. 33).

For Popper, of course, the considerations involved in factor (3) can only be pursuit norms since, on his view, we never *accept* theories even *after* they are tested. In this light, the restriction to "survival" in factor (3) seems too strong. We might well wish to pursue *H* on the grounds that we stand to learn something in the process of trying to refute it. This relaxation of factor (3) also seems to be more compatible with Popper's falsificationist viewpoint.

Now at first blush, the denial of a logic of theory generation seems a counterintuitive conclusion. Are scientists just guessing when they come up with hypotheses? One has visions of that mythical tribe of chimpanzees which, by bashing away for long enough on a stack of typewriters, would eventually generate not only the complete works of Shakespeare but also Einstein's general theory of relativity and the Watson-Crick hypothesis for the structure of DNA. The position that the hypothetico-deductive account adopts with respect to this issue is not the result of prejudice but is based on arguments which many philosophers of science still find convincing. Let us now examine these arguments to see precisely what they do or do not serve to establish.

The first argument I shall call the 'discovery machine' objection or the argument from the impossibility of a 'logic of invention'. Popper and Hempel have devoted much attention to criticising what Hempel calls the 'narrow inductivist conception of scientific inquiry' (1966, p. 11). This conception, which is often attributed (somewhat unfairly I think) to Bacon and Mill, maintains that scientists are in possession of a method which enables them to infer mechanically true scientific theories from an exhaustive collection of facts gathered without any theoretical preconceptions. The criticism consists in pointing out that we are not in possession of such an algorithm and that it is impossible, in principle, for there to be one for three basic reasons:

(1) Without a prior hypothesis to guide our investigation we have no idea which facts are relevant to our inquiry and in any case the set of such facts could never be exhaustive.

(2) Theories are always underdetermined by the available data, so that even if the theory which is generated did conform to all the known facts, there is no guarantee that the theory is true.

(3) The discovery of significant theories such as those associated with the names of Newton, Maxwell and Einstein involves an essential element of creativity and conceptual innovation which could never be performed by a machine following an algorithm.

These arguments do constitute a serious objection to the conception of a logic of theory generation as an inductivist discovery machine. But they do not, it seems to me, rule out other, less objectionable, conceptions of a logic of theory generation. In particular one might still wish to consider procedures that are algorithmic (or failing that, at least programmable) but which are applied to small-scale discoveries that do not involve any conceptual innovation and where, so to speak, all the relevant data *is* in at the start. Cases which may prove amenable to such procedures involve pattern recognition in sets of data, *e.g.*, Mendeleef's periodic table of the elements, Balmer's formula for the spectral lines of hydrogen, and Kepler's third law. Herbert Simon and his associates are currently working on Heuristic Search Algorithms which may, eventually, be able to match the human mind in the ability to detect simple regularities in complex data (Simon, 1973). One might be forgiven, however, for failing to see how such programs will shed any philosophical light on important, conceptually innovative discoveries.

Nothing in the discovery machine objection affects the third conception of a logic of theory generation as a classification and analysis of the inferences that scientists make in arguing to hypotheses plus an account of why they are reasonable ones. To merely assert that such a conception is untenable because of the foregoing argument is to beg the question. Of course the hypothetico-deductive theorists believe that they do have arguments against this third conception, and it is to these that I now turn.

The two kinds of argument that I shall consider are directed against what is often called 'psychologism'. They are

(1) the 'eureka moments' objection, and
(2) the thesis of nonisomorphism.[9]

The eureka moment phenomenon is familiar to anyone who has heard the anecdote about Newton watching an apple fall in his mother's orchard while pondering the mysteries of gravitation. There are also well-documented accounts by successful scientists of what Popper calls moments of 'poetic intuition' when the solution to a problem first presents itself to the conscious mind. James Watson, for example, relates how his 'pulse began to race' (1969, p. 118) as he played with his crude templates of the nucleotide bases and saw how the bases might be hydrogen bonded like-with-like.

'Eureka moments', then, are those ephemeral moments of insight, often attended by feelings of conviction or illumination, which sometimes accompany the dawning of important scientific discoveries, as well as, one suspects, many an abortive speculation. Watson's pulse rate, for example, soon slowed down once it was pointed out to him that he had chosen the wrong tautomeric forms of guanine and adenine (Watson, 1969, p. 120).

The objection to the possibility of a logic of discovery in the sense of a logic of theory generation consists in pointing out that such eureka moments constitute an ineliminable psychological element in many discoveries and, as such, will always resist logical analysis.[10]

A second and related type of objection falls under the heading of what I have called the thesis of nonisomorphism. It is pointed out that it is only rarely that the series of psychological events in the mind of the individual scientist leading up to the discovery of a theory can be represented in propositional form. Koestler and Polanyi, among others, have stressed the nonlinguistic character of much of creative thinking. The point is illustrated in a letter from Schrödinger to Reichenbach in which the path to theoretical discoveries is described as often proceeding through a "series of inferences which are deeply veiled by the darkness of instinctive guessing" (Reichenbach, 1944, p. 67). Since, then, a logic of discovery is presumably concerned with the logical relations between propositions, it cannot faithfully mirror the nonpropositional thought processes occurring in the mind of the individual who first made the discovery.

The response to these two objections is simply to deny that a logic of discovery is intended as a blow-by-blow account of the sequence of psychological events leading up to the formulation of an hypothesis. In this respect the logic of discovery should be compared with the logic of testing which is not intended as a mere historical narrative of what scientists are doing when they build pieces of apparatus, secure funding for experiments or take measurements. Philosophy of science is a normative, critical enterprise. As such, it is not interested in simply describing what scientists do—that is the task of the historian, sociologist or psychologist—rather, it is concerned with the cognitive justification of scientists' actions when they propose, test, adopt or reject theories.[11] This point is often expressed by saying that what philosophers of science are interested in is to provide *rational reconstructions* of what scientists do. If, then, there is to be a logic of discovery it will have to be a rational reconstruction of the period of theory generation within the context of justification. In order to see what promise this suggestion holds for tackling the problem of a logic of discovery, let us turn and examine the views of Hans Reichenbach to whom the terms 'rational reconstruction' and 'context of justification' are due.[12]

II: THE INDUCTIVE-PROBABILITY ACCOUNT

Reichenbach introduced the terms 'context of discovery' and 'context of justification' in the first chapter of *Experience and Prediction.* Reichenbach's aim in doing this was to draw a sharp line of demarcation between, on the one hand, the psychological description of thought processes as a temporal sequence of steps (that is, the context of discovery) and, on the other hand, the atemporal, logical relations between propositions (that is, the context of justification). Logic, for Reichenbach, is a *normative* enterprise concerned not with the description of inferences that people actually make but with their justification through the provision of a rational reconstruction. He remarked that 'the manner in which logical inferences are actually made is strange and obscure and rarely resembles the formal method of logic' (1957, p. 43). Philosophy of science, then, is solely concerned with the logic of justification since, for Reichenbach, this is what logic is; it is the logic of the context of justification. Strictly speaking, if we wish to remain faithful to Reichenbach's usage, it is impossible, as a matter of definition, for there to be a 'logic of discovery' in the sense of a 'logic of the context of discovery' since this latter context is *defined* as being purely a psychological matter.

Now Hanson interpreted Reichenbach's distinction between the two contexts as tantamount to the view that 'philosophy of science 'cannot be concerned with [reasons for suggesting hypotheses], but only with [reasons for accepting hypothesis]''.[13] And so Hanson included Reichenbach in the Popper-Hempel camp as one of those who denied that a logic of discovery was possible. In fact, however, this was far from Reichenbach's expressed intention. Hanson was guilty of interpreting Reichenbach's *logical* distinction as a *temporal* one. For Reichenbach believed that in *bona fide* cases of scientific discovery, a rational reconstruction was possible to an extent which philosophers (and even some scientists) were seldom prepared to admit. Reichenbach was just as much opposed to the 'happy guess' or 'mystic presentiment' view of the proposal of scientific hypotheses as was Hanson. In Reichenbach's view, despite what scientific researchers had to say about their discoveries in terms of 'natural hypotheses', 'simplicity' and 'mystic presentiments', the deep structure of scientific discovery is guided by the principle of induction (Reichenbach, 1938, p. 403).

Reichenbach made it quite plain that he was not claiming that the discoveries of men like Einstein were actually made by consciously performing chains of inductive inference, only that in these and other genuine cases of scientific discovery, the discovery could be rationally reconstructed within the context of justification. In the rational reconstruction the theory is made plausible or probable prior to testing by its inductive relation to antecedently known facts. Reichenbach asks:

> Why was Einstein's theory of gravitation a great discovery even before it was confirmed by astro-

nomical observations? Because Einstein saw—as his predecessors had not seen—that the known facts indicate such a theory; *i.e.*, that an inductive expansion of the known facts leads to the new theory. . . . (1938, p. 382)

The restriction to cases of *bona fide* discovery reflects Reichenbach's view that only in these cases are there inductive relations for the shrewd scientist to uncover and which serve to distinguish the status of his hypothesis from that of a mere guess. The logical situation after testing is no different in kind from that before testing except that the theory has attained a higher degree of inductive support. While the truth of the theory and the success of its predictions are not guaranteed by such inductive confirmation, Reichenbach did think he could show that the genuinely discovered theory represented the best posit, the hypothesis of greatest *inductive* simplicity, with respect to any given evidential base.

In the terms introduced earlier, what Reichenbach is offering as a logic of discovery is a logic of prior assessment and a logic of theory generation. The logic of prior assessment is a logic of prior plausibility, couched exclusively in terms of inductive probability. As such, it is coextensive with the logic of acceptance after testing.[14] The logic of theory generation consists solely of inductive inferences plus a heroic attempt to justify these inferences as reasonable ones in terms of the famous vindication approach.

Whatever the deficiencies of Reichenbach's conception of a logic of discovery, he should receive full credit for having had an honest and provocative stab at the problem. There is not time to develop my criticisms in detail. I shall just indicate briefly what I believe to be the shortcomings of Reichenbach's approach and why I think it ultimately fails.

(1) It is deficient as a logic of theory generation. It is difficult to maintain that inferences to theories such as Einstein's general theory of relativity, or even Newton's theory of universal gravitation, can be reduced to a concatenation of inductive inferences from known laws and data. One reason for this difficulty is the Duhemian observation that theories like Newton's usually entail the falsity of the laws like Kepler's which they are adduced to explain. The inferences in these cases are more like explanatory ones. For example, Newton's theory explains why the planets obey Kepler's laws to the extent that they do.

(2) It fails as a logic of prior assessment. I can see no way of assigning probabilities to hypotheses on a relative frequency interpretation of probability. Popper's criticisms of Reichenbach's attempt to do this (Popper, 1959, Section 80) are, I think, conclusive on this score.

The principal value of Reichenbach's analysis lies in its clear demonstration that the aspect of the logic of theory generation that concerns the philosopher of science is the justification of inferences to the theory via the provision of a rational reconstruction, and that this justification is achieved by appeal to the kind of appraisals made in the logic of prior assessment. I am convinced, however, that Reichenbach chose the wrong logic of prior assessment—the logic of probability, rather than the logic of pursuit.

III: THE ABDUCTIVE INFERENCE ACCOUNT

I shall now consider an attempt to approach the logic of discovery in terms of abductive inference. The major proponent of this approach was C. S. Peirce who coined the term 'abduction'. Recently his view have been championed in a somewhat modified form by N. R. Hanson.[15]

In 1910, Peirce drafted a letter to Paul Carus in which he confessed that in almost everything he had written before the turn of the century, he had more or less mixed up abductive inferences with inductive ones and that he now realized that abduction has nothing to do with probability (Peirce, 8.227). Before the transitional decade

1891–1901, Peirce regarded induction and abduction as inferences from data, either of which can play a role in theory generation, depending on the type of theory inferred (Burks, 1946; Goudge, 1950). Descriptive generalizations like Boyle's law are arrived at inductively; explanatory theories like the kinetic theory of gases are arrived at abductively. The difference between induction and abduction lies principally in the nature of the hypothesis which forms the conclusion.

After 1901 Peirce proposed that abduction, deduction, and induction are distinct types of inference corresponding to different stages of scientific inquiry. Deduction and induction are confined to the period of theory testing much as in the hypothetico-deductive account. Abduction, however, for Peirce, is now regarded as a form of ampliative inference peculiar to the period of theory generation. All hypotheses are arrived at abductively, regardless of their character.

Many commentators have experienced difficulty in finding a consistent interpretation of Peirce's conception of abduction, even when they have restricted their attention to his later writings.[16] This difficulty stems from the fact that Peirce seems to hold that:

(1) Abduction is the process of forming explanatory hypotheses and is the only logical operation which introduces new ideas.

(2) Hypotheses originate through instinctive guesses and flashes of insight.

(3) Abduction is the process of adopting an hypothesis 'on probation' as Peirce puts it. 'Adopting' here does not mean accepting the hypothesis as true, or even as inductively probable, but regarding the hypothesis as a working conjecture, a hopeful suggestion which is worth taking seriously enough to submit to detailed exploration and testing.

Thus, Peirce seems to be committed at the same time to the views that:

(1) Abduction is the logic of theory generation.

(2) Discoveries involve an ineliminable psychological component.

(3) Abduction is the logic of prior assessment.

From what was said earlier in response to the charges of psychologism, I do not think that there is any necessary incompatibility between (1) and (2). Thus, we are left with the task of reconciling (1) and (3). What kind of logic of theory generation did Peirce have in mind? And what kind of logic of prior assessment?

The reason that Peirce called abduction an inference is that he believed that it exhibits a definite logical form. He represented that form as:

> The surprising fact, *C*, is observed.
>
> But if *A* were true, *C* would be a matter of course.
>
> Hence, there is reason to suspect that *A* is true.
>
> (Peirce, 5.189)

As Peirce himself observed, the content of the hypothesis, *A*, is already contained in the premises. Hence, this mode of argument cannot literally represent the sequence of steps to the hypothesis in the first place. But then there is no reason why we should be interested in a mere narrative account. Since Peirce also rejected the idea of an algorithm for generating theories, this means that what Peirce's abductive inference schema was intended to represent is (a) a logic of theory generation in sense (B3), and (b) a logic of prior assessment. These two conceptions are not irreconcilable, since I have argued that it is the logic of prior assessment that provides the justification for the reasonableness of the inferences to the theory in our rational reconstruction. So, what kind of logic of prior assessment did Peirce intend to capture with his abductive inference? Though from its stated form it looks as if abduction is a logic of probability, I shall argue both that Peirce intended it as a logic of pursuit and that this is the more reasonable option.

Peter Achinstein (1970, 1971) has criticised abduction as a logic of probability on the grounds that it is far too permissive to establish its conclusion with any degree of plausibility. He illustrates the objection by means of the following kind of example. The surprising fact, *C,* is observed, that I am here today delivering this talk. The hypothesis, *A,* that I have been offered $1 million to speak, if true, would explain *C.* But this affords no grounds for believing that I am shortly to become a millionaire. In fact, from our background knowledge about universities, the hypothesis *A* is extremely improbable.

Achinstein's objection loses its force, however, when one realizes that abductive inference is not intended to establish that the hypothesis in question is probable or likely to be true, but only that it is reasonable to pursue, other things being equal. The categories of pursuit appraisal and probability appraisal are radically different. Peirce's schema for abductive inference should, accordingly, be modified along the following lines:

> The surprising fact, *C,* is observed.
> The hypothesis, *A,* is capable of explaining *C.*
> Hence, there are *prima facie* grounds for *pursuing A.*

The grounds are only *prima facie* because:

(1) There are almost certainly other hypotheses involved in our pursuit appraisal which may be capable of providing better explanations of *C.*

(2) There are other factors in the logic of pursuit. The ones specifically discussed by Peirce are simplicity, cost, and the implications of the pursuit of *A* for the rest of science. All these factors Peirce grouped together under the heading of the 'economy of research'. He wrote: 'the rules of scientific abduction ought to be based exclusively upon the economy of research' (7.220, n. 18).[17]

The justification for the categories of assessment in the logic of pursuit are pragmatic. We pursue explanatory theories because these are the kinds of theory we are interested in having. Explanatoriness is a stipulative requirement on our part, not necessarily a reliable indicator of truth.[18] Simplicity is not necessarily a sign that we have unlocked the mysteries of nature but a prudent preference for theories that are easier to work with and to test.

Peirce never quite relinquished the notion that abduction was also the key to the logic of probability. This was, I think, because of his belief that *via* abductive inferences, scientists had been remarkably successful at generating true theories. This struck Peirce as "quite the most surprising of all the wonders of the universe" (8.238). But Peirce was never able to justify abduction as leading to probable theories. The best that he could suggest as an explanation was the hypothesis that there is a peculiar affinity of the human mind with nature which leads us to guess correctly after only a small number of attempts. But, as Peirce himself conceded, this hypothesis is itself arrived at abductively. It cannot serve as an independent justification for the success of abduction.[19] Peirce never seems to have contemplated taking the step that I would recommend, namely, that of denying that we have as yet managed to give any convincing grounds for believing that our theories are true, close to the truth or probable, before they have been tested.

In conclusion, none of the objections examined in this paper establish the impossibility of a logic of discovery conceived as either a logic of prior appraisal or as a rational reconstruction of inferences to the theory. There are valuable insights to be gleaned from Peirce's treatment of abductive inference and his discussion of the economy of research. According to the interpretation developed here, the only workable conception of a logic of prior appraisal is of a logic of pursuit and it is the logic of pursuit which provides the justification for inferences to the theory in our rational reconstruction of the period of theory generation. And this is as it should be. The factors that justify our inferences to theories in the first place are the same as those that we use to decide which theory to pursue after they have been generated.[20]

NOTES

1. My suspicion is that providing the correct description of *X* is of primary relevance, not to the issue of whether *X* has been discovered, but to the question of whether *P* deserves the credit for having discovered *X*.

2. I also reject the placing of epistemic conditions on *T* before *T* can be said to have been discovered. Kordig (1978), for example, insists that 'real discoveries' have to be 'well-established' and 'justified' to count as genuine discoveries at all. On this account, *T* has been discovered by *P* only if *P* offers good reasons for believing that *T* is true or for 'accepting' *T*. While this may reflect the views of some scientists (consider, for example, Newton's objections to Hooke's claim to have discovered the inverse square law of gravitation), it seems an objectionably stringent requirement to impose on the meaning of the term 'discovery'. When, for example, on Kordig's account was Darwin's theory of evolution or Copernicus's theory of the heavens 'discovered' and by whom? A further disquieting consequence of Kordig's proposal is that it settles the issue of whether a logic of discovery is possible by purely verbal definition since all 'discoveries' (in his sense) necessarily involve the logic of justification or acceptance.

3. The Watson and Crick discovery is a typical example. Any adequate structural model of the DNA molecule had to be consistent with the known chemical composition of the compound, its water content, its chemical properties, the accepted quantum-mechanical laws of chemical binding and Chargaff's ratios. Furthermore, the model had to be compatible with the available X-ray photographs of the dry 'A' form and, later, the wet 'B' form furnished by Wilkins and Franklin. Also, since the molecule was assumed to be the carrier of all the genetic information in the cell nucleus, it had to be capable of containing a stable code for the formation of a very large number of sequences of amino acids arranged in an arbitrary order. Satisfying these contraints was no easy task. Both Linus Pauling and the Cambridge group tried and failed before Watson and Crick hit upon their final structure. By not calling the meeting of these antecedently known conditions a test, I by no means wish to underrate the ingenuity required to satisfy them. In the Watson and Crick case, to cite just one small example from their project, it was necessary to extend the theory of X-ray diffraction in order to ascertain what the pattern yielded by a helical molecule of specified dimensions would look like. According to the definition of a test adopted here, the Watson and Crick double helix model was first tested when the semi-conservative model of replication was confirmed by Meselson and Stahl and by Taylor in 1957, the mechanism of protein synthesis elucidated, and the genetic code cracked in the early 1960's. All of these developments required (among other things) derivations from the Watson and Crick hypothesis of results which were not previously known. As this example illustrates and as Laudan (1977) and Nickles (1978) have emphasized, the constraints surrounding the generation of a theory seldom, if ever, consist solely of *empirical* requirements. Nickles, this volume, discusses the further question of when it is rational to violate constraints.

4. One advantage of this terminology is that it avoids the awkwardness of speaking of a test of a theory which is still in the process of being formulated. (How can one 'test' something that does not yet exist?) But it also reflects my conviction along with Popper and the non-Bayesians that even after a theory has been articulated, compatibility with data already available is not a reliable test of that theory at all.

5. Gutting (1974) marks this distinction by the terms 'inventive discovery' and 'critical discovery'. Salmon (1967) similarly distinguishes between (1) thinking of the hypothesis, and (2) plausibility considerations. Both of these writers argue that the logic of discovery is necessarily restricted to the logic of prior assessment ('critical discovery' or 'plausibility considerations').

6. I have deliberately refrained from introducing the term 'plausible' here, since a plausible hypothesis is usually understood to be one that it is reasonable to believe to be true, at least provisionally. Interestingly, Peirce, in one of his last papers, the 'Note on the Doctrine of Chances' (1910), uses the term 'plausible' to mean what I have called 'worthy of pursuit'. Thus Peirce

describes a plausible theory as one that (a) has not yet been tested, (b) would explain more or less surprising phenomena if it were true, and (c) recommends itself for further examination (Buchler, 1955, p. 167). But Peirce goes on to add that if a theory is *highly* plausible, then this would 'justify us in a seriously inclining toward belief in it, as long as the phenomena be inexplicable otherwise' (Buchler, 1955, p. 167). While I think that Peirce has correctly described the circumstances under which we experience a strong psychological tendency to regard an untested theory as true, I wish to avoid the suggestion that this kind of 'inference to the best explanation' *justifies* the belief that the theory is true or objectively probable.

7. There are, of course, degenerate cases in which we can make probability judgments quite easily. For example, we can tell that H is false (and hence highly improbable) if we detect an inconsistency in it. Also, if H_2 is logically equivalent to H_1 plus some independent empirical propositions, then if the probability of H_1 is well-defined, we know that the probability of H_2 is lower than than of H_1, with respect to the same evidential base. It also seems reasonable to assume that H is likely to be false if it violates a wide range of empirical constraints, that is, if H is incompatible with a large body of 'well-established' laws and theories and if H gives no explanation of why these accepted results are erroneous. Something like this kind of reasoning occurs when scientists distinguish between a daring new theory and a mere 'crank' proposal. So I admit that we sometimes are justified in making *im*plausibility judgments and these can provide a basis for deciding the relative plausibility of new theories. But in the common sort of case in which competing hypotheses do not dramatically violate the existing constraints, I claim that there are no grounds for deciding their relative plausibility in the sense of probability or truth-likelihood. What is called 'making a plausibility judgment' in these cases is really based on pursuit-worthiness rather than truth-likeness or probability.

8. In actual scientific practice, of course, these three conditions will be relativized to the state of our knowledge: H is not known (or believed) to be self-contradictory; on the basis of current knowledge, H is not known (or believed) to have any false consequences or to be inconsistent with any statements that are known (or believed) to be true.

9. I have adopted the term 'eureka moments' from Shaffner (1974). Schaffner speculates that 'the subjectively significant 'aha!' or 'eureka!' experience that many discoverers report may be epiphenomenal as compared with the logical moves to the discovery, and may be quite irrelevant in a rational reconstruction of the discovery' (Schaffner, 1974, p. 383, n. 75). The thesis of nonisomorphism was suggested by Blackwell (1976).

10. In commenting on an earlier draft of this paper, Thomas Nickles pointed out that even if eureka moments are sufficient for a discovery, since these are not necessary (some discoveries are made without them), they do not count against the *possibility* of a logic of discovery.

11. Hanson, for example, has written: 'What leads to the initial formulation of H—the 'click', intuition, hunch, insight, perception, etc.—this *is* a matter of psychology' (1958, p. 200, n. 2).

12. Reichenbach actually borrowed the term 'rational reconstruction' from Carnap. See Reichenbach (1938, p. 5).

13. Hanson (1961, p. 20). Though Hanson presents the passage in single quotes as Reichenbach's own words, what Reichenbach actually wrote was: 'We pointed out in the beginning of our inquiry the distinction between the context of discovery and the context of justification. We emphasized that the epistemology cannot be concerned with the first but only with the latter' (1938, pp. 381–382). A comparison of Hanson's paraphrase with Reichenbach's original reveals that Hanson attributed to Reichenbach the identification of the context of discovery with 'reasons for suggesting hypotheses' and the context of justification with 'reasons for accepting hypotheses'. It is argued in this paper that Hanson's interpretation involved a misrepresentation of Reichenbach's actual views concerning the two contexts.

14. This is made explicit in Salmon's (1970) elaboration of Reichenbach's approach where the prior plausibility of a hypothesis is represented as a prior probability in Bayes's equation.

15. Blackwell (1969) contains a useful bibliography of Hanson's writings on the logic of discovery. For reasons of space, I have omitted any detailed discussion of Hanson's views.

16. See, for example, Fann (1970) and Frankfurt (1958).

17. Peirce (7.220, n. 18). Rescher (1976) has

recently emphasized the importance that Peirce placed on the economy of research for scientific method. Rescher relates the economy of research, however, to *inductive* reasoning—the problem of choosing between 'the welter of abductively eligible hypotheses'—rather than seeing it as an integral part of abductive inference as suggested here. I believe the latter account to be closer to Peirce's own views.

18. Popper appears to admit this in his (1957, pp. 34–35).

19. The circularity of this attempt at justifying abduction is noted by Fann (1970, pp. 37 and 51–54). In her recent paper on Peirce and Popper, Haack (1977) accuses Peirce of a *non sequitur* in trying to justify abduction in this manner. She argues that our success at selecting hypotheses from among the abductively eligible ones does not bear on the question of why the data supports a particular hypothesis that is abductively inferred from it (Haack, 1977, p. 72). If 'support' here means 'renders probable', then Haack's objection is sound. But as I have suggested in response to Achinstein's criticism, we should not regard abduction as an inference to a hypothesis but as an inference to the conclusion that a particular hypothesis should be pursued. On this interpretation, what (pragmatically) justifies an abductive inference is the fact that the hypothesis in question has methodological features which we desire in order to attempt to realize the aim of science. The abductive inference in this case is not different in kind from the inference we make in deciding which hypothesis to pursue when several competing hypotheses are compared, prior to their being tested.

20. I would like to thank Thomas Nickles and Stephen Wykstra for their helpful comments and suggestions concerning this paper.

BIBLIOGRAPHY

Achinstein, P.: 1970, 'Inference to Scientific Laws', in R. Steuwer (ed.), *Minnesota Studies in the Philosophy of Science,* Vol. V, Univ. of Minnesota Press, Minneapolis, pp. 87–111.

Achinstein, P.: 1971, *Law and Explanation,* Clarendon Press, Oxford.

Blackwell, R. J.: 1969, *Discovery in the Physical Sciences,* Univ. of Notre Dame, Indiana.

Blackwell, R. J. 1976, 'Scientific Discovery and the Laws of Logic', *The New Scholasticism* **50,** 333–344.

Braithwaite, R. B.: 1955, *Scientific Explanation,* Cambridge Univ. Press, Cambridge.

Buchler, J.: 1955, *Philosophical Writings of Peirce,* Dover, New York.

Burks, A. W.: 1946, 'Peirce's Theory of Abduction', *Philosophy of Science* **13,** 301–306.

Duhem, P.: 1955, *The Aim and Structure of Physical Theory,* Princeton Univ. Press, Princeton.

Einstein, A.: 1933, 'On the Method of Theoretical Physics', The Herbert Spencer Lecture, Oxford, reprinted in *The World As I See It,* Covici-Friede, New York, 1934.

Fann, K. T.: 1970, *Peirce's Theory of Abduction,* Martinus Nijhoff, The Hague.

Frankfurt, H. G.: 1958, 'Peirce's Notion of Abduction', *Journal of Philosophy* **55,** 593–597.

Goudge, T. A.: 1950, *The Thought of C. S. Peirce,* Univ. of Toronto Press, Toronto.

Gutting, G.: 1974, 'A Defense of the Logic of Discovery', *Philosophical Forum* **4,** 384–405.

Haack, S.: 1977, 'Two Fallibilists in Search of the Truth', *Aristotelian Society,* Supplementary Vol. **51,** 63–84.

Hanson, N. R.: 1958, *Patterns of Discovery,* Cambridge Univ. Press, Cambridge.

Hanson, N. R.: 1961, 'Is There a Logic of Scientific Discovery?', in H. Feigl and G. Maxwell (eds.), *Current Issues in the Philosophy of Science,* Holt, Rinehart and Winston, New York, pp. 20–35.

Hempel, C. G.: 1966, *Philosophy of Natural Science,* Prentice-Hall, Englewood Cliffs, New Jersey.

Kordig, C. R.: 1978, 'Discovery and Justification', *Philosophy of Science* **45,** 110–117.

Laudan, L.: 1977, *Progress and Its Problems,* Univ. of California Press, Berkeley.

Nickles, T.: 1978, 'Scientific Problems and Constraints', in P. D. Asquith and I. Hacking (eds.), *PSA 1978,* Volume 1, Philosophy of Science Association, East Lansing, Michigan, pp. 134–148.

Nickles, T.: This volume, 'Can Scientific Constraints Be Violated Rationally?' pp. 285–315.

Peirce, C. S.: 1931–1958, *Collected Papers of Charles Sanders Peirce,* C. Hartshorne *et al.* (eds.), Harvard Univ. Press, Cambridge.

Popper, K.: 1957, 'The Aim of Science', *Ratio* **1,** 24–35.

Popper, K.: 1959, *The Logic of Scientific Discovery,* Basic Books, New York.

Reichenbach, H.: 1938, *Experience and Prediction,* Univ. of Chicago Press, Chicago.

Reichenbach, H.: 1944, *Philosophic Foundations of Quantum Mechanics,* Univ. of California Press, Berkeley.

Reichenbach, H.: 1957, *The Philosophy of Space and Time,* Dover, New York.

Salmon, W.: 1967, *Foundations of Scientific Inference,* Univ. of Pittsburgh Press, Pittsburgh.

Salmon, W.: 1970, 'Bayes's Theorem and the History of Science', in R. Steuwer (ed.), *Minnesota Studies in the Philosophy of Science,* Vol. V. Univ. of Minnesota Press, Minneapolis, pp. 87–111.

Schaffner, K.: 1974, 'Logic of Discovery and Logic of Justification in Regulatory Genetics', *Studies in History and Philosophy of Science* **4,** 349–385.

Simon, H. A.: 1973, 'Does Scientific Discovery Have a Logic?', *Philosophy of Science* **40,** 471–480.

Watson, J. D.: 1969, *The Double Helix,* New American Library, New York.

PART IV Selected Problems of Particular Sciences

Space and Time: Geometry and Physics

INTRODUCTION

From the time Euclid systematized the knowledge of geometry around 300 B.C. until the nineteenth century, it was assumed that his geometrical system described space. But when non-Euclidean geometries were discovered in the nineteenth century, a question arose as to whether space might be non-Euclidean. A second question that arises from reflection on the first is to ask whether it makes any sense to ask whether space is Euclidean or not. These debates about the shape of space connected with discussions that had begun at the time of Newton and Leibniz concerning whether any sense could be attached to claims about the absolute position or velocity of a body. Does it make any sense that a particle is moving with regard to the underlying space, or must any statement with content specify that the particle is moving with respect to some other object—the earth or sun, for example?

Professor Poincaré eloquently defended the relativity of space, both regarding motion and shape. He traces the geometrical conception we have of space to our early experience, both individually and as a race. He even extends his argument to the three-dimensionality of space, attributing the appearance of three dimensionality to the nature of the human mind, although he describes the possibility of other kinds of creatures who might have a spatial conception with more or fewer dimensions. One point that he emphasizes, however, is that although we could in principle conceive of space as having a different geometry, the translation of our physics into that alternative geometry would prove so complicated that we would never choose to use such a system of description.

Subsequent events, namely, the development of the theory of general relativity, proved Poincaré wrong, for physicists found non-Euclidean geometries much more natural companions to relativity theory than Euclidean geometry. Professor Reichenbach reviews the history of geometry from the early Egyptian beginnings through the discovery of non-

Euclidean gometry. He argues that Poincaré's conventionalist view of geometry is incorrect because Poincaré did not consider the important interactions between geometry and physics. Reichenbach distinguishes between mathematical geometry, a pure subject that concerns only abstract objects in mathematical space, from physical geometry, which is intended to be the description of physical objects in physical space.

He points out that the application of the geometrical axioms to physical objects requires coordinating definitions that specify what is a point, a straight line, and so on. In particular, pure geometry may tell us that a straight line is the shortest distance between two points but unless we also choose to identify the path of a light ray with a geometrical straight line, we cannot conclude that the path of a light ray is the shortest distance between two physical points. He argues that Poincaré was partly right in that any of a number of choices of geometry is logically consistent with observations if one makes suitable compensating adjustments in phyics.

To summarize, it is the package of physics *plus geometry* that is compared to observation, and thus alternate geometries can be used, although only at the cost of changing physics. Where Reichenbach disagrees with Poincaré is that the overall package, including Euclidean geometry, proves to be so much less convenient, because of the complications in physics, that the preferable theory combines non-Euclidean geometry with relativistic physics.

Professor Salmon develops these themes at greater length. He also traces the development of geometry and of attitudes toward geometry, emphasizing the historical roots of Poincaré's attitude toward geometry in Kant's thought. He also exhibits some examples of how non-Euclidean geometry differs and answers one possible objection to the application of those geometries, namely, that they cannot be visualized. His main moral is that once more we have discovered that we can have no knowledge of the world from pure mathematics or pure thought alone.

Henri Poincaré

THE RELATIVITY OF SPACE

I.

It is impossible to picture empty space. All our efforts to imagine pure space from which the changing images of material objects are excluded can only result in a representation in which highly-coloured surfaces, for instance, are replaced by lines of slight colouration, and if we continued in this direction to the end, everything would disappear and end in nothing. Hence arises the irreducible relativity of space.

Whoever speaks of absolute space uses a word devoid of meaning. This is a truth that has been long proclaimed by all who have

---

Reprinted by permission from Henri Poincaré, *Science and Method,* Francis Maitland, trans., Dover Publications, Inc., 1952.

reflected on the question, but one which we are too often inclined to forget.

If I am at a definite point in Paris, at the Place du Panthéon, for instance, and I say, "I will come back *here* to-morrow;" if I am asked, "Do you mean that you will come back to the same point in space?" I should be tempted to answer yes. Yet I should be wrong, since between now and to-morrow the earth will have moved, carrying with it the Place du Panthéon, which will have travelled more than a million miles. And if I wished to speak more accurately, I should gain nothing, since this million of miles has been covered by our globe in its motion in relation to the sun, and the sun in its turn moves in relation to the Milky Way, and the Milky Way itself is no doubt in motion without our being able to recognize its velocity. So that we are, and shall always be, completely ignorant how far the Place du Panthéon moves in a day. In fact, what I meant to say was, "To-morrow I shall see once more the dome and pediment of the Panthéon," and if there was no Panthéon my sentence would have no meaning and space would disappear.

This is one of the most commonplace forms of the principle of the relativity of space, but there is another on which Delbeuf has laid particular stress. Suppose that in one night all the dimensions of the universe became a thousand times larger. The world will remain *similar* to itself, if we give the word *similitude* the meaning it has in the third book of Euclid. Only, what was formerly a metre long will now measure a kilometre, and what was a millimetre long will become a metre. The bed in which I went to sleep and my body itself will have grown in the same proportion. When I wake in the morning what will be my feeling in face of such an astonishing transformation? Well, I shall not notice anything at all. The most exact measures will be incapable of revealing anything of this tremendous change, since the yard-measures I shall use will have varied in exactly the same proportions as the objects I shall attempt to measure. In reality the change only exists for those who argue as if space were absolute. If I have argued for a moment as they do, it was only in order to make it clearer that their view implies a contradiction. In reality it would be better to say that as space is relative, nothing at all has happened, and that it is for that reason that we have noticed nothing.

Have we any right, therefore, to say that we know the distance between two points? No, since that distance could undergo enormous variations without our being able to perceive it, provided other distances varied in the same proportions. We saw just now that when I say I shall be here to-morrow, that does not mean that to-morrow I shall be at the point in space where I am to-day, but that to-morrow I shall be at the same distance from the Panthéon as I am to-day. And already this statement is not sufficient, and I ought to say that to-morrow and to-day my distance from the Panthéon will be equal to the same number of times the length of my body.

But that is not all. I imagined the dimensions of the world changing, but at least the world remaining always similar to itself. We can go much further than that, and one of the most surprising theories of modern physicists will furnish the occasion. According to a hypothesis of Lorentz and Fitzgerald,* all bodies carried forward in the earth's motion undergo a deformation. This deformation is, in truth, very slight, since all dimensions parallel with the earth's motion are diminished by a hundred-millionth, while dimensions perpendicular to this motion are not altered. But it matters little that it is slight; it is enough that it should exist for the conclusion I am soon going to draw from it. Besides, though I said that it is slight, I really know nothing about it. I have myself fallen a victim to the tenacious illusion that makes us believe that we think of an absolute space. I was thinking of the earth's motion on its elliptical orbit round

**Vide infra*, Book III. Chap. ii.

the sun, and I allowed 18 miles a second for its velocity. But its true velocity (I mean this time, not its absolute velocity, which has no sense, but its velocity in relation to the ether), this I do not know and have no means of knowing. It is, perhaps, 10 or 100 times as high, and then the deformation will be 100 or 10,000 times as great.

It is evident that we cannot demonstrate this deformation. Take a cube with sides a yard long. It is deformed on account of the earth's velocity; one of its sides, that parallel with the motion, becomes smaller, the others do not vary. If I wish to assure myself of this with the help of a yard-measure, I shall measure first one of the sides perpendicular to the motion, and satisfy myself that my measure fits this side exactly; and indeed neither one nor other of these lengths is altered, since they are both perpendicular to the motion. I then wish to measure the other side, that parallel with the motion; for this purpose I change the position of my measure, and turn it so as to apply it to this side. But the yard-measure, having changed its direction and having become parallel with the motion, has in its turn undergone the deformation, so that, though the side is no longer a yard long, it will still fit it exactly, and I shall be aware of nothing.

What, then, I shall be asked, is the use of the hypothesis of Lorentz and Fitzgerald if no experiment can enable us to verify it? The fact is that my statement has been incomplete. I have only spoken of measurements that can be made with a yard-measure, but we can also measure a distance by the time that light takes to traverse it, on condition that we admit that the velocity of light is constant, and independent of its direction. Lorentz could have accounted for the facts by supposing that the velocity of light is greater in the direction of the earth's motion than in the perpendicular direction. He preferred to admit that the velocity is the same in the two directions, but that bodies are smaller in the former than in the latter. If the surfaces of the waves of light had undergone the same deformations as material bodies, we should never have perceived the Lorentz-Fitzgerald deformation.

In the one case as in the other, there can be no question of absolute magnitude, but of the measurement of that magnitude by means of some instrument. This instrument may be a yard-measure or the path traversed by light. It is only the relation of the magnitude to the instrument that we measure, and if this relation is altered, we have no means of knowing whether it is the magnitude or the instrument that has changed.

But what I wish to make clear is, that in this deformation the world has not remained similar to itself. Squares have become rectangles or parallelograms, circles ellipses, and spheres ellipsoids. And yet we have no means of knowing whether this deformation is real.

It is clear that we might go much further. Instead of the Lorentz-Fitzgerald deformation, with its extremely simple laws, we might imagine a deformation of any kind whatever; bodies might be deformed in accordance with any laws, as complicated as we liked, and we should not perceive it, provided all bodies without exception were deformed in accordance with the same laws. When I say all bodies without exception, I include, of course, our own bodies and the rays of light emanating from the different objects.

If we look at the world in one of those mirrors of complicated form which deform objects in an odd way, the mutual relations of the different parts of the world are not altered; if, in fact, two real objects touch, their images likewise appear to touch. In truth, when we look in such a mirror we readily perceive the deformation, but it is because the real world exists beside its deformed image. And even if this real world were hidden from us, there is something which cannot be hidden, and that is ourselves. We cannot help seeing, or at least feeling, our body and our members which have not been deformed, and continue to act as measuring instruments. But if we imagine our body itself deformed, and in the same way as if it were seen in the mirror, these measuring instruments will fail us in their turn, and the deformation will no longer be able to be ascertained.

Imagine, in the same way, two universes which are the image one of the other. With each object P in the universe A, there corresponds, in the universe B, an object P^1 which is its image. The co-ordinates of this image P^1 are determinate functions of those of the object P; moreover, these functions may be of any kind whatever—I assume only that they are chosen once for all. Between the position of P and that of P^1 there is a constant relation; it matters little what that relation may be, it is enough that it should be constant.

Well, these two universes will be indistinguishable. I mean to say that the former will be for its inhabitants what the second is for its own. This would be true so long as the two universes remained foreign to one another. Suppose we are inhabitants of the universe A; we have constructed our science and particularly our geometry. During this time the inhabitants of the universe B have constructed a science, and as their world is the image of ours, their geometry will also be the image of ours, or, more accurately, it will be the same. But if one day a window were to open for us upon the universe B, we should feel contempt for them, and we should say, "These wretched people imagine that they have made a geometry, but what they so name is only a grotesque image of ours; their straight lines are all twisted, their circles are hunchbacked, and their spheres have capricious inequalities." We should have no suspicion that they were saying the same of us, and that no one will ever know which is right.

We see in how large a sense we must understand the relativity of space. Space is in reality amorphous, and it is only the things that are in it that give it a form. What are we to think, then, of that direct intuition we have of a straight line or of distance? We have so little the intuition of distance in itself that, in a single night, as we have said, a distance could become a thousand times greater without our being able to perceive it, if all other distances had undergone the same alteration. And in a night the universe B might even be substituted for the universe A without our having any means of knowing it, and then the straight lines of yesterday would have ceased to be straight, and we should not be aware of anything.

One part of space is not by itself and in the absolute sense of the word equal to another part of space, for if it is so for us, it will not be so for the inhabitants of the universe B, and they have precisely as much right to reject our opinion as we have to condemn theirs.

I have shown elsewhere what are the consequences of these facts from the point of view of the idea that we should construct non-Euclidian and other analogous geometries. I do not wish to return to this, and I will take a somewhat different point of view.

II.

If this intuition of distance, of direction, of the straight line, if, in a word, this direct intuition of space does not exist, whence comes it that we imagine we have it? If this is only an illusion, whence comes it that the illusion is so tenacious? This is what we must examine. There is no direct intuition of magnitude, as we have said, and we can only arrive at the relation of the magnitude to our measuring instruments. Accordingly we could not have constructed space if we had not had an instrument for measuring it. Well, that instrument to which we refer everything, which we use instinctively, is our own body. It is in reference to our own body that we locate exterior objects, and the only special relations of these objects that we can picture to ourselves are their relations with our body. It is our body that serves us, so to speak, as a system of axes of co-ordinates.

For instance, at a moment α the presence of an object A is revealed to me by the sense of sight; at another moment β the presence of another object B is revealed by another sense, that, for instance, of hearing or of touch. I judge that this object B occupies the same place as the object A. What does this mean? To begin with, it does not imply that these two objects occupy, at two different moments, the same point in an absolute space, which, even if it existed, would escape our knowledge, since between the moments α and β the solar system has been displaced and we cannot know what this displacement is. It means that these two objects occupy the same relative position in reference to our body.

But what is meant even by this? The impressions that have come to us from these objects have followed absolutely different paths—the optic nerve for the object A, and the acoustic nerve for the object B; they have nothing in common from the qualitative point of view. The representations we can form of these two objects are absolutely heterogeneous and irreducible one to the other. Only I know that, in order to reach the object A, I have only to extend my right arm in a certain way; even though I refrain from doing it, I represent to myself the muscular and other analogous sensations which accompany that extension, and that representation is associated with that of the object A.

Now I know equally that I can reach the object B by extending my right arm in the same way, an extension accompanied by the same train of muscular sensations. And I mean nothing else but this when I say that these two objects occupy the same position.

I know also that I could have reached the object A by another appropriate movement of the left arm, and I represent to myself the muscular sensations that would have accompanied the movement. And by the same movement of the left arm, accompanied by the same sensations, I could equally have reached the object B.

And this is very important, since it is in this way that I could defend myself against the dangers with which the object A or the object B might threaten me. With each of the blows that may strike us, nature has associated one or several parries which enable us to protect ourselves against them. The same parry may answer to several blows. It is thus, for instance, that the same movement of the right arm would have enabled us to defend ourselves at the moment α against the object A, and at the moment β against the object B. Similarly, the same blow may be parried in several ways, and we have said, for instance, that we could reach the object A equally well either by a certain movement of the right arm, or by a certain movement of the left.

All these parries have nothing in common with one another, except that they enable us to avoid the same blow, and it is that, and nothing but that, we mean when we say that they are movements ending in the same point in space. Similarly, these objects, of which we say that they occupy the same point in space, have nothing in common, except that the same parry can enable us to defend ourselves against them.

Or, if we prefer it, let us imagine innumerable telegraph wires, some centripetal and others centrifugal. The centripetal wires warn us of accidents that occur outside, the centrifugal wires have to provide the remedy. Connexions are established in such a way that when one of the centripetal wires is traversed by a current, this current acts on a central exchange, and so excites a current in one of the centrifugal wires, and matters are so arranged that several centripetal wires can act on the same centrifugal wire, if the same remedy is applicable to several evils, and that one centripetal wire can disturb several centrifugal wires, either simultaneously or one in default of the other, every time that the same evil can be cured by several remedies.

It is this complex system of associations, it is this distribution board, so to speak, that is our whole geometry, or, if you will, all that is distinctive in our geometry. What we call our intuition of a straight line or of distance is the consciousness we have of these associations and of their imperious character.

Whence this imperious character itself

comes, it is easy to understand. The older an association is, the more indestructible it will appear to us. But these associations are not, for the most part, conquests made by the individual, since we see traces of them in the newly-born infant; they are conquests made by the race. The more necessary these conquests were, the more quickly they must have been brought about by natural selection.

On this account those we have been speaking of must have been among the earliest, since without them the defence of the organism would have been impossible. As soon as the cells were no longer merely in juxtaposition, as soon as they were called upon to give mutual assistance to each other, some such mechanism as we have been describing must necessarily have been organized in order that the assistance should meet the danger without miscarrying.

When a frog's head has been cut off, and a drop of acid is placed at some point on its skin, it tries to rub off the acid with the nearest foot; and if that foot is cut off, it removes it with the other foot. Here we have, clearly, that double parry I spoke of just now, making it possible to oppose an evil by a second remedy if the first fails. It is this multiplicity of parries, and the resulting co-ordination, that is space.

We see to what depths of unconsciousness we have to descend to find the first traces of these spacial associations, since the lowest parts of the nervous system alone come into play. Once we have realized this, how can we be astonished at the resistance we oppose to any attempt to dissociate what has been so long associated? Now, it is this very resistance that we call the evidence of the truths of geometry. This evidence is nothing else than the repugnance we feel at breaking with very old habits with which we have always got on very well.

III.

The space thus created is only a small space that does not extend beyond what my arm can reach, and the intervention of memory is necessary to set back its limits. There are points that will always remain out of my reach, whatever effort I may make to stretch out my hand to them. If I were attached to the ground, like a sea-polype, for instance, which can only extend its tentacles, all these points would be outside space, since the sensations we might experience from the action of bodies placed there would not be associated with the idea of any movement enabling us to reach them, or with any appropriate parry. These sensations would not seem to us to have any spacial character, and we should not attempt to locate them.

But we are not fixed to the ground like the inferior animals. If the enemy is too far off, we can advance upon him first and extend our hand when we are near enough. This is still a parry, but a long-distance parry. Moreover, it is a complex parry, and into the representation we make of it there enter the representation of the muscular sensations caused by the movement of the legs, that of the muscular sensations caused by the final movement of the arm, that of the sensations of the semi-circular canals, etc. Besides, we have to make a representation, not of a complexus of simultaneous sensations, but of a complexus of successive sensations, following one another in a determined order, and it is for this reason that I said just now that the intervention of memory is necessary.

We must further observe that, to reach the same point, I can approach nearer the object to be attained, in order not to have to extend my hand so far. And how much more might be said? It is not one only, but a thousand parries I can oppose to the same danger. All these parries are formed of sensations that may have nothing in common, and yet we regard them as defining the same point in space, because they can answer to the same danger and are one and all of them associated with the notion of that danger. It is the possibility of parrying the same blow which makes the unity of these different parries, just as it is the possibility of being parried in the same way which makes the unity of the blows of such different kinds that can threaten us from the same point in space. It is this double unity that makes the individuality of each point in space, and in the notion of such a point there is nothing else but this.

The space I pictured in the preceding section, which I might call *restricted space,* was referred to axes of co-ordinates attached to my body. These axes were fixed, since my body did not move, and it was only my limbs that changed their position. What are the axes to which the *extended space* is naturally referred—that is to say, the new space I have just defined? We define a point by the succession of movements we require to make to reach it, starting from a certain initial position of the body. The axes are accordingly attached to this initial position of the body.

But the position I call initial may be arbitrarily chosen from among all the positions my body has successively occupied. If a more or less unconscious memory of these successive positions is necessary for the genesis of the notion of space, this memory can go back more or less into the past. Hence results a certain indeterminateness in the very definition of space, and it is precisely this indeterminateness which constitutes its relativity.

Absolute space exists no longer; there is only space relative to a certain initial position of the body. For a conscious being, fixed to the ground like the inferior animals, who would consequently only know restricted space, space would still be relative, since it would be referred to his body, but this being would not be conscious of the relativity, because the axes to which he referred this restricted space would not change. No doubt the rock to which he was chained would not be motionless, since it would be involved in the motion of our planet; for us, consequently, these axes would change every moment, but for him they would not change. We have the faculty of referring our extended space at one time to the position A of our body considered as initial, at another to the position B which it occupied some moments later, which we are free to consider in its turn as initial, and, accordingly, we make unconscious changes in the co-ordinates every moment. This faculty would fail our imaginary being, and, through not having travelled, he would think space absolute. Every moment his system of axes would be imposed on him; this system might change to any extent in reality, for him it would be always the same, since it would always be the *unique* system. It is not the same for us who possess, each moment, several systems between which we can choose at will, and on condition of going back by memory more or less into the past.

That is not all, for the restricted space would not be homogeneous. The different points of this space could not be regarded as equivalent, since some could only be reached at the cost of the greatest efforts, while others could be reached with ease. On the contrary, our extended space appears to us homogeneous, and we say that all its points are equivalent. What does this mean?

If we start from a certain position A, we can, starting from that position, effect certain movements M, characterized by a certain complexus of muscular sensations. But, starting from another position B, we can execute movements M^1 which will be characterized by the same muscular sensations. Then let a be the situation of a certain point in the body, the tip of the forefinger of the right hand, for instance, in the initial position A, and let b be the position of this same forefinger when, starting from that position A, we have executed the movements M. Then let a^1 be the situation of the forefinger in the position B, and b^1 its situation when, starting from the position B, we have executed the movements M^1.

Well, I am in the habit of saying that the points a and b are, in relation to each other, as the points a^1 and b^1, and that means simply that the two series of movements M and M^1 are accompanied by the same muscular sensations. And as I am conscious that, in passing from the position A to the position B, my body has remained capable of the same movements, I know that there is a point in space which is to the point a^1 what some point *b* is to the point *a*, so that the two points *a* and a^1 are equivalent. It is this that is called the homogeneity of space, and at the same time it is for this reason that space is relative, since its properties remain the same whether they are referred to the axes A or to the axes B. So that the relativity of space and its homogeneity are one and the same thing.

Now, if I wish to pass to the great space, which is no longer to serve for my individual use only, but in which I can lodge the universe, I shall arrive at it by an act of imagination. I shall imagine what a giant would experience who could reach the planets in a few steps, or, if we prefer, what I should feel myself in presence of a world in miniature, in which these planets would be replaced by little balls, while on one of these little balls there would move a Lilliputian that I should call myself. But this act of imagination would be impossible for me if I had not previously constructed my restricted space and my extended space for my personal use.

IV.

Now we come to the question why all these spaces have three dimensions. Let us refer to the "distribution board" spoken of above. We have, on the one side, a list of the different possible dangers—let us designate them as A1, A2, etc.—and, on the other side, the list of the different remedies, which I will call in the same way B1, B2, etc. Then we have connexions between the contact studs of the first list and those of the second in such a way that when, for instance, the alarm for danger A3 works, it sets in motion or may set in motion the relay corresponding to the parry B4.

As I spoke above of centripetal or centrifugal wires, I am afraid that all I have said may be taken, not as a simple comparison, but as a description of the nervous system. Such is not my thought, and that for several reasons. Firstly, I should not presume to pronounce an opinion on the structure of the nervous system which I do not know, while those who have studied it only do so with circumspection. Secondly, because, in spite of my incompetence, I fully realize that this scheme would be far too simple. And lastly, because, on my list of parries, there appear some that are very complex, which may even, in the case of extended space, as we have seen above, consist of several steps followed by a movement of the arm. It is not a question, then, of physical connexion between two real conductors, but of psychological association between two series of sensations.

If, for instance, A1 and A2 are both of them associated with the parry B1, and if A1 is similarly associated with B2, it will generally be the case that A2 and B2 will also be associated. If this fundamental law were not generally true, there would only be an immense confusion, and there would be nothing that could bear any resemblance to a conception of space or to a geometry. How, indeed, have we defined a point in space? We defined it in two ways: on the one hand, it is the whole of the alarms A which are in connexion with the same parry B; on the other, it is the whole of the parries B which are in connexion with the same alarm A. If our law were not true, we should be obliged to say that A1 and A2 correspond with the same point, since they are both in connexion with B1; but we should be equally obliged to say that they do not correspond with the same point, since A1 would be in connexion with B2, and this would not be true of A2—which would be a contradiction.

But from another aspect, if the law were rigorously and invariably true, space would be quite different from what it is. We should have well-defined categories, among which would be apportioned the alarms A on the one side and the parries B on the other. These categories would be exceedingly numerous, but they would be entirely separated one from the other. Space would be formed of points, very numerous but discrete; it would be *discontinuous*. There would be no reason for arranging these points in one order rather than another, nor, consequently, for attributing three dimensions to space.

But this is not the case. May I be permitted for a moment to use the language of those who know geometry already? It is necessary that I should do so, since it is the language best understood by those to whom I wish to make myself clear. When I wish to parry the blow, I try to reach the point whence the blow comes, but it is enough if I come fairly near it. Then the parry B1 may answer to A1, and to A2 if the point which corresponds with B1 is sufficiently close both to that which corresponds with A1 and to that which corresponds with A2. But it may happen that the point which corresponds with another parry B2 is near enough to the point corresponding with A1, and not near enough to the point corresponding with A2. And so the parry B2 may answer to A1 and not be able to answer to A2.

For those who do not yet know geometry, this may be translated simply by a modification of the law enunciated above. Then what happens is as follows. Two parries, B1 and B2, are associated with one alarm A1, and with a very great number of alarms that we will place in the same category as A1, and make to correspond with the same point in space. But we may find alarms A2 which are associated with B2 and not with B1, but on the other hand are associated with B3, which are not with A1, and so on in succession, so that we may write the sequence

B1, A1, B2, A2, B3, A3, B4, A4,

in which each term is associated with the succeeding and preceding terms, but not with those that are several places removed.

It is unnecessary to add that each of the terms of these sequences is not isolated, but forms part of a very numerous category of other alarms or other parries which has the same connexions as it, and may be regarded as belonging to the same point in space. Thus the fundamental law, though admitting of exceptions, remains almost always true. Only, in consequence of these exceptions, these categories, instead of being entirely separate, partially encroach upon each other and mutually overlap to a certain extent, so that space becomes continuous.

Furthermore, the order in which these categories must be arranged is no longer arbitrary, and a reference to the preceding sequence will make it clear that B2 must be placed between A1 and A2, and, consequently, between B1 and B3, and that it could not be placed, for instance, between B3 and B4.

Accordingly there is an order in which our categories range themselves naturally which corresponds with the points in space, and experience teaches us that this order presents itself in the form of a three-circuit distribution board, and it is for this reason that space has three dimensions.

V.

Thus the characteristic property of space, that of having three dimensions, is only a property of our distribution board, a property residing, so to speak, in the human intelligence. The destruction of some of these connexions, that is to say, of these associations of ideas, would be sufficient to give us a different distribution board, and that might be enough to endow space with a fourth dimension.

Some people will be astonished at such a result. The exterior world, they think, must surely count for something. If the number of dimensions comes from the way in which we are made, there might be thinking beings living in our world, but made differently from us, who would think that space has more or less than three dimensions. Has not M. de Cyon said that Japanese mice, having only two pairs of semicircular canals, think that space has two dimensions? Then will not this thinking being, if he is capable of constructing a physical system, make a system of two or four dimensions, which yet, in a sense, will be the same as ours, since it will be the description of the same world in another language?

It quite seems, indeed, that it would be possible to translate our physics into the language of geometry of four dimensions. Attempting such a translation would be giving oneself a great deal of trouble for little profit, and I will content myself with mentioning Hertz's mechanics, in which something of the kind may be seen. Yet it seems that the translation would always be less simple than the text, and that it would never lose the appearance of a translation, for the language of three dimensions seems the best suited to the description of our world, even though that description may be made, in case of necessity, in another idiom.

Besides, it is not by chance that our distribution board has been formed. There is a connexion between the alarm A1 and the parry B1, that is, a property residing in our intelligence. But why is there this connexion? It is because the parry B1 enables us effectively to defend ourselves against the danger A1, and that is a fact exterior to us, a property of the exterior world. Our distribution board, then, is only the translation of an assemblage of exterior facts; if it has three dimensions, it is because it has adapted itself to a world having certain properties, and the most important of these properties is that there exist natural solids which are clearly displaced in accordance with the laws we call laws of motion of unvarying solids. If, then, the language of three dimensions is that which enables us most easily to describe our world, we must not be surprised. This language is founded on our distribution board, and it is in order to enable us to live in this world that this board has been established.

I have said that we could conceive of thinking beings, living in our world, whose distribution board would have four dimensions, who would, consequently, think in hyperspace. It is not certain, however, that such beings, admitting that they were born, would be able to live and defend themselves against the thousand dangers by which they would be assailed.

VI.

A few remarks in conclusion. There is a striking contrast between the roughness of this primitive geometry which is reduced to what I call a distribution board, and the infinite precision of the geometry of geometricians. And yet the latter is the child of the former, but not of it alone; it required to be fertilized by the faculty we have of constructing mathematical concepts, such, for instance, as that of the group. It was necessary to find among these pure concepts the one that was best adapted to this rough space, whose genesis I have tried to explain in the preceding pages, the space which is common to us and the higher animals.

The evidence of certain geometrical postulates is only, as I have said, our unwillingness to give up very old habits. But these postulates are infinitely precise, while the habits have about them something essentially fluid. As soon as we wish to think, we are bound to have infinitely precise postulates, since this is the only means of avoiding contradiction. But among all the possible systems of postulates, there are some that we shall be unwilling to choose, because they do not accord sufficiently with our habits. However fluid and elastic these may be, they have a limit of elasticity.

It will be seen that though geometry is not an experimental science, it is a science born in connexion with experience; that we have created the space it studies, but adapting it to the world in which we live. We have chosen the most convenient space, but experience guided our choice. As the choice was unconscious, it appears to be imposed upon us. Some say that it is imposed by experience, and others that we are born with our space ready-made. After the preceding considerations, it will be seen what proportion of truth and of error there is in these two opinions.

In this progressive education which has

resulted in the construction of space, it is very difficult to determine what is the share of the individual and what of the race. To what extent could one of us, transported from his birth into an entirely different world, where, for instance, there existed bodies displaced in accordance with the laws of motion of non-Euclidian solids—to what extent, I say, would he be able to give up the ancestral space in order to build up an entirely new space?

The share of the race seems to preponderate largely, and yet if it is to it that we owe the rough space, the fluid space of which I spoke just now, the space of the higher animals, is it not to the unconscious experience of the individual that we owe the infinitely precise space of the geometrician? This is a question that is not easy of solution. I would mention, however, a fact which shows that the space bequeathed to us by our ancestors still preserves a certain plasticity. Certain hunters learn to shoot fish under the water, although the image of these fish is raised by refraction; and, moreover, they do it instinctively. Accordingly they have learnt to modify their ancient instinct of direction, or, if you will, to substitute for the association A1, B1, another association A1, B2, because experience has shown them that the former does not succeed.

Hans Reichenbach

THE NATURE OF GEOMETRY

Ever since the death of Kant in 1804 science has gone through a development, gradual at first and rapidly increasing in tempo, in which it abandoned all absolute truths and preconceived ideas. The principles which Kant had considered to be indispensable to science and nonanalytic in their nature have been recognized as holding only to a limited degree. Important laws of classical physics were found to apply only to phenomena occurring in our ordinary environment. For astronomical and for submicroscopic dimensions they had to be replaced by laws of the new physics, and this fact alone makes it obvious that they were empirical laws and not laws forced on us by reason itself. Let me illustrate this *disintegration of the synthetic a priori* by tracing the development of geometry.

The historical origin of geometry, which goes back to the Egyptians, supplies one of the many instances in which intellectual discoveries have grown from material needs. The annual floods of the Nile which fertilized the soil of Egypt brought trouble to landowners: the borderlines of their estates were destroyed every year and had to be reestablished by means of geometrical measurements. The geographical and social conditions of their country, therefore, compelled the Egyptians to invent the art of surveying. Geometry thus arose as an empirical science, whose laws were the results of observations. For instance, the Egyptians knew from practical experience that if they made a triangle the sides of which were respectively 3, 4, and 5 units long it would be a right triangle. The deductive proof for this result was provided much later by Pythagoras, whose famous theorem explains the Egyptian findings by the fact that the sum of the squares of 3 and 4 is equal to the square of 5.

Pythagoras' theorem illustrates the con-

Reprinted from Hans Reichenbach, *The Rise of Scientific Philosophy,* University of California Press, 1962.

tribution which the Greeks made to geometry: the discovery that geometry can be built up as a deductive system, in which every theorem is strictly derivable from the set of axioms (see p. 96). The construction of geometry in the form of an axiomatic system is forever connected with the name of Euclid. His logically ordered presentation of geometry has remained the program of every course in geometry and was used until recently as a text in our schools.

The axioms of Euclid's system appeared so natural and obvious that their truth seemed unquestionable. In this respect, Euclid's system confirmed earlier conceptions, developed before the principles of geometry acquired the form of an ordered system. Plato, who lived a generation before Euclid, was led by the apparent self-evidence of geometrical principles to his theory of ideas; and it was explained in Chapter 2 that the axioms of geometry were regarded by him as revealed to us through an act of vision, which showed geometrical relations as properties of ideal objects. The long line of development beginning with Plato, which did not essentially change this conception, terminated in the more precise though less poetical theory of Kant, according to which the axioms were synthetic a priori. Mathematicians more or less shared these views, but they were not so much interested in the philosophical discussion of the axioms as in the analysis of the mathematical relations holding between them. They tried to reduce the axioms to a minimum by showing some of them to be derivable from the others.

There was in particular one axiom, the axiom of parallels, which they disliked and attempted to eliminate. The axiom states that through a given point one and only one parallel can be drawn with respect to a given line; that is, there is one and only one line that does not ultimately intersect with a given line and yet lies in the same plane. We do not know why the mathematicians disliked this axiom, but we know of many attempts, dating back to antiquity, that were made with the intention of transforming this axiom into a theorem, that is, of deriving it from the other axioms. Mathematicians repeatedly believed that they had found a way of deriving the proposition about parallels from the other axioms. Invariably, however, these proofs have later been demonstrated to be fallacious. The mathematicians had unknowingly introduced some assumption which was not included in the other axioms but was of equal efficacy as the axiom of the parallels. The result of this development was, then, that there are equivalents of this axiom. But the mathematician had no more reason to accept these equivalents than to accept Euclid's axiom. For instance, an equivalent of the axiom of the parallels is the principle that the sum of the angles of a triangle is equal to two right angles. Euclid had derived this principle from his axiom, but it was shown that conversely the principle of the parallels is derivable when the principle of the angular sum is assumed as an axiom. What is an axiom in one system, thus becomes a theorem in another system, and vice versa.

The problem of parallels had occupied mathematicians for more than two thousand years before it found its solution. About twenty years after the death of Kant, a young Hungarian mathematician, John Bolyai (1802–1860), discovered that the axiom of parallels is not a necessary constituent of a geometry. He constructed a geometry in which the axiom of parallels was abandoned and replaced by the novel assumption that there exists more than one parallel to a given line through a given point. The same discovery was made about the same time by the Russian mathematician N. I. Lobachevski (1793–1856) and by the German mathematician K. F. Gauss (1777–1855). The geometries so constructed were called *non-Euclidean geometries.* A more general form of a non-Euclidean geometry, which includes systems in which there exist no parallel lines at all, was later developed by the German mathematician B. Riemann (1826–1866).

A non-Euclidean geometry contradicts Euclidean geometry—for instance, in a non-Euclidean triangle the sum of the angles is different from 180 degrees. Still, each non-Euclidean geometry is free from internal contradictions; it is a consistent system in the same sense that Euclid's geometry is consistent. A plurality of geometries thus replaces the unique system of Euclid. It is true that the Euclidean geometry is distinguished from all others by the fact that it is easily accessible to a visual presentation, whereas it seems impossible to visualize a geometry in which there is more than one parallel to a given line through a given point. But the mathematicians were not very much concerned about questions of visualization and regarded various geometrical systems as being of equal mathematical validity. In keeping with this somewhat detached attitude of the mathematician, I shall postpone the discussion of visualization until I have discussed some other problems.

The existence of a plurality of geometries demanded a new approach to the problem of the geometry of the physical world. As long as there was only one geometry, the Euclidean geometry, there was no question of the geometry of physical space. In the absence of an alternative, Euclid's geometry was naturally assumed to apply to physical reality. It was Kant's merit to emphasize more than others that the coincidence of mathematical and physical geometry calls for an explanation, and his theory of the synthetic a priori must be regarded as the great attempt of a philosopher to account for this coincidence. With the discovery of a plurality of geometries the situation changed completely. If the mathematician was offered a choice between geometries, there arose the question which of them was the geometry of the physical world. It was obvious that reason could not answer this question, that its answer was left to empirical observation.

The first to draw this conclusion was Gauss. After his discovery of non-Euclidean geometry he attempted to carry through an empirical test by means of which the geometry of the physical world was to be ascertained. For this purpose, Gauss measured the angles of a triangle the corners of which were marked by three mountain tops. The result of his measurements was carefully worded: it said that within the errors of observation the Euclidean principle was true, or, in other words, that if there was a deviation of the angular sum from 180 degrees, the inevitable errors of the observation made it impossible to prove its existence. If the world was non-Euclidean, it was controlled by a non-Euclidean geometry so slightly different from the Euclidean that discrimination between the two was impossible.

But Gauss' measurement requires some discussion. The problem of the geometry of physical space is more complicated than Gauss assumed and cannot be answered in so simple a way.

Assume for a moment that Gauss' result had been positive and that the angular sum of the triangle he measured had been different from 180 degrees. Would it follow that the geometry of the world is non-Euclidean?

There is a way to evade this consequence. Measuring angles between two distant objects is done by sighting the objects through lenses attached to a sextant or a similar instrument. Thus the light rays traveling from the objects to the sight device are used as defining the sides of the triangle. How do we know that the light rays move along straight lines? It would be possible to maintain that they do not, that their path is curved, and that Gauss' measurement did not refer to a triangle whose sides were straight lines. On this assumption the measurement was not conclusive.

Is there a way to test the new assumption? A straight line is the shortest distance between two points. If the path of the light ray is curved, it must be possible to connect the starting point with the end point by another line, which is shorter than the path of the light ray. Such a measurement could be made, in principle at least, with the help of measuring rods. The rods would have to be carried along the path of the light ray and then along several other lines of connection. If there is a shorter line of connection, it would thus be found by repeated trials.

Suppose that the test was carried through and it was negative, that is, we found the path of the light ray to be the shortest connection between the two points. Would this result, in combination with the previous measurement of the angular sum, prove the geometry to be Euclidean?

It is easily seen that the situation is as inconclusive as before. We questioned the behavior of light rays and checked it through measurements with solid rods. We now can question the behavior of the solid rods. The measurement of a distance is reliable only if the rod does not change its length while it is transported. We might assume that the rod transported along the path of the light ray was expanded by some unknown force; then the number of rods that can be deposited along the path is made smaller, and the numerical value found for the distance would be too small. We thus would believe the path of the light ray to be shorter than other paths, whereas in reality it is longer. Testing whether a line is the shortest distance thus depends on the behavior of measuring rods. How can we test whether a solid rod is really solid, that is, does not expand or contract?

We transport a solid rod from one place to a distant point. Is it still as long as before? In order to test its length we would have to employ a second rod. Assume that at the first place the two rods have equal length when one is put on top of the other; then one is transported to a different place. Do the two rods still have equal length? We cannot answer this question. In order to compare the rods, we should have either to transport the one rod back to the first place, or the other rod to the second place, since a comparison of length is possible only when one rod is on top of the other. In such a way we would find that they have equal length, too, when they are both at the second place. But there is no way of knowing whether two rods are equal when they are in different places.

The objection might be raised that there are other means of comparison. For instance, if a rod changes its length on transportation, we should discover the change if we compared the rod with the length of our arms. To eliminate this objection let us assume that the forces contracting or expanding transported bodies are universal, that is, that all physical objects, including human bodies, change their length in the same way. It is obvious that then no change would be observable.

The problem under consideration is the problem of congruence. It must be realized that there is no means of testing congruence. Suppose that during the night all physical objects, including our own bodies, became ten times as large. On awakening this morning we should be in no condition to test this assumption. In fact, we shall never be able to find it out. The consequences of such change are, in accordance with the conditions laid down, unobservable, and hence we can collect no evidence either for or against it. Perhaps we all are ten times as tall today as we were yesterday.

There is only one way to escape such ambiguities: to regard the question of congruence not as a matter of observation, but of definition. We must not say, "the two rods located at different places *are* equal", but we must say that we *call* these two rods equal. The transportation of solid rods defines congruence. This interpretation eliminates the unreasonable problems mentioned. It no longer makes sense to ask whether today we are ten times as tall as we were yesterday; we *call* our height of today equal to that of yesterday, and it has no meaning to ask whether it really is the same height. Definitions of this kind are called *coördinative definitions.* They coördinate a physical object, a solid rod, to the concept "equal length" and thus specify its denotation; this peculiarity explains the name.

Statements about the geometry of the physical world, herefore, have a meaning only after a coördinative definition of congruence is set up. If we change the coördinative definition of congruence, a different geometry will result. This fact is called the *relativity of geometry*. To illustrate the meaning of this result, assume again that Gauss' measurement had proved a deviation of the angular sum from 180 degrees and that measurements with solid rods had confirmed light rays to be the shortest distance: still there would be nothing to prevent us from regarding the geometry of our space as Euclidean. We then would say that the light rays are curved and the rods expanded; and we could figure out the amount of these distortions in such a way that the "corrected" congruence leads to a Euclidean geometry. The distortions may be regarded as the effect of forces which vary from place to place, but are alike for all bodies and light rays and thus are *universal forces*. The assumption of such forces means merely a change in the coördinative definition of congruence. This consideration shows that there is not just one geometrical description of the physical world, but that there exists a class of *equivalent descriptions;* each of these descriptions is true, and apparent differences between them concern, not their content, but only the languages in which they are formulated.

On first sight this result looks like a confirmation of Kant's theory of space. If every geometry can be applied to the physical world, it seems as though geometry does not express a property of the physical world and is merely a subjective addition by the human observer, who in this way establishes an order among the objects of his perception. Neo-Kantians have used this argument in defense of their philosophy; and it was used in a philosophical conception called *conventionalism,* introduced by the French mathematician Henri Poincaré, according to whom geometry is a matter of convention and there is no meaning in a statement which purports to describe the geometry of the physical world.

Closer investigation shows the argument to be untenable. Although every geometrical system can be used to describe the structure of the physical world, the geometrical system taken alone does not describe the structure completely. The description will be complete only if it includes a statement about the behavior of solid bodies and light rays. When we call two descriptions equivalent, or equally true, we refer to complete descriptions, in this sense. Among the equivalent descriptions, there will be one, and only one, in which solid bodies and light rays are not called "deformed" through universal forces. For this description I shall employ the name *normal system.* The question can now be asked which geometry leads to the normal system; and this geometry may be called the *natural geometry.* Obviously, the question as to the natural geometry, that is, the geometry for which solid bodies and light rays are not deformed, can be answered only through empirical investigation. In this sense, the question of the geometry of physical space is an empirical question.

The empirical meaning of geometry can be illustrated by reference to other relative concepts. If a New Yorker says, "Fifth Avenue is to the left of Fourth Avenue", this statement is neither true nor false unless he specifies the direction from which he looks at these streets. Only the complete statement "Fifth Avenue is to the left of Fourth Avenue seen from the South" is verifiable; and it is equivalent to the statement "Fifth Avenue is to the right of Fourth Avenue seen from the North." Relative concepts like "to the left of" or "to the right of" thus can very well be used in the formulation of empirical knowledge, but care must be taken that the formulation includes the point of reference. In the same sense, geometry is a relative concept. We can speak about the geometry of the physical world only after a coördinative definition of congruence has been given. But on that condition an empirical statement about the geometry of the physical world can be made. When we speak about physical geometry, it is therefore understood that some coördinative definition of congruence has been laid down.

Poincaré was right if he wanted to say that the choice of one from the class of equivalent descriptions is a matter of convention. But he was mistaken if he believed that the determination of natural geometry, in the sense defined, is a matter of convention. This geometry can only be ascertained empirically. It seems Poincaré believed erroneously that the "solid" rod and thus congruence can be defined only by the requirement that the resulting geometry must be Euclidean. Thus he argued that if measurements on triangles should lead to an angular sum different from 180 degrees, the physicist *must* introduce corrections for the paths of light rays and the lengths of solid rods because otherwise he could not say what he meant by equal length. But Poincaré overlooked the fact that such a requirement might compel the physicist to assume universal forces,* and that vice versa the definition of congruence can be given by the requirement that universal forces are to be excluded. By the use of this definition of congruence, an empirical statement about geometry can be made.

I should like to explain my criticism of Poincaré more fully, because recently Professor Einstein has undertaken a witty defense of conventionalism by depicting an imaginary conversation between Poincaré and me.† Since I believe that there can be no differences of opinion between mathematical philosophers if only opinions are clearly stated, I wish to state my conception in such a way that it might convince, if not Poincaré, yet Professor Einstein, for whose scientific work I certainly have as much admiration as he has so charmingly expressed for the work of Poincaré.

Assume that empirical observations are compatible with the following two descriptions:

CLASS I

(a) The geometry is Euclidean, but there are universal forces distorting light rays and measuring rods.

(b) The geometry is non-Euclidean, and there are no universal forces.

Poincaré is right when he argues that each of these descriptions can be assumed as true, and that it would be erroneous to discriminate between them. They are merely different languages describing the same state of affairs.

Now assume that in a different world, or in a different part of our world, empirical observations were made which are compatible with the following two descriptions:

CLASS II

(a) The geometry is Euclidean, and there are no universal forces.

(b) The geometry is non-Euclidean, but there are universal forces distorting light rays and measuring rods.

Once more Poincaré is right when he argues that these two descriptions are both true; they are equivalent descriptions.

But Poincaré would be mistaken if he were to argue that the two worlds I and II were the same. They are objectively different. Although for each world there is a class of equivalent descriptions, the different *classes* are not of equal truth value. Only one class can be true for a given kind of world; which class it is, only empirical observation can tell. Conventionalism sees

*The rule always to use Euclidean geometry for the ordering of geometrical observations can even lead to further complications, namely, to certain violations of the principle of causality. This will be the case if the space of physics is topologically different from Euclid's space, for instance, if it is finite. In such cases, at least one of Kant's a priori principles, either Euclidean geometry or causality, has to be abandoned. See the author's *Philosophie der Raum-Zeit-Lehre* (Berlin, 1928), p. 82.

†In P. A. Schilpp, *Albert Einstein, Philosopher-Scientist*, Evanston, 1949, pp. 677–679.

only the equivalence of the descriptions within one class, but stops short of recognizing the differences between the classes. The theory of equivalent descriptions enables us, however, to describe the world objectively by assigning empirical truth to only one class of descriptions, although within each class all descriptions are of equal truth value.

Instead of using classes of descriptions, it is convenient to single out, in each class, one description as the *normal system* and use it as a representative of the whole class. In this sense, we can select the description for which universal forces vanish as the normal system, calling it *natural geometry*. Incidentally, we cannot even prove that there must be a normal system; that in our world there is one, and only one, must be regarded as an empirical fact. (For instance, it might happen that the geometry of light rays differs from that of solid bodies.)

The theory of equivalent descriptions thus does not rule out an empirical meaning of geometry; it merely demands that we state the geometrical structure of the physical world by the addition of certain qualifications, namely, in the form of a statement about the natural geometry. In this sense Gauss' experiment presents important empirical evidence. The natural geometry of the space of our environment, within the exactness accessible to us, is Euclidean; or, in other words, the solid bodies and light rays of our environment behave according to the laws of Euclid. If Gauss' experiment had led to a different result, if it had revealed a measurable deviation from Euclidean relations, the natural geometry of our terrestrial environment would be different. In order to carry through a Euclidean geometry we then would have had to resort to the assumption of universal forces that distort light rays and transported bodies in a peculiar way. That the natural geometry of the world of our environment is Euclidean must be regarded as a fortunate empirical fact.

These formulations allow us to state the additions which Einstein made with respect to the problem of space. From his general theory of relativity he derived the conclusion that in astronomic dimensions the natural geometry of space is non-Euclidean. This result does not contradict Gauss' measurement according to which the geometry of terrestrial dimensions is Euclidean, because it is a general property of a non-Euclidean geometry that for small areas it is practically identical with the Euclidean geometry. Terrestrial dimensions are small as compared with astronomic dimensions. We are unable to observe the deviations from Euclidean geometry through terrestrial observations, because within these dimensions the deviations are too small. Gauss' measurement would have to be made with an exactness many thousands of times greater, in order to prove a deviation of the angular sum from 180 degrees. But such exactness is far beyond our reach and will presumably forever remain so. Only for larger triangles would the non-Euclidean character become measurable, since the angular deviation from 180 degrees grows with the size of the triangle. If we could measure the angles of a triangle whose corners were represented by three fixed stars, or better, by three galaxies, we would actually observe that the angular sum is more than 180 degrees. We shall have to wait for the establishment of cosmic travel before such a direct test can be made, since we would have to visit each of the three stars separately in order to be able to measure the three angles. So, we have to be satisfied by the use of indirect methods of inference, which even in the present status of our knowledge indicate that stellar geometry is non-Euclidean.

There is a further addition made by Einstein. According to his conception, the cause of the deviation from Euclidean geometry is to be found in the gravitational forces originating from the masses of the stars. In the neighborhood of a star the deviations are stronger than in interstellar space. Einstein has thus established a relation between geometry and gravitation. This amazing discovery, which was confirmed by measurements made during an eclipse of the sun and which had never before been anticipated, demonstrates anew the empirical character of physical space.

Space is not a form of order by means of which the human observer constructs his world—it is a system formulating the relations of order holding between transported solid bodies and light rays and thus expressing a very general feature of the physical world, which constitutes the basis for all other physical measurements. Space is not subjective, but real—that is the outcome of the development of modern mathematics and physics. Strangely enough, this long historical line leads ultimately back to the position held at its beginning: geometry began as an empirical science with the Egyptians, was made a deductive science by the Greeks, and finally was turned back into an empirical science after logical analysis of highest perfection had uncovered a plurality of geometries, one and only one of which is the geometry of the physical world.

This consideration shows that we have to distinguish between mathematical and physical geometry. Mathematically speaking, there exist many geometrical systems. Each of them is logically consistent, and that is all a mathematician can ask. He is interested not in the truth of the axioms, but in the implications between axioms and theorems: "if the axioms are true, then the theorem is true"—of this form are the geometrical statements made by the mathematician. But these implications are analytic; they are validated by deductive logic. The geometry of the mathematician is therefore of an analytic nature. Only when the implications are broken up, and axioms and theorems are asserted separately, does geometry lead to synthetic statements. The axioms then require an interpretation through coördinative definitions and thus become statements about physical objects; and geometry is thus made a system which is descriptive of the physical world. In that meaning, however, it is not a priori, but of an empirical nature. There is no synthetic a priori of geometry: either geometry is a priori, and then it is mathematical geometry and analytic—or geometry is synthetic, and then it is physical geometry and empirical. The evolution of geometry culminates in the disintegration of the synthetic a priori.

One question remains to be answered, the question of visualization. How can we ever visualize non-Euclidean relations in the way we can see the Euclidean relations? It may be true that by means of mathematical formulas we are able to deal with non-Euclidean geometries; but will they ever be as presentative as the Euclidean geometry, that is, will we be able to see their rules in our imagination in the way we see the Euclidean rules?

The foregoing analysis enables us to answer this question satisfactorily. Euclidean geometry is the geometry of our physical environment; no wonder that our visual conceptions have become adjusted to this environment and thus follow Euclidean rules. Should we ever live in an environment whose geometrical structure is noticeably different from Euclidean geometry, we would get adjusted to the new environment and learn to see non-Euclidean triangles and laws in the same way that we now see Euclidean structures. We would find it natural that the angles in a triangle add up to more than 180 degrees and would learn to estimate distances in terms of the congruence defined by the solid bodies of that world. To imagine geometrical relations visually means to imagine the experiences which we would have if we lived in a world where those relations hold. It was the physicist Helmholtz who gave this explanation of visualization. The philosopher had committed the mistake of regarding as a vision of ideas, or as laws of reason, what is actually the product of habit. It took more than two thousand years to uncover this fact; without the work of the mathematician and all its technicalities we would never have been able to break away from established habits and free our minds from alleged laws of reason.

The historical development of the problem of geometry is a striking illustration of the philosophical potentialities contained in the development of science. The philosopher who claimed to have uncovered the laws of reason rendered a bad service to the theory of knowledge: what he regarded as laws of reason was actually a conditioning of human imagination by the physical structure of the environment in which human beings live. The power of reason must be sought not in rules that reason dictates to our imagination, but in the ability to free ourselves from any kind of rules to which we have been conditioned through experience and tradition. It would never have been possible to overcome the compulsion of established habits by philosophical reflection alone. The versatility of the human mind could not become manifest before the scientist had shown ways of handling structures different from those for which an age-old tradition had trained our minds. On the path to philosophical insight the scientist is the trail blazer.

The philosophical aspect of geometry has at all times reflected itself in the basic trend of philosophy, and thus philosophy has been strongly influenced in its historical development by that of geometry. Philosophic rationalism, from Plato to Kant, had insisted that all knowledge should be constructed after the pattern of geometry. The rationalist philosopher had built up his argument on an interpretation of geometry which, for more than two thousand years, had remained unquestioned: on the conception that geometry is both a product of reason and descriptive of the physical world. Empiricist philosophers had fought in vain against this argument; the rationalist had the mathematician on his side, and the battle against his logic appeared hopeless. With the discovery of non-Euclidean geometries the situation was reversed. The mathematician discovered that what he could prove was merely the system of mathematical implications, of *if-then* relations leading from the axioms of geometry to its theorems. He no longer felt entitled to assert the axioms as true, and he left this assertion to the physicist. Mathematical geometry was thus reduced to analytic truth, and the synthetic part of geometry was surrendered to empirical science. The rationalist philosopher had lost his most powerful ally, and the path was free for empiricism.

Had these mathematical developments begun some two thousand years earlier, the history of philosophy would present a different picture. In fact, one of Euclid's disciples might very well have been a Bolyai and might have discovered the non-Euclidean geometry; the elements of this geometry can be developed with rather simple means of the kind available in Euclid's era. After all, the heliocentric system was discovered in that time, and Greek-Roman civilization had developed forms of abstract thought that rank with those of modern times. Such a mathematical development would have greatly changed the systems of the philosophers. Plato's doctrine of ideas would have been abandoned as lacking its basis in geometrical knowledge. The skeptics would have had no inducement to be more skeptical toward empirical knowledge than toward geometry and might have found the courage to teach a positive empiricism. The Middle Ages would have found no consistent rationalism which could be incorporated into theology. Spinoza would not have written his *Ethics Presented after the Geometrical Method,* and Kant would not have written his *Critique of Pure Reason.*

Or am I too optimistic? Can error be weeded out by teaching the truth? The psychological motives which led to philosophic rationalism are so strong that one might well assume they would have found other forms of expression. They might have pounced upon other productions of the mathematician and turned them into alleged evidence for a rationalist interpretation of the world. In fact, since Bolyai's discovery, more than a hundred years have passed and rationalism has not died out. Truth is not a sufficient weapon to outlaw error—or rather, the intellectual recognition of truth does not always endow the human mind with the strength to resist the deep-rooted emotional appeal of the search for certainty.

But truth is a powerful weapon, and it has at all times collected followers among the best. There is good evidence that the circle of its followers is growing larger and larger. And perhaps that is all that can be hoped for.

Wesley C. Salmon

THE TWIN SISTERS: PHILOSOPHY AND GEOMETRY

Almost everyone, nowadays, is aware of the intimate relations between the sciences and the various branches of mathematics. The mathematical character of modern physics is a commonplace, and the uses of mathematics in the biological and social sciences are widely publicized. You might think philosophy, in contrast to the sciences, would bear no especially significant relationship to mathematics—geometry in particular—but such a view would be quite mistaken. To be sure, the relationship is not of the same type as you find between mathematics and the empirical sciences, where some branch of mathematics is a useful, powerful, perhaps even indispensable, tool for the science in question. In the case of philosophy, the relationship goes far back, to a time long before there were any other sciences besides mathematics worthy of the name. It is a very deep relationship, representing one of the most profound and fruitful interactions between any two disciplines in the entire history of human thought. This is a rather strong statement, but I hope to justify it by sketching some of the high points of the story of this relationship.

Geometry and philosophy were born together at the same time, in the same place, and indeed, they had the same father. They are more like twin sisters than servant and master. The Greek philosopher, Thales of Miletus, who flourished around 600 B.C., made a trip to Egypt, where he learned something of the art of the surveyors, and of the geometrical knowledge they had accumulated. He brought this knowledge back to Greece, but he also introduced a significant change of viewpoint. Instead of regarding geometrical truths simply as rules of thumb, or practical guides furnished by experience, he regarded them as propositions which could actually be proved. This transformed geometry into a genuinely mathematical discipline. Thales is credited, for instance, with proving the theorem (among others) that the base angles of an isosceles triangle are equal to one another. At the same time, he is usually cited in the history books as the first important philosopher. We do not know too much about his philosophical work, for little of his writing has survived, but it appears that he believed that all things are composed of water and all things are full of gods. Mathematics seems to have had a more auspicious beginning than philosophy, but however that may be, they did come into being together with the work of Thales.

During the next three centuries both

Reprinted by permission from Wesley C. Salmon, *Space, Time and Motion: A Philosophical Introduction*, University of Minnesota Press, 1980.

geometry and philosophy flourished; for example, the mathematician Pythagoras and the philosophers Plato and Aristotle made their monumental contributions in this period. Pythagoras, incidentally, is famed as a philosopher as well as a mathematician; he stands with Descartes, Leibniz, and Russell as a philosopher whose contributions to mathematics would assure him immortality on that basis alone. By around 300 B.C. geometry was so well developed that Euclid was able to write his epoch-making work, *The Elements,* in which he reduced the whole of geometrical science to an axiomatic form in which all of the propositions (theorems) are deduced from a very small number of starting assumptions (axioms and postulates). This placed geometry in a unique position. It was by far the most thoroughly developed and most highly perfected science that existed in antiquity; indeed, I think it is fair to say, until the publication of Isaac Newton's *Principia* in 1686—just about two thousand years later—no other science existed which had an equal degree of development and perfection. The nearest rival is perhaps astronomy, but it was little more than a branch of applied geometry before Newton supplied physical explanations for the motions of the heavenly bodies.

Shortly before the time of Euclid, Plato established his Academy, a famous ancient school of philosophy. It is said that there was a sign over the door which read, "Let no one enter here unless he knows geometry." After Euclid had written *The Elements,* Plato's successors at the Academy are said to have changed the sign to read, "Let no one enter here unless he knows Euclid." Now, this might strike you as the kind of dark saying that philosophers are supposed to be famous for, but actually Plato had a very good reason for his attitude toward geometry. For him, geometry held the key to philosophical knowledge and truth—it held the key to the understanding of reality. Philosophers generally are interested in questions concerning the nature and the foundations of human knowledge. They want to know what knowledge is, how it is acquired, and on what basis it rests. Plato certainly was deeply interested in questions of this sort.

When Plato looked around for an example of human knowledge, geometry must have struck him as the outstanding candidate, for it was by far the most perfect instance of knowledge available at the time. It was evident to him that geometry was a science of pure reason; it was not the empirical science of the Egyptians who had learned by experience, for instance, that a triangle with sides of three, four, and five units, respectively, must contain a right angle. By taking a rope twelve units long, and marking off the units, they could use this knowledge to provide a practical method for constructing right angles—a task of considerable importance in surveying and building. The geometry Plato was thinking about was much like that of Euclid, where abstract propositions about abstract entities—perfect straight lines, perfect circles, perfect triangles—were proved by pure reason. To Plato this indicated that man's *reason* is capable of achieving knowledge of the most important sort; indeed, he believed the abstract geometrical entities to be more real than ordinary physical objects which exemplify these figures only imperfectly. Geometry thus constituted the best example of scientific knowledge, and in Plato's view it provided the keys both to the nature of human knowledge and to the nature of ultimate reality.

This attitude was by no means confined to Plato. Throughout hundreds of years, philosophers have held that pure reason does provide us with knowledge of the world, and that it is the best source of knowledge we have. It is far superior to our senses, which are quite capable of leading us astray. Thus, when René Descartes (1596–1650) ushered in the modern period of philosophy by wholesale doubting of the reliability of sense experience, he was not doing anything new. When, in his *Meditations,* he argued that we could never be absolutely sure of the pronouncement of the senses because we could never know for certain that we are not dreaming, Descartes was simply giving forceful expression to an idea that was familiar from before the time of Plato; namely, that our senses are subject to illusion, and they can therefore deceive us. And when he looked to geometry for a way out of the difficulty, arguing that geometrical properties (extension) represent the essence of matter, he was undertaking a program he was neither the first nor the last to attempt. He was also, incidentally, squarely in the tradition of Thales, for Descartes's invention of analytic geometry was an extraordinary mathematical contribution.

The doctrine that geometry provides useful knowledge of the physical world via pure reason was given its clearest formulation by the eighteenth-century philosopher, Immanuel Kant, who said that the propositions of geometry are *synthetic a priori* truths. By *a priori* he meant that they can be deduced by pure reason from postulates that are apparently self-evident; we do not have to perform observations and experiments in order to prove geometrical theorems. In addition, geometrical propositions are *synthetic*, which means that they provide information about the physical world in which we live—information which is clearly useful in such enterprises as surveying, navigation, architecture, engineering, and the natural sciences. In other words, geometry provides us knowledge of the actual structure of the space in which we live and move, and of the spatial relations among the objects that we meet with in everyday life.

This, then, is the picture of scientific knowledge that emerged from more than two millenia of contemplating the example provided by geometry. Knowledge that can be used to understand, predict, and control the happenings in our world can be established by purely logical demonstration. To be sure, the kinds of observations made by the Egyptian surveyors may *suggest* geometrical theorems to us, but they do not enter into the geometrical *proofs* in any way. Such observations are as irrelevant as the figures we draw on a chalkboard to the demonstrations of geometrical theorems. Empirical observations may serve a heuristic function, but they have no bearing upon the proofs.

Demonstrations must, of course, start somewhere, and this is where the *postulates* come in. The postulates are, so to speak, the basic premises for all of the geometrical deductions. This, naturally, leads us to question their status. These basic propositions, from which all of the theorems of geometry are supposed to follow, were long regarded as self-evident propositions. You can see that they must be true by just contemplating them—no one in his right mind could doubt them. Even John Stuart Mill, in his classic essay "On Liberty" (written several decades after the discovery of non-Euclidean geometry)—an essay devoted to defense of free discussion of all sorts of issues—can find no value in disputing the postulates of geometry! Thus, scientific knowledge is seen as a series of deductions from self-evident premises.

The view that the postulates of geometry are self-evident truths, though widely held, was not universally shared. As a matter of fact, we are not quite sure how Euclid stood on this point. He seems to have had no doubts about the first four of his postulates:

- P-1. A straight line can be drawn between any two points.
- P-2. A finite straight line can be extended continuously in a straight line.
- P-3. A circle can be drawn with any center and any radius.
- P-4. All right angles are equal to one another.

But there was, in addition, the infamous fifth postulate—the parallel postulate:

- P-5. Given a straight line and a point not on that line, there is one and only one line through that point parallel to the given line.

Euclid proved the first twenty-six propositions at the beginning of his *Elements* before he made any use at all of the fifth postulate. It looks a bit as if he wanted to prove all that he could without having to employ the parallel postulate; perhaps he regarded it as a little dubious, unlike the other four.

Whatever Euclid may have thought, subsequent geometers certainly regarded the fifth postulate as less self-evident than the other four. It was not that they doubted its truth; it was rather that the fifth postulate was more complicated than the others, so that its truth may not have been equally obvious. For more than two millenia mathematicians engaged in futile efforts to prove the parallel postulate. All were unsuccessful. Either they committed some logical fallacy, or they substituted an assumption which is just as difficult to justify as Euclid's fifth postulate. Some of the alternative assumptions which enable one to derive the parallel postulate are quite interesting. For example, it is sufficient to assume that the sum of the angles of the triangle is equal to two right angles, or that there are triangles of the same shape (similar triangles) but not of the same size, or that a line which is everywhere equidistant from a straight line is itself a straight line. Attempts to prove the parallel postulate on the basis of such assumptions are, of course, question begging, for these assumptions are equivalent to the parallel postulate itself.

One particular attempt to prove the fifth postulate deserves special mention. In 1733, Girolamo Saccheri, an Italian Jesuit, attempted to prove the parallel postulate by assuming it to be false, and then deducing an absurdity. This form of argument is known as *reductio ad absurdum,* and it is perfectly valid; in mathematics it is sometimes called "indirect proof." Saccheri's task split naturally into two parts. In the first place, one can deny the parallel postulate by maintaining that parallel lines do not exist at all. On the basis of this assumption Saccheri did succeed in deriving a contradiction, for the first four postulates *do* imply that there is at least one line through the given point parallel to the given line. In the second place, one can deny the parallel postulate by asserting that there is more than one line through the given point parallel to the given line. Saccheri satisfied himself that he had deduced a contradiction from this assumption as well, but in fact he did no such thing. What he actually succeeded in doing was deducing some interesting theorems of what later came to be known as non-Euclidean geometry. Unfortunately, he mistook one of the stranger ones for an absurdity. He was, therefore, the unwitting discoverer of non-Euclidean geometry, even though he died believing he had "cleansed Euclid of every blemish."

Saccheri did not know what he had done, but near the beginning of the nineteenth century three famous mathematicians, Carl Friedrich Gauss, Johann Bolyai, and Nikolai Ivanovich Lobachevski, working on the problem of the parallels, came to the conclusion that it is possible to assume that the parallel postulate of Euclid is false without getting into any absurdity or contradiction. In fact, they realized, it is possible to adopt Euclid's first four postulates while denying the fifth one (by asserting the existence of more than one parallel), and to develop a perfectly consistent non-Euclidean geometry on that basis. Gauss, who was quite possibly the greatest mathematician of all time, was the first to make the discovery, but he was so reluctant to engage in the kinds of senseless dispute he felt would result from publication that he did not publish his discovery until much later. In the meantime, about 1820, the same results were established more or less simultaneously, and certainly independently of one another, by Bolyai and Lobachevski. They denied Euclid's fifth postulate by saying that, instead of one parallel, there are many parallel lines, and on this basis they developed a new geometry.

A number of years later, around the middle of the nineteenth century, the mathematician Georg Friedrich Bernhard Riemann discovered that it is possible, if one tinkers a bit with the first four postulates, to develop another type of non-Euclidean geometry on the basis of a postulate that denies the existence of parallels altogether. He carried out this program successfully. Thus, at about the time Kant died, Gauss was secretly working out the details of a non-Euclidean geometry, and about twenty years later Bolyai and Lobachevski published their versions of such a geometry. Fifty years after Kant's death, Riemann elaborated a second form of non-Euclidean geometry. At this point, three distinct types of geometry are available: the geometry of Euclid (one parallel), the geometry of Bolyai and Lobachevski (many parallels), and the geometry of Riemann (no parallels). In addition, Riemann had worked out a generalized system of geometry in which the foregoing three fit as special cases.

In order to get an intuitive feel for these geometries, it will be a good idea to see how they can be realized. If we confine attention to the two-dimensional case, it is easy to see some of the important distinguishing characteristics of these types of geometry, for each is exemplified by a two-dimensional surface of a particular sort. As we all know, a two-dimensional surface like the surface of a chalkboard represents part of the two-dimensional Euclidean plane. Lines, circles, triangles and so on, in the ideal flat surface satisfy the relations laid down for two-dimensional Euclidean geometry.

A different sort of geometry is illustrated by the surface of a sphere. The sphere itself is a three-dimensional solid object in three-dimensional Euclidean space, but its surface is two-dimensional, as is indicated by the fact that points on the surface of the earth (which is approximately spherical) can be unambiguously located by two coordinates (longitude and latitude). If we look at this surface in the same way as we regard the surface of the chalkboard when we do ordinary plane Euclidean geometry, we can use it to illustrate one form of non-Euclidean geometry—namely, the Riemannian geometry of no parallels. In order to see this we must, first of all, be quite clear as to what we mean by a "straight line" on the surface of a sphere.

At first blush it looks as if there are no straight lines on the surface of the sphere; if we try to join two distinct points on the surface by a straight line, that line will not lie on the surface but will depart from it. For instance, to join the north and south poles of the earth (thinking of it as if it were a perfect sphere—which, of course, it is not) we get the axis of rotation which goes through the center of the earth and does not remain on its surface. But, by a suitable definition, we can introduce the concept of straight lines which do lie on the surface. The definition we need is simply the customary one according to which a straight line is the shortest distance between two points—but the line must remain on the surface. It cannot bore through the sphere, nor can it leave the sphere and go out into space—it must stay right on the surface. Defined in this way, the straight lines (often called "geodesics") on the sphere turn out to be the so-called "great circles."

Thinking again of the earth as a sphere, we see that one of these great circles is the equator, and all the lines of longitude are also great circles. Indeed, any plane that passes through the center of the sphere intersects the surface of the sphere in a great circle. You can test these claims by stretching a string between points on the surface of your globe. The lines of latitude, as you will see, are definitely not straight lines; they are circles which are *not* great circles. They are not shortest paths between the points that lie in them. Consequently, if you want to travel on the surface of the earth between two distant points on the same latitude, you take what is called a "great circle route," which is familiar in connection with intercontinental air travel. This route does not follow the same latitude, but rather a great circle. If, for instance, you want to fly from Los Angeles to Jerusalem, both cities being at roughly 30° north latitude, your great circle route would take you over Greenland at about 60° north latitude

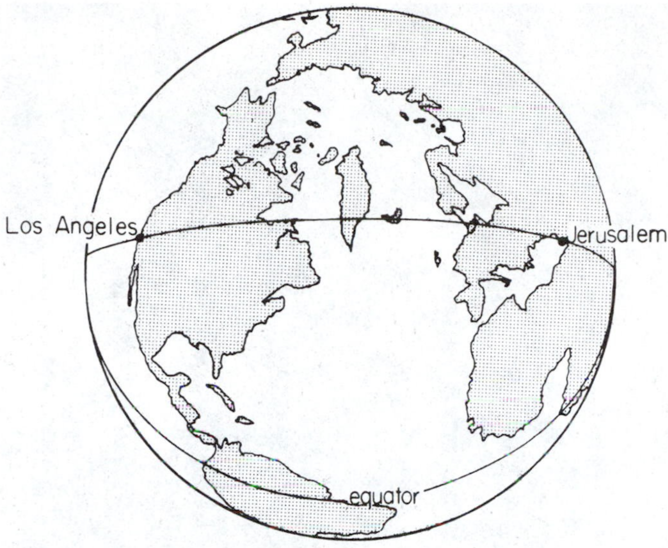

FIGURE 1

(halfway to the North Pole from 30°), but that is the shortest route (see Figure 1). If you have any doubt, test it with your stretched string.

Another way to see that lines of latitude are not shortest paths is to consider two opposite points on a line of latitude drawn close to the North Pole. It is obvious that to get from one point to the other you would not follow the circle around the pole; you would go straight across the pole instead.

With this understanding of what qualifies as a straight line, we can ask questions about parallel lines, triangles, and circles. From here on, we shall use the unqualified word "line" to refer to straight lines; for example, when we speak of parallel lines, these are to be understood to be *straight* lines that do not intersect one another. First of all, it should be obvious that there are no parallel lines on the surface of the sphere because all great circles intersect one another. The lines of longitude intersect at the two poles, the equator intersects the lines of longitude, and any other great circle that you draw is going to intersect the lines of longitude, the equator, and any other great circles that there might be. This is, indeed, a geometry of no parallels; given a straight line and a point not on that line, there is no straight line through that point which does not intersect the given line.

Circles and triangles on the surface of the sphere differ remarkably from circles and triangles in the Euclidean plane. Consider a large triangle ABC (see Figure 2) whose base line is along the equator and whose sides run from the equator to the north pole along lines of longitude. The lines of longitude intersect the equator in right angles, so that the two base angles at B and C are both right angles. The sum of the angles of triangle ABC is, consequently, greater than two right angles by the amount of the angle at A. We have all learned in Euclidean geometry that the sum of the angles of a triangle is two right angles, so here we have one of the basic distinctions between this particular type of geometry and the familiar Euclidean geometry we studied in high school. The sum of the angles of the triangle in Euclidean geometry is always 180°, whereas, in the non-Euclidean geometry of the sphere the angular sum is greater than 180°. Moreover, on the sphere the sum of the angles is not constant but depends

FIGURE 2

rather upon the size of the triangle. If you draw a very tiny triangle (DEF, Figure 2), the sum of the angles will be very nearly 180°; if it is small enough the angular sum will be indistinguishable for practical purposes from 180°.

Consider next a circle that is drawn on the surface of the sphere (Figure 3). The circle is simply a locus of points equidistant from a given center P. We can, of course, draw a straight line QR through the center; QR is a diameter. We can now ask about the relationship between the circumference and the diameter of this circle. Once again, in Euclidean geometry, we have learned that the ratio of the circumference to the diameter c/d is the number $\pi(= 3.1415 \ldots)$. In the geometry of the sphere, however, this ratio is smaller than π, roughly because the diameter QR has to bend a little bit to get from one side of the circle to the other. You can see this very clearly by considering the equator as a circle. As a matter of fact, it is a great circle, and so it is both a circle and a straight line, paradoxical as that might seem! Consider a diameter of the equator; it is simply a line of longitude. The line of longitude extending from the equator on one side, through the pole, to the equator on the other side, is obviously just half the length of the equator itself; consequently, in that case the ratio $c/d = 2$. Again, with the ratio c/d, as with the angular sum of the triangle, the value is not constant but depends upon the size of the figure. If we have a very large circle, the discrepancy between c/d and π is rather large; in a small circle the ratio will be very nearly equal to π. As the circle gets very, very tiny, the ratio becomes practically indistinguishable from π.

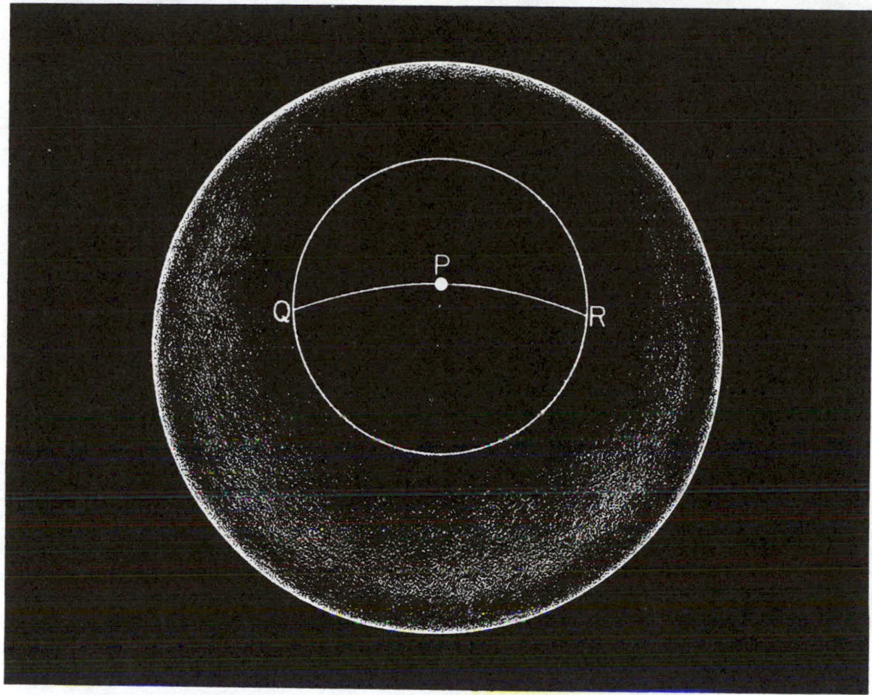

FIGURE 3.

The foregoing remarks have brought out a couple of the most outstanding features of the non-Euclidean geometry which is represented by the surface of the sphere. The geometry of many parallels, developed by Bolyai and Lobachevski, is exemplified by a surface which is known as a pseudo-sphere (Figure 4A). The saddle surface (Figure 4B) provides a useful approximate representation of that type of geometry. The important thing about surfaces like the pseudo-sphere and the saddle is that, unlike the sphere, if you look at one cross section of the surface (front to back) it curves down; if you look at a perpendicular cross section (side to side), it curves up. The sphere, of course, curves in the same direction regardless of which cross section you examine. This is the basis for saying that the surface of the sphere has positive curvature while the surface of the pseudo-sphere has negative curvature.

In dealing with the geometry of the saddle surface (or pseudo-sphere) we adopt the same meaning as before for "straight line"—namely, the shortest distance between two points. In Figure 5 we have a picture of a saddle surface with a straight line L and a point P above, with two straight lines through the point P. These two solid lines are both parallel to the original given straight line L, for no matter how far they are extended in either direction they will not intersect L. They approach L (in opposite directions), but they never meet L. Moreover, they both go through the point P, so here we see how the saddle surface realizes the condition of this type of non-Euclidean geometry: given a straight line and a point not on that line, there is more than one straight line through that point parallel to the given line.

In Figure 6 we have a triangle drawn upon the saddle surface. The triangle appears to be concave, but the sides of the triangle are indeed straight lines—that is, shortest paths on this non-Euclidean surface. You can see quite clearly that the sum

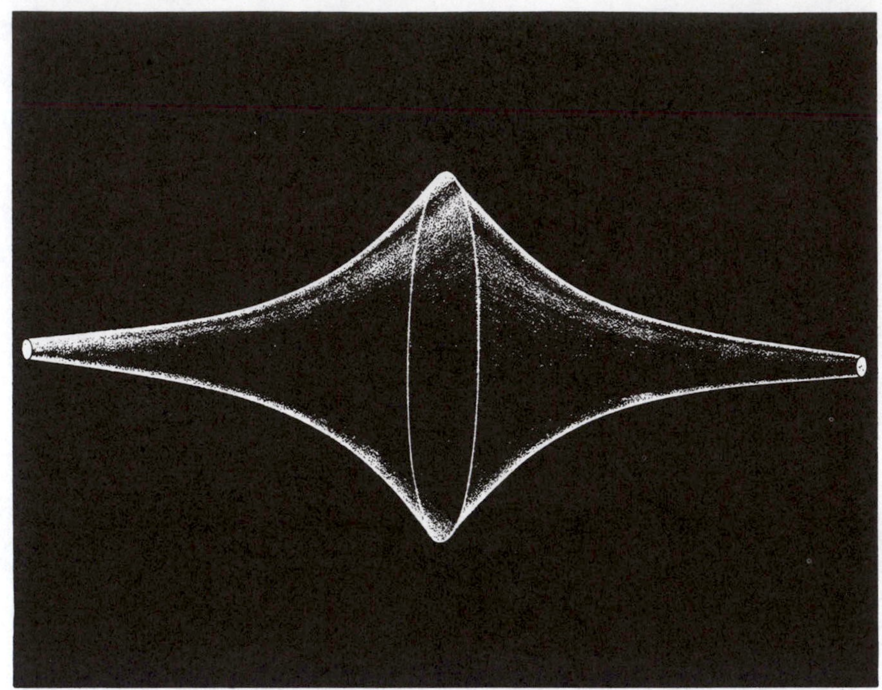

FIGURE 4A Pseudo-sphere.

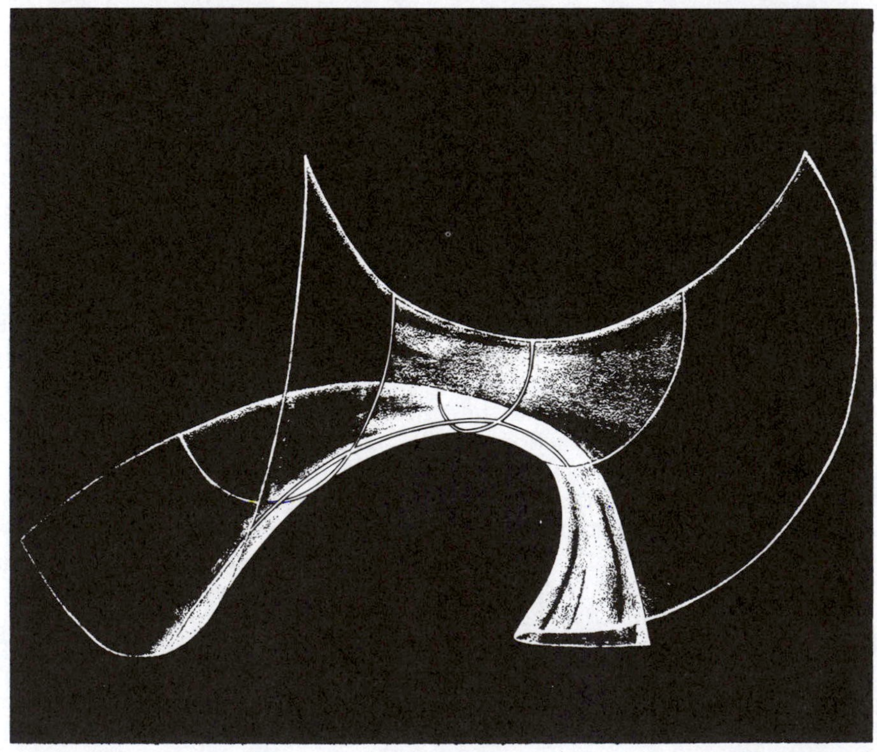

FIGURE 4B Saddle.

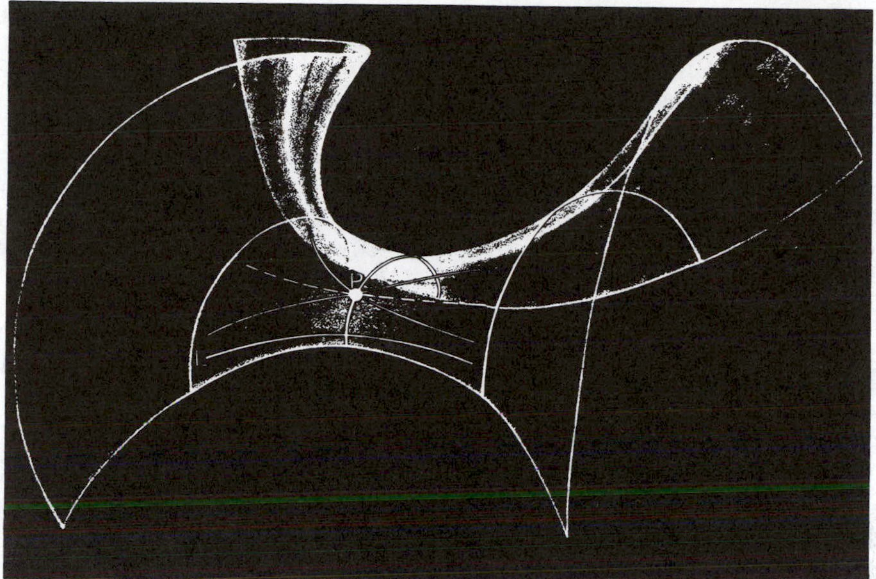

FIGURE 5

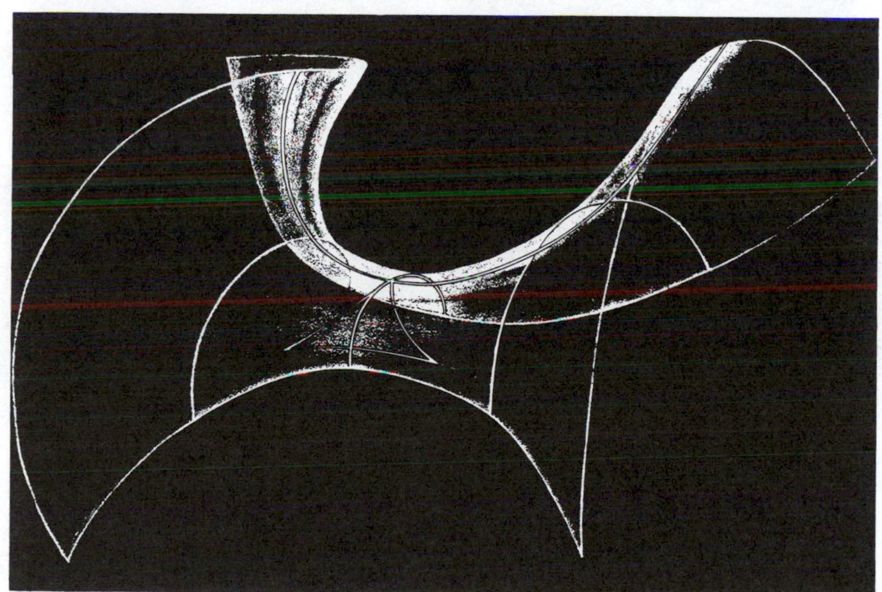

FIGURE 6

of the angles of that kind of triangle will be less than the sum of the angles on the flat Euclidean surface. Thus, on the saddle-type surface, the sum of the angles of a triangle is less than 180°. As in the case of the sphere, the sum of the angles is not constant, but depends upon the size of the triangle. The larger the triangle, the larger the discrepancy between the angular sum and 180°; the smaller the triangle, the smaller the discrepancy. In a very tiny triangle, the discrepancy is not noticeable at all.

Now, although it is hard to illustrate this point with a picture, it is nevertheless true that the ratio *c/d* of the circumference to the diameter of a circle in the saddle surface differs from π, the Euclidean value. In this case, *c/d* is greater than π, instead of less as in the case of the spherical surface. If you happen to have a saddle (or a Pringle "New-fangled Potato Chip") you might try verifying this fact using a string, a string compass, and a ruler. Again, the discrepancy depends upon the size of the figure.

We have now presented several major features of the three types of geometry; they are summarized in Table 1. But what sort of moral are we to draw from all of this? After all, for well over 2000 years there was one and only one geometry, and it was universally accepted as *the* true geometry. It evidently did not occur to anyone to suppose seriously that there were others, different from Euclid's geometry, which might be equally acceptable. Then, a mere century and a half ago, the non-Euclidean geometries were developed. The situation changed markedly. Now, there are several different kinds of geometry to choose among, and the question naturally arises as to which is correct. Each type can be elaborated in a purely formal and axiomatic fashion. Each can be formally generalized to three dimensions, just as Euclidean plane geometry can be generalized to Euclidean solid geometry. Furthermore, from a purely formal standpoint, all of these geometries are on a par. While it is impossible to provide an unconditional proof that the non-Euclidean geometries are free from contradiction, no such proof can be given for Euclidean geometry either. But it is possible to prove relative consistency: if any one of these geometries contains a contradiction, so does every other one. In particular, if non-Euclidean geometry is inconsistent, then so is Euclidean geometry. With this discovery the *logically* privileged status of Euclidean geometry was forever destroyed.

If there is no logical basis for selecting a particular geometry, the next question is obviously, "Which of these geometries correctly describes the physical space of our universe?" Gauss, himself, saw this question

TABLE 1.

	Riemann	Euclid	Bolyai-Lobachevski
Parallels	Zero	One	Many
Surface	Sphere	Plane	Pseudosphere (or saddle as a good approximation)
Curvature	Positive	Zero	Negative
Angular sum for triangles	>180° depends on size of triangle	=180° independent of size of triangle	<180° depends on size of triangle
Ratio of circumference to diameter of circle	<π depends on size of circle	=π independent of size of circle	>π depends on size of circle

clearly and performed an experiment in an attempt to answer it. He went to the tops of three mountains in the Alps, and he sighted from one mountaintop to another to lay out a triangle with the mountaintops as the three vertices. The sides of these triangles were defined by the paths of the light rays along which he sighted. This was *not* a triangle on the surface of the earth; the light rays go through space near the earth's surface, but the triangle touches the earth only at its three vertices. He made careful measurements of the angles of this triangle but, within the experimental accuracy available in that case, he could find no discrepancy between his results and 180°. We are not terribly surprised to learn that he was not able to find any disparity between triangles in actual space and Euclidean triangles, because his terrestrial triangle was, in terms of the total size of the universe, an exceedingly small one. We recall the point made above: in each type of non-Euclidean geometry the difference between the angular sum of a triangle and two right angles is a function of the size of the triangle—the smaller the triangle the smaller the difference. Even if physical space were non-Euclidean on a cosmic scale, the angular sum of Gauss's triangle would be very nearly two right angles. But whether the outcome is surprising or not, Gauss's experiment was philosophically important because of the direct way in which it confronts the problem of the geometrical structure of physical space.

Looking at Gauss's experiment, we may be tempted to suppose that the question of which geometry correctly describes physical space admits of a fairly straightforward empirical answer. If we make more and more precise measurements of larger and larger triangles, and if the results were always the same—no detectable difference from 180°—we would seem justified in saying that, within the accuracy of our methods, physical space reveals no departure from Euclidean relations. But we must consider what would happen if the results were different—if measurement of some triangles were to yield a distinct discrepancy from two right angles. Would we not be required to conclude that the geometry of physical space is non-Euclidean? Unfortunately, a simple unqualified affirmative answer is not warranted.

To illustrate this point, imagine what might have happened if Gauss had obtained a different outcome from the measurement of his Alpine triangle. Suppose, for example, that his very careful measurements had produced an angular sum distinctly below two right angles. What then? The normal reaction at this point would probably be to say, "Something has gone wrong; maybe this figure is not a triangle after all—perhaps the light rays along which we sighted did not travel in straight lines." If the figure had curved sides rather than straight ones, there would be no conflict with Euclidean geometry in saying that its angular sum is less than two right angles. It would then be necessary to investigate further and find out whether we were indeed measuring the angles of a triangle, or whether we were dealing with some curvilinear figure instead.

Let us see what might happen as we pursue this question. Suppose that somehow or other we take meter sticks and measure the paths of the light rays between the various mountaintops. And suppose we find that the paths of the light rays are in fact shorter, as measured by our meter sticks, than any other paths connecting these peaks. What then? It would still be open to someone to object, "After all, it could be that strange forces are affecting our measuring rods—perturbations which make them change their size or shape as they are moved from place to place." Again, the result does not prove conclusively that the sum of the angles of the triangle is less than 180°; it might be taken to prove instead that the figure in question is simply not a triangle—and that we must adjust our views about light rays and measuring rods accordingly.

Around the end of the nineteenth century, the French philosopher-scientist Henri Poincaré, following this line of thought, concluded that *any* apparent deviations from Euclidean geometry could always be explained away much as we have just indicated—that is, by saying that any disagreement with Euclidean spatial relations is a sure sign that something is wrong with our measuring rods or our experimental technique. In all such cases, we can account for the discrepancy from Euclidean geometry by means of some sort of perturbation affecting the measuring rods or other laboratory equipment.

This is, perhaps, the plausible thing to say. It means that we can always, if we wish, adjust our experimental results to fit Euclidean geometry; we can, if we wish, preserve Euclidean geometry at all costs. Perhaps that is what should be done in such situations. Kant would certainly have agreed. As already mentioned, Kant maintained that Euclidean geometry is synthetic a priori—necessarily true—and that all of our experiences *must* fit the Euclidean framework. Accordingly, any kind of apparent conflict between physical measurement and Euclidean geometry would have to be resolved in favor of Euclidean geometry.

It is a rather common error to suppose that the discovery of non-Euclidean geometries in and of itself constitutes a refutation of Kant's doctrine of the synthetic a priori character of Euclidean geometry. Even in the light of the relative consistency proof of Euclidean and non-Euclidean geometries, the existence of alternative types of geometry does not refute Kant's thesis. Kant did not maintain that alternative geometries are self-contradictory, for that would make Euclidean geometry analytic rather than synthetic. And the fact that it is logically possible for our physical measurements to yield results in apparent conflict with Euclidean geometry does not necessarily rob Euclidean geometry of its a priori status; we have just seen how Euclidean geometry can be maintained in the face of any apparent empirical refutation. The fact that all of these different geometries are on a par with respect to logical consistency does not mean that they are equally suitable for the description of the physical world. Kant's thesis is simply a denial of the possibility of using any geometry other than Euclid's for the visualization of spatial relations.

This view is extremely tempting, and it is widely held today by people who may not even recognize it as Kantian. Kant maintained that Euclidean geometry is a necessary form of spatial intuition. Stated less pretentiously, this view runs somewhat as follows: "Non-Euclidean geometry may be an amusing game for mathematicians to play, but you cannot really conceive of a non-Euclidean space. You cannot imagine what it would be like for our three-dimensional space to be non-Euclidean. To be sure, we can picture a two-dimensional non-Euclidean surface, but that is because we can stand outside of it in our three-dimensional Euclidean space and observe the curvature of the surface as it is embedded in three-dimensional space. But we cannot picture the curvature of our whole three-dimensional space because we cannot imagine stepping off into four-dimensional space." This common sort of statement seems to me to express Kant's view that Euclidean geometry is *the* necessary form of spatial intuition. While all of the geometries are on an equal footing from a logical standpoint, they are not on an equal footing epistemologically. Non-Euclidean geometries cannot be visualized, Kant says, and so they cannot be used for the description of spatial relations. *Physical* space cannot be conceived to be non-Euclidean. Euclidean geometry, according to Kant, enjoys an insuperable epistemological advantage over the other types of geometry precisely because we cannot visualize a three-dimensional non-Euclidean space.

In order to deal with this Kantian doctrine we must first try to determine whether it is true that we cannot visualize non-Euclidean spaces. Even though it may be possible to describe physical space in Euclidean terms, is it necessary to do so? Can we organize our external experiences in Euclidean space alone, or is it also possible to do so in a non-Euclidean framework? If we are to come to terms with such questions, we must become quite clear on what we are to mean by "visualization."

There are no special difficulties in visualizing two-dimensional spaces, whether they be Euclidean or non-Euclidean. We can literally see the surfaces or we can call them up in our imagination. In either case, they are seen from without as two-dimensional manifolds embedded in three-dimensional space. And the curvature of the non-Euclidean surfaces can readily be seen as a departure from the shape of the Euclidean plane. Let us call this *external visualization.*

It is easy to suppose that there is no particular problem in visualizing a Euclidean space of three dimensions, but clearly this is very different from the external visualization of a two-dimensional space. We cannot step outside of our three-dimensional space into a four-dimensional space in order to visualize externally *either* a three-dimensional Euclidean space *or* a three-dimensional non-Euclidean space. Instead, we must formulate an appropriate conception of *internal visualization* in order to understand what is involved in the visualization of a three-dimensional space of any variety. If we visualize three-dimensional Euclidean space, we do so from the standpoint of beings confined within that space; if we want to visualize three-dimensional non-Euclidean space, we must likewise do so from within. It was Hermann von Helmholtz, a nineteenth-century scientist and philosopher, who first saw this point and formulated a suitable concept of internal visualization. To visualize a space *internally,* he said, is simply *to imagine the kinds of experiences one would have if he were living in such a space.* We internally visualize a Euclidean space very easily because we are used to the kinds of experience people who are living in such a space have on a regular basis. But obviously this capability in no way entails an ability to view our three-dimensional universe from a vantage point in a four-dimensional super-space.

With Helmoltz's distinction between internal and external visualization clearly in mind, we can now see how it is possible to visualize three-dimensional non-Euclidean spaces. To make this point somewhat easier let us go back to the two-dimensional case. Imagine some two-dimensional creatures whose existence is confined entirely within a two-dimensional surface. Suppose that they can move around and make measurements (using two-dimensional instruments) in their two-dimensional space just as we can move about and make measurements in our three-dimensional space. They can no more leave their two-dimensional world to view it from a three-dimensional standpoint than can we leave our three-dimensional world to view it from a four-dimensional standpoint. Now, if it should happen that these two-dimensional beings inhabit the surface of a sphere, they would be unable to see the curvature as we can from our external vantage-point, but they could detect the curvature indirectly. If they were to lay out triangles and circles, measuring angular sums and diameters and circumferences, they would find that the angular sum is always greater than two right angles, and that the ratio of circumference to diameter is always less than π. They would find, moreover, that the amount of the discrepancy depends upon the size of the figure. In such a case, they would have detected the curvature of the spherical surface without having seen it externally. For two-dimensional beings to visualize internally this type of non-Euclidean space *is* to imagine getting the foregoing sorts of results in making measurements. The internal visualization of Euclidean and Lobachevskian spaces of two dimensions is completely analogous.

Similar considerations apply to the internal visualization of three-dimensional Euclidean and non-Euclidean spaces as well. We do not find it inconceivable that, in three-dimensional space, the measurement of the angles of triangles, regardless of their orientation in space, might result in a sum different from two right angles. Nor is it inconceivable that this result might vary with the size of the triangle. It is not inconceivable that measurements of circumferences and diameters of circles might yield a ratio other than π, whose value again depends upon the size of the figure. Analogous departures from Euclidean relations might also occur regarding the relation between the radius and the surface or the radius and volume of a sphere. These sorts of considerations show how, in principle, we could make a geometrical survey of our three-dimensional universe—in a manner quite analogous to our imaginary two-dimensional beings—in an effort to ascertain the geometrical structure of physical space. As yet, the empirical results are inconclusive, but we can imagine what they would have to be like to indicate any of the types of geometry we have discussed. According to Helmholtz's conception, the ability to visualize internally spaces characterized by the different types of geometry consists in this very ability to imagine sets of experiences of the sort just described.

It is extremely difficult, we must admit, to accept psychologically the import of Helmholtz's distinction between internal and external visualization, and to apply it to the problem of visualization of a three-dimensional non-Euclidean space. We find no difficulty in making the transition from *external* visualization of a two-dimensional Euclidean plane to the *internal* visualization of a three-dimensional Euclidean space, because all of our everyday spatial experiences since infancy have fit nicely into the Euclidean framework. If our universe has a non-Euclidean structure, it reveals itself only on a very large scale, and the deviation from Euclidean structure is too slight in moderate-sized regions to be noticed in ordinary experience. Because the transition from external to internal visualization is so easy in the Euclidean case, we may not even be aware of any distinction between the two types of visualization.

When, however, we turn to the problem of visualizing a non-Euclidean three-dimensional space, we may find the transition from the external visualization of the two-dimensional surface to the internal visualization of the three-dimensional space a severe psychological strain. We may feel, in fact, that there is something phony about translating the problem of "visualizing" or "imagining" a three-dimensional non-Euclidean space into the problem of internal visualization in Helmholtz's sense. It may seem simply impossible to imagine or picture a three-dimensional non-Euclidean space, no matter how successfully we can conceive intellectually the possibility of physical measurements that fit a non-Euclidean geometry. We must not forget, however, the psychological power of our lifelong conditioning to the Euclidean framework. If we had grown up in a world in which non-Euclidean relationships were a matter of daily experience, then it seems likely that the visualization of a three-dimensional non-Euclidean space would pose no more psychological difficulty than does the visualization of three-dimensional Euclidean space for us. If the experiences we must imagine (with some serious expenditure of intellectual effort) in order to visualize internally a three-dimensional non-Euclidean space were extremely familiar, this very same internal visualization would, it seems safe to say, be easy and psychologically satisfying. It is, therefore, a matter of psychological fact—not a priori necessity—that we have no problem internally visualizing a three-dimensional Euclidean space, while the visualization of a three-dimensional non-Euclidean space seems at first to present insuperable difficulties. Recognition of the distinction between internal and external visualization, and appreciation of the epistemological significance of that distinction, is a philosophical accomplishment of major proportions. Only after we have fully understood this point can we see that the problem of visualizing three-dimensional non-Euclidean space is, indeed, the problem of internal visualization in precisely the sense we have been discussing.

According to the foregoing results, we can visualize three-dimensional non-Euclidean spaces in precisely the same sense as we can visualize three-dimensional Euclidean spaces. Euclidean and non-Euclidean geometries are, therefore, on an equal footing from the standpoint of visualizability as well as from the purely formal standpoint. This conclusion, which goes far beyond mere logical parity, *does* constitute a serious blow to the Kantian doctrine that Euclidean geometry is synthetic a priori. Of course, it is always possible, as we have seen, to explain away any apparent non-Euclidean results by alleging the perturbation of measuring devices, but it is not necessary to do so, because we can visualize non-Euclidean spaces. To revert to Kantian terminology, non-Euclidean geometry thus emerges as a possible form of spatial intuition.

Although we have found that Euclidean and non-Euclidean geometries are formally and epistemologically admissible, there remains the question of which geometry correctly describes the physical space of our universe. At most we have shown that none of these geometries can be ruled out a priori; we have not really shown how to answer the a posteriori problem of which one to adopt. The question is a vexing one because, as we have seen, empirical results of measurement do not seem to offer unambiguous answers. Any measurement which would seem to provide a non-Euclidean result can be interpreted as fitting into a Euclidean framework, and any measurement which seems to provide a Euclidean result can be adapted to a non-Euclidean framework. Whatever results we get by using light rays, solid rods, and similar instruments for making measurements can be explained away, for we can always say that they are perturbed. We can always claim that light rays bend and that solid rods shrink and expand. Poincaré has made this point very forcefully.

Consider another fanciful example, which was put forth by Hans Reichenbach. Imagine that world A, shown in cross section in Figure 7, is a two-dimensional world consisting of a flat plane with a hump in the middle. Again, imagine that this world is inhabited by two-dimensional creatures who move about and make measurements in an attempt to survey it geometrically. In the peripheral regions of their space, they would find that it has Euclidean characteristics—the ratio of circumference to diameter of a circle is always π, the angles of the triangle always add up to 180°, and so on. In the central part, where the hump is located, they would find that their space has the geometrical properties of the surface of a sphere. In the region where the hump joins smoothly with the flat plane, they would find characteristics rather like those of the saddle surface. Moving about in their space and making such measurements, they

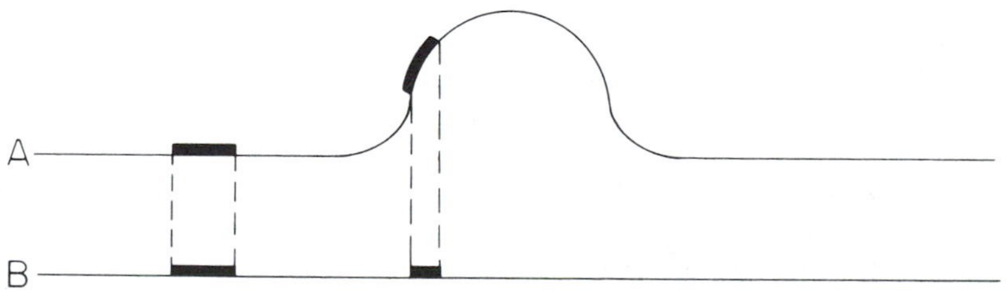

FIGURE 7

could find out that their space has precisely the kind of curvature just described, though they would be unable to form a three-dimensional mental image like ours.

Imagine, at the same time, beings who live in world B, located directly below world A, with measuring rods that behave in a peculiar fashion. Whenever these rods are moved from one place to another, they behave exactly as if they were vertical projections of measuring rods in world A above. Since *all* material bodies are assumed to expand and contract in the same manner, these beings would be quite unaware of such changes in their measuring rods. Obviously, the creatures in world B would get exactly the same results from their measurements as the inhabitants of world A. Now, this shows quite clearly that there are two interpretations available. A two-dimensional Poincaré living in world B might say, "Wait a minute—it is not necessarily true that our world is a flat surface with a hump in it. Maybe our world is one in which the measuring rods behave in a very odd manner, contracting and expanding in such a way that this world appears to have a hump when in fact it is perfectly flat." None of his countrymen in world B could refute him, as we saw in our discussion of Gauss's experiment. At the same time, precisely the same argument would be open to any inhabitant of world A as well, for the inhabitants of both worlds have precisely the same experiences. We see that, although these two worlds look very different from our Godlike external vantage point, from the standpoint of beings confined within these two worlds there is absolutely no way of distinguishing one from the other. Indeed, these are identical worlds! We have merely offered two equivalent ways of describing the same spatial facts. Henceforth we shall refer, not to two worlds, but rather to two descriptions, A and B, of the world of Figure 7.

It might be objected, however, that we have not yet come to grips with the basic problem—namely, the question of whether the measuring rods really do expand and contract as they are moved about, or whether they remain the same length. If we knew the answer to that question we would know whether the world really had a hump or not. But how can we find out? We do know that measuring rods expand and contract under the influence of certain kinds of forces, for example, as a result of changes of temperature. This kind of expansion and contraction is empirically detectable because it affects different materials in different ways. A copper rod, for example, expands a different amount and at a different rate from a glass rod. Forces whose influence depends upon the chemical constitution of the object affected are known as *differential forces;* their results are called *differential effects.* Whenever precise measurement is attempted, great care must be taken to correct for differential effects upon the instruments employed.

When we raise a question, however, about the expansion or contraction of measuring rods which are moved from place to place, we are assuming that such rods are free from the influence of any differential forces or that suitable corrections have been made for any differential effects. If these rods are subject to such expansion and contraction, it is a *universal effect* which depends only upon the position of the object. It is due to a *universal force* that affects all substances in the same manner and to the same degree. The fact that such effects are universal, not occurring differentially in different objects, guarantees that they will be empirically undetectable. To the question of whether two measuring rods located at different places are really the same length, or whether the same measuring rod keeps the same length as it is moved from place to place, there is no possible empirical answer. Consequently, Reichenbach argues, the question is not a factual question at all; it must be construed as a demand for a definition or convention. The concept which requires definition is *congruence;* we must define equality of spatial intervals when these intervals are located at different places. One natural definition of congruence results when we say that a solid rod remains congruent to itself—that is, it retains the same size—wherever it is located and however it is oriented in space (correcting, of course, for any differential effects). This is not, however, the only possible definition of congruence. Our two-dimensional Poincaré got rid of the hump by introducing a different definition of congruence.

Reichenbach's analysis shows, I think, that congruence cannot be discovered, but must be defined, *if* equality or inequality of length (or distance) is constituted by, and has no meaning apart from, the behavior of solid rods and other physical measuring devices. This means, in effect, that length is not a characteristic of space, but of the bodies that occupy space. Such a view differs fundamentally from Newton's account. According to Newton, absolute space is some sort of entity that is entirely distinct from, and not to be confused with, material bodies that may be used to measure space. Just as a table should not be identified with the ruler used to measure it, so also must absolute space be regarded as something separate from our measuring rods. Absolute space is taken by Newton to be a container for material objects, in no way depending upon them for its nature or existence. Hence, in a Newtonian framework, it is still appropriate to ask whether a measuring rod really expands or contracts when moved from place to place, since it is meaningful on the Newtonian view to ask whether it occupies the same amount of absolute space in one place as in the other. At one time, the two ends of the measuring rod are located at points P_1 and P_2; at a later time, the two ends of the same measuring rod are located at P_3 and P_4. The Newtonian question is this: is the quantity of absolute space in the interval P_1P_2 equal to the quantity of absolute space in the interval P_3P_4? According to Reichenbach's analysis, we are free to *define* the two as equal on the basis of the behavior of the measuring rods; according to Newton's approach we must *discover* whether the two amounts of space are equal. This issue resolves itself into the question of whether physical space has an intrinsic metrical structure. If it does, there is a true answer to the question of whether measuring rods change their size when they are transported to different locations. If space has such an intrinsic structure, then it also has a determinate intrinsic geometry. If no such intrinsic structure exists, it is open to us to define congruence, and different definitions of congruence may yield different geometries as correct descriptions of the same space, as we saw in connection with the world with a hump.

It was Riemann who first clearly posed the question of whether space has an intrinsic metric. His answer—the correct answer, I believe—was negative. Although his reasoning was somewhat faulty, due to the fact that Georg Cantor's set theory had not been developed at the time, it seems basically to be on the right track. I shall not elaborate in detail in this chapter for this problem is closely related to one of Zeno's paradoxes (not one of the famous paradoxes of motion, but a less familiar paradox of plurality), which we shall discuss at some length in the next chapter. But the general approach is this. A continuous finite line segment, regardless of length, contains an infinite number of points; indeed, it can be shown that any finite line segment contains precisely the same infinite number of points as any other. These points are, moreover, ordered in precisely the same way in any two such segments. This shows that the one line segment is *isomorphic* to the other (regardless of length)—that is, they have the same structure. Since all finite line segments are identical in internal structure, difference of length must somehow be imposed on an extrinsic basis. This is precisely where Reichenbach's definition of congruence comes in; it provides a basis for comparison of length by means of solid bodies which function as measuring rods.

It turns out, then, that the question of whether the measuring rod really changes its length when moved from place to place is without significance. Hence, the question of whether space is really Euclidean or non-Euclidean loses its significance as well. But this does not mean that whatever we might say about the geometrical structure of physical space is without significance. Nor does it mean that we are free to settle all questions about the geometry of space by means of definitions. Rather, the question of what geometry correctly describes physical space is an incomplete question unless a definition of congruence is provided. In the world of Figure 7, we may adopt the definition of congruence which says that the measuring rod does not change in length as it is moved about, and then determine by empirical measurements that the geometry is non-Euclidean. Or, we may adopt the definition of congruence which says that the measuring rods behave as stipulated by our two-dimensional Poincaré, and ascertain by empirical measurement that the geometry is Euclidean. These two different definitions of congruence yield two different descriptions of the world. To the question, "Which of these descriptions is correct?" the answer is that both are. They are *equivalent descriptions.* Every observable fact which tends to support the one also tends to support the other; any observable fact that would tend to refute the one would also tend to refute the other. The situation is somewhat analogous (though not completely so) to the question, "Is New York really to the right of Chicago?" This question obviously has no definite answer until we specify a vantage point. To a vacationer returning to Cleveland from Florida, the answer is affirmative; to an explorer headed toward the same city on his return from the North Pole, the answer is negative. But no one finds any difficulty in seeing why either a *yes* or a *no* can be correct, depending upon the way the question is specified.

It is equally obvious that the incompleteness of such questions does not mean that every conceivable answer is correct. It is incorrect to say that Chicago is to the left of New York from the standpoint of our returning explorer. It is equally incorrect to say that the geometry of the world of Figure 7 is Euclidean given that the measuring rods do not shrink or expand as they are moved about. This would be a correct description of some other world, but it is a false description of the world of Figure 7; such a world is pictured in Figure 8. This same world can correctly be described as non-Euclidean (having a hump) if congruence is defined in terms of the vertical projection shown in description B.

Given the fact that we have a choice between descriptions A and B of Figure 7, on what basis should we choose? Truth or falsity is not a ground for choice, since both are true descriptions of the world of Figure 7 (and both would be false descriptions of the world of Figure 8). The choice must be made on such grounds as simplicity, economy, and elegance. The outcome of the choice may be open to dispute. Some people might say that a definition of congruence which involves no "universal forces" is sim-

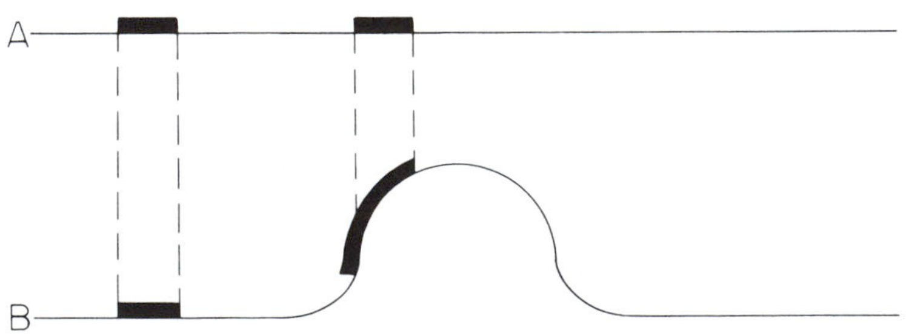

FIGURE 8

pler than one which has them; hence, description A is preferable. Others might say that Euclidean geometry is simpler than non-Euclidean geometry; hence, description B is preferable. Although Poincaré seems to have chosen the second alternative, Einstein (in the general theory of relativity) has chosen the first, and a majority of contemporary relativity physicists seem to agree. But the fundamental philosophical point is this: a description of physical space involves at least two elements, a specification of congruence relations and a geometry. When congruence has been defined, the question of which geometry applies is a determinate empirical question that can be settled in principle by physical measurement. One may instead choose a geometry by definition—we have discussed ways of preserving Euclidean geometry come what may. Such a stipulation, though not uniquely singling out a congruence relation, does severely limit the admissible candidates. Which ones are eligible must be ascertained by the empirical facts. Roughly speaking, the geometrical description of physical space involves two factors, a specification of congruence and a type of geometry, one of which can be chosen as a matter of convention. The remaining ingredient involves the ascertainment of empirical facts.

We may summarize the situation in the following way. There are various mathematical systems—systems of geometry—Euclidean and non-Euclidean. We want to try to apply these systems to the space in which we live. In order to do this we must say what we mean by the various terms that occur in these systems—"straight line," "intersection," "point," and most especially, "congruence."

If we are to make measurements, we must have measuring instruments of some sort. Obviously, we have to define a unit of measurement; we have to say whether we are going to measure in meters, centimeters, feet, yards, and so forth. Furthermore, having once defined the unit of measurement, we have to say what it means to talk about *equality of length*—that is to say, *congruence*—at different places. And that is nothing more than the question of what happens to the measuring rods as they are moved from place to place. Do they stay the same size? Do they change? This is not a question of fact; it is a question of definition. If we do not say what we mean by sameness of size, then there is no answer to the question of whether the world is Euclidean or non-Euclidean. We can say either because we are free to make different definitions regarding length and congruence. If, however, we pin down the concept of congruence by specifying physically what we mean by the length of a measuring rod as it is moved from place to place, then the question as to the geometry of our space becomes a straightforward, though difficult, empirical one. The function of definitions—often called *coordinating definitions*—is to provide a link between the abstract mathematical concepts that appear in the geometrical systems and the physical objects that populate the physical world. Without such coordinating definitions, there is no connection whatever between the mathematical system of the real world.

What are the philosophical conclusions to be drawn from this long story—a story beginning in 600 B.C. and tracing the relationships between philosophy and geometry for some 2500 years? Is the philosophical dream of Plato a realizable one? Can we, after all these centuries, come to the same conclusion as Plato did—that geometry provides us real knowledge of the physical world based solely upon reason? Is geometrical knowledge both synthetic and a priori as Kant claimed?

The answer to all these questions is, I think, negative. Although the discovery of non-Euclidean geometries does not, by itself, refute Kant's thesis, the existence of several alternative geometrical systems does force upon us a distinction between two ways of looking at geometry. In the first place, we can view a geometry as a mathematical system in which we lay down our postulates, whether they be the postulates of Euclidean or non-Euclidean geometry, and then make our deductions from them. From this standpoint we are not in the least concerned with whether the postulates are true; the only question is the logical relation between the postulates and the theorems that follow from them. Indeed, within this type of abstract mathematical system, the postulates and theorems are neither true nor false, for they contain primitive terms such as "point," "straight line," and "congruent" which, strictly speaking, have no meaning at all prior to the introduction of coordinating definitions which assign meanings to them. A system of this kind, containing uninterpreted primitive terms, is a system of *pure geometry;* as such, it is entirely a priori. This does not mean that the postulates (or theorems) are a priori truths. The *logical relation* between the postulates and the theorems is the a priori aspect of pure geometry. The proof of the theorems does not depend in any way upon experiment or experience. They are proved by deduction from the postulates. Such abstract systems of pure mathematics are not, however, synthetic; by themselves they convey no information whatever about the physical world. In the absence of coordinating definitions of the geometrical terms, they bear no relation to the physical world.

In the second place, we can look at geometry as a mathematical system to be used for the purpose of describing the world. When we ask what geometry describes the physical world, we are looking at geometry as *applied geometry.* And now, as we have seen, *if we make some appropriate coordinating definitions to begin with,* there is a definite answer to the question, "What geometry describes physical space?"—a question that may be hard to answer, but which is, nevertheless, a perfectly reasonable physical question. It becomes a reasonable question because our coordinating definitions associate such previously undefined geometrical concepts as straight line and congruence with such physical objects as light rays and measuring rods. This type of geometry—applied geometry—does give us knowledge of the physical world. It does inform us about the spatial structure of our universe. It is synthetic, but unfortunately, it is not a priori. Pure reason, unaided by the senses, cannot tell us how light rays and solid rods will behave; there is no way to tell a priori which geometry is the actual geometry of physical space. That can only be decided by empirical observations and physical experiments.

We find, consequently, that there is a type of geometry which is a priori—namely, pure geometry, an axiomatic system of pure mathematics. There is a type of geometry that is synthetic—namely, applied geometry, a system used to describe the physical world. Unfortunately, pure geometry is not applied geometry, and applied geometry is not pure geometry; there is no type of geometry which is both a priori and synthetic. The clear recognition of this fundamental logical distinction between pure and applied mathematics came about as a result of the development of non-Euclidean geometries. And a full appreciation of its significance requires a careful analysis of the nature and function of coordinating definitions.

Thus, the hope that geometry held out to philosophers for well over two millenia—the hope that pure reason could give us basic information about the nature of the world we live in—is seen ultimately to have been dashed, and by a development in the field of geometry itself. To know what kind of world we live in, to achieve scientific knowledge of it, we must look to the empirical sciences. Pure mathematics, by itself, will never yield such results. Mathematics has to be applied within the natural sciences, and in this context it turns out to be an enormously useful tool for our understanding of reality. But without the observation and experiment which form the indispensable core of the empirical sciences, pure mathematics can tell us nothing about the world we live in.

SUGGESTED READINGS

1. Adler, Irving. *A New Look at Geometry.* New York: The New American Library (Signet Y3225), 1966.

2. Aleksandrov, A. D., Kolmogorov, A. N., and Lavrent'ev, M. A., eds. *Mathematics: Its Content, Methods, and Meaning.* Cambridge, Mass.: The M.I.T. Press, 1963. Especially Chap. XVII, "Non-Euclidean Geometry," by A. D. Aleksandrov.

3. Bonola, Roberto. *Non-Euclidean Geometry.* New York: Dover Publications, Inc., 1955. Contains also Bolyai, "The Science of Absolute Space," and Lobachevski, "The Theory of Parallels."

4. Euclid. *The Thirteen Books of Euclid's Elements.* Sir Thomas L. Heath, trans. New York: Dover Publications, Inc., 1956. Contains extensive commentary on Euclid.

5. Grünbaum, Adolf. *Philosophic Problems of Space and Time.* New York: Alfred A. Knopf, 1963. 2nd ed., Boston: D. Reidel Publishing Co., 1974.

6. Hilbert, David. *Foundations of Geometry,* 2nd ed. La Salle, Ill.: Open Court, 1971.

7. Poincaré, Henri. *Science and Hypothesis.* New York: Dover Publications, Inc., 1952.

8. Reichenbach, Hans. *The Philosophy of Space and Time.* New York: Dover Publications, Inc., 1958.

9. Smart, J. J. C. *Problems of Space and Time.* New York: The Macmillan Co., 1964. An anthology of philosophical readings.

10. van Fraassen, Bas C. *An Introduction to the Philosophy of Time and Space.* New York: Random House, 1970.

11. Wolfe, Harold E. *Introduction to Non-Euclidean Geometry.* New York: Holt, Rinehart and Winston, Inc., 1945.

12. Young, J. W. A., ed. *Monographs on Topics of Modern Mathematics.* New York: Dover Publications, Inc., 1955. Article I, "The Foundations of Geometry" by Oswald Veblen provides a modern axiomatization of Euclidean geometry, and Article III, "Non-Euclidean Geometry" by Frederick S. Woods, presents non-Euclidean geometry axiomatically by suitable modifications of Veblen's axioms for Euclidean geometry.

The Philosophy of Biology

INTRODUCTION

One of the fundamental questions in the philosophy of biology is whether all living phenomena can ultimately be explained in physico-chemical terms or whether there is something ultimately different about living things, something which prevents their being explained in physico-chemical terms. The discussion of this issue, which is the topic of this section, will necessarily involve our returning to issues in the theory of explanation, issues which we discussed in Part II of this book.

In an early classic article reprinted in this book, Ernest Nagel argues that no good reasons have been given for rejecting the mechanistic view that living phenomena can be explained in physico-chemical terms. Nagel points out that such explanations have only been offered for a relatively small number of biological phenomena (remember: Nagel was writing in 1951) and that such explanations are not necessary for the fruitful study of living processes. Still, he criticizes the claim that such explanations could not in principle be offered.

After offering an account (a classical reductionist account) of what such an explanation would be like, Nagel considers three arguments offered by some biologists to show that such explanations are not possible. The arguments are: (1) Living organisms are hierarchically organized and cannot be explained in terms of the modes of behavior of physico-chemical parts; (2) living organisms have a unity which cannot be explained as the sum of the occurrences involving their physico-chemical parts; (3) living organisms cannot be explained in terms of their physico-chemical parts because we cannot synthesize them out of non-living materials. Nagel claims that none of these arguments are successful in proving that the mechanistic approach is incorrect.

One of the issues which Nagel does not address in the reprinted article is the question of whether biology is special and irreducible to physico-chemical explanations because it involves goal-directed (teleological) explanations. In other writings, Nagel (and some of his empiricist colleagues such as Richard B. Braithwaite and Carl G. Hempel) offered accounts according to which teleological explanations are reducible to accounts which emphasize the structural details by which the behavior in question helps ensure the proper working of the organism. A simple version of Hempel's analysis (as presented in Cummins's article) is this:

(a) at *t*, *s* functions adequately in a setting of kind *c*
(b) *s* functions adequately in a setting of kind *c* only if necessary condition *n* is satisfied
(c) if trait *i* were present in *s*, then n would be satisfied
(d) trait *i* is present in *s*

(a)–(c) are all claims which are in theory reducible to physico-chemical claims and they are useful claims for biologists to make. But (d) does *not* follow (deductively or inductively) from (a)–(c). Therefore, by the covering-law model of explanation, (a)–(c) do not explain (d).

Larry Wright objects to this empiricist analysis on the grounds that it overemphasizes the structure (in claims of type (c)) and misses the obviousness of the teleology. He prefers instead to work within an approach—due to Charles Taylor—which avoids such structural claims and emphasizes a more direct goal-oriented approach. Taylor's original scheme for analyzing claims that a behavior (*B*) occurs for a goal (*G*) is:

(i) *B* is necessary for *G* to obtain
(ii) That (i) is sufficient for *B* to occur

Wright believes that Taylor's approach can be strengthened by modifying it to read:

(i)* *B* tends to bring about *G*
(ii)* That (i*) causes *B* to occur

This enables Wright to argue that teleological explanations are a form of causal explanations (which meet Hempel's requirements) while insisting that their claims are both observational and yet postulated to explain behavior.

Wright's analysis, as different as it is from that offered by the empiricists, shares with the empiricist analysis two major assumptions:

(a) functional characterizations, if they do explain anything, explain the presence of the item whose function is being characterized
(b) to perform a function is to have effects which contribute to some larger system's performance of some activity.

It is these assumptions which are challenged by Robert Cummins in his essay. He feels that assumption (a) should have been rejected when the idea that living thing were there as a result of deliberate action was rejected. He argues, moreover, that the often-appealed-to claim that evolutionary advantage can take the place of deliberate action is false. He also argues that Wright has failed to justify his version of (a). Cummins concludes then that functional accounts explain the capacities of the containing system rather than the occurrence of the item whose function is being characterized. In the rest of his paper, Cummins goes on to analyze when it is appropriate to explain the capacities of a containing system by reference to the function of one of its components.

The essays by Baruch Brody and Michael Ruse argue for the importance of evolutionary theory in understanding the nature and role of teleological explanations. Ruse claims, contrary to Cummins, that evolution justifies teleology after the idea of nature as God's artifact is dropped. Brody argues for the view that functional characterizations explain in a Hempelian fashion the continued presence of the item whose function is being characterized, even though they cannot explain the origin of the item. Seeing what functional explanations can do and what they cannot do is essential to understanding the way in which modern teleology rests upon the truth of evolutionary theory.

In short, then, we have seen that questions of functional explanations in biology raise metaphysical questions about the relation between living and physico-chemical phenomena, historical questions about the implications of evolutionary theory, and methodological questions about whether all explanations really are Hempelian covering-law explanations.

Ernest Nagel

MECHANISTIC EXPLANATION AND ORGANISMIC BIOLOGY

Vitalism of the substantival type sponsored by Driesch and other biologists during the preceding and early part of the present century is now a dead issue in the philosophy of biology—an issue that has become quiescent less, perhaps, because of the methodological and philosophical criticism that has been levelled against the doctrine than because of the infertility of vitalism as a guide in biological research and because of the superior heuristic value of alternative approaches for the investigation of biological phenomena. Nevertheless, the historically influential Cartesian conception of biology as simply a chapter of physics continues to meet resistance; and outstanding biologists who find no merit in vitalism believe there are conclusive reasons for maintaining the irreducibility of biology to physics and for asserting the intrinsic autonomy of biological method. The standpoint from which this thesis is currently advanced commonly carries the label of "organismic biology"; and though the label covers a variety of special biological doctrines that are not all mutually comparible, those who fall under it

From *Philosophy and Phenomenological Research,* I, No. 3, March, 1951. Reprinted with permission of the author and the editor.

are united by the common conviction that biological phenomena cannot be understood adequately in terms of theories and explanations which are of the so-called "mechanistic type." It is the aim of the present paper to examine this claim.

It is, however, not always clear what thesis organismic biologists are rejecting when they declare that "mechanistic" explanations are not fully satisfactory in biology. In one familiar sense of "mechanistic," a theory is mechanistic if it employs only such concepts which are distinctive of the science of mechanics. It is doubtful, however, whether any professed mechanist in biology would today explicate his position in this manner. Physicists themselves have long since abandoned the seventeenth-century hope that a universal science of nature would be developed within the framework of the fundamental conceptions of mechanics. And no one today, it is safe to say, subscribes literally to the Cartesian program of reducing all the sciences to the science of mechanics and specifically to the mechanics of contact-action. On the other hand, it is not easy to state precisely what is the identifying mark of a mechanistic explanation if it is not to coincide with an explanation that falls within the science of mechanics. In a preliminary way, and for lack of anything better and clearer, I shall adopt in the present paper the criterion proposed long ago by Jacques Loeb, according to whom a mechanist in biology is one who believes that all living phenomena "can be unequivocally explained in physico-chemical terms," that is, in terms of theories that have been originally developed for domains of inquiry in which the distinction between the living and nonliving plays no role, and that by common consent are classified as belonging to physics and chemistry.

As will presently appear, this brief characterization of the mechanistic thesis in biology does not suffice to distinguish in certain important respects mechanists in biology from those who adopt the organismic standpoint; but the above indication will do for the moment. It does suffice to give point to one further preliminary remark which needs to be made before I turn to the central issue between mechanists and organismic biologists. It is an obvious commonplace, but one that must not be ignored if that issue is to be justly appraised, that there are large sectors of biological study in which physico-chemical explanations play no role at present and that a number of outstanding biological theories have been successfully exploited which are not physico-chemical in character. For example, a vast array of important information has been obtained concerning embryological processes, though no explanation of such regularities in physico-chemical terms is available; and neither the theory of evolution even in its current form, nor the gene theory of heredity is based on any definite physico-chemical assumptions concerning living processes. Accordingly, organismic biologists possess at least some grounds for their skepticism concerning the inevitability of the mechanistic standpoint; and just as a physicist may be warranted in holding that some given branch of physics (e.g., electromagnetic theory) is not reducible to some other branch (e.g., mechanics), so an organismic biologist may be warranted in holding an analogous view with respect to the relation of biology and physico-chemistry. If there is a genuine issue between mechanists and organismic biologists, it is not prima facie a pseudo-question.

However, organismic biologists are not content with making the obviously justified observation that only a relatively small sector of biological phenomena has thus far been explained in physico-chemical terms; they also maintain that *in principle* the mode of analysis associated with mechanistic explanations is inapplicable to some of the major problems of biology, and that therefore mechanistic biology cannot be adopted as the ultimate ideal in biological research. What are the grounds for this contention and how solid is the support which organismic biologists claim for their thesis?

The central theme of organismic biology

is that living creatures are not assemblages of tissues and organs functioning independently of one another, but are integrated structures of parts. Accordingly, living organisms must be studied as "wholes," and not as the mere "sums" of parts. Each part, it is maintained, has physico-chemical properties; but the interrelation of the parts involves a distinctive organization, so that the study of the physico-chemical properties of the parts taken in isolation of their membership in the organized whole which is the living body fails to provide an adequate understanding of the facts of biology. In consequence, the continuous adaptation of an organism to its environment and of its parts to one another so as to maintain its characteristic structure and activities cannot be described in terms of physical and chemical principles. Biology must employ categories and a vocabulary which are foreign to the sciences of the inorganic, and it must recognize modes and laws of behavior which are inexplicable in physico-chemical terms.

There is time to cite but one brief quotation from the writings of organismic biologists. I offer the following from E. S. Russell as a typical statement of this point of view:

> Any action of the whole organism would appear to be susceptible of analysis to an indefinite degree—and this is in general the aim of the physiologist, to analyze, to decompose into their elementary processes the broad activities and functions of the organism.
>
> But . . . by such a procedure something is lost, for the action of the whole has a certain unifiedness and completeness which is left out of account in the process of analysis. . . . In our conception of the organism we must . . . take into account the unifiedness and wholeness of its activities [especially since] the activities of the organism all have reference to one or other of three great ends [development, maintenance, and reproduction], and both the past and the future enter into their determination. . . .
>
> . . . It follows that the activities of the organism as a whole are to be regarded as of a different order from physico-chemical relations, both in themselves and for the purposes of our understanding. . . .
>
> . . . Bio-chemistry studies essentially the *conditions* of action of cells and organisms, while organismal biology attempts to study the actual modes of action of whole organisms, regarded as conditioned by, but irreducible to, the modes of action of lower unities. . . . (*Interpretation of Development and Heredity,* pp. 171–72, 187–88.)

Accordingly, while organismic biology rejects every form of substantial vitalism, it also rejects the possibility of physico-chemical explanation of vital phenomena. But does it, in point of fact, present a clear alternative to physico-chemical theories of living processes, and, if so, what types of explanatory theories does it recommend as worth exploring in biology?

(1) At first blush, the sole issue that seems to be raised by organismic biology is that commonly discussed under the heading of "emergence" in other branches of science, including the physical sciences; and, although other questions are involved in the organismic standpoint, I shall begin with this aspect of the question.

The crux of the doctrine of emergence, as I see it, is the determination of the conditions under which one science can be reduced to some other one, i.e., the formulation of the logical and empirical conditions which must be satisfied if the laws and other statements of one discipline can be subsumed under, or explained by, the theories and principles of a second discipline. Omitting details and refinements, the two conditions which seem to be necessary and sufficient for such a reduction are briefly as follows. Let S_1 be some science or group of sciences such as physics and chemistry, hereafter to be called the "primary discipline," to which a second science, S_2, for example biology, is to be reduced. Then, (i) every term which occurs in the statements of S_2 (e.g., terms like *cell, mytosis, heredity,* etc.) must be either explicitly definable with the help of the vocabulary specific to the primary discipline (e.g., with the help of expressions like *length, electric charge, osmosis*); or well-established empirical laws must be available with the help of which it is possible to state the sufficient conditions for the applications of all expressions in S_2, exclusively in terms of expressions occurring in the explanatory principles of S_1. For example, it must be possible to state the truth-conditions of a statement of the form *x is a cell* by means of sentences constructed exclusively out of the vocabulary belonging to the physico-chemical sciences. Though the label is not entirely appropriate, this first condition will be referred to as the condition of definability. (ii) Every statement in the secondary discipline, S_2, and especially those statements which formulate the laws established in S_2, must be derivable logically from some appropriate class of statements that can be established in the primary science, S_1—such classes of statements will include the fundamental theoretical assumptions of S_1. This second condition will be referred to as the condition of derivability.

It is evident that the second condition cannot be fulfilled unless the first one is, although the realization of the first condition does not entail the satisfaction of the second one. It is also quite beyond dispute that in the sense of reduction specified by these conditions biology has thus far not been reduced to physics and chemistry, since not even the first step in the process of reduction has been completed—for example, we are not yet in the position to specify exhaustively in physico-chemical terms the conditions for the occurrence of cellular division.

Accordingly, organismic biologists are on firm ground if what they maintain is that all biological phenomena are not explicable thus far physico-chemically, and that no physico-chemical theory can possibly explain such phenomena until the descriptive and theoretical terms of biology meet the condition of definability. On the other hand, nothing in the facts surveyed up to this point warrants the conclusion that biology is *in principle* irreducible to physico-chemistry. Whether biology is reducible to physio-chemistry is a question that only further experimental and logical research can settle; for the supposition that each of the two conditions for the reduction of biology to physico-chemistry may some day be satisfied involves no patent contradiction.

(2) There are, however, other though related considerations underlying the organismic claim that biology is intrinsically autonomous. A frequent argument used to support this claim is based on the fact that living organisms are hierarchically organized and that, in consequence, modes of behaviour characterizing the so-called "higher levels" of organization cannot be explained in terms of the structures and modes of behavior which parts of the organism exhibit on lower levels of the hierarchy.

There can, of course, be no serious dispute over the fact that organisms do exhibit structures of parts that have an obvious hierarchical organization. Living cells are structures of cellular parts (e.g., of the nucleus, cytoplasm, central bodies, etc.), each of which in turn appears to be composed of complex molecules; and, except in the case of unicellular organisms, cells are further organized into tissues, which in turn are elements of various organs that make up the individual organism. Nor is there any question but that parts of an organism which occupy a place at one level of its complex hierarchical organization stand in relations and exhibit activities which parts occupying positions at other levels of organization do not manifest: A cat can stalk and catch mice, but though its heart is involved in these activities, that organ cannot perform these feats; again, the heart can pump blood by contracting and expanding its muscular tissues, but no tissue is able to do this; and no tissue is able to divide by fission, though its constituent cells may have this power; and so on down the line. If such facts are taken in an obvious sense, they undoubtedly support the conclusion that behavior on higher levels of organization is not explained by merely citing the various behaviors of parts on lower levels of the hierarchy. Organismic biologists do not, of course, deny that the higher level behaviors occur only when the component parts of an organism are appropriately organized on the various levels of the hierarchy; but they appear to have reason on their side in maintaining that a knowledge of the behavior of these parts, when these latter are not component elements in their structured living organism, does not suffice as a premise for deducing anything about the behavior of the whole organism in which the parts do stand in certain specific and complex relations to one another.

But do these admitted facts establish the organismic thesis that mechanistic explanations are not adequate in biology? This does not appear to be the case, and for several reasons. It should be noted, in the first place, that various forms of hierarchical organization are exhibited by the materials of physics and chemistry, and not only by those of biology. On the basis of current theories of matter, we are compelled to regard atoms as structures of electric charges, molecules as organizations of atoms, solids and liquids as complex systems of molecules; and we must also recognize that the elements occupying positions at different levels of the indicated hierarchy generally exhibit traits and modes of activity that their component parts do not possess. Nonetheless, this fact has not stood in the way of establishing comprehensive theories for the more elementary physical particles, in terms of which it has been possible to explain some, if not all, of the physico-chemical properties exhibited by things having a more complex organization. We do not, to be sure, possess at the present time a comprehensive and unified theory which is competent to explain the whole range of physico-chemical phenomena at all levels of complexity. Whether such a theory will ever be achieved is certainly an open question. But even if such an inclusive theory were never achieved, the mere fact that we can now explain some features of relatively highly organized bodies on the basis of theories formulated in terms of relations between relatively more simply structured elements—for example, the specific heats of solids in terms of quantum theory or the changes in phase of compounds in terms of the thermodynamics of mixtures—should give us pause in accepting the conclusion that the mere fact of the hierarchical organization of biological materials precludes the possibility of a mechanistic explanation.

This observation leads to a second point. Organismic biologists do not deny that biological organisms are complex structures of physico-chemical processes, although like everyone else they do not claim to know in minute detail just what these processes are or just how the various physico-chemical elements (assumed as the ultimate parts of living creatures) are related to one another in a living organism. They do maintain, however, (or appear to maintain) that even if our knowledge in this respect were ideally complete, it would still be impossible to account for the characteristic behavior of biological organisms—their ability to maintain themselves, to develop, and to reproduce—in mechanistic terms. Thus, it has been claimed that even if we were able to describe in full detail in physico-chemical terms what is taking place when a fertilized egg segments, we would, nevertheless, be unable to explain mechanistically the fact of segmentation—in the language of E. S. Russell, we would then be able to state the physico-chemical *conditions* for the occurrence of segmentation, but we would still be unable to "explain the *course* which development takes." Now this claim seems to me to rest on a misunderstanding, if not on a confusion. It is entirely correct to maintain that a knowledge of the physico-chemical composition of a biological organism does not suffice to explain mechanistically its mode of action—anymore than an enumeration of the parts of a clock and a knowledge of their distribution and arrangement suffices to explain and predict the mode of behavior of the time piece. To do the latter one must *also* assume some theory or set of laws (e.g., the theory of mechanics) which formulates the way in which certain elementary objects behave when they occur in certain initial distributions and arrangements, and with the help of which we can calculate and predict the course of subsequent development of the mechanism. Now it may indeed be the case that our information at a given time may suffice to describe physico-chemically the constitution of a biological organism; nevertheless, the established physico-chemical theories may not be adequate, even when combined with a physico-chemical description of the initial state of the organism, for deducing just what the course of the latter's development will be. To put the point in terms of the distinction previously introduced, the condition of definability may be realized without the condition of derivability being fulfilled. But this fact must not be interpreted to mean that it is possible under any circumstances to give explanations without the use of some theoretical assumptions, or that because one body of physico-chemical theory is not competent to explain certain biological phenomena it is *in principle impossible* to construct and establish mechanistic theories which might do so.

(3) I must now examine the consideration which appears to constitute the main reason for the negative attitude of organismic biologists toward mechanistic explanations. Organismic biologists have placed great stress on what they call the "unifiedness," the "unity," the "completeness," or the "wholeness" of organic behavior; and, since they believe that biological organisms are complex systems of mutually determining and interdependent processes to which subordinate organs contribute in various ways, they have maintained that organic behavior cannot be analyzed into a set of independently determinable component behaviors of the parts of an organism, whose "sum" may be equated to the total behavior of the organism. On the other hand, they also maintain that mechanistic philosophies of organic action are "machine theories" of the organism, which assume the "additive point of view" with respect to biological phenomena. What distinguishes mechanistic theories from organismic ones, from this perspective, is that the former do while the latter do not regard an organism as a "machine," whose "parts" are separable and can be studied in isolation from their actual functioning in the whole living organism, so that the latter may then be understood and explained as an aggregate of such independent parts. Accordingly, the fundamental reasons for the dissatisfaction which organismic biologists feel toward mechanistic theories is the "additive point of view" that allegedly characterizes the latter. However, whether this argument has any merit can be decided only if the highly ambiguous and metaphorical notion of "sum" receives at least partial clarification; and it is to this phase of the question that I first briefly turn.

(i) As is well known, the word "sum" has a large variety of different uses, a number of which bear to each other certain formal analogies while others are so vague that nothing definite is conveyed by the word. There are well-defined senses of the term in various domains of pure mathematics, e.g., arithmetical sum, algebraic sum, vector sum, and the like; there are also definite uses established for the word in the natural sciences, e.g., sum of weights, sum of forces, sum of velocities, etc. But with notable exceptions; those who have employed it to distinguish wholes which are sums of their parts from wholes which supposedly are not, have not taken the trouble to indicate just what would be the sum of parts of a whole which allegedly is not equal to that whole.

I, therefore, wish to suggest a sense for the word "sum" which seems to me relevant to the claim of organismic biologists that the total behavior of an organism is not the sum of the behavior of its parts. That is, I wish to indicate more explicitly than organismic biologists have done—though I hasten to add that the proposed indication is only moderately more precise than is customary—what it is they are asserting when they maintain, for example, that the behavior of the kidneys in an animal body is more than the "sum" of the behaviors of the tissues, blood stream, blood vessels, and the rest of the parts of the body involved in the functioning of the kidneys.

Let me first state the suggestion in schematic, abstract form. Let T be a definite body of theory which is capable of explaining a certain indefinitely large class of statements concerning the simultaneous or successive occurrence of some set of properties $P_1, P_2, \ldots P_k$. Suppose further that it is possible with the help of the Theory T to explain the behavior of a set of individuals i with respect to their manifesting these properties P when these individuals form a closed system s_1 under circumstances C_1; and that it is also possible with the help of T to explain the behavior of another set of individuals j with respect to their manifesting these properties P when the individuals j form a closed system s_2 under circumstances C_2. Now assume that the two sets of individuals i and j form an enlarged closed system s_3 under circumstances C_3, in which they exhibit certain modes of behavior which are formulated in a set of laws L. Two cases may now be distinguished: (a) It may be possible to deduce the laws L from T conjoined with the relevant initial conditions which obtain in C_3; in this case, the behavior of the system s_3 may be said to be the sum of the behaviors of its parts s_1 and s_2; or (b) the laws L cannot be so deduced, in which case the behavior of the system s_3 may be said *not* to be the sum of the behaviors of its parts.

Two examples may help to make clearer what is here intended. The laws of mechanics enable us to explain the mechanical behaviors of a set of cogwheels when they occur in certain arrangements; those laws also enable us to explain the behavior of freely-falling bodies moving against some resisting forces, and also the behavior of compound pendula. But the laws of mechanics also explain the behavior of the system obtained by arranging cogs, weights, and pendulum in certain ways so as to form a clock; and, accordingly, the behavior of a clock can be regarded as the sum of the behavior of its parts. On the other hand, the kinetic theory of matter as developed during the nineteenth century was able to explain certain thermal properties of gases at various temperatures, including the relations between the specific heats of gases; but it was unable to explain the relations between the specific heats of solids—that is, it was unable to account for these relations theoretically when the state of aggregation of molecules is that of a solid rather than a gas. Accordingly, the thermal behavior of solids is not the sum of the behavior of its parts.

Whether the above proposal to interpret the distinction between wholes which are and those which are not the sums of their parts would be acceptable to organismic biologists, I do not know. But, while I am aware that the suggestion requires much elaboration and refinement to be an adequate tool of analysis, in broad outline it represents what seems to me to be the sole intellectual content of what organismic biologists have had to say in this connection. However, if the proposed interpretation of the distinction is accepted as reasonable, then one important consequence needs to be noted. For, on the above proposal, the distinction between wholes which are and those which are not sums of parts is clearly *relative to some assumed body of theory T;* and, accordingly, though a given whole may not be the sum of its parts relative to one theory, it may indeed be such a sum relative to another. Thus, though the thermal behavior of solids is not the sum of the behavior of its parts relative to the classical kinetic theory of matter, it is such a sum relative to modern quantum mechanics. To say, therefore, that the behavior of an organism is not the sum of the behavior of its parts, and that its total behavior cannot be understood adequately in physico-chemical terms even though the behavior of each of its parts is explicable mechanistically, can only mean that no body of general theory is now available from which statements about the total behavior of the organism are derivable. The assertion, even if true, does *not* mean that it is *in principle* impossible to explain such total behavior mechanistically, and it supplies no competent evidence for such a claim.

(ii) There is a second point related to the organismic emphasis on the "wholeness" of organic action upon which I wish to comment briefly. It is frequently overlooked, even by those who really know better, that no theory, whether in the physical sciences or elsewhere, can explain the operations of any concrete system, unless various restrictive or boundary conditions are placed on the generality of the theory and unless, also, specific initial conditions, relevantly formulated, are supplied for the application of the theory. For example, electrostatic theory is unable to specify the distribution of electric charges on the surface of a given body unless certain special information, not deducible from the fundamental equation of the theory (Poisson's equation) is supplied. This information must include statements concerning the shape and size of the body, whether it is a conductor or not, the distribution of other charges (if any) in the environment of the body, and the value of the dialectric constant of the medium in which the body is immersed.

But though this point is elementary, organismic biologists seem to me to neglect it quite often. They sometimes argue that though mechanistic explanations can be given for the behaviors of certain parts of organisms when these parts are studies in abstraction or isolation from the rest of the organism, such explanations are not possible if those parts are functioning conjointly and in mutual dependence as actual constituents of a living organism. This argument seems to me to have no force whatever. What it overlooks is that the initial and boundary conditions which must be supplied in explaining physico-chemically the behavior of an organic part acting in isolation are, in general, *not sufficient* for explaining mechanistically the conjoint functioning of such parts. For when these parts are assumed to be acting in mutual dependence, the environment of each part no longer continues to be what it was supposed to be when it was acting in isolation. Accordingly, a necessary requirement for the mechanistic explanation of the unified behavior of organisms is that boundary and initial conditions bearing on the actual relations of parts as parts of living organisms be stated in *physico-chemical* terms. Unless, therefore, appropriate data concerning the physico-chemical constitution and arrangement of the various parts of organisms are specified, it is not surprising that mechanistic explanations of the total behavior of organisms cannot be given. In point of fact, this requirement has not yet been fulfilled even in the case of the simplest forms of living organisms, for our ignorance concerning the detailed physico-chemical constitution of organic parts is profound. Moreover, even if we were to succeed in completing our knowledge in this respect—this would be equivalent to satisfying the condition of definability stated earlier—biological phenomena might still not be all explicable mechanistically: For this further step could be taken only if a comprehensive and independently warranted physico-chemical theory were available from which, together with the necessary boundary and initial conditions, the laws and other statements of biology are derivable. We have certainly failed thus far in finding mechanistic explanations for the total range of biological phenomena, and we may never succeed in doing so. But, though we continue to fail, then if this paper is not completely in error, the reasons for such failure are not the a priori arguments advanced by organismic biology.

(4) One final critical comment must be added. It is important to distinguish the question whether mechanistic explanations of biological phenomena are possible, from the quite different though related problem whether living organisms can be effectively synthesized in a laboratory out of nonliving materials. Many biologists apparently deny the first possibility because of their skepticism concerning the second, even when

their skepticism does not extend to the possibility of an artificial synthesis of every chemical compound that is normally produced by biological organisms. But the two questions are not related in a manner so intimate; and though it may never be possible to create living organisms by artificial means, it does not follow from this assumption that biological phenomena are incapable of being explained mechanistically. We do not possess the power to manufacture nebulae or solar systems, though we do have available physico-chemical theories in terms of which the behaviors of nebulae and solar systems are tolerably well understood; and, while modern physics and chemistry are beginning to supply explanations for the various properties of metals in terms of the electronic structure of their atoms, there is no compelling reason to suppose that we shall one day be able to manufacture gold by putting together artificially its subatomic constituents. And yet the general tenor, if not the explicit assertions, of some of the literature of organismic biology is that the possibility of mechanistic explanations in biology entails the possibility of taking apart and putting together in overt fashion the various parts of living organisms to reconstitute them as unified creatures. But in point of fact, the condition for achieving mechanistic explanations is quite different from that necessary for the artificial manufacture of living organisms. The former involves the construction of factually warranted *theories* of physico-chemical processes; the latter depends on the availability of certain physico-chemical substances and on the invention of effective techniques of control. It is no doubt unlikely that living organisms will ever be synthesized in the laboratory except with the help of mechanistic theories of organic processes—in the absence of such theories, the artificial creation of living things would at best be only a fortunate accident. But, however this may be, these conditions are logically independent of each other, and either might be realized without the other being satisfied.

(5) The central thesis of this paper is that none of the arguments advanced by organismic biologists establish the inherent impossibility of physico-chemical explanations of vital processes. Nevertheless, the stress which organismic biologists have placed on the facts of the hierarchical organization of living things and on the mutual dependence of their parts is not without value. For though organismic biology has not demonstrated what it proposes to prove, it has succeeded in making the heuristically valuable point that the explanation of biological processes in physico-chemical terms is not a necessary condition for the fruitful study of such processes. There is, in fact, no more good reason for dissatisfaction with a biological theory (e.g., modern genetics) because it is not explicable mechanistically than there is for dissatisfaction with a physical theory (e.g., electro-magnetism) because it is not reducible to some other branch of that discipline (e.g., to mechanics). And a wise strategy of research may, in fact, require that a given discipline be cultivated as an autonomous branch of science, at least during a certain period of its development, rather than as a mere appendage to some other and more inclusive discipline. The protest of organismic biology against the dogmatism frequently associated with mechanistic approaches to biology is salutary.

On the other hand, organismic biologists sometimes write as if any analysis of living processes into the behaviors of distinguishable parts of organisms entails a radical distortion of our understanding of such processes. Thus, Wildon Carr, one proponent of the organismic standpoint, proclaimed that "Life is individual; it exists only in living beings, and each living being is indivisible, a whole not constituted of parts." Such pronouncements exhibit a tendency that seems far more dangerous than is the dogmatism of intransigent mechanists. For it is beyond serious question that advances in biology occur only through the use of an abstractive method, which proceeds to study various aspects of organic behavior in relative isolation of other aspects. Organismic biologists proceed in this way, for they have no alternative. For example, in spite of his insistence on the indivisible unity of the organism, J. S. Haldane's work on respiration and the chemistry of the blood did not proceed by considering the body as a whole, but by studying the relations between the behavior of one part of the body (e.g., the quantity of carbon dioxide taken in by the lungs) and the behavior of another part (the chemical action of the red blood cells). Organismic biologists, like everyone else who contributes to the advance of science, must be selective in their procedure and must study the behavior of living organisms under specialized and isolating conditions—on pain of making the free but unenlightening use of expressions like "wholeness" and "unifiedness" substitutes for genuine knowledge.

Larry Wright

EXPLANATION AND TELEOLOGY

This paper develops and draws the consequences of an etiological analysis of goal-directedness modeled on one that functions centrally in Charles Taylor's work on action. The author first presents, criticizes, and modifies Taylor's formulation, and then shows his modified formulation accounts easily for much of the fine-structure of teleological concepts and conceptualizations. Throughout, the author is at pains to show that teleological explanations are orthodox from an empiricist's point of view: they require nothing novel methodologically.

1. INTRODUCTION

It is the central logical property of teleological characterizations that they explain what they characterize.* When we say '*A* in order that *B*', or '*A* for the sake of *B*', we ipso facto answer a question of the form 'Why *A*?' This is as true of functions as it is in the case of goal-directed behavior,[1] but here I am concerned only with the latter.[2] When I say the rabbit is running in order to escape from the dog, I am saying *why* the rabbit is behaving as it is. And since merely characterizing the behavior as fleeing implies that it takes place "in order to escape . . .," this characterization itself offers an explanation. If the rabbit is not running in order to escape, it is not fleeing. Similarly for stalking, evading, and anything of the form 'trying to *X*'. So whatever else a general account of teleological behavior must contain, it must at least provide the form of an explanation of this behavior; it must say something in general about why behavior of this sort takes place.

This view is implicit in the three classic "empiricist" analyses of teleological (goal-directed) behavior, authored by N. Wiener, *et al.* [10], R. B. Braithwaite [3], and Ernest Nagel [5]. For Wiener *et al.*, teleological behavior involves "a continuous feedback from the goal that modifies and guides the behaving object." Purposeful reactions are those "which are controlled by the error of

*The research for this project was carried out under a grant from the University of California Humanities Institute. This version of the paper has benefited significantly from suggestions and criticisms offered by the editors of *Philosophy of Science*.

Reprinted by permission from the author and *Philosophy of Science,* Vol. 39, no. 2, June 1972, pp. 204–218.

the reaction." On this interpretation, teleological behavior "becomes synonymous with behavior controlled by negative feedback . . ." ([10], p. 24). For Braithwaite, behavior is teleological if, starting from some specific initial condition, it can be represented as a standard (efficient) causal chain which attains or produces a specified final condition under a variety of environmental conditions. Nagel's analysis can be innocuously oversimplified by saying that behavior is teleological, with respect to a certain condition, if whenever there are changes in the system's state description which would, uncompensated for, take the system out of that condition, those changes naturally induce *other* changes which keep the system in the initial condition. (I have discussed these analyses in much more detail in [11].) In each case, to say that behavior is teleological is to say something about the causal laws or mechanisms producing the behavior, which is to say something explanatory of the behavior.

The *sort* of explanation offered by a teleological characterization on each of these accounts might be called 'etiological', that is causal[3] in the broad sense in which any explanation aimed at showing what produced, or brought about, the behavior is causal. This is, in a way, contentious. For there is a school of thought which takes teleological explanations to be sui generis, and a fortiori not to be understood in ordinary causal terms. And while some of the arguments from this perspective against the empiricist analyses have been telling, they have not, I think, been enough to justify calling off the search for an etiological account of teleological explanation. Three considerations urge this position upon us. First, an etiological analysis of teleological explanation would offer an escape from the eschatology and post hoc ergo propter hoc problems which have plagued the sui generist's accounts. Second, it would show teleological explanations to be explanations in a standard and clearly acceptable sense of the term, in no need of semantic apology or defense against pun. Third, and perhaps most important, if we accept human action explanations as a model of legitimate teleology, then other teleological explanations should be expected to share their salient logical features; and explanations in terms of reasons, motives, and intent are plainly etiological: they help us understand what brought about the behavior in question.[4] So it is not at all absurd to suppose that the causal/teleological contrast represents a distinction *among* different kinds of etiologies, as opposed to a contrast between etiologies and something else.

On the other hand, the empiricist attempts to carry out the program (those summarized above) have failed, legendarily by now, for a variety of reasons. Israel Scheffler was perhaps the trail blazer here with his catalog of "difficulties" in [7].[5] But it is crucial to notice that, quite apart from other considerations, it is precisely their preoccupation with the underlying structural detail which discomfits these analyses as adequate general characterizations of goal-directedness. This is so for two reasons. First, describing behavior teleologically—saying it is directed toward a certain goal—usually does not commit us to an exact function relating position or state description to time.[6] Usually, from the point of view of an underlying causal-deterministic account, a teleological characterization only restricts behavior to a certain range; it provides us certain broad constraints. Accordingly, within the appropriate range, determinism can fail without affecting the goal-directedness of the behavior. This alone prevents any deterministic substructure-unpacking from producing the desired result. But there is another consideration in virtue of which we should have expected analyses of the above sort to have been wide of the mark.

In a large number of cases, the goal-directedness of a bit of behavior is obvious on its face. Many of our teleological judgments are as reliable and intersubjective as the run of normal perceptual judgments. Occasionally there simply is no question about it: the rabbit is fleeing, the cat stalking, the squirrel building a nest. Certain complex behavior patterns seem to *demand* teleological characterization. It is because of this that we have any reason at all to think that there is anything to give a philosophical analysis of. The notion is interesting precisely because it functions so clearly in these contexts. Accordingly, we should view with suspicion any analysis which contends that goal-directedness consists in a relationship among parameters of which we are quite usually ignorant in the contexts of these reliable judgments. And in these cases we simply do not know the laws and state descriptions, the causal chain and variency, the underlying mechanism. Without further argument, then, an analysis in terms of coarser, more rudimentarily perceptual aspects of behavior would be vastly more plausible.

From this vantage, the analysis of teleological behavior offered by Charles Taylor in *The Explanation of Behaviour* [9] represents a big step in just the right direction. His analysis, while still etiological, offers us something much further removed from the details of an underlying mechanism than any of those that went before it. It is the first one to provide us with something which can legitimately be called the general *form* of the account behavior must have when it is teleological. For Taylor, behavior is goal-directed or teleological, that is, occurs for the sake of some end, simply if it "occurs because it is the type of event which brings about this end" (ibid., p. 9). And his elaboration of this position makes it clear that 'because' here is to have etiological force: it concerns what brings about the behavior. In the very next sentence he holds that to account for behavior teleologically is to "account for it by laws in terms of which an event's occurring is held to be dependent on that event's being required for some end." Then, later and more elaborately, he says,

> . . . (teleological) behavior is a function of the state of the system and (in the case of animate organisms) its environment; but the relevant feature of system and environment on which behaviour depends will be what the condition of both makes necessary if the end concerned is to be realized. (ibid.)

Teleological behavior then has what might be called a "requirement etiology": what brings it about is its being required. Moreover, in several places Taylor quite plausibly insists that it is often "perfectly observable" that a specific sort of behavior is what is required for a certain end. Often we can *see* what's required. So this seems to be an analysis of the sort we are looking for. These are much more the kind of features we should expect to find in an analysis if it is to account for the actual deployment of teleological concepts. In point of fact, Taylor's analysis is much more exciting than it at first appears. With a slight reworking and clarification it will allow us to give an empiricist defense of several things quite generally disparaged by empiricists: final causation, anthropomorphism and teleological explanation not reducible to an underlying, deterministic causal mechanism.

Now, there is some obvious conflict among Taylor's different formulations of his principle. He maintains that the difference between the "type" formulation and the "requirement" formulation is insignificant—merely apparent, superficial, should be ignored. This is questionable, and I will treat it in some detail below. But what is important to notice here is that these formulations are, in different contexts, very close to things we naturally say in embellishing and supporting our teleological characterizations in actual cases. Why did he do that? He *had* to do that if he was to avoid getting caught. Which is to say, he did it because it was required for a goal which he had. Alternatively, we sometimes say things like "He did it because that would get him around the obstacle." This is more like Taylor's first formulation: he did it because it was the type of thing which would bring about the end. Accordingly if we can use formulations of this kind as the basis of a philosophical analysis, we will have strong assurance at the outset that the analysis will offer us some insight into these macroscopic identificatory features of teleological behavior which are the source of our interest.

2. THE TAYLOR FORMULA

The formulation Taylor finally settles on in *The Explanation of Behaviour,* the one operative in his discussions of teleological explanation, is somewhat more exacting, and hence more contentious than any of those above. It is,

B occurs for the sake of *G* means:
(1) (i) *B* is necessary (required) for *G* to obtain;
(ii) *B*'s being necessary for *G* is sufficient for *B* to occur.[7]

This way of putting it certainly seems more precise, and perhaps more easily testable, than the others. And it might be urged that this formulation merely makes explicit the notion of functional dependency implicit in the others. But although Taylor sees very little difference among all these formulations, I will argue below that the differences are very important indeed. Some renderings get us into difficulties which others easily avoid.

Taylor is quite aware that on its most natural reading (1) is open to several objections, and he adds three qualifications about how it is to be understood. First, systems which are the objects of teleological descriptions obviously may have a specific end at one time (e.g., food) and not have it at another. So it is to be taken as implicit in this analysis that the event's being required is sufficient for its occurrence only when the end *is* an end for the system in question. To use Taylor's own example we can talk teleologically of a predator stalking his prey if and only if it is a sufficient condition for the predator's stalking that he have the goal of obtaining food and stalking is required to obtain it ([9], p. 9). Second, when he talks of sufficiency, or sufficient conditions, he is prepared to allow exceptions to the rule so long as they can "be cogently explained by interfering factors" (*ibid.,* p. 14, fn.). Accordingly, the predator's stalking may be interrupted by a hunter's trap, even though he still has the goal of a meal and stalking will still be required to get it; and this does not mean that we were wrong to talk teleologically in the first place. Third, Taylor is prepared to admit that in many cases there may be a number of different activities which would result in the achievement of a given end, and that these are normally selected among by appeal to some principle such as least effort. So he further qualifies (1) by including the selection criterion in the behavioral context so that we may "generally assume the selection has been made, and speak elliptically of 'the event required for' the goal or end" ([9], p. 9, fn.).

Unfortunately, even with these qualifications (1) is still enormously problematic. It still encounters several telling objections and a huge array of counter-examples. But since it is so promising on other grounds, I will attempt to modify Taylor's analysis to circumvent the objections while doing as little as possible to alter its insightful appreciation of the subject matter.

The most important modification to be made is to the (qualified) sufficient condition in the formula. As pointed out above, Taylor clearly sees that the subtlety of the relation between the requirement of a bit of behavior and its actual occurrence in a teleological system immediately defeats any attempt to carry out this analysis in terms of an unqualified sufficient condition. However, his simple expedient of adding a clause excepting cases of interference is not equal to the subtlety either. Even with this qualifying clause, there are four distinct objections to using sufficient conditions here.

First, and perhaps least importantly, it is very strained and artificial in some cases to describe as an appeal to interfering factors, the explanation of why a system failed to behave in a certain way in spite of the fact that such behavior was required for an end which it had. For example, it is possible that, instead of falling into a hunter's trap, our predator stops stalking because he is tired, even though still hungry and stalking is still required to get food. In this case, according to Taylor's formula, we would have to say that the predator's getting tired was an interfering factor. But if he stopped merely to rest, then presumably he wasn't even *trying* to stalk (at that moment); and in order to explain the interruption by *interference* he must at least be *trying*.

This case is really just one of a large number of difficult cases which arise out of the fact that teleological systems (e.g., organisms) almost always can be said to have more than one goal. A very general difficulty arises here because quite often a system's attention is directed toward only one of these goals at a time. Accordingly, this selective attention would have to be characterized as an interfering factor to explain why the behavior required for the other goals does not obtain. This is *in general* a peculiar thing to say because, *ex hypothesi,* the system is not even trying to accomplish those other goals. This terminological artificiality is not in itself that profound, philosophically speaking; but it is relevant because it is, as usual, symptomatic of a deeper, conceptual difficulty. This will appear below.

The second objection to characterizing the requirement of behavior as (qualifiedly) sufficient for its occurrence in a teleological system also primarily concerns the role of the interference clause. This clause seems to admit all sorts of bizarre accidents into the category of goal-directed activity, to the trivialization of Taylor's formula. Taylor attempts to avoid this discomfiture by including the term 'cogently' in his exception clause, but the criteria for the cogency of an explanation are not sufficiently straightforward to be applicable in most of the important cases. For instance, our favorite predator (suppose it's a feline), while resting between stalks, might be licking his paw, claws exposed, when a low-flying bird collides with the claws, killing itself and providing a meal for our still hungry friend. A very good case indeed can be made for saying that this rather disinterested exposure of the claws was precisely what was required to snare the unwary fowl. And if *anything* can count as interference, from having other goals to becoming tired, then it would appear to be impossible to adduce evidence against the claim that this disinterested claw exposure happens every time it is required to capture birds, barring interference, of course. Hence, Taylor's formula forces us to say, what is palpably false, that the predator was licking his paw in order to catch (or, for the sake of catching) the bird.

A third and even more important objection is closely analogous to the "multiple goals" difficulty Scheffler raises against Braithwaite. Let's return to the conceptually fertile predator, and this time suppose he is both hungry and thirsty. He does not see any water, but *does* see something which would pass for prey, and begins manifesting behavior which is, in some obvious sense, stalking. Before catching the prey, however, his stalking leads him to some water, and he pauses in his pursuit to drink. At this point Taylor's formula would require that we say the predator's behavior all along had been directed toward (i.e., for the sake of) obtaining water: Because he *was* thirsty; and he did *just* what was required to get water; and, barring interference of the sort Taylor allows, going to water always obtains when it is required to satisfy thirst.[8] Q.E.D. But this conflicts directly with the fact that the predator's behavior was obviously stalking; it was obviously *for the sake of* catching its prey. In fact, *ex hypothesi,* the predator couldn't even see the water prior to exhibiting the behavior in question.

Fourth, the (qualified) sufficiency criterion prevents Taylor's analysis from distinguishing between functions and goals in some cases in which the distinction between these two notions is clear and valuable. There are, for example, many rudimentary activities such as sleeping and eating, which are periodically required for the attainment of almost any goal an organism might have.

If an animal doesn't sleep for an extended period, it becomes less and less able to achieve even its most simple ends. Furthermore, barring abnormal conditions (i.e., interference), whenever an animal finds itself so tired that sleep is necessary for it to continue its everyday, goal-achieving activities, it sleeps. In other words, the requirement of sleep for these goals is (qualifiedly) sufficient to produce it. So on Taylor's analysis, we would have to say that the animal went to sleep in order to achieve all sorts of goals the next day. But this is absurd. A good case could be made for saying that the achievement of all those goals the next day is the *function* of sleep. But no case can be made for saying that this is its purpose in any other sense. In fact, we might plausibly maintain in a specific instance that the beast went to sleep merely because it was tired—not "for the sake of" *anything*. But even in cases in which we might urge that the induction of sleep was goal-directed—e.g., he went to sleep on purpose—the best candidate for the *goal* of going to sleep would be relief of the feeling of fatigue, nothing more long-term than that. To ascribe long-term ends-in-view to animals—and sometimes even men—when they eat, sleep, play, copulate, etc., is to indefensibly conflate goals and functions.[9]

3. A MODIFICATION OF THE FORMULA

All four of these objections are circumvented if instead of saying that the requirement of behavior *B* is (qualifiedly) sufficient for *B*, we change the formula to read: the requirement of *B* causes, or brings about, *B*. The difference between these formulas is subtle and may just have been overlooked by Taylor, who, I think, would regard them as synonymous. That there is a difference between them can be seen by examination of any of the recent sophisticated attempts to analyze the notion of causation.[10] And it is precisely these differences which admit the four objections above. I will not attempt an analysis of causation here, and it is not necessary.[11] But briefly, the two features of 'cause' which allow it to succeed where 'sufficient condition' failed are: First, 'sufficient condition' applies to a much wider range of relationships than does 'cause' (there are noncausal guarantees); and second, the ex post facto, diagnostic nature of cause eliminates the need for "cogency" criteria which Taylor must invoke. The necessity of this modification serves, at bottom, to once again illustrate the unworkability of the usual deterministic-inferential account of causation. The relationship we are searching for here is *obviously* causal; but just as clearly, even a modified and qualified sufficient condition does not capture it.

To demonstrate the power of this modification, let us look again at the above objections and notice the improvement effected by the new formula. First, the strained and artificial cases of interference simply vanish, because we have no need for an exception clause. The question of whether or not stalking is caused by (brought about by) the fact that it is required, doesn't even arise if no stalking behavior is being exhibited. All that must be established is that when stalking *does* obtain, it is produced/caused/brought about by its being required. And although this may sometimes be a complicated procedure, all of the philosophical and methodological principles are rather well-embodied in Mill's Methods. Obviously, this argument can be generalized to all cases of selective attention. The other three objections can be dispatched rather quickly. It is clear in these examples that: its requirement to catch the bird is *not* what *brought about* the predator's licking of his paw; the fact that walking in the direction he did was required to get water *was not* what *caused* the predator to stalk in that direction; its being required for the next day's activities was *not* what *produced* sleep. So the modified formula relieves us of the burden of egregiously misdescribing these cases.

One other modification must be made to

Taylor's formula. To insist that the behavior *B* be required for a given end before it can be characterized teleologically, is to insist on a condition which cannot be met in many normal purposive cases. Nothing a predator does by way of stalking is in any strong sense *required* for the goal of food, although the activity is clearly purposive. Prey may accidentally walk (or fly) into a predator's unsuspecting clutches, or some fairly unsophisticated trickery might result in the same end without stalking. Taylor's recognition of this is what leads him to add the "principle of selection" qualification mentioned above. But the condition is still too strong. As long as accidents are possible, an end may be satisfied without employing *anything* in the organism's repertoire of purposive activities. *None* of them is required, even if we include the selection criterion in the context.

Again, the way out is provided by introducing the notion of causing or bringing about. This is the terminology Taylor used in his earliest statement of the criterion but dropped it in the formulation he eventually settled on, either because he thought the two were equivalent, which they are not, or because he thought the former statement was too imprecise to be useful. However, if instead of saying that what causes *B* is the fact that it is (qualifiedly) *required,* we say that what causes *B* is the fact that it *brings about* the goal in question, we eliminate the difficulty of the previous paragraph. For the fact that the goal is sometimes achieved by accident is irrelevant to the claim that it is, sometimes, brought about by doing *B*. Of course (as Taylor probably recognized) this formulation is too tough: a teleological *B* does not *always* bring about its *G*. Taylor himself occasionally employs a weaker one: *B* is the *type of thing* which brings about *G*. In some cases a slightly different weak formulation is preferable: *B tends to*[12] bring about *G*. Satisfaction of any of these conditions would do. So the best way to state the general condition is disjunctively: *B* occurs for the sake of *G* if and only if *B* occurs because it brings about *G*, or because it is the type of thing which brings about *G*, or because it tends to bring about *G*, or, perhaps, because it is the type of thing which tends to bring about *G*.

Using the 'tends to' expression to represent the entire disjunction we can then easily formulate this modification of the Taylor formula thus:

S does *B* for the sake of *G* means:
(2) (i) *B* tends to bring about *G*.
(ii) *B* occurs because (i.e. is brought about by the fact that) it tends to bring about *G*.

This way of putting it may make it look frighteningly difficult to apply in concrete cases, and that may be what led Taylor to his more precise, albeit less defensible formulation. The question of application will occupy us shortly. For now it is important only to recognize that this embryonic looking statement of the formula in terms of results rather than requirements is the best that can be done. We had to exchange a rather tight statement employing a requirement-etiology for a much looser one employing a result or (better) *consequence*-etiology in order to account accurately for the most rudimentary properties of teleological systems. This is what the argument of the last few pages has shown. Behavior is teleological when it is being brought about by its tendency to produce a certain result, understanding 'tendency to produce' as including all the stricter formulations.[13]

4. TESTABILITY

One of Taylor's two main claims for his analysis is that it avoids the charge of empirical undecidability which functioned so centrally in the vitalist controversy earlier this century. He claims, as we saw, that whether or not something is required for something else is often the legitimate object of direct observation. Consequently, whether or not this requirement, barring interference, guarantees the occurrence of some specific behavior should be testable, empirically determinable. If so, the crucial question now becomes: does this property of the analysis survive my modification? Is my sloppier formulation empirically demonstrable?

Since even *speaking* of behavior teleologically presupposes the applicability of (2), we must begin such a demonstration with behavior described nonteleologically. Let me use the term 'physico-geometrical' to represent this sort of description. Now the behavior we are concerned with here—that to be explained, to be characterized teleologically—can from a physico-geometrical point of view be of a wide variety of different kinds. It can be a certain path or set of paths (pattern); it can be stationary gesticulation, or it can be some other configurational change or pattern of changes; and of course it can be the complete absence of motion and change; and the motion or motionlessness can be with respect to virtually anything in the environment. Such behavior is quite often referred to using a demonstrative pronoun in its presence, and nearly always requires and presumes an ostensive familiarity with it for the adequacy of any descriptive expression. "Why did he do that?" "What?" "What he just did, you know, go up to the edge, stop, and flail the water all frothy." "Oh, that. He was just trying to attract attention."

Whether a bit of physico-geometrically characterized behavior will have the property we are interested in—whether it tends to bring about *G*—will usually depend on some details of the environment. Whether a certain path will lead to prey or effect escape obviously depends, inter alia, on the nature and location of obstacles, and other agents. So, by altering these features of the environment, we can change precisely that property of the behavior on which the application of (2) depends. Hence, if we can show that the occurrence of a particular bit of behavior *depends on* the environment being such that the behavior has that property, we have empirically demonstrated that the formula applies: the behavior is directed towards *G*.

It is valuable to see how this works in a spectrum of cases. For the most transparent illustration, we could arrange an experimental environment such that from a certain initial state, there is only one bit of behavior open to "agent" *A* that would result in *G* and this would be *certain* to result in *G*. We could then vary the environment to make other bits of behavior meet this same condition. If in these circumstances the *B* meeting the condition were invariably exhibited by *A*, then it would be merely a standard experimental design problem to assure that it was the satisfaction of this requirement which was bringing about *B*. Using the standard tools of random variation and variable matching, we could assure ourselves that what was producing a particular *B* was that that particular *B* was the one which would bring about *G*. And since this is one of the things which satisfies the quasi-disjunctive condition (2) i, (2) has been empirically demonstrated to hold.

However, the stringent conditions imposed in this case are almost never met in practice. There are almost no circumstances in which there are not alternative methods and routines for achieving a particular goal. But the relaxation of the requirement to include these cases merely adds a complication: we just stipulate that the particular *B* exhibited always be one of the requisite set. In this case it is its membership in the set which brings about *B*, which is to say it occurs because it is the *sort of thing* which brings about *G*. This would satisfy (2) i as well. It may be wholly random which particular *B* within the potentially successful set occurs, or the occurrence of that particular member may be explicable in terms of some consideration such as Taylor's least effort principle. For our purposes it matters not. The applicability of (2) is indifferent to this further issue: once the preliminary test is passed, the formula applies. This shows that

(2) is consistent with a degree of indeterminacy in the regularities governing the constituents of *A* and its environment; and in this it represents something of an advance over Taylor's analysis.

Much more important is the fact that the other part of the requirement must be relaxed too. We must be able to allow for failures of perception and execution in nearly every case. There are all sorts of ways in which behavior can fail to achieve its evident goal and still be directed toward it. We can objectively and intersubjectively judge that the behavior exhibited was plausible, appropriate, the right sort of thing, etc., given *G*, even though the performance was flawed. This point is actually tricky and worth belaboring. The "sort of thing" which is "right," "plausible," and "appropriate" is specifiable physico-geometrically: a path, configurational change or pattern. On the other hand our *recognition* of the behavior as appropriate, even when characterized in these terms, may be considered wholly anthropomorphic: that's what I would have done, certainly reasonable, it couldn't have seen that it wouldn't work, etc. But this sort of anthropomorphism is not pernicious. As long as we are able to tell when the behavior has the requisite propriety with the reliability and intersubjectivity of standard perceptual judgements, our ascriptions can function in an experimental test; they are empirically demonstrable. And this we seem quite able to do in an enormous range of cases, both organismic and mechanistic. Jumping over the fence is, in some circumstances, *obviously* appropriate escape behavior, even if there is a net concealed on the other side ready to ensnare the would-be escapee. So if we discover that as the environment changes, the concomitant behavior is always[14] objectively appropriate, though sometimes unsuccessful in one way or another, the formula must still apply. This is directly responsible for the ". . . tends to . . ." language in (2). Behavior is appropriate vis-à-vis the achievement of *G* only if it moves things in the right direction: only if it tends to get the job done.

Interestingly, appropriate but unsuccessful behavior may well be the most central kind of teleological behavior, both conceptually and identificatorily; for it is the behavior of trying. And not only is trying one of the most emphatically teleological concepts, trying behavior constitutes the majority of that systematically complex behavior we are most reliable in identifying as teleological. The clearest cases of hunting, fleeing, and building consist largely of attempts—success is quite usually elusive. Furthermore, it is precisely trying behavior which functioned so centrally in earlier discussions under the headings of "plasticity" and "persistence" (see [3], [5], and [11]), which were there taken as the paradigms of objectively identifiable teleological behavior traits. What makes us say a predator is stalking—rather than writhing or undergoing spasms—is the systematic organization of the movements about the goal object, or about the obvious clues to the goal object, or about something which might be mistaken for a clue. It is this systematicity which makes the *direction* of the behavior so obvious. And the particular systematicity which gives direction to a bit of behavior is that which obtains when the behavior arises because it tends to produce a certain result, e.g., the apprehension of prey. This kind of systematicity just is plasticity and persistence of the sort Braithwaite and Nagel were concerned with, but misunderstood. If we were to conduct a giant Mill's Methods experiment to demonstrate that what was causing a certain kind of behavior in some context was that behavior's tendency to produce a certain result, the description of one successful outcome *could* be put: the behavior was plastic and persistent with respect to that result. Of course, details of such an experiment would in general be enormously difficult to specify from what I have called the physico-geometrical point of view. So, perhaps it is fortunate that it is at this end of the spectrum that we are most reliable perceptually. I take the ease with which it handles trying behavior to be another significant advantage of (2) over (1).

5. SOME CONSEQUENCES

We are now in a position to eradicate two of the most important difficulties that plague discussions of teleology: the charge of anthropomorphism, and the apparent circularity of teleological explanations. The first of these is perhaps the more notorious. It is often alleged that to use teleological terminology to describe the behavior of things other than humans is *simply* anthropomorphic. And often we do fall back on suspicious sounding reconstructions like, "well, that is how I would behave if I were there." But although individuals may commit anthromorphic mistakes, we have seen that there is nothing inherent in teleological concepts that demands such mistakes. In a number of cases we have examined, no anthropomorphism whatever is involved in justifying a teleological ascription. So the categorical charge is false. Still, in the more usual examples, teleological characterizations seem to involve something anthromorphic in the requisite propriety judgments. But we saw that these need only be objective, reliable judgments to serve adequately in (2); they need not be nonanthromorphic. Furthermore, teleological behavior is not *simply* appropriate behavior; it is appropriate behavior with a certain etiology. Establishing the etiology is what is central to the teleological characterization. So even here, the teleological ascription is not even wholly, much less simply anthromorphic. And what anthropomorphism there is is consistent with a reasonable empiricism: there is nothing essentially subjective or mysterious about it.

However, there is one further area in which the charge might seem to carry more weight. In those cases in which our teleological characterizations themselves are a matter of causal inspection of the phenomenon (much of higher organic behavior), what we claim to be perceiving is the etiology itself, not just the propriety of the behavior. We *see* that the squirrel is building a nest, the horse is going for the barn, not just that the behavior is appropriate for nest building or barn going. Here it might be maintained that we are simply transferring pattern recognition skills from human contexts to nonhuman. Wouldn't this be illicit, pernicious anthropomorphism?

Not necessarily. For here, as in perceptions generally, the reliability of our judgment is *everything:* to impugn a perceptual "skill" *merely* because it has anthropomorphic aspects, is to commit the genetic fallacy. We may quite legitimately rely upon anthropomorphic analogy to help us *recognize* complex behavior patterns. And a good case can be made for saying that we do employ some such tool in spotting the subtle differences in behavior which distinguish goal-directed behavior from behavior not directed at anything. For example, when we say that the motion of a cat's ear is *for the sake of* increased hearing acuity, as opposed to a fear or irritation *reaction,* we may well be "putting ourselves in its place" and characterizing the situations "from its point of view" and in terms of intentions, desires, hopes, and fears. The mistake (on both sides!) is only to confuse a recognition mnemonic with a justification. The anthropomorphic analogy may *explain our insight* into a complex phenomenon. The *justification of our teleological characterization,* however, has got to be that directive organization is the best etiological analysis or account of the movements; what is evoking them must be their tendency to produce a certain result. The correlation between this and our saying so is the *test* of our perceptual skill.[15] And the discussion of (2) in the previous section provides the practical framework in which such a test may be conducted. If there is a question of human perceptual arrogance here, this should help combat it.

Before going on to the second difficulty, a generally valuable consequence of the foregoing discussion should be drawn. There is a strong epistemological parallel here between the concept of "goal-directedness" and the class of concepts called "theoretical" in philosophical discussions of the observational confirmation of theories. Theoretical terms, in this context, are taken to refer to things which are in some strong sense unobservable. Hence, the *justification* for postulating them has to be that their existence *best explains* some phenomenon or other. Similarly, the justification for saying a certain bit of behavior is goal-directed, has to be that the best explanation of it is that it is being brought about by its tendency to produce a certain result. Furthermore, in some cases the goal-directedness of behavior is very difficult to detect (e.g., lower animals and plants). In these cases 'goal-directedness' would simply *be* a theory-term, being applied only after checking the consequence of its postulation.

This strong parallel may have been what tempted the vitalists to postulate an unobserved entity to direct the behavior of organisms toward a goal. The difficulty with that position is nearly classic, and I need not rehash it here. The real illumination provided by this parallel is on the observation/theory dispute. For it is clear that from a justificatory point of view, 'goal-directedness' *always* satisfies the epistemological criteria for being a theory-term: Saying something is goal-directed is *justified* if goal-directedness (i.e., our formula) *is the best account of* that something's behavior. But it is *also* clear that in some cases, the goal-directedness of behavior is obvious, palpable, instantly recognizable, and unmistakable to the normally sighted individual. Which is to say it is observable. This suggests that from at least one very important perspective, 'theory-term' and 'observation-term' do not mark an *epistemological* distinction.

The second difficulty mentioned above probably devolves to an uneasy feeling that there *must* be something circular about anything so involuted as the discussion of teleological systems on the last few pages. On one hand, I offer the plasticity and persistence of the behavior of a system toward a certain result as the empirical evidence demonstrating that the system has that result as a goal. On the other hand, I clearly want to be able to say we can *explain* the behavior of the system by appeal to its having a certain goal. But how can I explain the behavior of a system by appeal to its having a certain goal, when the evidence for its having that goal is a property of that behavior? Doesn't that really beg all the questions? Isn't that circular? I think that when the nature of such explanation is properly understood—i.e., in terms of the modified formula developed above—it is clear that whatever circularity there is, is not vicious. First of all, whether something *S* has goal G_i is clearly an empirical issue. One merely has to determine whether (2) is true when G_1 and *S*'s behavior are substituted in it. And this we saw was an empirical determination. But more importantly, (2) is true in this case only if *S*'s behavior occurs *because* it tends to bring about G_i.[16] This is part of what is *meant* by saying *S* has goal G_i. So just saying *S* has goal G_i is to offer an explanation of *S*'s behavior. This is why teleological explanations are *ascription*-explanations. What looks like circularity is merely the natural involutedness of ascription explanations: the content of the modified Taylor formula is implicit in the simple statement that *S has* goal *G*.

6. SUMMARY AND CONCLUSION

I have been trying to show that (2) represents the best way to put the insight of Charles Taylor's very perceptive discussion of goal-directedness. Part of this has consisted in showing (2) to satisfy a variety of desiderata: it accommodates the explanatoriness of teleological characterizations and explicates the notion of ascription explanations; it helps us understand why *trying* is so central a teleological concept; it squares teleological explanations with underlying causal indeterminacy; it accounts for the role of anthropomorphism; and it shows how teleological properties can be the legitimate object of direct observation. But there are two other features any analysis of the sort I am proposing should have. It should be able to make some sense of the forward orientation of teleological explanations, and it should explicate the contrast between teleological explanations and what might be called "merely" causal explanations. Let me close by showing that (2) can do both of these things.

One of the primary differences between (2) and (1) is that (2) offers consequence-etiologies as fundamental to teleology,[17] in place of the requirement-etiologies of (1). It is this which allows us to account for the forward orientation of teleological accounts of behavior: it is a direct result of their focus on consequences. When we say that teleological etiologies are consequence-etiologies, we are saying that the consequences of goal-directed behavior are involved in its etiology: it occurs *because* it has certain consequences. It occurs because it tends to achieve *G*. That, quite simply, is what the forward orientation consists in. And, as we saw, this involves nothing very heterodox empirically; e.g., in the experimental context, the causally relevant *manipulation* takes place before *X*, although the feature of that manipulation which determines whether *X* will occur (or be of a certain nature, etc.)—i.e., what *governs* the occurrence (etc.) of *X*—concerns what will ensue. So *X* occurs (etc.) *because* of what will ensue, but the statement of the cause is always appropriate put in the future tense: that things were such that *Y* will (tend to) ensue. And *this* statement concerns the state of affairs prior to *X*.

Not surprisingly, precisely this same feature reveals the distinction between teleological explanations and explanations we call "merely" causal. A merely causal explanation of *B* would provide an etiology in terms of the *antecedents* of *B*, not its consequences. The causal/teleological contrast is *among* etiologies, not between etiologies and something else. The form of a teleological explanation of *B* is: *B* does or tends to do thus and so, therefore *B*. A "merely" causal explanation of *B* would have the form: other sorts of things do or tend to do thus and so, therefore *B*. This may well represent an empiricistically defensible way of making sense of the Aristotelian distinction between efficient and final causes.

NOTES

1. Let me avail myself of Braithwaite's distinction between goal-directed behavior and goal-intended behavior, or action. My primary interest in this paper is in the goal-*directed* behavior of nonhuman things, although occasionally an argument will apply to human action as well. I take the legitimacy of teleological characterizations of human behavior to be much less controversial than that of teleological characterizations in other contexts.

2. I have argued the case for functions in [12].

3. I use the term 'etiological' (and, later, 'etiology') here because 'cause' (and 'causal') typically is used in this context to *contrast* with 'teleological', and I wish to keep that contrast. Although the term 'etiology' has impeccable etymological credentials for this use, I have adopted it mostly because its current employment in the field of medicine is very close to the job I wish it to do here.

4. If this is not obvious, compelling arguments to this effect have been made by Charles Taylor and Norman Malcolm (both anti-mechanists, interestingly) in [8], p. 57 ff., and [4], p. 59 ff.

5. For a detailed discussion of these and other dififculties, see [11].

6. Once again, see [11] for a more detailed discussion of this point.

7. On page 10 (ibid.) he says, ". . . in beings with a purpose an event's being required for a given end is a sufficient condition of its occurrence." Denis Noble also adopts this interpretation [7]. Since Taylor does not object to it in his reply, it would seem to have his implicit benediction.

8. This last proposition follows (for Taylor) from the reasonable presupposition that it is sometimes correct to say that the animal went to a watering place in order to get (for the sake of getting) a drink.

9. In case this distinction is in need of any defense see the discussions in [1], p. 145 ff., [2] and [12].

10. For example, see: *Causation in the Law,* H. L. A. Hart and A. M. Honore; Oxford: Clarendon Press, 1959, Part I.

11. The analysis could of course go on tracing logical connections forever. A good place to stop is with a concept whose application is clear in practice, in interesting cases.

12. Accept this locution at its pre-analytic face value for now. I will have occasion to explicate it shortly.

13. It might be contended that (2) is what Taylor meant to say. Just possibly, this is what he *thought* he said. He didn't. But I do not particularly wish to claim he didn't. I am most interested in making transparent just what interpretation *must* be placed on his analysis to account adequately for the topography of teleology. And in *this* I do claim some originality.

14. 'Always' is obviously stronger than necessary, but I am just trying to show *some* conditions under which the formula is empirically decidable, not to offer a general condition.

15. As in any other case, the evidence we have for our perceptual competence here may be colossally difficult to state in any detail; it may nevertheless be fairly obvious that the evidence is good. It may be obvious, e.g., that if our judgment were very far wrong, we immediately would be confronted with embarrassing consequences. We say the dog is chasing the rabbit, but he overtakes the rabbit and runs right by, and then he destroys himself by crashing full tilt into the first obstacle in his path.

16. Recall that I am construing the notion of "tending to bring about G" broadly enough to include the objective propriety of judgments discussed above in the context of unsuccessful (but) teleological behavior.

17. I have argued for exactly the same position with respect to functions in [12].

REFERENCES

1. Beckner, M. *The Biological Way of Thought.* New York: Columbia University Press, 1959.

2. Beckner, M. "Function and Teleology." *Journal of the History of Biology* 2 (1969).

3. Braithwaite, R. *Scientific Explanation.* New York: Harper, 1960.

4. Malcolm, N. "The Conceivability of Mechanism." *Philosophical Review* 77 (1968).

5. Nagel, E. *The Structure of Science.* New York: Harcourt, Brace and World, Inc., 1961 (Chapter 12).

6. Noble, D. "Charles Taylor on Teleological Explanation." *Analysis* 27 (1967).

7. Scheffler, I. "Thoughts on Teleology." *The British Journal for the Philosophy of Science* 9 (1959).

8. Taylor, C. "Explaining Action." *Inquiry* 13 (1970).

9. Taylor, C. *The Explanation of Behaviour.* London: Routledge and Kegan Paul, 1964.

10. Wiener, N.; Rosenblueth, A.; and Bigelow, J. "Behavior, Purpose, and Teleology." *Philosophy of Science* 10 (1943).

11. Wright, L. "The Case Against Teleological Reductionism." *The British Journal for the Philosophy of Science* 19 (1968).

12. Wright, L. "Functions," (forthcoming).

Robert Cummins

FUNCTIONAL ANALYSIS

I

A survey of the recent philosophical literature on the nature of functional analysis and explanation, beginning with the classic essays of Hempel in 1959 and Nagel in 1961, reveals that philosophical research on this topic has almost without exception proceeded under the following assumptions.[1]

(A) The point of functional characterization in science is to explain the presence of the item (organ, mechanism, process, or whatever) that is functionally characterized.

(B) For something to perform its function is for it to have certain effects on a containing system, which effects contribute to the performance of some activity of, or the maintenance of some condition in, that containing system.

Putting these two assumptions together, we have: a function-ascribing statement explains the presence of the functionally characterized item *i* in a system *s* by pointing out that *i* is present in *s* because it has certain effects on *s*. Give or take a nicety, this fusion of (A) and (B) constitutes the core of almost every recent attempt to give an account of functional analysis and explanation. Yet these assumptions are just that: assumptions. They have never been systematically defended; generally they are not defended at all. I think there are reasons to suspect that adherence to (A) and (B) has crippled the most serious attempts to analyze functional statements and explanation, as I will argue in sections II and III below. In section IV, I will briefly develop an alternative approach to the problem. This alternative is recommended largely by the fact that it emerges as the obvious approach once we take care to understand why accounts involving (A) and (B) go wrong.

Reprinted by permission from the author and *The Journal of Philosophy*, Vol. LXXII, no. 20 (November 20, 1975).

II

I begin this section with a critique of Hempel and Nagel. The objections are familiar for the most part, but it will be well to have them fresh in our minds as they form the backdrop against which I stage my attack on (A) and (B).

Hempel's treatment of functional analysis and explanation is a classic example of the fusion of (A) and (B). He begins by considering the following singular function-ascribing statement.

(1) The heartbeat in vertebrates has the function of circulating the blood through the organism.

He rejects the suggestion that "function" can *simply* be replaced by "effect" on the grounds that, although the heartbeat has the effect of producing heartsounds, this is not its function. Presuming (B) from the start, Hempel takes the problem to be how one effect—the having of which is the function of the heartbeat (circulation)—is to be distinguished from other effects of the heartbeat (e.g., heartsounds). His answer is that circulation, but not heartsounds, ensures a necessary condition for the "proper working of the organism." Thus, Hempel proposes (2) as an analysis of (1).

(2) The heartbeat in vertebrates has the effect of circulating the blood, and this ensures the satisfaction of certain conditions (supply of nutriment and removal of waste) which are necessary for the proper working of the organism.

As Hempel sees the matter, the main problem with this analysis is that functional statements so construed appear to have no explanatory force. Since he assumes (A), the problem for Hempel is to see whether (2) can be construed as a deductive nomological explanans for the presence of the heartbeat in vertebrates and, in general, to see whether statements having the form of (2) can be construed as deductive nomological explananda for the presence in a system of some trait or item that is functionally characterized.

Suppose, then, that we are interested in explaining the occurrence of a trait *i* in a system *s* (at a certain time *t*), and that the following functional analysis is offered:

(a) At *t*, *s* functions adequately in a setting of kind *c* (characterized by specific internal and external conditions).
(b) *s* functions adequately in a setting of kind *c* only if a certain necessary condition, *n*, is satisfied.
(c) If trait *i* were present in *s*, then, as an effect, condition *n* would be satisfied.
(d) Hence, at *t*, trait *i* is present in *s*.[2]

(d), of course, does not follow from (a)–(c), since some trait *i*′ different from *i* might well suffice for the satisfaction of condition *n*. The argument can be patched up by changing (c) to (c′): "condition *n* would be satisfied in *s* only if trait *i* were present in *s*," but Hempel rightly rejects this avenue on the grounds that instances of the resulting schema would typically be false. It is false, for example, that the heart is a necessary condition for circulation in vertebrates, since artificial pumps can be, and are, used to maintain the flow of blood. We are, thus, left with a dilemma. If the original schema is correct, then functional explanation is invalid. If the schema is revised so as to ensure the validity of the explanation, the explanation will typically be unsound, having a false third premise.

Ernest Nagel offers a defense of what is substantially Hempel's schema with (c) replaced by (c′).

> . . . a teleological statement of the form, "The function of *A* in a system *S* with organization *C* is to enable *S* in the environment *E* to engage in process *P*," can be formulated more explicitly by: every system *S* with organization *C* and in environment *E* engages in process P; if *S* with organization *C* and in environment *E* does not have *A*, then *S* does not engage in *P;* hence, *S* with organization *C* must have *A*.[3]

Thus he suggests that (3) is to be rendered as (4):

(3) The function of chlorophyll in plants is to enable them to perform photosynthesis.
(4) A necessary condition of the occurrence of photosynthesis in plants is the presence of chlorophyll.

So Nagel must face the second horn of Hempel's dilemma: (3) is presumably true, while (4) may well be false. Nagel is, of course, aware of this objection. His rather curious response is that, as far as we know, chlorophyll *is* necessary for photosynthesis in the green plants.[4] This may be so, but the response will not survive a change of example. Hearts are *not* necessary for circulation, artificial pumps having actually been incorporated into the circulatory systems of vertebrates in such a way as to preserve circulation and life.

A more promising defense of Nagel

might run as follows. While it is true that the presence of a working heart is not a necessary condition of circulation in vertebrates under all circumstances, still, under *normal* circumstances—most circumstances, in fact—a working heart is necessary for circulation. Thus, it is perhaps true that, at the present stage of evolution, a vertebrate that has not been tampered with surgically would exhibit circulation only if it were to contain a heart. If these circumstances are specifically included in the explanans, perhaps we can avoid Hempel's dilemma. Thus, instead of (4) we should have:

> (4) At the present stage of evolution, a necessary condition for circulation in vertebrates that have not been surgically tampered with is the operation of a heart (properly incorporated into the circulatory system).

(4′), in conjunction with statements asserting that a given vertebrate exhibits circulation and has not been surgically tampered with and is at the present stage of evolution, will logically imply that that vertebrate has a heart. It seems, then, that the Hempelian objection could be overcome if it were possible, given a true function-ascribing statement like (1) or (3), to specify "normal circumstances" in such a way as to make it true that, in those circumstances, the presence of the item in question is a necessary condition for the performance of the function ascribed to it.

This defense has some plausibility as long as we stick to the usual examples drawn from biology. But if we widen our view a bit, even within biology, I think it can be shown that this defense of Nagel's position will not suffice. Consider the kidneys. The function of the kidneys is to eliminate wastes from the blood. In particular, the function of my left kidney is to eliminate waste from my blood. Yet the presence of my left kidney is not, in normal circumstances, a necessary condition for the removal of the relevant wastes. Only if something seriously abnormal should befall my right kidney would the operation of my left kidney become necessary, and this only on the assumption that I am not hooked up to a kidney machine.[5]

A less obvious counter-example derives from the well-attested fact of hemispherical redundancy in the brain. No doubt it is in principle possible to specify conditions under which a particular duplicated mechanism would be necessary for normal functioning of the organism, but (a) in most cases we are not in a position to actually do this, though we are in a position to make well-confirmed statements about the functions of some of these mechanisms, and (b) these circumstances are by no means the normal circumstances. Indeed, given the fact that each individual nervous system develops somewhat differently owing to differing environmental factors, the circumstances in question might well be different for each individual, or for the same individual at different times.

Apparently Nagel was pursuing the wrong strategy in attempting to analyze functional ascriptions in terms of necessary conditions. Indeed, we are still faced with the dilemma noticed by Hempel: an analysis in terms of necessary conditions yields a valid but unsound explanatory schema; analysis in terms of sufficient conditions along the lines proposed by Hempel yields a schema with true premises, but validity is sacrificed.

Something has gone wrong, and it is not too difficult to locate the problem. An attempt to explain the presence of something by appeal to what it does—its function—is bound to leave unexplained why something else that does the same thing—a functional equivalent—isn't there instead. In itself, this is not a serious matter. But the accounts we have been considering assume that explanation is a species of deductive inference, and one cannot deduce hearts from circulation. This is what underlies the dilemma we have been considering. At best, one can deduce circulators from circulation. If we make this amendment, however, we are left with a functionally tainted analysis; 'the function of the heart is to circulate the blood' is rendered 'a blood circulator is a (necessary/sufficient) condition of circulation, and *the heart is a blood circulator*'. The expression in italics is surely as much in need of analysis as the analyzed expression. The problem, however, runs much deeper than the fact that the performance of a certain function does not determine how that function is performed. The problem is rather that to "explain" the presence of the heart in vertebrates by appeal to what the heart *does* is to "explain" its presence by appeal to factors which are causally irrelevant to its presence. Even if it were possible, as Nagel claimed, to *deduce* the presence of chlorophyll from the occurrence of photosynthesis, this would fail to explain the presence of chlorophyll in green plants in just the way deducing the presence and height of a building from the existence and length of its shadow would fail to explain why the building is there and has the height it does. This is not because all explanation is causal explanation: it is not. But to explain the presence of a naturally occurring structure or physical process—to explain why it is there, why such a thing exists in the place (system, context) it does—this does require specifying factors which causally determine the appearance of that structure or process.[6]

There is, of course, a sense in which the question "Why is *x* there?" is answered by giving *x*'s function. Consider the following exchange. X asks Y, "Why is that thing there (pointing to the gnomon of a sundial)?" Y answers, "Because it casts a shadow on the dial beneath, thereby indicating the time of day." It is exchanges of this sort that most philosophers have had in mind when they speak of functional explanation. But it seems to me that, although such exchanges do represent genuine explanations, the use of functional language in this sort of explanation is quite distinct from its explanatory use in science. In section IV below I will sketch what I think *is* the central explanatory use of functional language in science. Meanwhile, if I am right, the evident propriety of exchanges like that imagined between X and Y has led to premature acceptance of (A), hence to concentration on what is, from the point of view of scientific explanation, an irrelevant use of functional language. For it seems to me that the question "Why is *x* there?" can be answered by specifying *x*'s function only if *x* is or is part of an artifact. Y's answer, I think, explains the presence of the gnomon because it rationalizes the action of the agent who put it there by supplying a *reason* for putting it there. In general, when we are dealing with the result of a deliberate action, we may explain the result by explaining the action, and we may explain a deliberate action by supplying the agent's reason for doing it. Thus when we look at a sundial, we assume we *know* in a general way how the gnomon came to be there: someone deliberately put it there. But we may wish to know *why* it was put there. Specifying the gnomon's function allows us to formulate what we suppose to be the unknown agent's reason for putting it there, viz., a belief that it would cast a shadow such that . . . , and so on. When we do this, we are elaborating on what we assume is the crucial causal factor in determining the gnomon's presence, namely a certain deliberate action.

If this is on the right track, then the viability of the sort of explanation in question should depend on the assumption that the thing functionally characterized is there as the result of deliberate action. If that assumption is evidently false, specifying the thing's function will not answer the question. Suppose it emerges that the sundial is not, as such, an artifact. When the ancient building was ruined, a large stone fragment fell on a kind of zodiac mosaic and embedded itself there. Since no sign of the roof remains, Y has mistakenly supposed the thing was designed as a sundial. As it happens, the local people have been using the thing to tell time for centuries, so Y is right about the function of the thing X pointed to.[7] But it is simply false that the thing is there because it casts a shadow, for there is no agent who put it there "because it casts a shadow." Again, the function of a bowl-like depression in a huge stone may be to hold holy water, but we cannot explain why it is there by appeal to its function if we know it was left there by prehistoric glacial activity.

If this is right, then (A) will lead us to focus on a type of explanation which will not apply to natural systems: chlorophyll and hearts are not "there" as the result of any deliberate action; hence the essential presupposition of the explanatory move in question is missing. Once this becomes clear, to continue to insist that there *must* be *some* sense in which specifying the function of chlorophyll explains its presence is an act of desperation born of thinking there is no other explanatory use of functional characterization in science.

Why have philosophers identified functional explanation exclusively with the appeal to something's function in explaining why it is there? One reason, I suspect, is a failure to distinguish teleological explanation from functional explanation, perhaps because functional concepts do loom large in "explanations" having a teleological form. Someone who fails to make this distinction, but who senses that there is an important and legitimate use of functional characterization in scientific explanation, will see the problem as one of finding a legitimate explanatory role for functional characterization within the teleological form. Once we leave artifacts and go to natural systems, however, this approach is doomed to failure, as critics of teleology have seen for some time.

This mistake probably would have sorted itself out in time were it not the case that we do reason from the performance of a function to the presence of certain specific processes and structures, e.g., from photosynthesis to chlorophyll, or from coordinated activity to nerve tissue. This is perfectly legitimate reasoning: it is a species of inference to the best explanation. Our best (only) explanation of photosynthesis requires chlorophyll, and our best explanation of coordinated activity requires nerve tissue. But once we see what makes this reasoning legitimate, we see immediately that inference *to* an explanation has been mistaken for an explanation itself. Once this becomes clear, it becomes equally clear that (A) has matters reversed: given that photosynthesis is occurring in a particular plant, we may legitimately infer that chlorophyll is present in that plant precisely because chlorophyll enters into our best (only) explanation of photosynthesis, and given coordinated activity on the part of some animal, we may legitimately infer that nerve tissue is present precisely because nerve tissue enters into our best explanation of coordinated activity in animals.

To attempt to explain the heart's presence in vertebrates by appealing to its function in vertebrates is to attempt to explain the occurrence of hearts in vertebrates by appealing to factors which are causally irrelevant to its presence in vertebrates. This fact has given "functional explanation" a bad name. But it is (A) that deserves the blame. Once we see (A) as an undefended philosophical hypothesis about how to construe functional explanations rather than as a statement of the philosophical problem, the correct alternative is obvious: what we can and do explain by appeal to what something does is the behavior of a containing system.[8]

A much more promising suggestion in the light of these considerations is that (1) is appealed to in explaining *circulation.* If we reject (A) and adopt this suggestion, a simple deductive-nomological explanation with circulation as the explicandum turns out to be a sound argument.

(5) a. Vertebrates incorporating a beating heart in the usual way (in the way *s* does) exhibit circulation.
b. Vertebrate *s* incorporates a beating heart in the usual way.
c. Hence, *s*, exhibits circulation.

Though by no means flawless, (5) has several virtues, not the least of which is that it does not have biologists passing by an obvious application of evolution or genetics in favor of an invalid or unsound "functional" explanation of the presence of hearts. Also, the redundancy examples are easily handled, e.g., the removal of wastes is deduced in the kidney case.

The implausibility of (A) is obscured in examples taken from biology by the fact that there are two distinct uses of function statements in biology. Consider the following statements.

(a) The function of the contractile vacuole in protozoans is elimination of excess water from the organism.
(b) The function of the neurofibrils in the ciliates is coordination of the activity of the cilia.

These statements can be understood in either of two ways. (i) They are generally used in explaining how the organism in question comes to exhibit certain characteristics or behavior. Thus (a) explains how excess water, accumulated in the organism by osmosis, is eliminated from the organism; (b) explains how it happens that the activity of the cilia in paramecium, for instance, is coordinated. (ii) They may be used in explaining the continued survival of certain organisms incorporating structures of the sort in question by indicating the survival value which would accrue to such organisms in virtue of having structures of that sort. Thus (a) allows us to infer that incorporation of a contractile vacuole makes it possible for the organism to be surrounded by a semi-permeable membrane, allowing the passage of oxygen into, and the passage of wastes out of, the organism. Relatively free osmosis of this sort is obviously advantageous, and this is made possible by a structure which solves the excess water problem. Similarly, ciliates incorporating neurofibrils will be capable of fairly efficient locomotion, the survival value of which is obvious.[9]

The second sort of use occurs as part of an account which, if we are not careful, can easily be mistaken for an explanation of the presence of the sort of item functionally characterized, and this has perhaps encouraged philosophers to accept (A). For it might seem that natural selection provides the missing causal link between what something does in a certain type of organism and its presence in that type of organism. By performing their respective functions the contractile vacuole and the neurofibrils help species incorporating them to survive, and thereby contribute to their own continued presence in organisms of those species, and this might seem to explain the presence of those structures in the organisms incorporating them.

Plausible as this sounds, it involves a subtle yet fundamental misunderstanding of evolutionary theory. A clue to the mistake is found in the fact that the contractile vacuole occurs in marine protozoans which have no excess water problem but the reverse problem. Thus, the function and effect on survival of this structure is not the same in all protozoans. Yet the explanation of its presence in marine and fresh-water species is almost certainly the same. This fact reminds us that the processes actually responsible for the occurrence of contractile vacuoles in protozoans are totally insensitive to what that structure does. Failure to appreciate this point not only lends spurious plausibility to (A) as applied to biological examples, but seriously distorts our understanding of evolutionary theory. Whether an organism *o* incorporates *s* depends on whether *s* is "specified" by the genetic "plan" which *o* inherits and which, at a certain level of abstraction, is characteristic of *o*'s species. Alterations in the plan are not the effects of the presence or exercise of the structures the plan specifies. This is most obvious when the genetic change is the result of random mutation. Though not all genetic change is due to random mutation, some certainly is, and that fact is enough to show that specifying the function of a biological structure cannot, in general, explain the presence of that structure. If a plan is altered so that it specifies s' rather than s, then the organisms inheriting this plan will incorporate s'

regardless of the function or survival value of *s′* in those organisms. If the alteration is advantageous, the number of organisms inheriting that plan may increase, and, if it is disadvantageous, their number may decrease. But this typically has no effect on the plan, and therefore no effect on the occurrence of *s′* in the organisms in question.

One sometimes hears it said that natural selection is an instance of negative feedback. If this is meant to imply that the relative success or failure of organisms of a certain type can affect their inherited characteristics, it is simply a mistake: the characteristics of organisms which determine their relative success are determined by their genetic plan, and the characteristics of these plans are typically independent of the relative success of organisms having them. Of course, if *s* is very disadvantageous to organisms having a plan specifying *s*, then organisms having such plans may disappear altogether, and *s* will no longer occur. We could, therefore, think of natural selection as reacting on the *set* of plans generated by weeding out the bad plans: natural selection cannot alter a plan, but it can trim the set. Thus, we may be able to explain why a given plan is not a failure by appeal to the functions of the structures it specifies. Perhaps this is what some writers have had in mind. But this is not to explain why, e.g., contractile vacuoles occur in certain protozoans; it is to explain why the sort of protozoan incorporating contractile vacuoles occurs. Since we cannot appeal to the relative success or failure of these organisms to explain why their genetic plan specifies contractile vacuoles, we cannot appeal to the relative success or failure of these organisms to explain why they incorporate contractile vacuoles.

Once we are clear about the explanatory role of functions in evolutionary theory, it emerges that the function of an organ or process (or whatever) is appealed to in order to explain the biological capacities of the organism containing it, and from these capacities conclusions are drawn concerning the chances of survival for organisms of that type. For instance, appeal to the function of the contractile vacuole in certain protozoans explains how these organisms are able to keep from exploding in fresh-water. Thus, evolutionary biology does not provide support for (A), but for the idea instanced in (5): identifying the function of something helps to explain the capacities of a containing system.[10]

(A) misconstrues functional explanation by misidentifying what is explained. Let us abandon (A), then, in favor of the view that functions are appealed to in explaining the capacities of containing systems, and turn our attention to (B).

Whereas (A) is a thesis about functional explanation, (B) is a thesis about the analysis of function-ascribing statements. Perhaps when divorced from (A), as it is in (5), it will fare better than it does in the accounts of Hempel and Nagel.

III

In spite of the evident virtues of (5), (5a) has serious shortcomings as an analysis of (1). In fact it is subject to the same objection Hempel brings to the analysis which simply replaces 'function' by 'effect': vertebrates incorporating a working heart in the usual way exhibit the production of heartsounds, yet the production of heartsounds is not a function of hearts in vertebrates. The problem is that whereas the production of certain effects is essential to the heart's performing its function, there are some effects the production of which is irrelevant to the functioning of the heart. This problem is bound to infect any "selected effects" theory, i.e., any theory built on (B).

What is needed to establish a selected effects theory is a general formula which identifies the appropriate effects.[11] Both Hempel and Nagel attempt to solve this problem by identifying the function of something with just those effects which contribute to the maintenance of some special condition of, or the performance of some special activity of, some containing system.

If this sort of solution is to be viable, there must be some principled way of selecting the relevant activities or conditions of containing systems. For no matter which effects of something you happen to name, there will be some activity of the containing system to which just those effects contribute, or some condition of the containing system which is maintained with the help of just those effects. Heart activity, for example, keeps the circulatory system from being entirely quiet, and the appendix keeps people vulnerable to appendicitis.[12]

Hempel suggests that, in general, the crucial feature of a containing system, contribution to which is to count as the functioning of a contained part, is that the system be maintained in "adequate, or effective, or proper working order."[13] Hempel explicitly declines to discuss what constitutes proper working order, presumably because he rightly thinks that there are more serious problems with the analysis he is discussing than those introduced by this phrase. But it seems clear that for something to be in working order is just for it to be capable of performing its functions, and for it to be in adequate or effective or proper working order is just for it to be capable of performing its functions adequately or effectively or properly. Hempel seems to realize this himself, for in setting forth a deductive schema for functional explanation, he glosses the phrase in question as 'functions adequately'.[14] More generally, if we identify the function of something x with those effects of x which contribute to the performance of some activity a or to the maintenance of some condition c of a containing system s, then we must be prepared to say as well that a function of s is to perform a or to maintain c. This suggests the following formulation of "selected effects" theories.

(6) The function of an F in a G is f just in case (the capacity for) f is an effect of an F incorporated in a G in the usual way (or: in the way *this* F is incorporated in this G), and that effect contributes to the performance of a function of the containing G.

It seems that any theory based on (B)—what I have been calling "selected effects" theories—must ultimately amount to something like (6).[15] Yet (6) cannot be the whole story about functional ascriptions.

Suppose we follow (6) in rendering "The function of the contractile vacuole in protozoans is elimination of excess water from the organism." The result is (7).

(7) Elimination of excess water from the organism is an effect of a contractile vacuole incorporated in the usual way in a protozoan, and that effect contributes to the performance of a function of a protozoan.

In order to test (7) we should have to know a statement of the form "f is a function of a protozoan." Perhaps protozoans have no functions. If not, (7) is just a mistake. If they do, then presumably we shall have to appeal to (6) for an analysis of the statement attributing such a function and this will leave us with another unanalyzed functional ascription. Either we are launched on a regress, or the analysis breaks down at some level for lack of functions, or perhaps for lack of a plausible candidate for containing system. If we do not wish to simply acquiesce in the autonomy of functional ascriptions, it must be possible to analyze at least some functional ascriptions without appealing to functions of containing systems. If (6) can be shown to be the only plausible formulation of theories based on (B), then no such theory can be the whole story.

Our question, then, is whether a thing's function can plausibly be identified with those of its effects contributing to production of some activity of, or maintenance of some condition of, a containing system, where performance of the activity in question is not a function of the containing system. Let us begin by considering Hempel's suggestion that functions are to be identified with the production of effects contributing to the proper working order of a containing system. I claimed earlier that to say something is in proper working order is just to say that it properly performs its functions. This is fairly obvious in cases of artifacts or tools. To make a decision about which sort of behavior counts as working amounts to deciding about the thing's function. To say something is working, though not behaving or disposed to behave in a way having anything to do with its function, is to be open, at the very least, to the charge of arbitrariness.

When we are dealing with a living organism, or a society of living organisms, the situation is less clear. If we say, "The function of the contractile vacuole in protozoans is elimination of excess water from the organism," we do make reference to a containing organism, but not, apparently, to its function (if any). However, since contractile vacuoles do a number of things having nothing to do with their function, there must be some implicit principle of selection at work. Hempel's suggestion is that, in this context, to be in "proper working order" is simply to be alive and healthy. This works reasonably well for certain standard examples, e.g., (1) and (3): circulation does contribute to health and survival in vertebrates, and photosynthesis does contribute to health and survival in green plants.[16] But once again, the principle will not stand a change of example, even within the life sciences. First, there are cases in which proper functioning is actually inimical to health and life: functioning of the sex organs results in the death of individuals of many species (e.g., certain salmon). Second, a certain process in an organism may have effects which contribute to health and survival but which are not to be confused with the function of that process: secretion of adrenalin speeds metabolism and thereby contributes to elimination of harmful fat deposits in overweight humans, but this is not a function of adrenalin secretion in overweight humans.

A more plausible suggestion along these lines in the special context of evolutionary biology is this:

(8) The functions of a part or process in an organism are to be identified with those of its effects contributing to activities or conditions of the organism which sustain or increase the organism's capacity to contribute to survival of the species.

Give or take a nicety, (8) doubtless does capture a great many uses of functional language in biology. For instance, it correctly picks out elimination of excess water as the function of the contractile vacuole in fresh water protozoans only, and correctly identifies the function of sexual organs in species in which the exercise of these organs results in the death of the individual.[17]

In spite of these virtues, however, (8) is seriously misleading and extremely limited in applicability even within biology. Evidently, what contributes to an organism's capacity to maintain its species in one sort of environment may undermine that capacity in another. When this happens, we might say that the organ (or whatever) has lost its function. This is probably what we would say about the contractile vacuole if freshwater protozoans were successfully introduced into salt water, for in this case the capacity explained would no longer be exercised. But if the capacity explained by appeal to the function of a certain structure continued to be exercised in the new environment, though now to the individual's detriment, we would not say that that structure had lost its function. If, for some reason, flying ceased to contribute to the capacity of pigeons to maintain their species, or even undermined that capacity to some extent,[18] we would still say that a function of the wings in pigeons is to enable them to fly. Only if the wings ceased to function as wings, as in the penguins or ostriches, would we cease to analyze skeletal structure and the like functionally with an eye to explaining flight. Flight is a capacity which cries out for explanation in terms of anatomical functions regardless of its contribution to the capacity to maintain the species.

What this example shows is that functional analysis can properly be carried on in biology quite independently of evolutionary considerations: a complex capacity of an organism (or one of its parts or systems) may be explained by appeal to a functional analysis regardless of how it relates to the organism's capacity to maintain the species. At best, then, (8) picks out those effects which will be called functions when what is in the offing is an application of evolutionary theory. As we shall see in the next section, (8) is misleading as well in that it is not *which* effects are explained but the style of explanation that makes it appropriate to speak of functions. (8) simply identifies effects which, as it happens, are typically explained in that style.

We have not quite exhausted the lessons to be learned from (8). The plausibility of (8) rests on the plausibility of the claim that, for certain purposes, we may assume that a function of an organism is to contribute to the survival of its species. What (8) does, in effect, is identify a function of an important class of (uncontained) containing systems without providing an analysis of the claim that a function of an organism is to contribute to the survival of its species.

Of course, an advocate of (8) might insist that it is no part of his theory to claim that maintenance of the species is a function of an organism. But then the defense of (8) would have to be simply that it describes actual usage, i.e., that it is in fact effects contributing to an organism's capacity to maintain its species which evolutionary biologists single out as functions. Construed in this way, (8) would, at most, tell us *which* effects are picked out as functions; it would provide no hint as to *why* these effects are picked out *as functions.* We know why evolutionary biologists are interested in effects contributing to an organism's capacity to maintain its species, but why call them functions? This is precisely the sort of question a philosophical account of function-ascribing statements should answer. Either (8) is defended as an instance of (6)—maintenance of the species is declared a function of organisms—or it is defended as descriptive of usage. In neither case is any philosophical analysis provided. For in the first case (8) relies on an unanalyzed (and undefended) function-ascribing statement, and in the second it fails to give any hint as to the point of identifying certain effects as functions.

The failings of (8) are I think bound to cripple any theory which identifies a thing's functions with effects contributing to some antecedently specified type of condition or behavior of a containing system. If the theory is an instance of (6), it launches a regress or terminates in an unanalyzed functional ascription; if it is not an instance of (6), then it is bound to leave open the very question at issue, viz., why are the selected effects seen as functions?

IV

In this section I will sketch briefly an account of functional explanation which takes seriously the intuition that it is a genuinely distinctive style of explanation. The assumptions (A) and (B) form the core of approaches which seek to minimize the differences between functional explanations and explanations not formulated in functional terms. Such approaches have not given much attention to the characterization of the special explanatory strategy science employs in using functional language, for the problem as it was conceived in such approaches was to show that functional explanation is not really different in essentials from other kinds of scientific explanation. Once the problem is conceived in this way, one is almost certain to miss the distinctive features of functional explanation, and hence to miss the point of functional description. The account of this section reverses this tendency by placing primary emphasis on the kind of problem which is solved by appeal to functions.

1. Functions and Dispositions

Something may be capable of pumping even though it does not function as a pump (ever) and even though pumping is not its function. On the other hand, if something functions as a pump in a system *s*, or if the function of something in a system *s* is to pump, then it must be capable of pumping in *s*.[19] Thus, function-ascribing statements imply disposition statements; to attribute a function to something is, in part, to attribute a disposition to it. If the function of *x* in *s* is to ϕ, then *x* has a disposition to ϕ in *s*. For instance, if the function of the contractile vacuole in fresh-water protozoans is to eliminate excess water from the organism, then there must be circumstances under which the contractile vacuole would actually manifest a disposition to eliminate excess water from the protozoan which incorporates it.

To attribute a disposition *d* to an object *a* is to assert that the behavior of *a* is subject to (exhibits or would exhibit) a certain law-like regularity: to say *a* has *d* is to say that *a* would manifest *d* (shatter, dissolve) were any of a certain range of events to occur (*a* is put in water, *a* is struck sharply). The regularity associated with a disposition—call it the dispositional regularity—is a regularity which is special to the behavior of a certain kind of object and obtains in virtue of some special fact(s) about that kind of object. Not everything is water-soluble: such things behave in a special way in virtue of certain (structural) features special to water-soluble things. Thus it is that dispositions require explanation: if *x* has *d*, then *x* is subject to a regularity in behavior special to things having *d*, and such a fact needs to be explained.

To explain a dispositional regularity is to explain how manifestations of the disposition are brought about given the requisite precipitating conditions. In what follows I will describe two distinct strategies for accomplishing this. It is my contention that the appropriateness of function-ascribing statements corresponds to the appropriateness of the second of these two strategies. This, I think, explains the intuition that functional explanation is a special *kind* of explanation.

2. Two Explanatory Strategies[20]

(i) *The Instantiation strategy.* Since dispositions are properties, not events to explain a disposition requires explaining how it is instantiated. To explain an event, we cite its cause, and to explain an event type requires a recipe (law) for constructing causal explanations of its tokens. But dispositions, being properties, not events, are not explicable as effects. The *acquisition* of a property is an event, but explaining the acquisition of a property is quite distinct from explaining the property itself. One can explain why/how a thing became fragile without thereby explaining fragility, and one can explain why/how something changed properties—e.g., why something changed temperature—without thereby explaining the property that changed. To explain a property one must show how that property is instantiated in the things that have it.

Simple dispositions are explained by exhibiting their instantiations: water solubility is instantiated as a certain kind of molecular structure, temperature as (average) kinetic energy of molecules, flammability as a kind of subatomic structure (allowing for bonding with oxygen at relatively low temperatures). When we understand how a disposition is instantiated, we are in a position to understand why the dispositional regularity holds of the disposed objects.

Brian O'Shaughnessy has provided an example that allows a particularly simple illustration of this strategy.[21] Consider the disposition he calls elevancy: the tendency of an object to rise in water of its own accord. To explain elevancy, we must explain why freeing a submerged elevant object causes it to rise.[22] This we may do as follows. In every case, the ratio of an elevant object's mass to its nonpermeable volume is less than the density (mass per unit volume) of water: that is how elevancy is instantiated. Once we know this, we may apply Archimedes' Principle, which tells us that water exerts an upward force on a submerged object equal to the weight of the water displaced. In the case of an elevant object, this force evidently exceeds the weight of the object by some amount f. Freeing the object changes the net force on it from zero to a net force of magnitude f in the direction of the surface, and the object rises accordingly. Here we subsume the connection between freeings and risings under a general law connecting changes in net force with changes in motion by citing a feature of elevant objects which allows us (via Archimedes' Principle) to represent freeing them under water as an instance of introducing a net force in the direction of the surface.

(ii) *The analytical strategy.* Rather than deriving the dispositional regularity that specifies d (in a) from the facts of d's instantiation (in a), the analytical strategy proceeds by analyzing a disposition d of a into a number of other dispositions $d_1 \ldots d_n$ had by a or components of a such that programmed manifestation of the d_i results in or amounts to a manifestation of d.[23] The two strategies will fit together into a unified account if the analyzing dispositions (the d_i) can be made to yield to the instantiation strategy.

When the analytical strategy is in the offing one is apt to speak of capacities (or abilities) rather than of dispositions. This shift in terminology will put a more familiar face on the analytical strategy,[24] for we often explain capacities by analyzing them. Assembly-line production provides a transparent example of what I mean. Production is broken down into a number of distinct tasks. Each point on the line is responsible for a certain task, and it is the function of the components at that point to complete that task. If the line has the capacity to produce the product, it has it in virtue of the fact that the components have the capacities to perform their designated tasks, and in virtue of the fact that when these tasks are performed in a certain organized way—according to a certain program—the finished product results. Here we can explain the line's capacity to produce the product—i.e., explain how it is able to produce the product—by appeal to certain capacities of the components and their organization into an assembly line. Against this background we may pick out a certain capacity of an individual component the exercise of which is its function on the line. Of the many things it does and can do, its function on the line is doing whatever it is that we appeal to in explaining the capacity of the line as a whole. If the line produces several products—i.e., if it has several capacities—then, although a certain capacity c of a component is irrelevant to one capacity of the line, exercise of c by that component may be its function with respect to another capacity of the line as a whole.

Schematic diagrams in electronics provide another obvious illustration. Since each symbol represents any physical object whatever having a certain capacity, a schematic diagram of a complex device constitutes an analysis of the electronic capacities of the device as a whole into the capacities of its components. Such an analysis allows us to explain how the device as a whole exercises the analyzed capacity, for it allows us to see exercises of the analyzed capacity as programmed exercise of the analyzing capacities. In this case the "program" is given by the lines indicating how the components are hooked up. (Of course, the lines are themselves function-symbols.)

Functional analysis in biology is essentially similar. The biologically significant capacities of an entire organism are explained by analyzing the organism into a number of "systems"—the circulatory system, the digestive system, the nervous system, etc.—each of which has its characteristic capacities.[25] These capacities are in turn analyzed into capacities of component organs and structures. Ideally, this strategy is pressed until physiology takes over—i.e., until the analyzing capacities are amenable to the instantiation strategy. We can easily imagine biologists expressing their analyses in a form analogous to the schematic diagrams of electrical engineering, with special symbols for pumps, pipes, filters, and so on. Indeed, analyses of even simple cognitive capacities are typically expressed in flow-charts or programs, forms designed specifically to represent analyses of information-processing capabilities generally.

Perhaps the most extensive use of the analytical strategy in science occurs in psychology, for a large part of the psychologist's job is to explain how the complex behavioral capacities of organisms are acquired and how they are exercised. Both goals are greatly facilitated by analysis of the capacities in question, for then acquisition of the analyzed capacity resolves itself into acquisition of the analyzing capacities and the requisite organization, and the problem of performance resolves itself into the problem of how the analyzing capacities are exercised. This sort of strategy has dominated psychology ever since Watson attempted to explain such complex capacities as the ability to run a maze by analyzing the performance into a series of conditioned responses, the stimulus for each response being the previous response, or something encountered as the result of the previous response.[26] Acquisition of the complex capacity is resolved into a number of distinct cases of simple conditioning—i.e., the ability to learn the maze is resolved into the capacity for stimulus substitution, and the capacity to run the maze is resolved into abilities to respond in certain simple ways to simple stimuli. Watson's analysis proved to be of limited value, but the analytic strategy remains the dominant mode of explanation in behavioral psychology.[27]

3. Functions and Functional Analysis

In the context of an application of the analytical strategy, exercise of an analyzing capacity emerges as a function: it will be appropriate to say that x functions as a ϕ in s, or that the function of x in s is ϕ-ing, when we are speaking against the background of an analytical explanation of some capacity of s which appeals to the fact that x has a capacity to ϕ in s. It is appropriate to say that the heart functions as a pump against the background of an analysis of the circulatory system's capacity to transport food, oxygen, wastes, and so on, which appeals to the fact that the heart is capable of pumping. Since this is the usual background, it goes without saying, and this accounts for the fact that "The heart functions as a pump" sounds right, and "The heart functions as a noise-maker" sounds wrong, in some context-free sense. This effect is strengthened by the absence of any actual application of the analytical strategy which makes use of the fact that the heart makes noise.[28]

We can capture this implicit dependence on an analytical context by entering an explicit relativization in our regimented reconstruction of function-ascribing statements.

(9) x functions as a ϕ in s (or: the function of x in s is to ϕ) relative to an analytical account A of s's capacity to ψ just in case x is capable of ϕ-ing in s and A appropriately and adequately accounts for s's capacity to ψ by, in part, appealing to the capacity of x to ϕ in s.

Sometimes we explain a capacity of s by analyzing it into other capacities of s, as when we explain how someone ignorant of cookery is able to bake cakes by pointing out that he/she followed a recipe each instruction of which requires no special capacities for its execution. Here we don't speak of, e.g., stirring as a function of the cook, but rather of the function of stirring. Since stirring has different functions in different recipes, and at different points in the same recipe, a statement like "The function of stirring the mixture is to keep it from burning to the bottom of the pan" is implicitly relativized to a certain (perhaps somewhat vague) recipe.

To take account of this sort of case, we need a slightly different schema: where e is an activity or behavior of a system s (as a whole), the function of e in s is to ϕ relative to an analytical account A of s's capacity to ψ just in case A appropriately and adequately accounts for s's capacity to ψ by, in part, appealing to s's capacity to engage in e.

(9) explains the intuition behind the regress-ridden (6): functional ascriptions do require relativization to a "functional fact" about a containing system—i.e., to the fact that a certain capacity of a containing system is appropriately explained by appeal to a certain functional analysis. And, like (6), (9) makes no provision for speaking of the function of an organism except against a background analysis of a containing system (the hive, the corporation, the eco-system). Once we see that functions are appealed to in explaining the capacities of containing systems, and indeed that it is the applicability of a certain strategy for explaining these capacities that makes talk of functions appropriate, we see immediately why we do not speak of the functions of uncontained containers. What (6) fails to capture is the fact that uncontained containers can be functionally analyzed, and the way in which function-analytical explanation mediates the connection between functional ascriptions (x functions as a ϕ, the function of x is to ϕ) and the capacities of the containers.

4. Function-analytical explanation

If the account I have been sketching is to draw any distinctions, the availability and appropriateness of analytical explanations must be a nontrivial matter.[29] So let us examine an obviously trivial application of the analytical strategy with an eye to determining whether it can be dismissed on principled grounds.

(10) Each part of the mammalian circulatory system makes its own distinctive sound, and makes it continuously. These sounds combine to form the "circulatory noise" characteristic of all mammals. The mammalian circulatory system is capable of producing this sound at various volumes and various tempos. The heartbeat is responsible for the throbbing character of the sound, and it is the capacity of the heart to beat at various rates that explains the capacity of the circulatory system to produce a variously tempoed sound.

Everything in (10) is, presumably, true. The question is whether it allows us to say that the function of the heart is to produce a variously tempoed throbbing sound.[30] To answer this question we must, I think, get clear about the motivation for applying the analytical strategy. For my contention will be that the analytical strategy is most significantly applied in cases very unlike that envisaged in (10).

The explanatory interest of an analytical account is roughly proportional to (i) the extent to which the analyzing capacities are less sophisticated than the analyzed capacities, (ii) the extent to which the analyzing capacities are different in type from the analyzed capacities, and (iii) the relative sophistication of the program appealed to, i.e., the relative complexity of the organization of component parts/processes which is attributed to the system. (iii) is correlative with (i) and (ii): the greater the gap in sophistication and type between analyzing capacities and analyzed capacities, the more sophisticated the program must be to close the gap.

It is precisely the width of these gaps which, for instance, makes automata theory so interesting in its application to psychology. Automata theory supplies us with extremely powerful techniques for constructing diverse analyses of very sophisticated tasks into very unsophisticated tasks. This allows us to see how, in principle, a mechanism such as the brain, consisting of physiologically unsophisticated components (relatively speaking), can acquire very sophisticated capacities. It is the prospect of promoting the capacity to store ones and zeros into the capacity to solve problems of logic and recognize patterns that makes the analytical strategy so appealing in cognitive psychology.

As the program absorbs more and more of the explanatory burden, the physical facts underlying the analyzing capacities become less and less special to the analyzed system. This is why it is plausible to suppose that the capacity of a person and a machine to solve a certain problem might have substantially the same explanation, while it is not plausible to suppose that the capacities of a synthesizer and a bell to make similar sounds have substantially similar explanations. There is no work for a sophisticated hypothesis about the organization of various capacities to do in the case of the bell. Conversely, the less weight borne by the program, the less point to analysis. At this end of the scale we have cases like (10) in which the analyzed and analyzing capacities differ little if at all in type and sophistication. Here we could apply the instantiation strategy without significant loss, and thus talk of functions is comparatively strained and pointless. It must be admitted, however, that there is no black-white distinction here, but a case of more-or-less. As the role of organization becomes less and less significant, the analytical strategy becomes less and less appropriate, and talk of functions makes less and less sense. This may be philosophically disappointing, but there is no help for it.

CONCLUSION

Almost without exception, philosophical accounts of function-ascribing statements and of functional explanation have been crippled by adoption of assumptions (A) and (B). Though there has been widespread agreement that extant accounts are not satisfactory, (A) and (B) have escaped critical scrutiny, perhaps because they were thought of as somehow setting the problem rather than as part of proffered solutions. Once the problem is properly diagnosed, however, it becomes possible to give a more satisfactory and more illuminating account in terms of the explanatory strategy which provides the motivation and forms the context of function-ascribing statements. To ascribe a function to something is to ascribe a capacity to it which is singled out by its role in an analysis of some capacity of a containing system. When a capacity of a containing system is appropriately explained by analyzing it into a number of other capacities whose programmed exercise yields a manifestation of the analyzed capacity, the analyzing capacities emerge as functions. Since the appropriateness of this sort of explanatory strategy is a matter of degree, so is the appropriateness of function-ascribing statements.

NOTES

1. Cf. Carl Hempel, The logic of functional analysis, in *Aspects of Scientific Explanation,* New York, Free Press, 1965, reprinted from Llewellyn Gross, ed., *Symposium on Sociological Theory,* New York, Harper and Row, 1959; and Ernest Nagel, *The Structure of Science,* New York, Harcourt, Brace and World, 1961, chapter 12, section I. The assumptions, of course, predate Hempel's 1959 essay. See, for instance, Richard Braithwaite, *Scientific Explanation,* Cambridge, Cambridge University Press, 1955, chapter X; and Israel Scheffler, Thoughts on teleology, *British Journal for the Philosophy of Science,* 11 (1958). More recent examples include Francisco Ayala, Teleological explanations in evolutionary biology, *Philosophy of Science,* 37 (1970); Hugh Lehman, Functional explanations in biology, *Philosophy of Science,* 32 (1965); Richard Sorabji, Function, *Philosophical Quarterly,* 14 (1964); and Larry Wright, Functions, *Philosophical Review,* 82 (1973).

2. Hempel, p. 310.

3. Nagel, p. 403.

4. Ibid., p. 404.

5. It might be objected here that although it is the function of the kidneys to eliminate waste, that is not the function of a particular kidney unless operation of that kidney *is* necessary for removal of wastes. But suppose scientists had initially been aware of the existence of the left kidney only. Then, on the account being considered, anything they had said about the function of that organ would have been false, since, on that account, *it has no function in organisms having two kidneys!*

6. Even in the case of a designed artifact, it is at most the designer's *belief* that *x* will perform *f* in *s* which is causally relevant to *x*'s presence in *s*, not *x*'s actually performing *f* in *s*. The nearest I can come to describing a situation in which *x* performing *f* in *s* is causally relevant to *x*'s presence in *s* is this: the designer of *s* notices a thing like *x* performing *f* in a system like *s*, and this leads to belief that *x* will perform *f* in *s*, and this in turn leads the designer to put *x* in *s*.

7. *Is* casting a shadow the function of this fragment? Standard use may confer a function on something: if I standardly use a certain stone to sharpen knives, then that is its function, or if I standardly use a certain block of wood as a door stop, then the function of that block is to hold my door open. If non-artifacts *ever* have functions, appeals to those functions cannot explain their presence. The things functionally characterized in science are typically not artifacts.

8. A confused perception of this fact no doubt underlies (B), but the fact that (B) is nearly inseparable from (A) in the literature shows how confused this perception is.

9. Notice that the second use is parasitic on the first. It is only because the neurofibrils explain the coordinated activity of the cilia that we can assign a survival value to neurofibrils: the survival value of a structure *s* hangs on what capacities of the organism, if any, are explicable by appeal to the functioning of *s*.

10. In addition to the misunderstanding about evolutionary theory just discussed, biological examples have probably suggested (A) because biology was the *locus classicus* of teleological explanation. This has perhaps encouraged a confusion between the teleological *form* of explanation, incorporated in (A), with the explanatory role of functional ascriptions. Function-ascribing statements do occur in explanations having a teleological form, and when they do, their interest is vitiated by the incoherence of that form of explanation. It is the legitimate use of function-ascribing statements that needs examination, i.e., their contribution to non-teleological theories such as the theory of evolution.

11. Larry Wright (op. cit.) is aware of this problem but does not, to my mind, make much progress with it. Wright's analysis rules out "The function of the heart is to produce heartsounds," on the ground that the heart is not there because it produces heartsounds. I agree. But neither is it there because it pumps blood. Or if, as Wright maintains, there is a sense of "because" in which the heart *is* there because it pumps blood and not because it produces heartsounds, then this sense of "because" is as much in need of analysis as "function." Wright does not attempt to provide such an analysis, but depends on the fact that, in many cases, we are able to use the word in the required way. But we are also able to use "function" correctly in a variety of cases. Indeed, if Wright is right, the words are simply interchangeable with a little grammatical maneuvering. The problem is to make the conditions of correct use explicit. Failure to do this means that Wright's analysis provides no insight into the problem of how functional theories are confirmed, or whence they derive their explanatory force.

12. Surprisingly, when Nagel comes to formulate his general schema of functional attribution, he simply ignores this problem and thus leaves himself open to the trivialization just suggested. Cf. Nagel, p. 403.

13. Hempel, p. 306.

14. Ibid., p. 310.

15. Hugh Lehman (op. cit.) gives an analysis that appears to be essentially like (6).

16. Even these applications have their problems. Frankfurt and Poole, Functional explanations in biology, *British Journal for the Philosophy of Science,* 17 (1966), point out that heartsounds contribute to health and survival via their usefulness in diagnosis.

17. Michael Ruse has argued for a formulation like (8). See his Function statements in biology, *Philosophy of Science,* 38 (1971), and *The Philosophy of Biology,* London, Hutchinson, 1973.

18. Perhaps, in the absence of serious predators, with a readily available food supply, and with no need to migrate, flying simply wastes energy.

19. Throughout this section I am discounting appeals to the intentions of designers or users. *x* may be intended to prevent accidents without actually being capable of doing so. With reference to this intention, it *would* be proper in certain contexts to say, "*x*'s function is to prevent accidents, though it is not actually capable of doing so."

There can be no doubt that a thing's function is often identified with what it is typically or "standardly" used to do, or with what it was designed to do. But the sorts of things for which it is an important scientific problem to provide functional analyses—brains, organisms, societies, social institutions—either do not have designers or standard or regular uses at all, or it would be inappropriate to appeal to these in constructing and defending a scientific theory because the designer or use is not known—brains, devices dug up by archaeologists—or because there is some likelihood that real and intended functions diverge—social institutions, complex computers. Functional talk may have originated in contexts in which reference to intentions and purposes loomed large, but reference to intentions and purposes does not figure at all in the sort of functional analysis favored by contemporary natural scientists.

20. For a detailed discussion of the two explanatory strategies sketched here, see Cummins, *The Nature of Psychological Explanation*, Bradford Books/M.I.T. Press, Cambridge, 1983. In the original version of this paper, I called the two strategies the Subsumption Strategy and the Analytical Strategy. I have retained the latter term, but the former I have replaced. What I was calling the subsumption strategy in 1975 was simply a confusion; a conflation of causal subsumption of events, and the nomic derivation of a property via the facts of its instantiation. Since functions are dispositions and dispositions are properties, only the latter is relevant here.

21. Brian O'Shaughnessy, The powerlessness of dispositions, *Analysis*, October (1970). See also my discussion of this example in Dispositions, states and causes, *Analysis*, June (1974).

22. Also, we must explain why submerging a free elevant object causes it to rise, and why a free submerged object's becoming elevant causes it to rise. One of the convenient features of elevancy is that the same considerations dispose of all these cases. This does not hold generally: gentle rubbing, a sharp blow, or a sudden change in temperature may each cause a glass to manifest a disposition to shatter, but the explanations in each case are significantly different.

23. By "programmed" I simply mean organized in a way that could be specified in a program or flow chart: each instruction (box) specifies manifestation of one of the d_i such that if the program is executed (the chart followed), *a* manifests *d*.

24. Some might want to distinguish between dispositions and capacities, and argue that to ascribe a function to *x* is in part to ascribe a *capacity* to *x*, not a disposition as I have claimed. Certainly (1) is strained in a way (2) is not.

(1) Hearts are disposed to pump.
Hearts have a disposition to pump.
Sugar is capable of dissolving.
Sugar has a capacity to dissolve.

(2) Hearts are capable of pumping.
Hearts have a capacity to pump.
Sugar is disposed to dissolve.
Sugar has a disposition to dissolve.

25. Indeed, what makes something part of, e.g., the nervous system is that its capacities figure in an analysis of the capacity to respond to external stimuli, coordinate movement, etc. Thus there is no question that the glial cells are part of the brain, but there is some question whether they are part of the nervous system or merely auxiliary to it.

26. John B. Watson, *Behaviorism*, New York, W. W. Norton, 1930, chapters IX and XI.

27. Writers on the philosophy of psychology, especially Jerry Fodor, have grasped the connection between functional characterization and the analytical strategy in psychological theorizing but have not applied the lesson to the problem of functional explanation generally. The clearest statement occurs in J. A. Fodor, The appeal to tacit knowledge in psychological explanation, *Journal of Philosophy*, 65 (1968), 627–640.

28. It is sometimes suggested that heart-sounds do have a psychological function. In the context of an analysis of a psychological disposition appealing to the heart's noise-making capacity, "The heart functions as a noise-maker" (e.g., as a producer of regular thumps) would not even *sound* odd.

29. Of course, it might be that only arbitrary distinctions are to be drawn. Perhaps (9) describes usage, and usage is arbitrary, but I am unable to take this possibility seriously.

30. The issue is not whether (10) forces us, via (9), to say something false. Relative to *some* ana-

lytical explanation, it may be true that the function of the heart is to produce a variously tempoed throbbing. But the availability of (10) should not support such a claim.

B. A. Brody

THE REDUCTION OF TELEOLOGICAL SCIENCES

Consider the scientist whose ideology (scientific or otherwise) commits him to the view that all phenomena can be described and explained by use of the terms and theories of a suitably amplified science of physics.[1] What are the options open to him when he faces a phenomenon whose current description and/or explanation involves other terms and other theories?

One option is to subsume the current description and/or explanation under some physical descriptions and explanations. This involves the familiar process of connecting the non-physical terms (or their referents) with some physical terms (or their referents) and then deducing the non-physical theories (or something very close to them) from some physical theories. The other option is to deny the legitimacy of the current description and/or explanation. I take it that it is the former option which is chosen when some theory is reduced to a physical theory. There are those, however, who suppose that it is the latter option which is chosen in such cases. But what I have to say can be easily modified so as to fit in with their views, and so I need not consider that issue now.

What I would like to consider is a particular case of the latter type of move. In biology, psychology, and sociology, one finds functional (or teleological) explanations, explanations of some recurrent structure, activity, or behavior pattern in terms of its contribution to the preservation and/or development of some individual or group. The description of the phenomenon in question is, of course, offered in non-physical terms. But even if it could be reformulated in physical terms, the resulting explanations would still pose a problem for our scientist. After all, their logical structure seems to differ from the logical structure of explanations in physics, so it is just not clear how they could be subsumed under some physical explanations.

At this point, our scientist must feel the temptation of the second move. Naturally, if he can come up with an alternative satisfactory description and explanation of the phenomenon, he is done. But even if he can't he may still want to say that these explanations pose no problem for him because, on purely logical grounds, they can be shown to have no explanatory import. Of course, he cannot defend that claim *merely* by saying that they aren't of the same logical type as explanations in physics, for that would be to beg the question. But perhaps there are other arguments that he can offer, arguments already offered by C. G. Hempel in his "The Logic of Functional Analysis."[2]

In this paper, I shall try to show that Hempel's arguments fail. I shall also try to show, however, that teleological and functional explanations really pose no problem

Reprinted by permission from *American Philosophical Quarterly*, Vol. 12, no. 1, January 1975.

for our scientist because, on a proper analysis of their structure, they can be subsumed under physical explanations.

I

Hempel sets out the formal structure of functional explanations as follows:

(a) at time *t*, system *s* functions adequately in a setting of kind *c*
(b) *s* functions adequately in a setting of kind *c* only if a certain necessary condition *n* is satisfied
(c) if trait *i* were present in *s* then, as an effect, *n* would be satisfied
(d) therefore, at *t*, *i* is present in *s*.

And he sums up his argument against the explanatory import of this explanation as follows:

> . . . the information typically provided by a functional analysis of an item affords neither deductively nor inductively adequate grounds for expecting *i* rather than one of its alternatives. (P. 313.)

I want to note, in passing, that even if Hempel's account of the structure of functional explanations is essentially correct, he seems to be making its claim too strong. Consider, for example, the case of antelope horns; they function as a means of offense in struggles within the group as a means of defense against attacks by other animals. But it would be wrong to say that an antelope, in its normal environment, functions adequately only if it has horns. At best, one can say that the probability of its functioning adequately in that environment is much higher if it has these means of offense and defense. Put more formally, step (b) should be replaced with

(b) *s* has a higher probability of functioning adequately in a setting of kind *c* if a certain condition *n* is satisfied.

But as this point will not affect the argument, we can disregard it for now. We shall return, later on, to more important shortcomings in Hempel's analysis of the structure of functional explanations.

Granting Hempel's claim that the explanans in such an explanation does not offer grounds for expecting the presence of *i*, and certainly not for expecting *i* rather than one of its functional equivalents, *j*, *k*, etc., why should we suppose that this means that it has no explanatory import? Now, we should suppose this if we grant the assumption that an explanan has explanatory import only if it offers grounds for expecting that the explanandum is true. But why should we suppose that that assumption is true? Perhaps explanations in physics have that character, but, as we saw above, that is not the type of argument that our scientist can appeal to.

It should be noted that this assumption is one that Hempel makes in many places. Indeed, it seems to be one of the fundamental assumptions upon which the whole covering-law model for explanations rest. So it would be worth our while to look at it quite carefully.

What arguments does Hempel actually offer for this assumption? Strangely enough, despite its prominence in his thought, he never really argues for it. Indeed, he usually merely announces it.[3] But we can find in his articles several hints of possible arguments that someone might offer for this assumption.

At one point, in talking about a satisfactory explanation, he says:[4]

> The explanatory import of the whole argument lies in showing that the outcome described in the explanandum was to be expected in view of the antecedent circumstances and the several laws listed in the explanans.

This claim that the explanatory import lies in showing that the outcome was to be expected suggests the following argument: the explanans (in any scientific explanation) must offer grounds for expecting that the explanandum is true because explaining something *just is* offering grounds for expecting it to occur. Given that this is the nature of explanation, Hempel's assumption follows trivially.

The trouble with this argument is that there are, as Hempel himself recognizes elsewhere,[5] good reasons for rejecting this account of explanatory import, good reasons for rejecting the view that to explain something just is to offer good reasons for expecting that the phenomenon in question will occur. If this were the nature of explanation, then one could explain the rainfall by reference to the falling of the barometer and one could explain the height of the flagpole in light of the length of the shadow it casts and the position of the sun in the sky at the time that it casts the shadow. To be sure, it does not follow from the fact that explaining something is not identical with offering good reasons for expecting it that Hempel's assumption is wrong and that we can explain without offering good reasons for expecting that the explanandum is true. But it does follow that one cannot argue for Hempel's assumption from that view of the nature of explanation.

At one point, in explaining why Sizi's explanation of why Jupiter cannot have any moons would be a bad explanation even if Jupiter had no moons (Sizi's explanation was that Jupiter's having them would be incompatible with there being only seven openings in the head, etc.), Hempel says[6] that:

> The crucial defect of this argument is evident: the "facts" it adduces, even if accepted without question, are entirely irrelevant to the point at issue; they do not afford the slightest reasons for the assumption that Jupiter has no satelites . . . physical explanations meet the requirement of explanatory relevance: the explanatory information adduced affords good ground for believing that the phenomenon to be explained did, or does, indeed occur.

This suggests a second argument: it is clear that the relation of explanatory relevance must hold between the explanans and the explanandum. But it will only if the explanans provides good reasons for expecting that the explanandum is true, so Hempel's assumption is correct.

Clearly, however, this argument will also not do. While we can certainly agree that the explanans must be explanatorily relevant to the explanandum, what reasons does Hempel give us for supposing that this can happen only if the explanans provides us with good reasons for expecting that the explanandum is true? Perhaps this is true for physical explanations, perhaps it is not. We shall discuss that issue below. But what reason is there to suppose that this is true for explanations in general? There isn't even a hint of an argument for this additional claim.

In short, then, we have found no reason for agreeing with Hempel's assumption that the explanans must provide us with good reasons for expecting that the explanandum is true. But aren't we missing the crucial point? Isn't the crucial objection based upon the fact that the explanans provide us with no better reasons for expecting *i* rather than *j* or *k* or any of its other functional equivalents? This point can also be put as follows: in such cases of explanation, as in all other cases of explanation, we are concerned with understanding why the explanandum is true rather than some alternative to it. And the explanans, if it is to explain the explanandum, must differentiate between the explanandum and its alternatives. Now it clearly does that when it gives us good reasons for expecting that the explanandum is true (and that its alternatives are therefore false). But in a case like ours, where we could equally well substitute '*j*', '*k*', etc., for '*i*' in steps (c) and (d), the explanandum is not properly differentiated from its alternatives, and the explanation is not satisfactory. In short, then, isn't Hempel's objection really based upon this point about the existence of functional alternatives not ruled out by the explanans?

Perhaps so. But it seems to me that this argument against functional explanations is no better than the previous one. It too rests upon a dubious assumption, in this case, that the explanans must differentiate between the explanandum and its alternatives. As has been shown in recent work on statistical explanation,[7] this condition is not even satisfied by all explanations in the physical sciences. There are statistical explanations where the same explanans could be used to explain all the alternative explanandums, so it can hardly differentiate between an explanandum and its alternatives. Thus, to give but one example, one can explain the die coming up upon one by saying that it was a fair die that was tossed in an unbiased fashion and that the probability of its coming up upon one (or upon any of the heads) is therefore 1/6. But this same explanans could also explain the die coming up on two. Hempel would, of course, not accept this as an explanation because the explanans does not provide a high enough degree of probability for the explanandum, and therefore does not provide us with good reasons for expecting that the explanandum is true. But his rejection of such examples just shows how mistaken it is to work with his requirements on explanations.

It should be noted[8] that there are some cases of statistical explanation where the explanans does provide a high enough degree of probability for the explanandum, so Hempel's requirements laid down in his inductive-statistical model are satisfied, but does not differentiate between the explanandum and some of its alternatives. Thus, one can explain, even according to Hempel, the die coming upon one in 164 out of 496 throws. But the same explanans would also explain its coming upon one in 168 out of 996 throws by reference to the fact that it was a fair die tossed in an unbiased fashion; such a die has, after all, a reasonably high probability of coming upon one in 168 out of 996 throws. So it doesn't even follow from the fact that an explanation meets all of Hempel's requirements for statistical explanations that it meets the requirement that an explanans must differentiate between the explanandum and its alternatives.

I conclude, therefore, that Hempel has provided our scientist with no reasons that are independent of his ideology for rejecting functional explanations.

II

We shall now consider the possibility of our scientist claiming that functional explanations really do have the same logical structure as explanations in the physical sciences and that they therefore pose no logical problem for the program of subsuming all explanations under explanations from the physical sciences. I hope to show that this claim is correct.

Let us begin by looking at a biological example of a functional explanation, the explanation of the forward curving horn of the reedbuck antelope on the grounds that it provides it with an offensive and defensive weapon. Now such explanations occur in the context of evolutionary biology, and it is in that context that we shall examine them.[9] To make things easier, we shall assume for now that a significant evolutionary change, like the emergence of these horns, could take place instantaneously. We shall see later on that our account can be modified in a straightforward way to take into account the fact that it is far more likely that the emergence of such a significant trait is really the culmination of a large number of smaller evolutionary changes.

An evolutionary biologist would, keeping in mind the above-mentioned idealization, offer the following account: the first occurrence of a reedbuck antelope with such a horn is due to the randomly acting force of genetic mutations. It cannot be explained by reference to the advantages for the reedbuck in its environment for having these horns. To do this would involve illegitimate assumption about the "direction" and "purpose" of evolution. But once there are reedbucks with such horns, the fact that they have this function, the fact that they provide it with these offensive and defensive weapons, means that the reedbucks with these horns will probably produce a larger percentage of the next generation which will also do the same, until all reedbuck antelope have these horns. Moreover, unless some better-adapted competitor comes along, or unless the environment changes radically, it also means that there will probably continue to be reedbucks with these forward curving horns. To be sure, the same thing would have occured if the mutation had produced reedbucks with backward curving horns like the horns of the roan antelopes. But it didn't, so we don't have to consider that functional alternative.

The first thing to note is that there really are two different things being explained here, and that they are being explained in very different ways. The first thing being explained is the initial occurrence of the forward curving horns. It is explained as a result of the process of genetic mutation, a presumably random process.[10] This is a form of statistical explanation. To be sure, it is not a type of explanation that fits Hempel's model for statistical explanations, for the explanans (the set of laws governing the process of mutation together with the description of the forces that produced the mutation in this case) does not give a high degree of probability to the occurrence of just this mutation. But we have already called attention to the fact that there are even statistical explanations in physics that do no better. In short, then, the first part of this evolutionary explanatory account involves no functional elements, has the same logical structure as certain statistical explanations in physics, and poses no logical problem for our scientist's reductionist program.

The second thing being explained in this account is the proliferation of reedbucks with these forward curving horns and the persistence of this trait. It is, of course, only this part of the explanatory account that involves the function of the horns, and it is therefore this part of the explanatory account on which we must focus our attention.

Note that an important part of the explanans of this second phenomenon is derived from the first part of the explanatory account, viz., that genetic mutations have produced reedbuck antelopes with these types of horns but not with other types of horns and not with other types of weapons. This is, of course, the crucial point, because it helps us understand why this type of horn (rather than some functional equivalent) persisted in the reedbuck. Crudely speaking, it is the only one that persisted because it was the only one around.

There is a second point that should be noted. In the biologist's account, the main emphasis is on the contribution of the horns to differential reproduction rates within the species, and not, as in Hempel's formulation, on their contribution to the survival and flourishing of the particular organism.[11] In setting out our schematization of functional explanations in biology, we shall want to keep this point in mind.

I think that we are now in a position to see how this second functionalist part of the explanatory account can fit into the reductionist program that we are considering. After all, putting our two points together, we can see that the following is the proper schematization of functional explanations in biology:

(a*) At time t, species s is present in a setting of kind c.[12]

(b*) A member of species s in a setting of kind c will probably have a higher rate of reproduction if need n is satisfied than if it is not.

(c*) If trait i is present in a member of species s in a setting of kind c, then need n will be satisfied.

(d*) At some time sufficiently before t, mutations had produced members of species s with trait i, but at no time before or since then had they produced members of species s with some functional alternative.

Now the above explanans will, of course, have to be supplemented with the statistical laws about the ways in which adaptative traits spread and persist in populations, but it is clear that, when so supplemented, it provides us with a statistical explanation of the explanandum. Interestingly enough, the resulting statistical explanation, unlike the explanation of the first emergence of the trait, does satisfy all of the requirements of Hempel's model for statistical explanations.

In short, then, a proper analysis of the structure of functional explanations in their setting in evolutionary biology reveals that: (a) there are two phenomena to be explained, the initial occurrence of the trait and its domination of and persistence in the population; (b) the explanation of the first phenomenon, which involves no functional elements, is a type of statistical explanation found in the physical sciences, although not recognized in Hempel's model; (c) the explanation of the second phenomenon involves both functional elements and the fact that the trait has already appeared in the species, and, when properly analyzed, turns out to be of the ordinary statistical form; (d) in any case, neither of the explanations involve any special logical structure, and there is no logical reason why they cannot be subsumed under the laws of physics as part of some reductionist program.

There is one complication that we still have to attend to. It is highly unlikely that the evolution of the reedbuck horn was an instantaneous process, that the relevant mutation was from a reedbuck with no horn at all to one with the horn that we now know. There were undoubtedly intermediary steps in this process. How, if at all, does this affect the explanatory schema that we outlined above? And does this change any of our results about the reduction of functionalist explanations?

This complication seems to come to the following: given that the trait in question probably developed out of a more rudimentary version of it, it is probably the case that (d*) is false and that need *n* will have been satisfied, albeit in a less satisfactory fashion, by a more rudimentary version of that trait. But it is pretty clear how we can modify our account to deal with this complication. After all, it is still the case that the presence of trait *i* is more adaptive than the presence of its more rudimentary version, that it fulfills the need *n* to a greater degree. So our analysis should be modified as follows:

(a*) At time *t*, species *s* is present in a setting of kind *c*.

(b**) Members of species *s* in a setting of kind *c* will probably have rates of reproduction that are greater as need *n* is satisfied to a greater degree.

(c**) If trait *i* is present in a number of species *s* in a setting of kind *c*, then *n* will be satisfied to a degree *d*.

(d**) At some time sufficiently before *t*, mutations had produced members of species *s* with trait *i*, but at no time before or since have they produced members of species *s* with a functional alternative to *i* that satisfied *n* to a degree $\geqslant d$.

(e*) Therefore, the members of species *s* at time *t* have trait *i*.

In this analysis, as in the one it replaces, we find functional explanations in biology meeting all of Hempel's requirements for a statistical explanation. So their logical form poses no problem for our scientist's reductionist program.

III

Functional explanations are found in other sciences besides biology. Indeed, they are especially conspicuous in social anthropology, where the functionalist movement, the movement that concerns itself with providing functional explanations for societal phenomena, has been of great importance since the times of Malinowski and Radcliffe-Brown. They are also quite prevalent in sociology, and several important sociological theorists, especially Merton and Parsons, have written about their logic extensively. In this final section, we shall see whether the results of our analysis of functional explanations in biology can be used to shed light upon the nature of functional explanations in the social sciences.

The crux of our point about biology was that the biologist had two things to explain, the origin of the trait in the species and the spread of the trait through, and its persistence in, the species. Moreover, the functional explanation was only offered as an explanation of the latter phenomenon. Now it is obvious that the two analogous explanations in the social sciences are equally different and that the explanation of the origins of a given social structure in a given society is a different explanation from the explanation of its persistence in that society. The question that we must consider is whether the functional explanations in the social sciences are only offered as an explanation of the persistence of the social structures in question.

It is naturally difficult to be sure exactly what all the functionalists had in mind. But if one examines the writings of the functionalist who most addressed himself to this issue, Radcliffe-Brown, one rapidly sees that he would have agreed with this limitation on the scope of functional explanations; one sees that he certainly did not think that functional explanations explained the origins of anything.

In all of his methodological essays,[13] Radcliffe-Brown dealt extensively with the relation between historical investigations into the origins of social structures and functional accounts of these same structures. He was, to be sure, very suspicious of the role of the former in social anthropology because he felt that the resulting hypotheses would be, given the lack of records, far too conjectural. Thus, in discussing Frazer's theory of the origin of totemism, he said:[14]

> The methodological objection to this theory, and to all theories of the same type, is that there seems no possible way of verifying them . . . We are unable, by any means I can imagine, to prove that this is the way in which it actually did arise.

Nevertheless, he did not view functional analysis as offering an alternative explanation of the origins of the institution in question. Thus, after presenting his own account of the function of totemism, he wrote:[15]

> . . . it is possible to have a theory of totemism which, if substantiated, will help us to understand not only totemism but also many other things, without committing oneself to any hypothesis as to the historical origin or origins of totemism.

This theme is repeated again and again in all of his writings. Indeed, while he recognized that Durkhein had made fundamental contributions to the functionalist understanding of totemism, he criticized him for failing to understand that this was not also a contribution to understanding the origins of totemism. In light of this, we can see how Radcliffe-Brown was able to maintain that totemism served the same function in the different societies in which it was found while suspecting that different historical forces were responsible for the origin of totemism in different societies. In short, then, Radcliffe-Brown would certainly agree that functional explanations in social anthropology, like their counterparts in evolutionary biology, do not explain the origins of anything.

What then do functional explanations explain? Radcliffe-Brown was, strangely enough, very unclear on this point. On some occasion, he seems to suggest that they aren't intended as explanations at all, that they are only intended as fuller and deeper descriptions of the phenomenon in question. There is one remark of his that is, however, important for us in this context. In contrasting his approach to totemism with Frazer's he argued against Frazer as follows:[16]

> Moreover, the theory, and others like it, even if it explains how totemism at one time came into existence, does not explain how it succeeds in

continuing its existence. And that is a problem quite as important as the problem of origin.

So while Radcliffe-Brown may have held that functional explanations explain other things as well, he certainly seems to have held the view that they should explain the perseverance of social structures and institutions in a given society.

One finds a similar idea expressed by Merton. In writing about the importance of finding latent functions for social items, he says:[17]

> Operating with the concept of latent function, we are not too quick to conclude that if an activity of a group does not achieve its normal purpose, then its persistence can be described only as an instance of "inertia," "survival," or "manipulation by powerful subgroups in society."

This idea that we can use functional explanations to explain the survival of social items that are not fulfilling their manifest goal is, as Merton points out, very important in the functionalist tradition. After all, the early functionalists were especially concerned with attacking the views of people like Rivers who saw such a survival as being only "intelligible through its past history." The functionalists saw its survival as being due to the way it fulfilled its latent function.

Merton, it should be noted, sees such explanations as serving to account for the persistence of social items even when they are under explicit challenge. Thus, after offering a functional analysis of the political machine, he says:[18]

> It helps to explain why the periodic efforts at "political reform," "turning the rascals out," and "cleaning political house" are typically (though not necessarily) short-lived and ineffectual . . . unless the reform also involves a "re-forming" of the social and political structure such that the existing needs are satisfied by alternative structures or unless it involves a change which eliminates these needs altogether, the political machine will return to its integral place in the social scheme of things.

On the basis of this evidence (and a lot more of the same type that can be offered), we shall adopt the view that the primary thrust of functional explanations in the social sciences, as in biology, is toward explaining persistence rather than origins. Nevertheless, it would be mistaken to suppose that we can therefore also conclude that our model for functional analysis in biology can be carried over completely intact to the case of functional analysis in the social sciences. There are several crucial differences.

To begin with, the explanation of the origins will be different in the social sciences. In biology, the origin of the trait whose persistence is to be explained functionally is due to the randomly operating force of mutations, and the explanation of the origin is a statistical explanation. This is, of course, not the case in the social sciences. There, the origin of the social item whose persistence is to be explained functionally has to be explained historically. Moreover, while the function of the trait is, in the case of biology, irrelevant to the explanation of its origin, the function of the social item may be relevant to the explanation of its origin. After all, the function may be (or have been) a manifest function, and the originators of the social item may have brought it into existence precisely so that there will be something that fulfills that function. But in that case, the explanation of the origins will be a purposive explanation (one that involves the beliefs and motives of a variety of agents) and not a functional analysis of the type that we are concerned with.

To give an example, consider once more the political machine whose continued existence is due to the fact that it fulfills certain vital functions (e.g., to centralize political power so as to bring about the satisfaction of the needs of diverse sub-groups that would not otherwise be satisfied). Now, in the case of an ordinary political machine, its origins will have to be explained historically in terms of the ways in which its founders managed to build up power. But if some political reformers, in the course of their reform, create a functional alternative to the machine, a functional alternative whose continued existence would then be explained functionally, its origins would be explained in terms of the desires of these reformers to have a social structure that satisfies the function in question.

All of this means that our scientist, if he is to fully carry through his reductionist program, must be able to handle historical and purposive explanations as well as functional explanations. Whether their logical structure will pose a problem for him is, however, something that lies beyond the scope of this paper.

More importantly, there are differences between the logical structure of functional explanations of persistence in biology and functional explanations of persistence in the social sciences. These are due primarily to the mechanisms responsible for the persistence in these two cases. In biology, the mechanism is that of differential reproduction rates. Organisms with the trait in question have, on the whole, more descendants—who also have the trait—than members of the species that do not have that trait. Therefore, even if members of the species are born without the trait in question, the trait continues to persist in most members of the species. None of this is, of course, applicable to the cases in the social sciences. Indeed, the mechanism of persistence remains unclear at this point. This leads to a difference in the logical structure of the explanations since that is partially determined by the mechanism operating.

What then will be our schematism for functional explanations in the social sciences? The following seems appropriate:

(a#) At time *t*, society *s* is in a setting of kind *c*.

(b#) *s* has a much higher probability of functioning adequately in a setting of kind *c* if a certain condition *n* is satisfied.

(c#) If structure *i* is present in *s* then, as a result, *n* would be satisfied.

(d#) At some time before *t*, structure *i*, but none of its functional equivalents, had come into existence in *s*.

(e#) Therefore, structure *i* probably continues to be present in society *s* at time *t*.

As in the case of biology, the explanans needs supplementation. But there we knew which laws to add, viz., the laws about the ways in which adaptive traits spread in populations. Here, we are not so fortunate. We need some laws about the ways in which structures that meet conditions whose being met increase the probability of a society's functioning adequately probably persist in that society if no functional equivalents arise to replace them. And while functional analysis presupposes that there are such laws, at this time it cannot provide them. Functional explanations are then explanation-sketches rather than explanations.

Nevertheless, our essential point can still be made here. Given that we can include in our explanans the information that *i* (but none of its functional equivalents) has come into existence in *s* before *t*, our resulting functional explanation when filled out with some proper laws about persistence, will be a statistical explanation and as such, will pose no logical program for our scientist's reductionist program.[19]

NOTES

1. It is obvious that this program is not very precisely specified, for it is unclear which terms and theories are intrinsically nonphysical. But the results of this paper do not depend upon the way in which the program is defined more precisely, so we can disregard that issue here.

2. First published in L. Gross's *Symposium on Sociological Theory* (New York, 1959) and reprinted in his *Aspects of Scientific Explanation* (New York, 1965). All page references are to the reprint.

3. See, for example, his introduction of it on

pp. 367–368 of his *Aspects of Scientific Explanation (op. cit.)*.

4. *Ibid.*, p. 299.

5. See, for example, *ibid.*, p. 368.

6. *Philosophy of Natural Science* (Englewood Cliffs, 1966), p. 48.

7. See the papers collected in Wesley Salmon's *Statistical Explanation and Statistical Relevance* (Pittsburgh, 1971).

8. I owe this point to David Rosenthal.

9. Books on physiology, like W. B. Cannon's *The Wisdom of the Body* (New York, 1939), also use teleological and functional accounts, but their cases are all subsumable in the class of evolutionary functional accounts. The opposite is not, however, true, so we shall consider the wider content so as to cover all cases. On this point, see footnote 12.

10. If, of course, it turns out that scientific progress shows the process is not a random one, then perhaps this will mean that we may someday have a causal account of why the particular mutation occurred. But, at least for the moment, the explanation seems to be a non-Hempelian statistical explanation.

11. To be sure, in this case, the trait contributes to differential reproduction rates by contributing to the survival and flourishing of the particular organisms that have it. Still, the biologist is right in emphasizing the former because (a) it is what is involved in the mechanism for the persistence and domination of the trait and (b) in cases in which the two are unrelated, i.e., in cases (like attractive plumage which attracts members of the other sex) in which the trait contributes to differential reproduction rates without contributing to the survival and flourishing of the particular organism, the functional account still seems to be true. On these and related points, see ch. 14 of G. G. Simpson's *The Meaning of Evolution* (New Haven, 1949).

What if someone were to claim that the trait's contribution to the survival and flourishing of the particular organism has explanatory value independently of its contribution to differential reproduction rates? I think that it could be shown that such a claim would be mistaken. After all, it certainly does not explain the origin or persistence of the trait in the species, but neither can it be used to explain the origin or persistence of the trait in the individual. The origin of the trait in the individual is due to inheritance, and its persistence is due to its stability in the given situation. There seems then to be nothing for a trait's contribution to the survival and flourishing of the particular organism to explain except via its contribution to differential reproduction rates.

12. Actually, even this has to be modified to say that species *s* has been present in a setting of kind *c* for most of the time from the occurrence of the mutations until now. But we need not worry for now as to exactly how to state this additional modification.

13. The most important of which are collected in his *Method in Social Anthropology* (Chicago, 1958).

14. *Ibid.*, p. 20.

15. *Ibid.*, pp. 21–22.

16. *Ibid.*, p. 20.

17. R. K. Merton, *Social Theory and Social Structure* (New York, 1957), pp. 65 ff. (Italics are mine.)

18. *Ibid.*, p. 81.

19. I should like to thank David Rosenthal and several people at the University of North Carolina for helpful comments and questions.

Michael Ruse

TELEOLOGY REDUX

Although Mario Bunge is best known as a philosopher of physics—and deservedly so—he has in fact turned his keen attention to just about every other area of science. Truly he might be called the twentieth-century's answer to William Whewell![1] And like Whewell, Bunge has considered the biological sciences, and has things of interest to say about them. I think it is fair to conclude that, generally speaking, Bunge and I agree about biology. We are both suspicious of attempts to put biology apart from the physical sciences. In particular we both think that some sort of axiomatic or hypothetico-deductive ideal is appropriate in the biological sciences as well as in physics and chemistry [Bunge 1967; Ruse 1973].

But Bunge is too young and vigorous to bury with praise or agreement. I want therefore to turn to another aspect of biology, hoping thereby to provoke disagreement and to stir Bunge into action yet again. My concern in this paper will be with the problem of teleology. I think it is probably true to say that, inasmuch as the philosophy of biology can be said to have a "hot" topic, the nature of teleological explanation is that topic.[2] But what is meant here by "teleology"? Simply that in biology we find all kinds of funny language apparently making reference to, and indeed explaining in terms of, *ends*. Let me give just one example, and then I will explain why teleology has attracted philosophical attention.

The stegosaurus was a large herbivourous dinosaur of the Jurassic period. What made it very distinctive was the existence of a set of bony plates running along its back (see Figure 1). The question which paleontologists have long asked is, what "end" or "purpose" or "function" did they serve? What "problem" were they supposed to "solve?" A number of answers were put forward. Some said that the plates existed "in order to" make the stegosaurus seem bigger and more fearsome, thus frightening off predators. Some said that the plates existed "in order to" facilitate courtship recognition—with plates like that down one's backside it would be pretty difficult to make a mistake and spend one's time making romantic overtures to a member of the wrong species. And some said that the plates existed "in order to" aid heat regulation—the plates would act as sorts of cooling devices, radiating heat, thus enabling the non-sweating dinosaur to move about in the heat of the sun and get on with the business of living.

We see therefore the frankly teleological language of the biologist. And let us make no mistake about its value. By thinking in this way, biologists make great progress. Indeed, there is good reason to think that paleontologists have now solved the stegosaurus problem. The plates most probably existed for heat regulation—a conclusion based on their "design" for such regulation:

> The hypothesis that the dorsal plates of *Stegosaurus* were a heat-regulation device is based on the fact that the plates were porous and probably had a large supply of blood vessels, on their alternate placement to the left and right of the midline (suggesting cooling fins), on their large size over the most massive part of the body and on the constriction near their base, where they are closest to the heat source and would be inefficient heat radiators [Lewontin 1978, p. 218].

Reprinted by permission from J. Agassi and R. S. Cohen (eds.), *Scientific Philosophy Today,* 299–309.

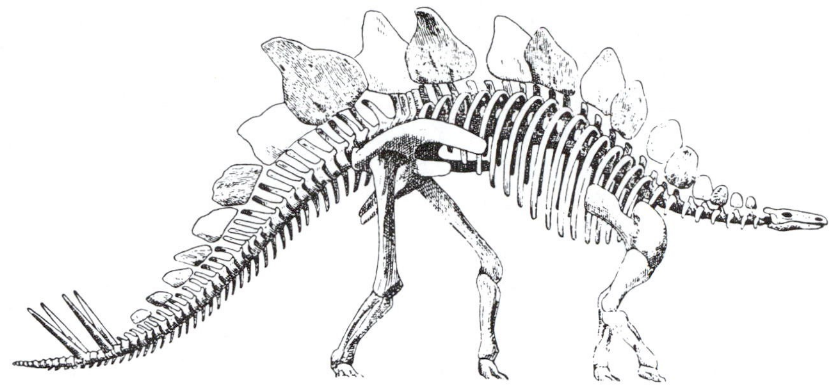

FIGURE 1. The stegosaurus. This skeleton in the American Museum of Natural History is 18 feet long. (From Lewontin, 1978)

And yet teleology is a problem. Not for biologists. They use it and they are happy to use it. But teleology is a problem for philosophers. If one looks at the physical sciences, rightly or wrongly still taken as the paradigm, one does not find the overt use of teleological language. It is true indeed that in the last century when Sir David Brewster—Scottish Man of Science—was asked what function the moon serves, he was happy to reply that it serves the end of lighting the way of nocturnal travellers [Brewster 1854]. But the fact that we today (at least in Ontario) find such an answer amusing, points to the incongruity of using teleology in physics. The moon does not exist "in order to" do anything. The very suggestion is not false, but absurd. It is like asking whether Tuesday is more tired than Monday. Tuesdays are not the sorts of things that can be tired, and moons are not the sorts of things that can have ends.

We see therefore that, philosophically speaking, teleology (or rather, the teleological way of thinking) is an interesting and important question. Physicists do not use it. Biologists do. What we need are answers to questions like, why biologists use a teleological way of thinking, how they can use it and get away with it, what they mean when they do use it, and whether or not biologists could stop using teleological language and still get as far in their biology? This last question might suggest a somewhat arrogant attitude—why on earth should biologists stop using teleological language? But I think it is true that biologists generally feel an urge to make their science as much like physics as possible. It is after all the paradigm science (not necessarily the nicest science), even though biologists will never admit to the position of physics. (As is well known, biologists tend to suffer from physics envy.)

In recent years there has been, as I have suggested, a great deal written about teleology, although non-philosophers should not be deceived by this productivity. Much that has been written stems not from the fact that the topic of teleology is important, which it is, or from the fact that the topic of teleology is difficult, which it also is. Rather, the popularity of the topic of teleology stems from the undeniable fact that the average philosopher of science wants as little to do with science as is humanly possible. Trained in one of the neo-scholastic education mills of North America, he has no contact with science, and, searching desperately for a thesis-topic, he has little inclination or pressure to make such contact.[3] Teleology therefore seems tailor-made. All one needs to know is the (false) fact that the heart beats in order to pump blood and that it does not beat in order to create heart sounds, and one is off and running. From then on one can drop all reference to things organic, and one can deal exclusively in such delightful philosophical phenomena as door knobs serving as paper-weights, sewing machines with knobs to be pressed when you want them to self-destruct, multi-purposed vibrators, or what have you.

But let me stop carping about my fellows and state my own position. This is that I think one can make some progress with the problem of teleology, but that this simply cannot be done in isolation from real science. One must stay close to science, both as it is now and perhaps even more importantly, as it was in the past. (I say this notwithstanding the fact that references to the history of science are about as popular amongst philosophers of science as are references to contemporary science.) I want to go further and doubt that an all-encompassing analysis of teleology can be offered. Or rather I want to suggest that if an all-encompassing analysis of teleology is possible, the way to get at it is not by trying to produce it immediately, but rather by producing analyses in the various branches of science, and then seeing if they can be put together. Following my own prescription therefore, what I want to do is look at biology and its history, and see how to unpack its teleological content.

If we look at the history of biology, we see that 1859 was a watershed. That was the year of the publication of Charles Darwin's *On the Origin of Species*. Literally, biology was revolutionized [Ruse 1979]. Before 1859 people believed that the organic world had been created miraculously by God. After 1859 people believed that the organic world was caused naturally by the action of unbroken law, and those who followed Darwin believed that the chief causal agent was natural selection working on random variation. From the point of view of teleology, 1859 and the publication of Darwin's *Origin* were crucial. And yet the *Origin's* importance is not straightforward. I am reminded of Conan Doyle's marvelous Sherlock Holmes story about the theft of the racehorse Silver Blaize. Holmes turned to Watson and said: "The most important fact my dear Watson, is the dog that barked in the night." "But the dog didn't bark in the night." "Precisely." And the most important fact about the arrival of the *Origin,* is that from the point of view of the teleology in biology, *it did not make the slightest bit of difference.* Before Darwin people cheerfully said that the eye existed in order to see. After Darwin people cheerfully said that the eye existed in order to see. The causes were different perhaps, but the teleology was not!

This being so, let us ask first about the teleology of pre-*Origin* biology. Perhaps then with this answered we can carry forward to modern biology. Now, the answer to questions about pre-*Origin* biology comes fairly readily. As every student of introductory philosophy knows well, people thought that they could speak teleologically about the organic world because they thought the organic world was teleological—it shows the ends put there by God for the benefit of man and other organisms. (Whether everything ultimately reduced to man's benefit was a nice point of natural theology.[4]) Thus, as Paley [1819] was happy to argue, we can speak of the eye having the function of seeing, because the eye does have that function—it was designed (literally) that way by God. The eye, if you please, is one of God's *artifacts.* And let it not be thought that only professional natural theologians thought that way. In a discussion in 1834 in Britain's leading science journal, the *Philosophical Transactions* of the Royal Society, of the adaptations of the kangaroo for feeding its young, we find Britain's leading zoologist, Richard Owen, calmly stating that such adaptations show "irrefragable evidence of creative forethought."

Of course, in an important sense, pre-Darwinians were using a metaphor or analogy. Literally they thought the eye was one of God's artifacts; but they were modelling it on human artifacts. We make the telescope, having our ends in mind. For this reason we can speak and think teleologically about the telescope. The eye in many respects seems very similar to the telescope. Therefore, it must have been made by God. ("Things which are so design-like just don't happen by chance or blind law.") So for this reason we can speak and think teleologically about the eye.

The key to pre-Darwinian biology, therefore, is the *artifact model.* Because the organic world seems *as if* it were designed, it was felt permissible and appropriate to speak of it as actually being designed. And coming straight across the *Origin* to post-Darwinian biology, the same point holds. Because the organic world seems *as if* it is designed, it is felt permissible and appropriate to speak of it as being designed. A molecule or the moon does not seem very much like an artifact, so we do not think such language appropriate in physics. The fins on the stegosaurus however are another matter. They look like turbo blades or the Heath Robinson contraptions that solar energy buffs are into. So why not talk and think that way? For Charles Darwin himself, incidentally, the continued use of teleological language came very naturally, because he had been brought up on Paley's *Natural Theology* [Darwin 1969].

None of this is to deny that pre- and post-*Origin* there is a difference. Pre-*Origin* the eye was caused by God. Post-*Origin* the eye was caused by natural selection. The teleology however remains the same.

> Much of evolutionary biology is the working out of an adaptationist program. Evolutionary biologists assume that each aspect of an organism's morphology, physiology and behavior has been molded by natural selection as a solution to a problem posed by the environment. The role of the evolutionary biologist is then to construct a plausible argument about how each part functions as an adaptive device. For example, functional anatomists study the structure of animal limbs and analyze their motions by time-lapse photography, comparing the action and the structure of the locomotor apparatus in different animals. Their interest is not, however, merely descriptive. Their work is informed by the adaptationist program, and their aim is to explain particular anatomical features by showing that they are well suited to the function they perform. Evolutionary ethologists and sociobiologists carry the adaptationist program into the realm of animal behavior, providing an adaptive explanation for differences among species in courting pattern, group size, aggressiveness, feeding behavior and so on. In each case they assume, like the functional anatomist, that the behavior is adaptive and that the goal of their analysis is to reveal the particular adaptation [Lewontin 1978, pp. 216–17].

Let me sum up now what I have tried to say or hint at so far. I argue that the teleology in modern biology is analogical. The organic world seems as if it is designed; therefore, we treat it as designed. The *artifact model* is the key to biological teleology. The phenomena of physics and chemistry do not seem as if designed. Therefore, we do not think teleological thought and language appropriate in those sciences. Of course, we do not today think biological phenomena really are designed—at least, not by the direct intervening agency of a miracle-working God. Rather, we think that natural selection working on random variation is the causal key. Therefore, if one wants to cash in a functional or teleological statement, "The *function* of the fins on the stegosaurus is to effect heat regulation," "The fins of the stegosaurus exist *in order that* its heat might be efficiently regulated," "The fins *solve the problem* (*serve the end*) of heat regulation," one must cash it out in terms of natural selection. The stegosaurus has fins because those of its ancestors which had fins survived and reproduced, and those that did not, did not. In short, the fins are an *adaption,* and to talk in functional language in biology is to refer to an *adaption.* If *x* has the function *y,* then *x* is an adaptation, and *y* is adaptive—it helps survival and reproduction.

This is the crux of what I want to say about biological teleology. In a sense it is all rather simple. But that does not mean it is not true. It makes sense of biology and of its history. Moreover, for once as a philosopher I am not prescribing what biologists ought to do. Rather, I am describing what they do do. For instance, in one of the most highly praised books of the past two decades, the evolutionist G. C. Williams commits himself to precisely the position on teleology I have just unpacked and endorsed.

> Whenever I believe that an effect is produced as the function of an adaptation perfected by natural selection to serve that function, I will use terms appropriate to human artifice and conscious design. The designation of something as the *means* or *mechanism* for a certain *goal* or *function* or *purpose* will imply that the machinery involved was fashioned by selection for the goal attributed to it. When I do not believe that such a relationship exists I will avoid such terms and use words appropriate to fortuitous relationships such as *cause* and *effect.*
>
> Thus, I would say that reproduction and dispersal are the goals or functions or purposes of apples and that the apple is a means or mechanism by which such goals are realized by apple trees. By contrast, the apple's contributions to Newtonian inspiration and the economy of Kalamazoo County are merely fortuitous effects and of no biological interest [Williams 1966, p. 9, his italics].

Enough by way of defence. Let me now conclude with four corollaries which follow from my main theorem.

First, one might ask why teleology is possible in biology? In one sense, I have already answered this question; in another sense, I do not have to answer it. My answer is that teleology is possible because the organic world is design-like. My answer does not depend on a correct answer to the further question: Why is the organic world design-like, or rather, why does natural selection make the organic world design-like? But having exempted myself from the responsibility of having to give a correct answer (!), what I would suggest is that the answer lies in a suggestion to be found in David Hume's *Dialogues Concerning Natural Religion,* namely, that irrespective of whether there is any similarity in the production of human artifacts and organic characteristics, the latter have to be design-like because if they were not, they simply would not work. The eye is like a telescope or some other human artifact for seeing, because if it were not like a telescope or related artifact, one would not see at all.

Second, is biological teleology real or hard-line teleology in the sense of explaining or trying to understand the past in terms of the future? And if it is, how do biologists avoid classic problems like that of the missing goal-object? (If one explains *x* in terms of future *y*, that is fine if *y* occurs; but what happens if *x* occurs, and then *y* does not?[5]) My position, which I shall more state than argue for here, is that the teleology is fairly hard-line. I would argue that when the biologist says *x* exists in order to *y*, or the function of *x* is *y* or (*y* ing), then the earlier *x* is being explained in terms of the later *y*. Consider (for support) the following two passages by G. G. Simpson, one of the world's leading evolutionists. He writes:

> In order to realize the new functions of a changed environment, an organism must, at the moment when the change or the occupation of a new environment begins, have at least some functions prospective with regard to the new environmental functions. This is merely a more technical way of saying that organisms can continue to live only under conditions to which they are already at least minimally adapted [Simpson 1953, p. 189].

> If a mutation, whether adaptive, non-adaptive, or inadaptive with respect to the adaptation of an ancestral population, does become fixed and spread by selection it is adaptive from the start with respect to the descending populations [Simpson 1953, pp. 194–5].

This, it seems to me, makes it all pretty clear that *x* is being explained in terms of the later (future) *y*.

But what then about the missing goal-object problem (and related problems)? I think these do not really arise (or only very rarely arise), because by the time that the biologist gets around to working out the function of *x*'s, the *y*'s (which were future to the *x*'s) are now past too—in short, the biologist knows that they occurred and were not missing! Assuming that the stegosaurus explanation is correct, one can as a paleontologist explain the plates in terms of (later) heat regulation, because the heat regulation is in our past also, and cannot therefore go missing. As far as today's organic world is concerned, the biologist is still working on organisms essentially as they have gone (or are going) past. If the biologist wants to state categorically that at this moment the function of *x* is to do *y* and *y* really is still future, then frankly I think he/she is taking a gamble. She/he might come up against the missing goal object problem. But of course, practically speaking, the biologist is on safe ground. The future, at least for this year's generation, is usually like the past—eyes go on seeing, and legs go on walking—and so by the time that anyone gets around to deciding that the future is not like the past, the future is not future any more.

Third, the above analysis raises the question of whether all of the teleology of biology has been located, and perhaps incidentally throws some light on a concept that keeps recurring in the philosophical literature on teleology, namely that of a *goal-directed system*.[6] Some philosophers have argued that all of the teleology of biology is analyzable in terms of goal-directedness, namely the ability to get back on target despite disruptions. This is clearly untrue—the assumption that the plates of the stegosaurus exist in order to effect heat-regulation says nothing about how the stegosaurus would react were something untoward to happen. But I think that goal-directedness is nevertheless important in biological teleology, and the above analysis shows how.

Consider: We get the teleology of functions in biology by analogy from human teleology, namely the teleology of human artifacts. Is there any other kind of human teleology? Clearly there is, namely our own teleology. We strive to achieve certain ends, just as we try to impregnate ends into our artifacts. I see no reason why we should not, and indeed would rather expect that we would, read this other kind of teleology into the organic world also. But how would we recognize this teleology, or rather what would lead us to impute such teleology to the non-human world? I suppose the existence of brains in other organisms would go part way, but I suspect that identifying goal-directed behavior is the really crucial factor. Suppose I am teleologically oriented towards something, say getting a Ph.D. The real mark that I have that Ph.D as a goal is the fact that I strive for it, despite all obstacles like unpleasant professors, irrelevant language exams, and so forth. Similarly, we see a teleology in the organic world, over and above the teleology of functions, when the wolf-pack shows a flexible strategy to bring down the much larger moose, and when the insect colony regroups and moves in the face of attack, and so forth. In other words, by analogy we see goal-directed behavior as teleological, of a kind different from that of function.

I am not sure however that we see all goal-directed behavior as teleological (and this is why supposed problem examples of goal-directedness drawn from the inorganic world do not worry me). As we get farther from human intelligence and its ability to conceive of and act towards goals, we see less teleology (of this second kind). Take something like sweating and shivering, which is generally conceded to fill the conditions for a goal-directed system. Is this teleological? A body capable of sweating and shivering is certainly teleological inasmuch as such processes are adaptations (which they are!). But is there more teleology (of the kind I find in the coordinating wolf pack)? I rather doubt it—no one thinks that any intelligence is involved in keeping the body at constant temperature through sweating and shivering.

I suspect however that there is a grey area about where this second form of teleology operates. We find it in dogs. We do not find it in trees. Do we find it in earth worms? In part, this is an empirical question. How goal-directed are earth worms capable of being—particularly, how goal-directed in terms of *behavior* are they capable of being? After that, the imputation of such teleology is all a bit arbitrary. But then, of course, this element of arbitrariness is characteristic of metaphor or analogy. Some people are prepared to push metaphors further than others. A grey area where there are disagreements confirms my position, rather than refutes it.

Fourth, and finally, let me mention reduction and eliminability (a point at which I suspect that Bunge and I might start to part company). Can we get rid of the teleology of biology? Would we want to? I suspect that we could, but that we would not want to. We could certainly talk always in terms of past causes (efficient causes) and drop all talk of function and so forth. But we could not then say as much. We might, for example, talk about the embryological development of the heart, talking in terms of cell-division and so forth, but we could not ask what the heart was for. Moreover, simply to switch to talk of "adaptation" would be no solution, because the forward-looking teleological aspect of biology still then remains. The whole way of looking at organisms in terms of design and artifacts has to be dropped. But why would one want to do this? Biology is not physics, nor should it have to be. If there were no good reason for biology's being different, then one would expect them to be the same. This is why Bunge and I feel justified in neglecting neo-vitalistic attempts to separate biology and physics on such pseudo-grounds as the purportedly peculiarly complex nature of biology. In pertinent respects, biology is no more complex, or unique, or what have you, than physics [Ruse 1977].

But, perhaps putting me apart from Bunge, when we come to teleology. I believe that we do have good reason for separating physics and biology, as they stand now. And the reason for this is that biology refers to phenomena that are design-like and physical phenomena are not, and biology as a science tries to capture and utilize this distinctive facet of the biological world. Moreover, interestingly, at the borderline if anything it is physics that gives. Where biology and physics meet, we do not get a total rejection of teleology. Rather, conversely teleology seeps in. At least, we find molecular biologists talking about such things as the need "to crack the genetic code." And if that is not to talk in terms of design, I do not know what is. In short, the teleology of biology is here to stay. Given its power in the hands of the evolutionist, for instance in finding out the purpose of the plates on the back of the stegosaurus, let us be thankful for this fact.

NOTES

1. I intend this as high praise.

2. Useful recent reviews include Woodfield [1976]; Wright [1976]; Nagel [1977].

3. I say "he" in this context deliberately. My experience is that female philosophers of science are far more sensitive to science than are their male counterparts.

4. See, for example, Buckland [1836].

5. This and related problems are discussed in Scheffler [1963].

6. Woodfield [1976] admirably digests and criticizes the literature on goal-directedness.

REFERENCES

Brewster, D.: 1854, *More Worlds than One: The Creed of the Philosopher and the Hope of the Christian*, Murray, London.

Buckland, W.: 1836, *Geology and Mineralogy: Bridgewater Treatise* 6, Pickering, London.

Bunge, M.: 1967, *Scientific Research*, Springer, New York.

Darwin, C.: 1859, *On the Origin of Species*, Murray, London.

Darwin, C.: 1969, *Autobiography*, N. Barlow (ed.), Norton, New York.

Lewontin, R. C.: 1978, "Adaptation," *Scientific American* 239, 212–30.

Nagel, E.: 1977, "Teleology Revisited," *J. Phil.* 74, 261–301.

Owen, R.: 1834, "On the Generation of the Marsupial Animals, with a Description of the Impregnated Uterus of the Kangaroo," *Phil. Trans.*, 333–64.

Paley, W.: 1819, *Natural Theology*, in *Collected Works*, Rivington, London, 1st published 1802.

Ruse, M.: 1973, *Philosophy of Biology*, Hutchinson, London.

Ruse, M.: 1977, "Is Biology Different from Physics?," in R. Colodny (ed.), *Logic, Laws, and Life*, University of Pittsburgh Press, Pittsburgh, pp. 89–127.

Ruse, M.: 1979, *The Darwinian Revolution: Science Red in Tooth and Claw*, University of Chicago Press, Chicago.

Scheffler, I.: 1963, *The Anatomy of Inquiry*, Knopf, New York.

Simpson, G. G.: 1953, *The Major Features of Evolution*, Columbia University Press. New York.

Williams, G. C.: 1966, *Adaptation and Natural Selection*, Princeton University Press, Princeton.

Woodfield, A.: 1976, *Teleology*, Cambridge University Press, Cambridge.

Wright, L.: 1976, *Teleological Explanations*, University of California Press, Berkeley.

BIBLIOGRAPHICAL ESSAY

This bibliography is intended as an introduction to, rather than a comprehensive survey of, the literature on the philosophy of science. The reader who wants further bibliographical information should consult the bibliographies of the books mentioned below.

The best introduction to the history of the philosophy of science is R. Blake, C. J. Ducasse, and E. Madden's *Theories of Scientific Method* (University of Washington Press, 1960). In connection with this book, the reader should consult the selections in the following historically oriented anthologies: B. A. Brody and N. Capaldi's *Science: Men, Methods and Goals* (W. A. Benjamin, 1968) and J. J. Kockelman's *Philosophy of Science: The Historical Background* (Free Press, 1968).

Several good texts introduce the reader to the classical positivist views in the philosophy of science. These include: C. G. Hempel, *Philosophy of Natural Science* (Prentice-Hall, 1966), E. Nagel, *The Structure of Science* (Harcourt, Brace, and World, 1961), A. Pap, *An Introduction to the Philosophy of Science* (Free Press, 1962), I. Scheffler, *The Anatomy of Inquiry* (Knopf, 1963), and J.J.C. Smart, *Between Science and Philosophy* (Random House, 1968). In connection with these texts, the reader may find it useful to look at some of the articles collected in A. Danto and S. Morgenbesser's *Philosophy of Science* (Appleton-Century-Crofts, 1953), H. Feigl and G. Maxwell's *Current Issues in the Philosophy of Science* (Holt, Rinehart and Winston, 1961), E. Madden's *The Structure of Scientific Thought* (Houghton Mifflin Co., 1960), and the many volumes of *Minnesota Studies in the Philosophy of Science*.

Most of the important current work in the philosophy of science appears in journals and collections of articles. Among the journals that specialize in the philosophy of science are *Philosophy of Science* and *British Journal for the Philosophy of Science* (*B.J.P.S.*).

The logical positivist conception of theories was presented in R. Braithwaite's *Scientific Explanation* (Cambridge University Press, 1953), Chapters 1–4; R. Carnap's "Testability and Meaning," *Philosophy of Science* (1936, 1937); and "The Methodological Character of Theoretical Concepts," *Minnesota Studies*, vol. I; C. G. Hempel's "The Theoretician's Dilemma," *Minnesota Studies*, vol. II; and E. Nagel's *Structure of Science*, Chapters 5 and 6. A classical critique of their view of the role of models was presented by M. Hesse in her book *Forces and Fields* (Nelson, 1961) and in her articles "Operational Definition and Analogy in Physical Theories," *B.J.P.S.* (1952), "The Role of Models in Physics," *B.J.P.S.* (1953), and "Theories, Dictionaries and Observation" *B.J.P.S.* (1958). The whole issue was evaluated in P. Achinstein's *Concepts of Science* (Johns Hopkins Press, 1968).

The logical positivist, covering-law models of explanation and prediction were defended in the many articles in C. G. Hempel's *Aspects of Scientific Explanation* (Free Press, 1965) and in Chapter 4 of E. Nagel's *The Structure of Science*. Their application to historical explanations was analyzed in P. Gardiner's *Theories of History* (Free Press, 1958) and W. Dray's *Philosophical Analysis and History* (Harper & Row, 1966).

A very basic but good introduction to the classical theory of confirmation is B. Skyrms's *Choice and Chance* (Dickenson, 1966). H. Kyburg's "Recent Work in Inductive Logic," *American Philosophical Quarterly* (1964), is a good summary of the classic work in this area. Most of the important classical articles in this area are collected in M. Foster and M. Martin's *Probability, Confirmation, and Simplicity* (Odyssey, 1966). Much of the discussion of Hempel's work is summarized in Part III of I. Scheffler's *The Anatomy of Inquiry*. Carnap's theory of degrees of confirmation is presented in his books *The Logical Foundations of Probability* (University of Chicago, 1950) and *The Continuum of Inductive Methods* (University of Chicago, 1952), and in "The Aim of Inductive Logic" in the first volume of *Logic, Methodology, and Philosophy of Science* (Stanford, 1962). The theory is discussed in several articles in P. Schilpp's *The Philosophy of Rudolf Carnap* (Open Court, 1963), where Carnap presented a reply to those criticisms.

Most of the authors whose recent essays dominate this second edition have produced volumes in which they develop their position at much greater length. Among the more crucial books are the following: Peter Achinstein's *Law and Explanation* (Oxford, 1971) and *The Nature of Explanation* (Oxford, 1983), in which he defends his views about explanations, laws, and causality; Nancy Cartwright's *How the Laws of Physics Lie* (Oxford, 1983), in which she develops her views about theories, causality, and explanation; Paul Feyerabend's *Against Method* (Humanities Press, 1974), in which he presents his views on explanation, reduction, causality, and scientific methods; Ronald Giere's *Explaining Science* (University of Chicago, 1988), in which he develops a philosophy of science based upon recent work in cognitive science; Clark Glymour's *Theory and Evidence* (Princeton, 1980), in which he elaborates upon his account of the confirmation of scientific theories; N. R. Hanson's *Patterns of Discovery* (University of Chicago, 1970), which is his classic defense of the idea of a logic of discovery; Thomas Kuhn's *The Structure of Scientific Revolutions*, 2nd ed. (University of Chicago, 1979) and *The Essential Tension* (University of Chicago, 1977), in which he presents his views about theory change and the meaning of scientific theories; Imre Lakatos's *The Methodology of Scientific Research Programs* (Cambridge, 1978), in which he presents his views about the confirmation and rejection of theories; Larry Laudan's *Progress and Its Problems* (University of California, 1977), in which he presents his views about the same topics; Wesley Salmon's *Scientific Explanation and the Causal Structure of the World* (Princeton, 1984), in which he develops his ideas about explanation, scientific laws, and causality; Dudley Shapere's *Reason and the Search for Knowledge* (Reidel, 1984), in which he develops his views about the meaning of scientific terms; Bas C. van Fraassen's *The Scientific Image* (Oxford, 1980), in which he develops his views about theories, explanation, and scientific realism.

All of this literature has produced many fine anthologies, which collect important and diverse responses to these ideas. Among the most important are: Paul Churchland and Clifford Hooker's *Images of Science: Essays on Realism and Empiricism* (University of Chicago, 1985); Ian Hacking's *Scientific Revolutions* (Oxford, 1981), Imre Lakatos and Alan Musgrave's *Criticism and the Growth of Knowledge* (Cambridge 1970), Jarrett Leplin's *Scientific Realism* (University of California, 1984), and Thomas Nickles's *Scientific Discovery* (Reidel, 1980).

The Poincaré-Reichenbach-Salmon debate has been continued in a number of important recent books, including Michael Friedman's *Foundations of Space-Time Theories* (Princeton, 1983), Adolf Grunbaum's *Philosophical Problems of Space and Time* (Knopf, 1963), and Larry Sklar's *Space, Time and Space-Time* (University of California, 1974).

Michael Ruse's *The Philosophy of Biology* (Addison Wesley, 1982) is a good introduction to many of the important issues in that area. An important anthology in the same area is Elliot Sober's *Conceptual Issues in Evolutionary Biology* (M.I.T., 1984).